Groundwater-Surface Water Interactions

Groundwater-Surface Water Interactions

Special Issue Editors

Jörg Lewandowski
Karin Meinikmann
Stefan Krause

MDPI • Basel • Beijing • Wuhan • Barcelona • Belgrade • Manchester • Tokyo • Cluj • Tianjin

Special Issue Editors

Jörg Lewandowski
Leibniz Institute of Freshwater
Ecology and Inland Fisheries
Germany
Humboldt University
Germany

Karin Meinikmann
Leibniz Institute of Freshwater Ecology
and Inland Fisheries
Germany
Julius Kühn-Institute
Germany

Stefan Krause
University of Birmingham
UK

Editorial Office
MDPI
St. Alban-Anlage 66
4052 Basel, Switzerland

This is a reprint of articles from the Special Issue published online in the open access journal *Water* (ISSN 2073-4441) (available at: https://www.mdpi.com/journal/water/special_issues/Groundwater-Surface_Water).

For citation purposes, cite each article independently as indicated on the article page online and as indicated below:

LastName, A.A.; LastName, B.B.; LastName, C.C. Article Title. *Journal Name* **Year**, *Article Number, Page Range.*

ISBN 978-3-03928-905-9 (Hbk)
ISBN 978-3-03928-906-6 (PDF)

Cover image courtesy of Jörg Lewandowski .

Contents

About the Special Issue Editors

Jörg Lewandowski is a senior scientist at the Leibniz-Institute of Freshwater Ecology and Inland Fisheries, as well as a lecturer at Humboldt University. His current research includes hydrological and biogeochemical processes involved in groundwater–surface water interactions at aquifer–stream (hyporheic zone) and aquifer–lake interfaces (lacustrine groundwater discharge). This includes topics of environmental and societal concern such as lake eutrophication and attenuation of trace organic compounds. Jörg's further ecohydrological research topics include lake restoration and bioturbation with a focus on *Chironomus plumosus*.

Karin Meinikmann Meinikmann has intensely investigated lacustrine groundwater discharge and implications for lake water quality and hydrology. She worked on hyporheic zone interactions in the context of pharmaceutical retention and degradation in wastewater-impacted rivers. At the Julius Kuehn Institute, her current focus is on plant protection products entering small standing waters via subsurface transport. This includes effects on aquatic biodiversity and the development of adequate monitoring schemes.

Stefan Krause is chair for ecohydrology and biogeochemistry at the University of Birmingham where he leads the Birmingham Water Council and holds an Invited Visiting Professorship at the University of Lyon. His research focus is on the fate and transport of emerging (e.g., engineered nanoparticles, pharmaceuticals, and microplastics) and legacy pollutants (e.g., chlorinated solvents) at ecohydrological interfaces. He develops advanced in situ environmental sensing technologies and integrated modelling approaches for identifying hotspots hot moment in environmental pollution. In collaboration with private sector partners, regulators, and decision makers, he is working toward developing sustainable solutions to wicked water problems that benefit society and the environment.

Editorial

Groundwater–Surface Water Interactions: Recent Advances and Interdisciplinary Challenges

Jörg Lewandowski [1,2,*], Karin Meinikmann [1,†] and Stefan Krause [3]

[1] Department Ecohydrology, Leibniz-Institute of Freshwater Ecology and Inland Fisheries, 12587 Berlin, Germany; Karin.Meinikmann@julius-kuehn.de
[2] Geography Department, Humboldt University of Berlin, 12489 Berlin, Germany
[3] School of Geography, Earth and Environmental Sciences, University of Birmingham, Birmingham B15 2TT, UK; S.Krause@bham.ac.uk
* Correspondence: lewe@igb-berlin.de; Tel.: +49-30-64181-668
† Current address: Julius Kühn-Institute, Institute for Ecological Chemistry, Plant Analysis and Stored Product Protection, 14195 Berlin, Germany.

Received: 14 January 2020; Accepted: 16 January 2020; Published: 19 January 2020

Abstract: The interactions of groundwater with surface waters such as streams, lakes, wetlands, or oceans are relevant for a wide range of reasons—for example, drinking water resources may rely on hydrologic fluxes between groundwater and surface water. However, nutrients and pollutants can also be transported across the interface and experience transformation, enrichment, or retention along the flow paths and cause impacts on the interconnected receptor systems. To maintain drinking water resources and ecosystem health, a mechanistic understanding of the underlying processes controlling the spatial patterns and temporal dynamics of groundwater–surface water interactions is crucial. This Special Issue provides an overview of current research advances and innovative approaches in the broad field of groundwater–surface water interactions. The 20 research articles and 1 communication of this Special Issue cover a wide range of thematic scopes, scales, and experimental and modelling methods across different disciplines (hydrology, aquatic ecology, biogeochemistry, environmental pollution) collaborating in research on groundwater–surface water interactions. The collection of research papers in this Special Issue also allows the identification of current knowledge gaps and reveals the challenges in establishing standardized measurement, observation, and assessment approaches. With regards to its relevance for environmental and water management and protection, the impact of groundwater–surface water interactions is still not fully understood and is often underestimated, which is not only due to a lack of awareness but also a lack of knowledge and experience regarding appropriate measurement and analysis approaches. This lack of knowledge exchange from research into management practice suggests that more efforts are needed to disseminate scientific results and methods to practitioners and policy makers.

Keywords: aquifer–stream interface; hyporheic zone; benthic zone; lacustrine groundwater discharge; submarine groundwater discharge; riparian corridors

---

1. Introduction

Recent years have seen a paradigm shift in understanding the importance of interactions between groundwater and surface water bodies. While for a long time surface waters and aquifers had been defined as discrete, separate entities, it is now understood that they are integral components of a surface–subsurface continuum [1]. This has fostered more holistic perspectives on a variety of ecosystem services at risk, such as the consideration of groundwater as a source of surface water pollution and ecosystem degradation (eutrophication, organic micropollutants, etc.) and vice versa (i.e., surface water as groundwater pollutant). The protection of potable water resources requires clean and safe water quality of both surface and subsurface waters.

This paradigm shift has triggered intense investigations of the water and mass transport processes across aquatic–terrestrial interfaces. Nevertheless, there is still a lack of mechanistic understanding and standardized methods to measure the processes involved. For example, it is well accepted that the reactive interface between surface water and the subsurface is of great importance for the quality and the quantity of exchange fluxes [2]. However, experimental and validated model-based evidence of the magnitude of the involved processes as well as of the underlying controls are scarce. One of the many reasons for this is that groundwater–surface water interactions integrate a large variety of scientific disciplines. Researchers from hydrology, biogeochemistry, microbiology, biology, physics, and chemistry work on the complex process interactions that require them to consider relevant aspects from other scientific fields. Additionally, interactions between surface and subsurface water take place in a range of different marine and freshwater systems, but the potential to transfer technologies and approaches as well as resulting knowledge and process understanding to other fields is not adequately exploited.

The aim of this Special Issue is to collectively present and integrate novel outcomes from interdisciplinary research on groundwater–surface water interactions. The presented studies cover a large variety of thematic areas, scales, and experimental and modelling-based methodologies and approaches, and thus promote interdisciplinary discussion and reveal knowledge gaps and future research needs.

2. Overview of the Special Issue Contributions

This Special Issue consists of 20 original research papers [3–23]. The majority of the contributions [3,5,8–11,15–17,19–23] focus on the interactions between rivers or streams and groundwater. Three studies [6,13,14] investigate the interactions between lakes and groundwater, while two studies [4,18] deal with the exchange between groundwater and oceanic water. Two additional contributions [7,12] show the relevance of groundwater exchange processes in wetlands. In the following we provide an overview that integrates the knowledge gains and innovations of the aforementioned studies and outlines the potential for knowledge and method exchange across study system boundaries.

2.1. Groundwater–River Interactions

2.1.1. Catchment-scale Hydrological Studies

The increasing awareness of the relevance of groundwater–surface water interactions leads to increased research into the quantification of water balances in systems with interacting river flow and aquifer dynamics. In this Special Issue, several studies aim at investigating the hydrological interactions between rivers and aquifers at a regional catchment scale. For example, Kelly et al. [8] used river gauges to quantify the baseflow indicator (BFI) in order to evaluate the relevance of groundwater dynamics and trends for an African river system. Vrzel et al. [19] applied different modelling approaches to groundwater and stream gauge data to disentangle and quantify the influence of precipitation and river water on an important aquifer system in Slovenia. Similarly, Parlov et al. [15] applied two- and three-component mixing models on stable isotope data to quantify the contributions of precipitation and surface water to aquifer recharge. Focusing on water fluxes crossing the interface in the other direction, Steiness et al. [16] aimed at localizing and quantifying the contributions of groundwater to stream flow. They therefore integrated a wide range of analysis methods (inter alia electrical resistivity tomography, ground penetrating radar, slug tests and hydraulic head measurements in catchment and riverbed, temperature measurements in the hyporheic zone, stable isotope measurements) in order to derive a spatially highly resolved picture of flow paths (a) from the catchment to the river and (b) in the hyporheic zone. Tang et al. [17] looked at the influence of groundwater as a significant proportion of stream discharge on temperature and chlorophyll-a dynamics. This study represents the only one in this Special Issue conducted in a karstic environment.

2.1.2. Hyporheic Zone Studies

Exchange fluxes across sediment–water interfaces are frequently studied in detail with a specific focus on the hyporheic zone (i.e., the zone of the stream–groundwater interface where groundwater and stream water mix). Lewandowski et al. [22] discuss the valuable ecosystem services provided by the hyporheic zone. Several studies focus on fluxes across the hyporheic zone by using temperature as a tracer. For example, Mojarrad et al. [11] used the temperature difference between groundwater and surface water to numerically model up- and downwelling areas in the hyporheic zone. Le Lay et al. [9] and Gilmore et al. [5] applied fiber-optic distributed temperature sensing (fo-DTS) for pattern identification as a prerequisite for optimized point measurements of groundwater exfiltration into a river. Le Lay et al. [10] used the diurnal atmospheric temperature signal propagating vertically through the hyporheic zone with a characteristic attenuation and phase shift to calculate vertical flow velocities. Yao et al. [23] used this approach to analyze the effects of a low-permeability sediment lens and surface discharge velocity on hyporheic flow. For this, they applied an innovative artificial rectangular sediment cuboid of a known heterogeneous sediment composition and placed it in a real streambed environment. Gilmore et al. [5] combined their fo-DTS measurements with traditional methods to calculate Darcy fluxes in order to minimize measurement efforts. Some of the research presented here used modelling tools to derive and improve the understanding of hyporheic flow mechanisms. For example, Broecker et al. [3] developed an innovative integral formulation for the sediment–water interface as an alternative to coupled modeling approaches that are frequently applied in groundwater–surface water modelling studies. Mojarrad et al. [11] used the empirical data from their study site to model the effects of sediment permeability and river discharge dynamics on hyporheic flow.

While most studies presented in this Special Issue focus on the investigation of exchange fluxes across groundwater–surface water interfaces, there are a few that focus additionally on biogeochemical processes at the interface. Ward et al. [20] used the resazurin–resorufin reactive tracer system to study aerobic microbial transformation at sediment–water interfaces and found that the resazurin-to-resorufin transformation rate was highest in youngest storage locations (i.e., that increasing residence times do not necessarily increase the reaction potential of solutes). Wolke et al. [21] systematically studied the impact of bed form movements on oxygen dynamics in the hyporheic zone. The advantage of such flume studies is the ability to conduct analyses under well-controlled environmental parameters. However, Wolke et al. [21] is the only study in this Special Issue using a lab flume for hyporheic research.

2.2. Groundwater–Lake Interactions

The discharge of groundwater to lakes is called lacustrine groundwater discharge (LGD) [24] The opposite flow direction (i.e., the recharge of the aquifer) has often been studied in the context of bank filtration as a method of drinking water production [25–27]. However, there are no studies on this topic within the Special Issue. Han et al. [6] show that the direction of interactions between aquifer and lake can change seasonally. They sampled precipitation, lake water, river water, and groundwater for chloride concentrations and water-stable isotopes to delineate discharge and recharge areas for the dry (cold) and the wet season in the complex hydrogeological environment of Lake Hulun. Nisbeth et al. [13,14] found that geogenic P might be a much more important P source in some instances than is generally assumed. Their study contributes to the still-overlooked problem of LGD-induced lake eutrophication [24,28–30].

2.3. Groundwater–Ocean Interactions

Submarine groundwater discharge (SGD) is the discharge of groundwater to oceans and their coastal areas [24]. Despite SGD having been studied much more intensively than LGD, there are still many knowledge gaps that need to be addressed, and there is still a need to develop appropriate measurement methods for SGD. Tirado-Conde et al. [18] compare the results of traditional seepage meter measurements with several analytical approaches of vertical sediment temperature profiles in a lagoon in Denmark. Duque et al. [4] applied stable isotopes of water to delineate flow paths and the origin of water in a coastal aquifer. This enabled them to draw valuable conclusions on nutrient transport and their fate along the vector from freshwater to saline environments.

2.4. Interactions of Groundwater with Wetlands

Wetlands often rely on intense interactions with groundwater [31], highlighting that in order to be successful and efficient, restoration efforts in groundwater-dependent terrestrial ecosystems need to consider these interactions. Two studies in this Special Issue provide valuable support for decision making in wetland management. Neff et al. [12] conceptually modelled the influence of depressional wetlands on groundwater flow at the landscape scale and found strong effects such as increased groundwater discharge and the generation of flow divides. From their results they developed an extensive guide for practitioners on how to determine the groundwater connectivity of wetlands. Harvey et al. [7] present an approach based on thermal infrared (TIR) to map groundwater influence in a wetland before and after restoration. They could show that TIR is a very promising tool for the establishment and monitoring of wetland restoration measures aiming at reestablishing groundwater connectivity.

3. Conclusions

The contributions to this Special Issue represent the immense variety of scopes and scientific disciplines that come together in researching the interactions between groundwater and surface water. The research questions addressed in this Special Issue cover, inter alia, the extensive spectrum of mechanistic process understanding of hydrological and biogeochemical interactions including the identification of drivers and controls as well as testing and improving methods and approaches for pattern identification and flux quantification. This Special Issue demonstrates how active and diverse the research community currently is, and proves the high relevance of groundwater–surface water interactions.

The research in this Special Issue covers a wide range of spatial scales, from point measurements to catchment approaches. While a large number of the contributions focused on the local-to-reach scale, the regional and catchment scales have also been studied. Several studies are dedicated to identifying the contribution of groundwater to the water balance of lakes, rivers, or watersheds. Two articles focus on recharge sources of aquifers in the Sava river catchment (Balkans). This indicates an increased awareness of the need to protect drinking water resources in the light of potential alterations of regional water balances in the future. However, there is still the need to improve the mechanistic understanding of processes at smaller scales, as indicated by the large proportion of contributions focusing on hyporheic zones.

Moreover, 15 out of 20 research papers in this Special Issue have a purely hydrological focus, while only five consider hydrological and biogeochemical interactions. In fact, only two contributions (both by Nisbeth et al. [13,14]) focus on mass transport between groundwater and surface water by determining groundwater-borne phosphorus loads to a lake. This could be due to the challenging complexity of groundwater–surface water exchange processes alone, because the investigation of related mass fluxes adds to further complexity. Given that exchange processes between groundwater and surface waters can facilitate the transport but also retention of nutrients and pollutants, the low number of studies here suggests that the scientific investigation of such cases is still difficult.

Furthermore, the large variety of methods and modelling approaches and their different applications might indicate the need for standardized approaches to specific recurring questions, especially at the applied level. Groundwater–surface water interactions will only be regularly considered in management and protection when simple and robust methods and analysis procedures are readily available. Therefore, we would like to call upon all scientists working in this field to reach out and actively share their experiences with practitioners and policy makers to foster the urgently needed attention for groundwater–surface water interaction in freshwater management and protection. In this context, Harvey et al. [7] provide an excellent example of how the application of thermal infrared (TIR) can effectively support wetland restoration planning and success monitoring. Additionally, Neff et al. [12] give well-developed hands-on advice on how to assess the groundwater connectivity of wetlands.

This Special Issue is another step forward in our understanding of surface water–groundwater interactions. The articles reveal not only the environmental and societal relevance of this interface but also identify many open questions and tasks that need to be addressed in future. Especially in the light of future challenges such as climate change, water scarcity, eutrophication, and the retention of pollutants such as pharmaceuticals and microplastics, there is an urgent need to continue our efforts to better understand the processes involved in groundwater–surface water exchange in order to be able to restore and sustain the ecosystem services provided by the groundwater–surface water interface.

Author Contributions: All authors contributed to recruiting and reviewing papers for this editorial. J.L. primarily led and coordinated this effort. J.L. and K.M. wrote a first draft of this editorial, and all authors contributed to reviewing and editing the editorial. All authors have read and agreed to the published version of the manuscript.

Funding: This work was funded by the European Union's Horizon 2020 research and innovation program under grant agreements No. 641939 (HypoTRAIN), No. 765553 (EuroFlow), and No. 734317 (HiFreq), and by the German Research Foundation's (DFG) graduate school "Urban Water Interfaces" under grant agreement GRK 2032/1.

Acknowledgments: Thanks to HypoTRAIN and Urban Water Interfaces for initiating the Special Issue, to all authors for their valuable contributions to the Special Issue, and to all reviewers for contributing to the development of the articles.

Conflicts of Interest: The authors declare no conflict of interest.

References

1. Winter, T.C.; Harvey, J.W.; Franke, O.L.; Alley, W.M. *Groundwater and Surface Water: A Single Resource, Circular 1139*; US Geological Survey: Denver, CO, USA, 1998; p. 79.

2. Krause, S.; Lewandowski, J.; Grimm, N.B.; Hannah, D.M.; Pinay, G.; McDonald, K.; Marti, E.; Argerich, A.; Pfister, L.; Klaus, J.; et al. Ecohydrological interfaces as hotspots of ecosystem processes. *Water Resour. Res.* **2017**, *53*, 6359–6376. [CrossRef]

3. Broecker, T.; Teuber, K.; Gollo, V.S.; Nützmann, G.; Lewandowski, J.; Hinkelmann, R. Integral flow modelling approach for surface water-groundwater interactions along a rippled streambed. *Water* **2019**, *11*, 1517. [CrossRef]

4. Duque, C.; Jessen, S.; Tirado-Conde, J.; Karan, S.; Engesgaard, P. Application of stable isotopes of water to study coupled submarine groundwater discharge and nutrient delivery. *Water* **2019**, *11*, 1842. [CrossRef]

5. Gilmore, T.E.; Johnson, M.; Korus, J.; Mittelstet, A.; Briggs, M.A.; Zlotnik, V.; Corcoran, S. Streambed flux measurement informed by distributed temperature sensing leads to a significantly different characterization of groundwater discharge. *Water* **2019**, *11*, 2312. [CrossRef]

6. Han, Z.; Shi, X.; Jia, K.; Sun, B.; Zhao, S.; Fu, C. Determining the discharge and recharge relationships between lake and groundwater in Lake Hulun using hydrogen and oxygen isotopes and chloride iIons. *Water* **2019**, *11*, 264. [CrossRef]

7. Harvey, M.C.; Hare, D.K.; Hackman, A.; Davenport, G.; Haynes, A.B.; Helton, A.; Lane, J.W., Jr.; Martin, A.; Briggs, M.A. Evaluation of stream and wetland restoration using UAS-based thermal infrared mapping. *Water* **2019**, *11*, 1568. [CrossRef]

8. Kelly, L.; Kalin, R.M.; Bertram, D.; Kanjaye, M.; Nkhata, M.; Sibande, H. Quantification of temporal variations in Base Flow Index using sporadic river data: Application to the Bua Catchment, Malawi. *Water* **2019**, *11*, 901. [CrossRef]

9. Le Lay, H.; Thomas, Z.; Rouault, F.; Pichelin, P.; Moatar, F. Characterization of diffuse groundwater inflows into streamwater (Part I: Spatial and temporal mapping framework based on fiber optic distributed temperature sensing). *Water* **2019**, *11*, 2389. [CrossRef]

10. Le Lay, H.; Thomas, Z.; Rouault, F.; Pichelin, P.; Moatar, F. Characterization of diffuse groundwater inflows into stream water (Part II: Quantifying groundwater inflows by coupling FO-DTS and vertical flow velocities). *Water* **2019**, *11*, 2430. [CrossRef]

11. Mojarrad, B.B.; Betterle, A.; Singh, T.; Olid, C.; Wörman, A. The effect of stream discharge on hyporheic exchange. *Water* **2019**, *11*, 1436. [CrossRef]

12. Neff, B.P.; Rosenberry, D.O.; Leibowitz, S.G.; Mushet, D.M.; Golden, H.E.; Rains, M.C.; Brooks, J.R.; Lane, C.R. A hydrologic landscapes perspective on groundwater connectivity of depressional wetlands. *Water* **2020**, *12*, 50. [CrossRef]

13. Nisbeth, C.S.; Kidmose, J.; Weckström, K.; Reitzel, K.; Odgaard, B.V.; Bennike, O.; Thorling, L.; McGowan, S.; Schomacker, A.; Kristensen, D.L.J.; et al. Dissolved inorganic geogenic phosphorus load to a groundwater-fed lake: Implications of terrestrial phosphorus cycling by groundwater. *Water* **2019**, *11*, 2213. [CrossRef]

14. Nisbeth, C.S.; Jessen, S.; Bennike, O.; Kidmose, J.; Reitzel, K. Role of groundwater-borne geogenic phosphorus for the internal P release in shallow lakes. *Water* **2019**, *11*, 1783. [CrossRef]

15. Parlov, J.; Kovač, Z.; Nakić, Z.; Barešić, J. Using water stable isotopes for identifying groundwater recharge sources of the unconfined alluvial Zagreb aquifer (Croatia). *Water* **2019**, *11*, 2177. [CrossRef]

16. Steiness, M.; Jessen, S.; Spitilli, M.; van't Veen, S.G.W.; Højberg, A.L.; Engesgaard, P. The role of management of stream–riparian zones on subsurface–surface flow components. *Water* **2019**, *11*, 1905. [CrossRef]

17. Tang, T.; Guo, S.; Tan, L.; Li, T.; Burrows, R.M.; Cai, Q. Temporal effects of groundwater on physical and biotic components of a karst stream. *Water* **2019**, *11*, 1299. [CrossRef]

18. Tirado-Conde, J.; Engesgaard, P.; Karan, S.; Müller, S.; Duque, C. Evaluation of temperature profiling and seepage meter methods for quantifying submarine groundwater discharge to coastal lagoons: Impacts of saltwater intrusion and the associated thermal regime. *Water* **2019**, *11*, 1648. [CrossRef]

19. Vrzel, J.; Ludwig, R.; Vižintin, G.; Ogrinc, N. An integrated approach for studying the hydrology of the Ljubljansko Polje Aquifer in Slovenia and its simulation. *Water* **2019**, *11*, 1753. [CrossRef]

20. Ward, A.S.; Kurz, M.J.; Schmadel, N.M.; Knapp, J.L.A.; Blaen, P.J.; Harman, C.J.; Drummond, J.D.; Hannah, D.M.; Krause, S.; Li, A.; et al. Solute transport and transformation in an intermittent, headwater mountain stream with diurnal discharge fluctuations. *Water* **2019**, *11*, 2208. [CrossRef]

21. Wolke, P.; Teitelbaum, Y.; Deng, C.; Lewandowski, J.; Arnon, S. Impact of bed form celerity on oxygen dynamics in the hyporheic zone. *Water* **2020**, *12*, 62. [CrossRef]

22. Lewandowski, J.; Arnon, S.; Banks, E.; Batelaan, O.; Betterle, A.; Broecker, T.; Coll, C.; Drummond, J.D.; Gaona Garcia, J.; Galloway, J.; et al. Is the hyporheic zone relevant beyond the scientific community? *Water* **2019**, *11*, 2230. [CrossRef]

23. Yao, C.; Lu, C.; Qin, W.; Lu, J. Field experiments of hyporheic flow affected by a clay lens. *Water* **2019**, *11*, 1613. [CrossRef]

24. Lewandowski, J.; Meinikmann, K.; Nützmann, G.; Rosenberry, D.O. Groundwater—The disregarded component in lake water and nutrient budgets. Part 2: Effects of groundwater on nutrients. *Hydrol. Process.* **2015**, *29*, 2922–2955. [CrossRef]

25. Tufenkji, N.; Ryan, J.N.; Elimelech, M. The promise of bank filtration. *Environ. Sci. Technol.* **2002**, *36*, 422A–428A. [CrossRef]

26. Sprenger, C.; Lorenzen, G.; Huelshoff, I.; Gruetzmacher, G.; Ronghang, M.; Pekdeger, A. Vulnerability of bank filtration systems to climate change. *Sci. Total Environ.* **2011**, *409*, 655–663. [CrossRef]

27. Hamann, E.; Stuyfzand, P.J.; Greskowiak, J.; Timmer, H.; Massmann, G. The fate of organic micropollutants during long-term/long-distance river bank filtration. *Sci. Total Environ.* **2016**, *545*, 629–640. [CrossRef]

28. Jarosiewicz, A.; Witek, Z. Where do nutrients in an inlet-less lake come from? The water and nutrient balance of a small mesotrophic lake. *Hydrobiologia* **2014**, *724*, 157–173. [CrossRef]

29. Meinikmann, K.; Hupfer, M.; Lewandowski, J. Phosphorus in groundwater discharge—A potential source for lake eutrophication. *J. Hydrol.* **2015**, *524*, 214–226. [CrossRef]

30. Schellenger, F.L.; Hellweger, F.L. Phosphorus loading from onsite wastewater systems to a lake (at long time scales). *Lake Reserv. Manag.* **2019**, *35*, 90–101. [CrossRef]
31. Krause, S.; Heathwaite, A.L.; Miller, F.; Hulme, P.; Crowe, A. Groundwater-dependent wetlands in the UK and Ireland: Controls, eco-hydrological functions and assessing the likelihood of damage from human activities. *Water Resour. Manag.* **2008**, *21*, 2015–2025. [CrossRef]

Article

Quantification of Temporal Variations in Base Flow Index Using Sporadic River Data: Application to the Bua Catchment, Malawi

Laura Kelly [1,*], **Robert M. Kalin** [1], **Douglas Bertram** [1], **Modesta Kanjaye** [2], **Macpherson Nkhata** [2] **and Hyde Sibande** [2]

[1] Department of Civil and Environmental Engineering, University of Strathclyde, Glasgow G1 1XJ, UK; robert.kalin@strath.ac.uk (R.M.K.); douglas.bertram@strath.ac.uk (D.B.)

[2] Ministry of Agriculture, Irrigation and Water Development, Government of Malawi, Tikwere House, City Centre, Private Bag 390, Lilongwe 3, Malawi; modesta.banda@gmail.com (M.K.); macpherson.nkhata@gmail.com (M.N.); hydesibande@yahoo.co.uk (H.S.)

* Correspondence: laura.kelly.100@strath.ac.uk

Received: 30 March 2019; Accepted: 26 April 2019; Published: 29 April 2019

Abstract: This study investigated how sporadic river datasets could be used to quantify temporal variations in the base flow index (BFI). The BFI represents the baseflow component of river flow which is often used as a proxy indicator for groundwater discharge to a river. The Bua catchment in Malawi was used as a case study, whereby the smoothed minima method was applied to river flow data from six gauges (ranging from 1953 to 2009) and the Mann-Kendall (MK) statistical test was used to identify trends in BFI. The results showed that baseflow plays an important role within the catchment. Average annual BFIs > 0.74 were found for gauges in the lower reaches of the catchment, in contrast to lower BFIs < 0.54 which were found for gauges in the higher reaches. Minimal difference between annual and wet season BFI was observed, however dry season BFI was >0.94 across all gauges indicating the importance of baseflow in maintaining any dry season flows. Long term trends were identified in the annual and wet season BFI, but no evidence of a trend was found in the dry season BFI. Sustainable management of the investigated catchment should, therefore, account for the temporal variations in baseflow, with special regard to water resources allocation within the region and consideration in future scheme appraisals aimed at developing water resources. Further, this demonstration of how to work with sporadic river data to investigate baseflow serves as an important example for other catchments faced with similar challenges.

Keywords: baseflow; base flow index; hydrograph; groundwater; Malawi

---

1. Introduction

Understanding temporal variations in baseflow are crucial for sustainable water resources management [1]. Baseflow is defined as the proportion of river flow derived from groundwater and other stored sources [2,3]. Other stored sources may include connected lakes, wetlands, melting snow, temporary storage in the banks of the river channel and slow-moving interflow [4]. Baseflow varies spatially and temporally influenced by several factors including geology, topography, climatic season and anthropogenic activities [5]. Baseflow can sustain river flows during prolonged periods of dry weather. Although dry season flows are significantly reduced and in some rivers approach zero flow, this water can be a vital life source for those who depend on it. Although globally pertinent, it is particularly crucial for semi-arid countries who experience long dry seasons each year [6]. Long term changes in baseflow can indicate unsustainable catchment management practices. Baseflow is thus a key consideration in many sustainable management approaches such as integrated water resources management (IWRM) and conjunctive water use. They are also a major focus of many worldwide initiatives including the United Nation Education Scientific and Cultural Organization (UNESCO) International Hydrological Programme [7]. Subsequently, it can be considered to underpin the United Nations Sustainable Development Goal (SDG) 6 'ensure availability and sustainable management of water and sanitation for all'.

There is a multitude of methods available to investigate baseflow which can be categorized into desk-based methods and field methods. Desk-based methods include hydrograph analysis (baseflow separation [8], frequency analysis [9] and recession analysis [10]), hydrogeological mapping [11], modelling [12] and mass balance [13]). Field methods, as described in Turner [14] include temperature profiling, seepage flux measurement, seepage meters, environmental tracers, artificial tracers, geophysics, remote sensing and ecological indicators. In some countries, however, investigation methods are limited to hydrograph analysis, specifically baseflow separation, which utilizes existing river flow data and provides estimates of baseflow without the need for complex modelling, detailed knowledge of soil characteristics or costly site investigations [15]. Such countries are usually those who experience long dry seasons each year and where baseflow knowledge is perhaps most pertinent. These are also countries often challenged by limited technical knowledge, lack of financial resources and experienced hydrological and hydrogeological staff.

Base flow index (BFI) is an important baseflow characteristic [16]. Originally developed as a parameter to index catchment geology and the ability of a catchment to store and release water, BFI is a numerical representation of the baseflow component of river flow [2]. BFI is calculated as the ratio of the flow under the baseflow hydrograph (the baseflow volume) to the flow under the river hydrograph (total flow volume) as presented in Equation (1) [17]. BFI is applied in hydrology and hydrogeology where it is used as a catchment descriptor in low flow studies [6], a groundwater availability indicator [18], and as a key engineering parameter for environmental flow requirements (EFR), which set a minimum flow required in a river to sustain its ecological health [19]. BFI is a popular means of providing a proxy indicator of groundwater discharge from the aquifer. [4,17,20]. A relative measure with no units, BFI ranges from near 0.0 to 1.0. A BFI close to 0.0 means a river has a low proportion of baseflow, an example would be a flashy river with relatively impermeable geology and little groundwater. A BFI close to 1.0 has a high proportion of baseflow, an example would be a stable river with relatively permeable geology and a lot of groundwater. [6,21]. In periods of dry weather, river flows can be significantly reduced, however, rivers with high BFI indicate that groundwater inflow is sustaining these reduced flows. Many countries and academics are now recognizing the importance of quantifying BFI including a global assessment based on over 3000 catchments worldwide [16], a national scale assessment in New Zealand [22], regional studies such as the Loss Plateau, China [23] and an experimental watershed in the Gulf Atlantic Coastal Plain, USA [4].

Equation (1) base flow index equation:

$$\text{Base Flow Index (BFI)} = \frac{\text{Baseflow volume}}{\text{Total flow volume}} \tag{1}$$

Baseflow is particularly important in Malawi, a semi-arid country known as the warm heart of Africa (Figure 1a). Malawi is rich in both groundwater and surface water resources in comparison to other African countries, however, these are unevenly distributed in time and space. Malawi experiences a distinct dry season each year with minimal to no rainfall. Many rivers still have some flow in the dry season, and it is presumed that they are sustained by baseflow from the region's superficial aquifers. However, anthropogenic activities such as over-abstraction of groundwater and deforestation are threatening flows in Malawi by negatively impacting baseflows. For example, sustained over-abstraction of groundwater can draw down the water table and result in reduced groundwater discharge to any connected rivers. Similarly, deforestation increases overland flows and leaves less water for infiltration and groundwater recharge. This can ultimately lead to reduced water available for groundwater discharge to connected rivers. Although, deforestation is widely reported in Malawi [9,24], there are no published studies confirming the over-abstraction of groundwater. The Ministry of Agriculture, Irrigation and Water Development has reported, based on internal assessments, a decline in groundwater levels and river flows which have resulted in the drying up of major rivers [25].

To date, few studies have been published which investigate baseflow and quantify BFI in Malawi. Preliminary work done by the South Africa FRIEND (flow regimes from international experimental and network data) programme produced an annual BFI map for South Africa which included Malawi however, the project has been inactive for a long time and the data that were collected are largely out of date [15]. More recently, a global BFI study reports estimates of annual BFI for Malawi [16] and the International Water Management Institute's tool; the Global Environmental Flow Information System also includes Malawi and provides estimates of annual baseflow [26]. Studies which are more site-specific, reporting annual BFI include Kumambala [27] who examined four stations along the Shire River in Southern Malawi and Ngongondo [18] who examined the Mulunguzi catchment. Only a few studies identify long term trends in baseflow; Ngongondo [18] identifies a trend in baseflow in the Mulungzui river showing a decline of approximately 50% from 1954 to 1998. In contrast, Kambombe et al. [28] identify an increase in baseflow in the Mulungzui catchment between 1970 and 1999. Kambombe et al. [28] also found a significant decreasing trend in baseflow of the Domasi, Likangala and Thondwe catchments during that period [28]. Further, baseflow is currently evaluated by the Surface Water Division of the Ministry of Agriculture, Irrigation and Water Development (MoAIWD) in Malawi, through use of their time series data management system 'HYDSTRA', however, focus appears to be mainly on annual baseflows. All these studies address BFI to a limited spatial and temporal coverage of flow data, with a focus on annual baseflow values. A gap in the research, therefore, exists to quantify seasonal and long-term trends in BFI for gauged catchments in Malawi.

This task is challenged by the lack of current data as river flow monitoring coverage has declined in Malawi since around 2010 and indeed is representative of sub-Saharan Africa [29]. Further, the data which is available is sporadic in nature, characterized by missing values.

This study demonstrates how to work with sporadic river flow data, using baseflow separation, to produce meaningful estimations on temporal variations in baseflow. We demonstrate this by using the Bua Catchment in Malawi as a case study, whereby the river data is considered representative of the wider Malawi. The objectives of this research were to (1) quantify the annual BFI; (2) quantify the seasonal BFI and (3) identify trends in the BFI. The results will provide important new insights on the behavior of baseflow in the catchment. It will also serve as an example to other catchments challenged by sporadic river data.

This study forms part of on-going research on baseflow in Malawi and has important implications for the sustainable management of water resources in the country. It offers support to the Government of Malawi in their journey towards SDG6 and as such, the research was conducted in a manner that will permit the exchange of knowledge with the water sector.

In Section 2 below, the study area is described in addition to the data and analysis methods. Specifically, the decision procedure for selection of the baseflow separation method and the implementation tool is described and the baseflow separation steps followed are provided. The results and discussion are discussed in Section 3, while the conclusions are summarized in Section 4.

2. Materials and Methods

2.1. Study Area

The Bua river originates on the western border of Malawi and flows in a northeasterly direction through Central Malawi to its outflow into Lake Malawi (Figure 1a). The Bua is joined by five major tributaries (Mphelele, Kasangadzi, Rusa, Ludzi and Namitete) and has numerous minor tributaries. It has a catchment area of 10,658 km^2 which is approximately 186 km in length and its width varies from approximately 87 km in the west to approximately 16 km is the east.

The catchment comprises three distinct hydrological zones; the flat plateau, steep slopes on the highland which rise from the plateau and the rift valley escarpment, and the lakeshore plain [30]. The plateau is generally at 1000–1100 m above sea level (masl). Towards the southwest are the Mchinji mountains which rise to over 1750 masl. Towards the west, where the river meets the lakeshore plain, the catchment drops rapidly through a series of steep slopes. High levels of sedimentation occur at the lakeshore plain as the gradient becomes gentle.

The Bua catchment is assigned Water Resource Area (WRA) 5 within the National Water Resources Master Plan (NWRMP) of Malawi [31]. WRA 5 is subdivided into four water resource units (WRUs) named 5C, 5D, 5E and 5F (Figure 1a). Both WRAs and WRUs are based on river basin boundaries. WRA 5 lies within the administrative districts of Mchinji, Kasungu, Nkhotakota, Lilongwe, Dowa and Ntchisi.

Land use in WRA 5, as shown in Figure 1c, mainly comprises cropland; arable agriculture of mainly maize crops and tobacco, and forest land; including Mchinji Forest Reserve and Kasungu National Park to the west, and Nkhotakota Game Reserve to the far east [32]. Wetlands or dambos are also scattered throughout the catchment. These wetlands become saturated in the wet season and provide a good source of water in the dry season [33]. The dambos are generally considered to drain the plateau area [30].

The climate of WRA 5 can be generally represented as sub-tropical [31]. The climate is divided into three weather variations; the warm wet season (1 November–30 April); the cool dry season (1 May–31 August); and the hot dry season (1 September–31 October), however, it's generally accepted to be bimodal referring to the wet season and the dry season [31]. Over 95% of the annual rainfall falls in the warm wet season or rainy season. The exact length of the wet season varies depending on the location within Malawi, reported to end in March in the south of the country, and April/May in the north [34]. No average annual rainfall or temperature values were available for the wet and dry season. The average annual rainfall for WRA 5 is 897 mm, with a range of 800–1000 mm [33]. The average annual temperature in WRA 5 ranges from 20 to 24 °C [33].

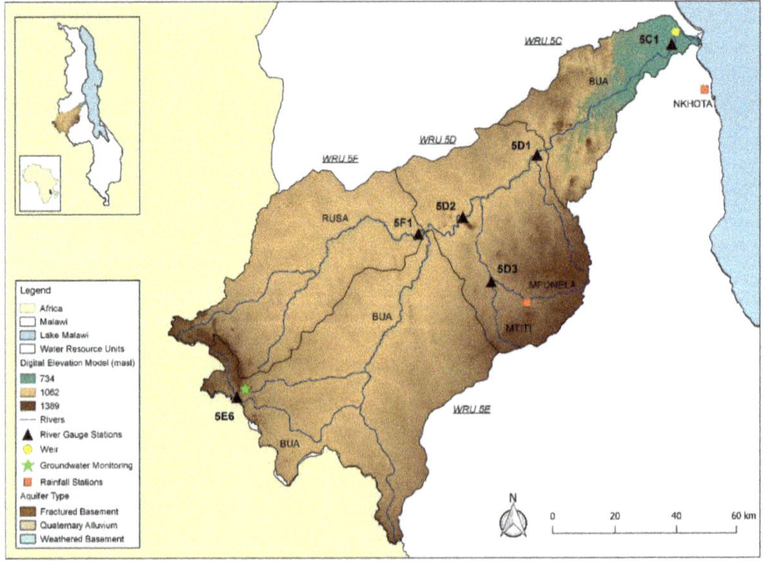

(a)

Figure 1. *Cont.*

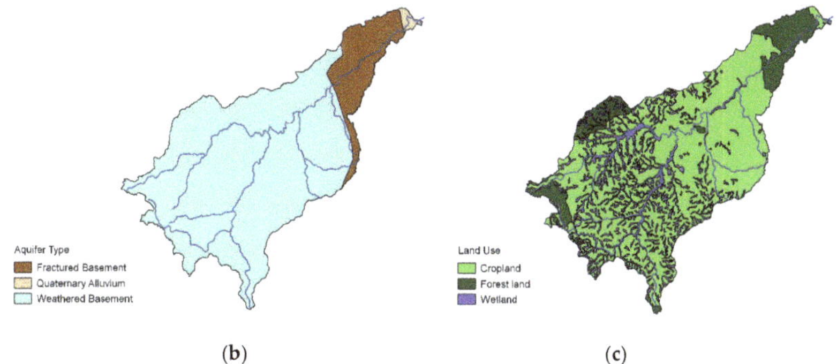

(b)   (c)

Figure 1. (**a**) Location of Malawi in Africa (insert), location of the Bua catchment in Malawi (insert) and digital elevation model of the Bua catchment (WRA 5) with rivers, river gauges, weir, rainfall stations and groundwater monitoring; (**b**) aquifer type map [35]; (**c**) land use map [32].

Malawi's groundwater occurrence is classified into three hydrogeological domains or aquifer types; (1) alluvial aquifers, (2) sedimentary aquifers and (3) basement aquifers. The sedimentary aquifers are subdivided into semi consolidated and consolidated aquifers, and the basement is subdivided into weathered and fractured aquifers [31]. Figure 1b shows the aquifer types in WRA 5. The basement aquifer considered one of Malawi's major aquifers underlays most of the Bua catchment [35]. In the west, the weathered basement is present over most of the plateau area and fractured basement occurs in the area of the Mchinji Forest Reserve. In the east lakeshore plain, the basement is overlain by the other major aquifer type, the alluvium aquifer with some pockets of the fractured basement also present [35]. There has been little published work on the hydrogeology and soils of the Bua catchment, however, a bulletin from 1983 presents details of the soil's patterns and the hydrogeological conditions of the weathered basement aquifer of a plateau area [30].

2.2. Data

This study focused on data from six river gauges within the Bua catchment (Table 1). Other gauges do exist, however, there was no data available for them. Four gauges monitor the main Bua river (5C1, 5D1, 5D2 and 5E6) which is a regulated river with a weir located downstream of gauge 5C1. Photos of the weir taken January 2019 and are provided in the Supplementary Material, Figure S1. The Rusa river, a major tributary of the Bua, is monitored by a fifth gauge, 5F1, and the Mtiti river (a tributary to the Kasangadzi river) is monitored by the final gauge, 5D3. Daily flow rate data were available for each gauge as follows; 5C1 (1957–2009), 5D1 (1958–2007), 5D2 (1953–2007), 5D3 (1958–2003), 5E6 (1970–2008) and 5F1 (1964–2005). Data coverage appears substantial ranging from 38–52 years, however, it is expected to have missing values throughout. Data were obtained from the Surface Water Division of the Department of Water Resources of Malawi.

Where possible, rainfall and groundwater data in the vicinity of the river gauges were also examined to provide support for the BFI analysis. Daily rainfall data for Nkhota station (18 km away from gauge 5C1 in a southeasterly direction) and Mponela station (13 km away from 5D3 in a southwesterly direction) were used. Stations are managed by the Department of Meteorological Services who provided the data. The rainfall data is of very good coverage with minimal missing values. Groundwater levels are monitored in WRA 5 via four monitoring boreholes, constructed around 2009/2010. The boreholes are managed by the Groundwater Division of the Department of Water Resources who provided the data. Only one of the monitoring boreholes, at Mchinji Water Office (GN196), had enough data coverage (2009–2013) to examine.

Table 1. Evaluation of baseflow separation tools against required criteria.

Require Criteria/Baseflow Separation Tools	Flow Screen R	FORTRAN BFI	SWAT	WEST Pro	BFlow	HYSEP	HydroClimATe	SAAS	RAP	WHAT	BFI+ 3.0	BFI Programme
Automated	Y	Y	Y	Y	Y	Y	Y	Y	Y	Y	Y	Y
Easily accessible	Y	N	Y	N	N	Y	Y	Y	Y	Y	Y	Y
Free to obtain and operate	Y	-	Y	-	-	Y	Y	Y	Y	Y	Y	Y
Requires minimal training to use	N	-	N	-	-	N	N	Y	N	Y	Y	Y
Can select seasonal periods	-	-	N	-	-	-	-	N	Y	N	N	Y

Where: Y = yes; N = No.

2.3. Decision Procedure for Selection of Baseflow Separation Method and Implementation Tool

Baseflow separation was selected to analyze the river data and determine BFI. Baseflow separation is categorized into graphical methods which are performed manually, and filtering methods which are automatically performed by a computer. [36]. There are a wide variety of filtering methods available, and a significant number of computer programs to implement the chosen method [37]. Although there is subjectivity involved in selecting an appropriate filtering method and an associated tool to implement it, merit holds in use of any of them as long as the use is consistent throughout the study [6,15,37]. The decision to select a filtering method and implementation tool is generally based on the criteria required for the study.

In this study, the selection of an appropriate implementation tool took precedence over the selection of a filtering method. The tool was required to meet certain criteria to allow the exchange of knowledge with the Government of Malawi. The tool needed to be automated, easily accessible, free to obtain and operate, require minimal training to use and capable of selecting seasonal periods from input data to quantify BFI. Several tools were evaluated against the required criteria including Flow Screen package for R [38], Formula Translation (FORTRAN) BFI program [39], Soil and Water Assessment Tool (SWAT) [40], Water Engineering Time Series PROcessing Tool (WEST PRO) [41], web-based BFlow [42], HYSEP [43], HydroClimATe: hydrologic and climate analysis toolkit [44], Streamflow Analysis and Assessment Software (SAAS) [45], River Analysis Package (RAP) [46], Web-based Hydrograph Analysis Tool (WHAT) [47], BFI + 3.0 of Hydro Office [48] and the BFI programme [6]. The evaluation assessment is presented in Table 1. As the BFI Programme [6] met all of the criteria it was selected for analysis.

The BFI programme is an excel based tool developed by Martin Morawietz at the Department of Geosciences in the University of Oslo, Norway. It was originally prepared for the textbook; Hydrological Drought-Processes and Estimation Methods for Streamflow and Groundwater [6]. It is free to download on the European Drought Centre website http://europeandroughtcentre.com/. The textbook provides working examples of how to use the tool. The tool implements the filtering method called the 'smoothed minima procedure' [21]. It uses smoothing and separation techniques to process a river hydrograph. Daily river flow data is partitioned into 5-day increments and the minimum flow in each period is identified [49]. Turing points are identified in the series of minimum flows and connected to draw the baseflow hydrograph. The precise details of the procedure are provided in the Low Flow Studies Report No 3 by the Institute of Hydrology [21] and by Wahl [39].

2.4. Baseflow Separation Steps

The raw river data were screened prior to baseflow separation to identify the periods of missing data. Before proceeding to analysis, there were two options available to deal with the missing data; (1) infill the missing data or (2) ignore the missing data and analyze only the raw data. Although there are merits to infilling data [50,51], most studies agree with the recommendation by Ladson et al. [52] that BFI should be determined from raw data only [20,23,53,54]. As such, this study did not infill data and analyzed the raw river flow data only. To do this, the flow data were prepared by dividing into periods of non-missing values [52].

The assessment periods selected were annual and seasonal periods defined by months. The annual period was taken as the hydrological year in Malawi as used by the Government of Malawi Water Resources Department and coincides with the start of the wet season and runs to the end of the dry season (1 November–31 October). The seasonal periods selected were the wet season defined as 1 November–30 April, and the dry season defined as 1 May–31 October. These periods are based on the weather variations recognized in Malawi and used in water resources assessments by the Water Resources Department [55] and the country's national irrigation master plan and investment framework [56].

The following steps were taken to perform the baseflow separation using the BFI programme:

(1) The baseflow separation was performed for each year of river data (1957–2009) producing a separate annual BFI value for each year where there was enough data in the period. It is commonly recommended in the literature to determine the long-term BFI which uses all the data successively [6,15], however here, it was not possible due to missing data. The mean annual BFI was therefore determined based on the individual years;

(2) The baseflow separation was performed for each season of data (1957–2009) in the same manner as the annual period described above;

(3) The total flow, baseflow and surface runoff flow from each baseflow separation were summed for each period;

(4) Descriptive statistics (average, maximum and minimum, standard deviation and coefficient of variation) were determined for the annual and seasonal periods.

2.5. Statistical Trend Analysis

The non-parametric Mann-Kendall (MK) statistical test [57,58] was used to identify if the BFI results had statistically significant increasing or decreasing trends. The test is prominently used in hydrology studies. For example, it is popular when identifying trends in streamflow [50,59–61], baseflow [50], BFI [4,62,63] and the vertical exchange fluxes between streambeds and connected aquifers [64]. It is also widely applied in identifying trends in rainfall [59,60]. Application of non-parametric testing is appropriate due to hydrological data not being normally distributed [61]. One of the main advantages of the MK test is that it is insensitive to missing data, which was a key challenge with the data in this study.

The hypothesis for the test, H0, was defined as 'there is no trend in the data', and the alternative hypothesis, Ha, was defined as 'there is a trend in the data'. If the *p*-value calculated was lower than the significance level, the H0 was rejected and the alternative Ha accepted, and a trend was indicated. If the *p*-value was greater than the significance level, no trend was indicated. The significance level is referred to as a Type 1 error and is the probability of rejecting the null hypothesis when it is true [61]. The direction of the trend was indicated by the test statistic, S, where a negative S value indicates a declining trend and a positive S value indicates an increasing trend. Details of the MK equations can be found in the literature [58].

The selection of the test parameters is important in statistical testing as they have a direct impact on the resulting trend. In this study, the following parameters were selected for the MK test; the 'exact p' method was used, the significance level was set to 0.01 (or 1%) and the equations were set to ignore missing data. Further, the 'normal' MK test was selected over the 'seasonal' MK test. Due to the decision not to infill data in this study, the BFI data was partitioned into annual and seasonal periods and as such the normal MK was applicable. If the data had been infilled, and there was no need to partition the data, the use of the seasonal MK test would have allowed comparison of the seasonal periods. The statistical programme XLSTAT, available at www.xlstat.com was used to perform the MK test [65].

3. Results and Discussion

3.1. Annual and Seasonal BFI Analysis Coverage

Annual and seasonal BFI was calculated for gauges 5C1, 5D1, 5D2, 5D3, 5E6 and 5F1. The results of the analysis for 5C1 are presented in Table 2 and the results of the other gauges are presented in Tables S1–S5.

Table 2. Results of the annual and seasonal BFI (base flow index) analysis (tabular) for the Bua River, gauge station 5C1, 1957–2009 (52 years).

Period	Annual BFI	Wet Season BFI	Dry Season BFI	Period	Annual BFI	Wet Season BFI	Dry Season BFI
1957/1958	-	-	0.94	1983/1984	-	-	-
1958/1959	0.66	0.65	0.85	1984/1985	-	-	-
1959/1960	0.53	0.48	0.96	1985/1986	-	0.80	-
1960/1961	-	0.44	-	1986/1987	0.81	0.80	0.99
1961/1962	0.83	0.81	0.91	1987/1988	0.62	0.58	0.95
1962/1963	-	-	0.99	1988/1989	-	-	-
1963/1964	0.77	0.75	0.98	1989/1990	0.77	0.75	0.92
1964/1965	0.79	0.77	0.96	1990/1991	0.76	0.74	0.97
1965/1966	-	0.69	-	1991/1992	0.43	0.41	0.87
1966/1967	0.48	0.40	0.94	1992/1993	-	0.50	-
1967/1968	0.58	0.54	0.83	1993/1994	-	-	0.95
1968/1969	-	-	0.81	1994/1995	0.60	0.60	0.91
1969/1970	-	-	-	1995/1996	0.54	0.53	0.84
1970/1971	-	-	-	1996/1997	0.76	0.75	0.89
1971/1972	-	0.64	-	1997/1998	0.90	0.90	0.87
1972/1973	-	0.47	-	1998/1999	0.76	0.74	0.92
1973/1974	0.68	0.62	0.94	1999/2000	0.75	0.73	0.87
1974/1975	0.72	0.72	0.99	2000/2001	-	-	0.95
1975/1976	0.69	0.61	0.95	2001/2002	0.94	0.88	0.98
1976/1977	0.81	0.77	0.99	2002/2003	-	0.85	-
1977/1978	-	-	0.91	2003/2004	-	-	0.99
1978/1979	0.80	0.76	0.99	2004/2005	0.84	0.82	0.92
1979/1980	-	0.65	-	2005/2006	0.90	0.82	0.98
1980/1981	0.75	0.71	0.99	2006/2007	0.87	0.81	0.96
1981/1982	-	-	-	2007/2008	0.92	0.87	0.99
1982/1983	-	0.64	-	2008/2009	0.88	0.81	0.99

- denotes no BFI determined due to insufficient data in that period.

As expected, the river data was characterized by missing values and this was seen across all datasets. This meant it was not possible to determine a BFI for all periods. To quantify the coverage of analysis, the number of periods for which a BFI was determined was counted and converted to a percentage based on the number of years of data (Table 3). For example, for 5C1, a BFI was determined for 30 full annual data periods, 39 wet seasons, and 37 dry seasons which equates to 58%, 75% and 71 % coverage for the respective periods. Data for each gauge ranged from 38 to 52 years and the percentage of coverage for each period (annual, wet and dry season) was consistently over 50%, with some periods as high as 80% coverage (Table 3). The results show, despite the sporadic nature of river flow data in Malawi, that such datasets can be analyzed to extract observations on baseflow. This is an important finding for Malawi and countries which hold similar datasets. They can begin to utilize such datasets and assess baseflow using minimal labor and financial resources.

Table 3. Percentage of data coverage in annual and seasonal BFI analysis for the gauges in WRA 5.

Gauge ID	River Name	Period of Data Coverage	No of Years of Available Data; No of Annual, Wet Season, Dry Season Periods with Data	Annual	Wet Season	Dry Season
5C1	Bua	1957–2009	52; 30, 39, 37	58%	75%	71%
5D1	Bua	1958–2007	49; 25, 29, 31	51%	59%	63%
5D2	Bua	1953–2005	52; 34, 42, 35	65%	81%	67%
5D3	Mtiti	1958–2003	45; 27, 30, 36	60%	67%	80%
5E6	Bua	1970–2008	38; 23, 27, 26	61%	61%	68%
5F1	Rusa	1964–2005	41; 24, 28, 27	59%	68%	66%

3.2. Average Annual BFI

Average annual BFI for the gauges were determined based on the BFI analysis results in Section 3.1. The results are presented in Figure 2 and Table 4.

Traditionally BFI has been determined on an annual basis. This study found high average annual BFIs for the gauges located on the lower elevation reaches of the Bua; 0.74 for 5C1, 0.75 for 5D1 and 0.76 for 5D2. This indicates that the river has a moderately high baseflow component of approximately 74–76% of the total annual river flow in the lower catchment. This finding is consistent with the annual BFI of 0.71 for 5C1 and 0.86 for 5D1 sourced from the HYDSTRA system in use by the Malawi Surface Water Division [33]. It also matches BFIs reported by Smith-Carington [30] of 0.85 (5D1) and 0.86 (5D2). Previous studies by UNESCO [15] and Beck et al. [16] reported similar annual BFI for Malawi in the range of 0.6 to 0.7 and 0.6 to 0.8 respectively. A moderately high baseflow was also found for 5F1 on the Rusa with a BFI of 0.80, or 80% of the total annual river flow which compares with a BFI of 0.81 from HYDSTRA.

In contrast, lower BFI values were found for the gauges located at higher elevations in the catchment. A BFI of 0.54 was found for 5E6, the highest gauged reach of the Bua. This doesn't match the BFI of 0.74 found from HYDSTRA. Finally, 5D3 on the Mtiti found a BFI of 0.48. There was no BFI available from HYDSTRA. Comparisons are provided for context only, it is important to bear in mind, that it's not generally recommended to compare BFIs across studies as different baseflow separation techniques and different data lengths will produce different baseflow volumes and this will affect the BFI [52]. Based on this study's annual average values, the Bua, the Rusa and the Mtiti rivers are considered perennial in nature with a stable flow regime.

3.3. Average Seasonal BFI (Wet and Dry Season)

Recent studies in BFI have sought to make seasonal adjustments, appreciating the variations that occur in baseflow both temporally and spatially and that annual BFI may not represent the true picture [66]. This study presents the first findings on seasonal BFI in the Bua catchment. Average seasonal BFI for the gauges was determined based on the BFI analysis results in Section 3.1. The results are presented in Figure 2 and Table 4.

For all gauges assessed, the results found minimal difference between the annual and the wet season BFI, however, in the dry season, all BFIs increased to over 0.80 (or 80% of the dry season flow was attributed to baseflow) as shown in Figure 2 and Table 4. For example, 5C1 had a BFI of 0.69 in the wet season increasing to 0.94 in the dry season. The increase in dry season BFI is indicative of the catchment geology. As mentioned in the literature, a high BFI indicates permeable catchment conditions whereby the catchment is storing water during the wet season and discharging it to the river during the dry season [6,17]. To support these BFI findings, it would have proved useful to compare river levels to groundwater levels near each gauging station. Unfortunately, however, of the groundwater data available there was none suitable for such a comparison.

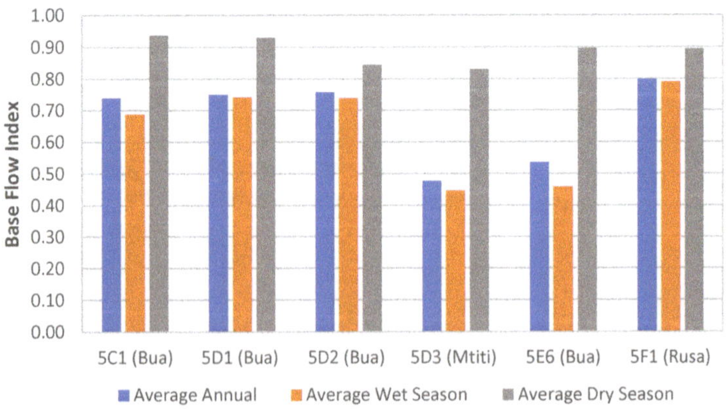

Figure 2. Results of annual and seasonal BFI analysis for the gauges in WRA 5 (graphical).

Table 4. Results of annual and seasonal BFI analysis for the gauges in WRA 5 (tabular).

Gauge ID (River)	5C1 (Bua)	5D1 (Bua)	5D2 (Bua)	5D3 (Mtiti)	5E6 (Bua)	5F1 (Rusa)
Data record	1957–2009	1958–2007	1953–2005	1958–2003	1970–2008	1964–2005
ANNUAL						
Average BFI	0.74	0.75	0.76	0.48	0.54	0.80
Minimum Average BFI	0.43	0.43	0.11	0.05	0.37	0.26
Maximum Average BFI	0.94	0.94	0.98	0.84	0.70	0.98
Standard Deviation	0.13	0.17	0.24	0.28	0.09	0.18
WET SEASON						
Average BFI	0.69	0.74	0.74	0.45	0.46	0.46
Minimum Average BFI	0.40	0.41	0.11	0.05	0.25	0.25
Maximum Average BFI	0.90	0.93	0.98	0.77	0.90	0.90
Standard Deviation	0.14	0.17	0.22	0.26	0.13	0.13
DRY SEASON						
Average BFI	0.94	0.93	0.84	0.83	0.90	0.89
Minimum Average BFI	0.83	0.55	0.55	0.00	0.47	0.61
Maximum Average BFI	0.99	1.00	1.00	1.00	0.98	1.00
Standard Deviation	0.05	0.11	0.11	0.23	0.12	0.10

Interestingly, from the wet season BFI results (Figure 2), there are two gauges which don't follow the high BFI seen in the other gauges; the 5D3 (Mtiti) and the 5E6 (Bua). The lower wet season BFI of these gauges can be attributed to the spatial variations in geology and topography which control baseflow. For example, both gauges are located at elevations of 1200masl, compared to the much lower elevations of 550–1000 masl for the other gauges. 5E6 is located on the headwaters of the Bua and drains the entirety of the Mchinji Forest Reserve (Figure S3) and 5D3 drains part of the Dowa Hills (Figure S4).

There is considerable variability seen across all gauges in the BFI within the annual and wet season periods shown by the minimum and maximum BFIs (Table 4). The coefficient of variation (CV) of the dry season BFI was low, compared to the annual and wet season BFI which was, as expected, much larger. For example, at gauge 5C1, the dry season CV was 6%, compared to the annual CV of 18%, and the wet season of 20%. This difference in variability highlights the varying behavior of baseflow. As mentioned in the literature, BFI is used in hydrology and hydrogeology in a range of applications [6,18,19]. Where there are variations between annual and seasonal values, as seen in these results, it is important to use the appropriate value as the use of an incorrect BFI could lead to inaccurate assessments. Several future scheme appraisals in Malawi would benefit from considering the seasonal BFI results of this study. For example, previous assessments for new investments in Malawi's water sector, which have taken account of EFRs and thus BFI values. The Water Resources Investment Strategy (WRIS) project, under the National Water development Program (NWDP), produced water resource assessments for the 17 WRAs in Malawi, including WRA 5 [33]. The project produced estimates for potential abstractable groundwater and sustainable surface water yield. Further, the National Irrigation Master Plan and Investment Framework (2014–2035), which sets out new investments for expansion of the irrigation sector in Malawi, is also centered around EFR, with one new dam proposed in the lower Bua catchment. It is presumed that these estimations have used annual BFI values which may lead to overestimation of available water resources. Seasonal variations are evidenced in this study and should be considered.

3.3.1. River Flow, Rainfall and Groundwater Patterns

Examining rainfall, river and groundwater patterns support the variation in wet and dry season BFI found above. For example, river flow and rainfall patterns for gauge 5C1 are shown in Figure 3. The baseflow separation divided the daily river flow into its daily baseflow and daily surface runoff components for each annual and seasonal period. Average monthly values for each flow component were determined for the years with no missing data; 30 in total. Figure 3 shows the average monthly flow volumes for the Bua and the average monthly rainfall volumes for Nkhota station. The observed river flow and rainfall patterns highlight the distinct wet and dry season pattern recognized in Malawi

Rainfall is high during the wet season (November–April) and in response the total river flow volume and the direct runoff increases. The baseflow also increases but to a much lesser extent. River flows start to decrease after the peak river discharge in March. During the dry season (May–October), rainfall and direct runoff are reduced to a minimum. However, the baseflow remains relatively stable and sustains the river. The ratio of baseflow to total river flow is much higher in the dry season than in the wet season, thus resulting in a higher BFI. This pattern is considered generally representative of the other gauges in the catchment.

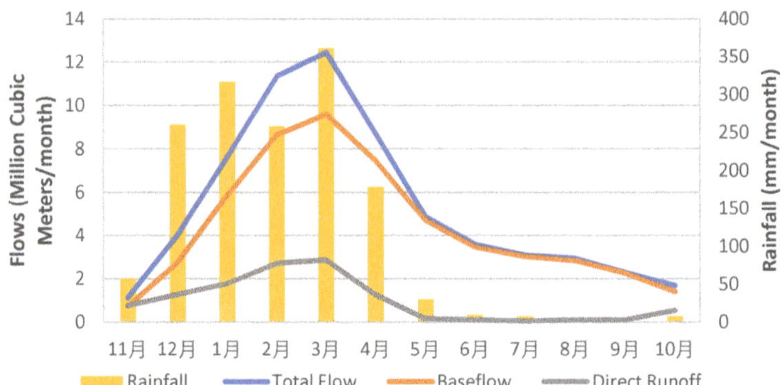

Figure 3. Average monthly flow volumes (total flow, baseflow, direct runoff), for the Bua River, Gauge 5C1, 1957–2009. Rainfall data for Nkhota station, 1960–2009.

There was not enough groundwater monitoring data available in the vicinity of gauge 5C1 for analysis. However, groundwater monitoring data at Mchinji Water Office (2009–2013), located 2 km from gauge 5E6 and at the same topographical elevation did have enough data. The data shows seasonal fluctuations in groundwater levels in line with the rainfall and river patterns above (Figure 4).

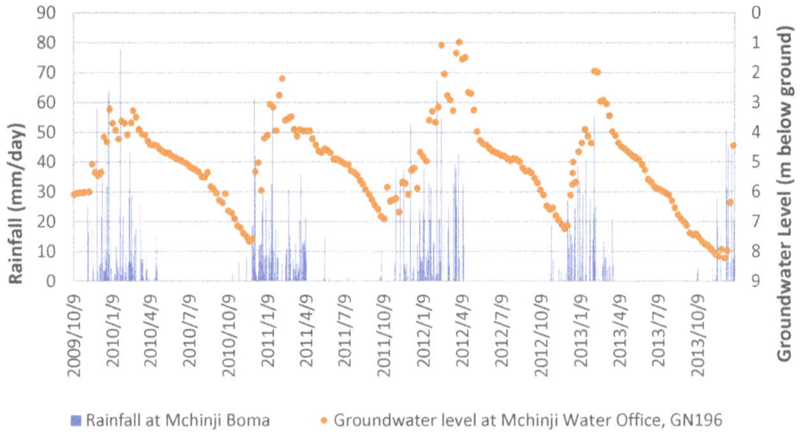

Figure 4. Daily rainfall (at Mchinji Boma) and sporadic groundwater levels (at Mchinji Water Office, GN196), located 2 km from the Bua River, Gauge 5E6, 2009–2013.

3.3.2. Comments on the Source of Baseflow

The baseflow separation approach used in this study assumes that baseflow is derived entirely from groundwater discharge from the aquifer, however, other stored sources can also contribute. The true source of baseflow is impossible to distinguish from baseflow separations alone and would require detailed site investigations to map each flow path [2]. It may be useful to provide some comments on the expected source of baseflow.

Based on the presence of aquifers identified through the literature and geological maps, we conceptualize that groundwater discharge from the local aquifers is the main contributor to baseflow during the wet and dry seasons. For example, an alluvium aquifer is present in the downstream reach of the Bua (5C1) and fractured basement dominates the entire upper catchment of the Bua (5E6) presenting good conditions for water to discharge to the river. Weathered basement aquifers underlay much of the middle reaches (5D1 and 5D2) and may contain pockets of perched aquifers. Further, interflow is expected to contribute to baseflow across all gauges during the wet season, though will not be a major source in the dry season. Finally, Dambos will also contribute to baseflow during the wet season and at the beginning of the dry season. Water is temporarily stored in the dambos and released slowly at the beginning of the dry season whereby it discharges to the river. Once the dambos have drained, the baseflow is maintained entirely from groundwater from the aquifers [30]. Dambos are present in much of the plateau area and the Rusa catchment (5F1) and have been previously identified as contributing to baseflow in the middle reaches of the Bua (5D1 and 5D2) [30].

3.4. Long Term Behavioral Changes in BFI—Statistical Trend Results

Detecting trends in BFI can help us understand the possible links between hydrological processes, anthropogenic activities and environmental changes. The MK test was used to identify increasing or decreasing statistically significant trends in the BFI results obtained in Section 3.1. The MK results are presented in Table 5. This study presents the first findings on detecting trends in BFI in the Bua catchment.

Table 5. Mann Kendall statistical results for BFI for gauges in WRA 5.

Gauge ID (River)	5C1 (Bua)	5D1 (Bua)	5D2 (Bua)	5D3 (Mtiti)	5E6 (Bua)	5F1 (Rusa)
Data record	1957–2009	1958–2007	1953–2005	1958–2003	1970–2008	1964–2005
ANNUAL						
MK Statistic 'S'	151	−166	−107	125	−90	−29
Trend (1% sig. level)	Increasing	Decreasing	No trend	Increasing	No trend	No trend
WET SEASON						
MK Statistic 'S'	241	−214	−188	161	−102	−50
Trend (1% sig. level)	Increasing	Decreasing	Decreasing	Increasing	No trend	No trend
DRY SEASON						
MK Statistic 'S'	62	−142	−82	16	4	−17
Trend (1% sig. level)	No trend	No trend	No trend	No trend	No trend	No trend

An increasing trend in BFI in the annual and wet season data was found at 5C1 (Bua) and 5D3 (Mtiti), however, no trend was found in the dry season data. Increases in baseflow have previously been linked to increases in groundwater levels as a result of prolonged increases in rainfall [3]. However, no trends in rainfall were detected in the annual, wet or dry season data from nearby rainfall stations; Nkhota station (close to 5C1) from 1960–2009, and Mponela station (close to 5D3) from 1960–2003 (Table S6). In contrast, a decreasing trend in BFI for the annual and wet season data was found for 5D1 (Bua) and 5D2 (Bua), however, no trend was found in dry season data. Decreases in BFI could be linked to prolonged over-abstraction of groundwater. Declining groundwater levels have been reported in Malawi; however, sparse monitoring of groundwater levels lends to lack of evidence of such trends. The natural vegetation of the plateau area was reported as Miombo woodland but had been cleared for cultivation in the 1980s which may have resulted in major changes to the hydrological cycle [30].

Interestingly, 5E6 (Bua) and 5F1 (Rusa) showed no trends in BFI for the annual, wet season or dry season data. The stability of the BFI here suggests that the systems are in balance, and the baseflow to the river has remained stable over the assessment period; 1970–2008 and 1964–2005 respectively. It may indicate minimal impact to groundwater levels in the area and a well-managed catchment. This is perhaps also true of 5E6 which drains the Mchinji Forest Reserve and can be expected to have minimal impacts from human activities.

These findings suggest that long term behavioral changes have occurred in the annual and wet season baseflow at several gauges in the Bua catchment as described above. Based on the tests being conducted at a significance level of 1%, there is a 1% risk of being wrong or a confidence level of 99% in the results. The trend results should, however, be interpreted with caution as further work is recommended to quantify the magnitude of the trends and examine potential drivers for such changes in baseflow behavior [61].

The above resultsprovide new evidence of temporal variations in baseflow in the Bua catchment. This will be of interest to the new National Water Resources Authority within the Malawi Government for catchment planning.

4. Conclusions

The main aim of this study was to demonstrate, using a case study, how to use sporadic river datasets to produce meaningful observations on temporal variations in baseflow. The findings can be summarized in terms of their contribution to knowledge.

4.1. Catchment Originality

This is the first study to quantify temporal variations in baseflow in the Bua catchment. Annually, average BFIs > 0.74 were found for gauges in the lower reaches of the catchment, with lower BFIs < 0.54 found for gauges in higher reaches. Seasonally, minimal difference was found between the annual and wet season BFI, however, baseflow increased in the dry season across all gauges with BFI all found to be >0.80. Long term trends were found in the annual and wet season BFI indicating behavioral changes in baseflow have occurred within the catchment. No trend was found in the dry season BFI. The source of baseflow is expected to be mainly groundwater discharge from the aquifers underlain the rivers, however, interflow and dambo storage may also play a role. An implication of these findings is that temporal variations in baseflow should be considered in future scheme appraisals in the catchment such as the proposed irrigation infrastructure. Further, the results should be included in catchment management plans set by the new National Water Resources Authority within the Malawi Government, to inform the seasonal allocation of water resources in the catchment.

4.2. Generic Relevance to the Reader and the Wider Research Community

Apart from the Bua catchment case study, this article serves as an important example for other gauged catchments in Malawi, and indeed other countries, which are required to assess variations in baseflow to underpin IWRM and SDG 6, but are faced with similar challenges of sporadic river data. Further research is now needed to quantify temporal variations in baseflow for all gauged catchments in Malawi. Our on-going baseflow research seeks to do this by using the approach demonstrated in this study.

Supplementary Materials: The following are available online at http://www.mdpi.com/2073-4441/11/5/901/s1; Supplementary Material word document containing the following Figures and Tables. **Figure S1**. Pictures of the weir located on the Bua river downstream of gauge 5C1, taken January 2019 by Oliver Phiri; **Figure S2**. Results of the annual and seasonal BFI analysis (graphical) for the Bua River, gauge station 5C1, 1957–2009 (52 years); **Figure S3**. River gauge 5E6 on the Bua river, draining Mchinji Forest Reserve, Google Earth Image, February 2019; **Figure S4**. River gauge 5D3 on the Mtiti river, draining part of the Dowa Hills, Google Earth Image, February 2019; **Table S1**. Results of the annual and seasonal BFI analysis (tabular) for the Bua river, gauge station 5D1, 1958–2007 (49 years); **Table S2**. Results of the annual and seasonal BFI analysis (tabular) for the Bua river, gauge station 5D2, 1953–2005 (52 years); **Table S3**. Results of the annual and seasonal BFI analysis (tabular) for the Mtiti river, gauge station 5D3, 1958–2003 (45 years); **Table S4**. Results of the annual and seasonal BFI analysis (tabular) for the Bua river, gauge station 5E6, 1970–2008 (38 years); **Table S5**. Results of the annual and seasonal BFI analysis (tabular) for the Rusa river, gauge station 5F1, 1964–2005 (41 years); **Table S6**. Mann Kendall statistical results for rainfall stations in WRA 5; Nkhota (1960–2009) and Mponela (1960–2003).

Author Contributions: Conceptualization, L.K. and R.M.K.; Formal analysis, L.K.; Funding acquisition, R.M.K.; Methodology, L.K.; Resources, M.K., M.N. and H.S.; Supervision, R.M.K. and D.B.; Validation, L.K., R.M.K. and D.B.; Visualization, L.K.; Writing—original draft, L.K.; Writing—review & editing, L.K., R.M.K., D.B., M.K. and M.N.

Funding: This research was funded by the Scottish Government under the Scottish Government Climate Justice Fund Water Futures Programme, research grant HN-CJF-03 awarded to the University of Strathclyde (R.M. Kalin).

Acknowledgments: The authors would like to gratefully acknowledge our partners the Malawi Government, in particular, the Surface Water Division and the Groundwater Division of the Department of Water Resources, and the Department of Meteorological Services for providing data for this study. A specific thank you to those who helped to facilitate and process the data requests; Jolamu Nkhokwe, Grey Muthali, Yobu Ezra Kachiwanda and Adams Chavula from the Department of Meteorological Services, Piasi Kaunda from the Surface Water Division.

Conflicts of Interest: The authors declare no conflict of interest. The funders had no role in the design of the study; in the collection, analyses, or interpretation of data; in the writing of the manuscript, or in the decision to publish the results.

References

1. Brodie, R.; Sundaram, B.; Tottenham, R.; Hostetler, S.; Ransley, T. *An Adaptive Management Framework for Connected Groundwater-Surface Water Resources in Australia*; Bureau Rural Sciences: Canberra, Australia, 2007.
2. International Commission on Groundwater. Surface Water and Groundwater Interaction. In *A Contribution to the International Hydrological Programme*; The United Nations Educational, Scientific and Cultural Organization (UNESCO): Paris, France, 1980.
3. Fetter, C.W. *Applied Hydrogeology*, 4th ed.; Lynch, P., Ed.; Prentice Hall Inc.: Upper Saddle River, NJ, USA, 2001.
4. Bosch, D.D.; Arnold, J.G.; Allen, P.G.; Lim, K.-J.; Park, Y.S. Temporal variations in baseflow for the Little River experimental watershed in South Georgia, USA. *J. Hydrol. Reg. Stud.* **2017**, *10*, 110–121. [CrossRef]
5. Smakhtin, V.U. Low flow hydrology: A review. *J. Hydrol.* **2001**, *240*, 147–186. [CrossRef]
6. Tallaksen, L.M.; Van Lanen, H.A. *Hydrological Drought: Processes and Estimation Methods for Streamflow and Groundwater*; Elsevier: Amsterdam, The Netherlands, 2004; Volume 48.
7. International Hydrological Programme of UNESCO. *Groundwater Resources Assessment under the Pressures of Humanity and Climate Changes GRAPHIC*; UNESCO: Paris, France, 2006.
8. Mei, Y.; Anagnostou, E.N. A hydrograph separation method based on information from rainfall and runoff records. *J. Hydrol.* **2015**, *523*, 636–649. [CrossRef]
9. Chimtengo, M.; Ngongondo, C.; Tumbare, M.; Monjerezi, M. Analysing changes in water availability to assess environmental water requirements in the Rivirivi River basin, Southern Malawi. *Phys. Chem. Earth Parts A/B/C* **2014**, *67*, 202–213. [CrossRef]
10. Tallaksen, L. A review of baseflow recession analysis. *J. Hydrol.* **1995**, *165*, 349–370. [CrossRef]
11. Bloomfield, J.; Allen, D.; Griffiths, K. Examining geological controls on baseflow index (BFI) using regression analysis: An illustration from the Thames Basin, UK. *J. Hydrol.* **2009**, *373*, 164–176. [CrossRef]
12. Rassam, D.W.; Werner, A. *Review of Groundwater-Surfacewater Interaction Modelling Approaches and Their Suitability for Australian Conditions*; eWater Cooperative Research Centre: Canberra, Australia, 2008.
13. Capesius, J.P.; Arnold, L.R. *Comparison of Two Methods for Estimating Base Flow in Selected Reaches of the South Platte River, Colorado*; US Geological Survey: Reston, VA, USA, 2012.
14. Turner, J.V. *Estimation and Prediction of the Exchange of Groundwater and Surface Water: Field Methodologies, eWater Technical Report*; eWater Cooperative Research Centre: Canberra, Australia, 2009.
15. *UNESCO Southern Africa FRIEND IHP-V Project 1.1 Technical Documents in Hydrology No.15*; UNESCO: Paris, France, 1997.
16. Beck, H.E.; van Dijk, A.I.J.M.; Miralles, D.G.; de Jeu, R.A.; Bruijnzeel, L.A.; McVicar, T.R.; Schellekens, J. Global patterns in base flow index and recession based on streamflow observations from 3394 catchments. *Water Resour. Res.* **2013**, *49*, 7843–7863. [CrossRef]
17. Gustard, A.; Bullock, A.; Dixon, J. *Low Flow Estimation in the United Kingdom*; Institute of Hydrology: Wallingford, UK, 1992.
18. Ngongondo, C.S. An analysis of long-term rainfall variability, trends and groundwater availability in the Mulunguzi river catchment area, Zomba mountain, Southern Malawi. *Quat. Int.* **2006**, *148*, 45–50. [CrossRef]
19. Hughes, D.A.; Hannart, P. A desktop model used to provide an initial estimate of the ecological instream flow requirements of rivers in South Africa. *J. Hydrol.* **2003**, *270*, 167–181. [CrossRef]
20. Esralew, R.A.; Lewis, J.M. *Trends in Base Flow, Total Flow, and Base-Flow Index of Selected Streams in and Near Oklahoma through 2008, Scientific Investigations Report 2010–5104*; U.S Department of the Interior, U.S Geological Survey: Reston, VA, USA, 2010.

21. Institute of Hydrology. *Low Flow Studies Report No 3*; Institute of Hydrology: Wallingford, UK, 1980.

22. Singh, S.K.; Pahlow, M.; Booker, D.J.; Shankar, U.; Chamorro, A. Towards baseflow index characterisation at national scale in New Zealand. *J. Hydrol.* **2019**, *568*, 646–657. [CrossRef]

23. Zhang, J.; Song, J.; Cheng, L.; Zheng, H.; Wang, Y.; Huai, B.; Sun, W.; Qi, S.; Zhao, P.; Wang, Y.; Li, Q. Baseflow estimation for catchments in the Loess Plateau, China. *J. Environ. Manag.* **2019**, *233*, 264–270. [CrossRef]

24. Hudak, A.T.; Wessman, C.A. Deforestation in Mwanza District, Malawi, from 1981 to 1992, as determined from Landsat MSS imagery. *Appl. Geogr.* **2000**, *20*, 155–175. [CrossRef]

25. Chitete, S. The Nation "Malawi Drying up". Available online: https://mwnation.com/malawi-drying-up/ (accessed on 16 April 2019).

26. Sood, A.; Smakhtin, V.; Eriyagama, N.; Villholth, K.G.; Liyanage, N.; Wada, Y.; Ebrahim, G.; Dickens, C. *Global Environmental Flow Information for the Sustainable Development Goals*; International Water Management Institute (IWMI): Colombo, Sri Lanka, 2017; Volume 168.

27. Kumambala, P.G. Sustainability of Water Resources Development for Malawi with Particular Emphasis on North and Central Malawi. Ph.D. Thesis, University of Glasgow, Glasgow, UK, 2010.

28. Kambombe, O.; Odongo, V.; Mutua, B.; Wambua, R. Impact of climate variability and land use change on streamflow in lake Chilwa basin, Malawi. *Int. J. Hydrol.* **2018**, *2*, 364–370.

29. Houghton-Carr, H.; Fry, M.; Wallingford, U. The decline of hydrological data collection for development of integrated water resource management tools in Southern Africa. *IAHS Publ.* **2006**, *308*, 51.

30. Smith-Carington, A. Hydrological bulletin for the Bua Catchment: Water resource unit number 5. In *Groundwater Section*; Department of Lands, Valuation and Water: Lilongwe, Malawi, 1983.

31. Government of Malawi, T. *National Water Resources Master Plan 2017. Main Report: Existing Situation*; Government of Malawi: Lilongwe, Malawi, 2017.

32. Government of Malawi. *Malawi Land Use Map*; Forestry Commission: Lilongwe, Malawi, 2018.

33. Government of Malawi. *Water Resources Investment Strategy. Component 1—Water Resources Assessment, Annex I(ii) for WRAs 5-10*; Government of Malawi: Lilongwe, Malawi, 2011.

34. Government of Malawi. *Final Report for consultancy services related to detailed design of the upgraded Kamuzu Barrage*; Extracts from Main Report Chapters 4–11 related to Hydrology–Hydraulics–Water Demand; Ministry of Water Development and Irrigation: Lilongwe, Malawi, 2013.

35. Government of Malawi. *Malawi Hydrogeological and Water Quality Map 2018*; Ministry of Agriculture, Irrigation and Water Development: Lilongwe, Malawi, 2018.

36. Brodie, R.; Sundaram, B.; Tottenham, R.; Hostetler, S.; Ransley, T. *An Overview of Tools for Assessing Groundwater-Surface Water Connectivity*; Bureau of Rural Sciences: Canberra, Australia, 2007; Volume 133.

37. Eckhardt, K. A comparison of baseflow indices, which were calculated with seven different baseflow separation methods. *J. Hydrol.* **2008**, *352*, 168–173. [CrossRef]

38. Dierauer, J.R.; Whitfield, P.H.; Allen, D.M. Assessing the suitability of hydrometric data for trend analysis: The "FlowScreen" package for R. *Can. Water Resour. J./Revue Canadienne Resour. Hydriques* **2017**, *42*, 269–275. [CrossRef]

39. Wahl, K.L.; Wahl, T.L. Determining the flow of comal springs at New Braunfels, Texas. *Unknown* **1995**, *95*, 16–17.

40. Arnold, J.G.; Moriasi, D.N.; Gassman, P.W.; Abbaspour, K.C.; White, M.J.; Srinivasan, R.; Santhi, C.; Harmel, R.; Van Griensven, A.; Van Liew, M.W. SWAT: Model use, calibration, and validation. *Trans. ASABE* **2012**, *55*, 1491–1508. [CrossRef]

41. Willems, P. A time series tool to support the multi-criteria performance evaluation of rainfall-runoff models. *Environ. Model. Softw.* **2009**, *24*, 311–321. [CrossRef]

42. Younghun Jung, Y.S.N.-I.W.; Lim, K.J. Web-Based BFlow System for the Assessment of Streamflow Characteristics at National Level. *Water* **2016**, *8*, 384. [CrossRef]

43. Sloto, R.A.; Crouse, M.Y. *HYSEP: A Computer Program for Streamflow Hydrograph Separation and Analysis*; US Department of the Interior, US Geological Survey: Reston, VA, USA, 1996.

44. Dickinson, J.E.; Hanson, R.T.; Predmore, S.K. *HydroClimATe: Hydrologic and Climatic Analysis Toolkit*; US Department of the Interior, US Geological Survey: Reston, VA, USA, 2014.

45. Metcalfe, R.A.; Schmidt, B. Streamflow Analysis and Assessment Software (SAAS) (V4.1). 2016. Available online: http://people.trentu.ca/~{}rmetcalfe/SAAS.html (accessed on 1 August 2018).

46. CRC for Catchment Hydrology. River Analysis Package (RAP) Brochure. 2003. Available online: https://toolkit.ewater.org.au/Tools/RAP (accessed on 2 August 2018).

47. Lim, K.J.; Engel, B.A.; Tang, Z.; Choi, J.; Kim, K.-S.; Muthukrishnan, S.; Tripathy, D. Automated Web GIS Based Hydrograph Analysis Tool, WHAT 1. *JAWRA J. Am. Water Resour. Assoc.* **2005**, *41*, 1407–1416. [CrossRef]

48. Gregor, B. *BFI+ 3.0 Users's Manual*; Department of Hydrogeology, Faculty of Natural Science, Comenius University: Bratislava, Slovakia, 2010; 21p.

49. Combalicer, E.; Lee, S.; Ahn, S.; Kim, D.; Im, S. Comparing groundwater recharge and base flow in the Bukmoongol small-forested watershed, Korea. *J. Earth Syst. Sci.* **2008**, *117*, 553–566. [CrossRef]

50. St. Jacques, J.-M.; Sauchyn, D.J. Increasing winter baseflow and mean annual streamflow from possible permafrost thawing in the Northwest Territories, Canada. *Geophys. Res. Lett.* **2009**, *36*. [CrossRef]

51. Harvey, C.L.; Dixon, H.; Hannaford, J. Developing best practice for infilling daily river flow data. In *Role of Hydrology in Managing Consequences of a Changing Global Environment, Proceedings of the BHS Third International Symposium, Newcastle, UK, 19–23 July 2010*; British Hydrological Society: London, UK, 2010; pp. 816–823.

52. Ladson, A.R.; Brown, R.; Neal, B.; Nathan, R. A standard approach to baseflow separation using the Lyne and Hollick filter. *Aust. J. Water Resour.* **2013**, *17*, 25–34. [CrossRef]

53. Hodgkins, G.A.; Dudley, R.W. Historical summer base flow and stormflow trends for New England rivers. *Water Resour. Res.* **2011**, *47*. [CrossRef]

54. Oki, D.S. *Trends in Streamflow Characteristics at Long-Term Gaging Stations, Hawaii, U.S Geological Survey Scientific Investigations Report 2004-5080*; US Department of the Interior, US Geological Survey: Reston, VA, USA, 2004.

55. Government of Malawi. *Water Resources Investment Strategy. Component 1—Water Resources Assessment. Annex II—Surface Water*; Government of Malawi: Lilongwe, Malawi, 2011.

56. Government of Malawi. *National Irrigation Master Plan and Investment Framework*; Ministry of Agriculture, Irrigation and Water Development: Lilongwe, Malawi, 2015.

57. Mann, H.B. Nonparametric tests against trend. *Econ. J. Econ. Soc.* **1945**, 245–259. [CrossRef]

58. Kendall, M.G. *Rank Correlation Methods*, 4th ed.; Griffin: London, UK, 1975.

59. Gumindoga, W.; Makurira, H.; Garedondo, B. Impacts of landcover changes on streamflows in the Middle Zambezi Catchment within Zimbabwe. *Proc. Int. Assoc. Hydrol. Sci.* **2018**, *378*, 43–50. [CrossRef]

60. Da Silva, R.M.; Santos, C.A.; Moreira, M.; Corte-Real, J.; Silva, V.C.; Medeiros, I.C. Rainfall and river flow trends using Mann-Kendall and Sen's slope estimator statistical tests in the Cobres River basin. *Nat. Hazards* **2015**, *77*, 1205–1221. [CrossRef]

61. Yue, S.; Pilon, P.; Cavadias, G. Power of the Mann-Kendall and Spearman's rho tests for detecting monotonic trends in hydrological series. *J. Hydrol.* **2002**, *259*, 254–271. [CrossRef]

62. Techamahasaranont, J.; Shrestha, S.; Babel, M.S.; Shrestha, R.P.; Jourdain, D. Spatial and temporal variation in the trends of hydrological response of forested watersheds in Thailand. *Environ. Earth Sci.* **2017**, *76*, 430. [CrossRef]

63. Zhang, X.S.; Amirthanathan, G.E.; Bari, M.A.; Laugesen, R.M.; Shin, D.; Kent, D.M.; MacDonald, A.M.; Turner, M.E.; Tuteja, N.K. How streamflow has changed across Australia since the 1950s: Evidence from the network of hydrologic reference stations. *Hydrol. Earth Syst. Sci.* **2016**, *20*, 3947. [CrossRef]

64. Anibas, C.; Schneidewind, U.; Vandersteen, G.; Joris, I.; Seuntjens, P.; Batelaan, O. From streambed temperature measurements to spatial-temporal flux quantification: Using the LPML method to study groundwater-surface water interaction. *Hydrol. Process.* **2016**, *30*, 203–216. [CrossRef]

65. *Addinsoft XLSTAT Statistical and Data Analysis Solution*; XLSTAT: Long Island, NY, USA, 2019.

66. Zhang, J.; Zhang, Y.; Song, J.; Cheng, L. Evaluating relative merits of four baseflow separation methods in Eastern Australia. *J. Hydrol.* **2017**, *549*, 252–263. [CrossRef]

Article

An Integrated Approach for Studying the Hydrology of the Ljubljansko Polje Aquifer in Slovenia and Its Simulation

Janja Vrzel [1,2,*], Ralf Ludwig [1], Goran Vižintin [3] and Nives Ogrinc [2,4]

[1] Department of Geography, Ludwig-Maximilians-Universität-München, Luisenstraße, 37, 80333 Munich, Germany
[2] Jožef Stefan International Postgraduate School, Jamova cesta 39, 1000 Ljubljana, Slovenia
[3] Department of Geotechnology, Mining and Environment, University of Ljubljana, Aškerčeva cesta 12, 1000 Ljubljana, Slovenia
[4] Department of Environmental Sciences, Jožef Stefan Institute, Jamova cesta 39, 1000 Ljubljana, Slovenia
* Correspondence: j.vrzel@iggf.geo.uni-muenchen.de

Received: 22 July 2019; Accepted: 17 August 2019; Published: 22 August 2019

Abstract: Groundwater and surface water are strongly connected. Therefore, understanding their interactions is important when studying the water balance of a complex aquatic system. This paper aims to present an integrated approach to study such processes, including a better understanding of the hydrological system behavior in the Ljubljansko polje (Slovenia). The study is based on multivariate statistical analyses of data collected over a long period, including the isotopic composition of groundwater, river water, and precipitation. The hydrology in the study domain was also simulated using a comprehensive modelling framework. Since boundary conditions are essential for simulating groundwater flow in a sensitive aquifer, a modelling system of rivers and channels (MIKE 11) and water flow and balance simulation model (WaSiM) were used to model river dynamics and the percolation of local precipitation, respectively. The results were then used as boundary conditions imposed on a transient state groundwater flow model performed in finite element subsurface flow simulation system (FEFLOW 6.2). Both the locations of recharge areas in the study domain and the calculated fluxes between the Sava River and the aquifer are graphically presented. The study revealed that a combination of the MIKE 11-FEFLOW-WaSiM tools offers a good solution for performing parallel simulations of groundwater and surface water dynamics.

Keywords: surface-groundwater interactions; hydrological modeling; aquifer responsiveness; mean residence time

---

1. Introduction

The European Union (EU) has attempted to add to the limited knowledge of the dynamic processes that connect surface water and groundwater [1]. The European Water Framework Directive and the related River Basin Management Plans for the period 2015–2021 and beyond, require, if needed, remedial measures to ensure a good groundwater chemical and quantitative status and prevent the deterioration of the status of all surface water and groundwater bodies. To achieve these aims, the EU encourages significant research activities aimed at improving our understanding of the complex water cycle, including the following: groundwater interactions at sensitive interfaces, identifying percolation areas and other sources of groundwater recharge, and the vulnerability of groundwater to different contaminants.

This paper presents a novel approach for simulating groundwater behavior at the reach-scale using multi-tools, in a system where surface-groundwater interactions are significant. For this purpose, FEFLOW was directly coupled with MIKE 11, and an indirect communication link between FEFLOW and WaSiM was established. Typically, the direct coupling means that models affect each other, but this is not the case for indirectly coupled models. Combining several tools is challenging because the behavior of a complex system cannot be expressed as the sum of its components or subsystems [2]. Unlike FEFLOW has been coupled with MIKE 11 many times [3,4], coupling FEFLOW to WaSiM is rarely reported in the literature [5], and, according to the authors' knowledge, this is the first study to include all three models.

An overview of other available modeling tools in hydrology is provided by Gunduz and Aral [6] as well as Barthel and Banzhaf [7]. Both publications report the limitations of existing modeling approaches, due to their diverse applications, complexity level, and local, regional, or even global geological, hydrological, and climatic characteristics in 2-dimensional or 3-dimensional space over different periods.

Krause and Bronstert [8] coupled WaSiM with MODFLOW to simulate groundwater-surface water interactions in North-eastern Germany. MODFLOW, which is often used in groundwater-focused studies [7], was also coupled with other models such as SWAT [9], LISFLOOD [10], and HEC-RAS [11]. Two examples of groundwater-focused modeling approaches that use FEFLOW were applied in the Zayandeh Rud catchment in Iran and in Covey Hill in Canada, where FEFLOW was coupled with SWAT and the Hydrological Evaluation of Landfill Performance (HELP), respectively [12,13]. Groundwater recharge through rainfall in Slovenia and the Far-North region of Cameroon was simulated with GROWA [14,15]. Lastly, the MODHMS surface water flow package (HydroGeoLogic, Inc., Reston, VA, USA) for MODFLOW, was used to model surface water-groundwater interactions [6].

Processes that occur at the interface between surface water and groundwater can dominate a system's evolution [2]. For this reason, fluxes between groundwater and surface water bodies, which are difficult to measure, have been the subject of local-scale and regional-scale studies [16–19]. This study is different because it deals exclusively with the connections between the river water level, percolation from local precipitation, and groundwater in an alluvial aquifer with an intergranular porosity. The Ljubljansko polje, located within the Sava River Basin (SRB), is ideal for testing such an approach since it is highly sensitive to changes in its two main groundwater sources: the Sava River water and local precipitation [20,21]. Understanding the hydrology of the Ljubljansko polje aquifer system is also important for local inhabitants because it represents the main source of drinking water.

2. Materials and Methods

2.1. The Ljubljansko Polje

The Ljubljansko polje is a ~71 km^2 basin located between latitudes 46.12° N, 46.08° N, and longitudes 14.43° E and 14.64° E in Central Slovenia (Figure 1). Its elevation ranges from 259.5 to 327.5 m and is separated from its adjacent geographic units (the Kamniško-Bistriško polje, the Kranjsko Sorško polje, and the Ljubljansko barje), by hills up to 450 m in height. The Sava River is the main river that flows in the basin together with its three main tributaries: the Ljubljanica, the Kamniška Bistrica, and the Gameljščica rivers.

The basin is filled with alluvial sediments with an intergranular porosity–Pleistocene and Holocene gravel, which may also contain silt and sand, with conglomerate lenses. Layers of clay or clay with gravel-stone are located only at specific locations, mainly in the southwestern part of the basin [22]. The sediments extend from a depth of 100 m in the central part of the basin to the surface (Šentjakob, Črnuče, Tacen and surrounding hills), where the impermeable bedrock outcrops.

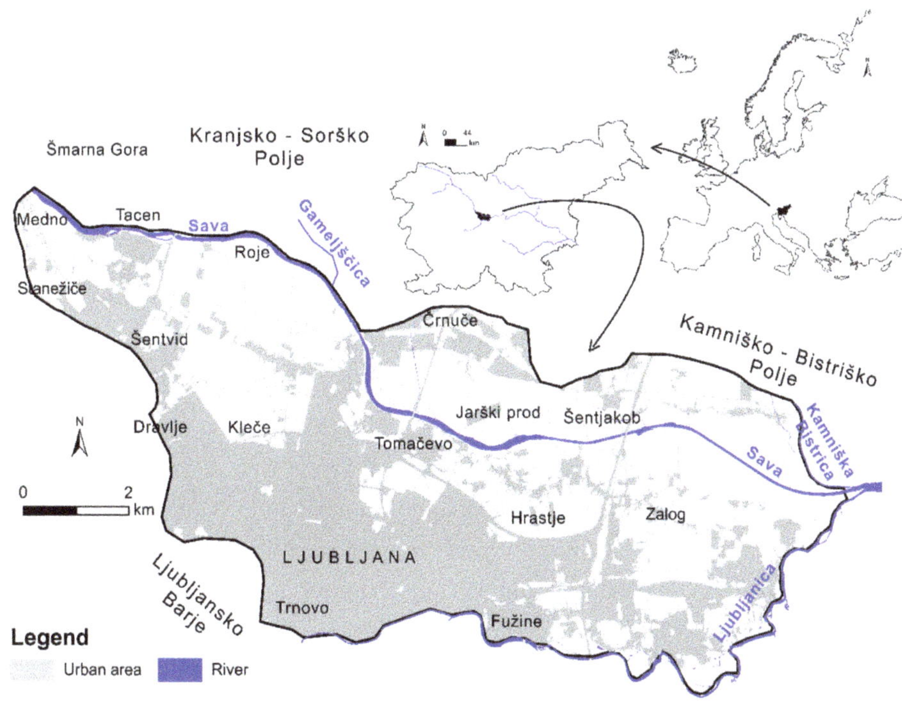

Figure 1. Map of the case study area–the Ljubljansko polje.

Local precipitation, leaching water from the Sava River, and the lateral underground inflows from neighboring groundwater bodies are the main sources of groundwater in the Ljubljansko polje aquifer [20,23,24]. The Ljubljansko polje is a small depression surrounded by impermeable bedrock outcrops, and it can be assumed that underground interactions of the system with neighboring groundwater bodies do not exist along its lateral boundary. Exceptions are narrow sections where bedrock sinks deeper under the surface, which enables underground interactions of the aquifer with the neighboring groundwater bodies. Although the underground interaction between the Ljubljansko polje and the Ljubljansko Barje is evident from the carbonate characteristics of the groundwater (Ca/Mg molar ratio), it has not been quantified [23].

A significant loss of groundwater occurs through water extraction, and, in addition to the Pivovarna Laško Union d.o.o. brewery, which extracts the groundwater, there are four pumping stations located at Kleče, Jarški prod, Hrastje, and Šentvid (Figure 2). Multivariate statistical methods were used as a basis for the conceptual model [20] to compare geochemical data with long-term datasets (precipitation levels, river discharge, hydraulic head, and groundwater pumping rate). The present hydrological model determines the upper boundary conditions of the two main groundwater sources: percolation and river water leaching, which differentiates it from the existing hydrological models that have been developed for the Ljubljansko polje [25,26].

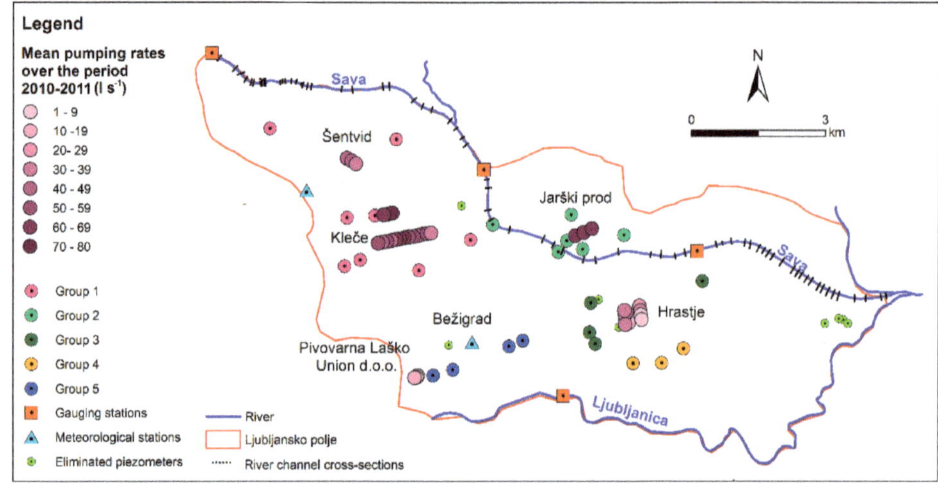

Figure 2. Locations of pumping stations and wells with mean pumping rates in L s^{-1} for 2010–2011. Locations of piezometers (five groups), three meteorological stations, and three gauging stations on the Sava River and one on the Ljubljanica River. The Sava River cross-sections used in MIKE 11.

2.2. Database

Hydraulic Head. Daily and monthly hydraulic head values from 52 and 25 piezometers, respectively, were provided by Javno podjetje VODOVOD KANALIZACIJA SNAGA (VOKA; URL: http://www.vo-ka.si). The observed hydraulic head was from 260.47 m to 299.08 m a.s.l. Only those piezometers with daily data and minimal gaps of available data were used in the simulation (Figure S1 in the supplementary material shows data availability). Locations of the piezometers are shown in Figure 2.

Groundwater Pumping. Daily rates of groundwater abstractions from the 32 wells for all four pumping stations were obtained from VOKA (Figure 2). Kleče is the main pumping station and produced five times more groundwater (53.0 L s^{-1}) than the other pumping stations during the study period (2010–2011).

River Discharge. Discharges observed in the Sava, and the Ljubljanica Rivers were used to simulate the spatially-explicit percolation (WaSiM model), while only the Sava River discharge was used in the transient state groundwater flow model (MIKE 11 and FEFLOW). The Ljubljanica River represented 20% to 30% of the Sava River discharge in 2010 and 2011 when the recorded discharge in the Sava River was 1021.66 m^3 s^{-1} but was not included in the groundwater flow model (GW flow model) because of its negligible effect on the aquifer due to the presence of a clogging layer [24]. The Sava River water level is the required input data for the steady-state groundwater model, and the data was obtained from Slovenian Environmental Agency (ARSO; URL: http://www.arso.gov.si) [27].

Underground Recharge/Discharge. The amounts of subsurface groundwater recharge/discharge were estimated during the calibration process of the steady-state GW flow model in FEFLOW.

Mean Residence Time (MRT). An estimation of MRT was made using the ^{3}H/^{3}He method for water samples collected on the 01.06.2010 at Kleče 8, Kleče 11, Kleče 12, Hrastje 3, Hrastje 8, Jarški prod 1, and Jarški prod 3 [20] (Table 3). It was not possible to estimate the MRT for Kleče due to the presence of "old" water. For this reason, the MRTs for Hrastje, and Jarški prod were used to calibrate the steady-state GW flow model.

Meteorological Data. Daily temperature (°C), precipitation (mm), global radiation (Wh m^{-2}), humidity (–), and wind speed (m s^{-1}) were obtained from ARSO for the period between 2003 to

2015 [28]. The inverse distance weighting (IDW) interpolation method was used for the horizontal interpolation of the meteorological data [29].

A significant flood event affected many urban areas in Slovenia in September 2010, including the southern part of the city Ljubljana, which belongs to the Ljubljansko Barje. An extremely high Sava River discharge, with a return period estimated at 10 years was also observed in Medno and Šentjakob [30]. The only year with a recorded higher precipitation than 2010 (1784 mm) was 1965 (1848 mm). In contrast, the lowest amount of precipitation (998 mm) was recorded in 2011 [31].

Soil Map. A precise soil texture map was provided by the TIS/ICPVO-Infrastructural Center for Pedology and Environmental Protection, Biotechnical Faculty, University of Ljubljana, Ljubljana 1999–2016 (URL: http://www.bf.uni-lj.si). Unfortunately, the soil texture type could not be defined in the built-up areas, which represents 39.60% of the study domain (artificial area in Table 1). Literature data for central Europe were used for model parameterization of the soil texture classes (Table 1) [32–34].

Table 1. Soil texture classes for the Ljubljansko polje (TIS/ICPVO).

Code	Soil Texture Type	Description	Percentage (%)
1	Artificial area	Data cannot be defined	39.60
2	Coarse	Sand/Loamy/Sand/Sandy loam	30.05
3	Coarse–Medium	Sand/Loamy sand/Sandy loam/Sandy clay loam/Clay loam/Loam/Silt loam	27.51
4	Medium	Sandy clay loam/Clay loam/Loam/Silt loam	2.77
5	Medium–Fine–Fine	Sandy clay loam/Clay loam/Loam/Silt loam/Silt/Silty clay loam/Silty clay/Sandy clay/Clay	0.06
6	Very fine	Silt/Silty clay loam/Silty clay/Sandy clay/Clay	0.01

Land Use Map. Land use data for the Slovenian part of the Sava River basin was taken from the CORINE Land Cover database (CLC2006) [35]. The original data have a spatial resolution of 25 m, which were downscaled to a resolution of 50 m for the Ljubljansko polje. Table 2 gives the land use classes for the Ljubljansko polje and is a combination of 2-level and 3-level in the CLC's 3-level hierarchical classification system.

Table 2. Land use classes for the Ljubljansko polje (CLC2006).

Code	Land Use Type	Percentage (%)
1	Heterogeneous agricultural areas	15.72
2	Arable land	22.70
3	Meadow	0.31
4	Mixed forest	4.74
5	Urban	30.44
6	Transitional woodland	0.92
7	Industrial area	19.30
8	Pastures	2.92
9	Water	2.97

Digital Elevation Model (DEM). The DEM at a spatial resolution of 5 m (YXZ format) was obtained from the Surveying and Mapping Authority of the Republic of Slovenia (URL: http://www.gu.gov.si/en/). The DEM was downscaled to a resolution of 50 m for the WaSiM model. LiDAR (light detection and ranging) data for the floodplain of the Sava River in the Ljubljansko polje were obtained from Savske elektrarne Ljubljana d.o.o. (URL: http://www.sel.si).

The River Channel Cross-sections. The Sava River channel cross-sections were measured in 2007, 2010, and 2013. Sixty-three cross sections were used in the MIKE 11 and FEFLOW models (Figure 2). The data (KOO format) was obtained from the Municipality of Ljubljana (URL: https://www.ljubljana.si)

A Counter Map of the Bedrock in the Ljubljansko Polje was obtained from VOKA, which had been prepared by local hydrologists. However, the map was updated and extended with information obtained from the latest drilling reports (115 in total), which were provided by VOKA and the Pivovarna Laško Union d.o.o. (URL: www.pivovarnalaskounion.si).

2.3. Data Quality Control

Any uncertainty in the hydro-chronological data, which appears due to a combination of measurement error, systematic error, natural variation, and inherent randomness [36], have been removed from the database by using the following methods.

1. Before applying multivariate statistical analysis, the database was reduced in size using linear correlation. The 47 piezometers, where the hydraulic heads were measured (2003–2015), were divided into five groups. Each group included piezometers with high correlation coefficients, $R^2 \geq 0.95$, due to similar patterns of hydraulic heads. These groups also have specific locations in the Ljubljansko polje (Figure 2). Only one piezometer from each group, with minimal data gaps in their sets, were selected as being representative and used for data analyses. Due to the high correlation ($R^2 > 0.99$) between the Sava River discharges observed at the three gauging stations, located at short distances from one another (~6.5 km), only discharges measured in Šentjakob were analyzed. The analysis also includes the Ljubljanica River discharge. Precipitation, which was recorded at the meteorological station in Ljubljana (Bežigrad, Figure 2), does not correlate with the discharges of the Sava and the Ljubljanica Rivers due to the remoteness of their springs where different climate conditions prevail.
2. Those piezometers, in which significantly different elevations of hydraulic heads from the majority of the piezometers were recorded, were excluded from the database (Figure S2).

2.4. System Responsiveness

Groundwater responsiveness to the Sava River discharge (river events) or local precipitation (precipitation events) was estimated using the moving window method. The method was used to cross-correlate the hydraulic heads, local precipitation levels, and river discharge. In this method, a defined time "window" (30 days) is moved over the daily data-set obtained for 2003–2015, where the distance moved is equal to the width of the "window." All data located within the "window" are statistically summarized [37].

2.5. Modeling Tools

Figure 3 shows the tools used in this study. Detailed descriptions of their setups, calibrations, and validations are provided in this chapter.

2.5.1. WaSiM

Percolation of local precipitation was simulated using the physically-based fully distributed hydrological model–Water Flow and Balance Simulation Model (WaSiM; URL: http://www.wasim.ch/en/). Depending on the general availability of data and the hydrological problem to be solved, several algorithms designed for simulating specific processes on various temporal and spatial scales are available. In addition, WaSiM uses the ASCII format, which allows easier and more optimal data exchange with other software packages [29].

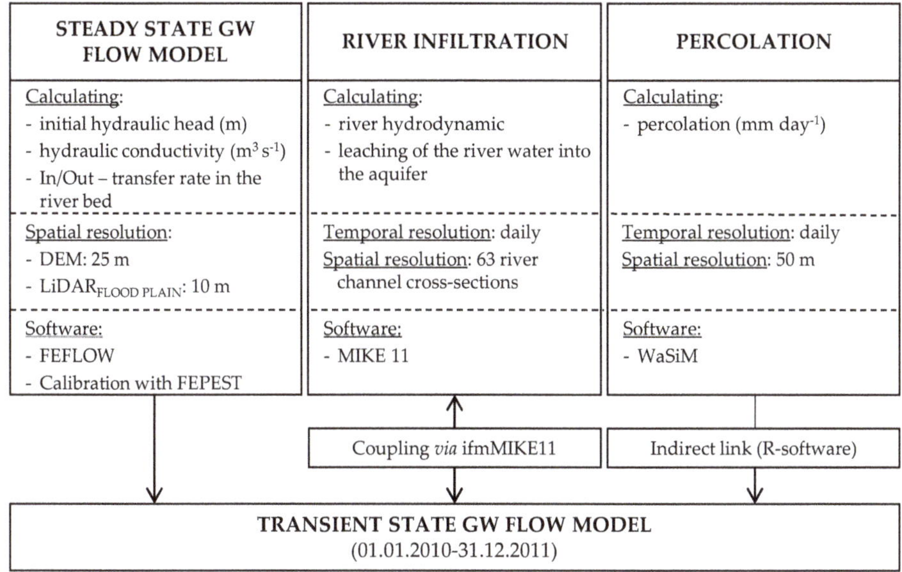

Figure 3. The hydrological modeling framework.

2.5.1.1. Setting Up the WaSiM Model

Input grids in WaSiM include topology, land use, soil features, aspect, and slope with a spatial resolution of 50 m. Climate conditions are defined by the station-based daily time sets of temperature (°C), precipitation (mm), global radiation (Wh m^{-2}), humidity (−), and wind speed (m s^{-1}) recorded in 2005 to 2015. The following parameters describe soil hydraulic properties for the four soil horizons in each soil texture class (Table 1): saturated hydraulic conductivity, saturated hydraulic conductivity recession with depth, saturated water content, residual water content, van Genuchten Parameter Alpha, van Genuchten Parameter n, Mualem Parameter in van-Genuchtens, number of numerical layers, and numerical layer thickness [29]. Land use classes (Table 2) for 12 Julian days over a year are parameterized by the following: albedo, leaf surface resistance, leaf area index, roughness length, vegetation-covered fraction, root depth, altitude correction, interception surface resistance, and soil surface resistance [29].

The hydrological model of the Slovenian part of the Sava River Basin (SRB), with a daily temporal resolution and a spatial resolution of 1 km, was developed, while only the Ljubljansko polje was later downscaled to a spatial resolution of 50 m. Results of the SRB model were applied as input data in the Ljubljansko polje model via the routing module, which is one of the modules in the WaSiM control file. The framework of the WaSiM model is shown in Figure 4. The SRB model calculates the lateral inflow of the surface water in the Ljubljansko polje from the upstream area of the SRB.

2.5.1.2. Calibration and Validation of the WaSiM Model

The trial-and-error method was used to calibrate the WaSiM model, which implies a manual parameter assessment through several calibrations run in order to determine realistic lower and upper limits of the adjustable parameters [38]. The parameters used for the calibration are given in Table 3. The reference evapotranspiration value was calculated using the Penman-Monteith method and is the amount of water that has evaporated from the referential plant and soil (Table 3). The reference surface is a hypothetical grass crop, in which the grass completely covers the soil, with a height of 0.12 m having a resistance of 70 s m^{-1} and an albedo of 0.23 [39].

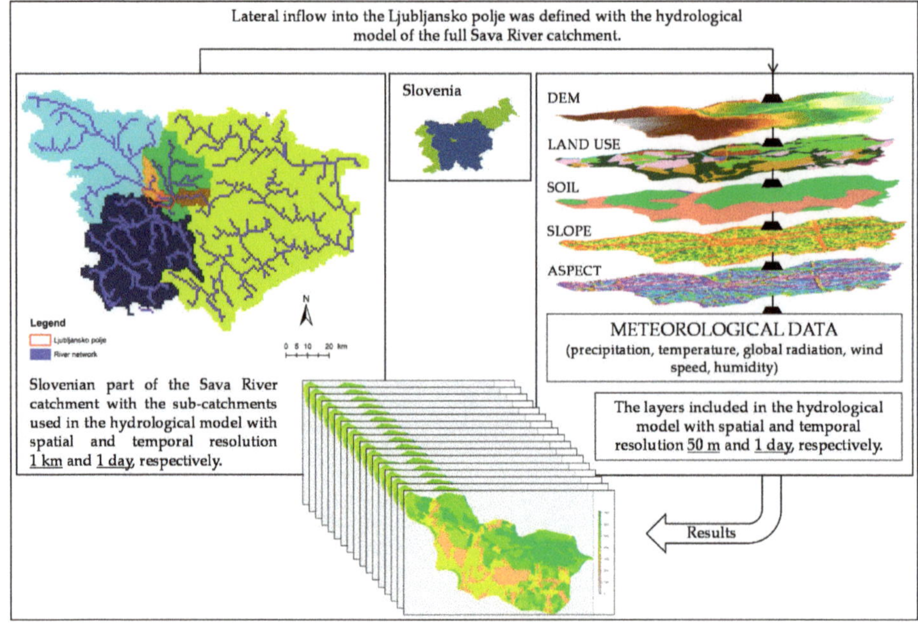

Figure 4. A framework of the modeling procedure with WaSiM.

Table 3. Parameters used for the calibration (C) and validation (V) of the WaSiM model during different periods.

Use	Module	Parameter	Period	Description
C	Evapotranspiration	Real ET (mm (dt)$^{-1}$)	1 January 2010–31 December 2014	Calculated real ET was compared with the referential ET in Bežigrad (46°3′56″ N, 14°30′45″ E, altitude 299 m), which were calculated by ARSO (Source: http://meteo.arso.gov.si).
V			1 January 2008–31 December 2009	
C	Unsaturated Zone	Soil moisture within the root zone (%)	1 January 2012–31 December 2012	Calculated soil moisture was compared with the observed soil moisture data in Kleče (46°5′11″ N, 14°29′56″ E, altitude 308 m) during the period January 2012–August 2012 [40].
C	Unsaturated Zone	Hydraulic head below the surface (m)	1 January 2010–31 December 2014	Trends of calculated percolation were compared with trends of the observed hydraulic head.
V			1 January 2009–31 December 2009	

Precipitation Events. Precipitation events that cause hydraulic head oscillations were used for the calibration of the WaSiM model within the domain of the Ljubljansko polje. These events were extracted from the full database in four steps by (1) calculating the two-day moving average for the hydraulic head, precipitation, river discharge, and groundwater abstraction, (2) extracting days when hydraulic heads, precipitation, and the river discharge were higher than the previous day, (3) extracting days with increased hydraulic heads and precipitation and decreased river discharge (↑hydraulic head|↑precipitation|↓the river discharge), and (4) plotting the extracted days versus hydraulic heads/precipitation/the river recharge/groundwater abstraction.

Due to a lack of available data, model validation was performed only for the year 2009 at an hourly time step using a grid size of 50 m. The warming up period for the model was three years.

2.5.2. MIKE 11

MIKE 11 from the MIKE Powered by the Danish Hydraulic Institute (DHI) is an implicit finite difference model that is capable of dealing with kinematic, diffusive, and dynamic Saint

Venant equations. It was chosen because it can solve one-dimensional unsteady flow problems [41]. Additionally, MIKE 11 handles the effect of riverbed morphology on the river water level and groundwater exchange, which is thought to be significant [42–45].

Setting Up of the MIKE 11 Model

MIKE 11 requires four input files, namely Network, Cross-section, Boundary, and Hydrodynamic (HD) Parameters. The river network and reference cross-sections are defined in the Network file. The Sava River channel in the study domain is 17.8 km long. The channel's cross-sections and their Manning's n coefficients are defined in the Cross-Section file. By comparing the natural situation, figures presented by Jarrett [46] and, with the help of local studies [47], the Manning's n coefficients were estimated to be between 0.027 and 0.030 in the steady domain. The Manning's n coefficient was later corrected for each cross-section for the whole calibration process of the transient state GW flow model.

Upstream and downstream of the model domain, the Inflow and Q-h boundary types, were defined, respectively. Daily discharges observed in Medno between 2010–2011 were used as the input data for the Inflow boundary type, while the auto calculation of the Q-h Table tool was used for the Q-h boundary type. The calculation is based on the Manning formula, with a river slope of 0.02 and a Manning's n parameter of 0.0285. In the HD Parameters file, the following parameters were defined: initial water level in the Sava River (observed at two gauging stations), groundwater leakage, and bed resistance. The Manning formula was chosen as the resistance formula, which uses 0.0324 as a global resistance number. The local groundwater leakage coefficients (0.0–1.0×10^{-6} s^{-1}) were defined for each cross-section for the calibration of the GW flow model.

2.5.3. FEFLOW

FEFLOW also belongs to the MIKE Powered by the DHI software family (URL: https://www.mikepoweredbydhi.com). It is an advanced finite element subsurface flow and transport modelling system, which supports several different file formats, but not WaSiM output-raster files [48]. Therefore, an indirect communication link between WaSiM and FEFLOW was established using open-source R-software (URL: https://www.r-project.org). Daily percolation data from a *.dat file was assigned to the in/outflow on the top/bottom parameter using the "assign material data to time stages" option.

FEFLOW was coupled to MIKE 11 via the FEFLOW Interface Manager (ifmMIKE11), owned by the MIKE Powered by the DHI. The ifmMIKE11 plug-in enables the user to define not only the surface water level at a single boundary node but also the water exchange area represented by the node and the nodal transfer rate [41,49].

However, the river water level, discharge, and leaching of the Sava River water into the aquifer, and groundwater into the river were simulated using coupled MIKE 11 with FEFLOW. The numerical stability of the stand-alone MIKE 11 model was tested initially because a numerically unstable model cannot be used with FEFLOW.

2.5.3.1. Development of the Physical Framework

A 2D unstructured triangle mesh with 110,249 elements was generated within the model domain. The mesh was refined around the central river line, the pumping wells, and the domain border. Direct estimation of the nodal distance method was used to calculate the relative distance between the pumping wells and their adjacent nodes [48]. The elements' volume ranged from 0.98 to 31,241.7 m^3, the maximal interior angle of the triangles was no smaller than 60°, and the Delaunay criterion violation is zero for all elements.

The 3D GW flow model is defined as being only a one-layer aquifer. An implementation of a multi-layer aquifer (described in Table S1) was not credible, due to limited data availability for the hydrogeological parameterization of all five potential layers. Nevertheless, to guarantee numerical stability, the vertical resolution of the model was refined by subdividing the layer into eight sub-layers. For the same reason, the layers were moved slightly lower from the river bed in the areas where an outcrop of bedrock appears. After making mesh corrections, these specific areas are located at an elevation of 275 m (Figure 5a, red areas). In this way, the continuity criterion was fulfilled [50]. The layer thickness varies from 0.5 m to 15.5 m (Figure 5b).

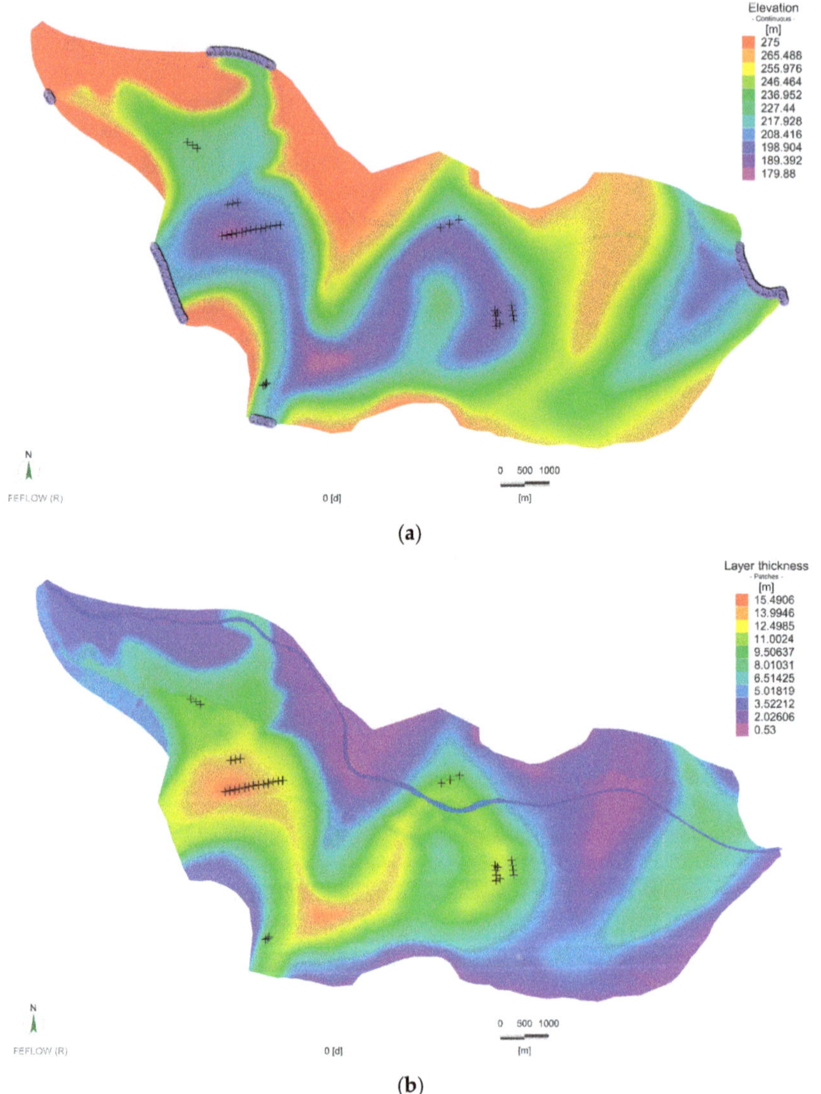

Figure 5. (**a**) Hydraulic head boundary condition (BC) and elevation of bedrock, and the (**b**) thickness of a sub-layer. Black crosses show locations of the Multilayer Well BC.

An interpolation of the model topography is based on the DEM at 25 m resolution (Figure 6a), while the interpolation of the Sava River floodplain or low terrace topography is based on more precise LiDAR data at a resolution of 10 m. The Sava's bed geometry was generated using the Sava River channel cross-sections data-sets. The distances between the cross-sections are approximately 20 m to 900 m along the flow path. Interpolation of the river bathymetry with the river channel cross-sections is essential because a high or low node elevation in a riverbed can produce unwanted or uncontrolled stream water-groundwater exchanges. The lithological composition of the aquifer, its thickness, and extension into the vertical direction downward to the bedrock was based on drilling reports and a bedrock counter map. The Ljubljanica River and outcrops of the bedrock were used as the main criteria for defining the lateral boundary of the model domain.

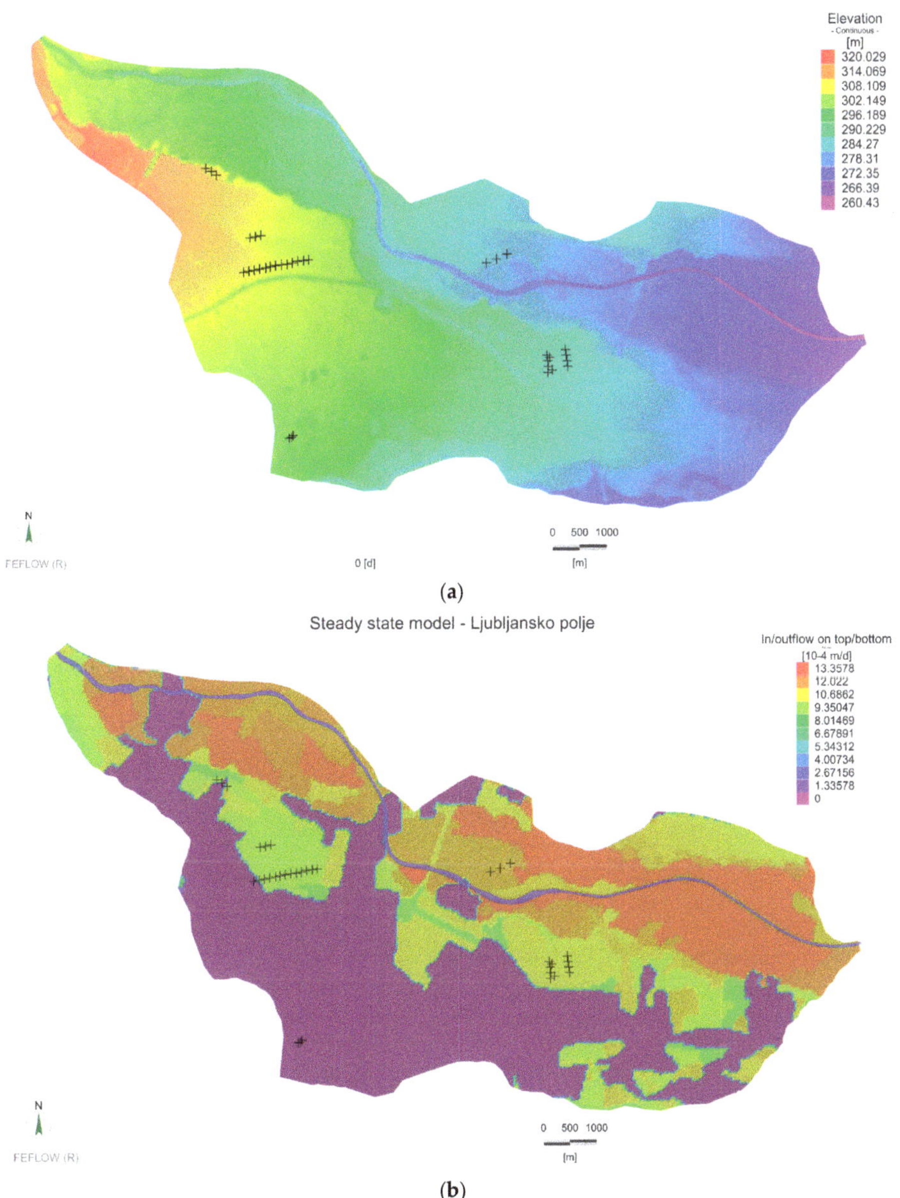

Figure 6. (**a**) Topography of the study domain, and (**b**) inflow/outflow on top/bottom BCs. The In/outflow on top/bottom BCs represents the mean percolation for the period 2010–2011 or the steady-state GW flow model. Black crosses show locations of the multilayer well DC.

The kriging method was used for the interpolation of the surface and bedrock, while the Sava River channel was interpolated using the 1D linear interpolation method. The kriging method is still the most accurate geostatistical interpolation method in FEFLOW since it takes into account both the distance and the degree of variation between known data points when estimating values in unknown areas [51]. It is important because the surface is interpolated from a grid based on a mesh of varying density. Every 10 neighboring points were used in the interpolation.

2.5.3.2. Setting Up the Steady-State and Transient State Groundwater Flow Models

The steady-state groundwater flow represents the underlying conditions in the Ljubljansko polje aquifer from 2010 to 2011. Its initial material properties (except in/outflow at the top/bottom) were estimated from consultants' reports. The initial values were later adjusted in the calibration process using a manual and an automatic trial-and-error method. Results of the steady-state GW flow model were then used for initializing the material parameters and boundary conditions (BC) in the transient state GW flow model.

Daily observed hydraulic heads, pumping ratios, and the Sava River water level were imported into FEFLOW through a time-series editor. The hydraulic heads are connected to the observation points in FEFLOW, pumping ratios to the Multilayer Well BC, and the Sava River water level to the Fluid Transfer BC. The latter is also coupled to MIKE 11 via the ifmMIKE11 plug-in. Percolation was assigned through the material property in/outflow on the top/bottom. The Hydraulic Head BC was defined along the outer model domain boundary sections where the bedrock goes deep under the surface (Figure 5a). In this case, interactions between the aquifer and neighboring water bodies are present. The outer boundaries along the bedrock outcrops and the base of the aquifer (the bedrock) are considered as non-flow boundaries.

Estimating the Hydraulic Head BC was demanding due to the limited understanding of hydrogeological processes between the Ljubljansko polje and neighboring groundwater bodies. To reduce the uncertainty of the model, the model's reactions to various hydraulic heads in the Hydraulic Head BC, from high to low values, were tested many times with the steady-state GW flow model. Many assumptions that were made in the parametrization and the design of the eastern part of the steady domain—where data availability is limited (hydraulic head, bedrock)—lead to a high outflow around the Kamniška Bistrica River and excluding the Hydraulic Head BC in Črnuče in the calibrated model.

Groundwater flows in the steady and transient state models are simulated using the standard (saturated) groundwater flow equation—Darcy's law, which can handle unconfined phreatic conditions in an aquifer [48]. Therefore, the top slice was defined as phreatic (fixed slice topping of an unconfined layer) and the bottom slice was defined as fixed, while the sub-slices depend on the top and bottom slices of the model. When the groundwater rises above the surface, the aquifer is treated as confined. The unconstrained head at the top of the model domain and the constrained boundary at the bottom of the model domain were selected as the head limits for the unconfined conditions. Residual water depth for the unconfined layers was estimated at 0.1 m.

2.5.3.3. Calibration and Validation of the FEFLOW Models

The material properties in the setup of the steady and transient state GW flow models (except percolation, an example of the percolation raster is in Figure 6b) were estimated during the calibration procedure, and include hydraulic conductivity for $K_{xx} = K_{yy}$ and K_{zz}, In and Out Transfer rates, and the specific yield. According to the authors' knowledge, the hydraulic conductivity of the alluvial sediments in the full modeling domain was not measured, which makes its estimation challenging.

The calibration and validation of the models were performed in the following five steps.

Step 1: The steady-state GW flow model was calibrated using manual and automatic trial-and-error history matching of the hydraulic heads in FEFLOW and FePEST, respectively. FePEST is a graphical user interface that links a software package for parameter estimation and uncertainty analysis of models (PEST) with FEFLOW [52].

Step 2: A simulation of the groundwater MRT was activated in FEFLOW. The initial MRT for the full aquifer is 4.99 years, which was calculated in FEFLOW, based on the ^{3}H/^{3}He results [20]. If the results after this step were unsatisfying, steps one and two were repeated.

The second step is a good indicator of the model's instability, either due to an inappropriate mesh discretization or parameterization. Nevertheless, the groundwater MRT has been widely used for evaluating and improving GW flow models in the past [53–56].

Step 3: The steady-state model was validated using mean hydraulic head elevations, the mean water levels of the Sava River, the mean percolation, and the mean volume of water extracted between 2013–2014.

Step 4: The steady-state GW flow model was upgraded to a transient state GW flow model, which was coupled to MIKE 11 via the ifmMIKE11 plug-in. The model was calibrated using daily hydraulic head records (1 April 2010–30 April 2010) in combination with manual and automatic trial-and-error history matching of the hydraulic heads and the Sava River discharge.

Step 5: Validation of the transient state model was performed for the period 1 January 2010–31 December 2011. Unfortunately, the initial results were unsatisfactory, and, as a result, several corrections to the model parameterization were made.

In FePEST (a graphical user interface for running FEFLOW models with Parameter Estimation code–PEST), a "pilot points" method was used for automatic trial-and-error history matching of the hydraulic heads and the MRT, and the kriging method for spatial interpolation of values from pilot points to the model domain [57]. Pilot points were located manually and had a higher density where a higher heterogeneous hydraulic conductivity was expected.

3. Results

3.1. WaSiM Model

Table S2 gives the obtained linear correlations for the observed and simulated real evapotranspiration and hydraulic heads. Soil moisture calculated by WaSiM and soil moisture observed by Pintar et al. [40] in the lysimeter in Kleče shows a good correlation (R^2 = 0.87) (Figure 7). Furthermore, a high Nash-Sutcliffe model efficiency (NSE = 0.87) between real evapotranspiration calculated using WaSiM and the referential evapotranspiration give adequate modeled values for the actual evapotranspiration in the Ljubljansko polje.

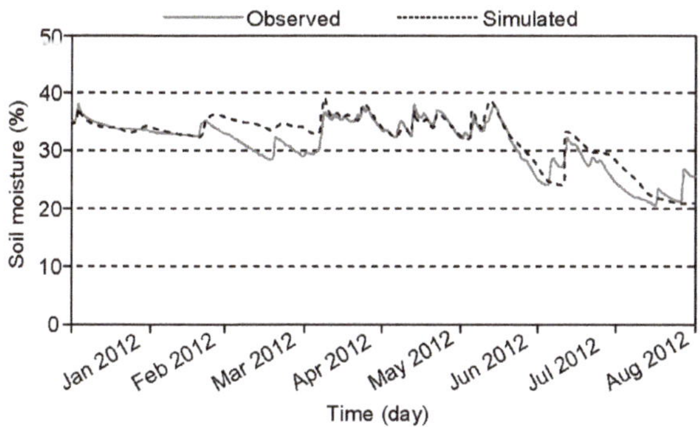

Figure 7. Comparison of the calculated and measured soil moisture [40].

The calibrated parameter set produced results that are in good agreement between the observed and the calculated real evapotranspiration and the hydraulic head for the validation period. The R^2 values for the real evapotranspiration and the well P-102 were 0.47 and 0.87 (Table S3), respectively.

The lowest correlation ($R^2 = 0.53$) was obtained between the actual and calculated hydraulic heads in well P-013.

3.2. Groundwater Flow Model (FEFLOW-WaSiM-ifmMIKE11)

3.2.1. The Steady-State Groundwater Flow Model

Simulated K_x and K_y in the steady-state model vary between 9.21×10^{-7} and 0.12 m s^{-1}, and K_z between 8.65×10^{-6} and 0.10 m s^{-1}. In and Out Transfer Rates in the Sava River bed vary between 5.0×10^{-3} and 2.9 day^{-1}. The FEFLOW model was stable when the hydraulic heads presented in Table 4 were used.

Table 4. Hydraulic head BC set up parameters. Hydraulic head BC in Črnuče was later excluded.

Location	Hydraulic Head BC
Trnovo	P-102 + 0.48 m
Dravlje	P-038 + (−2.00–2.20 m)
Roje	P-097 + (2.00–2.20 m)
Stanežiče	P−098 + 0.66 m
Outflow	P-017 + (−3.69–0.51 m)

3.2.2. The Transient State Groundwater Flow Model

Hydraulic conductivity ranged from 6.57×10^{-3} to $3.89 \times 10^{-2} \text{ m s}^{-1}$ and from 3.24×10^{-3} to $4.57 \times 10^{-3} \text{ m s}^{-1}$ for $K_x = K_y$ and K_z, respectively. The In and Out Transfer rates were $\leq 30.44 \text{ day}^{-1}$. All of these four parameters have values higher than that calculated from the calibration of the steady−state model. The simulated specific yield (Drain−/fillable porosity) ranged from 1.09×10^{-2} to 1.25×10^{-2}. The groundwater leakage was estimated at 8.00×10^{-5}, while the Manning's n coefficient was between 3.24×10^{-2} and 3.70×10^{-2}. Simulated leaching of the Sava River water into the aquifer (outflow) and the groundwater in the Sava River (inflow) can be as high as $20 \text{ m}^3 \text{ s}^{-1}$ or $0.26 \text{ m}^3 \text{ s}^{-1}$, respectively.

Tables in the supplementary material provide water balances for the Ljubljansko polje: (1) which were estimated in the past and were later summarized by Andjelov et al. [58] (see Table S4), (2) after calibration and validation of the steady−state model (Table S5), and (3) after validation of the transient models (Table S6).

4. Discussion

4.1. An Interpretation of Observed Data and an Estimation of the System Responsiveness

Results of a cross-correlation analysis (Figure 8) show a discrepancy in the time delay of groundwater responsiveness to the Sava River discharge dynamics (cross-corr. coef. 0.85–0.25). Posavec et al. [59] have observed a similar range of cross-correlation coefficients between groundwater and the Sava River. Nevertheless, diverse responsiveness of groundwater on local precipitation events at different locations may also be observed, despite the low cross-correlation coefficients (0.12 and 0.25). In general, the groundwater responsiveness responds more rapidly to river events than to precipitation events, due to the higher horizontal than vertical hydraulic conductivity. Post and von Asmuth [60] write that, besides the storage and transmissivity of an aquifer, the volume of an observation well, its screen length, the local permeability of the strata adjacent to the well, and atmospheric pressure also affect the groundwater responsiveness time. Group 2 responds to both of the primary groundwater sources within 24 h (Figure 8), while group 1 responds a month later. Additionally, the hydraulic head pattern in group 2 (Figure S3), which is located adjacent to the Sava River bed stands out from the other groups and has the strongest correlation with the Sava River discharge. Alternatively, the hydraulic head in group 3 correlates most strongly with local precipitation. The estimated time lag between groundwater and the Sava River in Zagreb, Croatia is in the same range as in this study [59].

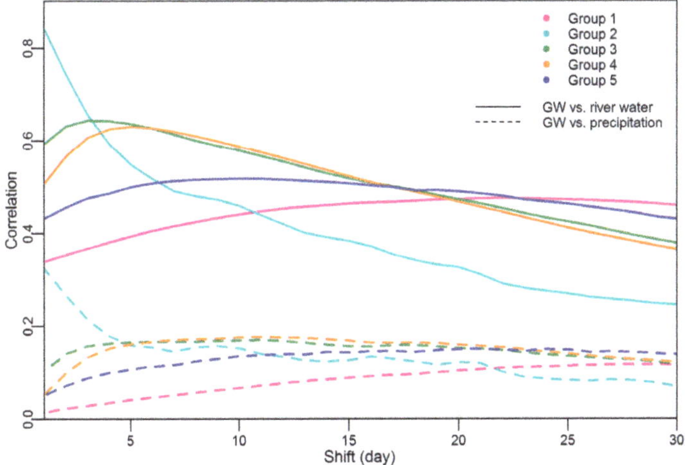

Figure 8. Cross-correlogram between the hydraulic heads and local precipitation levels (dash line)/Sava River discharge (full line) in different time lag data of the hydraulic head. Locations of groups are shown in Figure 2.

The distance between piezometers and the Sava River channel is the main factor in determining groundwater responsiveness. However, the Sava River is not the dominant groundwater source at a certain point. These observations agree with the study provided by Vrzel et al. [20], which is based on stable isotopes. Furthermore, response time is not equal to the groundwater MRT but is related exclusively to pressure changes. These results also show that running the transient state GW flow model at daily time steps is the most reasonable.

4.2. The WaSiM Model

A lower correlation between the observed and calculated hydraulic heads was expected because of the known hidden precipitation event signals that result from diverse groundwater responsiveness over the steady domain (Figure 8) and the presence of a second groundwater source known as the Sava River [20]. For instance, the influence of the Sava River water on the hydraulic head in the well P-013 and the longer MRT of the groundwater at Kleče [20] explains the lowest correlation ($R^2 = 0.53$) between the observed and calculated hydraulic heads in this well (Table S3). Built-up areas and locally presented layers with low hydraulic conductivity obstruct percolation in the aquifer. Additionally, the impact of groundwater abstraction on the hydraulic head was not taken into account in the WaSiM model. Nevertheless, $R^2 = 0.90$ for the well P-102 in the year 2013 (Table S2).

4.3. The Steady-State Groundwater Flow Model

The results in Table S5 show that simulated and estimated (Table S4) water balances are in the same range. A discrepancy between the observed and simulated underground recharge is likely related to the rough estimations of these fluxes in the limited number of published studies and the model. Despite the challenges in quantifying the fluxes of groundwater between aquifers as supported by Asmael et al. (2015) [61], this process deserves more attention in groundwater modeling, since Bouaziz et al. [62] revealed that inter-catchment groundwater flow processes affect water balance at a regional scale. Simulated groundwater leaching into the Sava River and simulated percolation are smaller than literature values (Tables S4 and S5), while underground recharge is not reported in the literature.

To ensure that the steady-state model provides a representative result, a comparison of the travel time of particles in the groundwater between sources and wells was completed as well as comparisons between the Sava River water fractions in groundwater mixtures (estimated by Vrzel et al. [20]) and the estimated system responsiveness (Figure 8). In Figure S4, shorter path lines of particles between the Sava River and Hrastje than those between the Sava River and Kleče can be seen and is the reason for the quick responsiveness of the system to the Sava River water in Hrastje than in Kleče, even though the fraction of the Sava River water is higher in Kleče than in Hrastje. Short paths of particles in Jarški prod explain why the Sava River water dominates in this area and why it has the shortest MRT [20] (Table 3).

The simulated MRT (Figure 9) reveals recharge areas of the aquifer (purple areas in Figure 9), which should receive the most attention for sustainable groundwater management [55]. The MRT, estimated using the $^3H/^3He$ data, is 8.4 years [20] (Table 3), while simulations show an even longer MRT at specific locations. Nevertheless, the old water (Figure 9: reddish areas) in the northern part of the basin and around Rožnik hill may not be accurate due to the influence of the model domain border–the boundary effect. Figure 9 explains that the five oscillation patterns of the hydraulic heads (Figure S3) and the diverse aquifer responsiveness over the steady domain (Figure 8) are related to the different recharge areas and groundwater flow directions. Additionally, calculated hydrological conductivity in the Ljubljansko polje agree with literature data and describe the poorly permeable layers in the area with lower simulated hydraulic conductivity [22].

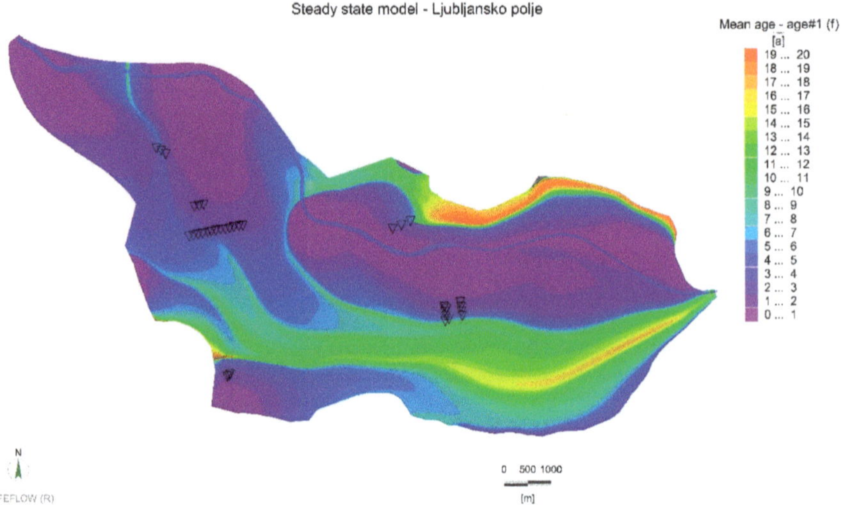

Figure 9. Simulated mean residence time. Black triangles show positions of pumping wells. The steady-state model gives a clear picture of groundwater flow recharge areas and BC. However, for understanding the system in time and under diverse climatic conditions, a transient state GW flow model is required.

4.4. The Transient State Groundwater Flow Model

A water balance for the transient state model (Table S6) was estimated for the 2 November 2010 high water level and the 1 July 2010 low water level in the Sava River. In general, water exchange rates are higher in the transient state model than in the steady-state model (Table S5), as well as in the water balance calculated from the literature (Table S4). Similarly, Ackerman et al. [63] observed higher exchange rates in the transient than in the steady-state model in the eastern Snake River Plain aquifer, USA. The simulated leaching of the Sava River water in the aquifer was higher when the Sava River water level was lower (1 July 2010). This observation may be explained by the higher pressure in the aquifer on the 2 November 2010, due to extremely wet conditions in September 2010 (Figure 10).

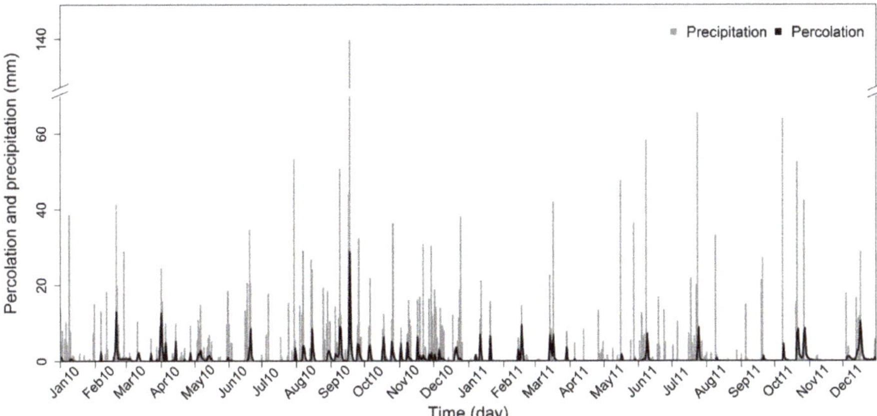

Figure 10. Daily precipitation observed in Bežigrad (mm) and percolation simulated in WaSiM (mm) for the full model domain–the Ljubljansko polje.

A comparison between the observed and simulated hydraulic heads after validation of the transient state model is shown in Figure 11. Their time series also match their maximal values, and only subtle deviations that are not critical were observed. These deviations are likely due to the use of only one model layer since multiple model layers can more realistically simulate the pressure in the aquifer, which is a fact confirmed by Zhou and Herath [64]. In order to be able to parameterize geological layers in more detail, geostatistical analyses [65] can be applied (besides obtaining new *in-situ* measurements). Realistic oscillations of the hydraulic head in the piezometers located near the Sava River (P-017) and a good agreement between the observed and simulated discharges and water levels in the Sava River (Figure S5) confirm that groundwater flow boundary conditions are well defined.

The simulated Sava River discharge was slightly lower than that observed, while their oscillation patterns match closely (Figure S5). The lower level of recharge downstream along the Sava River is another indicator of the leaching of the Sava River water into the aquifer (outflow). The outflow and inflow (a flux from the aquifer into the Sava River) have the opposite effect on the Sava River discharge. When the Sava River discharge increases, the outflow also increases, but the inflow decreases. Exact locations of the river water-groundwater interactions are presented in Figure 12 and Figure S6. In the literature, it is reported that groundwater drains into the Sava River only downstream of Šentjakob. However, model simulations indicate that this process already occurs upstream from Šentjakob (Figure 12).

Groundwater flow under different conditions, including high (2 November 2010) and low (1 July 2010) discharge in the Sava River, is shown in Figure S6. It should be noted that the data from 2 November 2010 is not representative of normal wet conditions. The system was still under the recovery phase after recent extremely wet conditions. However, a connection between the travelling time of particles in the groundwater between their sources and piezometers and the estimated system responsiveness can be observed (Figure S4).

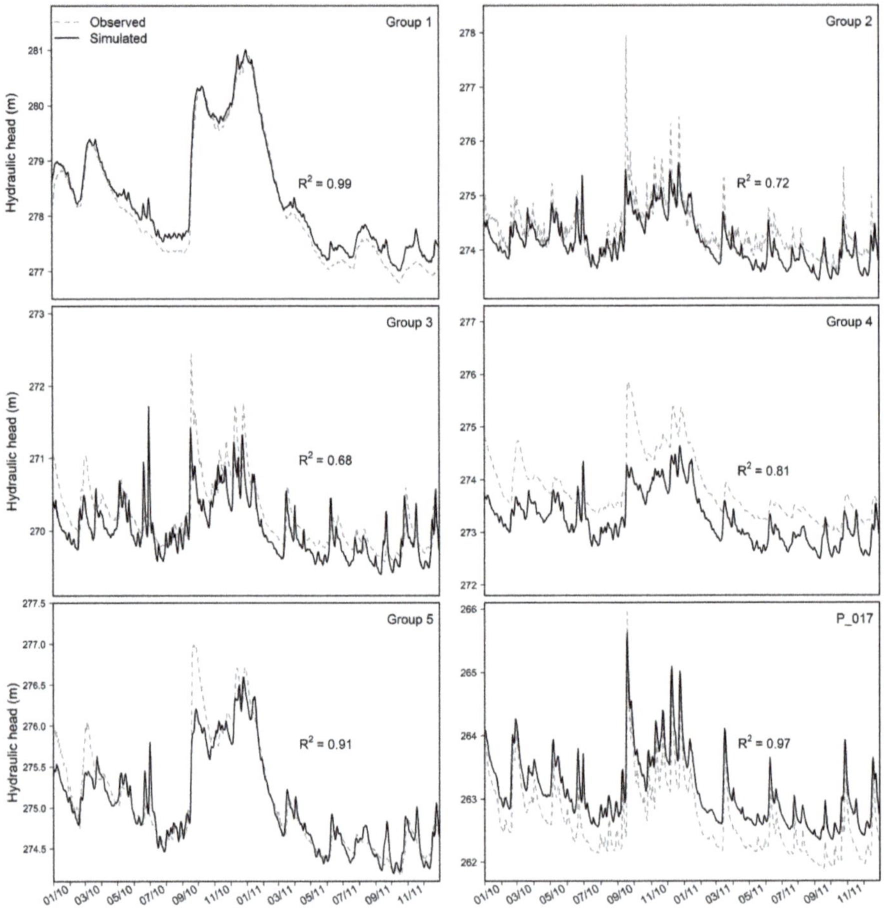

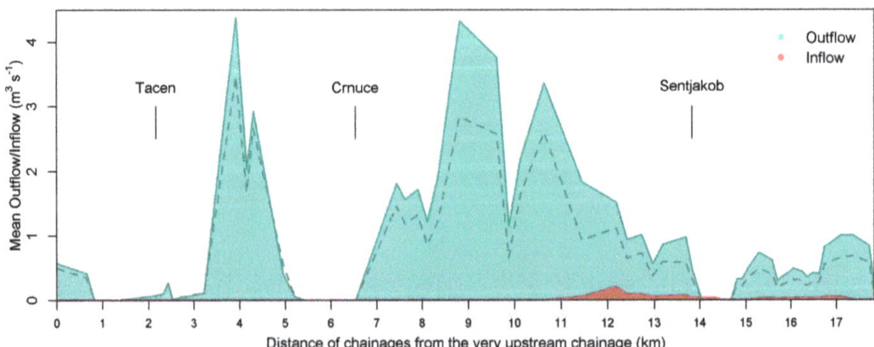

Figure 11. Observed and simulated hydraulic heads for the groups presented in Figure 2. Hydraulic heads were simulated in FEFLOW for the period 1 January 2010–31 December 2011.

Figure 12. The Sava River water-groundwater interactions areas, and mean seepage of river water into the aquifer (inflow) and vice versa (outflow) during 2010 (full line) and 2011 (dash line).

The travel time of the particles differs from the MRT estimated using the ^{3}H/^{3}He method and the simulated MRT (Figure 9). The reason is the different mathematical background of both tools. The calculation of particle travel time does not take into account the mixing of water from different sources, unlike in the simulation of the MRT. The logic behind the particle travel time calculation is similar to a piston flow by assuming that a parcel of water moves as a discrete volume along a flow path [66]. The results also show how Trnovo represents an important underground inflow of groundwater from the Ljubljansko Barje into the system.

Deviations also occur between groundwater flow paths in the steady and transient state models. For instance, Roje is the main recharge area for Hrastje in the transient state model, while Tomačevo is the main recharge area in the steady-state model. Figure 9 shows the mixing of groundwater around Hrastje from different recharge areas, which depends on hydrological conditions. The interpretation of the recharge area for groundwater in Hrastje might be ambiguous. However, the long MRT [20] (Table 3) and the small estimated Sava River water fractions in Hrastje [20] (Figure 6) suggest that the main recharge area for groundwater in Hrastje is in the North-Western part of the Ljubljansko polje.

According to Table 5, the observed and simulated values are most similar when the system contains more water. The highest discrepancy between the observed and simulated hydraulic heads is in the area around group 4 (Figure 11). One reason for this might be a lack of information about the depth of the bedrock. The dynamic bedrock elevation in this area was predicted based on limited information. A second reason might be an underestimation of the underground recharge in Trnovo.

Table 5. Statistics of the calibrated models after steps one, two, and five. Statistics for the calibrated transient state model (after step five) are presented for two days 2 November 2010 and 1 July 2010, when the Sava River recharge was high and low, respectively.

	Step 1 (Steady-State GW Flow Model)	Step 2 (Steady-State GW Flow Model + MRT)	Step 5 2 November 2010 (306 day)	Step 5 1 July 2010 (182 day)
$\bar{E}$	0.125	0.260	0.445	0.943
RMS	0.169	0.343	0.563	1.062
σ	0.173	0.351	0.575	1.085

5. Conclusions

An integrated approach using geochemical and multivariate statistical methods were used to study surface-groundwater interactions in the Ljubljansko polje aquifer system. The study involved collecting and processing a large amount of historical data and defining the isotopic composition of the groundwater. The main success of the work was the unravelling of information about groundwater behavior hidden in the data obtained from different sources and combining this into a hydrological model.

The paper presents the conceptualization of the study domain using chronological data and hydrological modeling. The chronological data of the hydraulic heads observed in many piezometers show five oscillation patterns that form five groups. The cross-correlation analyses revealed varying responsiveness of these five groups to the two main groundwater sources: the Sava River and local precipitation. Furthermore, the simulated mean residence time can explain the different patterns observed in the different recharge areas, and it identifies those areas that should receive special attention regarding groundwater protection.

Cross-correlation analyses reveal the importance of understanding the system in time, since the response to an increase in the high-water level of the Sava River or local precipitation can be fast, i.e., within 24 h. For this reason, a transient state model was developed to provide a reliable simulation of the complex interactions between groundwater and surface water that dominates the system's evolution (as is shown in the conceptual model) by combining multiple models. The groundwater flow was simulated using FEFLOW and MIKE 11, which were coupled using the ifmMIKE11 plug-in, for the simulation of the Sava River discharge, and fluxes between the river and groundwater water.

Percolation was simulated in WaSiM. For this reason, an indirect communication link was established between FEFLOW and WaSiM.

Solving hydrological issues related to sensitive hydrological systems can be time-consuming and complicated. However, the results of studies like this one provide valuable and unique information. It is even possible to quantify important components in the water balance of the basin under different conditions. This study also provided a complete picture of the full water cycle in the Ljubljansko polje. The GW flow model has great potential for many other applications such as (1) testing the impact of changing land use on the water cycle, including groundwater-surface water interactions, (2) projecting future hydrological conditions, and (3) as a model for simulating water tracers (e.g., isotopes) and pollutants (e.g., nutrients, pharmaceuticals, and toxic metals). The study results can also aid in the decision-making process regarding sustainable groundwater management in the Ljubljansko polje aquifer system, which is the main source of drinking water for the growing population of Ljubljana.

Supplementary Materials: The following are available online at http://www.mdpi.com/2073-4441/11/9/1753/s1. Figure S1: Distribution of observed hydraulic heads over time, Figure S2: Box plot of hydraulic heads observed in 50 piezometers over the period 2010–2011, Figure S3: Hydraulic head patterns of all five groups observed over the period 2010–2011, Figure S4: Travel time, backwards streamlines seeds (y), hydraulic head (m), and bedrock elevations (m), Figure S5: The upper two plots show outflows/inflows (m^3 s^{-1}) for random Sava River cross-sections simulated in MIKE 11 with ifmMIKE11 plug-in in FEFLOW Observed (full lines) and simulated (dash lines) discharges and water levels in the Sava River over the validated period 1 January 2010–31 December 2011, Figure S6: Groundwater flow under different conditions in the Sava River: (a) high discharge (left: upper legend) for 6 May 2010, (b) low discharge (right: lower legend) for 1 July 2010, Table S1: Lithological composition of the suggested model layers (ML) [67], Table S2: Linear correlations (R^2) between observed and calculated real evapotranspiration and trend of the hydraulic head after the calibration (period: January 2010–December 2014), Table S3: Correlations (R^2) between observed and calculated real evapotranspiration and trend of the hydraulic head after the validation (period: January 2009–December 2009), Table S4: Estimated components of a groundwater budget in the Ljubljansko polje (m^3 s^{-1}) [58], Table S5: Water balance after calibration and validation of the steady-state GW flow model for data observed over the period 2010–2011 and 2013–2014, respectively, Water balance after calibration for 2010–2011 data and included simulation of MRT (after step 2), Table S6: Water balance for a validated transient state model (after the step five) for two days 2 November 2010 (high water level in the Sava River) and 1 July 2010 (low water level in the Sava River).

Author Contributions: Conceptualization, J.V., N.O., and R.L. Methodology, J.V., N.O., and R.L. Software, J.V. Validation, J.V. Formal analysis, J.V. Investigation, J.V. Data curation, J.V. Writing—original draft preparation, J.V. Writing-review & editing, J.V., N.O., and R.L. Visualization, J.V. Supervision, N.O., R.L., and G.V.

Funding: The EU 7th Research Project–GLOBAQUA (Managing the effects of multiple stressors on aquatic ecosystem under water scarcity), grant number 603629-ENV-2013-6.2.1, funded this research. The research was partially conducted within the doctoral study of Janja Vrzel financed by the European Social Fund (KROP 2012) no. C2130-12-000070. The research activities were also performed within the national program P1-0143 and project L1-9191–Illicit drugs, alcohol and tobacco: wastewater based epidemiology, treatment efficiency, and vulnerability assessment of water catchments funded by the Slovenian Research Agency (ARRS).

Acknowledgments: The authors would like to thank all those institutes and companies who shared their data including the following: Environmental Agency of Slovenia, Holding Slovenske Elektrarne d.o.o., Institute for Water of the Republic of Slovenia (IzVRS), Javno Podjetje Vodovod-Kanalizacija Ljubljana (VOKA), Pivovarna Lašo Union d.o.o., TIS/ICPVO–Infrastructural Centre for Pedology and Environmental Protection, Biotechnical Faculty, University of Ljubljana, The City Municipality of Ljubljana, and The Surveying and Mapping Authority of the Republic of Slovenia. The authors would also like to thank MIKE Powered by DHI for the use of their software FEFLOW and the plug-in ifmMIKE11.

Conflicts of Interest: The authors declare no conflict of interest. The funders had no role in the design of the study, in the collection, analyses, or interpretation of the data, in the writing of the manuscript, or in the decision to publish the results.

References

1. Fleckenstein, J.H.; Krause, S.; Hannah, D.M.; Boano, F. Groundwater-surface water interactions: New methods and models to improve understanding of processes and dynamics. *Adv. Water Resour.* **2010**, *33*, 1291–1295. [CrossRef]

2. National Academic of Science, Engineering, and Medicine. *From Maps to Models: Augmenting the Nation's Geospatial Intelligence Capabilities*; The National Academies Press: Washington, DC, USA, 2016; p. 121.

3. Gyopari, M.C.; McAlister, D. *Wairarapa Valley Groundwater Resource Investigation: Upper Valley Catchment Hydrogeology and Modelling*; Technical Publication GW/EMI-T-10/74; Greater Wellington: Wellington, New Zealand, 2010; p. 108.

4. Monninkhoff, B.L.; Li, Z. Coupling FEFLOW and MIKE11 to optimise the flooding system of the Lower Havel polders in Germany. *Int. J. Water* **2009**, *5*, 163–180. [CrossRef]

5. Natkhin, M. Modellgestützte Analyse der Einflüsse von Veränderungen der Waldwirtschaft und des Klimas auf den Wasserhaushalt Grundwasserabhängiger Landschaftselemente. Master's Thesis, University of Potsdam, Potsdam, Germany, 2010. (In German).

6. Gunduz, O.; Aral, M. Surface Water—Groundwater Interactions: Integrated Modeling of a Coupled System. In *Handbook of Applied Hydrology*; Singh, V.P., Ed.; McGraw-Hill Inc.: New York, NY, USA, 2016; pp. 54.1–54.14.

7. Barthel, R.; Banzhaf, S. Groundwater and Surface Water Interaction at the Regional-Scale—A Review with Focus on Regional Integrated Models. *Water Resour. Manag.* **2016**, *30*, 1–32. [CrossRef]

8. Krause, S.; Bronstert, A. The impact of groundwater–surface water interactions on the water balance of a mesoscale lowland river catchment in northeastern Germany. *Hydrol. Process.* **2007**, *21*, 169–184. [CrossRef]

9. Wei, X.; Bailey, R.T.; Records, R.M.; Wible, T.C.; Arabi, M. Comprehensive simulation of nitrate transport in coupled surface-subsurface hydrologic systems using the linked SWAT-MODFLOW-RT3D model. *Environ. Model. Softw.* **2018**. [CrossRef]

10. Trichakis, I.; Burek, P.; De Roo, A.; Pistocchi, A. Towards a Pan-European Integrated Groundwater and Surface Water Model: Development and Applications. *Environ. Process.* **2017**, *17*, 81–93. [CrossRef]

11. Rodriguez, L.B.; Cello, P.A.; Vionnet, C.A.; Goodrich, D. Fully conservative coupling of HEC-RAS with MODFLOW to simulate stream–aquifer interactions in a drainage basin. *J. Hydrol.* **2008**, *353*, 129–142. [CrossRef]

12. Inter 3 Institute for Resource Management GmbH. Integrated Water Resources Management Zayandeh Rud: German-Iranian Research and Development Cooperation for a Better Future. Available online: https://www.inter3.de/fileadmin/user_upload/Downloads/Flyer_usw/download_IWRM-en.pdf (accessed on 16 August 2017).

13. Guay, C.; Nastev, M.; Paniconi, C.; Sulis, M. Comparison of Two Modeling Approaches for Groundwater–Surface Water Interactions. *Hydrol. Process.* **2013**, *27*, 2258–2270. [CrossRef]

14. Cheo, A.E.; Voigt, H.-J.; Wendland, F. Modeling groundwater recharge through rainfall in the Far-North region of Cameroon. *Groundw. Sustain. Dev.* **2017**, *5*, 118–130. [CrossRef]

15. Andjelov, M.; Mikulič, Z.; Tetzlaff, B.; Wendland, F.; Uhan, J. *Groundwater Recharge in Slovenia—Results of a Bilateral German-Slovenian Research Project*; Forschungszentrum Jülich GmbH Zentralbibliothek: Jülich, Germany, 2016; Volume 339.

16. Epting, J.; Huggenberger, P.; Radny, D.; Hammes, F.; Hollender, J.; Page, R.M.; Weber, S.; Bänninger, D.; Auckenthaler, A. Spatiotemporal scales of river-groundwater interaction—The role of local interaction processes and regional groundwater regimes. *Sci. Total Environ.* **2018**, *618*, 1224–1243. [CrossRef]

17. Feinstein, D. Since "Groundwater and Surface Water—A Single Resource": Some U.S. Geological Survey advances in modeling groundwater/surface-water interactions. *Acque Sotter. Ital. J. Groundw.* **2012**, *1*, 9–24. [CrossRef]

18. Brunner, P.; Simmons, C.T.; Cook, P.G. Spatial and Temporal Aspects of the Transition from Connection to Disconnection between Rivers, Lakes and Groundwater. *J. Hydrol.* **2009**, *376*, 159–169. [CrossRef]

19. Fleckenstein, J.H.; Niswonger, R.G.; Fogg, G.E. River-Aquifer Interactions, Geologic Heterogeneity, and Low-Flow Management. *Ground Water* **2006**, *44*, 837–852. [CrossRef] [PubMed]

20. Vrzel, J.; Solomon, D.K.; Blažeka, Ž.; Ogrinc, N. The study of the interactions between groundwater and Sava River water in the Ljubljansko polje aquifer system (Slovenia). *J. Hydrol.* **2018**, *556*, 384–396. [CrossRef]

21. Urbanc, J.; Jamnik, B. Isotope investigations of groundwater from Ljubljansko polje (Slovenia). *Geologia* **1998**, *41*, 355–364. [CrossRef]

22. Šram, D.; Brenčič, M.; Lapanje, A.; Janža, M. Perched Aquifers Spatial Model: A Case Study for Ljubljansko Polje (Central Slovenia). *Geologija* **2012**, *55*, 107–116. [CrossRef]

23. Cerar, S.; Urbanc, J. Carbonate Chemistry and Isotope Characteristics of Groundwater of Ljubljansko Polje and Ljubljansko Barje Aquifers in Slovenia. *Sci. World J.* **2013**, *2013*, 1–11. [CrossRef] [PubMed]

24. Auersperger, P.; Čenčur Curk, B.; Jamnik, B.; Janža, M.; Kus, J.; Prestor, J.; Urbanc, J. Dinamika Podzemne Vode. In *Podtalnica Ljubljanskega Polja*; Rejec Brancelj, I., Smrekar, A., Kladnik, D., Eds.; Založba ZRC: Ljubljana, Slovenia, 2005; pp. 39–61. (In Slovene)

25. All Work Package Leaders & the whole CC-WaterS Project Consortium. Climate Change and Impacts on Water Supply "CC-WaterS". Available online: http://www.ccwaters.eu/downloads/CC-WaterS_Project_Monography_final.pdf (accessed on 16 August 2017).

26. Janža, M.; Meglič, P.; Šram, D. *Numerical Hydrological Modelling*; Final Report P-II-30d/b-5/1-d; Geological Survey of Slovenia: Ljubljana, Slovenia, 2011; p. 66.

27. Slovenian Environment Agency. Dostop Do Podatkov o Vodah v Sloveniji (in Slovene). Available online: http://vode.arso.gov.si (accessed on 12 August 2017).

28. Slovenian Environment Agency, METEO. Available online: http://meteo.arso.gov.si (accessed on 12 August 2017). (In Slovene)

29. Schulla, J. *Model Description WaSiM (Water Balance Simulation Model)*; Technical Report; Hydrology Software Consulting J. Schulla: Zurich, Switzerland, 2015; p. 332.

30. Strojan, I.; Kobold, M.; Polajnar, J.; Šupek, M.; Pogačnik, N.; Jeromelj, M.; Petan, S.; Lalič, B.; Trček, R. Poplave v Dneh od 17. do 21. Septembra 2010. *Miš. Vod. Dan: Zbor. Ref.* **2010**, *21*, 1–11. (In Slovene)

31. Cegnar, T. Climate in Slovenia in 2011. *Ujma* **2012**, *26*, 20–32. (In Slovene)

32. Wessolek, G.; Kaupenjohann, M.; Renger, M. *Bodenphysikalische Kennwerte Und Berechnungsverfahren Für Die Praxis*; Facklam, M., Ed.; Technische Universitaet Berlin: Berlin, Germany, 2009. (In Germany)

33. Renger, M.; Bohne, K.; Facklam, M.; Harrach, T.; Riek, W.; Schäfer, W.; Wessolek, G.; Zacharias, S. *Ergebnisse und Vorschläge der DBG-Arbeitsgruppe "Kennwerte des Bodengefüges" zur Schätzung Bodenphysikalischer Kennwerte (in German)*; Deutsche Bodenkundliche Gesellschaft: Goettingen, Germany, 2008; p. 51.

34. Soil Moisture Classification. Available online: http://www.terragis.bees.unsw.edu.au/terraGIS_soil/sp_water-soil_moisture_classification.html (accessed on 3 May 2017).

35. CORINE Land Cover Database of the Year 2006. Available online: https://www.eea.europa.eu/data-and-maps/data/clc-2006-raster-4#tab-european-data (accessed on 5 December 2016).

36. Pignotti, G.; Rathjens, H.; Cibin, R.; Chaubey, I.; Crawford, M. Comparative Analysis of HRU and Grid-Based SWAT Models. *Water* **2017**, *9*, 272. [CrossRef]

37. Moving Windows. Available online: http://www.gitta.info/ContiSpatVar/en/html/SpatDependen_learningObject2.xhtml (accessed on 27 December 2017).

38. Refsgaard, J.C.; Storm, B. Construction, Calibration and Validation of Hydrological Models. In *Distributed Hydrological Modelling*; Abbott, M.B., Refsgaard, J.C., Eds.; Springer: Dordrecht, The Netherlands, 1990; Volume 22, pp. 41–54.

39. Povprečne Mesečne Vrednosti Evapotranspiracije v Obdobju 1971–2000 (in Slovene). Available online: http://meteo.arso.gov.si/met/sl/agromet/period/etp/ (accessed on 30 September 2018).

40. Pintar, M.; Korpar, P.; Zupanc, V. *Meritve na Lizimetrski Postaji v Klečah: Poročilo za Leto 2012*; University of Ljubljana: Ljubljana, Slovenia, 2013. (In Slovene)

41. Monninkhoff, B. *IfmMIKE11 2.1: User Manual*; DHI-WASY GmbH: Berlin, Germany, 2014; p. 132.

42. Diem, S.; Renard, P.; Schirmer, M. Assessing the effect of different river water level interpolation schemes on modeled groundwater residence times. *J. Hydrol.* **2014**, *510*, 393–402. [CrossRef]

43. Cardenas, M.B. Stream-aquifer interactions and hyporheic exchange in gaining and losing sinuous streams. *Water Resour. Res.* **2009**, *45*, 1–13. [CrossRef]

44. Cardenas, M.B.; Wilson, J.L.; Zlotnik, V.A. Impact of heterogeneity, bed forms, and stream curvature on subchannel hyporheic exchange. *Water Resour. Res.* **2004**, *40*, 1–13. [CrossRef]

45. Woessner, W.W. Stream and Fluvial Plain Ground Water Interactions: Rescaling Hydrogeologic Thought. *Ground Water* **2000**, *38*, 423–429. [CrossRef]

46. Jarrett, R.D. *Determination of Roughness Coefficients for Streams in Colorado*; Water-Resources Investigations 85–4004; U.S. Geological Survey: Lakewood, CO, USA, 1985; p. 54.

47. Steinman, F.; Banovec, P.; Gosar, L.; Rak, G.; Pogačnik, N.; Koželj, D.; Petkovšek, G.; Pipan, G.; Artač, M.; Šterk, M.; et al. *Interactive Visualization of Floodplain Areas for Disaster Response Support*; Final Report; Administration of the RS for Civil Protection and Disaster Relief, Ministry of Defence, Republic of Slovenia: Ljubljana, Slovenia, 2008; p. 258. (In Slovene)

48. Diersch, H.J.G. *FEFLOW Finite Element Modeling of Flow, Mass and Heat Transport in Porous and Fractured Media*; Springer: Berlin/Heidelberg, Germany, 2014.

49. Monninkhoff, L.; Hartnack, J. Improvements in the Coupling Interface between FEFLOW and MIKE11. 2009, pp. 1–12. Available online: https://pdfs.semanticscholar.org/aac7/a7ee00b41562869481555d7db99f2587c668.pdf?_ga=2.130108879.1923207088.1566342415-1995382604.1562081892 (accessed on 16 August 2017).

50. Kaiser, B.O.; Scheck-Wenderoth, M.; Cacace, M.; Lewerenz, B. Characterization of main heat transport processes in the Northeast German Basin: Constraints from 3-D numerical models. *Geochem. Geophys. Geosystems* **2011**, *12*, 1–17. [CrossRef]

51. Interpolation Methods. Available online: http://www.gisresources.com/types-interpolation-methods_3/ (accessed on 6 July 2017).

52. FePEST in FEFLOW 7.0, User Guide. Available online: https://www.mikepoweredbydhi.com/download/product-documentation (accessed on 16 August 2017).

53. Toews, M.; Daughney, C.J.; Cornaton, F.J.; Morgenstern, U.; Evison, R.D.; Jackson, B.M.; Petrus, K.; Mzila, D. Numerical simulation of transient groundwater age distributions assisting land and water management in the Middle Wairarapa Valley, New Zealand. *Water Resour. Res.* **2016**, *52*, 9430–9451. [CrossRef]

54. Gusyev, M.A.; Toews, M.; Morgenstern, U.; Stewart, M.; White, P.; Daughney, C.; Hadfield, J. Calibration of a transient transport model to tritium data in streams and simulation of groundwater ages in the western Lake Taupo catchment, New Zealand. *Hydrol. Earth Syst. Sci.* **2013**, *17*, 1217–1227. [CrossRef]

55. Sanford, W. Calibration of Models Using Groundwater Age. *Hydrogeol. J.* **2011**, *19*, 13–16. [CrossRef]

56. Sheets, R.A.; Bair, E.S.; Rowe, G.L. Use of 3 H/3 He Ages to evaluate and improve groundwater flow models in a complex buried-valley aquifer. *Water Resour. Res.* **1998**, *34*, 1077–1089. [CrossRef]

57. Doherty, J. Ground Water Model Calibration Using Pilot Points and Regularization. *Ground Water* **2003**, *41*, 170–177. [CrossRef]

58. Andjelov, M.; Bat, M.; Frantar, P.; Mikulič, Z.; Savić, V.; Uhan, J. Pregled Elementov Vodne Bilance. In *Podtalnica Ljubljanskega Polja*; Rejec Brancelj, I., Smrekar, A., Kladnik, D., Eds.; Založba ZRC: Ljubljana, Slovenia, 2005; pp. 27–38. (In Slovene)

59. Posavec, K.; Vukojević, P.; Ratkaj, M.; Bedeniković, T. Cross-Correlation Modelling of Surface Water—Groundwater Interaction Using the Excel Spreadsheet Application. *Rud. Geološko Naft. Zb.* **2017**, *32*, 25–32. [CrossRef]

60. Post, V.E.A.; Von Asmuth, J.R.; Asmuth, J.R. Review: Hydraulic head measurements—New technologies, classic pitfalls. *Hydrogeol. J.* **2013**, *21*, 737–750. [CrossRef]

61. Asmael, N.; Dupuy, A.; Huneau, F.; Hamid, S.; Le Coustumer, P. Groundwater Modeling as an Alternative Approach to Limited Data in the Northeastern Part of Mt. Hermon (Syria), to Develop a Preliminary Water Budget. *Water* **2015**, *7*, 3978–3996. [CrossRef]

62. Bouaziz, L.; Weerts, A.; Schellekens, J.; Sprokkereef, E.; Stam, J.; Savenije, H.; Hrachowitz, M. Redressing the balance: Quantifying net intercatchment groundwater flows. *Hydrol. Earth Syst. Sci.* **2018**, *22*, 6415–6434. [CrossRef]

63. Ackerman, D.J.; Rousseau, J.P.; Rattray, G.W.; Fisher, J.C. Steady-state and transient models of groundwater flow and advective transport, Eastern Snake River Plain aquifer, Idaho National Laboratory and vicinity, Idaho. *Sci. Investig. Rep.* **2010**. [CrossRef]

64. Zhou, Y.; Herath, H. Evaluation of alternative conceptual models for groundwater modelling. *Geosci. Front.* **2017**, *8*, 437–443. [CrossRef]

65. Guastaldi, E.; Carloni, A.; Pappalardo, G.; Nevini, J. Geostatistical Methods for Lithological Aquifer Characterization and Groundwater Flow Modeling of the Catania Plain Quaternary Aquifer (Italy). *J. Water Resour. Prot.* **2014**, *6*, 272–296. [CrossRef]

66. Anderson, M.P.; Woessner, W.W.; Hunt, R.J. *Applied Groundwater Modeling Simulation of Flow and Advective Transporte*, 2nd ed.; Elsevier Inc.: London, UK, 2015.

67. Vrzel, J.; Ogrinc, N.; Vižintin, G. Data Preparation for Groundwater Modelling—Ljubljansko Polje Aquifer System. *RMZ M&G* **2015**, *62*, 167–173.

 *water*

Article

Using Water Stable Isotopes for Identifying Groundwater Recharge Sources of the Unconfined Alluvial Zagreb Aquifer (Croatia)

Jelena Parlov [1], Zoran Kovač [1,*], Zoran Nakić [1] and Jadranka Barešić [2]

1   Faculty of Mining, Geology and Petroleum Engineering, University of Zagreb, 10 000 Zagreb, Croatia; jelena.parlov@rgn.hr (J.P.); zoran.nakic@rgn.hr (Z.N.)
2   Ruđer Bošković Institute, 10 000 Zagreb, Croatia; jbaresic@irb.hr
*   Correspondence: zoran.kovac@rgn.hr

Received: 11 September 2019; Accepted: 17 October 2019; Published: 19 October 2019

Abstract: The main purpose of this study was to understand the interactions between precipitation, surface water, and groundwater in the Zagreb aquifer system using water stable isotopes. The Zagreb aquifer is of the unconfined type and strongly hydraulically connected to the Sava River. As the groundwater is the main source of drinking water for one million inhabitants, it is essential to investigate each detail of the recharge processes of the aquifer to ensure adequate protection of the groundwater. Measuring the content of water stable isotopes in surface waters and groundwater enabled the creation of two- and three-component mixing models based on the isotopic mass balance for the purpose of the quantification of each recharge component. The mixing models gave ambiguous results. Observation wells equally distant from the Sava River did not have the same recharge component ratio. This indicated that there were more factors (in addition to the distance from the river) that were affecting groundwater recharge, and the properties of the unsaturated zone and surface cover data were therefore also taken into consideration. The thickness of the unsaturated zone and the characteristics of different soil types were identified as important factors in the recharge of the Zagreb aquifer. The areas with high thickness of the unsaturated zone and well-permeable soil had a very similar recharge component ratio to the areas with small thickness of the unsaturated zone but low-permeable soil.

Keywords: Zagreb aquifer; Sava River; groundwater–surface water interaction; water stable isotopes; soil properties

---

1. Introduction

Groundwater presents one of the most important water resources in the world. Water demand is continuously growing worldwide, and it is especially pronounced in arid regions [1]. Global groundwater extraction has reached about 1500 km^3 per year [2], which is higher than natural groundwater recharge [3]. Furthermore, a lot of alluvial aquifers have problems with groundwater depletion, which has become a global issue [4,5]. Recently, the problem has become more frequent in moderate climate areas, especially in the unconfined alluvial aquifers dependent on surface waters as a result of climate changes [6–8]. All these problems point to the necessity of detailed research of groundwater recharge mechanisms, which will enable sustainable management of groundwater resources, i.e., ensure sufficient quantities of potable water.

Groundwater–surface water interaction is generally investigated at the river scale and classified as connected or disconnected systems [9]. In connected systems, groundwater can discharge through the streambed to contribute to streamflow, and surface water can lose water, i.e., recharge the aquifer. Disconnected systems depend on the unsaturated zone beneath the surface water and can be completely disconnected (changes in the water table do not affect the infiltration rate from the river) and in a "transitional" state (changes in the water table possibly affect infiltration rate). However, a lot of alluvial aquifers that are connected with rivers have very dynamic relationships, where hydrologic conditions can generate very fast changes in groundwater flow direction and velocities. From that perspective, the isotopic composition of the different compounds can help to identify and quantify hydrological relationships. If hydrological conditions want to be evaluated on a local scale, where changes in the water isotopic composition can be very small, relevant hydrological relationships can sometimes be very difficult to define. In such aquifers, the characteristics of the unsaturated zone and soil type can have a significant impact on the isotope composition.

The Zagreb aquifer is a very important source of potable water for the Republic of Croatia, and it is designated as a part of the country's strategic water reserves. Although different isotopic research has been conducted, related to both groundwater quality and quantity [10–13], a distinct quantification of recharge of the Zagreb aquifer for both precipitation and infiltration from the Sava River has not been done before.

The main goal of this research was to evaluate groundwater–surface water interaction in the area of the Zagreb aquifer and to quantify the role of precipitation and the Sava River on the recharge of the aquifer. For this purpose, water stable isotopes ($\delta^{18}O$ and δ^2H) from surface water, groundwater, and precipitation were used. Stable isotopes of hydrogen (^1H and ^2H) and oxygen (^{16}O and ^{18}O) are conservative and can be used as tracers in different types of hydrological research [14–20]. Two- and three-component mixing models were used to quantify different recharge sources together with soil characteristics and hydrogeological properties of the saturated and unsaturated zones.

2. Study Area

The Zagreb aquifer covers approximately 300 km^2 and is located in the northwestern part of the Republic of Croatia (Figure 1). It presents the only source of potable water for approximately one million inhabitants of the City of Zagreb and part of the Zagreb County. The municipal water supply relies on the groundwater from the aquifer, which is abstracted at six currently operating well fields. During longer dry periods, smaller well fields located within the urban part of the City of Zagreb, which are usually not used by the water supply system, are put into operation. Total annual abstraction at Zagreb well fields in some years exceeds the annual groundwater recharge [21].

The Zagreb aquifer is an unconfined aquifer, which consists of two aquifer layers that are hydraulically connected. It consists of Quaternary sediments, which were deposited during the Middle and Upper Pleistocene and the Holocene. Holocene deposits belong to alluvial deposits, while Pleistocene deposits are lacustrine–marshy [22–24]. The main stratigraphic units have been identified according to microfloral and microfaunal analysis [25,26]. The boundary between Pleistocene and Holocene deposits has also been determined according to changes in the petrographic origin of the clasts in gravels and sands and changes in the sedimentary environment. It has been reported that, in the beginning of the Holocene, the Sava River started to flow and transport material from the Alps, which was dominantly carbonate, unlike Pleistocene deposits, which are mostly siliciclastic [22]. In general, Lower Pleistocene deposits are mainly composed of silty clays and clayey silts, with sporadic interbeds and lenses of gravelly sands. Lower and middle part of Middle Pleistocene deposits are made of sands, while clays and silts appear in the upper part. In Late Pleistocene deposits, frequent lateral changes in sands, gravels, silts, and clays can be found. Holocene deposits are mostly composed of sands and gravels [23].

Hydrogeologically, these deposits can be grouped into three units: unsaturated zone, shallow alluvial aquifer layer, and deeper lacustrine–marshy aquifer layer. The thickness of the unsaturated zone varies spatially and in time due to the hydrological conditions. It varies from 5 to 13 m during dry periods and low groundwater levels (Figure 2). In the area of the Zagreb aquifer, especially in the area where monitoring of groundwater quality is established (mostly according to the inflow areas of well fields), two main types of soils have developed: Fluvisols and Eutric Cambisols on Holocene deposits.

Apart from these types of soils, Pseudogley on plateau, Pseudogley–gley, Gleyic Fluvisols, Gleysol, Pseudogley, and Vertic Gleysols can also be found in small areas [27,28]. Hydraulic properties of the Fluvisols are generally permeable, with the exception of dry periods [29]. On the other hand, the Eutric Cambisols developed on Holocene deposits seem to be of low permeability, where percolation to the aquifer is very low [30]. Hydraulic conductivity values of the unsaturated zone vary between 1.26 m/day and 1015 m/day [11]. It must be emphasized that, in some parts of the Zagreb aquifer, the unsaturated zone is damaged by human influence (setting foundations for buildings, for example), especially on the left bank of the Sava River, where urban areas prevail.

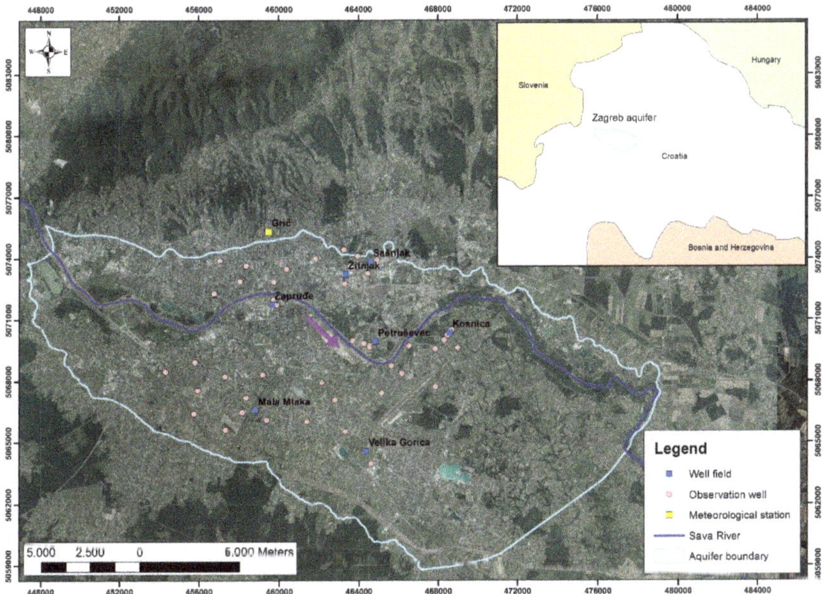

Figure 1. The location of the Zagreb aquifer system.

The shallow aquifer layer is in direct contact with the Sava River, with general groundwater flow direction from W/NW to E/SE. The Sava River presents the main source of recharge, and its influence on changes in groundwater levels is more pronounced in the vicinity of the river itself [31]. It has been shown that the Sava River drains the aquifer in some areas during low and medium water levels, while surface water infiltrates into the aquifer during high water levels. Furthermore, it has been found that a no-flow boundary prevails in the north of the Zagreb aquifer. Inflow boundaries were determined in the south and west and outflow boundary in the east. Those boundaries were determined based on the exploration of equipotential maps in different hydrological conditions, i.e., high, medium, and low water levels [32]. The thickness of the shallow aquifer varies from 5 to 40 m [33], while hydraulic conductivities can go up to 3000 m/day [11].

The deeper aquifer layer is characterized by frequent vertical and lateral alterations of gravel, sand, and clays [11], while its thickness goes up to 60 m in the eastern part [33]. Although shallow and deeper aquifer layers are hydraulically connected, geochemical stratification along the depth has been recognized. It has been shown that CaMg–HCO_3 hydrogeochemical facies prevail in the shallow aquifer, while higher concentrations of sodium can be found in the deeper aquifer, which results from longer retention in the underground, resulting in CaMgNa–HCO_3 hydrogeochemical facies [12,34].

Problems with groundwater quantity and quality have been observed. Groundwater levels are declining 1–2 m per decade, while permanent groundwater reserves have decreased by about 4% in the period from 1976 to 2006 throughout the aquifer [21]. It has been shown that the main reasons for groundwater decrease are related to deepening of the Sava riverbed, excessive pumping, and upstream construction of dykes along the Sava River [32]. Human influence on the recharge component of the Zagreb aquifer is evident. Probably the most important one is related to the extensive riverbed erosion as a consequence of the upstream Sava River regulation and gravel extraction from the river. Furthermore, due to occasional flooding, embankments were built along the Sava River, which stopped the potential infiltration to groundwater in some areas. Climate change also probably resulted in prolonged periods of drought. However, the extent of the impact of climate change on groundwater resources is not known and should be investigated in detail in future research in this area. Regarding quality, five main groups of contaminants have been identified, mostly related to agricultural activity, industrial development, leakage from sewage system, and septic tanks. They are nitrates, pesticides, potentially toxic metals, chlorinated aliphatics, and pharmaceuticals [11]. Most of the recent research has been related to nitrate origin and trends. Although elevated concentrations have been observed, results have shown that nitrate trends are decreasing in most parts of the aquifer, while nitrate origin is mainly organic [13,35–37].

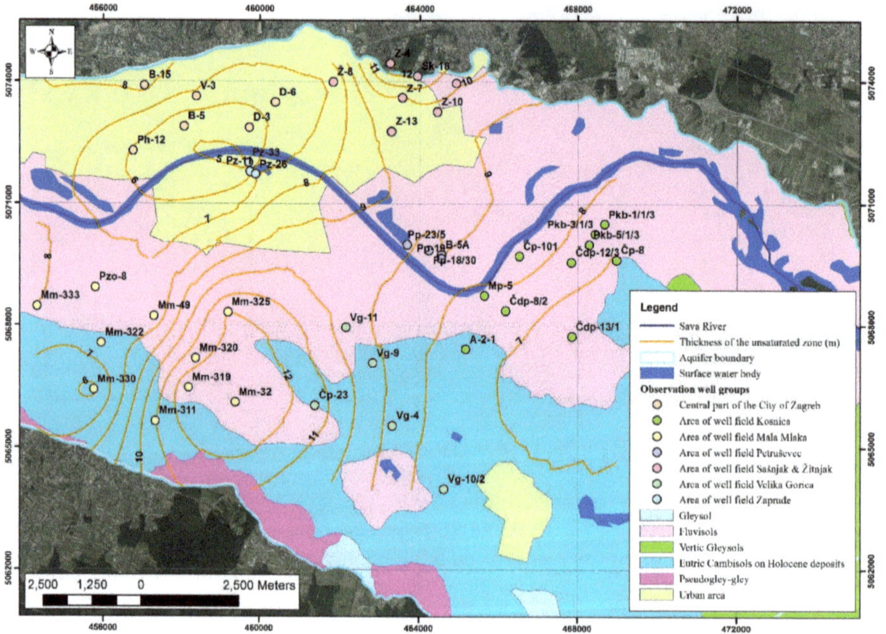

Figure 2. Maximum thickness of the unsaturated zone and soil types of the Zagreb aquifer.

3. Data and Methods

Water stable isotopes from groundwater ($\delta^{18}O$ and δ^2H) were measured in the area of well field Zaprude and Petruševec [10,11]; in the area of future well field Kosnica within the International Atomic Energy Agency (IAEA)'s technical cooperation (TC) project CRO7001 [38]; in the inflow areas of well field Sašnjak–Žitnjak, Mala Mlaka, and Velika Gorica; and in the urban part of the City of Zagreb [13] (Figure 1). Water stable isotopes from the Sava River were measured in the area of well field Petruševec [39] and in the area of well field Kosnica within IAEA TC project CRO7001 [38]. Stable isotope data of water from precipitation were used from Global Network of Isotopes in Precipitation (GNIP) [40,41] for meteorological station Grič, located in the urban part of the City of Zagreb. All $\delta^{18}O$ and δ^2H values are expressed in ‰ notation relative to VSMOW (Vienna Standard Mean Ocean Water), while deuterium excess (d-excess) has been defined according to Dansgaard [42] with a global mean slope of 8.

The mixing of groundwater and surface water is not uniform in all parts of the aquifer; it is spatially and temporary variable depending on the strength of the impact of each individual recharge source in different hydrological conditions. Their quantitative relationship can be derived from two- or three-end member mass balance equations [16,19,38,43–46], where the sum of end member contributions are usually expressed as fractions (f) equal to 1. The two-component model is based on the assumption of the presence of only two recharge sources: precipitation and the Sava River. Quantification of the Sava River and precipitation to the Zagreb aquifer recharge was estimated based on the two mass balance equations written as follows:

$$f_{sw} + f_p = 1 \tag{1}$$

$$f_{sw} \times \delta^{18}O_{sw} + f_p \times \delta^{18}O_p = \delta^{18}O_s \tag{2}$$

where f_{sw} and f_p are the Sava River and precipitation fractions, respectively; $\delta^{18}O_s$ is the isotopic composition of oxygen in the observation well; and $\delta^{18}O_{sw}$ and $\delta^{18}O_p$ are the isotopic compositions of oxygen in the surface water and precipitation, respectively.

The three-component mixing model was used for testing and quantifying the change in the isotopic composition of the groundwater that takes place within the aquifer but not as a consequence of infiltration of the precipitation or the Sava River. It is often the case that a third end member contributes to mixing. Here, the third mixing component is groundwater itself (horizontal inflow of groundwater from neighboring parts of the aquifer in relation to the observation well), and the three mass balance equations are written as follows:

$$f_{sw} + f_p + f_{gw} = 1 \tag{3}$$

$$f_{sw} \times \delta^{18}O_{sw} + f_p \times \delta^{18}O_p + f_{gw} \times \delta^{18}O_{gw} = \delta^{18}O_s \tag{4}$$

$$f_{sw} \times \delta^{18}O^2_{sw} + f_p \times \delta^{18}O^2_p + f_{gw} \times \delta^{18}O^2_{gw} = \delta^{18}O^2_s \tag{5}$$

where f_{sw}, f_p, and f_{gw} are the Sava River, precipitation, and groundwater fractions, respectively; $\delta^{18}O_s$ is the isotopic composition of oxygen in the observation well; and $\delta^{18}O_{sw}$, $\delta^{18}O_p$, and $\delta^{18}O_{gw}$ are the isotopic compositions of oxygen in the surface water, precipitation, and groundwater, respectively.

The average values of $\delta^{18}O$, δ^2H, number of analysis (n), and calculated d-excess from the observation wells included in the analysis, together with their affiliation to location area, soil type, and defined groundwater levels in low and high waters, are shown in Table 1. In sum, there are 45 observation wells, which are mostly spatially located in the inflow areas of different well fields (well field Kosnica, Mala Mlaka, Petruševec, Sašnjak and Žitnjak, and Zapruđe), while the rest are located in the central part of the City of Zagreb (CPCZ) (Figure 2). Furthermore, there are 28 observation wells on the right bank of the Sava River, while there are 17 observation wells on the left bank. The groundwater from the observation wells were sampled at different time periods, 4–12 times during 2010, 2011, 2015, and 2016 (sampling was done in accordance with the national monitoring plan). The average values of δ^2H in the observation wells varied from −64.32‰ to −60.40‰, while average values of $\delta^{18}O$ varied from −9.46‰ to −8.92‰. The d-excess varied from 9.32‰ to 13.12‰. Isotopic composition of the Sava River resulted in δ^2H values from −69.88‰ to −55.61‰, with the average value of −62.84‰, while $\delta^{18}O$ values varied from −10.33‰ to −8.54‰, with the average value of −9.47‰. Long-term average for $\delta^{18}O$ and δ^2H for the meteorological station Zagreb–Grič, based on the data from GNIP, was −8.28‰ and −60.0‰, respectively. In order to provide more detailed interpretation, groundwater stable isotope data was spatially grouped at the level of inflow areas related to different well fields but also based on the areas that belong to different types of soil. In the two- and three-component mixing models, long-term average of $\delta^{18}O$ for precipitation (−8.28‰; $n = 89$) from Zagreb–Grič meteorological station was used, while the average value of $\delta^{18}O$ from all available analysis was used for the Sava River (−9.47‰; $n = 56$). In the three-component mixing models, the average value of $\delta^{18}O$ from all observation wells was used (−9.16‰).

In addition to the isotopic data, different hydrogeological and pedological data were used. The Zagreb aquifer is of the unconfined type, and it is essential to determine the fluctuation of the groundwater table in order to determine the thickness of the unsaturated zone. Analysis of the minimum and maximum groundwater levels were done for the period from 2000 to 2014. The groundwater levels are measured continuously on one-third of the observation wells, but because the only purpose of many observation wells is water quality monitoring, they are not continually monitored. For these objects, maximum and minimum groundwater level data were taken from the water table contour maps created for wet and dry periods (Figure 3) [47]. In Table 1, groundwater levels for low and high waters characteristic for the area of each investigated observation well are shown. The difference between maximum and minimum groundwater levels varied from 0.8 to 5.09 m. Due to a large difference in the fluctuation of the groundwater level at certain points, only minimum groundwater levels, i.e., the maximum thickness of the unsaturated zone, were observed (Figure 2).

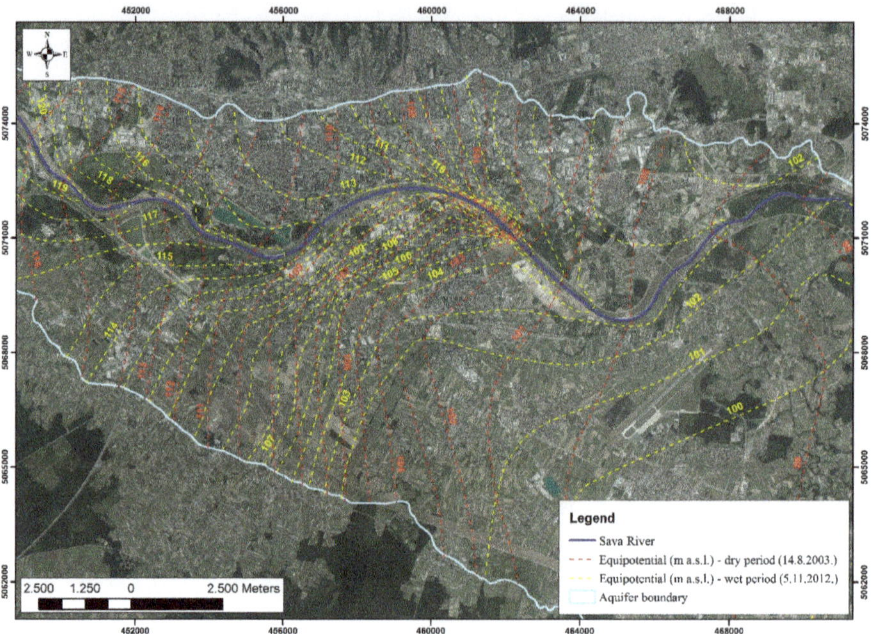

Figure 3. The water table contour map of the Zagreb aquifer during dry and wet periods (equipotentials in meters above sea level (m a.s.l.)) [47].

All investigated observation wells are also grouped according to their affiliation to different soil types; 8 observation wells are located in the Eutric Cambisol, 21 in the Fluvisols, and 16 in the urban part of the City of Zagreb (Table 1). Soil hydraulic properties were estimated based on the data from the research that was done in the soil profile in the area of the future well field Kosnica [29], which is located in the Fluvisols, and a soil profile that is located in the area of well field Velika Gorica [30], which is located in the Eutric Cambisols on Holocene deposits. Granulometric composition of soil profiles from these types of soils is given in Table 2. The soil profile in the Fluvisol has much more sand, while higher percentages of silt and clay in the Eutric Cambisols developed on Holocene deposits suggest much lower permeability. It has been shown that the Fluvisols are generally permeable, except in dry periods when water content is very low [29], while very low water content can be found in the Eutric Cambisols, which results in very small unsaturated hydraulic conductivities [30]. These results suggest that the possible influence of precipitation on the recharge of the Zagreb aquifer could be diminished in areas where the Eutric Cambisol prevails.

All maps were made in ArcMap 10.1 for Desktop, while georeferenced orthophoto image was obtained from geoportal of the Croatian Geodetic Administration. All maps are presented in the official coordinate system of the Republic of Croatia (HTRS96/TM).

Table 1. Hydrogeological, pedological, and water stable isotope data (average values of $\delta^{18}O$, δ^2H, and d-excess) used in the analysis (minimum and maximum water levels registered between 2000 and 2014).

Observation Well	Inflow Area According to Figure 2	Soil Type	Minimum Water Level (m a.s.l.)	Maximum Water Level (m a.s.l.)	δ^2H (‰)	$\delta^{18}O$ (‰)	d-Excess (‰)	n
D-6	CPCZ	Urban	106.6	107.9	−62.3	−9.1	10.2	4
D-3	CPCZ	Urban	108.0	109.2	−62.6	−9.1	10.1	4
Ph-12	CPCZ	Urban	110.4	112.8	−62.6	−9.1	10.3	4
B-5	CPCZ	Urban	109.5	111.8	−63.2	−9.1	9.7	4
V-3	CPCZ	Urban	109.2	110.3	−62.7	−9.2	10.6	4
B-15	CPCZ	Urban	110.4	111.2	−63.3	−9.2	10.4	4
Čp-8	Kosnica	Fluvisols	98.6	103.5	−60.7	−9.1	12.0	12
Pkb-1/1/3	Kosnica	Fluvisols	99.0	102.1	−61.2	−9.1	11.8	12
Čdp-13/1	Kosnica	Fluvisols	98.9	103.2	−61.0	−9.2	12.3	12
Čdp-12/3	Kosnica	Fluvisols	98.8	103.9	−61.3	−9.2	12.2	12
Pkb-3/1/3	Kosnica	Fluvisols	99.1	102.0	−61.4	−9.2	12.2	12
Pkb-5/1/3	Kosnica	Fluvisols	98.6	101.9	−62.3	−9.2	11.6	12
Mp-5	Kosnica	Fluvisols	100.0	103.1	−62.0	−9.2	11.9	12
Čp-101	Kosnica	Fluvisols	99.6	103.1	−61.9	−9.3	12.2	12
A-2-1	Kosnica	Fluvisols	99.9	101.5	−62.3	−9.3	12.1	12
Čdp-8/2	Kosnica	Fluvisols	99.3	104.1	−62.6	−9.3	11.8	12
Mm-49	Mala Mlaka	Fluvisols	106.4	108.1	−60.0	−8.9	10.5	4
Mm-325	Mala Mlaka	Fluvisols	102.0	104.9	−60.5	−9.0	11.2	4
Mm-311	Mala Mlaka	Eutric Cambisols	101.5	105.0	−61.0	−9.0	11.1	4
Pzo-8	Mala Mlaka	Fluvisols	109.0	111.0	−62.1	−9.1	10.5	4
Mm-333	Mala Mlaka	Fluvisols	109.0	111.0	−61.9	−9.1	10.7	4
Mm-320	Mala Mlaka	Fluvisols	102.5	105.4	−62.2	−9.1	10.9	4
Mm-322	Mala Mlaka	Eutric Cambisols	107.9	109.5	−62.2	−9.2	11.3	4
Mm-330	Mala Mlaka	Eutric Cambisols	108.8	111.0	−63.1	−9.3	10.1	4
Mm-319	Mala Mlaka	Fluvisols	101.1	104.6	−63.1	−9.3	11.4	4
Mm-32	Mala Mlaka	Fluvisols	100.4	103.8	−63.7	−9.3	11.1	4
Pp-18/30	Petruševec	Fluvisols	100.8	103.9	−61.0	−9.1	11.7	12
Pp-19	Petruševec	Fluvisols	100.5	103.8	−60.7	−9.2	12.9	12
B-5A	Petruševec	Fluvisols	99.8	103.8	−61.0	−9.2	12.8	11
Pp-23/5	Petruševec	Fluvisols	100.9	104.7	−61.2	−9.1	11.3	12
Ž-8	Sašnjak and Žitnjak	Urban	104.5	107.6	−62.0	−8.9	9.3	4
Z-7	Sašnjak and Žitnjak	Urban	101.7	104.1	−62.1	−9.0	9.6	4
Z-4	Sašnjak and Žitnjak	Urban	101.1	104.6	−62.0	−9.0	9.8	4
Sk-18	Sašnjak and Žitnjak	Urban	100.4	104.1	−61.8	−9.0	10.2	4
Sk-16/2	Sašnjak and Žitnjak	Urban	99.9	103.6	−62.0	−9.1	10.8	4

Table 1. *Cont.*

Observation Well	Inflow Area According to Figure 2	Soil Type	Minimum Water Level (m a.s.l.)	Maximum Water Level (m a.s.l.)	δ^2H (‰)	δ^{18}O (‰)	d-Excess (‰)	n
Z-10	Sašnjak and Žitnjak	Urban	99.9	103.5	−61.9	−9.3	12.1	4
Z-13	Sašnjak and Žitnjak	Urban	101.0	103.4	−61.8	−9.4	13.1	4
Vg-4	Velika Gorica	Eutric Cambisols	99.5	102.3	−62.1	−9.0	9.7	4
Vg-11	Velika Gorica	Fluvisols	101.0	104.1	−62.5	−9.3	11.5	4
Vg-10/2	Velika Gorica	Eutric Cambisols	99.6	102.2	−64.3	−9.3	10.3	4
Vg-9	Velika Gorica	Eutric Cambisols	100.4	103.2	−63.5	−9.4	11.4	4
Čp-23	Velika Gorica	Eutric Cambisols	99.4	102.0	−64.1	−9.5	11.6	4
Pz-11	Zapruđe	Urban	106.4	110.2	−62.2	−9.3	11.8	12
Pz-26	Zapruđe	Urban	105.5	109.8	−61.0	−9.2	12.3	12
Pz-33	Zapruđe	Urban	106.9	110.8	−60.4	−9.1	12.6	12

Table 2. Granulometric composition of the most common types of soils in the area of the Zagreb aquifer (modified according to [29,30]).

Type of the Soil	Soil Horizon	Depth (m)	Sand (%)	Silt (%)	Clay (%)
Fluvisols	A	0–0.19	24.56	65.27	10.17
	AC-C	0.19–0.68	13.79	76.69	9.52
	2C/Cl	0.68–1.1	56.33	38.23	5.44
	3Cl	1.1–1.4	43.29	47.43	9.28
	4Cl/Cr	1.4.–1.9	37.21	50.50	12.29
	5Cr	1.9–2.1	55.62	38.45	5.93
Eutric Cambisols	A	0–0.2	2.42	87.43	10.15
	Bw	0.2–0.4	1.8	89.54	8.65
		0.4–0.6	1.46	87.59	10.95
		0.6–0.8	2.72	84.97	12.32
	C	0.8–1	4.77	82.00	13.23
		1–1.2	3.84	89.78	6.37

4. Results and Discussion

The local meteoric water line (LMWL) for Zagreb is given in Figure 4, together with the LMWL for Ljubljana (because the Sava River is formed by the Sava Dolinka and the Sava Bohinjka headwaters in northwest Slovenia) as well as the average values of the observation wells and the Sava River. The water isotopic composition of the Sava River and groundwater from the Zagreb aquifer is more similar to the precipitation in the area of the City of Ljubljana than the precipitation in the area of the City of Zagreb, which is in agreement with previous results [10–13,38].

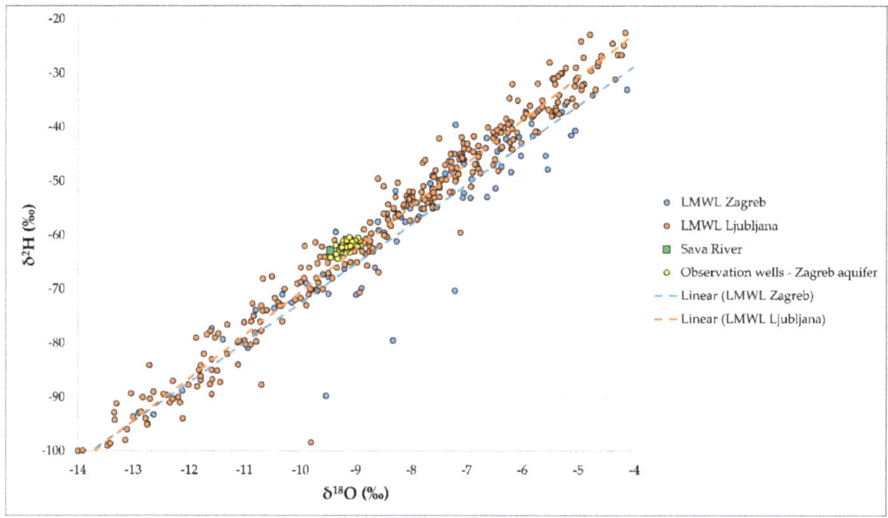

Figure 4. Comparison of the water isotopic composition of precipitation from Zagreb and Ljubljana with average values of observation wells from the Zagreb aquifer and the Sava River.

The d-excess values for groundwater were analyzed only for observation wells with at least 12 data measurements. The chosen objects are spatially representative for the entire aquifer and statistically more reliable. The d-excess values in groundwater samples varied from 9.07‰ to 14.21‰, while they varied from 10.49‰ to 14.69‰ in samples of the Sava River. Deuterium excess values in samples of the local precipitation varied a wide range from −13.38‰ to 18.18‰ (Figure 5). Groundwater and river water had similar isotopic signature as opposed to local precipitation.

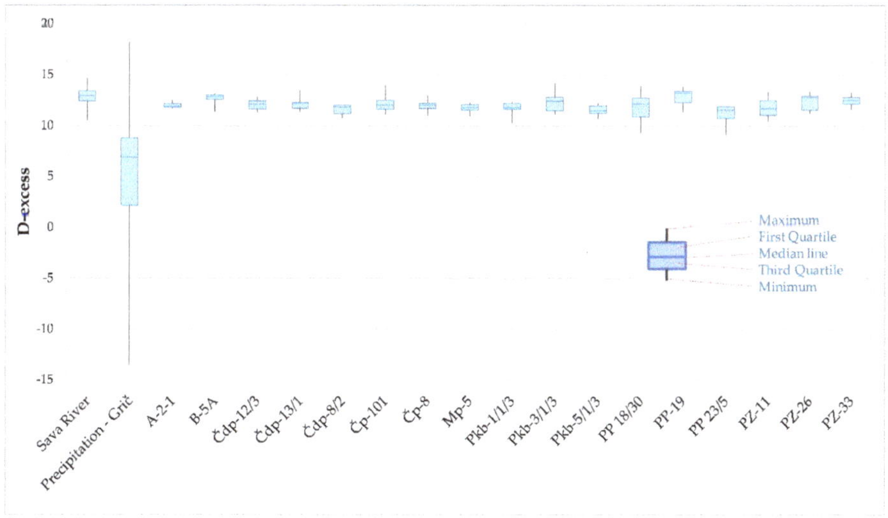

Figure 5. Average deuterium excess (d-excess) values in samples of the Sava River, precipitation, and groundwater.

The results from both mixing models, given in Table 3, show that the Sava River presents the most important source of recharge for the Zagreb aquifer. When observation wells were evaluated separately, in the two-component mixing model, contribution of the Sava River to aquifer recharge varied from 53.36% to 99.16%, with an average of 73.90%. Recharge from precipitation varied from 0.84% to 46.64%, with an average of 26.10%, and it was lower than the Sava River recharge for all observation wells.

Table 3. Results of two- and three-component mixing models.

Observation Well	Two-Component Mixing Model		Three-Component Mixing Model		
	Recharge from Precipitation (%)	Recharge from the Sava River (%)	Recharge from Precipitation (%)	Recharge from the Sava River (%)	Recharge from Groundwater (%)
D-6	34.03	65.97	32.89	62.73	4.38
D-3	33.19	66.81	32.04	63.53	4.43
Ph-12	30.88	69.12	29.69	65.72	4.59
B-5	30.04	69.96	28.83	66.52	4.65
V-3	26.68	73.32	25.42	69.72	4.86
B-15	22.27	77.73	20.93	73.92	5.15
Čp-8	31.93	68.07	30.76	64.72	4.52
Pkb-1/1/3	29.72	70.28	28.50	66.83	4.67
Čdp-13/1	26.47	73.53	25.20	69.92	4.88
Čdp-12/3	23.74	76.26	22.42	72.52	5.06
Pkb-3/1/3	22.13	77.87	20.79	74.05	5.16
Pkb-5/1/3	19.96	80.04	18.58	76.12	5.30
Mp-5	19.12	80.88	17.72	76.92	5.36
Čp-101	16.95	83.05	15.51	78.98	5.51
A-2-1	14.29	85.71	12.80	81.52	5.68
Čdp-8/2	14.92	85.08	13.45	80.92	5.63
Mm-49	45.38	54.62	44.43	51.94	3.63
Mm-325	42.65	57.35	41.65	54.53	3.82
Mm-311	38.66	61.34	37.59	58.33	4.08
Pzo-8	33.82	66.18	32.67	62.93	4.40
Mm-333	33.61	66.39	32.46	63.13	4.41
Mm-320	28.15	71.85	26.91	68.32	4.77
Mm-322	24.37	75.63	23.06	71.92	5.02
Mm-330	17.44	82.56	16.01	78.52	5.47
Mm-319	12.82	87.18	11.31	82.92	5.77
Mm-32	10.50	89.50	8.96	85.12	5.92
Pp-18/30	31.93	68.07	30.76	64.72	4.52
Pp-19	22.66	77.34	21.32	73.55	5.13
B-5A	20.10	79.90	18.72	75.98	5.30
Pp-23/5	34.03	65.97	32.89	62.73	4.38
Ž-8	46.64	53.36	45.71	50.73	3.56
Z-7	42.44	57.56	41.44	54.73	3.83
Z-4	41.18	58.82	40.16	55.93	3.91
Sk-18	39.08	60.92	38.02	57.93	4.05
Sk-16/2	31.09	68.91	29.90	65.51	4.57
Z-10	18.49	81.51	17.08	77.52	5.40
Z-13	8.61	91.39	7.04	86.91	6.05
Vg-4	41.60	58.40	40.58	55.53	3.88
Vg-11	18.07	81.93	16.65	77.92	5.43
Vg-10/2	11.55	88.45	10.03	84.12	5.85
Vg-9	9.24	90.76	7.68	86.32	6.00
Čp-23	0.84	99.16	-0.87	94.32	6.55
Pz-11	18.07	81.93	16.65	77.92	5.43
Pz-26	25.70	74.30	24.42	70.66	4.92
Pz-33	29.41	70.59	28.19	67.12	4.69

The three-component mixing model, when the fraction of groundwater was introduced, resulted in no significant difference with regard to the two-component mixing model (Table 3). The fraction of precipitation in aquifer recharge varied from −0.87% to 45.71%, with an average of 24.82%. Recharge of the Zagreb aquifer from the surface water of the Sava River varied from 50.73% to 94.32%, with an average of 70.28%. Changes in the groundwater isotopic composition were negligible, with fractions between 3.55% and 6.55% and the average value of 4.90%, suggesting that only a small amount of water comes from neighboring groundwater bodies. It is possible that a certain amount of water comes from the adjacent aquifer in the west but also from the S, SW boundary when low waters are occurring, which can be seen in the equipotential map (Figure 3). The difference between the average values of fractions in two- and three-component mass balance models regarding recharge from the Sava River was 3.62%, while it was 1.28% regarding recharge from precipitation. Both models gave very similar results. For further analysis, results from the two-component mixing model were used. It has to be emphasized that, in this case, the three-component mixing model resulted in one negative value for recharge from precipitation (Čp-23, Table 3). Uncertainty of the results is generally related to the limited number of available data measured during different time periods within several projects.

Although the results from the mixing models were as expected, i.e., the dominance of the Sava River as the main source of recharge was confirmed and consistent with previous research [10–13,38], some anomalies were identified. A few observation wells that are located far away from the Sava River have a higher river water fraction than those that are located near the Sava River. In order to explore this issue in more detail, data from the observation wells were mainly grouped according to the inflow areas of different well fields (Table 4). In most cases, the largest fractions in aquifer recharge were river water and seen in the inflow areas of well fields that are located near the Sava River (well fields Kosnica, Zapruđe, and Petruševec). In the urban part of the City of Zagreb, observation wells have very similar river water fraction values, while in the inflow areas of well fields Sašnjak and Žitnjak, the Sava River's influence is less pronounced, which can be expected due to their slightly longer distances from the Sava River. However, some results from the observation wells located in the inflow areas of well fields Velika Gorica and Mala Mlaka suggested that the Sava River's influence is the highest in the southern part of the Zagreb aquifer, near its edges (Figure 6).

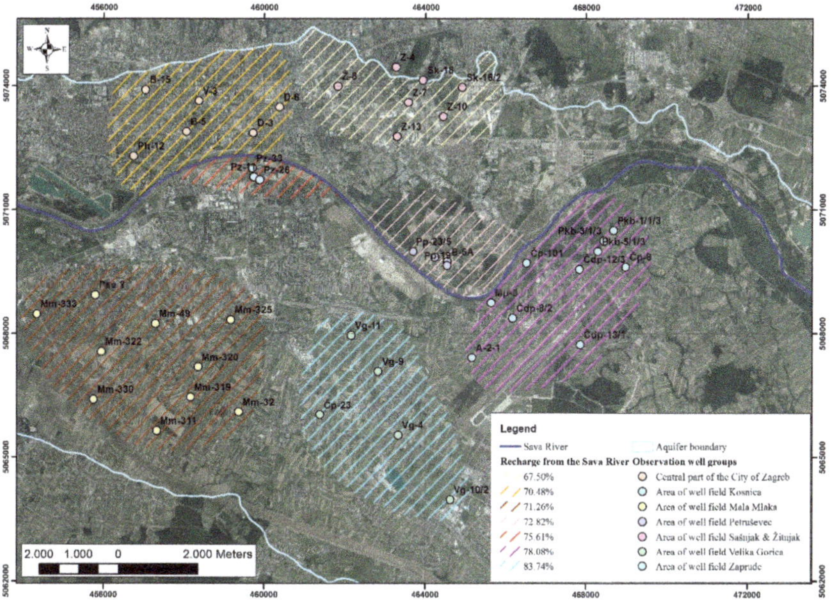

Figure 6. Spatial recharge fractions of the Zagreb aquifer from the Sava River.

Table 4. Recharge fractions in the spatially different areas of the Zagreb aquifer.

Grouped Observation Wells	Recharge from the Precipitation (%)	Recharge from the Sava River (%)
Central part of the City of Zagreb	29.52	70.48
Area of well field Kosnica	21.92	78.08
Area of well field Mala Mlaka	28.74	71.26
Area of well field Petruševec	27.18	72.82
Area of well fields Sašnjak and Žitnjak	32.50	67.50
Area of well field Velika Gorica	16.26	83.74
Area of well field Zapruđe	24.39	75.61

In order to clarify this issue, data of soil property and thickness of the unsaturated zone were analyzed. In the urban part of the City of Zagreb, the unsaturated zone, especially the upper part (first few meters of depth), is mainly disintegrated by human influence. Almost all observation wells located on the left bank of the Sava River are placed in the urban area.

Of all the 45 observation wells, 21 are located in the Fluvisols, eight of them in the Eutric Cambisols, and 16 in the urban area (Figure 2, Table 5). Most of them are located on the right bank of the Sava River. Although the observation wells located in the Fluvisols are closer to the Sava River, they show about 6% lower influence of the Sava River on the recharge of the Zagreb aquifer, which is probably the result of a more permeable soil type (Table 5). In the Eutric Cambisols, a higher fraction of finer particles, i.e., silts and clays, can be found, which reduces the amount of infiltration through the unsaturated zone into the Zagreb aquifer.

Table 5. Average recharge in different soil types in the area of the Zagreb aquifer.

Soil Type	Number of Observation Wells	Recharge from the Precipitation (%)	Recharge from the Sava River (%)
Fluvisols	21	25.65	74.35
Eutric Cambisols	8	19.75	80.25
Urban area	16	29.86	70.14

The thickness ranges in different types of soil were very similar. The thickness of the unsaturated zone varied from 5.7 to 11.5 m within the Eutric Cambisols soil type, from 6.1 to 12.9 m within the Fluvisols, and from 5 to 13 m within the urban part of the Zagreb aquifer. However, the minimum and maximum values of the precipitation fractions in the Zagreb aquifer recharge, calculated based on the data from the observation wells in different types of surface cover (Table 3), showed the importance of observing the characteristics of an unsaturated zone and its role in the estimation of groundwater recharge. Figure 7 illustrates how the soil type and thickness of the unsaturated zone affect the recharge of the aquifer from precipitation. The Eutric Cambisols soil type is less permeable than the Fluvisol and, combined with the large thickness, it results in extremely low infiltration of precipitation. Similarly, in spite of the greater permeability of the Fluvisols, large thickness of the unsaturated zone significantly reduces infiltration from the surface. However, detailed inspection of soil permeability should be done in future research. In the urban area, the soil and the unsaturated zone are mostly disturbed, making it difficult to interpret the results. The occurrence of low and high values of recharge from precipitation confirms emphasized heterogeneity caused by human activity in that area.

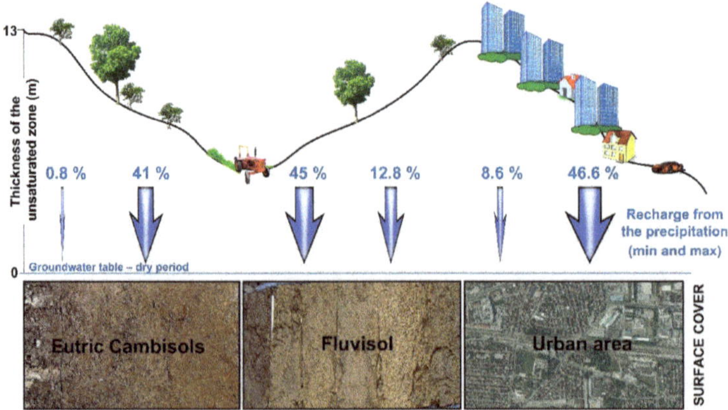

Figure 7. The conceptual illustration of groundwater recharge of the Zagreb aquifer system through the unsaturated zone with different thickness and surface covers in two-component mixing model.

5. Conclusions

The main aim of this study was to determine and quantify the main recharge components of the unconfined alluvial aquifer, which is strongly hydraulically connected with the river, using water stable isotopes. The final outcomes can be summarized as follows:

- Two- and three-component mixing models based on the isotopic mass balance are efficient tools for the quantification of groundwater recharge fractions. In the research area, these models showed that the Sava River, although spatially varying, is the dominating source of groundwater recharge.
- The thickness of the unsaturated zone and soil permeability has large influence on the aquifer recharge. Evaluation of soil permeability can be crucial in quantification of the shallow alluvial aquifer recharge and data interpretation.
- Water stable isotope composition of groundwater, precipitation, and the Sava River water confirmed strong connection between the alluvial part of the Zagreb aquifer and the Sava River.
- The Sava River presents the main recharge factor of the Zagreb aquifer. In general, more than 70% of the Zagreb aquifer recharge is related to the Sava River.
- Future research in the area of the Zagreb aquifer should be focused on detailed inspection of soil permeability in order to define the difference in percolation through different types of soil.
- The water management of the Zagreb aquifer should be focused on groundwater level stabilization by regulation of the Sava River.
- The identified pattern of groundwater–surface water interaction in the study area will enable the development of better monitoring networks as well as determination of aquifer areas that need better protection.

Author Contributions: J.P. participated in the evaluation of the water isotope and unsaturated zone analyses and contributed to writing of the article. Z.K. carried out data analysis and data interpretation and took the lead in writing the article. Z.N. contributed in writing the article and data interpretation. J.B. contributed in data processing of water isotope composition and related interpretation.

Funding: This research received no external funding.

Acknowledgments: The authors would like to thank the International Atomic Energy Agency for providing technical support within the IAEA TC project CRO7001 (Isotope Investigation of the Groundwater–Surface Water Interaction at the Well Field Kosnica in the Area of the City of Zagreb). The publication process was supported by the Development Fund of the Faculty of Mining, Geology, and Petroleum Engineering at the University of Zagreb.

References

1. Chen, Y.N.; Li, Z.; Fan, Y.T.; Wang, H.J.; Deng, H.J. Progress and prospects of climate change impacts on hydrology in the arid region of northwest China. *Environ. Res.* **2015**, *139*, 11–19. [CrossRef] [PubMed]
2. Döll, P.; Hoffmann-Dobrev, H.; Portmann, F.T.; Siebert, S.; Eicker, A.; Rodell, M.; Strassberg, G.; Scanlon, B. Impact of water withdrawals from groundwater and surface water on continental water storage variations. *J. Geodyn.* **2012**, *59*, 143–156. [CrossRef]
3. Joshi, S.K.; Rai, S.P.; Sinha, R.; Gupta, S.; Densmore, A.L.; Rawat, Y.S.; Shekhar, S. Tracing groundwater recharge sources in the northwestern Indian alluvial aquifer using water isotopes ($\delta^{18}O$, $\delta^{2}H$ and ^{3}H). *J. Hydrol.* **2018**, *559*, 835–847. [CrossRef]
4. Wada, Y.; van Beek, L.P.H.; Bierkens, M.F.P. Non-sustainable groundwater sustaining irrigation: A global assessment. *Water Resour. Res.* **2012**, *48*, 6. [CrossRef]
5. Gleeson, T.; Befus, K.M.; Jasechko, S.; Luijendijk, E.; Cardenas, M.B. The global volume and distribution of modern groundwater. *Nat. Geosci.* **2015**, *9*, 161–167. [CrossRef]
6. Kumar, C.P. Climate Change and Its Impact on Groundwater Resources. *Int. J. Eng. Sci.* **2012**, *1*, 43–60.
7. Gampe, D.; Nikulin, G.; Ludwig, R. Using an ensemble of regional climate models to assess climate change impacts on water scarcity in European river basins. *Sci. Total Environ.* **2016**, *573*, 1503–1518. [CrossRef]

8. Vrzel, J.; Ludwig, R.; Gampe, D.; Ogrinc, N. Hydrological system behaviour of an alluvial aquifer under climate change. *Sci. Total Environ.* **2019**, *649*, 1179–1188. [CrossRef]

9. Brunner, P.; Cook, P.G.; Simmons, C.T. Disconnected surface water and groundwater: From theory to practice. *Ground Water* **2010**, 1–8. [CrossRef]

10. Parlov, J.; Nakić, Z.; Posavec, K.; Bačani, A. Origin and dynamics of aquifer recharge in Zagreb area. In Proceedings of the Fifth International Scientific Conference on Water, Climate and Environment 2012, Balwois, Ohrid, Macedonia, 28 May–2 June 2012; ISBN 978-608-4510-10-9.

11. Nakić, Z.; Ružičić, S.; Posavec, K.; Mileusnić, M.; Parlov, J.; Bačani, A.; Durn, G. Conceptual model for groundwater status and risk assessment—Case study of the Zagreb aquifer system. *Geol. Croat.* **2013**, *66*, 55–76. [CrossRef]

12. Marković, T.; Brkić, Ž.; Larva, O. Using hydrochemical data and modelling to enhance the knowledge of groundwater flow and quality in an alluvial aquifer of Zagreb, Croatia. *Sci. Total Environ.* **2013**, *458–460*, 508–516. [CrossRef]

13. Kovač, Z.; Nakić, Z.; Barešić, J.; Parlov, J. Nitrate Origin in the Zagreb Aquifer System. *Geofluids* **2018**, *15*. [CrossRef]

14. Deshpande, R.D.; Gupta, S.K. Oxygen and hydrogen isotopes in hydrological cycle: New data from IWIN national pro-gramme. *Proc. Indian Natl. Sci. Acad.* **2012**, *78*, 321–331.

15. Achyuthan, H.; Deshpande, R.D.; Rao, M.S.; Kumar, B.; Nallathambi, T.; Shashikumar, K.; Ramesh, R.; Ramachandran, P.; Maurya, A.S.; Gupta, S.K. Spatio-temporal mixing inferred from water isotopes and salinity of surface waters in the Bay of Bengal. *Mar. Chem.* **2013**, *149*, 51–62. [CrossRef]

16. Shahul Hameed, A.S.; Resmi, T.R.; Suraj, S.; Unnikrishnan Warrier, C.; Sudheesh, M.; Deshpande, R.D. Isotopic characterization and mass balance reveals groundwater recharge pattern in Chaliyar river basin, Kerala, India. *J. Hydrol. Reg. Stud.* **2015**, *4*, 48–58. [CrossRef]

17. Fan, Y.; Chen, Y.; He, Q.; Li, W.; Wang, Y. Isotopic Characterization of River Waters and Water Source Identification in an Inland River, Central Asia. *Water* **2016**, *8*, 286. [CrossRef]

18. González-Trinidad, T.; Pacheco-Guerrero, A.; Júnez-Ferreira, H.; Bautista-Capetillo, C.; Hernández-Antonio, A. Identifying Groundwater Recharge Sites through Environmental Stable Isotopes in an Alluvial Aquifer. *Water* **2017**, *9*, 569. [CrossRef]

19. Vrzel, J.; Kip Solomon, D.; Blažeka, Ž.; Ogrinc, N. The study of the interactions between groundwater and Sava River water in the Ljubljansko polje aquifer system (Slovenia). *J. Hydrol.* **2018**, *556*, 384–396. [CrossRef]

20. Ala-aho, P.; Soulsby, C.; Pokrovsky, O.S.; Kirpotin, S.N.; Karlsson, J.; Serikova, S.; Vorobyev, S.N.; Manasypov, R.M.; Loiko, S.; Tetzlaff, D. Using stable isotopes to assess surface water source dynamics and hydrological connectivity in a high-latitude wetland and permafrost influenced landscape. *J. Hydrol.* **2018**, *556*, 279–293. [CrossRef]

21. Bačani, A.; Posavec, K.; Parlov, J. Groundwater quantity in the Zagreb aquifer. In *XXXVIII IAH Congress Groundwater Quality Sustainability*; Zuber, A., Kania, J., Kmiecik, E., Eds.; University of Silesia: Krakow, Poland, 2010; pp. 87–92.

22. Velić, J.; Saftić, B. Subsurface Spreading and Facies Characteristics of Middle Peistocene Deposits between Zaprešić and Samobor. *Geološki Vjesn.* **1991**, *44*, 69–82.

23. Velić, J.; Durn, G. Alternating Lacustrine-Marsh Sedimentation and Subaerial Exposure Phases during Quaternary: Prečko, Zagreb, Croatia. *Geol. Croat.* **1993**, *46*, 71–90. [CrossRef]

24. Velić, J.; Saftić, B.; Malvić, T. Lithologic composition and stratigraphy of Quaternary sediments in the area of the "Jakuševec" Waste Depository (Zagreb, Northern Croatia). *Geol. Croat.* **1999**, *52*, 119–130. [CrossRef]

25. Sokač, A. Pleistocene ostracode fauna of the Pannonian Basin in Croatia. *Paleontol. Jugosl.* **1978**, *20*, 1–51.

26. Hernitz, Z.; Kovačević, S.; Velić, J.; Urli, M. An example of complex geological and geophysical explorations of Quaternary deposits in the soroundings of Prevlaka. *Geološki Vjesn.* **1981**, *33*, 11–34. (In Croatian)

27. Bogunović, M.; Vidaček, Ž.; Husnjak, S.; Sraka, M.; Petošić, D. Inventory of Soils in Croatia. *Agric. Conspec. Sci.* **1998**, *63*, 105–112.

28. Sollitto, D.; Romić, M.; Castrignano, A.; Romić, D.; Bakić, H. Assessing heavy metal contamination in soils of the Zagreb region (Northwest Croatia) using multivariate geostatistics. *Catena* **2010**, *80*, 182–194. [CrossRef]

29. Ružičić, S.; Kovač, Z.; Nakić, Z.; Kireta, D. Fluvisol permeability estimation using soil water content variability. *Geofizika* **2017**, *34*, 141–155. [CrossRef]

30. Ružičić, S.; Kovač, Z.; Tumara, D. Physical and chemical properties in relation to soil permeability in the area of the Velika Gorica well field. *Min.-Geol.-Pet. Eng. Bull.* **2018**, *33*, 73–81. [CrossRef]

31. Posavec, K.; Vukojević, P.; Ratkaj, M.; Bedeniković, T. Cross-correlation modelling of surface water-groundwater interaction using the Excel spreadsheet application. *Min.-Geol.-Pet. Eng. Bull.* **2017**, *32*, 25–32. [CrossRef]

32. Posavec, K. Identification and prediction of minimum ground water levels of Zagreb alluvial aquifer using recession curve models. Ph.D. Thesis, University of Zagreb, Zagreb, Croatia, 2006; p. 89. unpublished. (In Croatian)

33. Nakić, Z.; Posavec, K.; Parlov, J.; Bačani, A. Development of the conceptual model of the Zagreb aquifer system. In *The Geology in Digital Age, Proceedings of the 17th Meeting of the Association of European Geological Societies (MAEGS 17), Belgrade, Serbia, 17–18 September 2011*; The Serbian Geological Society: Belgrade, Serbia, 2011.

34. Vlahović, T.; Bačani, A.; Posavec, K. Hydrogeochemical stratification of the unconfined Samobor aquifer (Zagreb, Croatia). *Environ. Geol.* **2009**, *57*, 1707–1722. [CrossRef]

35. Kovač, Z.; Cvetković, M.; Parlov, J. Gaussian simulation of nitrate concentration distribution in the Zagreb aquifer. *J. Maps* **2017**, *13*, 727–732. [CrossRef]

36. Kovač, Z.; Nakić, Z.; Pavlić, K. Influence of groundwater quality indicators on nitrate concentrations in the Zagreb aquifer system. *Geol. Croat.* **2017**, *70*, 93–103. [CrossRef]

37. Kovač, Z.; Nakić, Z.; Špoljarić, D.; Stanek, D.; Bačani, A. Estimation of nitrate trends in the groundwater of the Zagreb aquifer. *Geosciences* **2018**, *8*, 5. [CrossRef]

38. Parlov, J.; Kovač, Z.; Barešić, J. The study of the interactions between Sava River and Zagreb aquifer system (Croatia) using water stable isotopes. In Proceedings of the 16th International Symposium on Water-Rock Interaction and the 13th International Symposium on Applied Isotope Geochemistry, Tomsk, Russia, 21–26 July 2019.

39. Horvatinčić, N.; Barešić, J.; Krajcar Bronić, I.; Obelić, B.; Kármán, K.; Fórisz, I. Study of the bank filtered groundwater system of the Sava River at Zagreb (Croatia) using isotope analyses. *Cent. Eur. Geol.* **2011**, *54*, 121–127. [CrossRef]

40. IAEA/WMO. Global Network of Isotopes in Precipitation. The GNIP Database. 2019. Available online: http://www.iaea.org/water (accessed on 21 May 2019).

41. Vreča, P.; Krajcar Bronić, I.; Horvatinčić, N.; Barešić, J. Isotopic characteristics of precipitation in Slovenia and Croatia: Comparison of continental and maritime stations. *J. Hydrol.* **2006**, *330*, 457–469. [CrossRef]

42. Dansgaard, W. Stable isotopes in precipitation. *Tellus* **1964**, *16*, 436–468. [CrossRef]

43. Clark, I.D.; Fritz, P. Tracing the hydrological cycle. In *Environmental Isotopes in Hydrogeology*; CRC Press: Boca Raton, FL, USA, 1997; pp. 35–60.

44. Peng, T.R.; Wang, C.H.; Lai, T.C.; Ho, F.S.K. Using hydrogen, oxygen, and tritium isotopes to identify the hydrological factors contributing to landslides in a mountainous area, central Taiwan. *Environ. Geol.* **2007**, *52*, 1617–1629. [CrossRef]

45. Peng, T.R.; Wang, C.H.; Hsu, S.M.; Wang, G.S.; Su, T.W.; Lee, J.F. Identification of groundwater sources of a local-scale creep slope: Using environmental stable isotopes as tracers. *J. Hydrol.* **2010**, *381*, 151–157. [CrossRef]

46. Clark, I. *Groundwater Geochemistry and Isotopes*; CRC Press, Taylor & Francis Group: Boca Raton, FL, USA, 2015; p. 421.

47. Kovač, Z. Nitrate origin in groundwater of the Zagreb alluvial aquifer. Ph.D. Thesis, University of Zagreb, Zagreb, Croatia, 2017; p. 255. (In Croatian)

 *water*

Article

The Role of Management of Stream–Riparian Zones on Subsurface–Surface Flow Components

Mads Steiness [1],*, Søren Jessen [1], Mattia Spitilli [2], Sofie G. W. van't Veen [3], Anker Lajer Højberg [4] and Peter Engesgaard [1]

[1] Department of Geosciences and Natural Resource Management, University of Copenhagen,
     1350 Copenhagen, Denmark; sj@ign.ku.dk (S.J.); pe@ign.ku.dk (P.E.)
[2] DICAM—Department of Civil Engineering, Chemistry, Environmental and Materials Engineering,
     University of Bologna, 40136 Bologna, Italy; mattiaspitilli@hotmail.it
[3] Department of Bioscience, Aarhus University, 8600 Silkeborg, Denmark; svv@bios.au.dk
[4] Geological Survey of Denmark and Greenland, Department of Hydrology, 1350 Copenhagen, Denmark;
     alh@geus.dk
* Correspondence: mast@ign.ku.dk; Tel.: +45-35332289

Received: 28 June 2019; Accepted: 10 September 2019; Published: 12 September 2019

Abstract: A managed riparian lowland in a glacial landscape (Holtum catchment, Denmark) was studied to quantify the relative importance of subsurface and surface flow to the recipient stream. The hydrogeological characterization combined geoelectrical methods, lithological logs, and piezometric heads with monthly flow measurements of springs, a ditch, and a drain, to determine seasonality and thereby infer flow paths. In addition, groundwater discharge through the streambed was estimated using temperature and water-stable isotopes as tracers. The lowland received large groundwater inputs with minimal seasonal variations from adjacent upland aquifers. This resulted in significant amounts of groundwater-fed surface flow to the stream, via man-made preferential flow paths comprising ditches, drainage systems, and a pond, and via two natural springs. Roughly, two thirds of the stream gain was due to surface flow to the stream, mainly via anthropogenic alterations. In contrast, direct groundwater discharge through the streambed accounted for only 4% of the stream flow gain, although bank seepage (not measured) to the straightened and deepened stream potentially accounted for an additional 17%. Comparison to analogous natural flow systems in the catchment substantiate the impact of anthropogenic alterations of riparian lowlands for the subsurface and surface flow components to their streams.

Keywords: riparian zone; groundwater; surface flow; springs; management

1. Introduction

Non-point source contamination of streams with nutrients, applied for agricultural production, stresses natural environmental systems. In Denmark, despite a nearly 50% reduction in nitrate leaching from the root zone [1], additional reductions are required to meet the goal of the EU Water Framework Directive. Riparian lowlands may serve as natural buffers, which may attenuate the movement of nutrients to streams [2]. Management of riparian lowlands are therefore often recommended as an integral part of watershed management practices in agricultural areas to reduce nutrient loading to streams [3–5]. Riparian lowlands used for agricultural purposes are frequently drained, but how much such drainage increases or decreases nutrient retention is uncertain. To effectively use riparian lowlands as buffer zones, or to implement new mitigation measures in the riparian lowlands, it is imperative to first understand the flow paths. This includes how discharge of groundwater from regional aquifers is distributed among e.g., direct groundwater discharge through the streambed, through drains to the stream, or from groundwater-fed seeps (focused and/or diffusive) leading to overland flow, which is then routed to the stream.

Hill [6] proposed different conceptual flow distribution models, where the hydrological controls were mainly determined by the size of the connected regional aquifer system and the thickness of the riparian aquifer. Several studies have since then, qualitatively confirmed these models. For example, studies [7,8] have demonstrated that flow across the lowland occurred as subsurface flow and groundwater-fed surface flow. Furthermore, Shabaga and Hill [8] demonstrated by dye tracer studies that surface flow took place as a combination of water discharging through macro-pores (soil pipes) and diffusively. Preferential flow through soil pipes in peat deposits has been shown to transmit large fluxes of water and with the possibility of causing springs resulting in overland flow [9].

Measuring the individual flow paths is not a trivial task and thus not often done. Brüsch and Nilsson [10] attempted to measure total surface runoff (i.e., the sum of all overland flow) from a riparian zone and found an average surface runoff of 0.3 L s^{-1}, compared with stream flow of 24–38 L s^{-1}. Furthermore, diffuse discharge to the surface only accounted for 3% of the runoff, so most of this occurred through focused seeps. Shabaga and Hill [8] measured surface flow directly to a river at one site. Johansen et al. [11] measured flow in three springs and a network of ditches. In none of these studies, the direct seepage through the streambed was measured. It was therefore not possible to quantify the relative role of subsurface flow to overland flow. Langhoff et al. [12] attempted to measure all flow paths. Direct groundwater discharge to the streams was measured by use of seepage meters at eight sites. Bank seepage and overland flow was estimated or measured as well. At three of the sites, more than 50% of the discharge was composed of direct groundwater discharge through the streambed; at one site, surface flow contributed all the stream gain. From their study, it is evident that surface flow is a non-negligible flow path with a potential to represent a significant part of the overall stream–gain–water balance. Frederiksen et al. [13] performed a similar, but less intense, study to evaluate methods for estimating groundwater discharge at different scales. Here they compared the relative contribution of groundwater discharge measured by seepage meters, tile drainage, and overland flow to that of stream flow gain from differential gauging and hydrograph separation. Their study catchment could be divided into two separate units; an upstream reach, which was tile drainage dominated and a lower reach, which was groundwater dominated. In the latter, groundwater discharge was >60% of stream flow gain.

The objectives and overall methodology of this study were to: (1) investigate the connection between the groundwater system and the recipient stream in a managed riparian lowland, based on frequent monitoring of major water flux components of the stream–gain–water balance throughout a year, (2) quantify the relative importance of subsurface and surface flow to the stream, and (3) discuss these results in relation to similar studies in the same catchment on of how management (or mismanagement) of a riparian lowland may change flow paths. The hydrogeology of the riparian lowland and adjacent upland was characterized extensively using geoelectrical methods, hand-drillings, water-stable isotopes, piezometric head observations, along with monthly monitoring of flow in several springs, a ditch, and through the streambed.

2. Study Site

As study site, a riparian lowland located in Holtum catchment (central Jutland, western Denmark) was chosen (Figure 1). From north to south through Holtum catchment runs the Main Stationary Line (MSL) of the late Weichselian glaciation [14]. West of the MSL outwash plain sediments dominate, while clayey tills are frequent east of the MSL. Land use in the catchment is mainly agriculture (49%), followed by forest (31%), urban (15%), and wetland and surface-water areas (5%). The annual precipitation in the catchment is 984 ± 55 mm year^{-1} (mean ± 1 standard deviation), and actual evapotranspiration is 510 ± 45 mm year^{-1} [15].

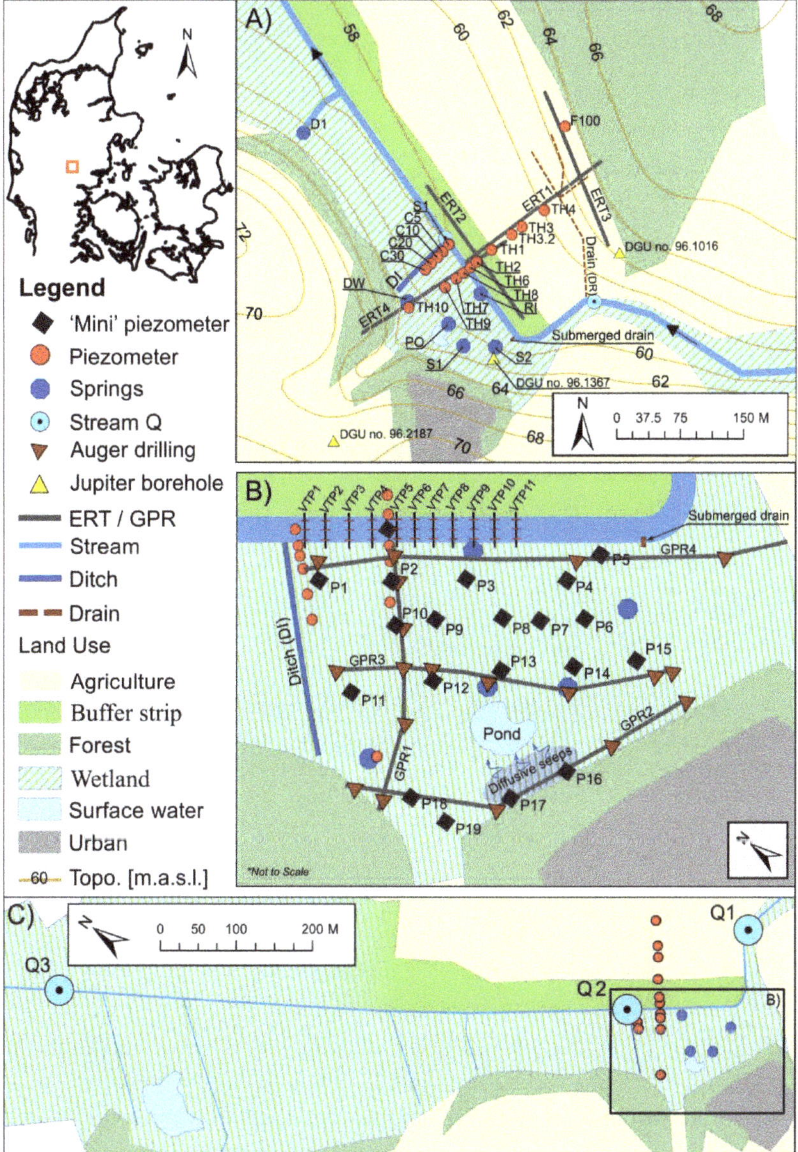

Figure 1. Field site in Holtum catchment in the central part of Jutland, Denmark. (**A**) The south side of the stream is a wetland functioning as a cow pasture. The north side is an approx. 35 m wide buffer strip of grass and an agricultural field extending to the northern valley hillslope. On the south/wetland side of the stream, six overland flow systems are located: two springs S1 and S2, overflowing drainage well (DW), an overflow pipe from the pond (PO), a ditch (DI) and a small rivulet (RI). The DI and RI discharge directly to the stream. One submerged drain has been found. One pipe draining the north/agricultural side to the stream has been found. Piezometer network, three wells from the national database (JUPITER), and location of Electrical Resistivity Tomography (ERT) lines (grey lines) are shown; (**B**) Mini-piezometer network (P1–P19), Ground-Penetrating Radar (GPR) lines (grey lines), and Vertical Temperature Profiles (VTP). This area is shown as the rectangle in C; (**C**) Location of stream discharge measurements (Q).

The riparian lowland is located along a section of Holtum stream running in the valley of an old sub-glacial stream trench, and hence the riparian aquifer comprises sandy outwash deposits. Hillslopes north and south of the lowland area comprise clayey tills. Beneath these Quaternary deposits sits a regional Miocene micaceous sand aquifer. The stream is a lowland second-order gaining stream with a width of approx. three meters at the field site. The stream is perennial and has been straightened and likely deepened at the field site, a very frequent management practice along streams in agricultural landscapes. The height of the stream bank is about 1–1.3 m. The catchment boundary lies about 1 km southwest and 0.4 km northeast of the study site (distances perpendicular to the stream). Land use southwest of the riparian lowland is agriculture all the way to the catchment boundary. A forest plantation, mainly with pine trees, is located to the northeast of the riparian lowland and here extends to the catchment boundary.

Figure 1A,B displays the observation points, land use, and drainage features, and defines the naming used henceforth in this paper. South of the stream, the riparian lowland consists of an 85 m wide wetland functioning as a cow pasture. The wetland is bordered by a steep hillslope to the south rising 12 m. The hillslope is cut by a small gorge in the center. Several old drainage wells and two springs (S1 and S2, Figure 1) are present in the wetland. One of the drainage wells (DW) is constantly overflowing. Springs and the drainage well contribute to diffusive overland flow. The wetland's drainage system has not been well maintained and it is likely that drains connecting to the drainage wells are not properly functioning. A submerged drainpipe in the stream in the eastern part of the wetland exists, also indicated in Figure 1A,B. An elevated small pond is present in the middle of the wetland. The banks of the pond are to the north (towards the stream) and northeast constructed from earthen levees. The pond is drained through an overflow pipe (Pond Outlet, PO) also contributing to overland flow. As a result, a small rivulet (RI) runs from the wetland into the stream. To the west, a nearly overgrown ditch (DI) is deep enough to reach sand beneath the foot of the clayey till hillslope; this forms a spring at the southern end of DI. The water captured by DI discharges directly into the stream. Several similar ditches exist downstream of the field site (between Q2 and Q3, Figure 1C), with discharge mainly consisting of groundwater. S1, S2, DW, PO, RI and DI are collectively termed "springs" from here on as they represent a visible and measurable flow. Furthermore, all "springs" one way or another represent water emerging to the surface in a, at least apparently, spring-like fashion. Along the foot of the hillslope, seeps have been observed maintaining wet surface areas. The wetland area is therefore mostly very wet, with areas fully saturated throughout the year, and consists of the main part of saturated peat and mud.

In 2015, a transect consisting of ten piezometers with screens at varying depths was established [16]. Some of these piezometers are shown in Figure 1A (along the ditch; S1, C5, C10, C20, C30).

On the north side of the stream, land use consisted of an approx. 35 m wide buffer strip of grass functioning as a cow pasture and a 145 m wide field for crop production extending from the buffer strip to the northern valley hillslope. Field crops have alternated between corn and oats. A field drain (DR), leading directly into the stream (not submerged and therefore easily measurable), drains a part of the agricultural area which is prone to waterlogging due to a depression in the surface topography.

3. Materials and Methods

3.1. Hydrogeological Characterization

Electrical Resistivity Tomography (ERT) surveys were done on both sides of the stream similar to surveys carried out in comparable settings [17,18]. A SYSCAL PRO resistivity meter with a Wenner configuration was used. The ERT1 and ERT2 profiles were measured in November 2015; the ERT3 profile was measured in March 2016 and the ERT4 profile in March 2017. All profiles (locations shown in Figure 1A) were measured using 96 electrodes with 2 m electrode spacing; equaling a length of each profile of 192 m and a maximum penetration depth of ~35 m in the center of the profiles. The location and elevation (meters above sea level; MASL) of each electrode were measured with a Differential

Global Positioning System (DGPS) using a Trimble® R8s GNSS receiver with a Trimble TSC3 controller (Trimble Inc., Sunnyvale, CA, USA). The inversion of the measured data was done with Res2DInv by Geotomo software (Res2DInv ver. 3.57; Geotomo Software, Penang, Malaysia), using a least squared inversion technique to produce the tomographic profiles.

Common offset Ground-Penetrating Radar (GPR) surveys were conducted across four transects in the wetland south of the stream (Figure 1B). The GPR survey was carried out to identify the thickness and spatial geometry of the peat in the wetland. GPR has been frequently used to identify the peat-layer thickness in wetlands [19,20]. A PulseEKKO® PRO SmartTow GPR system (Sensors & Softwarer Inc., Mississauga, Canada) was used with a 250 MHz transducer, with a separation of 0.4 m between the transmitter and receiver. A step size of 0.05 m was used and a stacking of 256 pulses, collected at each position along the GPR transects, with the receiver set with a sample time window of 400 ns. No topographical corrections were made since the ground surface is relatively flat in the surveyed area. The profile length was measured using an odometer. The collected data was post-processed in Ekko_View Deluxe (Sensors & Software Inc., Mississauga, Canada). A high pass (DEWOW) filter was applied to reduce low frequency noise and The Spherical Exponential Calibrated Compensation (SEC2) gain was used to compensate for spherical spreading loses and dissipation of the radar signal.

Eighteen shallow boreholes were hand-drilled, using an Eijkelkamp auger (Eijkelkamp Soil & Water, Geisbeek, the Netherlands), in the wetland to determine the peat-layer thickness and to ground truth depth of reflectors and obtain lithological logs from sediments retracted from the boreholes.

3.2. Hydrology

In addition to the existing piezometers, 14 new piezometers were installed in a northeast southwest transect perpendicular to the stream. Galvanized steel pipes (Ø2.54 cm) equipped with a 9 cm long screen, were hammered into the subsurface using a pneumatic hammer. Universal Transverse Mercator (UTM) coordinates of the piezometers including elevation of top of well and ground surface were measured with a DGPS.

Naming of piezometers, presented in Figure 1, were done in chronological order of installation. Installation of piezometers TH1d, TH2d, TH3, and TH4 took place in November 2015. Piezometers TH3 and TH4 had to be removed in March 2016 to allow the farmer to access the field. In September 2016, TH3 was re-established as TH3.2 and removed again in March 2017.

The transect was extended in March 2016 with the addition of TH1s and TH2s, next to the existing TH1d and TH2d. On the same occasion, TH6 was also installed approx. 1.5 m from the stream. On the opposite side of the stream, in the wetland, Installation of TH8, approx. 1 m from the stream, took place in January 2017. Two piezometers TH7d and TH7s, forming a nest, were installed in October 2016 and January 2017 respectively. Installation of TH10 near the hillslope, and another nest (consisting of TH9s and TH9d) took place in March 2017. Hence, four piezometer nests were installed at the field site, two on each side of the stream; nested piezometers are referred to as TH1, TH2, TH7, and TH9.

Nine of the piezometers were slug-tested using the falling head method, five of which was slug-tested at multiple screen depths. The slug-tests were conducted three consecutive times and analyzed using the Hvorslev method for unconfined aquifers.

Daily values of precipitation for 2017 were obtained from Danish Meteorological Institute's Nørre Snede station (UTM-E: 524034, UTM-N: 6202165) approx. 4 km east of the field site.

3.3. Discharge Measurements in Stream and Springs

Stream discharge and flows from the springs were measured once every month and the monthly measurement was repeated twice. It would have been better if we had continuous stream discharge data, but, on the other hand, it was not possible for us to measure continuous flow from the springs. Therefore, the two types of measurements go in tandem and provide a snapshot of the monthly stream flow gain and spring flows. This also means that our water balance is the average of these monthly stream flow gains and spring flows.

Stream discharge was measured using an OTT Acoustic Digital Current meter (ADC) (OTT Hydromet, Kempten, Germany). The ADC uses a 6 MHz acoustic sensor for point flow velocity measurements, and corrects the measurement for water depth (water above the sensor) and temperature. Stream discharge was measured using the mean-section method at three locations; immediately upstream of the field site (Q1), right after the ditch draining the wetland (Q2) and downstream of the field site (Q3) (Figure 1C). The distances along the stream are 250 m and 750 m between Q1 and Q2, and Q2 and Q3, respectively. Measurements were performed monthly (except in July 2017) from January 2017 to January 2018.

"Spring" discharge in the wetland (Figure 1A,B) was measured using a cutthroat flume (Baski Inc.; Denver, CO USA). Using a spirit level to keep the base of the flume horizontal in both the longitudinal (direction of flow) and transverse directions. The banks on either side of the flume were dammed to make sure water would not flow around the flume. After installation, the flume was left for up to half an hour before measurements were taken, to allow flow to equilibrate and to ensure free flow conditions had been reached. Experimental work [21] have established a relationship between the head (h_a), upstream of the flume throat, and discharge rate (Q), given different sizes of cutthroat flumes under free flow conditions:

$$Q_{free\,flow} = C_f h_a^{n_1} \tag{1}$$

where $Q_{free\,flow}$ is the flow rate in $L^3 \, t^{-1}$; C_f is the free flow coefficient, which is a function of flume length, given by a flume length coefficient, and throat width of the flume $\left(C_f = K_l \cdot W^{1.025}\right)$; and n_1 is the free flow exponent.

3.4. Direct Seepage through the Streambed: 1D Vertical Temperature Profiling

The mean near-surface groundwater temperature for Danish conditions and within the catchment is relatively constant at ~8 °C [17,22,23]. This feature allows the use of natural temperature variations as tracer to determine the flux between groundwater and stream. Vertical groundwater fluxes were hence calculated from measurements of Vertical Temperature Profiles (VTPs) and thermal conductivity of the streambed.

The measurements were carried out in September 2016, where the contrast between groundwater and stream water was relatively high after the summer period and flow and heat transport conditions were assumed to be at steady state. The same method has previously been used to estimate stream- and groundwater exchange in Holtum stream [16,17,22]. Notice that we here only have one estimate (September) of the streambed seepage. We quickly found that streambed seepage was much smaller than flows from the springs, so we did not repeat this monthly. This is quite a time-consuming process.

Three VTPs, one near each of the banks and one in the center of the stream, were measured for 11 transects along a 90 m stream section (Figure 1B). Each VTP was recorded after a 10-min equilibration time [24] at 0, 0.025, 0.05, 0.075, 0.1, 0.15, 0.2, 0.3, 0.4, and 0.5 m depth below the streambed surface. A steady-state analytical solution [25] for a 1-dimensional conduction–convection system was used to estimate the vertical groundwater fluxes.

$$T(z) = T_s + \left(T_g - T_s\right) \frac{exp\left(\frac{N_{pe}z}{L} - 1\right)}{exp\left(N_{pe} - 1\right)} \tag{2}$$

where $T(z)$ is the temperature (°C) of the streambed sediment at depth z (m); T_s is the temperature (°C) of the stream water at the streambed (measured at depth 0); T_g is the groundwater temperature (°C) at depth L (m) and N_{pe} is the Peclet number expressing the ratio caused by convection and conduction. For this, study T_g and L are defined as 8.3 °C and 4 m.

$$N_{pe} = \frac{q_z \cdot \rho_f \cdot C_f \cdot L}{\kappa_e} \tag{3}$$

where q_z is the vertical Darcy flux (m s^{-1}); $\rho_f \cdot C_f$ is the volumetric heat capacity of the water (J m^{-3} °C^{-1}) and κ_e is the thermal conductivity of the sediment $\left(\text{J m}^{-1}\text{ s}^{-1}\text{ °C}^{-1}\right)$. The accuracy of the temperature measurements have been reported to be ±0.2 °C [16,24].

The thermal conductivity (κ_e) of the top 3–10 cm of the streambed sediment was measured in the field using a KD2 pro with the SH-1 sensor (Decagon Devices, Pullman, WA, USA). The instrument has an accuracy of ±10% at thermal conductivities between 0.002 and 2 W m^{-1} °C^{-1}. The sensor is equipped with a heater needle spaced 6 mm apart from a temperature sensor needle. A read time of 2 min was used, during which a heat pulse was generated for the first half of the read time and then data (temperature) was recorded to compute thermal properties for the full read time. Prior to initiation of the heat pulse, the instrument allows 30 s for temperature equilibration of the sensor. Measurements were conducted next to each VTP measurement, yielding the bulk thermal conductivity and a measurement error for each location. Previous studies have used the same instrument to measure in situ κ_e [24] and κ_e of cores collected in Holtum catchment [26]. Both studies found that the thermal conductivity can assume a wide range of values. In general, the thermal conductivity of a porous material depends on the sediment mineral composition, grain size, pore space, and water content hereof (in this case fully saturated). Thermal properties of selected materials are available in the literature [27].

3.5. Water Sampling and Analysis

Water samples were collected for water-stable isotopes (^{2}H and ^{18}O) and water quality, an- and cations, alkalinity, Fe^{2+} and H$_2$S. In this paper, only the water-stable isotopes will be reported. Samples from springs in the wetland (see Figure 1A,B) were collected monthly between January 2017 and January 2018 (excluding July 2017). Grab samples of stream water and the drain on the crop side were collected for the same period.

Groundwater samples from the piezometers were collected on several occasion. During installation of piezometers screened deeper than 2 m below ground level (MBGL) (TH1d, TH1s, TH2d, TH2s, TH7d, TH7d and TH9d) multilevel water sampling was conducted by taking samples at several depths. The piezometers were further sampled 2–5 times between January 2015 and December 2017.

Groundwater samples were retrieved using a peristaltic pump. Following three purges, water was pumped through a flow cell equipped with electrodes for dissolved oxygen (DO; WTW FDO 925 IDS electrode), pH (Hach PHC101 electrode), Electrical Conductivity (EC) and temperature (both recorded using an Hach CDC401 electrode) connected to WTW Oxi 3310 IDS and HACH HQ30 flexi instruments. Sampling commenced when values were stable. Samples were taken before water entered the flow cell, using a syringe and a three-way valve.

All samples of stable hydrogen and oxygen water isotopes were filtered with a 0.20 μm Minisart cellulose acetate syringe filter by Sartorius (Göttingen, Germany), into a 1.5 mL glass vial in the field and then stored in a refrigerator. The samples were analyzed at the Geological Survey of Denmark and Greenland (GEUS) on a PICARRO L2120-i (PICARO Inc., Santa Clara, CA, USA) using cavity ring-down spectroscopy (CRDS). All values for isotopes are expressed in per-mill (‰) with the δ-notation indicating the deviation from the VSMOW (Vienna Standard Mean Ocean Water) standard:

$$\delta\text{-value }(‰) = \left(\frac{R_{Sample} - R_{VSMOW}}{R_{VSMOW}}\right) \times 1000‰ \tag{4}$$

where R is the ratio of the heavy to the light isotope.

In June 2017, a campaign was undertaken to investigate the spatial patterns of the water-stable isotopes of groundwater directly underneath the peat cover of the wetland. Nineteen point samples were obtained using stainless steel mini-piezometers with an inner diameter of 0.5 cm and a 5 cm screen length. The mini-piezometer was pushed vertically down to ~1 m below the surface by hand or with the use of a hammer, and withdrawn again after sampling. Sampling followed the same procedure as described above. Likewise, a mini-piezometer was later (November 2018) used to sample water 1 m beneath the streambed.

4. Results

4.1. Hydrogeological Characterization

Figure 2 shows the geology as interpreted by comparing resistivity cross sections to four borehole logs available from the Danish National Well Database (Jupiter), in which boreholes are denoted with a unique borehole number (DGU no.). The logs are from DGU Nos. 96.2187, 96.1367 and 96.1026, the locations of which are shown in Figure 1A. DGU No. 96.1076 is located about 425 m north of the end of ERT1.

A zone near the surface with resistivity values above 600 Ωm is visible to the southwest (from 0 to about 38 m along the profile). This is most likely due to the steep increase in slope, which facilitates the presence of a deeper unsaturated zone. Just below the unsaturated zone is a ~4 m thick clay lens with low resistivity values (30–70 Ωm), which could be part of a more extensive clay till layer seen in borehole 96.2187 from 67.5 to 64 MASL. A saturated gravel or sand layer (100–400 Ωm) was found underneath, again in accordance with 96.2187 with gravelly outwash deposits from 64 to 54.5 MASL. The upland aquifer depth to the south (southwest), considering only the unconfined top unit of permeable sediments, is 15.5 m thick.

The geology described in borehole 96.1367 consists of a peat layer of 1 m thickness followed by 14 m sand. Data from ERT4 is in reasonable agreement with the lithology of the borehole. Resistivities in the top of the profile are below 30 Ωm representing peat. Below, resistivities are in the range of 70–400 Ωm corresponding to saturated gravelly or coarse sand. The peat has been formed in a small depression in the top-center of the profile. The peat layer has a length of about 40 m and a thickness ranging from less than 0.5 m up to approx. 3 m. A gravelly sand layer with varying thickness of ~5–8 m is found below the peat followed by a sand layer (resistivity values ~70 Ωm) extending to the bottom of the profile. At the southern edges and deepest part of profile ERT4, a clay layer is likely present, since resistivity values are <50 Ωm. Borehole 96.2187 shows a micaceous clay layer at a depth of 54.5–26 MASL and it may be that this clay layer has a trough-like bathymetry extending beneath the whole stream valley (but not captured with ERT).

On the crop side of the stream (ERT1), the sandy aquifer is unconfined with resistivity values ranging from 70 to 400 Ωm. The aquifer is covered by approx. 0.5–1 m sandy soil with high resistivity values (>400 Ωm), where the soil is unsaturated. Smaller areas near the surface along the profile have resistivity values ranging from ~90–190 Ωm indicating saturated soils all the way to the surface. These are areas prone to water logging and are drained (DR).

At the far northern end of profile ERT1, resistivity values are high associated with increase in thickness of the unsaturated zone—or indicative of high resistivity sediments such as gravel. The geology of borehole 96.1026 consists of sand and gravel down to 60.3 MASL, then from 60.3 to 53 MASL a layer of sandy clay is present and below this is a clay layer (53–39 MASL). This is also indicated by the resistivity values found at the foot of the hillslope for ERT1, with sand and gravelly deposits on top of a moraine till/clay. The aquifer depth to the north (northeast), down to the micaceous clay, is about 13 m thick. This is comparable to the thickness found to the south of the stream.

The hydraulic conductivities (K) range from ~0.1 to 50 m day^{-1} with a mean of 14 m day^{-1}. These K-values (Figure 2) are all representative for sandy aquifer materials and are similar to those found for a sandy riparian aquifer in the same catchment [17]. High K-values are found at depths of 2–6 MBGL for TH7, where also the ERT4 profile indicated the presence of gravelly or coarse sand. A similar agreement is with TH2 and TH1, both with higher hydraulic conductivities from 2–6 MBGL and 5–12 MBGL, respectively, in areas with coarse sands as indicated by ERT1. In addition, 11 out of the 18 hand-drilled boreholes were described as having coarse sand and some with pieces of gravel-sized particles.

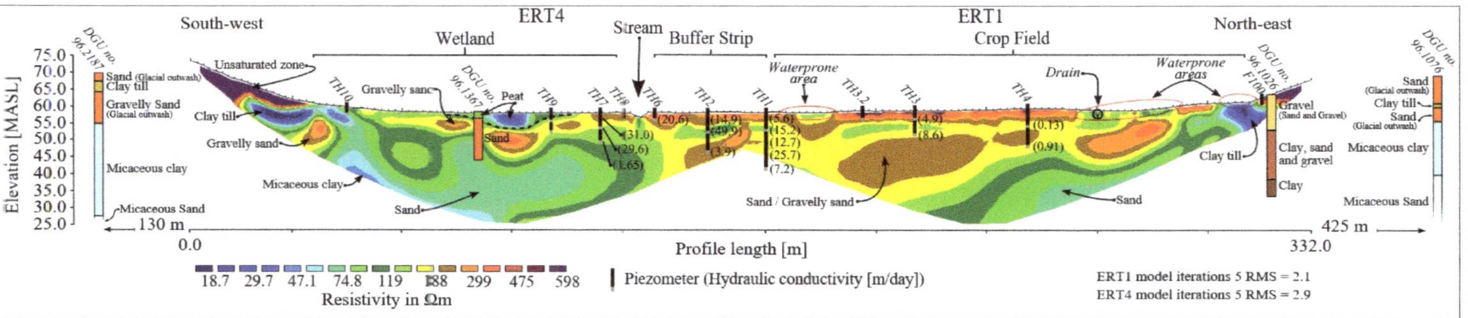

Figure 2. Resistivity profiles ERT4 and ERT1. Measured hydraulic conductivities from slug-tests performed in the piezometers at several depths are shown in m day^{-1}. Data from four boreholes from the national well database (Jupiter, GEUS), DGU. No. 96.2187, DGU. No. 96.1367, DGU. No. 96.1026 and DGU. No. 96.1076 are imposed on to the profiles.

A joint interpretation of ERT4, the GPR survey, and shallow borehole logs is shown in Figure 3, which displays the resulting thickness of the peat that overlies the sandy aquifer in the wetland. Peat thicknesses varies between 0.1 and 1.5 m. Springs are in elongate bands with peat thickness >0.5 m which runs parallel to the stream.

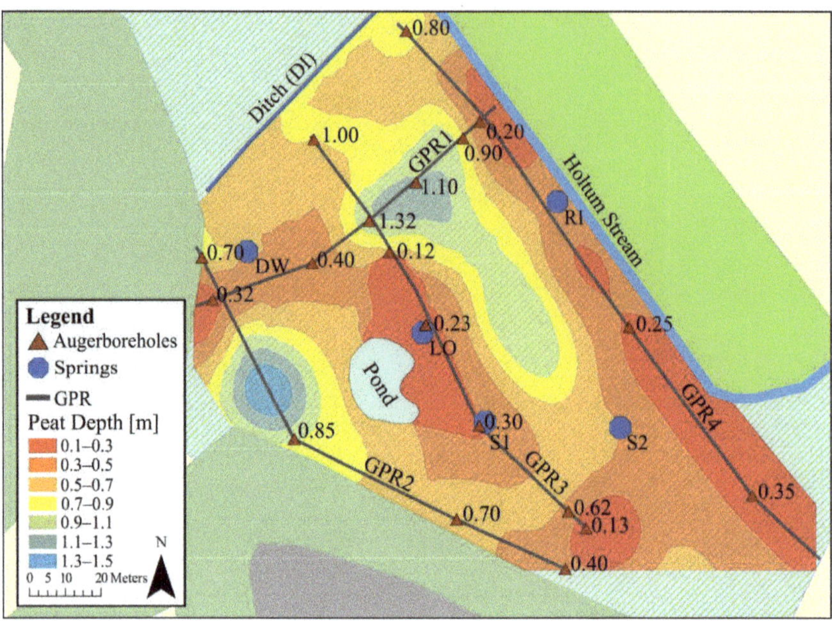

Figure 3. Estimated peat-layer thickness in the wetland. Interpolation of GPR data and shallow auger boreholes (data from June 2017). The depth (thickness of peat or soil layer) to the underlying sand is noted next to the auger boreholes (m). Springs are shown for reference.

4.2. Hydrology

Figure 4 displays the seasonal fluctuations in precipitation, groundwater hydraulic heads, and stream stage. The stream stage and hydraulic heads followed the fluctuations of precipitation (Figure 4A,B). June had 114 mm of rainfall, which resulted in increased hydraulic heads in all piezometers within the same month. The same was seen for October with 136 mm of rainfall.

In all piezometers and throughout the monitoring period, hydraulic heads were higher than the stream stage (Figure 4). The corresponding average gradients between the most northern (F100) and southernmost (TH10) piezometers (Figure 4C) and the stream were 0.01 and 0.02, respectively (positive towards to the stream).

The two nested piezometers TH1d and TH1s, on the crop side of the stream and screened at 16 and 6 MBGL, respectively, showed a significant downward gradient. On average, the head observed in TH1s was 0.7 m higher than that of TH1d, and TH1s was the only piezometer with an artesian head at the field site. The nested piezometers TH2s and TH2d, located closer to the stream and screened at 5.5 and 12 MBGL, respectively, showed vertical head difference from 0.05 to 0.07 m over the monitoring period, yielding a persistent upward hydraulic gradient of about 0.01.

To the other side of the stream, the TH7d screened at nine MBGL always had higher hydraulic heads than the shallower TH7s piezometer (4 MBGL), varying from 0.03 to 0.08 m giving upward hydraulic gradients of between 0.006 and 0.016. TH9d and TH9s that are also located in the wetland, but further away from the stream, screened at 5.5 and 1.5 MBGL, respectively. Here, the hydraulic heads were generally higher for the shallow TH9s, yielding downward gradients ranging from 0.001 to 0.004.

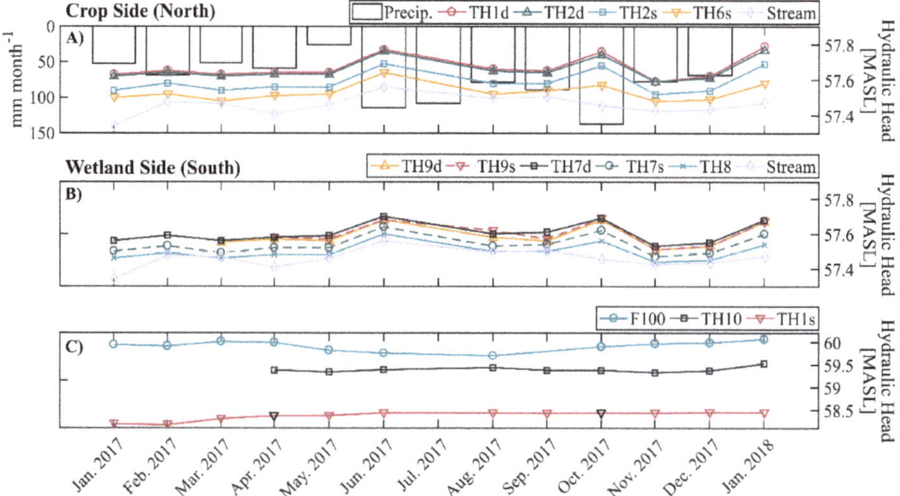

Figure 4. Monthly precipitation, hydraulic heads, and stream stage from January 2017 through January 2018. (**A**) Piezometers on the crop side (north) of the stream and precipitation; (**B**) Piezometers on the wetland side of the stream (south) and stream stage; (**C**) Piezometers TH10 and F100, respectively the most southerly and northerly piezometer.

The difference between maximum and minimum heads ranges from 0.16 to 0.37 m with an average of 0.21 m. The largest fluctuations were observed in F100 with -0.21 to 0.16 m of fluctuations from the observed average value of 59.91 MASL. Piezometer TH10 also had slightly higher fluctuations, when compared to piezometers further away from the edges of the riparian lowland, with -0.06 to 0.14 m of fluctuations from the average head of 59.39 MASL.

In general, the temporal fluctuations throughout the monitoring period on both sides of the stream were identical (Figure 4). Although the magnitude of the fluctuations in observed heads were slightly larger in the piezometers on the crop side, ranging from 0.16 m to 0.24 m, not including F100 furthest from the stream. Fluctuations were more dampened in the observed hydraulic head in the wetland ranging from 0.16 to 0.18 m, not including TH10 at the hillslope.

4.3. Stream and Spring Discharge

4.3.1. Stream Discharge

Measured stream discharges (Q_s) at the three locations are shown in Figure 5. Mainly for August, October (in 2017), and January (2018), the higher discharges can be attributed to higher intensity of precipitation (Figure 4A)—stream discharges were highest in August for all three locations. The average Q_s, excluding measurements affected by precipitation, for Q1, Q2, and Q3 were 0.239 m^3 s^{-1}, 0.257 m^3 s^{-1}, and 0.421 m^3 s^{-1}, respectively.

The differences between Q1 and Q2 monthly are barely noticeable in Figure 5. The stream section between the two locations is, however, gaining for six out of the 11 months. There was no difference in discharge in February. Discharge measurement associated with rainfall event, was also when the observed largest gaining (April and October 2017) and losing conditions (August 2017 and Jan 2018) occurred in the stream between Q1 and Q2. The significant decrease in discharge between Q1 and Q2 (losing conditions, -0.22 m^3 s^{-1}) seen in August, is certainly because Q2 was measured last and after a rainfall had stopped—whereas Q1 and Q3 was measured during rainfall. Excluding the month where discharge can be coupled with rainfall, the average gain between Q1 and Q2 yields 5.1×10^{-5} m^3 s^{-1} m^{-1} with a standard deviation of 7.6×10^{-5} m^3 s^{-1} m^{-1}. Hence, it is not strictly possible to determine whether the stream section between Q1 and Q2 is gaining or loosing. For later comparison, stream gain per meter between Q1 and Q2 would correspond to a total gain of 4.58×10^{-3} m^3 s^{-1} along a 90 m stream reach, corresponding to where the VTP measurements was conducted.

Between Q1 and Q3, the stream was gaining and had an average gain of 1.9×10^{-4} m^3 s^{-1} m^{-1} (standard deviation (SD) 5.4×10^{-5} m^3 s^{-1} m^{-1}) which is considerable higher than between Q1 and Q2. Seasonal variations are more visible at Q3 with lower discharge during late spring to late summer (May through September) and with higher discharge in the winter.

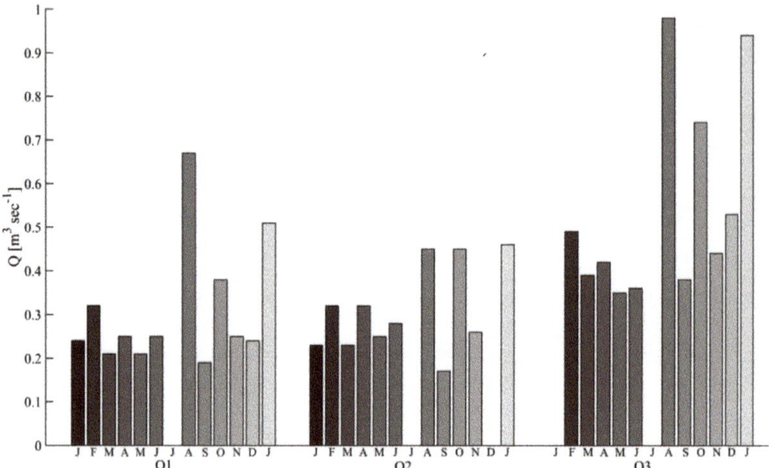

Figure 5. Monthly measured stream discharge (Q_s) at Q1, Q2, and Q3 from January 2017 through January 2018, except July 2017, December 2017 for Q2, and January 2017 for Q3. High discharge in April, August, and October 2017, and January 2018 can be related to precipitation events. Month of measurement is given below the x-axis starting at January 2017 (dark grey) to January 2018 (light grey) (J, F, M, A, M, J, J, A, S, O, N, D, J).

4.3.2. Wetland Springs Discharge

Measured discharges at the six springs are shown in Figure 6. Discharge ranged from 6.1×10^{-5} m^3 s^{-1} at S2 in December 2017 to 1.2×10^{-3} m^3 s^{-1} at S1 in February 2017. S1 had the highest average discharge of 9.4×10^{-4} m^3 s^{-1} followed by the rivulet (RI) with an average discharge of 6.7×10^{-4} m^3 s^{-1}.

The seasonality of the discharge at RI, S1 and PO was especially similar (Figure 6). For all three locations, discharge was highest in late winter to early spring (February–April). Discharge decreased until September, with the lowest discharges. In early fall discharge increased again, until the last measurements in January 2018. The similarity in the seasonality of PO and RI is expected since the PO drains the pond into the bulrush-covered central part of the wetland. Immediately on the opposite side of the bulrush (towards the stream), the rivulet RI is formed and discharges into the stream. RI therefore likely receives water from the pond. However, discharge measured at RI is larger than PO (Figure 6), indicating that RI receives water not only from PO. This may include water from other springs upstream of RI, being S1, DW, and/or S2 or from diffusive groundwater upwelling in the central part of the wetland.

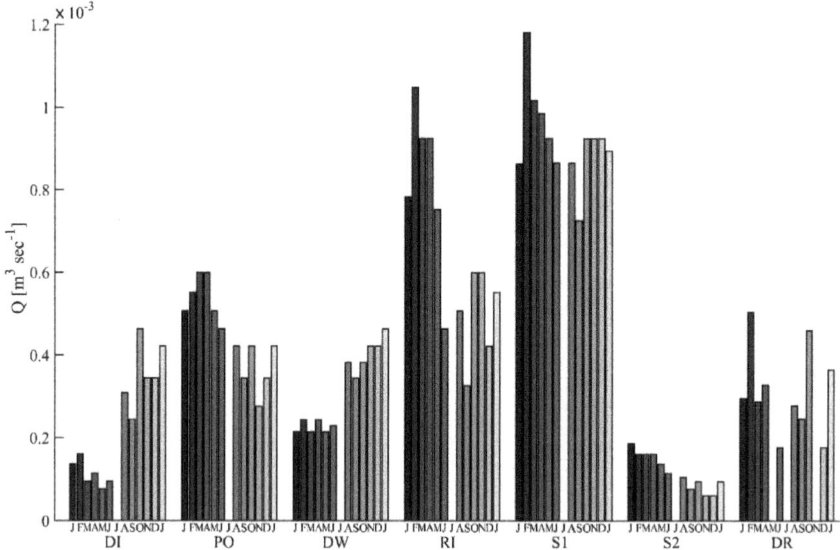

Figure 6. Measured discharge from six locations in the wetland on the south side of the stream: Ditch (DI), Pond Outlet (PO), Drainage Well (DW), Rivulet (RI), Spring 1 (S1), Spring 2 (S2) and the drain (DR) draining the crop field. Month of measurement is indicated below the x-axis starting at January 2017 (dark grey) to January 2018 (light grey).

The ditch (DI) and drainage well (DW) display a sudden increase in discharge in August and continuing into the fall. Possibly, part of the water from DW runs as surface flow to DI and thus added to the groundwater emerging at the south end of DI. S2 showed an overall decrease in discharge over the monitoring period, as well as the lowest average discharge of the springs (1.2×10^{-4} m^3 s^{-1}). Drain discharge from the crop field north of the stream showed an average discharge of 3.17×10^{-4} m^3 s^{-1} corresponding to a contribution of about 7% of the stream gain of 4.58×10^{-3} m^3 s^{-1} for a 90 m stream section.

If all springs in the wetland were assumed to discharge to the stream individually, the average combined spring discharge would amount to 2.7×10^{-3} m^3 s^{-1}, corresponding to 59% of the average stream gain. In September 2016, when discharge through the streambed was measured (see below), the total discharge from all springs amounted to 2.5×10^{-3} m^3 s^{-1}, which is very similar to the average spring discharge measured throughout the full monitoring period. DI and RI are the only two springs visibly discharging directly into the stream and may receive water from one or more of the other springs. Therefore, these springs two (DI and RI) can be considered separately. Their combined discharge only accounts for about one third of the combined discharge of DW, PO, S1, and S2. Furthermore, the combined discharge from DI and RI equals 0.9×10^{-3} m^3 s^{-1} accounting for 20% of the gain in the stream.

4.4. Streambed and Bank Fluxes

For each VTP location, the vertical groundwater Darcy flux (q) was estimated using the in situ, locally determined thermal conductivity, which ranged from 0.44 to 2.70 W m^{-1} °C^{-1} (avg. 1.88 W m^{-1} °C^{-1}) (Figure 7A). Estimated fluxes through the streambed are presented in Figure 7B, with VTP1 located furthest downstream and VTP11 furthest upstream (locations shown in Figure 1B).

Estimated q values always indicated upwards flow to the stream. The mean estimated q was 7.02×10^{-7} m s^{-1}. The maximum q of 1.6×10^{-6} m s^{-1} was measured at VTP2, i.e., at the north bank and the center of the stream. Here, VTP2 had the highest observed κ_e, reflected in the highest average q across the stream of 1.57×10^{-6}. The lowest flux of 1.35×10^{-7} m s^{-1} was estimated for the north bank profile of VTP7, which also had the lowest observed κ_e. Accordingly, q varied by a factor of five.

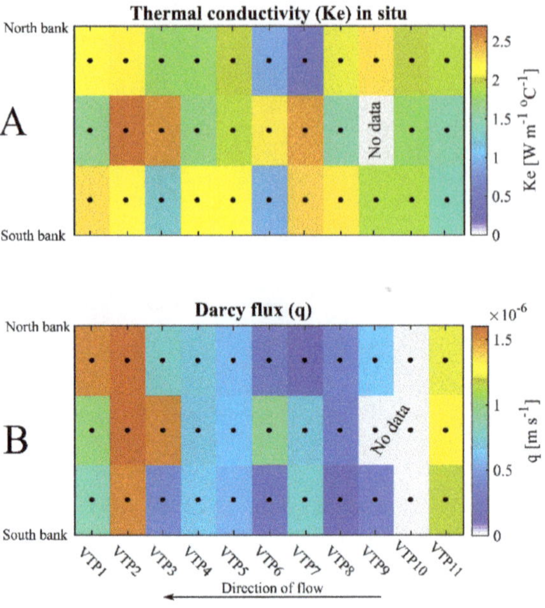

Figure 7. Thermal conductivities (**A**) and derived vertical streambed fluxes (**B**) through the ~90 m × 3 m streambed (Figure 1B).

Assuming a stream width of 3 m along the 90 m stream segment surveyed with VTP, the total direct groundwater discharge to the stream amounts to 2.08×10^{-4} m^3 s^{-1}. The groundwater discharge to the stream only represents 4% of the stream gain. Adding the combined average discharge of all springs (59%) together with the groundwater discharge (assuming half comes from the south and the other half from the north), this accounts for 61% of the gain in discharge in the stream gain that comes from the southern wetland section.

For comparison, Darcy's Law can be applied to calculate the bank seepage to the stream, using the average stream stage, and average hydraulic head and distance to the stream's bank for piezometers TH6 (north) and TH8 (south). The distance used is the piezometer distance to the center of the stream minus 1.5 m, assuming the stream is 3 m wide. The K-value measured in TH6 (20.6 m day^{-1}) is used on the north bank, while the measured K-value in TH7 at 2 MBGL of 31 m day^{-1} is used on the southern side. The flow area used for both banks is the 90 m stream reach times a water depth of 0.35 m. The potential bank seepage to the stream from the northern bank is then 3.85×10^{-4} m^3 s^{-1} and through the southern bank 3.97×10^{-4} m^3 s^{-1}. Potentially, this would amount to nearly 17% of the total stream gain, twice the amount, from each individual bank, than the flux estimated through the stream bottom using VTP-data.

4.5. Water-Stable Isotopes

Stable oxygen and hydrogen isotope composition of precipitation for a 2-year period from a nearby station (Voulund, UTM-E: 510000, UTM-N: 6210000) has previously been reported [28]. The δ^{18}O values of rainfall at Voulund had a weighted average of −7.69‰ and ranged from −5.5 to −10.0‰. δ^2H values had a weighted average of −51.5‰ and ranged from −38 to −73‰. Stable isotope concentrations, from all piezometers and for all samples collected in the monitoring period, yielded an average δ^{18}O value of −7.79‰, ranging from −8.27‰ to −7.20‰. δ^2H had an average of −51.5‰ ranging from −47.3‰ to −54.0‰.

Figure 8 shows the δ^{18}O value of springs, the stream and the drain (DR) on the crop side of the stream as a function of time. About 15% of the samples had δ^{18}O concentrations outside the δ^{18}O mean groundwater signal ± 1 standard deviation, indicated in Figure 8 by grey shading. The springs include three surface-water bodies with free-flowing water (DI, RI and PO) where the isotope signal can be affected due to evaporation or precipitation. Indeed, DI and RI had δ^{18}O values outside the range of the groundwater; RI was more was more enriched in May 2017 and the same was the case for DI in January, February, and April 2017. Both DI and RI, and the stream samples were more depleted in January. Direct precipitation, upwelling groundwater, evaporation, and their relative magnitude, determines the δ^{18}O and δ^2H composition at these three locations in the wetland.

DI δ^{18}O values were variable the first four months after which they became steadier until the last measurement in January 2018. DI had the two most enriched δ^{18}O values measured (January and April) and had a δ^{18}O range of −8.67‰ to −6.63‰, which is the largest range for all the springs. The rivulet (RI) had the second largest range (−8.12‰ to −7.43‰) with a similar pattern to that of DI. The δ^{18}O range of the PO draining the pond was −7.98‰ to −7.63‰, which could indicate enough residence time in the pond, to buffer changes due to variations in groundwater δ^{18}O input.

The δ^{18}O of the stream (ST) and the drain (DR) followed the same pattern with the same peaks and troughs. Water from the drain was, for all months except for March, enriched compared to the stream. S1, S2 and DW had the lowest fluctuations with a δ^{18}O range of −7.99‰ to −7.48‰, −7.82‰ to −7.56‰, and −8.04‰ to −7.56‰, respectively. This can also be seen from the low standard deviations (legend, Figure 8). The δ^{18}O ranges of S1, S2, and DW were all lower than what were measured for the groundwater at the field site.

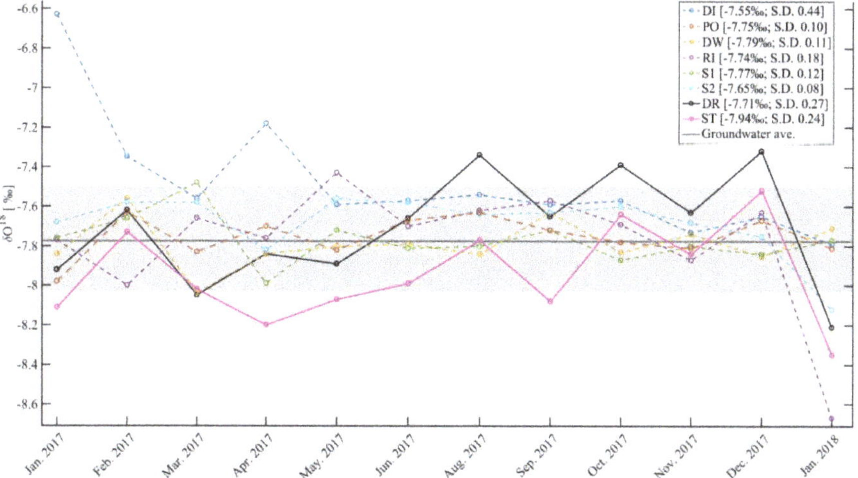

Figure 8. Monthly measured δ^{18}O from springs, the stream and drain (on the crop side) the greyed area indicates the water-stable isotope signal measured in the piezometers in March and November 2017 (mean ± 1 standard deviation). The legend shows the median concentration and standard deviation (SD).

Samples collected of the shallow groundwater in June 2017 under the wetland, using mini-piezometers and analyzed for δ^{18}O are shown in Figure 9A. Average δ^{18}O was −7.69‰ ranging from −8.06‰ to −7.11‰ and average δ^2H was −51.06‰ ranging from −48.27‰ to −52.84‰. The spatial patterns of the measured δ^{18}O show enriched signals in the center of the wetland—whereas depleted signals, compared to the average δ^{18}O signal reported by Müller et al. [28], were mainly found along the hillslope to the southwest and 10–15 m from the stream (except for one measurement with a $\delta^{18}O$ of −7.11‰). Water from DW, PO, and S1 discharge to the surface of the wetland and flows into the center of the wetland, where the peat has its maximum thickness. In this area samples from six of the mini-piezometers showed slight enrichment (−7.37 to −7.66‰, yellow area in Figure 9A).

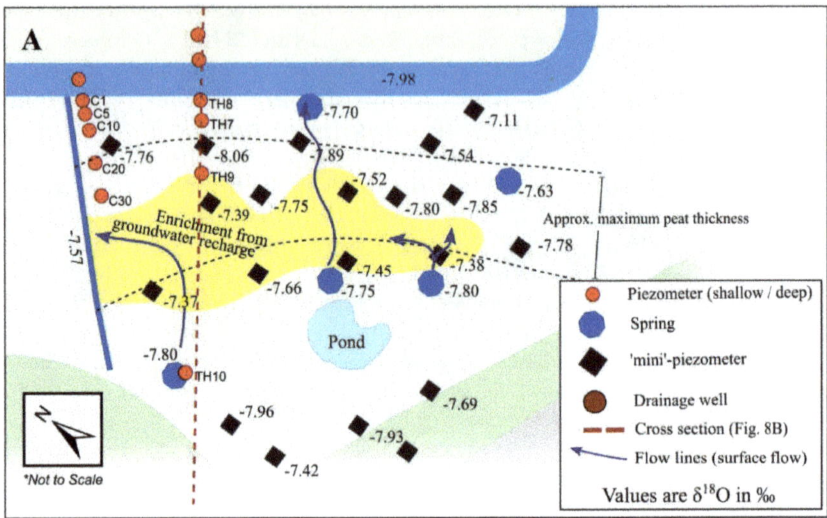

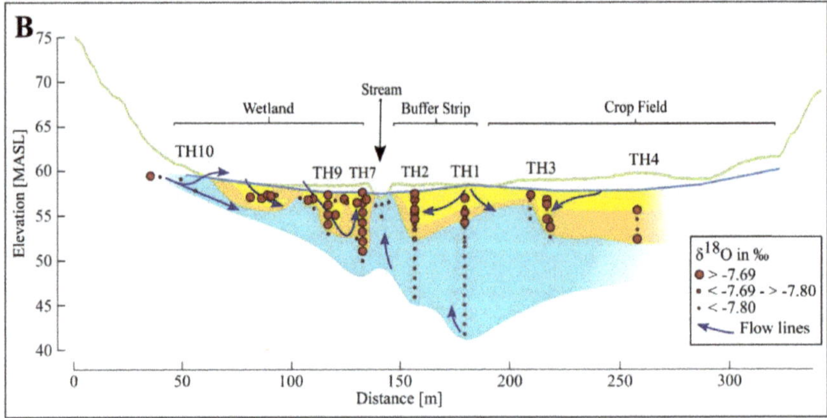

Figure 9. Measured $\delta^{18}O$ signals from piezometers and mini-piezometers. (**A**) Samples from the shallow groundwater under the peat layer in the wetland, June 2017; (**B**) Cross-section of field site based on all samples taken from piezometers and mini-piezometers at different times. The measured $\delta^{18}O$ have been divided into three broad groups and hand interpolated. The grouping corresponds broadly to the three set intervals of the dots: yellow $\delta^{18}O$ >−7.69‰, orange <−7.69‰ to >−7.80‰, and blue <−7.80‰.

From the piezometric head observation, it was established that the groundwater table in the wetland was below the ground surface, throughout the monitoring period. This suggests that water, originating from precipitation and the springs, on top of the peat represents a perched water table resulting in a surface-water-saturated soil. In addition, a downward gradient was present between piezometers TH9s and TH9d. The enriched isotope signals in the center of the wetland could therefore indicate water enriched from exposure to evaporation percolating through the peat and recharging the groundwater system.

A cross-section of the field site (Figure 9B) showing all samples collected for $\delta^{18}O$ in both piezometers and mini-piezometers have been divided in to three broad sub-groups and interpolated by hand. A plume of enriched water is seen near the surface along the entire cross-section, except for directly under the stream where $\delta^{18}O$ in groundwater is depleted. Samples taken under the streambed ($-8.2‰$) and from TH6 ($-7.87‰$ SD 0.17‰) and TH8 ($-7.87‰$ SD 0.05‰) right next to the stream showed consistently depleted signals. Depleted isotopic groundwater also enters the wetland from the southwest and either flows to the stream as groundwater or initially as surface flow. The surface flow is enriched and the observations of an enriched signal from the mini-piezometers indicate that water re-infiltrates. This also fits with the observed downward gradient between the nested piezometers TH9s and TH9d. Piezometer TH9d, sampled every meter at the time of installation, had enriched values of $\delta^{18}O$ in the top of the well, but samples became more depleted with depth. $\delta^{18}O$ concentrations from TH7d, sampled with depth in October 2016, were slightly more enriched and variable with depth. Downward seepage through the peat upstream TH7 followed by the apparent upward flow due to upward gradient between TH7d and TH7s, maybe the cause of a more uniform distribution.

Samples taken from TH10 at the hillslope (March and November 2017) also showed depleted $\delta^{18}O$ ($-7.91‰$ and $-8.02‰$, respectively). The depleted $\delta^{18}O$ values measured along the hillslope in mini-piezometers and in TH10 are similar to those measured in the two deepest piezometers; TH1d (16 MBGL) and TH2d (12 MBGL) on the crop side. Stable isotopes were sampled five times from TH1d, November 2015, February and September of 2016, and March and November of 2017, with an average $\delta^{18}O$ of $-8.02‰$ and a $\delta^{18}O$ minimum of $-8.15‰$ and maximum of $-7.82‰$. TH2d was sampled on four occasions in November 2015, February and September of 2016 and in March 2017 with an average $\delta^{18}O$ of $-8.11‰$ with maximum and minimum of $-7.96‰$ and $-8.26‰$, respectively. Provided the isotope signals measured in TH1d and TH2d are representative of well-mixed groundwater, and not event-based signals, it would appear that a groundwater source with a similar signature is entering the wetland from the southwest. Some of this water bypasses the peat and flow to the stream through the 5–8 m thick gravelly/coarse sand. Deeper groundwater may also flow upwards to the stream as indicated by the depleted $\delta^{18}O$ signals in the deeper parts of TH1 and TH2.

Piezometer TH3 had enriched $\delta^{18}O$ values with depth indicating a downward flux (recharge area); the same is seen with TH4. From TH1d and TH2d $\delta^{18}O$ signals became depleted with depth and from five MBGL and seven MBGL $\delta^{18}O$ signals became steadier. The $\delta^{18}O$ data from the crop side therefore indicates vertical downward seepage. In addition, upward leakage from the bottom of the profile cannot be ruled out.

5. Discussion

The main objective of the study has been to understand and quantify the heterogeneity in water fluxes to riparian zones and streams in a lowland catchment and specifically address the relative importance of groundwater-fed surface flow to subsurface flow directly to the stream. We will here try to discuss our results in relation to existing conceptual models of riparian zones e.g., [6,12,29,30]. For example, Vidon and Hill [30] proposed a conceptual model, where aquifer thickness and slope are two characteristics determining the link between riparian zones and the upland. Several studies [6,12,29] focus on how flow paths in typical riparian zones are defined by the landscape type, riparian zone–upland link, riparian aquifer thickness, and permeability of a confining peat layer. However, only one of these studies includes the geomorphology of the stream (width, sinuosity) [12].

Most conceptual flow path models emphasize different parts of the link between stream–riparian zone regional aquifer, but with no guidelines to evaluate the relative contribution of subsurface and groundwater-fed surface flow. The size of the sub-catchment to discharge station Q_2 at our field site is approximately 28 km^2 with roughly 23 km of stream network. Precipitation and actual evapotranspiration average 984 mm year^{-1} and 510 mm year^{-1}, respectively, giving an average recharge of 474 mm/year [15]. Average stream flow Q_2 was measured in 2017 to ~10^7 m^3 year^{-1} corresponding to a runoff of 377 mm/year. Thus, roughly 20% of recharge is not captured by the stream network, but flows out of the sub-catchment via groundwater. If it is assumed that all the stream flow originates as base flow (direct groundwater discharge) then one can compute the average flux. Assuming the average width is ~1.5 m then the 23 km network has a streambed area of 34000 m^2. This gives an average streambed flux of 1.1 m day^{-1}. Clearly, actual rates in this type of catchments are orders of magnitude lower; we measured 0.06 m day^{-1}. Accordingly, a lot of water is fed into the stream valley and only a small portion is captured by the stream directly though the streambed. Flow to the stream therefore takes alternative pathways.

5.1. Hydraulic Connectivity

Features such as upland linkage, topography, and hydrogeological setting, both adjacent to and in the riparian lowland, are known to influence the flow distribution in riparian lowlands [29,30]. Landscape characteristics such as slope of riparian zone and depth of upland permeable sediments influence the hydraulic connectivity, water table fluctuation and groundwater flow direction [30]. In the study area, both to the north and south, relatively thick unconfined aquifers with a gently (north, ~2.5%) too steeply (south, ~7%) sloping topography are present. Vidon and Hill [30] have shown that such landscape and hydrogeological characteristics would give the southern riparian lowland a permanent upland aquifer linkage with a constant flow direction, a continuous large flux, and small water table variations. For the northern riparian lowland, the upland–riparian link to the upland aquifer would still be permanent, but with small to moderate fluxes and moderate water table variations. This is confirmed by our measurements. For example, the largest fluctuations in hydraulic head was found near the edges of the riparian lowland, but notably with smaller fluctuations on the south side (TH10 with hydraulic head variation of ±0.2 m) compared with the north side (F100 ±0.37 m). Nevertheless, generally, observations showed small variations in the groundwater table, indicating a permanent link to an upland aquifer. This ensures a constant groundwater input and maintains a year-round hydraulic gradient towards the stream. The largest fluctuations observed were to the north, indicating a linkage to a smaller upland aquifer. Considering that the upland aquifer thickness is slightly less and the length to the upland boundary is shorter compared to the southern side of the stream, this could explain the slightly larger fluctuations. To the south, the connectivity of the riparian lowland appears to maintain a more continuous groundwater flux giving less seasonal variation near the hillslope.

The lack of identifying a confining layer at the bottom of the riparian aquifer suggests the potential of deeper groundwater flow paths bringing deeper and more regional water to the riparian zone from further away in the catchment. The δ^{18}O signal (Figure 9) confirms this, where measurements below the streambed were close to the meteoric average.

5.2. Heterogeneity in Flow Paths to Stream

Conceptual models of flow paths in riparian zones as a function of the link to an upland aquifer and the thickness of the riparian aquifer have been proposed [6]. This model have since been extended to include groundwater-fed surface flow [8]. A link to larger upland aquifer results in large groundwater inputs and small water table fluctuations (see above) [28]. With large volumes of water transferred to the riparian aquifer and with a limited area of discharge more water is routed to the surface as seeps and springs and reach the stream via overland flow.

The studied riparian aquifer consists of sand. On the southern side of the stream, a semi-confining peat layer is present on top of the sand. The depth of the aquifer increases from ~15 m near the hillslope to >30 m in the central part (a deeper confining layer of the riparian aquifer has not been identified from boreholes or the ERT profiles). The large thickness and the permanent link to a more regional aquifer fits well with other conceptual models [6,8] where a significant part of the regional flow to the riparian zone is routed to the stream via seeps and springs—despite the considerable thickness of the aquifer.

This was best observed on the south side of the stream. For example, the observed diffusive seepage occurring in the area above the pond. Furthermore, the presence of springs and the more extensive need for drainage (ditch, DW) is evidence of a larger groundwater to surface-water component. This is not the case on the north side and may be explained by a link to a much smaller regional aquifer. Here flow to the stream is mainly governed by groundwater, except for the part captured by the drain near the depression in the landscape.

Like in other studies there is an exchange of water between flow paths (e.g., [8,31]). TH9s and TH9d on the southern side of the stream show a small downward gradient. TH9s is screened just below the peat layer and TH9d is screened in the coarse sand deposited under the peat located approx. 2–6 MBGL (Figure 2). The more conductive coarse sand could produce a downward gradient; generating a preferential flow path to the stream under the peat. Surface flow could therefore infiltrate again (Figure 9), flow towards the stream and upwards due to the upward gradient at TH7s and TH7d (TH7s is screened in the high conductive sand and TH7d screened below in a less conductive material) and the presence of the stream. The enriched $\delta^{18}O$ plume near TH9 and TH7 confirms this.

Part of the groundwater will pass through the aquifer to the stream from both the north and the south (Figure 9). The individual fluxes estimated by VTP across a 90 m stream reach vary with close to an order of magnitude. Groundwater discharge has been demonstrated by other local studies, [16,17,22] and elsewhere [32,33] to vary substantially in time and space both across and along a stream reach. Compared to these local studies further downstream of Holtum stream, the estimated streambed fluxes in this study is at the lower end (average of 7×10^{-7} m s^{-1}). Two local studies [16,34] reported mean fluxes of 4–5×10^{-6} m s^{-1} approximately a factor 6 higher on average than the fluxes measured at our site. This corresponds nicely with the observations made in one local study [16] where it was found that stream flow in areas downstream of our site was less rain-event controlled. The difference between our site and the other sites in Holtum is that our site is, or rather was, managed to some degree and now appears with overgrown ditches, overflowing drainage wells, and a possible broken drainage system.

In September 2016, the overland flow component was a factor of 12 higher than discharge estimated through the streambed. This clearly indicates that groundwater-fed surface flow is the most important flow component at the site. This ratio may change over the season. We do not have the data to quantify this as discharge through the streambed was only measured once.

Langhoff et al. [12] proposed a coefficient C called "the apparent relative seepage width" to indicate the proportion of water entering the stream through the streambed relative to overland flow. C is defined as the width of the wet zone of the flood plain (b_r) divided by the effective width of the stream (Sb_s), i.e., C = $b_r/(Sb_s)$. Here, S is the stream sinuosity (which for Holtum stream is S = 1, i.e., a straight stream) and b_s is the width of the stream. High values of C correspond to relative low contribution to stream flow gain from direct seepage through the streambed and vice versa. At our field site, the wet zone is on the south side (b_r 85 m) and with a stream width of b_s = 3 m, C can be computed as ~28. This is similar to Site 3 studied in Langhoff et al. [12], who also found that the majority of the water flowed to the stream as overland flow (actually 110%). It thus appears that the C ratio can be an efficient way of quantifying heterogeneity in flow paths, because it takes into account the stream valley morphology and wetness.

5.3. Water Balance

The gain in stream discharge between Q1 and Q2 was estimated to be on average 5.09×10^{-5} m^3 s^{-1} m^{-1}. Higher discharges were attributed to higher intensity of precipitation, for where the largest gaining and losing conditions in the stream also was prevalent—these (four) measurements was not included in the average stream gain. The stream gain had a standard deviation of 7.6×10^{-5} meaning that the stream between Q1 and Q2 can experience both losing and gaining conditions.

However, only two of the discharge measurements represented losing conditions and the stream section between Q1 and Q3 can be classified as gaining. The ADC is expected to have an uncertainty of 5% of the measured discharge [35]. This limits the minimum distance between two measurement stations when using the ADC, as the accretion of a discharge should be higher or less than the uncertainty—the average accretion for our measurements was 6%, excluding months where discharge measurements was known to be affected by precipitation. It would have been better to have continuous (daily) measurements of stream discharge, as this would have reduced the uncertainty on the stream gain. However, as explained above, it was not possible to at the same time measure continuous (daily) spring flows. This means that the water balance is to be regarded as the average of monthly snapshots of stream flow gains, spring flows, and only one measurement of direct seepage through the streambed (which turned out to be minor).

The estimated gain between Q1 and Q2 was significantly less than that estimated from Q1 to Q3. This could be explained by the large portion of arable land with drain tiles to the south and the riparian lowland with several ditches and a small tributary between stations Q2 and Q3 (Figure 1C). The dynamics of the stream stage followed the fluctuations of precipitation (flashy) and that of the hydraulic heads. This might reflect the riparian zone with conductive layers; therefore, it quickly drains and returns to its state of equilibrium. The discharge measurement of August, where discharge at Q2 was measured after a rainfall event, showed a significant decrease in discharge compared to the measurements at Q1 and Q3 made during the event. The stream therefore responds quickly to rainfall events. This is supported by the findings of Poulsen et al. [16] who estimated that the contribution to stream discharge from rainfall events for the same sub-catchment as studied here could be as high as 65%.

The average estimated flux though the streambed can only account for 4% of the stream gain. Langhoff et al. [12] measured bank seepage to a stream that had been deepened and found that bank seepage could contribute of up to 25% of the stream gain. With a similar setting at the field site, where the streambed is situated at a depth below the peat layer to the south, it is likely that bank seepage plays a moderate to important role. It was calculated based on data from the two nearest piezometers that bank seepage potentially could contribute with 17% of the stream gain.

The drain located on crop field discharges directly into the stream contributing about 7% to the stream gain. The most prominent contribution to the stream gain is water discharging through the springs in the wetland to the south. The estimated contribution of measured spring discharge to the stream gain was from 20% and up to 59% (depending on whether they contribute individually or not).

With a potential of close to 66% of the stream gain coming from focused sources a contribution of 4% and 17% from direct groundwater discharge to the stream seem insignificant. With up to 87% of the stream gain accounted for the water balance is not closed. Diffusive overbank flow, unmeasured discharge from (one known) submerged drains or highly focused bank seepage could account for the missing 13%. These numbers are similar to those found in the study by Langhoff et al. [12]. Here the measured or unmeasured bank seepage and overland flow are of similar magnitude. The results emphasize the important role of groundwater-fed overland flow [8].

We cannot exclude the possibility that some of the springs are unnatural in that they could be the result of a broken drainpipe system. We know of the existence of the (overflowing) drainage well (DW) and the submerged drain outlet to the stream. This point to the potential role of an unmanaged drainage system of the hydrology of a wetland system. It has not been possible to obtain information about the drainage system, but the landscape with a steep increase in terrain to the south suggests that a drain network is present only in the wetland and does not drain the sandy/gravelly agricultural fields located further south (see e.g., Figure 2 with well DGU 92.2187). The drains thus capture groundwater that probably otherwise would have resulted in groundwater-fed overland flow. The human alteration of the system (channelization, drain pipes, ditches) has potentially increased the lateral connectivity of the stream and the terrestrial landscape [36]. As mentioned above, the fluxes through the streambed is roughly six times lower compared to other more natural (unmanaged) systems downstream of our site.

Our results are also different to those reported by Frederiksen et al. [13] for a groundwater-dominated stream west of our site. Here overland flow was negligible, and groundwater was the main flow path (>60% of stream flow gain). A drainage system (broken or not, ditch) thus have re-organized the flow paths and most groundwater enters the stream by groundwater-fed surface flow than groundwater-fed stream flow. This will of course also have implications for nitrate transport and removal, which is currently being studied. With that said, some springs may be entirely natural and could be due to a thin peat cover (e.g., S1 and S2). For example, Johansen et al. [11] found that peat thickness controlled vertical seepage to a rich fen.

6. Conclusions

From the present study, it can be concluded that management for agricultural purposes have changed the natural flow paths through the riparian lowland to the stream. In the present study we have addressed the connection of a groundwater system and a riparian lowland to specifically determine the relative importance of subsurface and surface flow to a headwater stream.

The investigated riparian lowland had a permanent upland linkage ensuring continuous large groundwater inputs, small groundwater table fluctuations, and maintaining a constant direction of flow towards the stream. The permanent upland connectivity and aquifer thickness result in large groundwater inputs. Thus, not all the groundwater from the upland aquifer can be accommodated as seepage only through the streambed of the relatively small headwater stream. In our case we were able to demonstrate that the direct discharge through the streambed was approximately a factor of six lower compared to other riparian–stream systems in the same catchment (more natural riparian zones with no or little human alterations). In other words, the highly managed wetland system captures a high proportion of the regional groundwater discharging to the riparian zone (~2/3 of the regional flow bypasses the subsurface). This can explain the need for anthropogenic alterations of the area in the past—changes needed to make the land usable for agricultural production or as, in this case, a cow pasture. The anthropogenic changes include ditches, drainage wells, and the man-made pond. Most of the "springs" are therefore man-made preferential flow paths. The straightening of the stream has likely made these changes even more needed.

Such anthropogenic changes are common for many riparian lowlands situated in agricultural lowland catchments. Many of these are easily detectable from maps (stream straightening, ditches) and of course by site visits. To determine the relative importance of groundwater-fed overland flow to direct stream seepage is of course more complicated as this requires detailed field monitoring. This just highlights the importance of measuring all flow components in a water budget for a riparian–stream system. In the end, anthropogenic alterations will have consequences for the effectiveness of N-removal in the subsurface as the main part of groundwater is routed to the stream via drains, ditches and diffuse surface flow/rivulet with short residence times and little contact with subsurface reactive materials.

Author Contributions: Conceptualization, M.S. (Mads Steiness), P.E., and S.J.; Methodology, M.S. (Mads Steiness), S.G.W.v.V.; Formal analysis, M.S. (Mads Steiness) and M.S. (Mattia Spitilli); Investigation, M.S. (Mads Steiness), S.G.W.v.V., and M.S. (Mattia Spitilli); Writing—original draft preparation, M.S. (Mads Steiness); Writing—review and editing, M.S. (Mads Steiness), P.E., S.J., A.L.H., S.G.W.v.V. and M.S. (Mattia Spitilli); Visualization, M.S. (Mads Steiness); Supervision, P.E., A.L.H., S.J.; Project administration, A.L.H.; Funding acquisition, A.L.H.

Funding: This research was part of the TReNDS project and funded by Innovation Fund Denmark (grant no. 4106-00027B).

Acknowledgments: We would like to thank family Hauge for granting us access to the field site. We would also like to thank the people who have participated in field trips and assisted in collecting the data: Fransesca Parnanzone, Iris Tobelaim, Joel Tirado-Conde, and Wenjing Qin. We also thank the associate editor and two reviewers for insightful comments and suggestions to improve the manuscript.

Conflicts of Interest: The authors declare no conflict of interest.

References

1. Kronvang, B.; Andersen, H.E.; Børgesen, C.; Dalgaard, T.; Larsen, S.E.; Bøgestrand, J.; Blicher-Mathiasen, G. Effects of policy measures implemented in Denmark on nitrogen pollution of the aquatic environment. *Environ. Sci. Policy* **2008**, *11*, 144–152. [CrossRef]
2. Hoffmann, C.C.; Kronvang, B.; Audet, J. Evaluation of nutrient retention in four restored Danish riparian wetlands. *Hydrobiologia* **2011**, *674*, 5–24. [CrossRef]
3. Lowrance, R.; Altier, L.S.; Newbold, J.D.; Schnabel, R.R.; Groffman, P.M.; Denver, J.M.; Correll, D.L.; Gilliam, J.W.; Robinson, J.L.; Brinsfield, R.B.; et al. Water quality functions of riparian forest buffers in chesapeake bay watersheds. *Environ. Manag.* **1997**, *21*, 687–712. [CrossRef]
4. Peterjohn, W.T.; Correll, D.L. Nutrient dynamics in an agricultural watershed: Observations on the role of a Riparian forest. *Ecology* **1984**, *65*, 1466–1475. [CrossRef]
5. Vidon, P.; Allan, C.; Burns, D.; Duval, T.P.; Gurwick, N.; Inamdar, S.; Lowrance, R.; Okay, J.; Scott, D.; Sebestyen, S. Hot spots and hot moments in riparian zones: Potential for improved water quality management. *J. Am. Water Res. Assoc.* **2010**, *46*, 278–298. [CrossRef]
6. Hill, A. Nitrate Removal in Stream Riparian Zones. *J. Environ. Qual.* **1996**, *25*, 743–755. [CrossRef]
7. Devito, K.J.; Fitzgerald, D.; Hill, A.R.; Aravena, R. Nitrate Dynamics in Relation to Lithology and Hydrologic Flow Path in a River Riparian Zone. *J. Environ. Qual.* **2000**, *29*, 1075–1084. [CrossRef]
8. Shabaga, J.A.; Hill, A.R. Groundwater-fed surface flow path hydrodynamics and nitrate removal in three riparian zones in southern Ontario, Canada. *J. Hydrol.* **2010**, *388*, 52–64. [CrossRef]
9. Hill, A.R. The impact of pipe flow in riparian peat deposits on nitrate transport and removal. *Hydrol. Processes* **2012**, *26*, 3135–3146. [CrossRef]
10. Brüsch, W.; Nilsson, B. Nitrate transformation and water movement in a wetland area. *Hydrobiologia* **1993**, *251*, 103–111. [CrossRef]
11. Johansen, O.M.; Pedersen, M.L.; Jensen, J.B. Effect of groundwater abstraction on fen ecosystems. *J. Hydrol.* **2011**, *402*, 357–366. [CrossRef]
12. Langhoff, J.H.; Rasmussen, K.R.; Christensen, S. Quantification and regionalization of groundwater–surface water interaction along an alluvial stream. *J. Hydrol.* **2006**, *320*, 342–358. [CrossRef]
13. Frederiksen, R.R.; Christensen, S.; Rasmussen, K.R. Estimating groundwater discharge to a lowland alluvial stream using methods at point-, reach-, and catchment-scale. *J. Hydrol.* **2018**, *564*, 836–845. [CrossRef]
14. Houmark-Nielsen, M. The last interglacial-glacial cycle in Denmark. *Quatern. Int.* **1989**, *3–4*, 31–39. [CrossRef]
15. Sebok, E.; Refsgaard, J.C.; Warmink, J.J.; Stisen, S.; Jensen, K.H. Using expert elicitation to quantify catchment water balances and their uncertainties. *Water Resources Res.* **2016**, *52*, 5111–5133. [CrossRef]
16. Poulsen, J.R.; Sebok, E.; Duque, C.; Tetzlaff, D.; Engesgaard, P.K. Detecting groundwater discharge dynamics from point-to-catchment scale in a lowland stream: Combining hydraulic and tracer methods. *Hydrol. Earth Syst. Sci.* **2015**, *19*, 1871–1886. [CrossRef]
17. Karan, S.; Engesgaard, P.; Looms, M.C.; Laier, T.; Kazmierczak, J. Groundwater flow and mixing in a wetland–stream system: Field study and numerical modeling. *J. Hydrol.* **2013**, *488*, 73–83. [CrossRef]
18. Jensen, J.K.; Engesgaard, P.; Johnsen, A.R.; Marti, V.; Nilsson, B. Hydrological mediated denitrification in groundwater below a seasonal flooded restored riparian zone. *Water Resources Res.* **2017**, *53*, 2074–2094. [CrossRef]
19. Lowry, C.S.; Fratta, D.; Anderson, M.P. Ground penetrating radar and spring formation in a groundwater dominated peat wetland. *J. Hydrol.* **2009**, *373*, 68–79. [CrossRef]
20. Van Bellen, S.; Dallaire, P.-L.; Garneau, M.; Bergeron, Y. Quantifying spatial and temporal Holocene carbon accumulation in ombrotrophic peatlands of the Eastmain region, Quebec, Canada. *Glob. Biogeochem. Cycles* **2011**, *25*. [CrossRef]
21. Skogerboe, G.V.; Bennett, R.S.; Walker, W.R. Generalized discharge relation for cutthroat flumes. *J. Irrigat. Drain. Div.* **1972**, *98*, 569–583.
22. Jensen, J.K.; Engesgaard, P. Nonuniform Groundwater Discharge across a Streambed: Heat as a Tracer. *Vadose Zone J.* **2011**, *10*, 99–109. [CrossRef]
23. Kidmose, J.; Engesgaard, P.; Nilsson, B.; Laier, T.; Looms, M. Spatial Distribution of Seepage at a Flow-Through Lake: Lake Hampen, Western Denmark. *Vadose Zone J.* **2011**, *10*, 110–124. [CrossRef]

24. Duque, C.; Müller, S.; Sebok, E.; Haider, K.; Engesgaard, P. Estimating groundwater discharge to surface waters using heat as a tracer in low flux environments: The role of thermal conductivity. *Hydrol. Processes* **2016**, *30*, 383–395. [CrossRef]

25. Bredehoeft, J.D.; Papadolopulos, I.S. Rates of Vertical Groundwater Movement Estimated from the Earth's Thermal Profile. *Water Resources Res.* **1965**, *1*, 325–328. [CrossRef]

26. Sebok, E.; Müller, S. The effect of sediment thermal conductivity on vertical groundwater flux estimates. *Hydrol. Earth Syst. Sci. Discuss.* **2018**. [CrossRef]

27. Stonestrom, D.A.; Blasch, K.W. *Determining Temperature and Thermal Properties for Heat-Based Studies of Surface-Water Ground-Water Interactions: Appendix A of Heat as a Tool for Studying the Movement of Ground Water Near Streams (Cir1260)*; Stonestrom, D.A., Constantz, J., Eds.; U.S. Geological Survey: Reston, VA, USA, 2003; pp. 73–80.

28. Müller, S.; Stumpp, C.; Sørensen, J.H.; Jessen, S. Spatiotemporal variation of stable isotopic composition in precipitation: Post-condensational effects in a humid area. *Hydrol. Processes* **2017**, *31*, 3146–3159. [CrossRef]

29. Dahl, M.; Nilsson, B.; Langhoff, J.H.; Refsgaard, J.C. Review of classification systems and new multi-scale typology of groundwater–surface water interaction. *J. Hydrol.* **2007**, *344*, 1–16. [CrossRef]

30. Vidon, P.G.F.; Hill, A.R. Landscape controls on the hydrology of stream riparian zones. *J. Hydrol.* **2004**, *292*, 210–228. [CrossRef]

31. Cey, E.E.; Rudolph, D.L.; Parkin, G. Role of the riparian zone in controlling the distribution and fate of agricultural nitrogen near a small stream in southern Ontario. *J. Contamin. Hydrol.* **1999**, *37*, 45–67. [CrossRef]

32. Anibas, C.; Buis, K.; Verhoeven, R.; Meire, P.; Batelaan, O. A simple thermal mapping method for seasonal spatial patterns of groundwater–surface water interaction. *J. Hydrol.* **2011**, *397*, 93–104. [CrossRef]

33. Schmidt, C.; Bayer-Raich, M.; Schirmer, M. Characterization of spatial heterogeneity of groundwater-stream water interactions using multiple depth streambed temperature measurements at the reach scale. *Hydrol. Earth Syst. Sci.* **2006**, *10*, 849–859. [CrossRef]

34. Karan, S.; Sebok, E.; Engesgaard, P. Air/water/sediment temperature contrasts in small streams to identify groundwater seepage locations. *Hydrol. Processes* **2017**, *31*, 1258–1270. [CrossRef]

35. Turnepseed, D.P.; Sauer, V.B. *Discharge Measurements at Gaging Stations*; U.S. Geological Survey Techniques and Methods Book 3; U.S. Geological Survey: Reston, VA, USA, 2010.

36. Covino, T. Hydrologic connectivity as a framework for understanding biogeochemical flux through watersheds and along fluvial networks. *Geomorphology* **2017**, *277*, 133–144. [CrossRef]

Article

Temporal Effects of Groundwater on Physical and Biotic Components of a Karst Stream

Tao Tang [1], Shuhan Guo [1,2], Lu Tan [1], Tao Li [2,3], Ryan M. Burrows [4] and Qinghua Cai [1,*]

[1] State Key Laboratory of Freshwater Ecology and Biotechnology, Institute of Hydrobiology, Chinese Academy of Sciences, Wuhan 430072, China; tangtao@ihb.ac.cn (T.T.); guoshuhan@ihb.ac.cn (S.G.); tanlu@ihb.ac.cn (L.T.)

[2] University of Chinese Academy of Sciences, Beijing 100049, China; taoli_st@rcees.ac.cn

[3] State Key Laboratory of Urban and Regional Ecology, Research Center for Eco-Environmental Sciences, Chinese Academy of Sciences, Beijing 100085, China

[4] Australian Rivers Institute, Griffith University, Brisbane 4111, Australia; r.burrows@griffith.edu.au

* Correspondence: qhcai@ihb.ac.cn; Tel.: +86-027-68780865

Received: 11 May 2019; Accepted: 19 June 2019; Published: 21 June 2019

Abstract: Although most lotic ecosystems are groundwater dependent, our knowledge on the relatively long-term ecological effects of groundwater discharge on downstream reaches remains limited. We surveyed four connected reaches of a Chinese karst stream network for 72 consecutive months, with one reach, named Hong Shi Zi (HSZ), evidently affected by groundwater. We tested whether, compared with other reaches, HSZ had (1) milder water temperature and flow regimes, and (2) weaker influences of water temperature and flow on benthic algal biomass represented by chlorophyll *a* (Chl. *a*) concentrations. We found that the maximum monthly mean water temperature in HSZ was 0.6 °C lower than of the adjacent upstream reach, and the minimum monthly mean water temperature was 1.0 °C higher than of the adjacent downstream reach. HSZ had the smallest coefficient of variation (CV) for water temperature but the largest CV for discharge. Water temperature and discharge displayed a significant 12-month periodicity in all reaches not directly groundwater influenced. Only water temperature displayed such periodicity in HSZ. Water temperature was an important predictor of temporal variation in Chl. *a* in all reaches, but its influence was weakest in HSZ. Our findings demonstrate that longer survey data can provide insight into groundwater–surface water interactions.

Keywords: groundwater–surface water interactions; time series; water temperature regime; flow regime; chlorophyll *a*; wavelet analysis; convergent cross-mapping

1. Introduction

Most streams and rivers are groundwater-influenced systems [1,2]. Groundwater commonly upwells into surface waters at the hyporheic zone [3,4], which influences the physical and biotic characteristics of surface waters substantially [5–8]. The importance of groundwater to surface lotic ecosystems has been documented extensively, with most of the work focusing on distinct functions of the upwelling zone. Upwelling zones have been observed to have more favorable conditions for many organisms, including stable temperature and flow conditions as well as higher concentrations of dissolved nutrients [9,10]. Consequently, upwelling zones generally harbor higher biodiversity and more specialized biotic assemblages compared with other lotic environments [11,12]. The upwelling zone also serves as a refuge for lotic organisms to survive adverse environmental conditions such as extreme temperatures, floods, and stream drying [13,14]. A handful of studies have estimated the ecological impact of groundwater discharge for organisms in downstream reaches; however, findings are equivocal. Some studies demonstrate that important ecological characteristics, such as primary production and biodiversity, of downstream reaches benefit from groundwater inputs [15]. Further, coldwater fish have been confirmed to use groundwater-fed reaches as a thermal refuge during warming periods [16,17]. In contrast, other studies did not find significant groundwater effects for aquatic organisms [18,19].

Study duration is likely a major reason underlying variation in the reported reach-scale physical and biotic effects attributed to groundwater discharge. Most studies have evaluated groundwater effects over short time scales (usually never exceeding two years) and without the adequate characterization of the changes in seasonal patterns and processes [2,16,18,20]. Since the influence of groundwater for lotic ecosystems may vary remarkably among seasons and years [21,22], inconsistent results may be expected if study designs do not adequately account for this temporal variation. Moreover, short-term studies often cannot establish accurate or meaningful quantitative relationships between measured variables due to limited data. This limitation hampers mechanical exploration of pathways describing how groundwater influences surface waters. Therefore, studies of longer duration and with a more intensive survey effort are needed for exploring long-term and quantitative effects of groundwater discharge on downstream ecosystems, which will be valuable for a better understanding of groundwater–surface water interactions.

Short-term studies may also fail to accurately characterize important groundwater effects on stream physicochemical conditions. For instance, stable water temperature and flow conditions are key features of groundwater-influenced stream reaches, and these conditions may mitigate temporal fluctuations of water temperature and flow in downstream reaches [23]. However, due to short durations, many studies only measure changes in the magnitude of downstream water temperature or flow (i.e., the incremental deviation from the average conditions caused by groundwater inputs) [24,25]. Importantly, other widely used regime descriptors for any continuous phenomena, including variability (temporal fluctuations across regular time intervals), frequency (the number of events occurring at a certain magnitude or exceedance), timing (dates that particular events occur), and predictability (the reliability of event recurrence), have been poorly documented [26,27]. Consequently, it is unclear if or how groundwater can change other aspects of temperature and flow regimes other than the response magnitude in downstream reaches. Such knowledge has ecological value because the magnitude, variability, frequency, timing, and predictability are all important aspects influencing lotic ecological structure and function [28,29].

In the present study, four connected reaches in a Chinese Karst stream were surveyed continuously for 72 months, one of which is known to receive direct groundwater inputs. We monitored water temperature, discharge, and benthic algal biomass (represented by Chlorophyll *a* [Chl. *a*] concentrations) to evaluate how groundwater discharge affects these important physical and biotic components of lotic ecosystems. Benthic algae are a high-quality food source in streams and can be the main energy source to the aquatic food web [30,31]. Water temperature and discharge are commonly influenced by groundwater discharge [9,10] and are also major drivers of the temporal dynamics of benthic algal biomass [32,33]. Our main objective was to characterize the water temperature and discharge regimes for each reach. Further, we assessed reach-specific relationships between benthic Chl. *a* concentrations and water temperature as well as discharge. We tested two hypotheses: the reach most affected by groundwater will have (1) milder water temperature and flow regimes and (2) weaker influences of water temperature and flow on benthic algal biomass dynamics. Support for these two hypotheses will provide evidence that groundwater has an important influence on temporal dynamics of adjacent downstream reaches in the karst stream.

2. Materials and Methods

2.1. Study Region

We conducted the study on the Nan River, located in Shengnongjia (31,015′–31,075′ N, 109,056′–110,058′ E), the western mountainous region in Hubei province of China. Elevation in this region varies from 420 to 3015 m (with a mean elevation >1500 m), resulting in an altitudinal climate gradient. The annual mean air temperature ranges from 7.1 °C in the western region to 14.5 °C in the eastern region, with the coldest and warmest months in January and July, respectively [34]. The annual precipitation commonly varies between 800 and 2500 mm, in which more than 85% of rainfall occurs in April to October with December to February accounting for <8% [34]. Karst terrain is a typical geomorphological characteristic in Shengnongjia where ~30% of bedrock is limestone. Due to the high solubility of this type of rock, there are enormous underground rivers, karst caves, swallow holes, groove, and springs in this region [34,35].

We surveyed four reaches in the headwaters of Nan River (Figure 1). The streambed substrate is dominated by gravels, pebbles, and cobbles, with bedrock of limestone [34]. Sheng Nong Yuan (SNY) and Da Long Tan (DLT) are two first-order reaches with elevation of 2320 m and 2220 m, respectively (Table 1). SNY is located in the mainstream, while DLT is in an adjacent tributary (Figure 1). Swallow holes are common along the 1-km-long stream channel just downstream of the confluence of DLT tributary (Figure 1). Surface water seeps into swallow holes, leading to reduced flow in the reaches immediately downstream. The downwelling reaches (i.e., sections with swallow holes), which were not surveyed in this study, have been observed to dry up from late December to February when rainfall is minimal (personal observation). Groundwater recharges to the surface channel through discrete springs in reaches downstream of the swallow holes and ensure sufficient base flow to maintain surface water during dry periods. Hong Shi Zi (HSZ) is a second-order downstream reach approximately 2 km downstream from the confluence point of DLT with an elevation of 1950 m. Several visible springs can be observed along this reach; indicating that this reach is influenced by groundwater discharge. Ji Zi Gou (JZG) is another second-order reach about 4 km downstream HSZ, located at an elevation of 1820 m. All the four study reaches have permanent surface flow. The DLT tributary reach had the narrowest wetted width and the shallowest water depth during the study (Table 1). As for the three mainstream reaches, mean wetted width and water depth both increased longitudinally among reaches from SNY to JZG. All the four reaches had closed riparian canopy and low degree of human disturbance because of effective protection of this region.

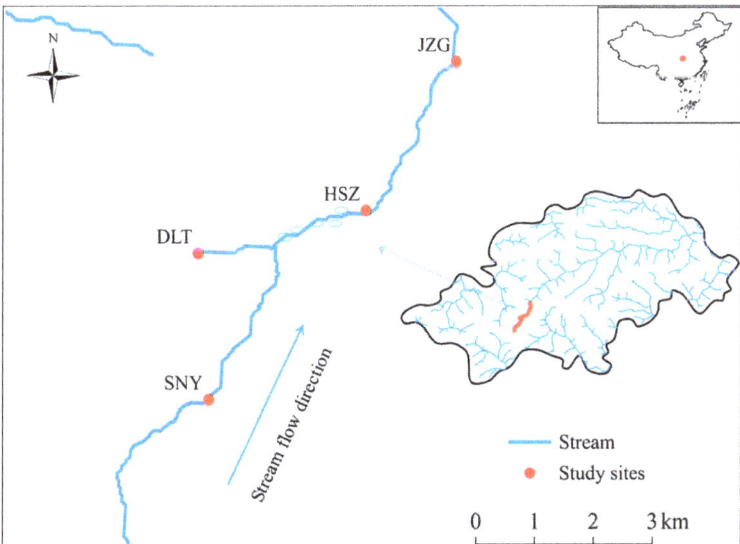

Figure 1. Locations of the 4 stream reaches in the headwaters of Nan River. The ellipse circles illustrate stream channel where discrete swallow holes and springs were observed. SNY = Sheng Nong Yaun; DLT = Da Long Tan; HSZ = Hong Shi Zi; JZG = Ji Zi Gou.

Table 1. Channel morphology characteristics for the four study reaches.

	Sheng Nong Yuan	Da Long Tan	Hong Shi Zi	Ji Zi Gou
Abbreviation	SNY	DLT	HSZ	JZG
Latitude (°N)	31.4722	31.4935	31.4997	31.5215
Longitude (°E)	110.2983	110.2967	110.3222	110.3361
Elevation (m a.s.l.)	2320	2220	1950	1820
Stream order	1	1	2	2
Wetted width (m, mean ± SD)	3.03 ± 2.34	2.81 ± 0.87	5.82 ± 3.63	6.05 ± 3.07
Water depth (cm, mean± SD)	16.5 ± 7.4	11.5 ± 3.8	18.1 ± 7.8	20.5 ± 5.0

2.2. Data Collection and Preparation

The four reaches were surveyed monthly from July 2011 to June 2017; however, data was not collected in April and July of 2015 due to large floods. We monitored stream discharge, air temperature, and water temperature for each reach on all sampling occasions. Mean reach discharge was calculated from measurements at three randomly selected cross-sections along a 50-m reach. At every 50-cm interval along each cross-section, flow velocity was measured using a Global Water Flow Probe Hand-Held Flowmeter (Xylem Inc., White Plains, NY, USA) and water depth was recorded using a tape. The wetted width of each cross section was also measured with a tape. Discharge ($m^3\ s^{-1}$) was calculated using the velocity-area method for each cross-section [36]. Air and water temperature were measured and averaged from the same three cross-sections with an YSI Pro Plus multimeter (YSI Inc., Yellow Springs, OH, USA). Although we did not sample all the reaches at the same time on each occasion, fieldwork in all reaches was completed within a 2 to 3-hour period to minimize variation due to time of sampling.

Benthic algae were sampled at the same aforementioned three cross-sections for each reach during each occasion. Five stones of 10–20 cm diameter were randomly selected at each cross-section, with a total of 15 stones for each reach. We defined the sampling area for each stone with a circular lid of 2.7 cm radius. The stone surface within the lid was vigorously scrubbed using a nylon brush and rinsed 3–4 times with distilled water. All subsamples from the same cross-section were combined into one composited sample and the volume was recorded [37]. Algal sample was passed through a glass fiber filter (Whatman GF/F) in the field and kept dark and cool. In the laboratory chlorophyll *a* was extracted with 90% acetone and absorbance values at 630, 645, 665 and 750 nm were measured using a Shimadzu UV-1601 spectrophotometer (Shimadzu Corp., Kyto, Japan), following a standard analysis procedure [38]. Benthic Chl. *a* concentration for each cross-section was calculated as mg per stone surface area ($mg\ m^{-2}$). Chl. *a* concentration for each reach was averaged from the three cross-section replicates.

2.3. Statistical Analysis

Although our data did not allow us to directly quantify variation in groundwater-surface water exchange among the study reaches and over time, we made use of long-term (six years) measurements to infer spatial and temporal variation in the relative influence of groundwater. This approach has been employed elsewhere [7,16,19,25]. We then relate this understanding of the relative variation in groundwater influence to the observed spatial and temporal variation in the physical and biotic parameters that we assessed.

We performed bivariate linear regression by using water temperature as the response variable and air temperature the predictor to investigate the degree of groundwater influence for each reach. Since monthly water temperature data is not generally autocorrelated, simple linear regression is effective in modeling such a relationship [28]. We expect that the reach most influenced by groundwater (i.e., HSZ) would have a greater intercept and a lesser slope for the regression equation than reaches not influenced by groundwater [39].

We calculated nine metrics characterizing stream water temperature and discharge regimes during the study period for each reach. These metrics represent the water temperature and discharge regime magnitude, variability, frequency, and timing. Magnitude was described with the total mean value, the maximum monthly mean value (Mean_max), and the minimum monthly mean value (Mean_min) of water temperature and discharge. Variability was represented by the range and coefficient of variation (CV) of the two variables. We counted the number of months with water temperature >10 °C for delineating temperature frequency because 10 °C is an important temperature threshold for many aquatic metabolic processes [40,41]. For instance, algal biomass was greatest when water temperature was at or below 10 °C in several forested streams [41]. We also selected 3 °C as another threshold for calculating water temperature frequency because we observed that the study reaches displayed substantial differences among the reaches when temperature was lower than this value. For discharge frequency, we counted number of months with discharge <0.02 or >0.20 m^3 s^{-1} which represents approximately the 25th and 75th percentiles of the distribution of discharge across all reaches. For the timing metrics, we identified specific months when the maximum (M_mean_max) and minimum (M_mean_min) monthly mean water temperature or discharge was reached. We also calculated all the aforementioned nine metrics for Chl. *a*, because algal biomass also varies temporally [42].

We used wavelet analysis to describe regime periodicities in water temperature, discharge, and Chl. *a* for each reach. Wavelet analysis is especially effective in detecting and differentiating scale-specific dynamics from noisy, complex and usually nonstationary time series data that violates assumptions of many other statistical methods [43,44]. This method performs a local time-scale decomposition (i.e., local wavelet transform) of the time series by estimating its spectral characteristics as a function of time [43]. Significance of local wavelet spectrum was tested by comparing against a background red noise spectrum (representing low-frequency/long-period cycles of the time series with autocorrelated residuals) simulated with Monte Carlo procedure [45,46]. Wavelet analysis was conducted by using 'wt' function in R package 'biwavelet', with continuous Morlet wavelet transforming as the base function due to its good balance in time and frequency localization [45]. To further determine which periodicities were most important throughout the study duration, we calculated time-averaged wavelet spectrum for temperature, discharge, and Chl. *a* with 'analyze.wavelet' function in package 'WaveletComp'. For each site, there was a total of 3–4 missing data during the study period. Since missing data are not allowed in wavelet analysis, for each time series, we interpolated missing values with the mean of all available data from the same month adjusted for that year's mean value using the decomposition method described by Cloern and Jassby [42,47].

We conducted convergent cross-mapping (CCM) to identify influences of water temperature and discharge on temporal fluctuations of Chl. *a*. This novel nonparametric method has been described in detail elsewhere [48–50], so we provide only a brief introduction here. Based on state space reconstruction, CCM measures the skill of cross-mapping (i.e., prediction) between two variables (e.g., x and y) as test for causality. Following Takens' Theorem, it is assumed that if x influences y, then y would contain unique information from x. Therefore, historical values of x can be reconstructed from y alone. Conversely, x does not contain information from y, hence will be incapable of cross-mapping y [50]. To accomplish the test, Simplex projection with a time delay embedding from y is first used to predict historical values of x [48]. Then, Pearson's correlation coefficient (ρ) between predicted x and measured x values is calculated to indicate the ability of y cross-mapping x. The ability of x cross-mapping y is also estimated with similar procedures. It is determined that x influences y if ρ value for y cross-mapping x is higher than that of x cross-mapping y, and vice versa [51]. Significance of the result was tested with Ebisuzaki phase-shift null model. This randomizing procedure removes dynamic correlations between time series, but preserves most of short-term behaviors [52]. A CCM result is considered significant if the ρ value for x cross-mapping y exceeds the range of the 5th and 95th percentiles of corresponding estimates for the null model. In the present study, we estimated reach-specific effects of stream water temperature and discharge on Chl. *a*. If temporal dynamics of Chl. *a* was driven by water temperature or discharge, Chl. *a* would be more effective in cross-mapping the other two variables than using the other two variables to cross map Chl. *a*. We applied simplex projection to calculate the optimal embedding dimension for individual time series, and set library size varying from 1 to 60 in steps of 6 with 100 bootstrapped library samples. CCM was performed with time lags of ±3 months to account for possible time delayed effects [53]. For the time lag with the highest cross map skill, we further tested the significance of the results by generating a null distribution with 100 surrogate data. All the time series were first-differenced to remove any trends, and normalized to zero mean and unit variance for allowing time series comparisons [48]. CCM was performed with package 'rEDM'.

Additionally, to investigate whether contrasting seasonal hydrological conditions may influence water temperature, discharge, and Chl. *a*, we calculated the mean values (± standard deviation) of these parameters for each reach during a typically low-flow (December to February) and high-flow (May to July) period in this region [34]. We assessed whether these parameters differed among reaches within seasons using Kruskal–Wallis test by ranks, with pairwise comparisons by Mann–Whitney tests with Bonferroni correction.

3. Results

3.1. Effects of Groundwater on Stream Water Temperature and Discharge Regimes

Reach-specific regression relationships between air and stream water temperature were detected (Table 2). HSZ had the highest intercept of 4.162 and the shallowest slope of 0.352 for the relationship between air and stream water temperature. HSZ also had a distinct water temperature regime when compared with other reaches (Table 3). Mean water temperature increased from SNY to JZG as elevation decreased. However, the maximum monthly mean water temperature (Mean_max) in HSZ was 0.6 °C lower than the adjacent higher elevation reach of DLT (Table 3). Further, the minimum monthly mean water temperature (Mean_min) was 1.0 °C higher than the adjacent lower elevation reach of JZG (Table 3). Moreover, HSZ had a narrower temperature range than that in DLT and had the smallest coefficient of variation (CV) among all the four reaches (Table 3). In accordance with low temperature variability, there was only two months in HSZ when water temperature was <3 °C (Table 3). In comparison, for JZG, where mean water temperature was 1.5 °C higher compared to HSZ, mean monthly water temperature was <3 °C on eight occasions (Table 3). The frequency for months with water temperature >10 °C in HSZ was similar to that recorded in DLT (Table 3). Although influenced by groundwater, HSZ had same timings for the maximum and minimum monthly mean values of water temperature as that in other reaches. All reaches also had the same temperature periodicity that ranged from 10 to 14 months, with the 12-month periodicity having the most significant wavelet spectrum power across the study duration (Table 4, Figure 2). That is, stream water temperature had an annual cycle that peaked in August and was lowest in January (Figure 2). During the high flow period, the mean water temperature in HSZ was similar to that in the higher elevation reach of DLT (~10 °C for both reaches). Moreover, HSZ had a higher mean water temperature than that in the lower elevation reach of JZG (3.9 vs. 3.5 °C) during the low flow period, despite the difference was not significant (Appendix A Figure A1).

The discharge regime in HSZ was also different from that recorded in the other three reaches. The discharge magnitude and variation range generally increased with stream size (Table 3). The mean and range of discharge in the largest reach (JZG) was approximately 6 and 5.5 times greater, respectively, than that recorded in the smallest reach (DLT) (Table 3). However, other regime metrics did not display such stream size-related trends. HSZ had the highest discharge CV, and the highest frequency of months with discharge lower than 0.02 m^3 s^{-1} (Table 3). Discharge displayed a significant 12-month periodicity in the all reaches across the study duration with the peak in May and trough in January or February (Table 4, Figure 3). However, significance of such periodicity (indicated by black line circles) varied among years and among reaches (Figure 3). Besides the annual periodicity, discharge also displayed a significant 2–6 months periodicity in 2013 across all reaches (Figure 3).

During the low flow period, mean discharge in HSZ was 0.03 m^3 s^{-1}, which was lower than the sum of mean discharge of the two upstream reaches (SNY: 0.03 m^3 s^{-1}, DLT: 0.02 m^3 s^{-1}). In contrast, this pattern was not evident during the high flow period (Appendix A Figure A1).

Table 2. The Y intercept, slope, and adjusted R^2 (R^2_{adj}) for the regression between stream water temperature and air temperature from July 2011 to June 2017 for each study reach. All regression models are significant with $p < 0.001$. SNY = Sheng Nong Yaun; DLT = Da Long Tan; HSZ = Hong Shi Zi; JZG = Ji Zi Gou.

Stream Reach	Regression Coefficient		R^2_{adj}
	Y Intercept	Slope	
SNY	2.296	0.373	0.807
DLT	1.834	0.453	0.830
HSZ	4.162	0.352	0.861
JZG	3.327	0.476	0.850

Table 3. Values of stream water temperature, discharge, and benthic chlorophyll *a* concentration (Chl. *a*) regime metrics for each study reach during the study. Mean_max = the maximum monthly mean value; Mean_min = the minimum monthly mean value; M_mean_max = the month with the maximum monthly mean value; M_mean_min = the month with the minimum monthly mean value. CV: coefficient of variation. SNY = Sheng Nong Yaun; DLT = Da Long Tan; HSZ = Hong Shi Zi; JZG = Ji Zi Gou.

Regime	Descriptors	Metrics	Stream Reach			
			SNY	DLT	HSZ	JZG
Water temperature	Magnitude	Mean (°C)	5.9	6.7	7.6	9.1
		Mean_max (°C)	9.6	12.0	11.4	14.7
		Mean_min (°C)	1.2	0.9	3.7	2.7
	Variability	Range (°C)	10.3	13.4	11.9	17.3
		CV (%)	52.4	59.5	37.9	48.6
	Frequency	Months < 3 °C	19	19	2	8
		Months > 10 °C	1	17	19	33
	Timing	M_mean_max	Aug	Aug	Aug	Aug
		M_mean_min	Jan	Jan	Jan	Jan
Discharge	Magnitude	Mean (m^3 s^{-1})	0.145	0.080	0.196	0.482
		Mean_max (m^3 s^{-1})	0.527	0.227	0.700	1.274
		Mean_min (m^3 s^{-1})	0.012	0.016	0.013	0.096
	Variability	Range (m^3 s^{-1})	1.717	0.553	3.251	4.329
		CV (%)	163.3	108.0	234.4	155.8
	Frequency	Months < 0.02 m^3 s^{-1}	15	14	19	0
		Months > 0.20 m^3 s^{-1}	12	5	14	38
	Timing	M_mean_max	May	May	May	May
		Mmean_min	Feb	Jan	Jan	Jan
Chl. *a*	Magnitude	Mean (mg m^{-2})	5.60	3.56	5.28	7.58
		Mean_max (mg m^{-2})	9.46	6.70	7.98	20.42
		Mean_min (mg m^{-2})	2.68	1.44	2.43	2.24
	Variability	Range (mg m^{-2})	16.83	13.81	18.38	33.82
		CV (%)	66.0	83.8	82.7	112.4
	Frequency	Months < 5 mg m^{-2}	32	54	43	38
		Months > 10 mg m^{-2}	8	3	11	19
	Timing	M_mean_max	Dec	Feb	Dec	Feb
		M_mean_min	July	July	May	Apr

Table 4. Significant ($p < 0.05$) periodicities (Unit: month) in stream water temperature (WT), discharge, and benthic chlorophyll *a* concentration (Chl. *a*) across July 2011 to June 2017 identified by comparing time-averaged wavelet spectrum to a null red-noise power spectrum by wavelet analysis for each reach. Numbers in parentheses represent the periods having the highest average wavelet power. -: no significant period is detected. SNY = Sheng Nong Yaun; DLT = Da Long Tan; HSZ = Hong Shi Zi; JZG = Ji Zi Gou.

	SNY	DLT	HSZ	JZG
WT	10–14 (12)	10–14 (12)	10–14 (12)	10–14 (12)
Discharge	12	11–13 (12)	12	12
Chl. *a*	10–13 (12)	11–13 (12)	-	10–13 (12)

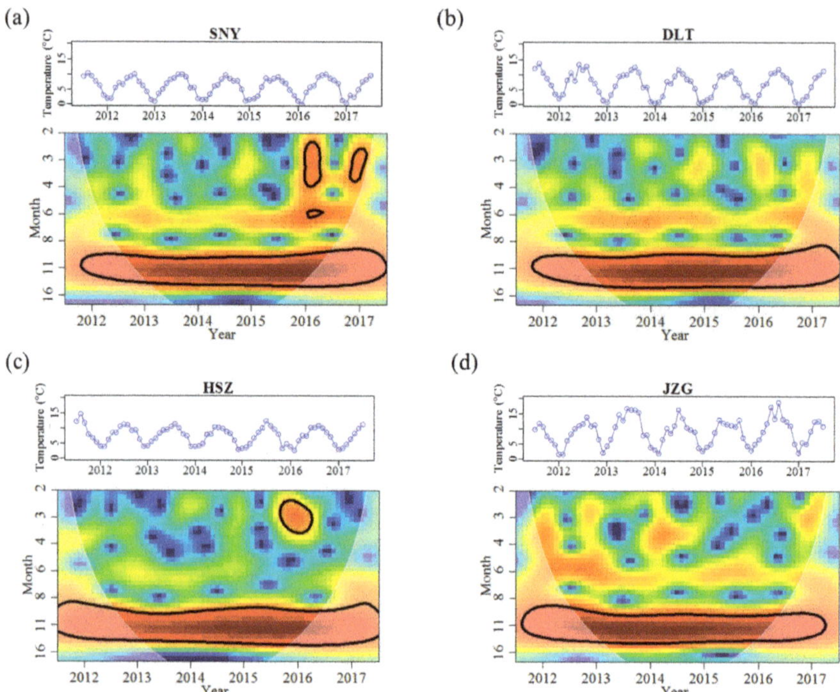

Figure 2. Continuous wavelet power spectra showing the strength of the periodicities of stream water temperature changes between July 2011 and June 2017 for each stream reach. The colors code for power intensity from dark blue (low intensity) to dark red (high intensity), and solid black line circles indicate significant ($p < 0.05$) coherent time–frequency regions. The shaded area is the time–frequency region affected by edge-effects and was not included in the analyses. The original time series are demonstrated above each wavelet plot. SNY = Sheng Nong Yaun; DLT = Da Long Tan; HSZ = Hong Shi Zi; JZG = Ji Zi Gou.

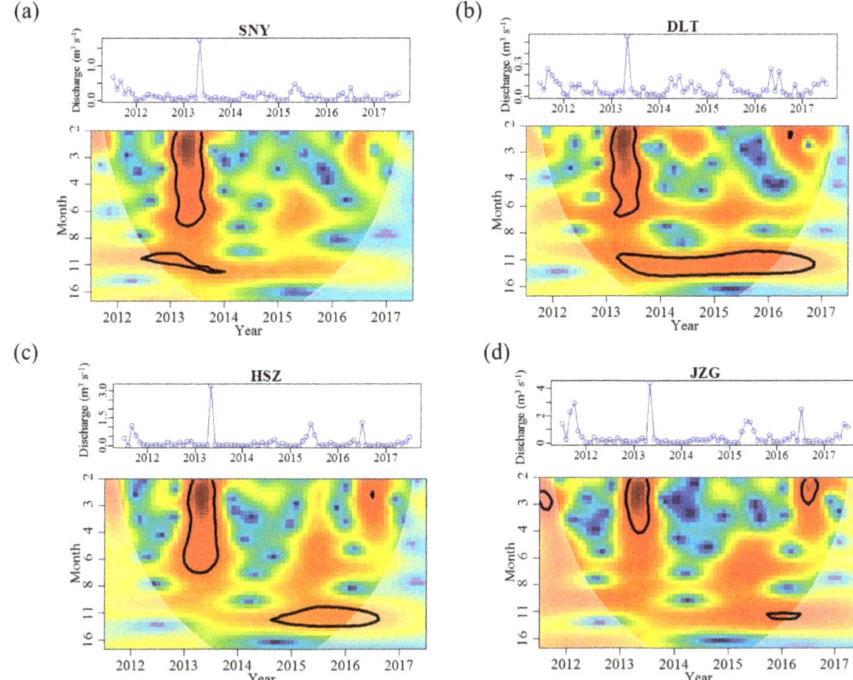

Figure 3. Continuous wavelet power spectra showing the periodicity of discharge from July 2011 to June 2017 for each stream reach. Further information on wavelet spectra are described in Figure 2. The original time series are demonstrated above each wavelet plot, with different value ranges to illustrate reach-specific variation. SNY = Sheng Nong Yuan; DLT = Da Long Tan; HSZ = Hong Shi Zi; JZG = Ji Zi Gou.

3.2. Effects of Groundwater on Associations between Chl. a Dynamics, Water Temperature, and Discharge

Across all stream reaches, the range of Chl. *a* concentrations and frequency of months with Chl. *a* concentration higher than 10 mg m^{-2} increased as elevation decreased (Table 3). The temporal dynamics of Chl. *a* concentrations in HSZ was, however, largely different than in the reaches less affected by groundwater (Table 3). Specifically, Chl. *a* magnitude was lower in HSZ when compared with the smaller upstream reach, SNY (Table 3). The CV of Chl. *a* concentrations in HSZ was similar to that in DLT (Table 3). The 12-month periodicity in Chl. *a* concentrations was significant for all reaches except for HSZ (Table 4, Figure 4). Moreover, HSZ had the highest mean Chl. *a* concentrations among the four reaches during the high flow period (Appendix A Figure A1); however, differences were not significant in this test (Kruskal–Wallis chi-squared = 5.092, df = 3, *p*-value = 0.165). During the low flow period, the mean Chl. *a* concentration in HSZ (7.17 mg m^{-2}) was slightly higher than that in the tributary reach of DLT (5.59 mg m^{-2}), and was lower than the other two mainstream reaches. These differences were also not significant (Appendix A Figure A1).

By using CCM, we found that water temperature had a significant influence on Chl. *a* dynamics in SNY, DLT, and JZG, but there was variation in the coupling of these variables through time and also the strength of this coupling (Figure 5; Appendix A Table A1; Appendix A Figure A2). The influence that water temperature had on Chl. *a* was tightly coupled in time except for in DLT where there was a 2-month delay in this coupling (Appendix A Table A1). By comparison, water temperature had the weakest and nonsignificant influence on Chl. *a* in HSZ as it had the lowest ρ value within the range of the 5th and 95th percentiles of the null model (Figure 5; Appendix A Table A1). In contrast to water temperature, discharge had a weak influence on Chl. *a* dynamics in all reaches, represented by low ρ values (Figure 5; Appendix A Table A1; Appendix A Figure A2). Furthermore, discharge did not affect Chl. *a* dynamics more significantly than that in the null model for all reaches (Figure 5).

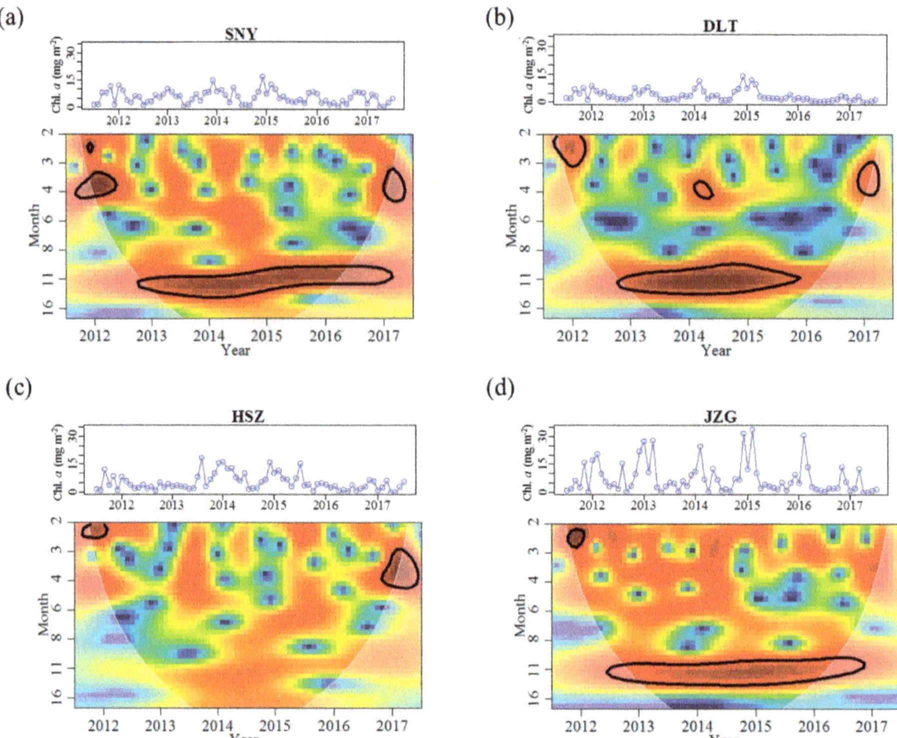

Figure 4. Continuous wavelet power spectra showing the periodicity of benthic algal chlorophyll *a* concentrations (Chl. *a*) from July 2011 to June 2017 for each study reach. Further information on wavelet spectra are described in Figure 2. The original time series are demonstrated above each wavelet plot. SNY = Sheng Nong Yaun; DLT = Da Long Tan; HSZ = Hong Shi Zi; JZG = Ji Zi Gou.

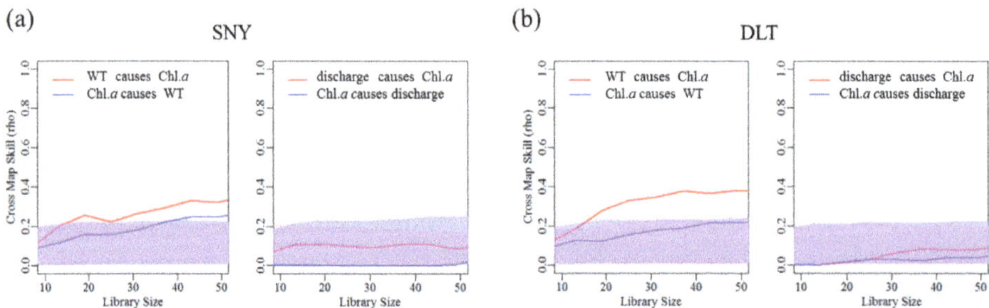

Figure 5. *Cont.*

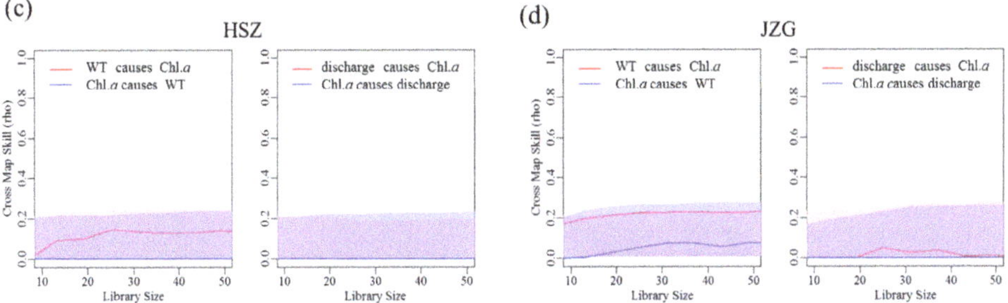

Figure 5. Convergent cross-mapping (CCM) between benthic chlorophyll *a* concentration (Chl. *a*) versus stream water temperature (WT) (left) and discharge (right) using time series from July 2011 to June 2017 for each study reach. The y-axis represents the Cross Map Skill (rho or Pearson's correlation coefficient) and the x-axis represents the library size (different lengths of data subsampled randomly from the time series) used for modeling. The time lags having the highest cross map skill in CCM were used in the final CCM for significance testing. Shaded areas are the 95% confidence intervals (5th to 95th percentiles) of 100 surrogate time series from the null model (light red area = 95% confidence intervals of the null model for the red line; light blue area = 95% confidence intervals of the null model for the blue line; overlap=light purple). rho improving with increasing library size indicates causality, and rho beyond the 95% confidence intervals of the null model implies significant causality. For JZG, the red line is beyond the light shaded area, indicating that WT has significant effects on Chl. *a*. SNY = Sheng Nong Yaun; DLT = Da Long Tan; HSZ = Hong Shi Zi; JZG = Ji Zi Gou.

4. Discussion

Differences in stream regime descriptors, including parameter magnitude, variability, frequency, timing, and predictability, among adjacent reaches can provide ecologically valuable information on how groundwater influences surface waters. We found that groundwater discharge can have a large influence on the temporal patterns of many stream regime metrics that describe variation in water temperature and discharge. Further, we found that changes in stream physical conditions due to groundwater discharge can alter the influence of these physical conditions for driving temporal variation in benthic algal biomass.

The most groundwater-influenced reach, HSZ, had a distinct air–water temperature relationship and a more stable thermal regime when compared with the other three reaches, confirming predictions and observations that this reach was substantially influenced by groundwater. In particular, HSZ had a lower maximum monthly mean water temperature than the adjacent higher elevation reach and a higher minimum monthly mean water temperature than the adjacent lower elevation reach. The mean water temperature in HSZ was also lower than that in higher elevation reach during the high flow period, and was higher than that in the lower elevation reach during the low flow period. Further, HSZ had the lowest frequency of months with temperatures < 3 °C among all reaches. These results indicate that stream water in HSZ was cooler in summer and warmer in winter compared to the other reaches. The smallest CV of water temperature and the shallowest slope for air–water temperature relationship in HSZ imply a slower heating and cooling rate (i.e., a temperature buffering effect) than the other reaches during the annual temperature cycle [20]. Such buffering effect has been commonly observed on daily temperature magnitude [54–56]. Our findings suggest that this phenomenon could also occur at longer time scales. The same timings for the maximum and minimum monthly mean water temperature across all reaches suggest that groundwater did not alter the timing of the metrics for water temperature in HSZ.

Although discharge increased with stream size, HSZ had the highest CV and the highest frequency for low-flow events. This result is related to the fact that groundwater was the sole source of flow for this reach when upstream reaches, which contain visible swallow holes, often ceased to flow between December to February. In fact, we found that mean discharge in HSZ (0.196 m^3 s^{-1}) was a little smaller than the sum of flows in the two upstream reaches of SNY and DLT (0.225 m^3 s^{-1}) in the whole study duration, and this difference was even more evident during the low flow period. It is likely that a proportion of downwelling flow had longer flow paths than the distance of the study reaches, longer subsurface residence times than we could detect in this study, or had lateral path(s) beyond the watershed boundary [57]. Therefore, although discharge magnitude increased longitudinally, karst topographic features likely induced a higher variability of flow dynamics in HSZ than in other reaches. Overall, the different regime metrics of discharge in HSZ compared with other reaches provides evidence that this reach is significantly influenced by groundwater discharge.

HSZ had a lower Chl. *a* magnitude and a higher low-concentration frequency than that in the upstream reach of SNY. This result seems unexpected because benthic algal biomass often increases with stream size and temperature [32]. Further, spring stream flow is generally rich in dissolved and organic nutrients that can promote downstream algal biomass [18,58]. One possible reason is that higher predation pressure by lotic grazers in HSZ reduced algal biomass because more stable temperature and flow conditions were also beneficial to grazer populations [59]. A comparison of the differences in nutrient concentrations between HSZ and nearby reaches and an evaluation of intensity of predation pressure and its influences on algal biomass in HSZ would be valuable topics for future research. Benthic algal biomass in HSZ did not display a significant annual cycle as that in other reaches. This phenomenon has previously been attributed to reduced temporal variation in physical conditions such as temperature, conductivity, and flow [60]. Our findings provide further support for this notion because water temperature and flow in HSZ had milder temporal regimes and their influences on temporal dynamics of Chl. *a* were weaker than reaches less influenced by groundwater. We found that stream discharge did not have a significant influence on the temporal dynamics of Chl. *a* at any study reach. A possible explanation for this is that stream discharge mainly affected benthic algal assemblages via indirect paths. Discharge was relatively low in most times during the study period and flow velocity may not have been high enough to reduce algal colonization or growth. Therefore, flow may have been most important for mediating the transfer of nutrients for autotrophic assimilation [31]. While floods might have negatively impacted lotic assemblages, the effect of these events may have been temporary and/or perhaps not detectible or important over log time periods, as in our study.

We used long-term data (six years) to gain a broad understanding of the physical and biotic influences of groundwater in our study stream. However, we were unable to adequately explore how groundwater–surface water exchange during contrasting hydrological conditions may influence important physical and biotic components of our stream system, due to the monthly sampling frequency of our study design. Further research with a more frequent and intensive survey effort would provide more valuable information on how the important physical and biotic components of streams that we assessed change in response to intra- and inter-annual variation in groundwater–surface water exchange.

In conclusion, water temperature in the most groundwater-influenced stream reach had the smallest variability (i.e., CV) and the lowest frequency for low-temperature months, when compared with other reaches. Groundwater inputs thus induced milder physical habitat conditions. These milder physical habitat conditions were associated with weaker temporal patterns for benthic algal biomass. That is, benthic algae biomass in this reach did not display a significant annual cycle as that in other reaches. Our findings demonstrate that groundwater inputs may alter regime metrics other than magnitude for physical components in downstream reach, and further induce substantial changes in temporal dynamics of important biotic components. Therefore, longer and more intensive studies will greatly improve our understanding of how groundwater effects stream ecosystems.

Author Contributions: Conceptualization, T.T. and Q.C.; Formal analysis, T.T.; Funding acquisition, T.T. and Q.C.; Investigation, S.G. and L.T.; Methodology, T.T.; Writing—original draft, T.T.; Writing—review & editing, T.L. and R.M.B.; and all authors reviewed the manuscript before submission.

Funding: This research was funded by the National Natural Science Foundation of China (No. 31470510), the National Key Research and Development Program of China (No. 2016YFC0503601, No. 2016YFC0502106), and the Special Project of Science and Technology Basic Work of Ministry of Science and Technology of China (No. 2014FY120200).

Acknowledgments: We thank Xiaoyu Dong, Tingting Sun, Ting Tang, Ze Ren, Fengzhi He, and Yang Li for assistance with fieldwork. We are grateful to Andrew Boulton and Naicheng Wu for valuable comments on the draft.

Conflicts of Interest: The authors declare no conflicts of interest.

Appendix A

Table A1. Convergent cross-mapping detecting the best cross-map skill (ρ) and time lag indicating the causal effects of stream water temperature (WT) or discharge on dynamics of benthic chlorophyll a concentration (Chl. *a*) from July 2011 to June 2017 for each reach. -: No causality is detected. SNY = Sheng Nong Yaun; DLT = Da Long Tan; HSZ = Hong Shi Zi; JZG = Ji Zi Gou.

Stream Reach	WT Causing Chl. *a*		Discharge Causing Chl. *a*	
	ρ	Time Lag (Month)	ρ	Time Lag (Month)
SNY	0.267	0	0.101	2
DLT	0.309	2	0.103	2
HSZ	0.135	0	-	
JZG	0.230	0	0.089	2

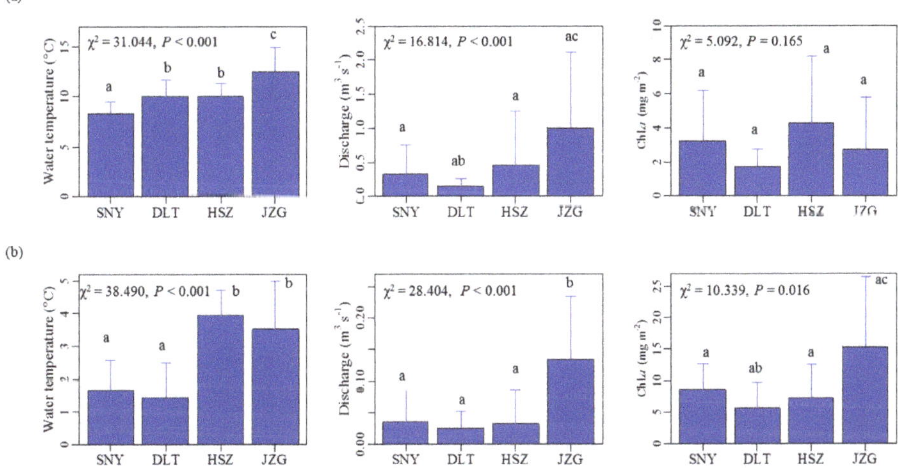

Figure A1. Plots demonstrating among-reach differences in the mean (± standard deviation) for water temperature, discharge, and chlorophyll a concentration (Chl. a) during the high-flow period (a, May to July) and the low-flow period (b, December to February), respectively. The χ^2 value and the *p*-value are the results of Kruskal–Wallis tests among the four reaches. The lower-case letters above whiskers are the results of pairwise comparisons, with the same letters indicating no difference at $p \leq 0.05$. SNY = Sheng Nong Yaun; DLT = Da Long Tan; HSZ = Hong Shi Zi; JZG = Ji Zi Gou.

SNY:

DLT:

HSZ:

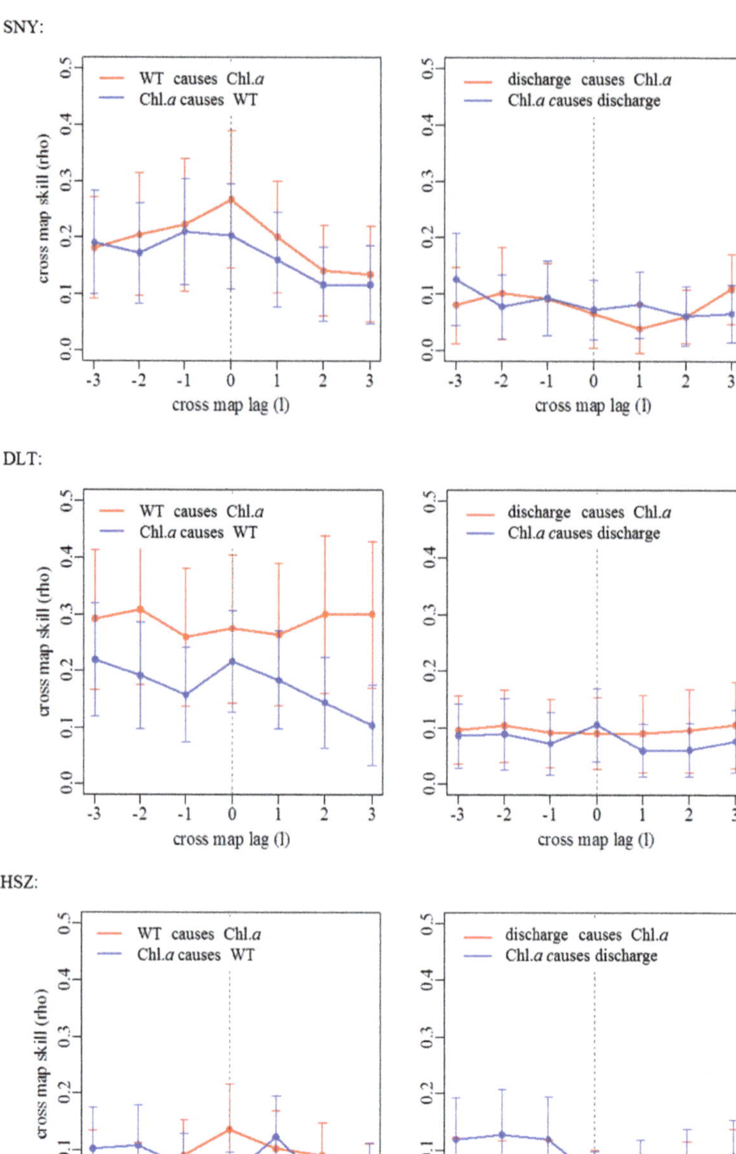

Figure A2. *Cont.*

JZG:

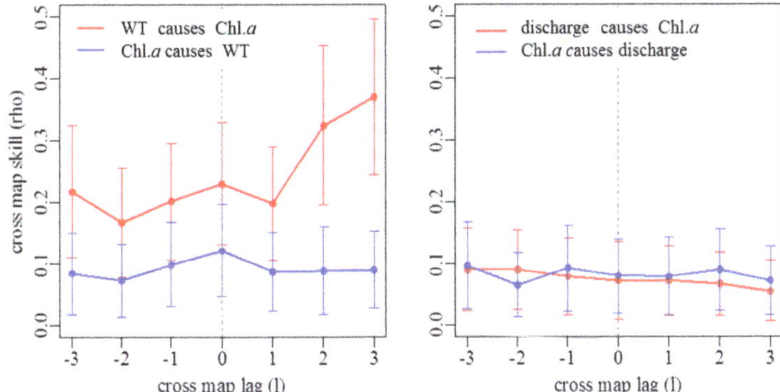

Figure A2. Extended convergent cross-mapping detecting causality and time lags between chlorophyll a concentrations (Chl. *a*) and water temperature (left) or discharge (right) using time series from July 2011 to June 2017 for each study reach. SNY = Sheng Nong Yaun; DLT = Da Long Tan; HSZ = Hong Shi Zi; JZG = Ji Zi Gou.

References

1. Hayashi, M.; Rosenberry, D.O. Effects of ground water exchange on the hydrology and ecology of surface water. *Groundwater* **2002**, *40*, 309–316. [CrossRef]
2. Wondzell, S.M. Groundwater–surface-water interactions: Perspectives on the development of the science over the last 20 years. *Freshw. Sci.* **2015**, *34*, 368–376. [CrossRef]
3. Boulton, A.J.; Hancock, P.J. Rivers as groundwater-dependent ecosystems: A review of degrees of dependency, riverine processes and management implications. *Aust. J. Bot.* **2006**, *54*, 133–144. [CrossRef]
4. Pipan, T. *Epikarst—A Promising Habitat: Copepod Fauna, Its Diversity and Ecology: A Case Study From Slovenia (Europe)*; ZRC Publishing: Postojna–Ljubljana, Slovenia, 2005.
5. Boulton, A.J.; Findlay, S.; Marmonier, P.; Stanley, E.H.; Valett, H.M. The functional significance of the hyporheic zone in streams and rivers. *Annu. Rev. Ecol. Syst.* **1998**, *29*, 59–81. [CrossRef]
6. Burrows, R.M.; Rutlidge, H.; Valdez, D.G.; Venarsky, M.; Bond, N.R.; Andersen, M.S.; Fry, B.; Eberhard, S.M.; Kennard, M.J. Groundwater supports intermittent-stream food webs. *Freshw. Sci.* **2018**, *37*, 42–53. [CrossRef]
7. Robinson, C.T.; Doering, M. Spatial patterns in macroinvertebrate assemblages in surface-flowing waters of a glacially-influenced floodplain. *Aquat. Sci.* **2013**, *75*, 373–384. [CrossRef]
8. Malard, F.; Tockner, K.; Dole-Olivier, M.J.; Ward, J. A landscape perspective of surface-subsurface hydrological exchanges in river corridors. *Freshw. Biol.* **2002**, *47*, 621–640. [CrossRef]
9. Findlay, S. Importance of surface-subsurface exchange in stream ecosystems: The hyporheic zone. *Limnol. Oceanogr.* **1995**, *40*, 159–164. [CrossRef]
10. Robertson, A.; Wood, P. Ecology of the hyporheic zone: Origins, current knowledge and future directions. *Fund. Appl. Limnol.* **2010**, *176*, 279–289. [CrossRef]
11. Boulton, A.J.; Datry, T.; Kasahara, T.; Mutz, M.; Stanford, J.A. Ecology and management of the hyporheic zone: Stream-groundwater interactions of running waters and their floodplains. *J. N. Am. Benthol. Soc.* **2010**, *29*, 26–40. [CrossRef]
12. Mathers, K.L.; Hill, M.J.; Wood, P.J. Benthic and hyporheic macroinvertebrate distribution within the heads and tails of riffles during baseflow conditions. *Hydrobiologia* **2017**, *794*, 17–30. [CrossRef]
13. Dole-Olivier, M.-J. The hyporheic refuge hypothesis reconsidered: A review of hydrological aspects. *Mar. Freshw. Res.* **2011**, *62*, 1281–1302. [CrossRef]
14. Kawanishi, R.; Inoue, M.; Dohi, R.; Fujii, A.; Miyake, Y. The role of the hyporheic zone for a benthic fish in an intermittent river: A refuge, not a graveyard. *Aquat. Sci.* **2013**, *75*, 425–431. [CrossRef]
15. Mejia, F.H.; Baxter, C.V.; Berntsen, E.K.; Fremier, A.K. Linking groundwater-surface water exchange to food production and salmonid growth. *Can. J. Fish. Aquat. Sci.* **2016**, *73*, 1650–1660. [CrossRef]

16. Olsen, D.A.; Young, R.G. Significance of river-aquifer interactions for reach-scale thermal patterns and trout growth potential in the Motueka River, New Zealand. *Hydrogeol. J.* **2009**, *17*, 175–183. [CrossRef]

17. Torgersen, C.E.; Price, D.M.; Li, H.W.; McIntosh, B.A. Multiscale thermal refugia and stream habitat associations of chinook salmon in northeastern Oregon. *Ecol. Appl.* **1999**, *9*, 301–319. [CrossRef]

18. Pepin, D.M.; Hauer, F.R. Benthic responses to groundwater–surface water exchange in 2 alluvial rivers in northwestern Montana. *J. N. Am. Benthol. Soc.* **2002**, *21*, 370–383. [CrossRef]

19. Wright, K.K.; Baxter, C.V.; Li, J.L. Restricted hyporheic exchange in an alluvial river system: Implications for theory and management. *J. N. Am. Benthol. Soc.* **2005**, *24*, 447–460. [CrossRef]

20. Arrigoni, A.S.; Poole, G.C.; Mertes, L.A.K.; O'Daniel, S.J.; Woessner, W.W.; Thomas, S.A. Buffered, lagged, or cooled? Disentangling hyporheic influences on temperature cycles in stream channels. *Water Resour. Res.* **2008**, *44*, W09418. [CrossRef]

21. Kath, J.; Harrison, E.; Kefford, B.J.; Moore, L.; Wood, P.J.; Schäfer, R.B.; Dyer, F. Looking beneath the surface: Using hydrogeology and traits to explain flow variability effects on stream macroinvertebrates. *Ecohydrology* **2016**, *9*, 1480–1495. [CrossRef]

22. Stubbington, R. The hyporheic zone as an invertebrate refuge: A review of variability in space, time, taxa and behaviour. *Mar. Freshw. Res.* **2012**, *63*, 293–311. [CrossRef]

23. Kløve, B.; Ala-aho, P.; Bertrand, G.; Boukalova, Z.; Ertürk, A.; Goldscheider, N.; Ilmonen, J.; Karakaya, N.; Kupfersberger, H.; Kvœrner, J.; et al. Groundwater dependent ecosystems. Part I: Hydroecological status and trends. *Environ. Sci. Policy* **2011**, *14*, 770–781. [CrossRef]

24. Coulter, A.A.; Galarowicz, T.L. Fish assemblage and environmental differences upstream and downstream of a cave: A potential reset mechanism. *Environ. Biol. Fishes* **2015**, *98*, 1223–1231. [CrossRef]

25. Stubbington, R.; Wood, P.J.; Reid, I.; Gunn, J. Benthic and hyporheic invertebrate community responses to seasonal flow recession in a groundwater-dominated stream. *Ecohydrology* **2011**, *4*, 500–511. [CrossRef]

26. Arismendi, I.; Johnson, S.L.; Dunham, J.B.; Haggerty, R. Descriptors of natural thermal regimes in streams and their responsiveness to change in the Pacific Northwest of North America. *Freshw. Biol.* **2013**, *58*, 880–894. [CrossRef]

27. Olden, J.D.; Naiman, R.J. Incorporating thermal regimes into environmental flows assessments: Modifying dam operations to restore freshwater ecosystem integrity. *Freshw. Biol.* **2010**, *55*, 86–107. [CrossRef]

28. Caissie, D. The thermal regime of rivers: A review. *Freshw. Biol.* **2006**, *51*, 1389–1406. [CrossRef]

29. Woodward, G.; Bonada, N.; Brown, L.E.; Death, R.G.; Durance, I.; Gray, C.; Hladyz, S.; Ledger, M.E.; Milner, A.M.; Ormerod, S.J. The effects of climatic fluctuations and extreme events on running water ecosystems. *Philos. Trans. R. Soc. B* **2016**, *371*, 20150274. [CrossRef]

30. Lau, D.C.P.; Goedkoop, W.; Vrede, T. Cross-ecosystem differences in lipid composition and growth limitation of a benthic generalist consumer. *Limnol. Oceanogr.* **2013**, *58*, 1149–1164. [CrossRef]

31. Brett, M.T.; Bunn, S.E.; Chandra, S.; Galloway, A.W.E.; Guo, F.; Kainz, M.J.; Kankaala, P.; Lau, D.C.P.; Moulton, T.P.; Power, M.E.; et al. How important are terrestrial organic carbon inputs for secondary production in freshwater ecosystems? *Freshw. Biol.* **2017**, *62*, 833–853. [CrossRef]

32. DeNicola, D. Periphyton responses to temperature at different ecological levels. In *Algal Ecology: Freshwater Benthic Ecosystems*; Stevenson, R.J., Bothwell, M.L., Lowe, R.L., Thorp, J.H., Eds.; Academic Press: San Diego, CA, USA, 1996; pp. 149–181. ISBN 0-12-668450-2.

33. Biggs, B.J.F.; Nikora, V.I.; Snelder, T.H. Linking scales of flow variability to lotic ecosystem structure and function. *River Res. Appl.* **2005**, *21*, 283–298. [CrossRef]

34. Zhu, Z.; Song, C. *Scientific Survey of Shennongjia Nature Reserve*; China Forestry Publishing House: Beijing, China, 1999; ISBN 7-5038-2335-8. (In Chinese)

35. Sweeting, M.M. *Karst in China: Its Geomorphology and Environment*; Springer-Verlag: Berlin/Heidelberg, Germany, 1995; 265p, ISBN 3642795226.

36. Gore, J.A.; Banning, J. Chapter 3—Discharge measurements and streamflow analysis. In *Methods in Stream Ecology*, 3rd ed.; Hauer, F.R., Lamberti, G.A., Eds.; Academic Press: Boston, MA, USA, 2017; Volume 1, pp. 49–70.

37. Tang, T.; Jia, X.; Jiang, W.; Cai, Q. Multi-scale temporal dynamics of epilithic algal assemblages: Evidence from a Chinese subtropical mountain river network. *Hydrobiologia* **2016**, *770*, 289–299. [CrossRef]

38. Bureau, C.E.P. *Standard Methods for Examination of Water and Wastewater*, 4th ed.; Chinese Environmental Science Press: Beijing, China, 2004; ISBN 7-80163-400-4. (In Chinese)

39. O'Driscoll, M.A.; DeWalle, D.R. Stream-air temperature relations to classify stream-ground water interactions in a karst setting, central Pennsylvania, USA. *J. Hydrol.* **2006**, *329*, 140–153. [CrossRef]

40. Harig, A.L.; Fausch, K.D. Minimum habitat requirements for establishing translocated cutthroat trout populations. *Ecol. Appl.* **2002**, *12*, 535–551. [CrossRef]

41. Rodman, A.R.; Scott, J.T. Comparing two periphyton collection methods commonly used for stream bioassessment and the development of numeric nutrient standards. *Environ. Monit. Assess.* **2017**, *189*, 360. [CrossRef]

42. Winder, M.; Cloern, J.E. The annual cycles of phytoplankton biomass. *Philos. Trans. R. Soc. B* **2010**, *365*, 3215–3226. [CrossRef] [PubMed]

43. Cazelles, B.; Chavez, M.; Berteaux, D.; Ménard, F.; Vik, J.O.; Jenouvrier, S.; Stenseth, N.C. Wavelet analysis of ecological time series. *Oecologia* **2008**, *156*, 287–304. [CrossRef]

44. Keitt, T.H.; Fischer, J. Detection of scale-specific community dynamics using wavelets. *Ecology* **2006**, *87*, 2895–2904. [CrossRef]

45. Torrence, C.; Compo, G.P. A practical guide to wavelet analysis. *Bull. Am. Meteorol. Soc.* **1998**, *79*, 61–78. [CrossRef]

46. Vasseur, D.A.; Yodzis, P. The color of environmental noise. *Ecology* **2004**, *85*, 1146–1152. [CrossRef]

47. Cloern, J.E.; Jassby, A.D. Patterns and scales of phytoplankton variability in estuarine–coastal ecosystems. *Estuaries coasts* **2010**, *33*, 230–241. [CrossRef]

48. Chang, C.-W.; Ushio, M.; Hsieh, C. Empirical dynamic modeling for beginners. *Ecol. Res.* **2017**, *32*, 785–796. [CrossRef]

49. Deyle, E.R.; Fogarty, M.; Hsieh, C.; Kaufman, L.; MacCall, A.D.; Munch, S.B.; Perretti, C.T.; Ye, H.; Sugihara, G. Predicting climate effects on Pacific sardine. *Proc. Natl. Acad. Sci. USA* **2013**, *110*, 6430–6435. [CrossRef] [PubMed]

50. Sugihara, G.; May, R.; Ye, H.; Hsieh, C.; Deyle, E.; Fogarty, M.; Munch, S. Detecting causality in complex ecosystems. *Science* **2012**, *338*, 496–500. [CrossRef] [PubMed]

51. Ye, H.; Sugihara, G. Information leverage in interconnected ecosystems: Overcoming the curse of dimensionality. *Science* **2016**, *353*, 922–925. [CrossRef] [PubMed]

52. Van Nes, E.H.; Scheffer, M.; Brovkin, V.; Lenton, T.M.; Ye, H.; Deyle, E.; Sugihara, G. Causal feedbacks in climate change. *Nat. Clim. Chang.* **2015**, *5*, 445–448. [CrossRef]

53. Ye, H.; Deyle, E.R.; Gilarranz, L.J.; Sugihara, G. Distinguishing time-delayed causal interactions using convergent cross mapping. *Sci. Rep.* **2015**, *5*, 14750. [CrossRef]

54. Burkholder, B.K.; Grant, G.E.; Haggerty, R.; Khangaonkar, T.; Wampler, P.J. Influence of hyporheic flow and geomorphology on temperature of a large, gravel-bed river, Clackamas River, Oregon, USA. *Hydrol. Process.* **2008**, *22*, 941–953. [CrossRef]

55. Hannah, D.M.; Malcolm, I.A.; Bradley, C. Seasonal hyporheic temperature dynamics over riffle bedforms. *Hydrol. Process.* **2009**, *23*, 2178–2194. [CrossRef]

56. Story, A.; Moore, R.; Macdonald, J. Stream temperatures in two shaded reaches below cutblocks and logging roads: Downstream cooling linked to subsurface hydrology. *Can. J. For. Res.* **2003**, *33*, 1383–1396. [CrossRef]

57. Ford, D.; Williams, P.D. *Karst Hydrogeology and Geomorphology*; John Wiley & Sons Ltd.: Chichester, UK, 2007; ISBN 978-0-470-84996-5.

58. Wyatt, K.H.; Hauer, F.R.; Pessoney, G.F. Benthic algal response to hyporheic-surface water exchange in an alluvial river. *Hydrobiologia* **2008**, *607*, 151–161. [CrossRef]

59. Álvarez, M.; Pardo, I. Factors controlling epilithon biomass in a temporary, karstic stream: The interaction between substratum and grazing. *J. N. Am. Benthol. Soc.* **2007**, *26*, 207–220. [CrossRef]

60. Godwin, C.M.; Carrick, H.J. Spatio-temporal variation of periphyton biomass and accumulation in a temperate spring-fed stream. *Aquat. Ecol.* **2008**, *42*, 583–595. [CrossRef]

Communication

Is the Hyporheic Zone Relevant beyond the Scientific Community?

Jörg Lewandowski [1,2,*], Shai Arnon [3], Eddie Banks [4], Okke Batelaan [4], Andrea Betterle [5,6], Tabea Broecker [7], Claudia Coll [8], Jennifer D. Drummond [9], Jaime Gaona Garcia [1,10,11], Jason Galloway [1,2], Jesus Gomez-Velez [12], Robert C. Grabowski [13], Skuyler P. Herzog [14], Reinhard Hinkelmann [7], Anja Höhne [1,15], Juliane Hollender [5], Marcus A. Horn [16,17], Anna Jaeger [1,2], Stefan Krause [9], Adrian Löchner Prats [18], Chiara Magliozzi [13,19], Karin Meinikmann [1,20], Brian Babak Mojarrad [21], Birgit Maria Mueller [1,22], Ignacio Peralta-Maraver [23], Andrea L. Popp [5,24], Malte Posselt [8], Anke Putschew [22], Michael Radke [25], Muhammad Raza [26,27], Joakim Riml [21], Anne Robertson [23], Cyrus Rutere [16], Jonas L. Schaper [1,22], Mario Schirmer [5], Hanna Schulz [1,2], Margaret Shanafield [4], Tanu Singh [9], Adam S. Ward [14], Philipp Wolke [1,28], Anders Wörman [21] and Liwen Wu [1,2]

1   Department Ecohydrology, Leibniz-Institute of Freshwater Ecology and Inland Fisheries, 12587 Berlin, Germany; jgaona@obsebre.es (J.G.G.); jason.austin.galloway@student.hu-berlin.de (J.G.); anja.hoehne@posteo.de (A.H.); anna.jaeger@igb-berlin.de (A.J.); karin.meinikmann@julius-kuehn.de (K.M.); b.mueller@igb-berlin.de (B.M.M.); schaper@igb-berlin.de (J.L.S.); h.schulz@igb-berlin.de (H.S.); wolke@igb-berlin.de (P.W.); liwen.wu@igb-berlin.de (L.W.)

2   Geography Department, Humboldt University of Berlin, 12489 Berlin, Germany

3   Zuckerberg Institute for Water Research, The Jacob Blaustein Institutes for Desert Research, Ben-Gurion University of the Negev, Midreshet Ben-Gurion 84990, Israel; sarnon@bgu.ac.il

4   National Centre for Groundwater Research and Training (NCGRT), College of Science & Engineering, Flinders University, Adelaide, SA 5001, Australia; eddie.banks@flinders.edu.au (E.B.); okke.batelaan@flinders.edu.au (O.B.); margaret.shanafield@flinders.edu.au (M.S.)

5   Eawag, Swiss Federal Institute of Aquatic Science and Technology, 8600 Dübendorf, Switzerland; andrea.betterle@eawag.ch (A.B.); juliane.hollender@eawag.ch (J.H.); andrea.popp@eawag.ch (A.L.P.); mario.schirmer@eawag.ch (M.S.)

6   AAWA-Autorità di distretto idrografico delle Alpi Orientali, 38122 Trento, Italy

7   Chair of Water Resources Management and Modeling of Hydrosystems, Technische Universität Berlin, 10623 Berlin, Germany; tabea.broecker@uwi.tu-berlin.de (T.B.); reinhard.hinkelmann@wahyd.tu-berlin.de (R.H.)

8   Department of Environmental Science and Analytical Chemistry (ACES), Stockholm University, 11418 Stockholm, Sweden; claudia.coll@aces.su.se (C.C.); malte.posselt@aces.su.se (M.P.)

9   School of Geography, Earth and Environmental Sciences, University of Birmingham, Birmingham B15 2TT, UK; j.drummond@bham.ac.uk (J.D.D.); s.krause@bham.ac.uk (S.K.); txs523@bham.ac.uk (T.S.)

10   Biology, Chemistry and Pharmacy Department, Free University Berlin, 14195 Berlin, Germany

11   Civil and Environmental Engineer Department, University of Trento, 38123 Trento, Italy

12   Department of Civil and Environmental Engineering, Vanderbilt University, Nashville, TN 37205, USA; jesus.gomezvelez@vanderbilt.edu

13   School of Water, Energy and Environment, Cranfield University, Cranfield MK43 0AL, UK; r.c.grabowski@cranfield.ac.uk (R.C.G.); chiara.magliozzi@isti.cnr.it (C.M.)

14   O'Neill School of Public and Environmental Affairs, Indiana University, Bloomington, IN 47405, USA; skuyherz@iu.edu (S.P.H.); adamward@indiana.edu (A.S.W.)

15   School of Earth Sciences, University of Western Australia, Crawley, Western Australia 6009, Australia

16   Department of Ecological Microbiology, University of Bayreuth, 95440 Bayreuth, Germany; marcus.horn@uni-bayreuth.de (M.A.H.); cyrusrnjeru@gmail.com (C.R.)

17   Institute of Microbiology, Leibniz University of Hannover, 30419 Hannover, Germany

18   Naturalea Conservació, SL, 08211 Castellar del Vallès, Spain; adrianlochner@naturalea.eu

19   Istituto di Scienza e Tecnologie dell'Informazione (ISTI) National Research Council (CNR), Area della Ricerca CNR di Pisa, 56124 Pisa, Italy

20   Julius Kühn-Institute, Institute for Ecological Chemistry, Plant Analysis and Stored Product Protection, 14195 Berlin, Germany

21 Department of Sustainable Development, Environmental Science and Engineering, KTH Royal Institute of Technology, 10044 Stockholm, Sweden; mojarrad@kth.se (B.B.M.); riml@kth.se (J.R.); worman@kth.se (A.W.)
22 Department of Environmental Science and Technology, Chair Water Quality Engineering, Technische Universität Berlin, Straße des 17. Juni 135, 10623 Berlin, Germany; anke.putschew@tu-berlin.de
23 Department of Life Sciences, University of Roehampton, London SW15 4JD, UK; nacho.peralta@roehampton.ac.uk (I.P.-M.); a.robertson@roehampton.ac.uk (A.R.)
24 Department of Environmental Systems Science, ETH Zurich, 8000 Zurich, Switzerland
25 Institute for Hygiene and Environment, Free and Hanseatic City of Hamburg, 20539 Hamburg, Germany; michael.radke@hu.hamburg.de
26 IWW Water Centre, 45476 Mülheim an der Ruhr, Germany; raza@geo.tu-darmstadt.de
27 Institute of Applied Geosciences, Technische Universität Darmstadt, 64287 Darmstadt, Germany
28 Department of Earth Science, Free University Berlin, 12249 Berlin, Germany
* Correspondence: lewe@igb-berlin.de; Tel.: +49-30-64181-668

Received: 29 September 2019; Accepted: 21 October 2019; Published: 25 October 2019

Abstract: Rivers are important ecosystems under continuous anthropogenic stresses. The hyporheic zone is a ubiquitous, reactive interface between the main channel and its surrounding sediments along the river network. We elaborate on the main physical, biological, and biogeochemical drivers and processes within the hyporheic zone that have been studied by multiple scientific disciplines for almost half a century. These previous efforts have shown that the hyporheic zone is a modulator for most metabolic stream processes and serves as a refuge and habitat for a diverse range of aquatic organisms. It also exerts a major control on river water quality by increasing the contact time with reactive environments, which in turn results in retention and transformation of nutrients, trace organic compounds, fine suspended particles, and microplastics, among others. The paper showcases the critical importance of hyporheic zones, both from a scientific and an applied perspective, and their role in ecosystem services to answer the question of the manuscript title. It identifies major research gaps in our understanding of hyporheic processes. In conclusion, we highlight the potential of hyporheic restoration to efficiently manage and reactivate ecosystem functions and services in river corridors.

Keywords: hyporheic zone; hyporheic exchange flow; surface water–groundwater exchange; ecosystem services; nutrient turnover; refuge; hyporheos; removal of trace organic compounds; emerging pollutants; self-purification capacity

---

1. Introduction

The "hyporheic zone" (HZ) is a unique habitat that is located at the interface of surface water and groundwater within river corridors. While the term hyporheic zone is sometimes used as a synonym for the streambed, it is more accurately the zone in which surface water and groundwater mix. The HZ is an interfacial zone important to many key stream processes and organisms. Because of the large surface area of sediment grains within the streambed and the high activity of microbes living in the HZ, it plays a key role as a reactive zone, transforming pollutants and natural solutes, as well as providing a habitat for benthic communities [1].

The term hyporheic zone was originally proposed by Orghidan in 1955 in Romanian, who described this interface as a discrete streambed compartment hosting a distinctive community [2]. Today, HZ research encompasses fields such as ecology, hydrology, hydrogeology, microbiology, geomorphology, biogeochemistry, environmental engineering, and conservation [3]. Therefore, a general definition and delineation of the HZ covering all disciplines is extraordinarily challenging [4]. Definitions of the HZ differ between disciplines, and sometimes even within the same discipline [5,6]. In ecology, it is generally assumed that the HZ is located just below the surface layer of the streambed (also known as the benthic zone) and that its thickness typically oscillates in the centimeter range. In hydrology, and especially in modelling studies, the HZ is defined as the zone that contains all the flowpaths that begin and end at the sediment–water interface, whereas in biogeochemistry, it is defined as a zone where surface water and groundwater mix and where at least a certain percentage (e.g., 10%) of surface water is present [5]. The depth to which the HZ extends can vary over time because fluctuations of surface water level, surface water flow velocity, groundwater table level, and water temperature impact subsurface flow paths. In contrast to the lower boundary, the upper boundary of the HZ is clearly determined by the sediment surface. A comprehensive discussion and comparison of these definitions can be found in Gooseff [5], Gomez-Velez et al. [7], and Ward [6]. Here, we use the definition that the HZ comprises (1) saturated, porous streambed sediments (2) with a characteristic hyporheic community, either with (3) flowpaths originating from and returning to surface water or (4) a mixture of groundwater and at least 10% of surface water, and (5) with hyporheic residence times on time scales relevant for the processes of interest [6]. The flow of water into, in, out, or across the HZ is termed hyporheic exchange flow, or equally hyporheic exchange flux, both abbreviated as HEF [8]. In our definition, HEF is a specific type of surface water–groundwater exchange, but the terms HEF and surface water–groundwater exchange are not interchangeable. While some authors have used HEF to describe the general exchange between surface water and groundwater [9,10], a HZ may not always exist. For example, in river sections with strong up- or downwelling flow, the HZ could be minimized or vanish, but there would still be fluxes within the saturated, porous streambed sediments [11]. Thus, surface water–groundwater exchange is a broad term describing exchange between the aquifer and river, while HEF is a specific exchange under the prerequisite that a HZ is present.

Since 1955, there has been a steady increase in HZ research and several key papers have been published. For example, Brunke and Gonser [12] reviewed the ecological significance of exchange processes between groundwater and rivers and discussed human impacts and alterations of natural exchange processes, such as reduced connectivity due to colmation by fine particle loads or organic and toxic contamination of surface water. Hancock [13] also reviewed human impacts on HZs and their ecosystem services and suggested that the HZ should be considered in river management. Boulton et al. [14] focused more on transport processes and biogeochemical turnover in the HZ itself. Their review includes the relevant mechanisms, the fate of major chemical compounds, and involved organisms. Fischer et al. [15] investigated hyporheic processes from a microbial perspective and highlighted the importance of the activity and composition of the microbial communities for biochemical reactions in the HZ. Due to its significance for carbon and nitrogen cycling, they called the HZ "the river's liver". Krause et al. [16] published a review of HZ functions and discussed how to advance HZ process understanding across disciplinary boundaries. This was further elaborated by Krause et al. [17], who discussed the high biogeochemical activity of the HZ. The review by Boano et al. [1], focusing on modelling water, heat, and dissolved and sediment transport processes, directed research towards the scale and magnitude of HZ fluxes [18], while Magliozzi et al. [11] summarized the five main drivers (i.e., hydrological, topographical, hydrogeological, ecological, and anthropogenic) at catchment, valley, and reach scales that control spatial and temporal HEF variations.

Ward [6] stated that our understanding of coupled, interacting hyporheic processes is still quite limited and that there is an urgent need for cross-site comparisons that consider hydrological, ecological, and biogeochemical processes. Recent research has deepened our understanding of the ecological importance of the HZ and the response of communities to hydrological extremes. For example, Stubbington [19] and Dole-Olivier [20] discuss HZ function as a potential refuge for benthic invertebrates, especially during floods, low flows, and drying events. Other authors have investigated interactions between ecology and chemical processes; for example, Peralta-Maraver et al. [21] focused on the hierarchical interplay of hydrology, community ecology, and fate of nutrients, as well as pollutants in the HZ. Methodological advances have increased the precision and resolution of measurements of HEF, which has been essential for biogeochemical research. For example, Anderson [22], Rau et al. [23], and Ren et al. [24] reviewed the use of heat as a tracer to study HEF. Knapp et al. [25] outlined the application of the "smart" tracer system resazurin–resorufin to study HEF and biogeochemical turnover in the HZ, as described in more detail below. Kalbus et al. [26] and Brunner et al. [27] gave an overview of the manifold measurement and modelling techniques for HEF processes. Further research has investigated biogeochemical processing in the HZ. While earlier studies focused on nutrients [28,29] and mining-derived pollutants, such as metals [30], recent papers have begun to investigate emerging pollutants, such as microplastics [31], pesticides [32], organic stormwater contaminants [33], and pharmaceuticals [34,35]. Even though there has been so much HZ research published in recent decades, it is unclear whether the hyporheic zone is of any relevance beyond the scientific community. This will be addressed with the present manuscript.

Despite the advancements in our scientific understanding of the HZ, further work is needed to link hydrological, ecological, and biogeochemical processes to develop a conceptual framework of the HZ and its associated ecosystem services [36,37]. Such ecosystem services are defined as "the benefits people obtain from ecosystems" by the Millennium Ecosystem Assessment [38], which also categorizes the services according to four main aspects: provisioning (e.g., food), regulating (e.g., water quality), supporting (e.g., nutrient cycling), and cultural services (e.g., recreation). From the history of HZ research, it is clear that the HZ provides ecosystem services, for example by supporting fish spawning and by serving as a "bioreactor", improving water quality. Ecosystem services provided by HZs are one option to show the relevance of the HZ beyond the research community. The ecosystem services framework has been criticized as overly anthropocentric and reductionist in its consideration of nature in purely monetary terms. Despite these shortcomings, we use ecosystem services provided by the HZ to illustrate the relevance of the HZ in a broader context, but avoid any monetary quantification in this review.

The aims of the present paper are (i) to provide a brief overview of recent developments in HZ research, (ii) to identify major research and knowledge gaps, and (iii) to show the relevance of the HZ beyond research focusing on HZ processes. Therefore, we identify ecosystem services provided by HZs and discuss the HZ's impact on the adjacent compartments, as well as on entire ecosystems.

2. Hyporheic Zone Drivers and Processes

2.1. Physical Drivers of Hyporheic Exchange Flows

HEFs (Figure 1) are driven by pressure gradients created by local streambed topographic variations and modulated by subsurface sediment architecture, combined with large-scale geomorphological and hydrogeological characteristics of the river network and adjacent aquifer systems, which can critically impact the spatial variability of HEF patterns [9,39–41]. At the scale of the HZ, very small water level fluctuations drive changes in the hydraulic gradients across streambed bedform structures.

The hydrogeology (i.e., the location, hydraulic conductivity, recharge, and discharge zones of local to regional aquifers) governs the overall spatio-temporal fluxes of surface water–groundwater exchange, and therefore the general gaining or losing character of rivers and river sections [42]. As discussed in the introduction, the HZ is the zone where flowpaths originate from and return to surface water. This zone may be compressed or absent in gaining and losing sections of streams [8,43], because gaining and losing flows that do not originate from the stream or terminate in the stream, respectively, are considered surface water–groundwater exchange flows, but not HEFs. Of course, gain or loss flows can still be highly relevant from an ecohydrological perspective. The absolute gaining or losing of water from streams is a dynamic feature which can vary spatially and temporally. The fragmentation of coherent gaining or losing zones at the streambed interface strongly depends on the regional groundwater contribution [41]. In gaining streams, local regional groundwater systems discharge groundwater through the HZ into surface waters. This is common in humid climates, where rivers drain groundwater systems. However, in (semi-)arid climates, losing rivers are predominant (i.e., groundwater pressures below streams are often lower than the stream water pressure, resulting in infiltration of stream water) [44].

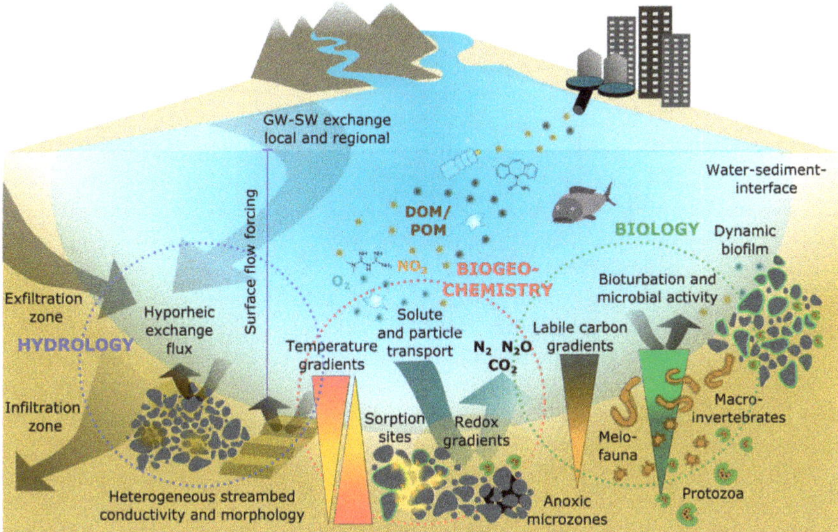

Figure 1. Conceptual model of the major hyporheic zone drivers and processes, as discussed in Section 2 of the present review. Dashed circles indicate the separation of disciplines in current hyporheic research, despite the high system complexity and manifold interconnections of hyporheic processes. GW-SW exchange is groundwater-surface water exchange; DOM and POM are dissolved or particulate organic matter, respectively.

HEFs can also be induced by hydrodynamic pressure gradients along the stream bed arising from the flowing surface water [45,46]. On a rugged streamed surface, the surface water flow field produces a heterogeneous pressure distribution acting on the exchange between surface and subsurface water. In this setting, the streamed morphology and the overlying flow field control the spatial patterns of pressure gradients and boundary shear stress along the surface–subsurface water interface. The pressure distribution on the streamed surface significantly differs between ripples and dunes. Therefore, determining the stream water level (reflecting the hydrostatic pressure), as well as the hydrodynamic pressure arising from inertia effects along the streamed surface, is crucial for HEF investigations.

At large scales, the stream water surface follows the streamed topography very closely, but as scales decrease the water surface tends to be smoother in comparison to the bed surface. This impedes direct use of the streamed topography to estimate the hydrostatic head, especially at small spatial scales, since very fine topographic features are often not reflected as similar features at the water surface. Recent investigations have suggested a scale (wavelength)-dependent ratio between stream water and bed surface fluctuations, thus providing a way to estimate the hydrostatic head distribution based on the streamed topography [40,41]. Pressure gradients might also be caused by flow of stream water over and around obstacles in the water body, such as woody debris [47–49]. Hydrodynamic pressure gradients and turbulent momentum transfer into the streamed sediments can control HEFs that are generally characterized by surficial flowpaths and short residence times. In particular, hyporheic exchange triggered by turbulent momentum transfer and by shear stresses at the sediment–water interface can be relevant, especially in the case of permeable sediments with large grain sizes [50,51]. This is due to the fact that increasing sediment permeability results in higher HEFs. Moreover, a larger grain size increases shear stress at the water–sediment interface and turbulence intensity of the boundary layer, which control the HEFs.

Other processes that can influence pressure gradients include the subtle differences in (hydrostatic) water levels across in-stream geomorphological features (e.g., HEF through a gravel island in the stream channel due to different water levels around the island). Another example is intra-meander groundwater flow, commonly triggered along tortuous rivers by longitudinal gradients of river stage [52]. Moreover, flow through streambed riffles is commonly driven by changing water levels along pool–riffle sequences of the stream. Besides pressure gradients, hydraulic conductivity of the sediment determines the intensity of HEFs, and the spatial heterogeneity of hydraulic conductivity might result in uneven distributions of HEF patterns [53–55]. In fact, according to the description of flow in saturated porous media (i.e., Darcy's law), HEFs tend to increase linearly with the hydraulic conductivity of the sediments, which, in turn, can span several orders of magnitude. As a result, given the short autocorrelation length of the hydraulic conductivity field in many environmental contexts, HEFs can display a lot of spatial heterogeneity, even at short scales [56,57].

Numerous well-known methods for measuring surface water–groundwater exchange exist [26,27,58]. Measurement techniques based on seepage meters, mini-piezometers, and thermal sensors have strongly improved our capacity to estimate HEFs. Vertical and longitudinal surveys of radon-222 (naturally occurring radioactive gas tracer with a half-life of 3.8 days) can also be used to quantify HEF along stream reaches [59,60]. Tracking surface water–groundwater interactions using temperature as a tracer is a particularly advantageous approach because it is easy to measure, the costs of temperature sensors are relatively low, and natural temperature differences at interfaces are common. Moreover, temperature depth profiles in the HZ are used to calculate HEFs [61–65]. Active heat pulse sensors have been developed to determine dynamic 3D flow fields in the near subsurface and quantify HEFs [66–68]. These overcome the limiting factors of streambed heat tracer studies, which use vertical, ambient temperature profiles and a 1D analytical solution of the heat diffusion–advection equation [22], ignoring horizontal flow components. Among the multiple thermal techniques, fiber-optic distributed temperature sensing (FO-DTS), pioneered by Selker et al. [69], Lowry et al. [70], and Tyler et al. [71], enables the spatio-temporal identification of the patterns of surface water–groundwater interactions. Subsequent research has showed the potential of this technique for estimating vertical exchanges by measuring thermal profiles with a higher depth resolution [72], or obtaining spatial flux patterns within the shallow streambed at relatively high resolution [73,74]. Combining FO-DTS with other techniques, such as thermal infrared reflectometry (TIR) [75] or geophysics [74,76,77] can provide insights into the impacts that the hydrogeological characteristics of the subsurface material can have on defining exchange patterns (i.e., to inform about the existence of different components of groundwater-surface water interactions, such as groundwater discharge, interflow, and local downwelling, depending on the hydrogeology) [74].

The physical drivers of HEF (channel slope, bedform geometry, flowpath length, sediment properties, and hydraulic head) and the transient hydraulic forcing control the residence times of water in the HZ, and thus, the development of potential hot spots for biogeochemical reactions [7,78,79]. For example, fine organic matter in the hyporheic zone reduces the hydraulic conductivity of the sediment, and thus increases hyporheic residence times. Consequently, there is a longer reaction time in the HZ and the redox pattern along flow paths will change. Under similar driving forces, HZs develop more easily under higher aspect ratios (the ratio between bedform height and width) and steeper channel slopes. This is important from an ecological perspective, as it is not only the amount of HEF but also the length and depth of the flowpaths in the HZ and the water residence time that play critical roles in modulating biogeochemical processes in the HZ [80].

On the reach scale, HEFs are often modelled using transient storage models [1]. This approach is subject to a variety of assumptions. For instance, when using transient storage models, it is commonly assumed that transient storage in the investigated river reach is primarily caused by HEFs [35,81]. This assumption only holds true if morphological features that cause surface water transient storage—such as pool–riffle sequences or side pools, where the flow velocity is orders of magnitudes lower than the advective surface water velocity—are largely absent in the investigated stream reach [82]. Furthermore, the most commonly used transient storage models predefine a characteristic shape for the residence time distribution (RTD) in the HZ (i.e., parametric transient storage models). In the simplest case, the transient storage zone is described as a well-mixed compartment that mathematically transfers to an exponentially shaped RTD [83]. Further research has subsequently shown that in some streams, log-normal [84] and truncated power law [85] parameterization of residence times in the transient storage zone fit measured breakthrough curves more accurately, particularly in regard to longer residence times. However, assuming a characteristic shape of the RTD might be the biggest shortcoming of all parametric transient storage models, as the RTD, in fact, integrates over all stream processes, and thus, should directly be used to infer insights on physical, chemical, and biological processes [86–88]. Therefore, the flexibility of a shape-free or non-parametric deconvolution approach can also capture non-traditional features of measured BTCs, such as multiple peaks [88–90]. However, having higher flexibility consequently results in a higher degree of freedom, which makes the optimization procedure challenging when estimating hydrological model parameters [88]. Nevertheless, only the shape-free approach allows an accurate estimation of the hyporheic residence time, and likewise, its significance on water quality in riverine systems.

2.2. Hyporheic Assemblages and Biological Processes

The high surface area of the sedimentary matrix in the HZ is an important habitat for a wide range of organisms (Figure 1). A diverse assemblage of biofilms grow attached to sediment grains and cover the cavities of the pore space [91]. Furthermore, protists, meiofauna, and macroinvertebrates (see Table 1 for definition) occupy the interstitial spaces among sediment particles in the HZ, swimming in the pore space or digging into the sediment. Hyporheic assemblages (the "hyporheos") [92] play a critical role in the ecological functioning of the HZ. Hyporheic biofilms are mainly composed of diverse consortia of archaea and bacteria embedded in a matrix of extracellular polymeric substances, including polysaccharides [91]. Consequently, hyporheic biofilms present a high diversity of operational units [93] and metabolic capabilities [94], and are hot spots of enzymatic activity [95]. Hyporheic biofilms have the ability to degrade and even to consume a broad range of dissolved compounds (including nutrients, pollutants, and trace organic compounds) [91], supporting the water purifying capacity of the stream (as a hyporheic bioreactor, discussed below). Thus, hyporheic biofilms are considered crucial components of the global biogeochemical fluxes of carbon, nitrogen, and phosphorous [96]. Biofilms degrade large quantities of organic matter, releasing carbon dioxide to the atmosphere [1], and also denitrify nitrate, emitting nitrous oxide and nitrogen gas to the atmosphere [97]. The type, relative abundance, and distribution and interactions of microorganisms result in dynamic microbial communities inhabiting both nutrient-rich, reducing HZ sediments, as well as nutrient-poor, oxidizing HZ sediments [98]. Protists, meiofauna, and invertebrates inhabiting the streambed sediments also come into play, boosting biofilm activity by grazing and bioturbating the hyporheic sediments [21].

Table 1. Definition of different groups inhabiting the streambed. Note these are paraphyletic groups.

Group	Definition
Biofilms	Unicellular consortia of prokaryotes (archaea and bacteria), fungi, and algae (in the top sediment layers) embedded in a porous extracellular matrix.
Protozoa	Eukaryotic single cell free-living organisms such as flagellates, ciliates, and amoeba.
Meiofauna	Eumetazoa invertebrates whose body size generally ranges between 0.45 µm and 500 µm.
Macroinvertebrates	Eumetazoa invertebrates whose body size is generally greater than 500 µm.

The HZ is not only colonized by the hyporheos, but also by organisms from adjacent environments, such as stygobites from groundwater or excavating benthic biota [99]. In addition, the HZ may act as a refuge for benthic organisms escaping from a variety of perturbations and the pressures of biotic interactions [100] (Section 3.4). Therefore, categorically stating that the hyporheos forms a discrete community (as an ecological entity) could be ambiguous and imprecise. Indeed, ecological communities, such as the hyporheos, are often inadequately described and quantified in ecological research [101]. However, Peralta-Maraver et al. [4] recently demonstrated that the hyporheos can be clearly distinguished from benthos as a discrete community with ecological integrity. Moreover, Peralta-Maraver et al. [4] also showed that the demarcation between both communities and the extension of their biotopes were quite dynamic as a result of the vertical hydrodynamic conditions and time. Vertical variability, direction, and magnitude of surface water–groundwater exchange, as well as heat and solute gradients, control the ecology in the HZ and in the benthic zone [14,102–107]. The concentration of solutes such as oxygen plays an important role in the ecology of the hyporheic zone (e.g., downwelling zones typically have higher concentrations of organic matter and oxygen, and possess a greater diversity and abundance of the hyporheos) [4]. These results reflect the ecotonal nature of the HZ [14] and also draw attention to the importance of ecological processes and services that the hyporheos sustains. To this end, hyporheic conditions favor the occurrence in the hyporheos of specific functional traits (e.g., body size and form) that reflect adaptation to the surrounding environment and that are dissimilar to those present in the benthos [108]. The consequences of trait diversification in the HZ extend beyond adaptive species shifts to exploit productive habitats, and suggest benefits to river restoration by enhancing functional interactions among different ecological niches [108].

Defining natural system boundaries is a critical aspect when assessing ecological processes and services [109,110]. Peralta-Maraver et al. [111] demonstrated that the streambed compartment and the biological features of the benthos and hyporheos drive the rate of coarse organic carbon (litter) decomposition, and thus play a crucial role in the wider functioning of the streambed ecosystem as a bioreactor. In the benthos, the biomass of metazoa was the primary predictor of decomposition, whereas in the hyporheic zone the protozoa were the strongest predictors [111]. Previous studies [112,113] showed that total mineralization of allochthonous organic carbon was actually higher in the benthic zone, while the HZ fulfils the role of an allochthonous organic carbon sink. The whole community of both compartments from biofilms to macroinvertebrates plays a vital and distinctive role in the ecological functioning of the hyporheic bioreactor. More importantly, the reduction in the supply of resources (e.g., oxygen) with depth and under upwelling conditions exerts a greater selective pressure on large body size classes [4]. As a result, streambed assemblages become more size-structured as environmental constraints increase, resulting in a reduction of the metabolic capacity of the hyporheic bioreactor [114].

2.3. Biogeochemical Processes

Most stream metabolic processes (including nutrient turnover, degradation of contaminants, removal of trace organic compounds, and other redox-related processes) occur not in the overlying water body but in the HZ (and the benthic zone), predominantly due to the presence of diverse microbial biofilms [25]. That is why they are called biogeochemical processes and not simply chemical or geochemical processes. Surface water that enters the HZ drives oxygen, nutrients, and other chemical compounds into the HZ, which in turn drives mineralization of organic matter along the flowpath through the HZ (Figure 1). Consequently, a redox zonation develops in the HZ. First, oxygen is consumed, and once oxygen is depleted nitrate is consumed, allowing for nitrate removal in the presence of non-limiting amounts of electron donors. This is followed by reduction and dissolution of iron(oxy)hydroxides. Subsequently, sulfate is reduced to sulfide and finally methane is produced. Archaea-driven methanogenesis is a widely spread anaerobic respiration mechanism in the HZ [115]. In a study by Jones et al. [116], methanogenesis accounted for all the respiration in anoxic sediments and up to 0.6% in oxic sediments. Paired with observations of denitrification in bulk-oxic sediments [80,117,118], these studies suggest that anoxic microzones with redox potentials below sulphate respiration occur, even in oxygenated sediments.

As discussed, bacterial (and archea) consortia mainly dominate hyporheic biofilms. This results in the coexistence of diverse operational taxonomic units and metabolic capabilities [91], such as chemolithotropy, where bacteria obtain energy from the oxidation of sulfide, sulfur, metal, ammonium, or nitrite, among others, to fuel their metabolism. For example, nitrification, in which ammonium is converted by bacteria into nitrate, is a chemolithotrophic process occurring in the HZ. Nitrification has a major impact on the predominant form and abundance of nitrogen found in different parts of the HZ [119]. Moreover, even in cases of low nitrifier abundance and productivity, nitrification can significantly impact oxygen dynamics, accounting for up to 50% of the biological oxygen demand in the HZ [98]. In anoxic zones, denitrification is an important process where anaerobic respiration using nitrate occurs. The nitrate may be supplied by infiltrating water, but in most cases coupled nitrification–denitrification reactions occur as water enters anoxic microzones within the aerobic HZ [120]. The contribution of denitrification to total organic matter decomposition in streambed sediments has been reported to account for up to 50% of total carbon mineralization [121], although denitrification rates can vary dramatically between sites and even at different locations within a single stream cross-section due to sediment composition heterogeneity [81]. Denitrification rates are higher under higher availability of nitrate and dissolved organic carbon (DOC), higher hyporheic exchange rates determining substrate transport and oxygen concentrations, greater abundance of denitrifying microbes, the occurrence of anoxic microzones, as well as higher surface area of granular material [81]. Bacteria also use Fe^{2+} and Mn^{2+}, H_2, or reduced sulfur compounds as electron donors under anoxic conditions [98]. HZs receiving reduced groundwater are, therefore, predictably colonized by iron-oxidizing bacteria, since Fe^{2+} is a major groundwater constituent. In most natural waters, Mn^{2+} occurs in low concentrations and manganese oxidizers are present in low abundance. Moreover, since iron and manganese oxidation yield low energy and their oxidation is predominantly performed by heterotrophic bacteria, these ions do not contribute considerably to hyporheic productivity, except under extremely low organic matter concentrations and high metal concentrations [98].

Redox gradients in HZs can be quite steep at millimeter or even submillimeter scales, and are primarily controlled by microbial activity. Other chemical compounds are released or degraded depending on the redox zonation. For example, the pharmaceutical gabapentin is degraded under oxic and suboxic redox conditions (i.e., when nitrate is still present) but is stable under anoxic conditions [122]. In contrast, phosphate is released under anoxic conditions because it is bound to iron(oxy)hydroxides that are dissolved under reducing conditions [123]. The reactivity of the HZ is highly sensitive to the interactions with groundwater, since surface water and groundwater can differ considerably in temperature and chemical composition [124,125].

The reactivity of the HZ can be determined by measuring the turnover of individual electron donors and acceptors, such as organic carbon compounds, nitrate, oxygen, or iron redox species. Alternatively, Haggerty et al. [126,127] proposed the resazurin–resorufin system to approximate the microbial reactivity of transient storage zones. It uses the reactive tracer resazurin and its transformation product resorufin to determine metabolic activity of surface and subsurface storage zones [25]. Such a system provides a good estimate of microbial activity in a rather short period of time relative to measurements of turnover rates of individual electron donors and acceptors.

3. Relevance of the Hyporheic Zone

Due to physical, biological, and biogeochemical characteristics and processes, the HZ plays a crucial role in nutrient turnover, removal of TrOCs, and particle retention in streams (Figure 2). Furthermore, hyporheic sediments constitute an important habitat and refuge for aquatic organisms, as well as a reservoir of biodiversity. Often these functions are called ecosystem services. The classic definition of ecosystem services is based on a rather anthropocentric view of ecosystems as resources for society; we, however, also want to emphasize the use of these services for the ecosystem itself and for organisms depending on the ecosystem. Furthermore, the concept of ecosystem service has been introduced to make the services monetarily quantifiable and comparable in our economic world. In the present manuscript, we look at and discuss the conceptual benefits of ecosystem services without attempting any monetary quantification. Thus, we look at the major ecosystem services provided by the HZ to answer the question of whether the HZ is relevant beyond the scientific community. Generally, interactions between groundwater and surface water play a fundamental role in the functioning of riverine and riparian ecosystems, and therefore underpin numerous ecosystem services. In the context of sustainable river basin management, it is crucial to understand and quantify HZ processes to understand the benefits they confer, and to restore functions and services where necessary [128]. Here, we highlight the role of the HZ in regulating and supporting ecosystem services.

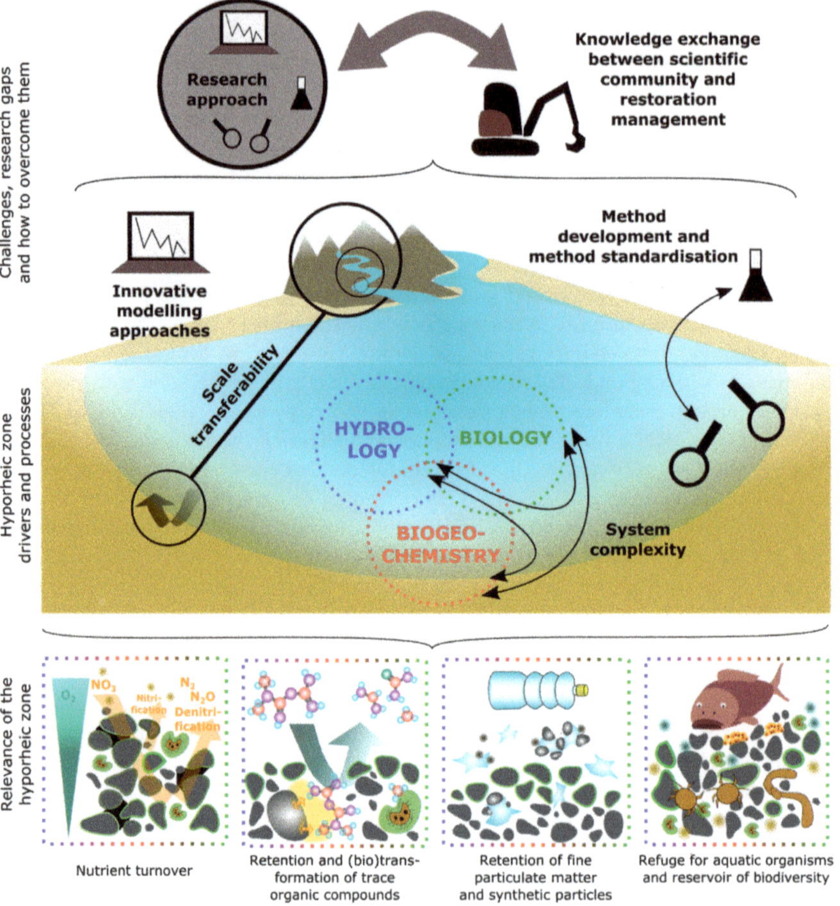

Figure 2. Graphic illustration of the relevance of the hyporheic zone based on different ecosystem services (lower row of panels in the figure; discussed in Section 3), major hyporheic research disciplines (central panel in the figure; see details in Figure 1, also discussed in Section 2), and challenges and research gaps of HZ research (shown as bold black in the figure; discussed in Section 4).

3.1. Nutrient Turnover

HZs are characterized by steep redox gradients, intense microbial diversity, and high turnover rates (Figure 1). Therefore, they are sometimes considered (hydrodynamically driven) bioreactors [21,129]. Physico-chemical controlled sorption on the large surface area of the sediment matrix removes various compounds from the pore water. Sorption reactions and filtration of particles might be followed by degradation or desorption. HZs are often hot spots of phosphorus (P), nitrogen (N), and carbon (C) turnover [80,117,119].

In other words, the HZ is both a sink and a source of nutrients. Phosphate and ammonium in the pore water in the HZ might originate from the mineralization of organic matter. Additionally, phosphate might be released by reductive dissolution of iron-bound phosphorus or weathering of bedrock [123]. Phosphate can only be released if it was previously uptaken, sorbed, or imported into the HZ as particulate matter, either due to gravity or filtration by HEFs [123]. Lapworth and Goody [130] found for a chalk stream that colloidal and particulate matter regulating bioavailable forms of P might also be formed in the HZ, perhaps through co-precipitation with $CaCO_3$. Depending on the biogeochemical milieu and the sediment composition, HZs might be sinks for dissolved phosphate. For example, Butturini and Sabater [131] measured the nutrient retention efficiency in the HZ of a sandy Mediterranean stream, finding an uptake length of 3.3 cm for ammonium and 37 cm for phosphate.

Denitrification and nitrification are mainly driven by nitrate, dissolved organic carbon, and oxygen concentrations, which is why the HZ has a high potential to regulate the fate of nitrate in streams. While inorganic nitrogen can be removed in the HZ by various processes, such as sorption or assimilation, the microbial process of denitrification is the major removal mechanism. At the same time, formation of nitrate by oxidation of ammonium (nitrification) can counteract the effect of denitrification. Nitrification particularly occurs in the surficial sections of the benthic biofilm [132]. Zarnetske et al. [80] investigated, in a gravel-bar-inducing HEF, the transition of nitrification to denitrification for a range of residence times. While short residence times at the head of the gravel bar caused net nitrification, longer residence times of more than 6.9 h, in this case, led to net denitrification, particularly at its tail. Formation of N_2 across all residence times showed that denitrification likely occurred in oxygen-reduced microsites, even where nitrification was the predominant process. The study demonstrated that whether the HZ is a net sink or a net source of nitrate depends to a large extent on the distribution of residence times. In addition, the quality of dissolved organic carbon influences denitrification [117]. Addition of acetate led to an increased denitrification rate in the HZ, showing that the nitrate removal process is limited by transport of labile dissolved organic carbon to the HZ.

Current river management measures focus not only on restoring flora and fauna but also aim to improve N removal by increasing hyporheic connectivity and optimizing hyporheic residence times [133]. To maximize the reaction yield, hyporheic water must interact with reactive sediments and biofilms for a period comparable with the relevant reaction timescale(s). Damköhler numbers can provide a quantitative estimate of the effectiveness of reactions in the HZ [81,134]. For first order approximations of attenuation rates, the hyporheic Damköhler number (Da_{HZ}) is the product of the hydraulic retention time and the reaction rate coefficient. A Da_{HZ} of 1 means that the amount of water treated by the HZ and the completeness of such treatment are balanced. In contrast, very small Da_{HZ} are indicative of reaction limitation, in which hyporheic retention times are much shorter compared to reaction timescales, so attenuation reactions do not have time to proceed effectively. In the opposite case of transport limitation ($Da_{HZ} >> 1$), attenuation reactions run to completion relatively early along a flowpath, meaning the remaining retention time is not beneficial in terms of water quality. The small HEF in the case of transport limitation implies that only a limited percentage of the stream water is affected by the biogeochemical reactions in the HZ.

Although the capacity to reduce nitrate loads by hyporheic restoration in individual stream reaches might be small [135,136], the cumulative nitrate removal capacity over longer reaches or stream networks can be significant under favorable environmental conditions. Morén et al. [137] showed in a nationwide simulation that small agricultural streams have high potential to reduce the terrestrial N load in Sweden, and thus, highlighted that hyporheic restoration can be seen as one action strategy of several in a spatially differentiated remediation plan [138] working towards the goals of the water framework directive. Harvey et al. [133] recently introduced the reaction significance factor and showed that intermediate levels of hyporheic connectivity, rather than the highest or lowest levels, are the most efficient ones in removing nitrogen from river networks. Moreover, it has been shown that the hydrological regime of a stream can substantially affect the reach-scale turnover of nutrients [139]. Streamflow dynamics and river morphology jointly control river stage and flow velocity, and in turn, the extent of HEFs. As a consequence, temporal variability of river flow conditions mediated by landscape, geomorphological, and climatic features at the catchment scale is of paramount importance for the temporal dynamics of biogeochemical reaction yields.

3.2. Retention and (Bio)transformation of Trace Organic Compounds

Trace organic compounds (TrOCs) of anthropogenic origin occur in very low concentrations (μg L^{-1} to ng L^{-1}) in freshwaters, but might exhibit ecotoxicological effects, such as endocrine disruption, oxidative stress, growth inhibition, or altered behavior, even in the ng L^{-1} range [140–143]. For the HZ, especially polar, and thus highly mobile and persistent or pseudo-persistent (continuous release in the environment) TrOCs, are of relevance. Examples include industrial chemicals, pesticides, washing and cleaning agents, personal care products, pharmaceuticals, artificial sweeteners, and substances leaching from facade or surface sealing. TrOCs typically enter surface waters via wastewater treatment plant effluents, since wastewater is most commonly only purified in treatment plants with respect to organic matter, phosphorus, and nitrogen [144], while many TrOCs are not, or only partially, degraded in wastewater treatment plants [145]. Furthermore, urban drainage, combined sewer overflow [146], or leaking water from sewer systems [147], onsite wastewater treatment systems [148], and landfills [149] reach the aquatic environment without undergoing human-induced purification processes. This means that a significant amount of TrOCs are discharged into rivers and streams [150]. In cities with partially closed water cycles or with water supply wells downstream of TrOC sources, TrOCs might end up in drinking water supply systems [151].

Biotransformation, sorption, dispersion, photolysis, and volatilization influence the fate and attenuation of TrOCs in aquatic systems [152] (Figure 2). All these processes are affected by the physico-chemical properties of the compounds (i.e., octanol water partitioning coefficient log Kow, functional groups, ionization), as well as by physical and biological parameters of the river and the sediment (i.e., river flow rate, hydraulic conductivity, turbidity, dissolved oxygen concentration, pH, temperature, the structure of microbial communities, the hydraulic regime, and the extent of hyporheic exchange in the river) [153]. Several studies on bank filtration for drinking water production strongly indicate that natural attenuation of TrOCs is most pronounced in the first few meters of infiltration (i.e., within the highly reactive HZ) [12,154–158], making this zone very important in contributing to this ecosystem service.

As described for nutrient retention, hyporheic connectivity, water, and solute residence times are relevant for TrOC removal as well. The specific biogeochemistry, especially redox conditions and carbon availability, along a flowpath plays a major role in the turnover efficiency and reactivity of TrOCs in the HZ. TrOC reactivity (i.e., the rate of chemical transformation of a given TrOC) along a hyporheic flowpath was found to be a function of the ambient redox conditions, the availability of biodegradable organic carbon, and the structure and diversity of the microbial community. Burke et al. [159] examined the fate of ten TrOCs in sediment cores taken from a bank filtration site under varying redox conditions caused by different temperatures. Although they observed compound-specific behavior, especially related to redox sensitivity of compounds, they generally showed that reactivity was highest under warm/oxic conditions, lower in cold/oxic conditions, and lowest in warm manganese-reducing conditions. Moreover, Schaper et al. [122] found that several TrOCs are preferentially transformed under oxic and suboxic conditions. Schaper et al. [160] investigated attenuation of 28 TrOCs along a hyporheic flowpath in an urban lowland river in situ. They observed differences in turnover rates and differences in retardation coefficients caused by reversible sorption between consecutive sections of the flowpath (0–10, 10–30, and 30–40 cm). Most compounds showed highest transformation in the first 10 cm of the flowpath, although the oxic zone reached down to 30 cm. They attributed the spatial difference in transformation within the same redox zone to the higher availability of biodegradable dissolved organic matter in the first 10 cm, which led to higher microbial activity in the shallow HZ. Not only the activity, but also the diversity and composition of TrOC-transforming bacteria communities can affect the reaction rate coefficients. In a mesocosm study investigating TrOC half-lives in recirculating flumes, three different levels of sediment bacterial diversity were compared. Higher bacterial diversity significantly increased degradation of both the artificial sweetener acesulfame and the anti-epileptic drug carbamazepine [161]. In addition, the microscopic trophic interactions between bacterial biofilms and microscopic grazers (i.e., flagellates and ciliates) in the pore space can also determine the performance of the HZ during processing of TrOCs. Moderate levels of grazing might even have a positive effect on bacterial activity due to the selective consumption of less active bacteria [162], via the predator effect on bacteria dispersal in the medium (as a consequence of swimming around and grazing). Grazing on biofilms may result in better exposure of TrOCs to potential degraders [163], predation-induced recycling of nutrients (microbial loop) [162], and the increase of the absorption surface [21]. Nevertheless, this hypothesis has not been tested yet.

In contrast to most TrOCs that are either best degraded under oxic or suboxic conditions or which are degraded independent of the redox potential, very few persistent compounds and transformation products are preferentially degraded under anoxic conditions in saturated sediments. El-Athman et al. [164–166] showed that the very persistent triiodinated benzoic acid unit of the iodinated X-ray contrast media can be deiodinated in reducing environments using cobalamines (e.g., vitamin B12 as an electron shuttle). Under aerobic conditions, the apparent removal via transformation of iodinated X-ray contrast media is just based on side chain alterations, but not on deiodination. It is expected that alternating redox conditions may enhance the degree of transformation; for example, by opening the aromatic structure of the triiodo benzoic acid derivatives, which might be possible after removal of the relatively large iodine atoms under reduced conditions followed by a further transformation under aerobic conditions.

Many TrOCs, such as pharmaceuticals from urban drainage, but also phosphorus in runoff from agriculture or heavy metals, can be removed in the HZ through physico-chemical controlled adsorption on the large inner surfaces of the sediments, often followed by other transforming or degrading reactions. The degree of retention of solutes subject to reversible sorption is directly proportional to HEFs and the solute residence time in the HZ, but also depends on the equilibrium partition coefficient of the sorption reaction [34,167]. Sorption of TrOCs is also impacted by biogeochemical conditions in the HZ. For example, pH and organic matter in the HZ affect the sorption of charged and ionizable pharmaceuticals [168]. Pharmaceutical compounds with specific functional groups, such as ibuprofen and sulfamethoxazole, have the capability to transform into anionic species with increasing pH, while becoming neutral at lower pH [169]; thus, sorption rates increase at lower pH and high organic matter content. Similar mechanisms arise by adsorption of solutes on colloidal particles that are transported in flowing water, subjected to HEF, and subsequently clogged in the pores of the sediment matrix of the HZ [170,171].

Often, biodegradation of TrOCs does not result in complete mineralization, but in a myriad set of different transformation products (TPs). It has been demonstrated that some TPs are more toxic [172] and more stable [173,174] than their parent compounds. Usually, much less ecotoxicological data exist about TPs than their parent compounds. For example, in a bench-scaled flume experiment designed to study hyporheic processes [175], TPs were detected, which were only formed in the sediment but released to the water, hinting at secondary contamination with TPs [176]. In particular, valsartan acid, a rather stable TP of the common blood pressure medication valsartan, forms in situ in the HZ [160,177] and significantly increases its concentrations in the surface water of wastewater receiving streams [178].

3.3. Retention of Fine Particulate Matter and Synthetic Particles

Fine particles in the stream water column are problematic if they occur in high concentrations or are carrying sorbed contaminants [179]. Hyporheic filtration can reduce the concentration of fine particles despite low settling velocities [180,181]. Therefore, the HZ and HEF are highly relevant for particle removal [182]. Increased sediment loads that often result from agricultural run-off, particularly fine sediments <2 mm in size, can lead to colmation (clogging) in the benthic zone and HZ. This results in reduced hydraulic connectivity and concomitant changes in the hyporheos [183–185]. The functional traits of hyporheos also change along a colmation gradient [184,186]. The ecosystem services delivered by the HZ also diminish under colmation (e.g., survival of salmonid eggs and embryos is much lower under colmated conditions) [187,188]. While colmation has a negative impact on the biological community in the HZ, the removal of particulate matter by filtration is positive from the perspective of surface water quality [180] (Figure 2).

Previous research on deposition and retention of fine suspended particles in streams focused on clay and organic matter particles [182,189]. The impacts of microplastics and nanomaterials in streams are still unclear. While research on marine microplastics is advancing rapidly, there is still a lack of data and knowledge on microplastic abundance and fate in freshwater systems. Few studies have been conducted in rivers, with most of them focusing on larger systems and urban environments [190,191]. The observed microplastic abundances in riverine water, as well as hyporheic sediment samples, were generally high, and sometimes an order of magnitude above levels reported for marine environments [192].

The major sources of microplastics in rivers are sludge from wastewater treatment plants (WWTP) that is spread over agriculture fields and illegally dumped plastic [193]. WWTP effluents contain fewer particles than sludge, but due to their continuous input into the environment they can be a significant source to streams [194,195]. Road runoff and stormwater runoff from urban areas also source microplastics [196]. Since most rivers in central Europe receive WWTP effluent, road, and stormwater runoff, it can be assumed that most, if not all, rivers and their HZs are contaminated by microplastics.

Laboratory experiments indicate that heteroaggregates with suspended solids can be formed, which is supported by modelling approaches [197,198]. Aggregation, fouling, and particle size distribution appear to affect sedimentation behavior. Several authors highlight the uneven distribution of sampled microplastics in water and sediment along river corridors [191,192,198,199]. Hotspots with considerable concentrations were commonly found in the HZ. This result coincides with a recent experimental study showing that microplastics are transported similarly to naturally occurring allochthonous particles [200], which are known to develop hotspots in river corridors. Finally, deposit feeders seem to affect microplastic transport into the HZ [201] and flood events can partially remobilize microplastics retained in river sediments [202].

3.4. Refuge for Aquatic Organisms and Reservoir of Biodiversity

The HZ is considered a refuge for aquatic organisms, especially during adverse environmental conditions, such as floods, droughts, and heat waves [19,20,203] (Figure 2). Disturbances are generally reduced in the HZ compared to the benthic zone due to its capacity for retaining water during drying periods and its greater stability during floods [99]. In addition, the HZ is a refuge for invertebrates and fish during their early stages of development due to the reduced predator pressure [92]. For example, the HZ is critical in salmonid life histories; salmonids bury their eggs in the HZ of gravel bed streams, and the developing embryos remain there until emerging as free-swimming fish some months later [204]. The use of the HZ as a refuge is not exclusive to invertebrates or fish. Biofilms, composed of consortia of bacterial strains, use hyporheic sediments as a refuge during dry events, surviving in the deeper wetted sections and recolonising the sediment matrix when interstitial pore spaces become re-filled with water [205]. This is especially relevant to the role of streambed biofilms on the functioning of the hyporheic bioreactor during pollutant breakdown and nutrient cycling.

Still, the importance of the HZ as a refuge is debated and some studies contradict this idea [99]. Nevertheless, the importance of the HZ as a refuge becomes evident in intermittent systems [21]. These systems are common all around the world, sustaining and supporting diverse communities that are well adapted to persist in the HZ during dry conditions [206]. Streambed communities in Mediterranean streams and rivers are the typical example of well-adapted organisms to natural intermittency that make use of the HZ as a refuge [207].

4. Challenges, Research Gaps, and How to Overcome Them

Despite a great amount of HZ research during the past two decades, there are still major research gaps (Figure 2) that need to be addressed to progress our understanding of HZ processes and functioning. Further research closing the gaps is necessary to improve our understanding of HZs, to protect HZs and ecosystem services provided by them, and finally to conduct adequate and efficient management measures to restore ecosystem functions and ecosystem services.

4.1. System Complexity

Our knowledge about coupled physical, chemical, and biological processes is still limited [6]. Until now, most studies have focused on 1–3 hyporheic processes, which limits our ability to characterize and understand interactions [6]. Individual studies address, for example, the hyporheos, HEFs, hyporheic biogeochemistry, the fate of pharmaceuticals in streambeds, the geomorphology of streams, subsurface hydrogeology, microplastic abundance in streams, or ecological effects of river management measures and damming. Therefore, the water–sediment interface is also an interface of different scientific disciplines. Each discipline has its own methods, definitions, rules, and standards, and often methods vary even between different research groups. In addition, most research groups work only on specific catchments, making it difficult to synthesize the results and translate them to a bigger picture. Thus, there is a need for large studies bringing together different disciplines and research groups to simultaneously investigate various hypotheses. Examples of such efforts are the joint experiments of the project HypoTRAIN [161] and the field campaigns of the project, "Where Rivers, Groundwater, and Disciplines Meet: A Hyporheic Research Network", funded by the Leverhulme Trust [49,208].

Simplifications are common and necessary in HZ research because of the complexity of this system and the involved processes. Nevertheless, much care is required to assure that common simplifications do not result in a systematic bias. For example, bedforms occurring in many streams are dynamic features that form, change shape, migrate, and erode by the force of flowing water. As long as there is sufficient flow velocity in the overlying water body, these bedforms migrate downstream. However, nearly all flume studies have investigated stagnant bedforms and their impacts on HEF. The researchers only rarely considered bed movement, even though the relevance of bed movement for hyporheic exchange has been long recognized [45,46], and can lead, for example, to overestimation of nitrate removal [209]. Another example of simplification of complexity is the representation of redox potentials in HZs. There is little empirical evidence on the spatio-temporal extent of the redox zones, how these zones change in response to dynamic hydrologic conditions, and their impacts on nutrients. Furthermore, in many streams the oxygen concentration of the surface water varies dramatically between day and night due to photosynthesis during daytime and respiration of organic matter during nighttime, as already shown in 1956 by Odum [210], as well as in studies by Mulholland et al. [211], Roberts et al. [212], and Rajwa-Kuligiewicz et al. [213]. However, impacts of fluctuating surface water oxygen concentrations on the extent of the oxic zone in the streambed have rarely been studied (although one example is Brandt et al. [214]). Fieldwork is typically only conducted during the day, potentially missing important diurnal variations in processes.

4.2. Scale Transferability

Most field investigations have scale limitations, meaning they are either very localized point measurements in heterogeneous HZs or lack sufficient resolution to draw conclusions about local conditions. Tracer tests remain the primary method for exploring HZ residence times and even nutrient cycling. However, they typically only give a spatially averaged indication of HZ residence times and other characteristics. An improved process understanding is usually difficult or impossible based on large scale investigations. Few studies combine tracer tests with high resolution sampling within the study area. For example, Zarnetske et al. [117] conducted a $\delta^{15}NO_3$ and chloride tracer test in a stream in Oregon, United States, in which they sampled detailed solute and nutrient concentrations at many locations within a gravel bar. Schaper et al. [35] combined plot- and reach-scale investigations to identify the relevance of hyporheic removal of TrOCs for their removal in the whole stream.

Furthermore, temporal dynamics are often insufficiently studied because common methods are often labor- and cost-intensive, resulting in one-time investigations. In addition, it is necessary to know spatial and random variability to ensure that observed differences are indeed due to time variance and not due to spatial or random differences. Thus, we need improvements in sensing and modelling of space–time variability of processes in the HZ. Yet, it has not been satisfactorily assessed how temporal dynamics affect ecological processes in the benthic zone and the HZ. Temporal dynamics (from daily- to seasonal-scale) affect hydrological exchange between the benthic zone and the HZ [26,215] and the organization of the streambed biota [216,217]. Hence, it is reasonable to expect that the location of the boundary between benthos and hyporheos is time-dependent. Peralta-Maraver et al. [4] discussed, in their one-month study, that the line of demarcation between benthos and hyporheos tended to be relatively persistent, although their study was not replicated through time. On the one hand, future surveys might assess the integrity of benthos and hyporheos at a daily scale by increasing the frequency of sampling times. Recalling the daily variation of the surface water level, a reasonable strategy would be repeating sample collection during the maximum and minimum water stage level. On the other hand, assessing the seasonal variation in the boundary between both communities implies repeating the same protocols during different seasons.

As described above (2.1), hydrostatic and hydrodynamic heads along uneven streambeds have long been known as drivers of hyporheic flow [1,27,45] that pose a hydromechanical (transport) limitation on nutrient biogeochemistry [39,218,219] and impact the regional groundwater discharge patterns [41]. However, there is still uncertainty connected to the importance of these drivers over a wide range of temporal and spatial scales [220,221]. A comprehensive description of the spatio-temporal variability of physical and biogeochemical exchange processes at the surface water–groundwater interface is, thus, necessary to upscale and understand the ecosystem services provided by hyporheic processes along river networks.

4.3. Research Approach

Researchers focus typically either on field investigations, mesocosm studies, batch scale experiments, or modelling. Each type of study has its advantages: Field investigations represent reality best but spatial and temporal variation of manifold environmental factors, for example the complexity of the streambed hinders individual process understanding, manipulation experiments, and generalizability. Batch-scale experiments are further from reality but are easy to conduct, allow manipulation and control of all environmental factors, and therefore are useful for investigating individual processes. Flume studies are in between field investigations and batch experiments. Some environmental factors (e.g., discharge, temperature, sunlight, precipitation) can be controlled, they are closer to reality than batch experiments, and allow the study of coupled processes, however the effort is much higher. Modelling is used to estimate non-measurable process variables, such as turnover rates and process interactions, and is usually based on measured data. Combining several types of studies at the same site and from multidisciplinary perspectives reduces the shortcomings of single methods, and thus, adds invaluable insight into processes in the HZ [161]. Boano et al. [1] suggested that detailed measurements within the HZ improve our understanding of solute transport and residence times using tracer studies. However, the logistical and economic efforts to gather in situ measurements within the HZ are relatively high, and also the implementation of such data in the model approaches themselves is challenging [222]. Therefore, novel numerical models that include both complex transport and reaction models but also appropriately conceptualize the HZ are needed to interpret those in situ measurements.

4.4. Method Development and Method Standardization

As mentioned above (4.1), methods differ between disciplines, but also between different research groups of the same scientific discipline. For a comparison of results obtained at different sites, it is necessary to measure the same basic parameters with comparable methods. For that purpose, a general method standardization and harmonization is an essential prerequisite. Thus, there is a need for research projects focusing on method development, on clearly defined protocols, as well as on intercomparison studies. In this context, it is very important to develop simple and cheap methods and easy to follow protocols so that they can be applied reliably and with manageable effort by scientists with different disciplinary backgrounds. Standardization of metadata and system characterization will allow comparison of different study sites and result in a big step forward in HZ research.

Furthermore, there is an urgent need for novel and innovative methods to improve process understanding. For example, it is particularly important to develop methods of flowpath identification to be able to sample along those flowpaths. Only when flowpath geometries are known is it possible to study biogeochemical processing of water parcels moving through the HZ [66].

Generally, microbial respiration and activity measurements are restricted to a daily temporal resolution or to methods involving the extraction of sediment cores, which makes in situ investigations on a temporal scale difficult. There is a need for in situ measurements that can assess microbial activities on a sub-daily scale to cover sub-daily fluctuations of environmental conditions and their effect on the microbial biota in hyporheic sediments. Similarly, new microbiology tools, such as high-throughput sequencing can help to characterize microbial communities in the HZ, and in particular, identify microorganisms that cannot be cultured. Further development of these methods may help to elucidate the complex microbiological interactions that regulate TrOC and nutrient turnover.

Novel methods such as highly dynamic sampling of pore water [223], time-integrated passive sampling techniques [177], and isotopic techniques are under development and will allow better insights into hyporheic processes. Peter et al. [33] injected visible dye into shallow HZ sediments in a known downwelling location and pinpointed the area where the labeled flowpath re-emerged to the stream. Piezometers and seepage meters were installed at these locations and used to collect paired influent–effluent samples and determine the water treatment occurring along the flowpath.

4.5. Innovative Modelling Approaches

Integration of theoretical advances into modelling studies is urgently required. Although the understanding of HZ processes has improved over the past decades, most numerical models still use different model concepts for groundwater and turbulent open channel flow, assuming hydrostatic pressure distributions, and couple these models [124,224–232]. Until now, only a few fully integrated surface water–groundwater flow models exist, such as HydroGeoSphere [233] or the solver porousInter by Oxtoby et al. [234]. For the porousInter solver, an extended version of the Navier–Stokes equation is applied for surface water and groundwater, including porosities as well as an additional drag term for the application within the sediment and different turbulence models [235,236]. Similar to the transformation from separate investigations to an integral consideration of processes within groundwater and surface water, we expect a change towards more integral numerical approaches for a better resolution of processes at the stream and aquifer interface. As such approaches require a very high spatial and temporal resolution, upscaling methods must be developed to come from small scales to the river reach scale. Commercial numerical software (such as Comsol Multiphysics®) are nowadays attempting to explicitly couple different physics within and across modelling domains. In surface water–groundwater interactions, the possibility of coupling, for example, thermodynamic and turbulent or laminar flows could open promising avenues to reveal the effect of feedback processes in the overall functioning of the HZ.

Classic modelling of the fate and transport of TrOC in rivers often struggles to capture the spatial and temporal variability in river conditions. Most models do not consider the specific processes in the HZ, especially at larger scales [237–239]. Similarly to hydraulic modelling, these approaches also have high data demands regarding spatial and temporal resolution. In situ monitoring data and characterization of partitioning and degradation rates are not available for the great majority of case studies. The degradation rates, in particular, are obtained from laboratory experiments or in silico (as a function of structure); the extrapolation to river conditions carries large uncertainties [240]. Recent modelling of TrOC in the Rhine River basin showed that degradation rates in small- and medium-sized streams, where the HZ is usually most important, are being underestimated [237].

The relevance of the HZ for nutrient turnover, removal of trace organic compounds, and pollutant dynamics in streams can be assessed by quantifying the relative contribution of hyporheic removal to in-stream (i.e., reach-scale removal of a reactive compound). The relative contribution is not only a function of the hyporheic reactivity (i.e., the turnover rate or reaction rate of a given compound in the HZ), but also of the physical exchange characteristics. The relative hyporheic contribution has been assessed by simultaneously quantifying hyporheic reactivity and the reach-scale relative removal of a reactive compound, as well as the characteristics of the HEF on the reach-scale, for instance via transient storage models [81,122].

For many compounds, such as pharmaceuticals and nitrate, hyporheic reactivity is, furthermore, dependent on redox conditions in the HZ. To properly assess the relative contribution of the HZ to overall in-stream removal of these compounds, it could, therefore, be more appropriate to not only quantify the overall residence time in the HZ but also to disentangle reach-scale exposure times to different redox conditions in streambed sediments [241]. The hyporheic turnover length (i.e., the distance that is required for streamflow to be entirely exchanged with the HZ) increases with river discharge [18]. Also, in higher-order streams, lateral hyporheic flowpaths (e.g., through meander bends) along which redox conditions are likely to become anoxic, gaining importance relative to short, vertical flowpaths [9]. It is, thus, reasonable to assume that the relative contribution of the HZ to in-stream compound removal decreases with river discharge, particularly for compounds that are preferably removed in oxic and suboxic sections of the HZ.

First, future research should aim to develop a more versatile model approach to account for complex stream transport processes, inform about the actual RTD, and consider that compound reactivity along a hyporheic flowpath varies as a function of depth and residence time in the transient storage zone. Natural tracers, such as radon-222 [59,60,242], could be combined with conservative tracer tests to reduce uncertainties inherent in the quantification of HEFs. In addition, smart tracers, such as the resazurin–resorufin system [25], could be used to disentangle hyporheic and surface water contributions to transient storage, and may provide the means to quantify reach-scale exposure times to oxic redox conditions in the HZ. Secondly, in order to inform river management and hyporheic restoration efforts, future studies should aim to experimentally investigate the relative contribution of the HZ to overall in-stream removal of reactive compounds in various lotic systems and river networks that differ with respect to their hydrological conditions.

4.6. Knowledge Exchange Between the Scientific Community and Restoration Practitioners

River restoration typically focuses on the preservation or creation of habitats that have been lost by human alteration to rivers and floodplains to increase ecological diversity, biomass, presence of target species, or flood retention potential. HEF and streambed biogeochemical processes are rarely considered, other than in the context of the restoration of spawning grounds for salmonids. However, there is considerable potential to use catchment, river habitat, and flow restoration techniques to improve HEF and adjust residence times by acting on the physical drivers of HEF or hydrogeological factors influencing HEF (e.g., hydraulic conductivity, channel bedforms, hydraulic head; see Section 2.1), for instance through nature-based solutions, such as large, woody debris, to promote habitat structure and nutrient spiraling [47].

At river-basin-scale, improved land management practices can reduce fine sediment generation and delivery to the river (e.g., cover crops, buffer strips), which deposits on, and ingresses into, the riverbed, clogging coarse bed sediments [182,243]. In-channel measures can be used to narrow the channel, increase river flow velocities, and induce scour (e.g., willow spilling, flow deflectors, and natural riparian vegetation growth) to flush fine sediment from the riverbed, which increases hydraulic conductivity and promotes the formation of bedforms [244].

Many stream restoration practices can reliably improve hyporheic connectivity, but do not explicitly control hyporheic residence times, with overall water quality improvements depending on whether transport timescales align with reaction rates of interest [135,245,246]. However, most restoration interventions are not targeting the development of optimal flowpath lengths or biogeochemical conditions to yield specific biogeochemical reactions of interest [247]. The efficiency of river restoration structures can likely be improved by attending to design variables that can alter HZ functioning. Example variables include controlling hydraulic gradients (e.g., the height of a step [218]), manipulating hydraulic conductivities (e.g., with sediment coarsening [248]), changing flowpath geometries (e.g., with baffle walls [54,249] or hyporheic caps and increased HZ depth [33]), or shielding a structure from groundwater upwelling or downwelling (e.g., with a liner). Future research should focus on tailoring river restoration practices to deliver specific regulating ecosystem services, while also recognizing that in-channel structures alone are not able to overcome catchment-scale degradation of these services [250]. Additionally, despite extensive research demonstrating potential effects of restoration measures on river hydrodynamics, little is known about their effects on the hyporheos. Experimental studies on large wood, commonly used in restoration design, and structure-induced HEF [251] have confirmed that there is both a taxonomic and functional effect on the local benthos and hyporheos [108,252].

Despite the potential for restoration to impact HEF, there is little evidence as yet of an improvement in river water quality with restoration [253]. There are several reasons for this: (i) the objective of most restoration projects has been habitat creation not water quality; (ii) there are numerous factors influencing water quality in rivers (e.g., diffuse pollution from urban and rural environments or treated wastewater); and (iii) water quality and ecological monitoring is infrequently conducted at the appropriate scale and sufficient duration prior to and after restoration. However, new research should support river managers and restoration practitioners in incorporating HEF in their restoration goals. For example, Magliozzi et al. [254] developed a framework to integrate existing environmental data to prioritize catchments, sub-catchments, and river reaches for HEF restoration.

Thus, as river restoration aims to address physical habitat degradation to improve biodiversity (i.e., species and ecosystem diversity), targeting biological responses of hyporheos communities would be a logical direction for a holistic approach to river functioning. These results suggest that there is an increasing emphasis on addressing the HZ into site-specific restoration design to optimize ecohydrological understanding of aquatic ecosystems and explore new methods to target retention of local priority pollutants. Finally, while hyporheic structures have primarily been considered in river restoration, there is potential to utilize HZ treatment processes in stormwater, wastewater, and agricultural contexts. For example, engineered HZ could be used in artificial or heavily modified channels, such as stormwater drainages, canals, channels that convey treated wastewater to receiving water bodies, and in irrigation return flow ditches. However, more research is needed to show whether hyporheic processing is an efficient water quality management technique and to integrate it with existing management.

5. Conclusions

Coming back to the title of the present manuscript, we conclude: Yes, the HZ can be highly relevant beyond the scientific community. Several important ecosystem services (e.g., nutrient turnover, TrOC transformation, filtering of fine particles, refuge for aquatic organisms, and reservoir of biodiversity) provided by streams are based on HZ processes and are relevant at the catchment level. Thus, HZ research can support sustainable management practices of water resources. Nevertheless, it is also clear that the restoration of hyporheic functions is only one piece in a comprehensive river management system. For example, there is a need to reduce nutrient emissions to aquifers and surface water bodies. A well-functioning HZ can help to improve water quality by a further reduction of remaining nutrient loads, but it cannot be the sole management measure compensating for high nutrient emissions. Similarly, emissions of TrOCs, such as pharmaceuticals, need to be reduced in the first place by developing easily degradable pharmaceuticals, responsible use, reduced release of the compounds, and the implementation of advanced treatment steps in wastewater treatment plants. Subsequently, the HZ may reduce the remaining TrOC loads. In this way, future exposure of aquatic organisms and humans to TrOCs and their potentially adverse effects can be avoided. For example, the widespread use of antibiotics for both human and animal treatment is attracting rising attention because of the undesirable consequences that an increased bacterial resistance of pathogens can have on life. The massive release of antibiotics in surface water bodies will likely increase the attention on self-purification processes of river networks and on the potential for HZs to reduce contaminant loads in the forthcoming years.

It is clear that even though our knowledge of the HZ has improved over the last 70 years, there are still more open questions than answers. HZ research is extraordinarily challenging because this interface is also a place where different disciplines meet. In addition, temporal fluctuations of the overlying water body and spatial variability of the underlying aquifer render HZ research extremely challenging for process understanding and upscaling. There is a need for novel methods and method standardization, joint investigations, studies that avoid systematic simplifications, and data-model integration. Furthermore, interdisciplinary approaches combining expertise obtained through large-scale field surveys with carefully designed experiments are needed to acquire a fully mechanistic understanding of the ecosystem services provided by the HZ and predict its functioning given the upcoming global change. Specifically, based on the state of the science, as described in detail in the present paper, focused research on HEF processes is still needed:

- to understand geomorphic-climatic controls that underlie spatial patterns of streamflow dynamics to quantify hydrologically critical drivers of HEF across different scales [255,256]. A proper description of the spatial variability of hydrological processes would help clarify how the ecosystem services provided by HEFs can be extended and upscaled to entire river networks.
- to enlighten the role of HEF, hyporheic sediments, and processes in cycling of microplastics, as HEF has the potential to retain large amounts of microplastics.
- to develop methods of incorporating stream restoration structures into site-specific designs that optimize retention of local priority pollutants.
- to clarify the relative contribution of the HZ to overall in-stream removal of reactive compounds in various stream systems differing with respect to their hydrological characteristics.

Author Contributions: Conceptualization, J.L., O.B., T.B., S.P.H., R.H., J.H., S.K., K.M. and B.M.M.; figures, A.J. and C.C.; writing—original draft preparation, J.L., S.A., E.B., O.B., A.B., C.C., J.D.D., J.G.G., S.P.H., A.H., J.H., A.J., A.L.P. (Adrian Löchner Prats), C.M., K.M., B.B.M., B.M.M., I.P.-M., A.P., M.P., A.L.P. (Andrea L. Popp), J.R., A.R., C.R., J.L.S., M.S. (Mario Schirmer), H.S., M.S. (Margaret Shanafield), T.S., A.S.W., P.W., A.W. and L.W.; writing—review and editing, J.L., S.A., E.B., O.B., A.B., T.B., C.C., J.D.D., J.G., J.G.-V., R.C.G., S.P.H., R.H., A.H., J.H., M.A.H., A.J., C.M., B.B.M., B.M.M., I.P.-M., A.L.P., M.P., M.R. (Michael Radke), M.R. (Muhammad Raza), J.R., A.R., C.R., J.L.S., M.S. and A.W.

Funding: This research was funded by the European Union's Horizon 2020 research and innovation program under grant agreements No. 641939 (HypoTRAIN), No. 765553 (EuroFlow), and No. 734317 (HiFreq), and by the German Research Foundation's (DFG) graduate school "Urban Water Interfaces" under grant agreement GRK 2032/1.

Acknowledgments: We thank two anonymous reviewers for their input to our manuscript.

References

1. Boano, F.; Harvey, J.W.; Marion, A.; Packman, A.I.; Revelli, R.; Ridolfi, L.; Wörman, A. Hyporheic flow and transport processes: Mechanisms, models, and biogeochemical implications. *Rev. Geophys.* **2014**, *52*, 603–679. [CrossRef]

2. Orghidan, T. A new habitat of subsurface waters: The hyporheic biotope. *Fundam. Appl. Limnol. Arch. Hydrobiol.* **2010**, *176*, 291–302. [CrossRef]

3. Dahm, C.N.; Valett, H.M.; Baxter, C.V.; Woessner, W.W. Hyporheic zones. In *Methods in Stream Ecology*, 2nd ed.; Hauer, F.R., Lamberti, G., Eds.; Academic Press: San Diego, CA, USA, 2007; pp. 119–236.

4. Peralta-Maraver, I.; Galloway, J.; Posselt, M.; Arnon, S.; Reiss, J.; Lewandowski, J.; Robertson, A.L. Environmental filtering and community delineation in the streambed ecotone. *Sci. Rep.* **2018**, *8*, 15871. [CrossRef] [PubMed]

5. Gooseff, M.N. Defining hyporheic zones–advancing our conceptual and operational definitions of where stream water and groundwater meet. *Geogr. Compass* **2010**, *4*, 945–955. [CrossRef]

6. Ward, A.S. The evolution and state of interdisciplinary hyporheic research. *Wiley Interdiscip. Rev. Water* **2016**, *3*, 83–103. [CrossRef]

7. Gomez-Velez, J.D.; Wilson, J.L.; Cardenas, M.B.; Harvey, J.W. Flow and residence times of dynamic river bank storage and sinuosity-driven hyporheic exchange. *Water Resour. Res.* **2017**, *53*, 8572–8595. [CrossRef]

8. Fox, A.; Boano, F.; Arnon, S. Impact of losing and gaining streamflow conditions on hyporheic exchange fluxes induced by dune-shaped bed forms. *Water Resour. Res.* **2014**, *50*, 1895–1907. [CrossRef]

9. Gomez-Velez, J.D.; Harvey, J.; Cardenas, M.B.; Kiel, B. Denitrification in the Mississippi River network controlled by flow through river bedforms. *Nat. Geosci.* **2015**, *8*, 941–945. [CrossRef]

10. Boulton, A.J.; Hancock, P.J. Rivers as groundwater-dependent ecosystems: A review of degrees of dependency, riverine processes and management implications. *Aust. J. Bot.* **2006**, *54*, 133–144. [CrossRef]

11. Magliozzi, C.; Grabowski, R.C.; Packman, A.I.; Krause, S. Toward a conceptual framework of hyporheic exchange across spatial scales. *Hydrol. Earth Syst. Sci.* **2018**, *22*, 6163–6185. [CrossRef]

12. Brunke, M.; Gonser, T. The ecological significance of exchange processes between rivers and groundwater. *Freshw. Biol.* **1997**, *37*, 1–33. [CrossRef]

13. Hancock, P.J. Human impacts on the stream-groundwater exchange zone. *Environ. Manag.* **2002**, *29*, 763–781. [CrossRef] [PubMed]

14. Boulton, A.J.; Findlay, S.; Marmonier, P.; Stanley, E.H.; Valett, H.M. The functional significance of the hyporheic zone in streams and rivers. *Annu. Rev. Ecol. Evol. Syst.* **1998**, *29*, 59–81. [CrossRef]

15. Fischer, H.; Kloep, F.; Wilzcek, S.; Pusch, M.T. A river's liver–microbial processes within the hyporheic zone of a large lowland river. *Biogeochemistry* **2005**, *76*, 349–371. [CrossRef]

16. Krause, S.; Hannah, D.M.; Fleckenstein, J.H.; Heppell, C.M.; Kaeser, D.; Pickup, R.; Pinay, G.; Robertson, A.L.; Wood, P.J. Inter-disciplinary perspectives on processes in the hyporheic zone. *Ecohydrology* **2011**, *4*, 481–499. [CrossRef]

17. Krause, S.; Lewandowski, J.; Grimm, N.B.; Hannah, D.M.; Pinay, G.; McDonald, K.; Marti, E.; Argerich, A.; Pfister, L.; Klaus, J.; et al. Ecohydrological interfaces as hot spots of ecosystem processes. *Water Resour. Res.* **2017**, *53*, 6359–6376. [CrossRef]

18. Cranswick, R.H.; Cook, P.G. Scales and magnitude of hyporheic, river-aquifer and bank storage exchange fluxes. *Hydrol. Process.* **2015**, *29*, 3084–3097. [CrossRef]

19. Stubbington, R. The hyporheic zone as an invertebrate refuge: A review of variability in space, time, taxa and behavior. *Mar. Freshw. Res.* **2012**, *63*, 293–311. [CrossRef]

20. Dole-Olivier, M.J. The hyporheic refuge hypothesis reconsidered: A review of hydrological aspects. *Mar. Freshw. Res.* **2011**, *62*, 1281–1302. [CrossRef]

21. Peralta-Maraver, I.; Reiss, J.; Robertson, A.L. Interplay of hydrology, community ecology and pollutant attenuation in the hyporheic zone. *Sci. Total Environ.* **2018**, *610*, 267–275. [CrossRef]

22. Anderson, M.P. Heat as a ground water tracer. *Ground Water* **2005**, *43*, 951–968. [CrossRef] [PubMed]

23. Rau, G.C.; Andersen, M.S.; McCallum, A.M.; Roshan, H.; Acworth, R.I. Heat as a tracer to quantify water flow in near-surface sediments. *Earth Sci. Rev.* **2014**, *129*, 40–58. [CrossRef]

24. Ren, J.; Cheng, J.Q.; Yang, J.; Zhou, Y.J. A review on using heat as a tool for studying groundwater-surface water interactions. *Environ. Earth Sci.* **2018**, *77*, 756. [CrossRef]

25. Knapp, J.L.A.; González-Pinzón, R.; Haggerty, R. The resazurin-resorufin system: Insights from a decade of "Smart" tracer development for hydrologic applications. *Water Resour. Res.* **2018**, *54*, 6877–6889. [CrossRef]

26. Kalbus, E.; Reinstorf, F.; Schirmer, M. Measuring methods for groundwater–surface water interactions: A review. *Hydrol. Earth Syst. Sci.* **2006**, *10*, 873–887. [CrossRef]

27. Brunner, P.; Therrien, R.; Renard, P.; Simmons, C.T.; Franssen, H.J.H. Advances in understanding river-groundwater interactions. *Rev. Geophys.* **2017**, *55*, 818–854. [CrossRef]

28. Hill, A.R. Nitrate removal in stream riparian zones. *J. Environ. Qual.* **1996**, *25*, 743–755. [CrossRef]

29. Birgand, F.; Skaggs, R.W.; Chescheir, G.M.; Gilliam, J.W. Nitrogen removal in streams of agricultural catchments -A literature review. *Crit. Rev. Environ. Sci. Technol.* **2007**, *37*, 381–487. [CrossRef]

30. Gandy, C.J.; Smith, J.W.N.; Jarvis, A.P. Attenuation of mining-derived pollutants in the hyporheic zone: A review. *Sci. Total Environ.* **2007**, *273*, 435–446. [CrossRef]

31. Klein, S.; Worch, E.; Knepper, T.P. Occurrence and spatial distribution of microplastics in river shore sediments of the Rhine-Main area in Germany. *Environ. Sci. Technol.* **2015**, *49*, 6070–6076. [CrossRef]

32. Boutron, O.; Margoum, C.; Chovelon, J.-M.; Guillemain, C.; Gouy, V. Effect of the submergence, the bed form geometry, and the speed of the surface water flow on the mitigation of pesticides in agricultural ditches. *Water Resour. Res.* **2011**, *47*, 1–13. [CrossRef]

33. Peter, K.T.; Herzog, S.; Tian, Z.Y.; Wu, C.; McCray, J.E.; Lynch, K.; Kolodziej, E.P. Evaluating emerging organic contaminant removal in an engineered hyporheic zone using high resolution mass spectrometry. *Water Res.* **2019**, *150*, 140–152. [CrossRef] [PubMed]

34. Riml, J.; Wörman, A.; Kunkel, U.; Radke, M. Evaluating the fate of six common pharmaceuticals using a reactive transport model: Insights from a stream tracer test. *Sci. Total Environ.* **2013**, *458–460*, 344–354. [CrossRef] [PubMed]

35. Schaper, J.L.; Posselt, M.; McCallum, J.L.; Banks, E.W.; Hoehne, A.; Meinikmann, K.; Shanafield, M.A.; Batelaan, O.; Lewandowski, J. Hyporheic exchange controls fate of trace organic compounds in an urban stream. *Environ. Sci. Technol.* **2018**, *52*, 12285–12294. [CrossRef] [PubMed]

36. Conant, B.; Robinson, C.E.; Hinton, M.J.; Russell, H.A. A framework for conceptualizing groundwater-surface water interactions and identifying potential impacts on water quality, water quantity, and ecosystems. *J. Hydrol.* **2019**, *574*, 609–627. [CrossRef]

37. Ward, A.S.; Packman, A.I. Advancing our predictive understanding of river corridor exchange. *Wiley Interdiscip. Rev. Water* **2018**, *6*, e1327. [CrossRef]

38. Millennium Ecosystem Assessment. *Ecosystems and Human Well-Being: Synthesis*; Island Press: Washington, DC, USA, 2005; p. 160.

39. Gomez-Velez, J.D.; Harvey, J.W. A hydrogeomorphic river network model predicts where and why hyporheic exchange is important in large basins. *Geophys. Res. Lett.* **2014**, *41*, 6403–6412. [CrossRef]

40. Caruso, A.; Ridolfi, L.; Boano, F. Impact of watershed topography on hyporheic exchange. *Adv. Water Resour.* **2016**, *94*, 400–411. [CrossRef]

41. Mojarrad, B.B.; Riml, J.; Wörman, A.; Laudon, H. Fragmentation of the hyporheic zone due to regional groundwater circulation. *Water Resour. Res.* **2019**, *55*, 1242–1262. [CrossRef]

42. Kurth, A.-M.; Dawes, N.; Selker, J.; Schirmer, M. Autonomous distributed temperature sensing for long-term heated applications in remote areas. *Geosci. Instrum. Meth.* **2013**, *2*, 71–77. [CrossRef]

43. Boano, F.; Revelli, R.; Ridolfi, L. Quantifying the impact of groundwater discharge on the surface-subsurface exchange. *Hydrol. Process.* **2009**, *23*, 2108–2116. [CrossRef]

44. Shanafield, M.; Cook, P.G. Transmission losses, infiltration and groundwater recharge through ephemeral and intermittent streambeds: A review of applied methods. *J. Hydrol.* **2014**, *511*, 518–529. [CrossRef]

45. Elliott, A.H.; Brooks, N.H. Transfer of nonsorbing solutes to a streambed with bed forms: Theory. *Water Resour. Res.* **1997**, *33*, 123–136. [CrossRef]

46. Elliott, A.H.; Brooks, N.H. Transfer of nonsorbing solutes to a streambed with bed forms: Laboratory experiments. *Water Resour. Res.* **1997**, *33*, 137–151. [CrossRef]

47. Krause, S.; Klaar, M.; Hannah, D.; Mant, J.; Bridgeman, J.; Trimmer, M.; Manning-Jones, S. The potential of large woody debris to alter biogeochemical processes and ecosystem services in lowland rivers. *Wiley Interdiscip. Rev. Water* **2014**, *1*, 263–275. [CrossRef]

48. Dixon, S.J.; Sear, D.A. The influence of geomorphology on large wood dynamics in a low gradient headwater stream. *Water Resour. Res.* **2014**, *50*, 9194–9210. [CrossRef]

49. Blaen, P.J.; Kurz, M.J.; Drummond, J.D.; Knapp, J.L.A.; Mendoza-Lera, C.; Schmadel, N.M.; Klaar, M.J.; Jäger, A.; Folegot, S.; Lee-Cullin, J.; et al. Woody debris is related to reach-scale hotspots of lowland stream ecosystem respiration under baseflow conditions. *Ecohydrology* **2018**, e1952. [CrossRef]

50. Packman, A.; Salehin, M.; Zaramella, M. Hyporheic exchange with gravel beds: Basic hydrodynamic interactions and bedform-induced advective flows. *J. Hydraul. Eng.* **2004**, *130*, 1647–1656. [CrossRef]

51. Roche, K.R.; Blois, G.; Best, J.L.; Christensen, K.T.; Aubeneau, A.F.; Packman, A.I. Turbulence links momentum and solute exchange in coarse-grained streambeds. *Water Resour. Res.* **2018**, *54*, 3225–3242. [CrossRef]

52. Wroblicky, G.; Campana, M.; Valett, H.; Dahm, C. Seasonal variation in surface-subsurface water exchange and lateral hyporheic area of two stream-aquifer systems. *Water Resour. Res.* **1998**, *34*, 317–328. [CrossRef]

53. Salehin, M.; Packman, A.I.; Paradis, M. Hyporheic exchange with heterogeneous streambeds: Laboratory experiments and modelling. *Water Resour. Res.* **2004**, *40*, W11504. [CrossRef]

54. Vaux, W.G. Intragravel flow and interchange of water in a streambed. *Fish. Bull. NOAA* **1968**, *66*, 479–489.
55. Pryshlak, T.T.; Sawyer, A.H.; Stonedahl, S.H.; Soltanian, M.R. Multiscale hyporheic exchange through strongly heterogeneous sediments. *Water Resour. Res.* **2015**, *51*, 9127–9140. [CrossRef]
56. Kalbus, E.; Schmidt, C.; Molson, J.W.; Reinstorf, F.; Schirmer, M. Influence of aquifer and streambed heterogeneity on the distribution of groundwater discharge. *Hydrol. Earth Syst. Sci.* **2009**, *13*, 69–77. [CrossRef]
57. Raghavendra Naganna, S.; Chandra Deka, P. Variability of streambed hydraulic conductivity in an intermittent stream reach regulated by vented dams: A case study. *J. Hydrol.* **2018**, *562*, 477–491. [CrossRef]
58. Schirmer, M.; Leschik, S.; Musolff, A. Current research in urban hydrogeology—A review. *Adv. Water Resour.* **2013**, *51*, 280–291. [CrossRef]
59. Bourke, S.A.; Cook, P.G.; Shanafield, M.; Dogramaci, S.; Clark, J.F. Characterisation of hyporheic exchange in a losing stream using radon-222. *J. Hydrol.* **2014**, *519*, 94–105. [CrossRef]
60. Cranswick, R.H.; Cook, P.G.; Lamontagne, S. Hyporheic zone exchange fluxes and residence times inferred from riverbed temperature and radon data. *J. Hydrol.* **2014**, *519*, 1870–1881. [CrossRef]
61. Schmidt, C.; Bayer-Raich, M.; Schirmer, M. Characterization of spatial heterogeneity of groundwater-stream water interactions using multiple depth streambed temperature measurements at the reach scale. *Hydrol. Earth Syst. Sci.* **2006**, *10*, 849–859. [CrossRef]
62. Gordon, R.P.; Lautz, L.K.; Briggs, M.A.; McKenzie, J.M. Automated calculation of vertical pore-water flux from field temperature time series using the VFLUX method and computer program. *J. Hydrol.* **2012**, *420*, 142–158. [CrossRef]
63. Mojarrad, B.B.; Betterle, A.; Singh, T.; Olid, C.; Wörman, A. The effect of stream discharge on hyporheic exchange. *Water* **2019**, *11*, 1436. [CrossRef]
64. Vandersteen, G.; Schneidewind, U.; Anibas, C.; Schmidt, C.; Seuntjens, P.; Batelaan, O. Determining groundwater-surface water exchange from temperature-time series: Combining a local polynomial method with a maximum likelihood estimator. *Water Resour. Res.* **2015**, *51*, 922–939. [CrossRef]
65. Schneidewind, U.; van Berkel, M.; Anibas, C.; Vandersteen, G.; Schmidt, C.; Joris, I.; Seuntjens, P.; Batelaan, O.; Zwart, H.J. LPMLE3: A novel 1-D approach to study water flow in streambeds using heat as a tracer. *Water Resour. Res.* **2016**, *52*, 6596–6610. [CrossRef]
66. Lewandowski, J.; Angermann, L.; Nützmann, G.; Fleckenstein, J.H. A heat pulse technique for the determination of small-scale flow directions and flow velocities in the streambed of sand-bed streams. *Hydrol. Process.* **2011**, *25*, 3244–3255. [CrossRef]
67. Angermann, L.; Krause, S.; Lewandowski, J. Application of heat pulse injections for investigating shallow hyporheic flow in a lowland river. *Water Resour. Res.* **2012**, *48*, W00P02. [CrossRef]
68. Banks, E.W.; Shanafield, M.A.; Noorduijn, S.; McCallum, J.; Lewandowski, J.; Batelaan, O. Active heat pulse sensing of 3D-flow fields in streambeds. *Hydrol. Earth Syst. Sci.* **2018**, *22*, 1917–1929. [CrossRef]
69. Selker, J.; Van de Giesen, N.; Westhoff, M.; Luxemburg, W.; Parlange, M.B. Fiber optics opens window on stream dynamics. *Geophys. Res. Lett.* **2006**, *33*, L24401. [CrossRef]
70. Lowry, C.S.; Walker, J.F.; Hunt, R.J.; Anderson, M.P. Identifying spatial variability of groundwater discharge in a wetland stream using a distributed temperature sensor. *Water Resour. Res.* **2007**, *43*, W10408. [CrossRef]
71. Tyler, S.W.; Selker, J.S.; Hausner, M.B.; Hatch, C.E.; Torgersen, T.; Thodal, C.E.; Schladow, S.G. Environmental temperature sensing using Raman spectra DTS fiber-optic methods. *Water Resour. Res.* **2009**, *45*, W00D23. [CrossRef]
72. Vogt, T.; Schneider, P.; Hahn-Woernle, L.; Cirpka, O.A. Estimation of seepage rates in a losing stream by means of fiber-optic high-resolution vertical temperature profiling. *J. Hydrol.* **2010**, *380*, 154–164. [CrossRef]
73. Shanafield, M.; McCallum, J.L.; Cook, P.G.; Noorduijn, S. Using basic metrics to analyze high-resolution temperature data in the subsurface. *Hydrogeol. J.* **2017**, *25*, 1501–1508. [CrossRef]
74. Gaona, J.; Meinikmann, K.; Lewandowski, J. Identification of groundwater exfiltration, interflow discharge, and hyporheic exchange flows by fibre optic distributed temperature sensing supported by electromagnetic induction geophysics. *Hydrol. Process.* **2019**, *33*, 1390–1402. [CrossRef]
75. Hare, D.K.; Briggs, M.A.; Rosenberry, D.O.; Boutt, D.F.; Lane, J.W. A comparison of thermal infrared to fiber-optic distributed temperature sensing for evaluation of groundwater discharge to surface water. *J. Hydrol.* **2015**, *530*, 153–166. [CrossRef]

76. Ward, A.S.; Gooseff, M.N.; Singha, K. Imaging hyporheic zone solute transport using electrical resistivity. *Hydrol. Process.* **2010**, *24*, 948–953. [CrossRef]

77. McLachlan, P.J.; Chambers, J.E.; Uhlemann, S.S.; Binley, A. Geophysical characterisation of the groundwater-surface water interface. *Adv. Water Resour.* **2017**, *109*, 302–319. [CrossRef]

78. Wu, L.; Singh, T.; Gomez-Velez, J.; Nützmann, G.; Wörman, A.; Krause, S.; Lewandowski, J. Impact of dynamically changing discharge on hyporheic exchange processes under gaining and losing groundwater conditions. *Water Resour. Res.* **2018**, *54*, 10076–10093. [CrossRef]

79. Singh, T.; Wu, L.; Gomez-Velez, J.D.; Lewandowski, J.; Hannah, D.M.; Krause, S. Dynamic hyporheic zones: Exploring the role of peak flow events on bedform-induced hyporheic exchange. *Water Resour. Res.* **2019**, *55*, 218–235. [CrossRef]

80. Zarnetske, J.P.; Haggerty, R.; Wondzell, S.M.; Baker, M.A. Dynamics of nitrate production and removal as a function of residence time in the hyporheic zone. *J. Geophys. Res.* **2011**, *116*, G01025. [CrossRef]

81. Harvey, J.W.; Bohlke, J.K.; Voytek, M.A.; Scott, D.; Tobias, C.R. Hyporheic zone denitrification: Controls on effective reaction depth and contribution to whole-stream mass balance. *Water Resour. Res.* **2013**, *49*, 6298–6316. [CrossRef]

82. Johnson, Z.C.; Warwick, J.J.; Schumer, R. Factors affecting hyporheic and surface transient storage in a western U.S. river. *J. Hydrol.* **2014**, *510*, 325–339. [CrossRef]

83. Bencala, K.E.; Walters, R.A. Simulation of solute transport in a mountain pool-and-riffle stream: A transient storage model. *Water Resour. Res.* **1983**, *19*, 718–724. [CrossRef]

84. Wörman, A.; Packman, A.I.; Johansson, H.; Jonsson, K. Effect of flow-induced exchange in hyporheic zones on longitudinal transport of solutes in streams and rivers. *Water Resour. Res.* **2002**, *38*, 2-1–2-15. [CrossRef]

85. Gooseff, M.N.; Wondzell, S.M.; Haggerty, R.; Anderson, J. Comparing transient storage modeling and residence time distribution (RTD) analysis in geomorphically varied reaches in the Lookout Creek basin, Oregon, USA. *Adv. Water Resour.* **2003**, *26*, 925–937. [CrossRef]

86. Jury, W.A. Simulation of solute transport using a transfer function model. *Water Resour. Res.* **1982**, *18*, 363–368. [CrossRef]

87. Sardin, M.; Schweich, D.; Leu, F.J.; Van Genuchten, M.T. Modeling the nonequilibrium transport of linearly interacting solutes in porous media: A review. *Water Resour. Res.* **1991**, *27*, 2287–2307. [CrossRef]

88. Knapp, J.L.A.; Cirpka, O.A. Determination of hyporheic travel time distributions and other parameters from concurrent conservative and reactive tracer tests by local-in-global optimization. *Water Resour. Res.* **2017**, *53*, 4984–5001. [CrossRef]

89. Liao, Z.; Cirpka, O.A. Shape-free inference of hyporheic traveltime distributions from synthetic conservative and "smart" tracer tests in streams. *Water Resour. Res.* **2011**, *47*, W07510. [CrossRef]

90. Liao, Z.; Lemke, D.; Osenbrück, K.; Cirpka, O.A. Modeling and inverting reactive stream tracers undergoing two-site sorption and decay in the hyporheic zone. *Water Resour. Res.* **2013**, *49*, 3406–3422. [CrossRef]

91. Battin, T.J.; Besemer, K.; Bengtsson, M.M.; Romani, A.M.; Packmann, A.I. The ecology and biogeochemistry of stream biofilms. *Nat. Rev. Microbiol.* **2016**, *14*, 251–263. [CrossRef]

92. Williams, D.D.; Hynes, H.B.N. The occurrence of benthos deep in the substratum of a stream. *Freshw. Biol.* **1974**, *4*, 233–256. [CrossRef]

93. Zeglin, H.L. Stream microbial diversity responds to environmental changes: Review and synthesis of existing research. *Front. Microbiol.* **2015**, *6*, 454. [CrossRef] [PubMed]

94. Singer, G.; Besemer, K.; Schmitt-Kopplin, P.; Hödl, I.; Battin, T.J. Physical heterogeneity increases biofilm resource use and its molecular diversity in stream mesocosms. *PLoS ONE* **2010**, *5*, e9988. [CrossRef] [PubMed]

95. Romaní, A.M.; Fund, K.; Artigas, J.; Schwartz, T.; Sabater, S.; Obst, U. Relevance of polymeric matrix enzymes during biofilm formation. *Microb. Ecol.* **2008**, *56*, 427–436. [CrossRef] [PubMed]

96. Battin, T.J.; Kaplan, L.A.; Findlay, S.; Hopkinson, C.S.; Marti, E.; Packman, A.I.; Newbold, J.D.; Sabater, F. Biophysical controls on organic carbon fluxes in fluvial networks. *Nat. Geosci.* **2008**, *1*, 95–100. [CrossRef]

97. Mulholland, P.J.; Helton, A.M.; Poole, G.C.; Hall, R.O.; Hamilton, S.K.; Peterson, B.J.; Tank, J.L.; Ashkenas, L.R.; Cooper, L.W.; Dahm, C.N.; et al. Stream denitrification across biomes and its response to anthropogenic nitrate loading. *Nature* **2008**, *452*, 202–205. [CrossRef]

98. Storey, R.G.; Fulthorpe, R.R.; Williams, D.D. Perspectives and predictions on the microbial ecology of the hyporheic zone. *Freshw. Biol.* **1999**, *41*, 119–130. [CrossRef]

99. Robertson, A.L.; Wood, P.J. Ecology of the hyporheic zone: Origins, current knowledge and future directions. *Fundam. Appl. Limnol Arch. Hydrobiol.* **2010**, *176*, 279–289. [CrossRef]

100. Wood, P.J.; Boulton, A.J.; Little, S.; Stubbington, R. Is the hyporheic zone a refugium for aquatic macroinvertebrates during severe low flow conditions? *Fundam. Appl. Limnol./Arch. Hydrobiol.* **2010**, *176*, 377–390. [CrossRef]

101. Eichhorn, M. *Natural Systems: The Organisation of Life*; John Wiley and Sons: Chichester, UK, 2016; p. 359.

102. Dole-Olivier, M.J. Surface water-groundwater exchanges in three dimensions on a backwater of the Rhône River. *Freshw. Biol.* **1998**, *40*, 93–109. [CrossRef]

103. Fraser, B.G.; Williams, D.D. Seasonal boundary dynamics of a groundwater/surface-water ecotone. *Ecology* **1998**, *79*, 2019–2031. [CrossRef]

104. Miyake, Y.; Nakano, S. Effects of substratum stability on diversity of stream invertebrates during baseflow at two spatial scales. *Freshw. Biol.* **2002**, *47*, 219–230. [CrossRef]

105. Sliva, L.; Williams, D.D. Responses of hyporheic meiofauna to habitat manipulation. *Hydrobiologia* **2005**, *548*, 217–232. [CrossRef]

106. Davy-Bowker, J.; Sweeting, W.; Wright, N.; Clarke, R.T.; Arnott, S. The distribution of benthic and hyporheic macroinvertebrates from the heads and tails of riffles. *Hydrobiologia* **2006**, *563*, 109–123. [CrossRef]

107. Andrushchyshyn, O.P.; Wilson, K.P.; Williams, D.D. Ciliate communities in shallow groundwater: Seasonal and spatial characteristics. *Freshw. Biol.* **2007**, *52*, 1745–1761. [CrossRef]

108. Magliozzi, C.; Usseglio-Polatera, P.; Meyer, A.; Grabowski, R.C. Functional traits of hyporheic and benthic invertebrates reveal importance of wood-driven geomorphological processes in rivers. *Funct. Ecol.* **2019**, *33*, 1758–1770. [CrossRef]

109. Smock, L.A.; Gladden, J.E.; Riekenberg, J.L.; Smith, L.C.; Black, C.R. Lotic macroinvertebrate production in three dimensions: Channel surface, hyporheic, and floodplain environments. *Ecology* **1992**, *73*, 875–886. [CrossRef]

110. Post, D.M.; Doyle, M.W.; Sabo, J.L.; Finlay, J.C. The problem of boundaries in defining ecosystems: A potential landmine for uniting geomorphology and ecology. *Geomorphology* **2007**, *89*, 111–126. [CrossRef]

111. Peralta-Maraver, I.; Perkins, D.M.; Thompson, M.S.; Fussmann, K.; Reiss, J.; Robertson, A.L. Comparing biotic drivers of litter breakdown across streams compartments. *J. Anim. Ecol.* **2019**, *88*, 1146–1157. [CrossRef]

112. Smith, J.J.; Lake, P.S. The breakdown of buried and surface–placed leaf litter in an upland stream. *Hydrobiologia* **1993**, *271*, 141–148. [CrossRef]

113. Cornut, J.; Elger, A.; Lambrigot, D.; Marmonier, P.; Chauvet, E. Early stages of leaf decomposition are mediated by aquatic fungi in the hyporheic zone of woodland streams. *Freshw. Biol.* **2010**, *55*, 2541–2556. [CrossRef]

114. Peralta-Maraver, I.; Robertson, A.L.; Perkins, D.M. Depth and vertical hydrodynamics constrain the size structure of a lowland streambed community. *Biol. Lett.* **2019**, *15*, 20190317. [CrossRef] [PubMed]

115. Stumm, W.; Morgan, J.J. Redox conditions in natural waters. In *Aquatic Chemistry: Chemical Equilibria and Rates in Natural Waters*, 3rd ed.; Schnoor, J.L., Zehnder, A., Eds.; Wiley: New York, NY, USA, 1996; pp. 464–489. [CrossRef]

116. Jones, J.B.; Holmes, R.M.; Fisher, S.G.; Grimm, N.B.; Greene, D.M. Methanogenesis in Arizona, USA dryland streams. *Biogeochemistry* **1995**, *31*, 155–173. [CrossRef]

117. Zarnetske, J.P.; Haggerty, R.; Wondzell, S.M.; Baker, M.A. Labile dissolved organic carbon supply limits hyporheic denitrification. *J. Geophys. Res.* **2011**, *116*, G04036. [CrossRef]

118. Briggs, M.A.; Day-Lewis, F.D.; Zarnetske, J.P.; Harvey, J.W. A physical explanation for the development of redox microzones in hyporheic flow. *Geophys. Res. Lett.* **2015**, *42*, 4402–4410. [CrossRef]

119. Triska, F.J.; Duff, J.H.; Avanzino, R.J. The role of water exchange between a stream channel and its hyporheic zone in nitrogen cycling at the terrestrial-aquatic interface. *Hydrobiologia* **1993**, *251*, 167–184. [CrossRef]

120. Triska, F.J.; Duff, J.H.; Avanzino, R.J. Patterns of hydrological exchange and nutrient transformation in the hyporheic zone of a gravel-bottom stream: Examining terrestrial-aquatic linkages. *Freshw. Biol.* **1993**, *29*, 259–274. [CrossRef]

121. Christensen, P.B.; Nielsen, L.P.; Sørensen, J.; Revsbech, N.P. Denitrification in nitrate-rich streams: Diurnal and seasonal variation related to benthic oxygen metabolism. *Limnol. Oceanogr.* **1990**, *35*, 640–651. [CrossRef]

122. Schaper, J.L.; Seher, W.; Nützmann, G.; Putschew, A.; Jekel, M.; Lewandowski, J. The fate of polar trace organic compounds in the hyporheic zone. *Water Res.* **2018**, *140*, 158–166. [CrossRef]

123. Lewandowski, J.; Nützmann, G. Nutrient retention and release in a floodplain's aquifer and in the hyporheic zone of a lowland river. *Ecol. Eng.* **2010**, *36*, 1156–1166. [CrossRef]

124. Trauth, N.; Schmidt, C.; Vieweg, M.; Maier, U.; Fleckenstein, J.H. Hyporheic transport and biogeochemical reactions in pool-riffle systems under varying ambient groundwater flow conditions. *J. Geophys. Res. Biogeosci.* **2014**, *119*, 910–928. [CrossRef]

125. De Falco, N.; Boano, F.; Bogler, A.; Bar-Zeev, E.; Arnon, S. Influence of stream-subsurface exchange flux and bacterial biofilms on oxygen consumption under nutrient-rich conditions. *J. Geophys. Res. Biogeosci.* **2018**, *123*, 2021–2034. [CrossRef]

126. Haggerty, R.; Argerich, A.; Marti, E. Development of a "smart" tracer for the assessment of microbiological activity and sediment-water interaction in natural waters: The resazurin-resorufin system. *Water Resour. Res.* **2008**, *44*, W00D01. [CrossRef]

127. Haggerty, R.; Marti, E.; Argerich, A.; von Schiller, D.; Grimm, N.B. Resazurin as a "smart" tracer for quantifying metabolically active transient storage in stream ecosystems. *J. Geophys. Res. Biogeosci.* **2009**, *114*, G03014. [CrossRef]

128. Schirmer, M.; Luster, J.; Linde, N.; Perona, P.; Mitchell, E.A.D.; Barry, D.A.; Hollender, J.; Cirpka, O.A.; Schneider, P.; Vogt, T.; et al. Morphological, hydrological, biogeochemical and ecological changes and challenges in river restoration–The Thur River case study. *Hydrol. Earth Syst. Sci.* **2014**, *18*, 1–14. [CrossRef]

129. Lewandowski, J.; Putschew, A.; Schwesig, D.; Neumann, C.; Radke, M. Fate of organic micropollutants in the hyporheic zone of a eutrophic lowland stream: Results of a preliminary field study. *Sci. Total Environ.* **2011**, *409*, 1824–1835. [CrossRef]

130. Lapworth, D.J.; Gooddy, D.C.; Jarvie, H.P. Understanding phosphorus mobility and bioavailability in the hyporheic zone of a chalk stream. *Water Air Soil Pollut.* **2010**, *218*, 213–226. [CrossRef]

131. Butturini, A.; Sabater, F. Importance of transient storage zones for ammonium and phosphate retention in a sandy-bottom Mediterranean stream. *Freshw. Biol.* **1999**, *41*, 593–603. [CrossRef]

132. Arnon, S.; Yanuka, K.; Nejidat, A. Impact of overlying water velocity on ammonium uptake by benthic biofilms. *Hydrol. Process.* **2013**, *27*, 570–578. [CrossRef]

133. Harvey, J.J.; Gomez-Velez, J.; Schmadel, N.; Scott, D.; Boyer, E.; Alexander, R.; Eng, K.; Golden, H.; Kettner, A.; Konrad, C.; et al. How Hydrologic connectivity regulates water quality in river corridors. *J. Am. Water Resour. Assoc.* **2019**, *55*, 369–381. [CrossRef]

134. Ocampo, C.J.; Oldham, C.E.; Sivapalan, M. Nitrate attenuation in agricultural catchments: Shifting balances between transport and reaction. *Water Resour. Res.* **2006**, *42*, W01408. [CrossRef]

135. Gordon, R.P.; Lautz, L.K.; Daniluk, T.L. Spatial patterns of hyporheic exchange and biogeochemical cycling around cross-vane restoration structures: Implications for stream restoration design. *Water Resour. Res.* **2013**, *49*, 2040–2055. [CrossRef]

136. Hester, E.T.; Hammond, B.; Scott, D.T. Effects of inset floodplains and hyporheic exchange induced by in-stream structures on nitrate removal in a headwater stream. *Ecol. Eng.* **2016**, *97*, 452–464. [CrossRef]

137. Morén, I.; Riml, J.; Wörman, A. *In-Stream Water Management Strategies for Reducing Nutrient Loads to the Baltic Sea*; BONUS Soils2Sea Deliverable 4.4; KTH Royal Institute of Technology: Stockholm, Sweden, March 2018; Available online: www.Soils2Sea.eu (accessed on 23 October 2019).

138. Refsgaard, J.C.; Chubarenko, B.; Donnely, C.; de Jonge, H.; Olesen, J.E.; Steljes, N.; Wachniew, P.; Wörman, A. Spatially differentiated regulation measures-can it save the Baltic Sea from excessive N-loads? *Ambio* **2019**. [CrossRef] [PubMed]

139. Botter, G.; Basu, N.B.; Zanardo, S.; Rao, P.S.C.; Rinaldo, A. Stochastic modeling of nutrient losses in streams: Interactions of climatic, hydrologic and biogeochemical controls. *Water Resour. Res.* **2010**, *46*, W08509. [CrossRef]

140. Kidd, K.A.; Blanchfield, P.J.; Mills, K.H.; Palace, V.P.; Evans, R.E.; Lazorchak, J.M.; Flick, R.W. Collapse of a fish population after exposure to a synthetic estrogen. *Proc. Natl. Acad. Sci. USA* **2007**, *104*, 8897–8901. [CrossRef] [PubMed]

141. Galus, M.; Rangarajan, S.; Lai, A.; Shaya, L.; Balshine, S.; Wilson, J.Y. Effects of chronic, parental pharmaceutical exposure on zebrafish (*Danio rerio*) offspring. *Aquat. Toxicol.* **2014**, *151*, 124–134. [CrossRef]

142. Cruz-Rojas, C.; SanJuan-Reyes, N.; Fuentes-Benites, M.P.A.G.; Dublan-García, O.; Galar-Martínez, M.; Islas-Flores, H.; Gómez-Oliván, L.M. Acesulfame potassium: Its ecotoxicity measured through oxidative stress biomarkers in common carp (Cyprinus carpio). *Sci. Total Environ.* **2019**, *647*, 772–784. [CrossRef]

143. Kubec, J.; Hossain, S.; Grabicová, K.; Randák, T.; Kouba, A.; Grabic, R.; Roje, S.; Buřič, M. Oxazepam alters the behavior of crayfish at diluted concentrations, venlafaxine does not. *Water* **2019**, *11*, 196. [CrossRef]

144. Luo, Y.; Guo, W.; Ngo, H.H.; Nghiem, L.D.; Hai, F.I.; Zhang, J.; Liang, S.; Wang, X.C. A review on the occurrence of micropollutants in the aquatic environment and their fate and removal during wastewater treatment. *Sci. Total Environ.* **2014**, *473–474*, 619–641. [CrossRef]

145. Melvin, S.; Leusch, F. Removal of trace organic contaminants from domestic wastewater: A meta-analysis comparison of sewage treatment technologies. *Environ. Int.* **2016**, *92–93*, 183–188. [CrossRef]

146. Phillips, P.; Chalmers, A.; Gray, J.; Kolpin, D.; Foreman, W.; Wall, G. Combined Sewer Overflows: An Environmental Source of Hormones and Wastewater Micropollutants. *Environ. Sci. Technol.* **2012**, *46*, 563–579. [CrossRef] [PubMed]

147. Roehrdanz, P.R.; Feraud, M.; Lee, D.G.; Means, J.C.; Snyder, S.A.; Holden, P.A. Spatial models of sewer pipe leakage predict the occurrence of wastewater indicators in shallow urban groundwater. *Environ. Sci. Technol.* **2017**, *51*, 1213–1223. [CrossRef] [PubMed]

148. Schaider, L.A.; Ackerman, J.M.; Rudel, R.A. Septic systems as sources of organic wastewater compounds in domestic drinking water wells in a shallow sand and gravel aquifer. *Sci. Total Environ.* **2016**, *547*, 470–481. [CrossRef] [PubMed]

149. Eggen, T.; Moeder, M.; Arukwe, A. Municipal landfill leachates: A significant source for new and emerging pollutants. *Sci. Total Environ.* **2010**, *408*, 5147–5157. [CrossRef]

150. Masoner, J.R.; Kolpin, D.W.; Cozzarelli, I.M.; Barber, L.B.; Burden, D.S.; Foreman, W.T.; Forshay, K.J.; Furlong, E.T.; Groves, J.F.; Hladik, M.L.; et al. Urban stormwater: An overlooked pathway of extensive mixed contaminants to surface and groundwaters in the United States. *Environ. Sci. Technol.* **2019**, *53*, 10070–10081. [CrossRef]

151. Jekel, M.; Ruhl, A.S.; Meinel, F.; Zietzschmann, F.; Lima, S.P.; Baur, N.; Wenzel, M.; Gnirß, R.; Sperlich, A.; Dünnbier, U.; et al. Anthropogenic organic micro-pollutants and pathogens in the urban water cycle: Assessment, barriers and risk communication (ASKURIS). *Environ. Sci. Eur.* **2013**, *25*, 1–8. [CrossRef]

152. Schwarzenbach, R.P.; Escher, B.I.; Fenner, K.; Hofstetter, T.B.; Johnson, C.A.; Von Gunten, U.; Wehrli, B. The challenge of micropollutants in aquatic systems. *Science* **2006**, *313*, 1072–1077. [CrossRef]

153. Boulton, A.J.; Datry, T.; Kasahara, T.; Mutz, M.; Stanford, J.A. Ecology and management of the hyporheic zone: Stream-groundwater interactions of running waters and their floodplains. *Freshw. Sci.* **2010**, *29*, 26–40. [CrossRef]

154. Heberer, T.; Massmann, G.; Fanck, B.; Taute, T.; Dunnbier, U. Behaviour and redox sensitivity of antimicrobial residues during bank filtration. *Chemosphere* **2008**, *73*, 451–460. [CrossRef]

155. Huntscha, S.; Singer, H.P.; McArdell, C.S.; Frank, C.E.; Hollender, J. Multiresidue analysis of 88 polar organic micropollutants in ground, surface and wastewater using online mixed-bed multilayer solid-phase extraction coupled to high performance liquid chromatography-tandem mass spectrometry. *J. Chromatogr. A* **2012**, *1268*, 74–83. [CrossRef]

156. Lawrence, J.E.; Skold, M.E.; Hussain, F.A.; Silverman, D.R.; Resh, V.H.; Sedlak, D.L.; Luthy, R.G.; McCray, J.E. Hyporheic zone in urban streams: A review and opportunities for enhancing water quality and improving aquatic habitat by active management. *Environ. Eng. Sci.* **2013**, *30*, 480–501. [CrossRef]

157. Regnery, J.; Barringer, J.; Wing, A.D.; Hoppe-Jones, C.; Teerlink, J.; Drewes, J.E. Start-up performance of a full-scale riverbank filtration site regarding removal of DOC, nutrients, and trace organic chemicals. *Chemosphere* **2015**, *127*, 136–142. [CrossRef] [PubMed]

158. Hollender, J.; Rothardt, J.; Radny, D.; Loos, M.; Epting, J.; Huggenberger, P.; Borer, P.; Singer, H. Target and Non-target Screening of micropollutants at riverbank filtration sites with short residence times using LC-HRMS/MS. *Water Res. X* **2018**, *1*, 100007. [CrossRef]

159. Burke, V.; Greskowiak, J.; Asmuß, T.; Bremermann, R.; Taute, T.; Massmann, G. Temperature dependent redox zonation and attenuation of wastewater-derived organic micropollutants in the hyporheic zone. *Sci. Total Environ.* **2014**, *482–483*, 53–61. [CrossRef]

160. Schaper, J.L.; Posselt, M.; Bouchez, C.; Jaeger, A.; Nützmann, G.; Putschew, A.; Singer, G.; Lewandowski, J. Fate of trace organic compounds in the hyporheic zone: Influence of retardation, the benthic bio-layer and organic carbon. *Environ. Sci. Technol.* **2019**, *53*, 4224–4234. [CrossRef]

161. Jaeger, A.; Coll, C.; Posselt, M.; Rutere, C.; Mechelke, J.; Betterle, A.; Raza, M.; Meinikmann, K.; Melutens, A.; Portmann, A.; et al. Using recirculating flumes and a response surface model to investigate the role of

hyporheic exchange and bacterial diversity on micropollutant half lives. *Environ. Sci. Process. Impacts* **2019**. accepted. [CrossRef]

162. Shapiro, O.H.; Kushmaro, A.; Brenner, A. Bacteriophage predation regulates microbial abundance and diversity in a full-scale bioreactor treating industrial wastewater. *ISME J.* **2010**, *4*, 327–336. [CrossRef]

163. Otto, S.; Harms, H.; Wick, L.Y. Effects of predation and dispersal on bacterial abundance and contaminant biodegradation. *FEMS Microbiol. Ecol.* **2017**, *93*, fiw241. [CrossRef]

164. El-Athman, F.; Adrian, L.; Jekel, M.; Putschew, A. Deiodination in the presence of *Dehalococcoides mccartyi* strain CBDB1: Comparison of the native enzyme and co-factor vitamin B_{12}. *Environ. Sci. Pollut. Res.* **2019**. [CrossRef]

165. El-Athman, F.; Jekel, M.; Putschew, A. Reaction kinetics of corrinoid-mediated deiodination of iodinated X-ray contrast media and other iodinated organic compounds. *Chemosphere* **2019**, *234*, 971–977. [CrossRef]

166. El-Athman, F.; Adrian, L.; Jekel, M.; Putschew, A. Abiotic reductive deiodination of iodinated organic compounds and X-ray contrast media catalyzed by free corrinoids. *Chemosphere* **2019**, *221*, 212–218. [CrossRef] [PubMed]

167. Jonsson, K.; Johansson, H.; Wörman, A. Sorption behaviour and long-term retention of reactive solutes in the hyporheic zone of streams. *J. Environ. Eng.* **2004**, *130*, 573–584. [CrossRef]

168. Hebig, K.H.; Groza, L.G.; Sabourin, M.J.; Scheytt, T.J.; Ptacek, C.J. Transport behavior of the pharmaceutical compounds carbamazepine, sulfamethoxazole, gemfibrozil, ibuprofen, and naproxen, and the lifestyle drug caffeine, in saturated laboratory columns. *Sci. Total Environ.* **2017**, *590*, 708–719. [CrossRef] [PubMed]

169. Nghiem, L.D.; Hawkes, S. Effects of membrane fouling on the nanofiltration of trace organic contaminants. *Desalination* **2009**, *236*, 273–281. [CrossRef]

170. Packman, A.I.; Brooks, N.H.; Morgan, J.J. A physicochemical model for colloid exchange between a stream and a sand streambed with bed forms. *Water Resour. Res.* **2000**, *36*, 2351–2361. [CrossRef]

171. Packman, A.I.; Brooks, N.H.; Morgan, J.J. Kaolinite exchange between a stream and streambed: Laboratory experiments and validation of a colloid transport model. *Water Resour. Res.* **2000**, *36*, 2363–2372. [CrossRef]

172. Celiz, M.D.; Tso, J.; Aga, D.S. Pharmaceutical metabolites in the environment: Analytical challenges and ecological risks. *Environ. Toxicol. Chem.* **2009**, *28*, 2473–2484. [CrossRef]

173. Henning, N.; Kunkel, U.; Wick, A.; Ternes, T.A. Biotransformation of gabapentin in surface water matrices under different redox conditions and the occurrence of one major TP in the aquatic environment. *Water Res.* **2018**, *137*, 290–300. [CrossRef]

174. Nödler, K.; Hillebrand, O.; Idzik, K.; Strathmann, M.; Schiperski, F.; Zirlewagen, J.; Licha, T. Occurrence and fate of the angiotensin II receptor antagonist transformation product valsartan acid in the water cycle—A comparative study with selected β-blockers and the persistent anthropogenic wastewater indicators carbamazepine and acesulfame. *Water Res.* **2013**, *47*, 6650–6659. [CrossRef]

175. Li, Z.; Sobek, A.; Radke, M. Flume experiments to investigate the environmental fate of pharmaceuticals and their transformation products in streams. *Environ. Sci. Technol.* **2015**, *49*, 6009–6017. [CrossRef]

176. Writer, J.H.; Antweiler, R.C.; Ferrer, I.; Ryan, J.N.; Thurman, E.M. In-stream attenuation of neuro-active pharmaceuticals and their metabolites. *Environ. Sci. Technol.* **2013**, *47*, 9781–9790. [CrossRef] [PubMed]

177. Mechelke, J.; Vermeirssen, E.L.M.; Hollender, J. Passive sampling of organic contaminants across the water-sediment interface of an urban stream. *Water Res.* **2019**, *165*, 114966. [CrossRef] [PubMed]

178. Jaeger, A.; Posselt, M.; Betterle, A.; Schaper, J.; Mechelke, J.; Coll, C.; Lewandowski, J. Spatial and temporal variability in attenuation of polar trace organic compounds in an urban lowland stream. *Environ. Sci. Technol.* **2019**, *53*, 2383–2395. [CrossRef] [PubMed]

179. Ren, J.H.; Packman, A.I. Stream-subsurface exchange of zinc in the presence of silica and kaolinite colloids. *Environ. Sci. Technol.* **2004**, *38*, 6571–6581. [CrossRef] [PubMed]

180. Karwan, D.L.; Saiers, J.E. Hyporheic exchange and streambed filtration of suspended particles. *Water Resour. Res.* **2012**, *48*, W01519. [CrossRef]

181. Fox, A.; Packman, A.I.; Boano, F.; Phillips, C.B.; Arnon, S. Interactions between suspended kaolinite Deposition and hyporheic exchange flux under losing and gaining flow conditions. *Geophys. Res. Lett.* **2018**, *45*, 4077–4085. [CrossRef]

182. Wharton, G.; Mohajeri, S.H.; Righetti, M. The pernicious problem of streambed colmation: A multi-disciplinary reflection on the mechanisms, causes, impacts, and management challenges. *Wiley Interdiscip. Rev. Water* **2017**, e1231. [CrossRef]

183. Jones, I.; Growns, I.; Arnold, A.; McCall, S.; Bowes, M. The effects of increased flow and fine sediment on hyporheic invertebrates and nutrients in stream mesocosms. *Freshw. Biol.* **2015**, *60*, 813–826. [CrossRef]

184. Descloux, S.; Datry, T.; Marmonier, P. Benthic and hyporheic invertebrate assemblages along a gradient of increasing streambed colmation by fine sediment. *Aquat. Sci.* **2013**, *75*, 493–507. [CrossRef]

185. Mathers, K.L.; Millett, J.; Robertson, A.L.; Stubbington, R.; Wood, P.J. Faunal response to benthic and hyporheic sedimentation varies with direction of vertical hydrological exchange. *Freshw. Biol.* **2014**, *59*, 2278–2289. [CrossRef]

186. Dunscombe, M.; Robertson, A.; Peralta Maraver, I.; Shaw, P. Community structure and functioning below the stream bed across contrasting geologies. *Sci. Total Environ.* **2018**, *630*, 1028–1035. [CrossRef] [PubMed]

187. Kemp, P.; Sear, D.A.; Collins, A.L.; Naden, P.S.; Jones, J.I. The impacts of fine sediment on riverine fish. *Hydrol. Process.* **2011**, *25*, 1800–1821. [CrossRef]

188. Sear, D.A.; Jones, J.I.; Collins, A.L.; Hulin, A.; Burke, N.; Bateman, S.; Pattison, I.; Naden, P.S. Does fine sediment source as well as quantity affect salmonid embryo mortality and development? *Sci. Total Environ.* **2016**, *541*, 957–968. [CrossRef] [PubMed]

189. Larsen, L.; Harvey, J.; Skalak, K.; Goodman, M. Fluorescence-based source tracking of organic sediment in restored and unrestored urban streams. *Limnol. Oceanogr.* **2015**, *60*, 1439–1461. [CrossRef]

190. Wagner, M.; Scherer, C.; Alvarez-Muñoz, D.; Brennholt, N.; Bourrain, X.; Buchinger, S.; Fries, E.; Grosbois, C.; Klasmeier, J.; Marti, T. Microplastics in freshwater ecosystems: What we know and what we need to know. *Environ. Sci. Eur.* **2014**, *26*, 12. [CrossRef]

191. Dris, R.; Imhof, H.; Sanchez, W.; Gasperi, J.; Galgani, F.; Tassin, B.; Laforsch, C. Beyond the ocean: Contamination of freshwater ecosystems with (micro-) plastic particles. *Environ. Chem.* **2015**, *12*, 539–550. [CrossRef]

192. Mani, T.; Hauk, A.; Walter, U.; Burkhardt-Holm, P. Microplastics profile along the Rhine River. *Sci. Rep.* **2015**, *5*, 17988. [CrossRef]

193. Alimi, O.S.; Hernandez, L.M.; Tufenkji, N. Microplastics and nanoplastics in aquatic environments: Aggregation, deposition, and enhanced contaminant transport. *Environ. Sci. Technol.* **2018**, *52*, 1704–1724. [CrossRef]

194. Browne, M.A.; Crump, P.; Niven, S.J.; Teuten, E.; Tonkin, A.; Galloway, T.; Thompson, R. Accumulation of microplastic on shorelines worldwide: Sources and sinks. *Environ. Sci. Technol.* **2011**, *45*, 9175–9179. [CrossRef]

195. Magnusson, K.; Norén, F. *Screening of Microplastic Particles In and Down-Stream a Wastewater Treatment Plant*; Report C55; Swedish Environmental Research Institute: Stockholm, Sweeden, 2014; p. 22. Available online: http://www.diva-portal.org/smash/get/diva2:773505/FULLTEXT01.pdf (accessed on 23 October 2019).

196. Vogelsang, C.; Lusher, A.L.; Dadkhah, M.E.; Sundvor, I.; Umar, M.; Ranneklev, S.B.; Eidsvoll, D.; Meland, S. *Microplastics in Road Dust–Characteristics, Pathways and Measures*; Research Report 7361-2019; Institute of Transport Economics, Norwegian Centre for Transport Research: Oslo, Norway, 2019; p. 174. Available online: https://niva.brage.unit.no/niva-xmlui/handle/11250/2493537 (accessed on 23 October 2019).

197. Kooi, M.; Besseling, E.; Kroeze, C.; Van Wezel, A.P.; Koelmans, A.A. Modeling the fate and transport of plastic debris in freshwaters: Review and guidance. In *Freshwater Microplastics*; Springer: Cham, Switzerland, 2018; pp. 125–152. [CrossRef]

198. Besseling, E.; Quik, J.T.; Sun, M.; Koelmans, A.A. Fate of nano-and microplastic in freshwater systems: A modeling study. *Environ. Pollut.* **2017**, *220*, 540–548. [CrossRef]

199. Castañeda, R.A.; Avlijas, S.; Simard, M.A.; Ricciardi, A. Microplastic pollution in St. Lawrence river sediments. *Can. J. Fish. Aquat. Sci.* **2014**, *71*, 1767–1771. [CrossRef]

200. Hoellein, T.J.; Shogren, A.J.; Tank, J.L.; Risteca, P.; Kelly, J.J. Microplastic deposition velocity in streams follows patterns for naturally occurring allochthonous particles. *Sci. Rep.* **2019**, *9*, 3740. [CrossRef] [PubMed]

201. Nel, H.A.; Dalu, T.; Wasserman, R.J. Sinks and sources: Assessing microplastic abundance in river sediment and deposit feeders in an Austral temperate urban river system. *Sci. Total Environ.* **2018**, *612*, 950–956. [CrossRef]

202. Hurley, R.; Woodward, J.; Rothwell, J.J. Microplastic contamination of river beds significantly reduced by catchment-wide flooding. *Nat. Geosci* **2018**, *11*, 251–257. [CrossRef]

203. Maazouzi, C.; Galassi, D.; Claret, C.; Cellot, B.; Fiers, F.; Martin, D.; Marmonier, P.; Dole-Olivier, M.J. Do benthic invertebrates use hyporheic refuges during streambed drying? A manipulative field experiment in nested hyporheic flowpaths. *Ecohydrology* **2017**, *10*, e1865. [CrossRef]

204. Malcolm, I.A.; Soulsby, C.; Youngson, A.F.; Hannah, D.M.; McLaren, I.S.; Thorne, A. Hydrological influences on hyporheic water quality: Implications for salmon egg survival. *Hydrol. Process.* **2004**, *18*, 1543–1560. [CrossRef]

205. Febria, C.M.; Beddoes, P.; Fulthorpe, R.R.; Williams, D.D. Bacterial community dynamics in the hyporheic zone of an intermittent stream. *ISME J.* **2012**, *6*, 1078–1088. [CrossRef]

206. Datry, T.; Larned, S.T.; Tockner, K. Intermittent rivers: A challenge for freshwater ecology. *BioScience* **2014**, *64*, 229–235. [CrossRef]

207. Bonada, N.; Doledec, S.; Statzner, B. Taxonomic and biological trait differences of stream macroinvertebrate communities between mediterranean and temperate regions: Implications for future climatic scenarios. *Glob. Chang. Biol.* **2007**, *13*, 1658–1671. [CrossRef]

208. Ward, A.S.; Zarnetske, J.P.; Baranov, V.; Blaen, P.J.; Brekenfeld, N.; Chu, R.; Derelle, R.; Drummond, J.; Fleckenstein, J.; Garayburu-Caruso, V.; et al. Co-located contemporaneous mapping of morphological, hydrological, chemical, and biological conditions in a 5th order mountain stream network, Oregon, USA. *Earth Syst. Sci.* **2019**. [CrossRef]

209. Zheng, L.; Cardenas, M.B.; Wang, L.; Mohrig, D. Ripple effects: Bedform morphodynamics cascading into hyporheic zone biogeochemistry. *Water Resour. Res.* **2019**, *55*, 7320–7342. [CrossRef]

210. Odum, H.T. Primary production in flowing waters. *Limnol. Oceanogr.* **1956**, *1*, 102–117. [CrossRef]

211. Mulholland, P.J.; Houser, J.N.; Maloney, K.O. Stream diurnal dissolved oxygen profiles as indicators of in-stream metabolism and disturbance effects: Fort Benning as a case study. *Ecol. Indic.* **2005**, *5*, 243–252. [CrossRef]

212. Roberts, B.J.; Mulholland, P.J.; Hill, W.R. Multiple scales of temporal variability in ecosystem metabolism rates: Results from 2 years of continuous monitoring in a forested headwater stream. *Ecosystems* **2007**, *10*, 588–606. [CrossRef]

213. Rajwa-Kuligiewicz, A.; Bialik, R.J.; Rowiński, P.M. Dissolved oxygen and water temperature dynamics in lowland rivers over various timescales. *J. Hydrol. Hydromech.* **2015**, *63*, 353–363. [CrossRef]

214. Brandt, T.; Vieweg, M.; Laube, G.; Schima, R.; Goblirsch, T.; Fleckenstein, J.H.; Schmidt, C. Automated in situ oxygen profiling at aquatic–terrestrial interfaces. *Environ. Sci. Technol.* **2017**, *51*, 9970–9978. [CrossRef]

215. Gilbert, J.; Dole-Oliver, M.J.; Marmonier, P.; Vervier, P. Surface water-groundwater ecotones. In *The Ecology and Management of Aquatic–Terrestrial Ecotones*; Naiman, R.J., Décamps, H., Eds.; Unesco: Paris, France, 1990; pp. 199–225.

216. Townsend, C.R.; Scarsbrook, M.R.; Dolédec, S. The intermediate disturbance hypothesis, refugia, and biodiversity in streams. *Limnol. Oceanogr.* **1997**, *42*, 938–949. [CrossRef]

217. Robertson, A.L. Lotic meiofaunal community dynamics: Colonisation, resilience and persistence in a spatially and temporally heterogeneous environment. *Freshw. Biol.* **2000**, *44*, 135–147. [CrossRef]

218. Morén, I.; Wörman, A.; Riml, J. Design of remediation actions for nutrient mitigation in the hyporheic zone. *Water Resour. Res.* **2017**, *53*, 8872–8899. [CrossRef]

219. Grant, S.B.; Azizian, M.; Cook, P.; Boano, F.; Rippy, M.A. Factoring stream turbulence into global assessments of nitrogen pollution. *Science* **2018**, *359*, 1266–1269. [CrossRef]

220. Wörman, A.; Packman, A.I.; Marklund, L.; Harvey, J.W.; Stone, S.H. Fractal topography and subsurface water flows from fluvial bedforms to the continental shield. *Geophys. Res. Lett.* **2007**, *34*, L07402. [CrossRef]

221. Stonedahl, S.H.; Harvey, J.W.; Wörman, A.; Salehin, M.; Packman, A.I. A multiscale model for integrating hyporheic exchange from ripples to meanders. *Water Resour. Res.* **2010**, *46*, W12539. [CrossRef]

222. Bridge, J.W. *High Resolution In-Situ Monitoring of Hyporheic Zone Biogeochemistry*; Science Report SC030155/SR3; Environment Agency: Bristol, UK, 2005; p. 51. Available online: https://assets.publishing.service.gov.uk/government/uploads/system/uploads/attachment_data/file/291498/scho0605bjco-e-e.pdf (accessed on 23 October 2019).

223. Posselt, M.; Jaeger, A.; Schaper, J.L.; Radke, M.; Benskin, J.P. Determination of polar organic micropollutants in surface and pore water by high-resolution sampling-direct injection-ultra high performance liquid chromatography-tandem mass spectrometry. *Environ. Sci. Process. Impacts* **2018**, *20*, 1716–1727. [CrossRef] [PubMed]

224. Saenger, N.; Kitanidis, P.K.; Street, R. A numerical study of surface-subsurface exchange processes at a riffle-pool pair in the Lahn River, Germany. *Water Resour. Res.* **2005**, *41*, 12424. [CrossRef]

225. Cardenas, M.B.; Wilson, J.L. Hydrodynamics of coupled flow above and below a sediment-water interface with triangular bed-forms. *Adv. Water Resour.* **2007**, *30*, 301–313. [CrossRef]

226. Cardenas, M.B.; Wilson, J.L. Effects of current-bed form induced fluid flow on the thermal regime of sediments. *Water Resour. Res.* **2007**, *43*, W08431. [CrossRef]

227. Cardenas, M.B.; Wilson, J.L. Dunes, turbulent eddies, and interfacial exchange with permeable sediments. *Water Resour. Res.* **2007**, *43*, W08412. [CrossRef]

228. Bardini, L.; Boano, F.; Cardenas, M.B.; Revelli, R.; Ridolfi, L. Nutrient cycling in bedform induced hyporheic zones. *Geochim. Cosmochim. Acta* **2012**, *84*, 47–61. [CrossRef]

229. Trauth, N.; Schmidt, C.; Maier, U.; Vieweg, M.; Fleckenstein, J.H. Coupled 3-D stream flow and hyporheic flow model under varying stream and ambient groundwater flow conditions in a pool-riffle system. *Water Resour. Res.* **2013**, *49*, 5834–5850. [CrossRef]

230. Chen, X.B.; Cardenas, M.B.; Chen, L. Three-dimensional versus two-dimensional bed form-induced hyporheic exchange. *Water Resour. Res.* **2015**, *51*, 2923–2936. [CrossRef]

231. Chen, X.; Cardenas, M.B.; Chen, L. Hyporheic exchange driven by three-dimensional sandy bed forms: Sensitivity to and prediction from bed form geometry. *Water Resour. Res.* **2018**, *54*. [CrossRef]

232. Ren, J.; Wang, X.; Zhou, Y.; Chen, B.; Men, L. An analysis of the factors affecting hyporheic exchange based on numerical modeling. *Water* **2019**, *11*, 665. [CrossRef]

233. Brunner, P.; Simmons, C.T. HydroGeoSphere: A fully integrated, physically based hydrological model. *Groundwater* **2012**, *50*, 170–176. [CrossRef]

234. Oxtoby, O.; Heyns, J.; Suliman, R. A finite-volume solver for two-fluid flow in heterogeneous porous media based on OpenFOAM. Paper presented at the Open Source CFD International Conference, Hamburg, Germany, 24–25 October 2013.

235. Broecker, T.; Teuber, K.; Elsesser, W.; Hinkelmann, R. Multiphase modeling of hydrosystems using OpenFOAM. In *Advances in Hydroinformatics*; Springer: Singapore, 2018; pp. 1013–1029. [CrossRef]

236. Broecker, T.; Teuber, K.; Sobhi Gollo, V.; Nützmann, G.; Hinkelmann, R. Integral flow modelling approach for surface water-groundwater interaction along a rippled streambed. *Water* **2019**, *11*, 1517. [CrossRef]

237. Honti, M.; Bischoff, F.; Moser, A.; Stamm, C.; Baranya, S.; Fenner, K. Relating Degradation of Pharmaceutical Active Ingredients in a Stream Network to Degradation in Water-Sediment Simulation Tests. *Water Resour. Res.* **2018**, *54*, 9207–9223. [CrossRef]

238. Lindim, C.; van Gils, J.; Cousins, I.T. A large-scale model for simulating the fate & transport of organic contaminants in river basins. *Chemosphere* **2016**, *144*, 803–810. [CrossRef]

239. Feijtel, T.; Boeije, G.; Matthies, M.; Young, A.; Morris, G.; Gandolfi, C.; Hansen, B.; Fox, K.; Holt, M.; Koch, V.; et al. Development of a geography-referenced regional exposure assessment tool for European rivers-great-er contribution to great-er #1. *Chemosphere* **1997**, *34*, 2351–2373. [CrossRef]

240. Coll, C.; Lindim, C.; Sobek, A.; Sohn, M.D.; MacLeod, M. Prospects for finding Junge variability-lifetime relationships for micropollutants in the Danube river. *Environ. Sci. Process. Impacts* **2019**. [CrossRef]

241. Frei, S.; Peiffer, S. Exposure times rather than residence times control redox transformation efficiencies in riparian wetlands. *J. Hydrol.* **2016**, *543*, 182–196. [CrossRef]

242. Pittroff, M.; Frei, S.; Gilfedder, B.S. Quantifying nitrate and oxygen reduction rates in the hyporheic zone using ^{222}Rn to upscale biogeochemical turnover in rivers. *Water Resour. Res.* **2017**, *53*, 563–579. [CrossRef]

243. Grabowski, R.C.; Gurnell, A.M. Diagnosing problems of fine sediment delivery and transfer in a lowland catchment. *Aquat. Sci.* **2016**, *78*, 95–106. [CrossRef]

244. Grabowski, R.C.; Gurnell, A.M.; Burgess-Gamble, L.; England, J.; Holland, D.; Klaar, M.J.; Morrissey, I.; Uttley, C.; Wharton, G. The current state of the use of large wood in river restoration and management. *Water Environ. J.* **2019**, *33*, 366–377. [CrossRef]

245. Azinheira, D.L.; Scott, D.T.; Hession, W.; Hester, E.T. Comparison of effects of inset floodplains and hyporheic exchange induced by in-stream structures on solute retention. *Water Resour. Res.* **2014**, *50*, 6168–6190. [CrossRef]

246. Lammers, R.W.; Bledsoe, B.P. What role does stream restoration play in nutrient management? *Crit. Rev. Environ. Sci. Technol.* **2017**, *47*, 335–371. [CrossRef]

247. Herzog, S.P.; Higgins, C.P.; Singha, K.; McCray, J.E. Performance of engineered streambeds for inducing hyporheic transient storage and attenuation of resazurin. *Environ. Sci. Technol.* **2018**, *52*, 10627–10636. [CrossRef] [PubMed]
248. Ward, A.S.; Morgan, J.A.; White, J.R.; Royer, T.V. Streambed restoration to remove fine sediment alters reach-scale transient storage in a low-gradient fifth-order river, Indiana, USA. *Hydrol. Process.* **2018**, *32*, 1786–1800. [CrossRef]
249. Herzog, S.P.; Higgins, C.P.; McCray, J.E. Engineered streambeds for induced hyporheic flow: Enhanced removal of nutrients, pathogens, and metals from urban streams. *J. Environ. Eng.* **2015**, *142*, 04015053. [CrossRef]
250. Bernhardt, E.S.; Palmer, M.A. Restoring streams in an urbanizing world. *Freshw. Biol.* **2007**, *52*, 738–751. [CrossRef]
251. Mutz, M.; Kalbus, E.; Meinecke, S. Effect of instream wood on vertical water flux in low-energy sand bed flume experiments. *Water Resour. Res.* **2007**, *43*, 1–10. [CrossRef]
252. Thompson, M.S.A.; Brooks, S.J.; Sayer, C.D.; Woodward, G.; Axmacher, J.C.; Perkins, D.M.; Gray, C. Large woody debris "rewilding" rapidly restores biodiversity in riverine food webs. *J. Appl. Ecol.* **2018**, *55*, 895–904. [CrossRef]
253. Palmer, M.A.; Menninger, H.L.; Bernhardt, E. River restoration, habitat heterogeneity and biodiversity: A failure of theory or practice? *Freshw. Biol.* **2010**, *55*, 205–222. [CrossRef]
254. Magliozzi, C.; Coro, G.; Grabowski, R.C.; Packman, A.I.; Krause, S. A multiscale statistical method to identify potential areas of hyporheic exchange for river restoration planning. *Environ. Model. Softw.* **2019**, *111*, 311–323. [CrossRef]
255. Betterle, A.; Schirmer, M.; Botter, G. Characterizing the spatial correlation of daily streamflows. *Water Resour. Res.* **2017**, *53*, 1646–1663. [CrossRef]
256. Betterle, A.; Schirmer, M.; Botter, G. Flow dynamics at the continental scale: Streamflow correlation and hydrological similarity. *Hydrol. Process.* **2018**, *33*, 627–646. [CrossRef]

Article

The Effect of Stream Discharge on Hyporheic Exchange

Brian Babak Mojarrad [1,*], Andrea Betterle [2,3], Tanu Singh [4], Carolina Olid [5] and Anders Wörman [1]

[1] Department of Sustainable Development, Environmental Science and Engineering, KTH Royal Institute of Technology, 114 28 Stockholm, Sweden
[2] Department of Water Resources and Drinking Water, EAWAG Swiss Federal Institute of Aquatic Science and Technology, 8600 Dübendorf, Switzerland
[3] Department ICEA and International Center for Hydrology "Dino Tonini", University of Padova, 35100 Padua, Italy
[4] School of Geography, Earth and Environmental Sciences, University of Birmingham, Birmingham B45 0AJ, UK
[5] Climate Impacts Research Centre, Department of Ecology and Environmental Science, Umeå University, 98107 Abisko, Sweden
* Correspondence: mojarrad@kth.se; Tel.: +46-76-775-2405

Received: 24 June 2019; Accepted: 9 July 2019; Published: 12 July 2019

Abstract: Streambed morphology, streamflow dynamics, and the heterogeneity of streambed sediments critically controls the interaction between surface water and groundwater. The present study investigated the impact of different flow regimes on hyporheic exchange in a boreal stream in northern Sweden using experimental and numerical approaches. Low-, base-, and high-flow discharges were simulated by regulating the streamflow upstream in the study area, and temperature was used as the natural tracer to monitor the impact of the different flow discharges on hyporheic exchange fluxes in stretches of stream featuring gaining and losing conditions. A numerical model was developed using geomorphological and hydrological properties of the stream and was then used to perform a detailed analysis of the subsurface water flow. Additionally, the impact of heterogeneity in sediment permeability on hyporheic exchange fluxes was investigated. Both the experimental and modelling results show that temporally increasing flow resulted in a larger (deeper) extent of the hyporheic zone as well as longer hyporheic flow residence times. However, the result of the numerical analysis is strongly controlled by heterogeneity in sediment permeability. In particular, for homogeneous sediments, the fragmentation of upwelling length substantially varies with streamflow dynamics due to the contribution of deeper fluxes.

Keywords: hyporheic zone; transient flow discharge; groundwater-surface water interaction; experimental-modeling study; temperature measurement; depth decaying permeability

---

1. Introduction

In streams, groundwater–surface water interactions include hyporheic exchanges occurring at the sediment–water interface (SWI). Hyporheic exchanges trigger fluxes of water, solutes, oxygen, nutrients, and organic matter between surface water and sediment pore water [1,2]. The delivery of these compounds is in turn critical for the development of biotic communities in riverbed sediments [3]. Hyporheic exchange influences the residence times of water and solutes, controlling biogeochemical transformations and eventually influencing the functioning of the stream ecosystem [4,5]. These interactions occur at multiple spatial scales with both small-scale and large-scale fluxes determining the resultant spatial distribution of flow paths along the river corridor. Previous studies investigated the complex mechanism underlying exchanges such as transience in streamflow, topographical features, groundwater inflows/outflows, channel gradient, and heterogeneity in sediment properties [6–9].

Many of the aforementioned drivers and controlling factors of hyporheic exchange are spatially and temporally variable. The variability of streamflow in time and space results in complex hyporheic processes which demand comprehensive and systematic investigation. The study of steady-state streamflow discharges does not take into consideration temporal variations in the pressure distributions along the SWI. Natural river–aquifer systems are exposed to precipitation inputs, evapotranspiration, snow melt, outflow from dam operations, and wastewater treatment plants [10,11]. This leads to complex streamflow dynamics, which affect the hyporheic flow field, as well as the delivery of waterborne solutes. In addition to the pressure distribution on the SWI, subsurface heterogeneity has a substantial impact on the subsurface flow field [12,13]. Heterogeneity in hydraulic conductivity influences the magnitude of subsurface flow velocity, as well as the size of the hyporheic zone [14,15]. Previous studies have shown that hydraulic conductivity decays with depth following an exponential function that can be estimated based on field measurements [16–18]. Spatial heterogeneities of the hyporheic zone influence stream ecosystems [19] and, in turn, the fate and transport of contaminants in the subsurface region [20–22].

Variations in flow discharge and spatio–temporal temperature patterns are key aspects of river–aquifer systems. Various techniques have been applied to monitor the subsurface exchange fluxes of such systems [23]. Previous studies have well established that heat advected by water flow could be used as a natural tracer to identify and measure vertical hyporheic exchange [24–26]. The transportation of heat in porous media is governed by conduction (through a saturated bulk medium), convection (carried by water fluxes through pore spaces), and thermal dispersion (due to tortuous flow paths in the porous medium). The main advantage of using heat as a tracer over hydrometric methods is due to the independence of streambed thermal properties on sediment texture. This results in the variation in heat being less than the variation in hydraulic conductivity [27]. Nowadays, extensive and accurate temperature measurements are possible due to advances in high-resolution temperature data using fiber-optic technology [28–30]. Improved monitoring techniques using temperature loggers and temperature lances for shallow heat and flow dynamics has additionally improved our understanding of water quality in river–aquifer systems [31].

Various drivers and controls for hyporheic exchange have been extensively studied, particularly for steady-state discharge conditions. However, recently, transience in flow discharge in hyporheic modeling has attracted increasing attention [32–38]. Studies have highlighted the importance of storm-induced groundwater fluctuations for the volume of hyporheic exchange [35], residence time distributions [39], and other controlling factors, such as stream water velocity, hillslope lag, the amplitude of the hillslope, cross-valley and down-valley slopes on hyporheic flow paths [36,40]. Trauth and Fleckenstein [37] showed the importance of discharge events with different magnitudes and durations on the mean age of water and solutes in the hyporheic zone and their effect on the rates of aerobic and anaerobic respiration. Additionally, Sawyer et al. [10] conducted an experiment in the Colorado River, USA, and tried to estimate the influence of dam operation on hyporheic fluxes using high-resolution temporal data. Their experimental results indicated that the thermal, geochemical, and hydrological dynamics of the hyporheic zone were substantially affected by dam operation. Wroblicky et al. [41] studied the variation in the planimetric areal dimension of the hyporheic zone due to seasonal changes by developing numerical models for two first-order rivers. They showed that hyporheic depth is decreased in high flow and is sensitive to the hydraulic conductivity of the subsurface region.

The goal of this study is to investigate the impact of different stream discharge intensities (i.e., base-, low-, and high-flow) on the properties of hyporheic fluxes by combining experimental and numerical approaches. Unlike previous studies [10,37,41], here we implemented a numerical model and experimental approaches for the same region, which facilitates the interpretation of the results. The numerical model was constructed using morphological and hydrological data for a 1500 m long stream reach, and the experimental analysis was based on point temperature measurement at different streambed depths using temperature lances in stream reaches characterized by upwelling and downwelling. A vertical gradient in permeability is included in the model, and the role of the underlying decay function is investigated through the modeling results. In particular, we evaluate the magnitude of hyporheic exchange fluxes and the maximum depth of hyporheic fluxes, as well as the fragmentation of upwelling fluxes at the streambed interface under base-, low-, and high-flow discharges.

2. Materials and Methods

2.1. Study Site

The study is based on an experiment which was conducted in August 2017 in Krycklan Catchment, Sweden. Krycklan Catchment is a research catchment located approximately 50 km northwest of Umeå, Sweden (64°14′ N, 19°46′ E) [42]. The catchment has an area of 67.9 km^2 and an elevation range of 114–405 m a.s.l. Hydrological parameters were measured in Krycklan Catchment at 15 stations located in different parts of the catchment [43]. Krycklan Catchment is characterized by a cold and humid climate, featuring deep snow cover during the entire winter [44]. The 30-year (1981–2007) mean annual precipitation is 614 mm/year, 30–50% of which falls as snow [45]. The Quaternary deposits mostly consist of glacial till with a thickness of up to tens of meters [42].

The current study focused on a stretch of river with a length of approximately 1500 m, delimited by two hydrological stations, namely C5 and C6, in the upstream and downstream regions, respectively. Lake Stortjärn is located 100 m upstream of station C5. The lake has an area of 4 ha and is fed by a mire system. The river stretch can be considered as the gaining stream in which the stream water source is dominated by Lake Stortjärn. However, there are several parts of the stream segment with losing condition which is highly controlled by topography. The subcatchment area draining the stretch of river between stations C5 and C6 is mostly covered by forest containing Scots pine (*Pinus sylvestris*), Norway spruce (*Picea abies*), and birch (*Betula pubescens*). Peat dominates the riparian zones of the river stretch in which organic content increases in riverbanks. Streambed sediment type varies along the river, and includes dense decomposed organic matter, sand, and cobble/boulder [44]. The width of the stream channel between stations C5 and C6 varies between 0.3 and 2 m, and the average slope of the stream is approximately 3% [46]. Figure 1 shows the subcatchment of the Krycklan Catchment Study (KCS) area in which the experiment was carried out.

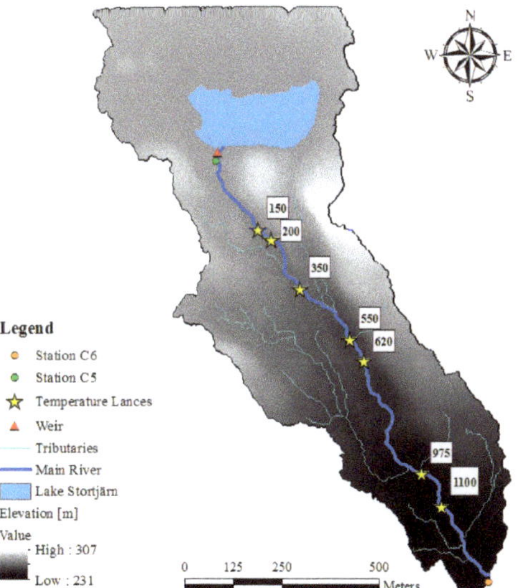

Figure 1. Map showing the experimental subcatchment and its topography, the main river (dark blue color), tributaries (cyan color), Lake Stortjarn (solid light blue region), and hydrological stations C5 and C6. Also shown are the locations of the V-notch weir (red flag) and temperature lances (yellow stars).

2.2. Field Measurement

During the experiment, the outflow from Lake Stortjärn into the stream discharge was regulated to simulate high and low downstream flow discharges. The regulation was conducted by closing a gate of a V-notch weir located approximately 50 m downstream of the lake outlet (Figure 1). The V-notch weir was completely closed during the period of 07–24 August 2017 so that no water entered the stream from Lake Stortjärn. Water was pumped downstream of the (closed) weir to simulate a high-flow pulse on 19 August (07:22 until 18:17) and 21 August (06:15 until 09:07) of 2017.

Leach et al. [44] delineated a set of upwelling and downwelling zones between stations C5 and C6. Vertical temperature profiles were recorded at seven locations along the stream reach. Starting from upstream, vertical temperature profiles were recorded at distances of 150, 200, 350, 550, 620, 975, and 1100 m from the outflow of Lake Stortjärn (Figure 1). Temperatures were recorded by means of Multi-Level Temperature Sticks (MLTS; UIT, Dresden, Germany) [47]. The MLTS sticks are polyoxymethylene probes featuring eight TSIC-506 thermometer sensors located at various distances from the head of the lance (i.e., the upper part), namely -60, -35, -25, -20, -15, -10, -5, and 0 cm. At the location 975 m from the outflow of the lake, a longer temperature stick was deployed with sensors located at -83, -65, -48, -31, -23, -17, -13, and 0 cm from the head of the lance. Each temperature sensor recorded at a temporal resolution of 5 min. The measuring range of the temperature sensors was between 5 and 45 °C, with a resolution of 0.04 °C. The typical accuracy of the sensors was 0.07 °C. Temperature sticks were hammered into the streambed on 3 August.

Stream discharge was continuously recorded at both the upstream (C5) and downstream (C6) stations at a temporal resolution of 1 min. Five different times of year were chosen in order to capture the intra-annual variability of the hydrological response of the catchment, namely, 22 May 2017 (snow melt flow), 27 June 2017 (summer base flow), 18 August 2017 (low flow), 19 August 2017 (high flow), and 30 August 2017 (autumn base flow). The stream discharge along the main channel was estimated for each of the aforementioned flow discharges (i.e., snow melt, summer, low, high, and autumn flows) at a spatial resolution of 50 m, accounting for the progressive downstream drainage area (Table 1 and Figure S1 in the Supplementary Materials). Moreover, the mean water depth of the river was measured at a spatial resolution of 50 m for each of the five mentioned flow discharges (Table 1). It should be mentioned that the mean water depth for summer flow was only measured up to 400 m from the upstream station.

2.3. Field Data Analysis

A power-law relationship was established between stream discharge estimates (Table 1) and measured water depths (Table 1) at a spatial resolution of 50 m along the stream. The power-law relationship provides the means for estimation of water depth with the spatial resolution of 50 m for low-, high-, and base-flow condition using the measured discharge intensities. Figure 2 (and Table S1 in the Supplementary Materials) shows the rating curves that were calculated for each specific location along the stream. The rating curves were used to determine the water depths which subsequently were used as the boundary condition on the streambed interface of the numerical model.

The experimental period (03–21 August) was divided into subperiods reflecting base-, low-, and high-flow discharges (Table 2). The mean stream discharge at a spatial resolution of 50 m along the main channel was estimated for each of the three aforementioned flow discharges accounting for the progressive downstream drainage area. Finally, the corresponding water depth for each flow discharge was estimated using the estimated discharge and correlation functions shown in Figure 2 at a spatial resolution of 50 m (Table S2 in Supplementary Materials). These water depths were later used (Section 2.3) to address the top boundary condition of the numerical model for base-, low-, and high-flow discharges.

Table 1. Estimated stream discharge values based on measured streamflow data and drainage area (Q), and measured mean water depth (d) for different flow discharges at 50 m spatial resolution.

Distance from Upstream Station [m]	Drainage Area [m^2]	Snow Melt Flow (22 May 2017)		Summer Base Flow (27 June 2017)		Low Flow (18 August 2017)		High Flow (19 August 2017)		Autumn Base Flow (30 August 2017)	
		Q [L/s]	d [cm]	Q [L/s]	d [cm]	Q [L/s]	d [cm]	Q [L/s]	d [cm]	Q [L/s]	d [cm]
0	234,446	54.48	31	2.63	11	0.19	3	25.58	17	7.36	9
50	236,031	54.54	31	2.65	11	0.19	3	25.59	17	7.42	9
100	243,262	54.81	31	2.76	11	0.21	2	25.63	11	7.48	7
150	250,838	55.09	32	2.86	15	0.23	18	25.67	24	7.55	26
200	258,334	55.37	28	2.97	17	0.26	21	25.71	35	7.61	25
250	262,731	55.53	32	3.03	9	0.27	8	25.74	26	7.67	14
300	265,497	55.63	33	3.07	19	0.27	7	25.76	16	7.73	13
350	267,926	55.72	27	3.11	7	0.28	6	25.77	8	7.79	10
400	291,324	56.59	27	3.44	11	0.35	9	25.90	12	7.86	6
450	293,516	56.67	29	3.47	[illegible]	0.35	7	25.91	19	7.92	14
500	296,214	56.77	21	3.51	[illegible]	0.36	5	25.93	16	7.98	10
550	337,040	58.29	20	4.09	[illegible]	0.47	26	26.16	34	8.04	22
600	343,055	58.51	42	4.17	[illegible]	0.49	15	26.19	24	8.11	16
650	357,882	59.07	32	4.39	[illegible]	0.53	8	26.28	19	8.17	13
700	373439	59.64	29	4.61	[illegible]	0.57	19	26.37	33	8.23	27
750	375,921	59.74	27	4.64	[illegible]	0.58	14	26.38	20	8.29	11
800	379,197	59.86	35	4.69	[illegible]	0.58	5	26.40	15	8.35	5
850	396,899	60.52	22	4.94	[illegible]	0.63	5	26.50	16	8.42	17
900	405,145	60.82	15	5.06	[illegible]	0.66	9	26.54	16	8.48	12
950	411,322	61.05	26	5.14	[illegible]	0.67	4	26.58	19	8.54	14
1000	413,347	61.13	29	5.17	[illegible]	0.68	24	26.59	29	8.60	29
1050	535,591	65.67	31	6.91	[illegible]	1.01	6	27.28	17	8.66	10
1100	558,355	66.51	21	7.24	[illegible]	1.07	4	27.41	9	8.73	8
1150	567,113	66.84	14	7.36	[illegible]	1.10	2	27.46	9	8.79	5
1200	568,492	66.89	28	7.38	[illegible]	1.10	2	27.47	20	8.85	4
1250	573,211	67.07	30	7.45	[illegible]	1.11	14	27.49	23	8.91	20
1300	578,673	67.27	35	7.52	[illegible]	1.13	19	27.52	32	8.98	26
1437	648,703	69.87	35	8.52	[illegible]	1.32	19	27.92	32	9.19	26

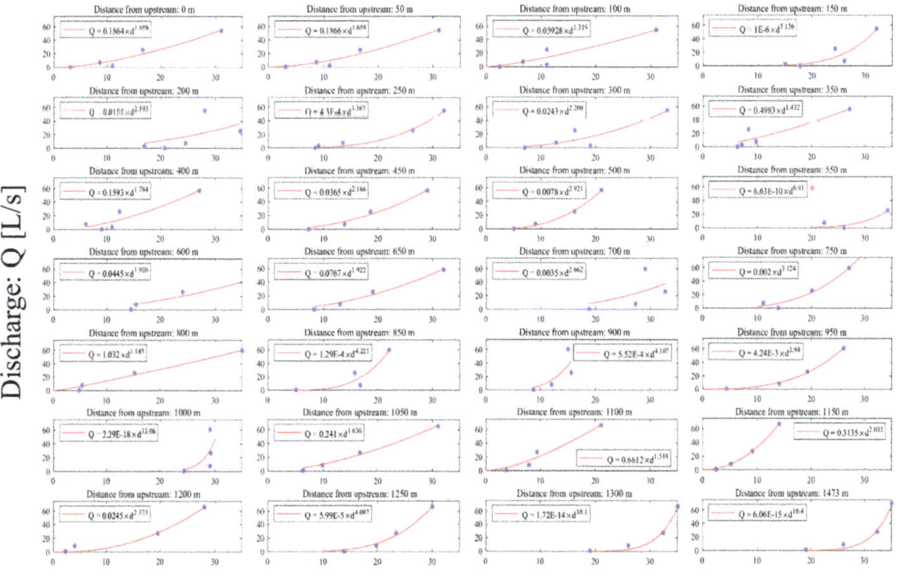

Figure 2. Relationships between discharge and water depth at a spatial resolution of 50 m (one of the discharge values at a distance of 550 m from the upstream station (shown with a magenta star) was not considered in the analysis due to its unrealistic value, which might be due to measurement error).

Table 2. Mean upstream and downstream stream discharge values for different flow discharges.

Flow Discharge	Time Period	Mean Upstream Discharge [L/s]	Mean Downstream Discharge [L/s]
Base flow	03–07 August	8.71	10.8
Low flow	07–19 August	0.22	1.56
High flow 1	19 August	25.59	27.96
High flow 2	21 August	25.57	30.09

2.4. Modeling Framework

The numerical model was developed using the COMSOL Multiphysics® software to simulate a two-dimensional longitudinal transect along the center of the stream network. The Brinkman–Darcy equation (Equation (1)) was used with the continuity equation to describe the subsurface flow in the hyporheic zone near the streambed surface:

$$\rho\left(\frac{1}{\varepsilon}\frac{\partial \mathbf{u}}{\partial t} + \frac{1}{\varepsilon^2}(\mathbf{u}\cdot\nabla\mathbf{u})\right) + \nabla p - \mu_e\nabla^2\mathbf{u} + \frac{\mu}{k}\mathbf{u} = 0 \tag{1}$$

where ρ [kg/m^3] is the water density, ε [–] is the Brinkman porosity, $\mathbf{u}$ [m/s] is the subsurface velocity vector, t [s] is the time, p [pa] is the water pressure, μ [pa.s] is the water's dynamic viscosity, μ_e [pa.s] is the water's effective viscosity, k [m^2] is the intrinsic permeability of the sediment, and $\left(\frac{\mu}{k}\mathbf{u}\right)$ is the Darcy term.

In the present study, the subsurface flow model includes Quaternary deposits, in which the top surface was quantified using a digital elevation model (DEM) file representing the stream morphology with a resolution of 50 cm. The depth of the Quaternary deposits varies along the stream, and is up to 17 m [48].

In order to investigate the effect of hydrogeological properties on the hyporheic flow fields, flow simulations were performed using two different sediment permeability scenarios: (a) constant permeability; and (b) depth-decaying permeability. The first permeability scenario was performed using a constant intrinsic permeability ($k = 10^{-9}$ [m^2] representing the streambed sediment [18]) for the entire subsurface region. The second permeability scenario was performed by applying a decaying intrinsic permeability in the top meter of the subsurface region (starting with $k = 10^{-9}$ [m^2] for the top surface decaying to $k = 10^{-12}$ [m^2] at a depth of 1 m) and a constant intrinsic permeability ($k = 10^{-12}$ [m^2] refers to the glacial sediment soil type) for the rest of the subsurface region (deeper than 1 m from the top surface). The depth-decaying function for the soil permeability was defined based on the findings of Morén et al. [18]. They performed an experiment in a small stream in Sweden (Tullstorps Brook) and measured the hydraulic conductivity at two depths (i.e., 3 and 7 cm) every 100 m for a reach of 1500 m. The permeability $k(z)$ of the top meter of the stream was described according to [17]:

$$k(z) = k_0 e^{cz} \tag{2}$$

where k_0 [m^2] is the intrinsic permeability at the streambed interface, z [m] is the depth from the streambed, and c [m^{-1}] is an empirical decay coefficient. An exponential function was fitted to the measured data of Morén et al. [18] with a lower hydraulic conductivity limit of $K = 10^{-6}$ [m/s] (corresponding to an intrinsic permeability of $k = 10^{-12}$ [m^2]) for a depth of 1 m from the streambed surface (Figure S2, Supplementary Materials).

The numerical model was constructed using a computational mesh with a non-uniform size and an increasing resolution at depths ranging between 0.03 and 14 m. The mesh size formation in longitudinal direction (along the stream) depends on streambed morphology fluctuation. The more morphology fluctuation, the lower mesh size.

A pressure boundary condition was assigned based on the surface flow discharge. In particular, a hydrostatic pressure was applied at the sediment interface based on the overlying water depths estimated as described in Section 2.3 (Table S2 in the Supplementary Materials). The spatially varying head boundary condition at the streambed surface was calculated by: $p(x, y = y_{topography}) = \rho g d_i(x)$, where $p(x, y = y_{topography})$ [pa] is the pressure distribution at the streambed topography surface, g [m/s^2] is the gravitational acceleration, and $d_i(x)$ [m] is the water depth for each flow discharge. The water depth was quantified using the corresponding mean discharge value for each flow discharge (see Section 2.3). Furthermore, an open boundary was assumed for the upstream and downstream sides, whereas a no-flow boundary was considered for the bottom surface of the model. Each model was run to a steady state flow condition based on hydrostatic head boundary condition on the top surface.

The flow velocities at the streambed interface, as well as trajectories and residence times of 1000 uniformly distributed water particles at the streambed water interface, were evaluated during the three types of flow discharge (i.e., base-, low-, and high-flow discharges). It should be noted that the movement of the streambed sediment was not accounted for in the numerical model. The hyporheic exchange depth, hyporheic flux residence times, and the fragmentation of upwelling length were considered as the metrics. Fragmentation quantifies the heterogeneity in the spatial distribution of the zones characterized by gaining and losing behavior [15]. In particular, fragmentation is defined as the size distribution of spatial coherent upward and downward lengths. Here, the frequency of occurrence of different lengths of coherent upwelling and downwelling regions at the streambed interface was represented by a cumulative distribution function (CDF). An increase in the fragmentation of coherent flow regions results in a shift towards shorter (upwelling/downwelling) lengths in the CDF plot. It should be mentioned that this definition is independent of the magnitude of the fluxes.

3. Results

3.1. Water Temperature

Time series of the measured water temperature at the seven monitoring locations for the different flow discharges (i.e., base-, low-, and high-flow) are presented in Figure 3. Each time series corresponds to the temperature recorded by sensors at increasing depth. Note that the period of record (i.e., the horizontal axis for each column in Figure 3) differs between the three flow regimes, with the low-flow and high-flow discharges having the longest and shortest records, respectively. Since no full diurnal cycles are captured during high-flow discharge (the recorded time period is about 10 h), the variability of the temperature signals is smaller compared to the high- and base-flow discharges. On the other hand, low-flow discharge displays larger temperature fluctuations at shallower depths compared to the base-flow discharge. Temperature lances located at distances of 150, 200, and 350 m from the upstream station recorded a relatively wider range of temperature for the base-flow discharge than the high-flow discharge; conversely, the difference in temperature distribution was negligible between the different flow discharges for temperature lances located at 550, 620, 975, and 1100 m. During high-flow discharge, all temperature time series clearly highlight the moment when the flood wave reaches each location (temperature increments are delayed at further downstream locations).

Figure 4 shows the temperature envelopes versus depth for all monitoring locations during the three different flow discharges. The envelopes represent the median and the interquartile range of the temperature signals. The statistics were evaluated for each sensor (i.e., for each depth) during different flow discharges. Median and quartiles were then linearly interpolated between the sensors. The envelopes effectively show the temperature pattern with increasing depth as well as the vertical variability of the temperature signal. Steep gradients of median temperature suggest an abrupt transition between higher surface-water temperatures and lower groundwater temperatures. This generally indicates cold upwelling fluxes approaching the surface of the streambed and, consequently, suggests the existence of shallow hyporheic zones. Strong temperature gradients with increasing depth are generally associated with decreased spatial variability in the temperature signal (i.e., a narrow interquartile range). This is a consequence of the relatively constant temperature of groundwater throughout the year (in the range of 8–12 °C), in contrast to the strong temporal dynamics of surface water temperature. The results indicate that low-flow discharge has the lowest temperature range (8–12 °C) and the base-flow discharge has the highest temperature range (9–16 °C). Additionally, the variability of the temperature envelope with depth for different flow discharges shows a similar behavior at distances of 150, 200, and 350 m from the upstream station, while the behavior at distances of 550, 620, 975, and 1100 m are similar to each other but different from those of the other flow group. It should be noted that, in case of base- and low-flow discharges, daily cycles in surface water temperature are mostly responsible for the variability in temperature envelopes (represented by the quartile range in Figure 4). On the other hand, the observed variability in temperatures during high-flow discharge is a consequence of the abrupt increment of temperature that is experienced at all motoring locations at the arrival of the flood wave.

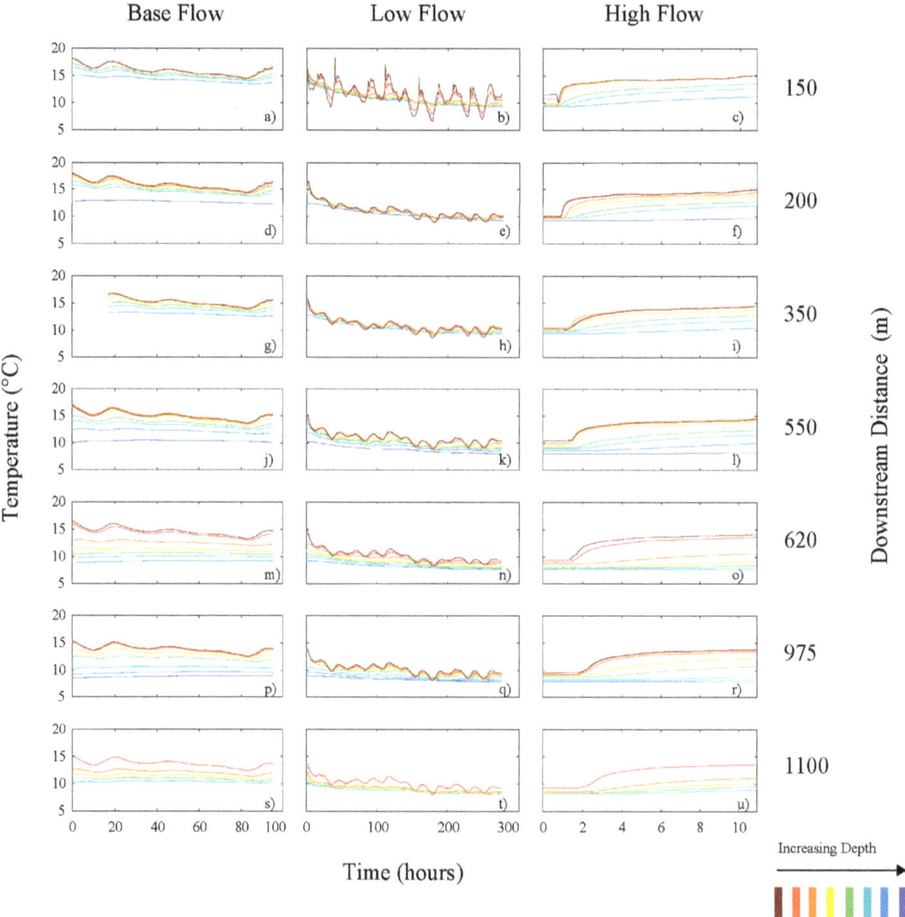

Figure 3. Water temperature time series measured at different locations along the stream network for different flow discharges. Each color represents the temperature recorded by sensors at increasing depth. Colors range from brown to blue as depth increases. Temperatures were not properly recorded during the first 20 h by the temperature stick located 350 m from the upstream station (panel g), and these were therefore neglected. High-flow discharge corresponds to high flow 1 period of Table 2.

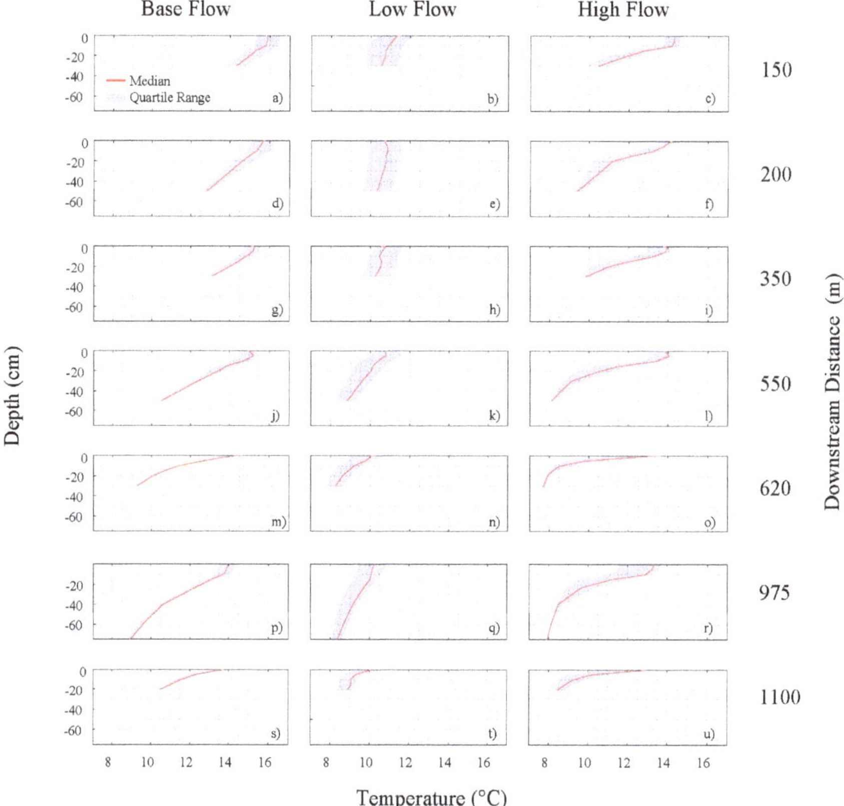

Figure 4. Vertical envelopes of temperature dynamics during base-, low-, and high-flow discharges at different monitoring locations. The envelopes indicate the interquartile range (shaded area) and the median (red line).

3.2. Modeling Results

The hyporheic flow field was quantified for the study reach. The spatial distribution of the maximum depth of the hyporheic fluxes was quantified based on the deepest point of the streamlines ($D_{HF,max}$). Figure 5 shows the calculated maximum depths of the hyporheic fluxes for the case of constant permeability (Figure 5a) and decaying permeability in the top meter of the flow domain (Figure 5b). The results indicate that $D_{HF,max}$ varied during the three flow discharges. In particular, the minimum $D_{HF,max}$ is highly affected by stream water flow intensity (the range of $D_{HF,max}$ is affected by the order of 10 and 100 in constant and decaying permeability scenarios, respectively). High-flow discharges have lower values of $D_{HF,max}$ than the base-flow discharge, whereas the minimum $D_{HF,max}$ for the base flow is higher than that for low-flow discharge, regardless of the permeability scenario applied. The median value of $D_{HF,max}$ varies slightly between different flow discharges, with the low-flow discharge having the highest median value of $D_{HF,max}$ among all the flow discharges, i.e., 4 m and 0.03 m in constant and decaying permeability scenarios, respectively (Figure S3 in the Supplementary Materials). However, in the case of decaying permeability, the median value of $D_{HF,max}$ (Figure 5b) is in the range of centimeters which is substantially lower than in the constant permeability case (Figure 5a) that is in a range of meters. Additionally, the logarithmic ranges of interquartile of the maximum depth of the hyporheic fluxes in the constant permeability case (Figure 5a) are smaller than those in the decaying permeability scenario (Figure 5b) for all flow discharges.

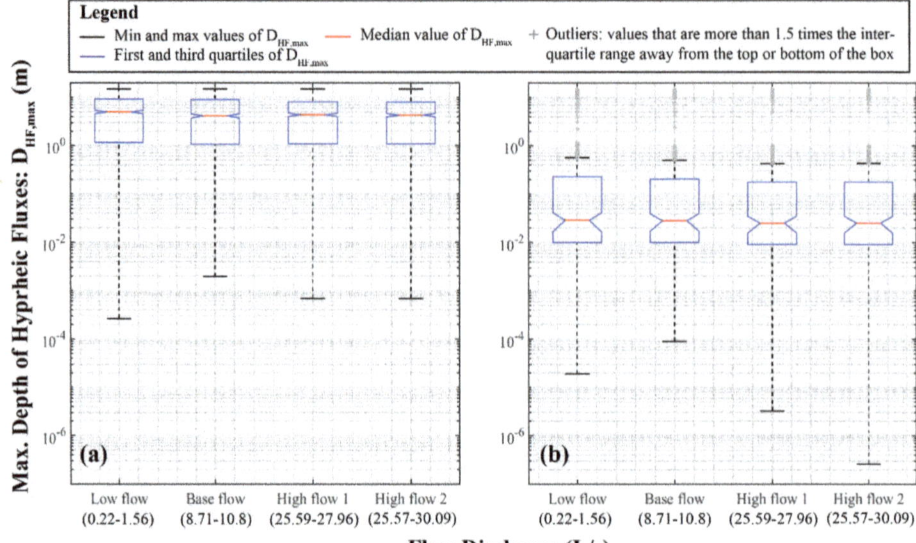

Figure 5. Boxplots showing the maximum depth of hyporheic fluxes under various flow discharges assuming (**a**) constant intrinsic permeability ($k = 10^{-9}$ [m^2]) for the entire subsurface region and (**b**) a decaying intrinsic permeability (starting from $k(z = 0) = 10^{-9}$ [m^2] at the surface–subsurface water interface and decaying exponentially to $k(z = -1) = 10^{-12}$ [m^2] at one meter depth). In the case of a vertically varying permeability, a constant permeability ($k = 10^{-12}$) was used for depths larger than one meter. The second row of the horizontal axis (i.e., numbers), are the ranges of stream flow discharge along the stream for each flow regime. $D_{HF,max}$: deepest point of the streamlines.

The second metric that was analyzed was the residence time of each released particle at the streambed interface. In this study, the residence time is defined as the time taken for the water to enter along a streamline through the streambed and return to the streambed surface again. Figure 6 presents the distribution of residence times for 1000 water parcels that were uniformly distributed at the streambed surface. In both permeability cases (i.e., decay and no decay), the high flow discharges have a larger range of residence times. The minimum residence time varies among different streamflow discharges, with increasing flow discharge intensity resulting in the decrease of the minimum hyporheic flow residence time. The influence of permeability is evident in the interquartile ranges of residence time. A vertical gradient of permeability induces a larger interquartile range of residence times (i.e., a larger variability in residence times). Additionally, the decaying of permeability results in a higher median value of residence time compared to the homogeneous case, i.e., in the order of 10 (Figure 6a,b).

Finally, the velocity field at the streambed interface was analyzed with a spatial resolution of 50 cm in order to investigate the distribution of coherent upwelling zones (i.e., lengths of upwelling zones) for different flow discharges. The CDF of the length of the coherent upwelling was computed. The results indicate that applying a decaying permeability function strongly influences the fragmentation of hyporheic flows (Figure 7). Considering a constant permeability for the whole subsurface (Figure 7a) results in significant variation in the fragmentation length between different flow intensities, with the high-flow discharges being less fragmented (larger upwelling/downwelling lengths) than the other flow discharges. In this case, for almost 90% of the lengths, the base flow and low flow are the most fragmented flows. On the other hand, assuming a depth-decaying permeability induces a similar fragmentation of coherent upwelling length for all the flow discharges (Figure 7b). However, in this case, the hyporheic exchange during low-flow discharge is less fragmented at lengths larger than 5 m compared to the other flow discharges.

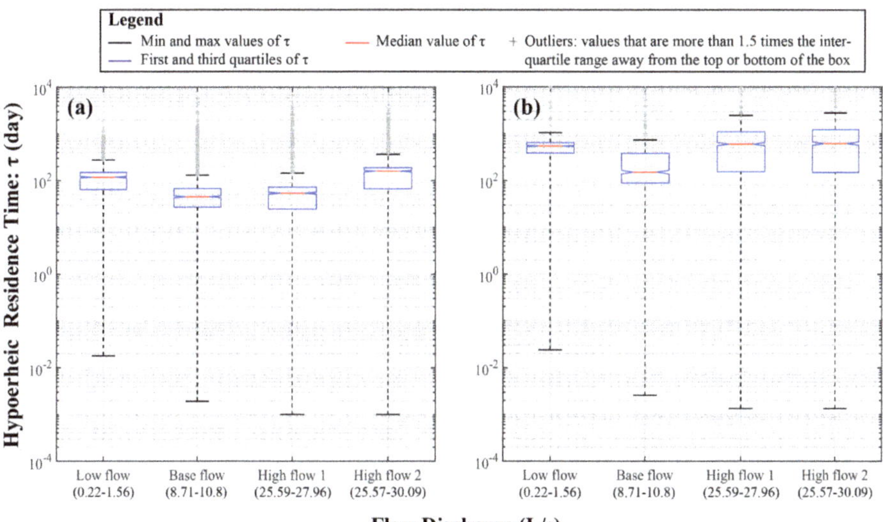

Figure 6. Box and whisker plots of hyporheic fluxes residence time under various flow discharges assuming (**a**) constant intrinsic permeability ($k = 10^{-9}$ [m^2]) for the entire subsurface region and (**b**) decaying intrinsic permeability (starting from $k(z = 0) = 10^{-9}$ [m^2] at the surface–subsurface water interface and decaying exponentially to $k(z = -1) = 10^{-12}$ [m^2] down to a depth of 1 m). In the case of vertically varying permeability, a constant permeability ($k = 10^{-12}$) was used for depths larger than 1 m. The second row of the horizontal axis (i.e., numbers), are the ranges of stream flow discharge along the stream for each flow regime. τ: residence time of the particles released at the streambed interface.

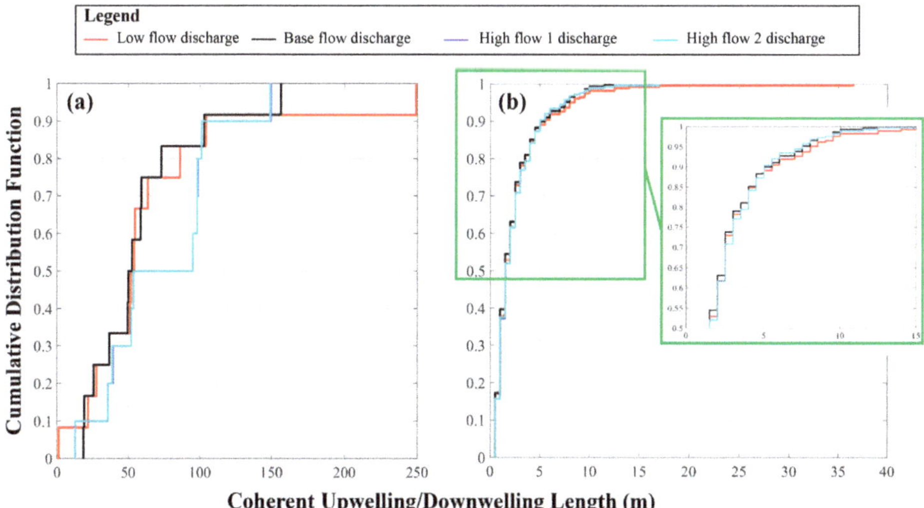

Figure 7. Cumulative distribution function of the length distribution of spatially coherent upwelling/downwelling stretches at the streambed interface under various flow discharges, assuming (**a**) constant intrinsic permeability ($k = 10^{-9}$ [m^2]) for the entire subsurface region, and (**b**) a decaying intrinsic permeability.

4. Discussion

A comprehensive understanding of the impacts of flow discharge on hyporheic exchange fluxes is essential to highlight the effect of streamflow dynamics on the extent of hyporheic exchange. Here, we combined field experiments with numerical modeling to estimate the variation in hyporheic depth and flow velocity field in a boreal stream.

4.1. Temperature Variation in the Streambed Sediment

The experiment of the present study was performed in summer, when groundwater is colder than stream water. Hence, the diurnal fluctuation of stream water temperature can be used to investigate the hyporheic exchange and identify gaining and losing reaches [49]. During summer, the water temperature of the hyporheic zone tends to decrease as the depth of the streambed increases [50]. At the same time, the stream water temperature signal is progressively damped at larger depths. These effects result from mixing between deep cold subsurface water and warm surface water mediated by the thermal dispersion associated with both thermal diffusion and hydrodynamic dispersion [25]. However, this general trend displays significant variations between monitoring locations as well as during different flow discharges (Figure 3). The temperature time series indicates areas of upwelling and of downwelling in the experimental reach. Strong downwelling conditions can be observed both in large and relatively erratic temperature records (e.g., at a distance of 150 m from the outflow of Lake Stortjärn for the base-flow discharge) and by observing the surface temperature dynamics at larger depths (e.g., at a distance of 150, 200, and 350 m from the outflow of Lake Stortjärn for low-flow discharges). A milder downwelling condition is suggested when the recorded temperature at greater depths is significantly larger than the average groundwater temperature (at a distance of 200 and 350 m from the outflow of Lake Stortjärn for the base-flow discharge), which suggests advective downward fluxes. The increase in temperature at larger depth during the high-flow discharge also suggests mild downwelling fluxes (at a distance of 350 m from the outflow of Lake Stortjärn for high-flow discharge). On the other hand, the relatively large difference in average temperature with increasing depth suggests areas of upwelling (at a distance of 550, 620, 975, and 1100 m from the outflow of Lake Stortjärn).

During low-flow discharge, the sediments in downwelling reaches are likely to experience reduced saturation, facilitating thermal diffusion towards deeper soils (due to lower thermal inertia) [51]. However, the lack of significant surface water fluxes during low-flow discharge hinders the interpretation of temperature data. Thermal diffusion can play a major role in sediment heat transfer, and its relation with soils having heterogeneous saturation can significantly complicate the analysis of temperature data. In general, the similarity of the temporal variation in temperature and the difference in temperature at different depths indicate upwelling and downwelling conditions in the stretches. A higher temperature at depth indicates a more robust downwelling condition, whereas a larger difference in temperature at depth shows upward fluxes.

The temperature envelopes shown in Figure 4 tend to confirm upwelling and downwelling stretches result as is suggested by temperature time series. Moreover, they can be used to further interpret the spatial patterns of hyporheic exchanges under the different flow discharges.

Under the base-flow discharge, the locations 150 and 200 m from the outflow of Lake Stortjärn (and to some extent 350 m) display clear signs of downwelling conditions. The gradient in the median temperature is lower than in the remaining sites, and the temperature is variable at greater depths. This suggests a significant convective heat transport following surface water flows towards deeper soil layers. Conversely, the strong vertical temperature gradients, together with the narrow interquartile ranges of temperature, confirm upwelling conditions at the locations 620 and 1100 m from the outflow of Lake Stortjärn (Figure 4m,s). Additionally, moderate upwelling conditions and the development of a shallow hyporheic zone are suggested by the small variability of temperature with depth observed at locations 550 and 975 m from the outflow of Lake Stortjärn.

As already mentioned, the temperature data under low-flow discharge is more complex than for the other flow discharges. In the low-flow discharge condition, the hyporheic flow at downwelling locations does not only include direct downwelling from the surface water, but can also include the upwelling of hyporheic water (e.g., water that has infiltrated upstream). On the other hand, the signature of groundwater upwelling might be more easily detectable, since large groundwater circulation patterns can be less affected by streamflow discharges. At low-flow discharge, temperature envelopes can be divided into two groups: the first includes the locations 150, 200, and 350 m from the outflow of Lake Stortjärn, and the second includes the remaining downstream sites. In the first three upstream locations, the median temperature remains almost constant as the depth in the streambed sediments increases. Furthermore, the temperature variability (represented by the interquartile range) remains stable across depth, while featuring extremely large values. The undisturbed propagation (i.e., small damping) of the temperature signal at larger depths suggests a reduced thermal inertia of the sediments during low-flow discharges. This may be related to unsaturated conditions established in the sediments during this phase. The establishment of unsaturated conditions further suggests the absence of significant groundwater flow in these locations.

The temperature envelopes at the remaining downstream locations (i.e., 550, 620, 975, and 1100 m from the outflow of Lake Stortjärn) display a different pattern. As depth increases, temperature decreases, both in terms of median temperature and interquartile range. Moreover, the temperatures measured at the surface are lower compared to those at the first three upstream locations. This is likely a consequence of groundwater upwelling.

Moreover, during the high-flow discharge, a significant increase in surface temperature is observed at all locations. At locations 150, 200, and 350 m from the outflow of Lake Stortjärn, the vertical temperature gradient increases compared to the base-flow discharge, suggesting the development of a hyporheic zone with a reduced extent. The inflection of the median temperature profile at sites 550 and 975 m from the outflow of Lake Stortjärn (Figure 4l,r) suggest a shallow (i.e., 10–20 cm in depth) hyporheic zone. The enhanced exchange of surface water with sediments is further suggested by the increased variability in the temperature profile compared with the base-flow discharge. On the other hand, the large temperature gradient at locations 620 and 1100 m from the outflow of Lake Stortjärn suggests significant upwelling fluxes, possibly reinforced by pressure gradients triggered by high flow discharges.

Our experimental findings regarding the regions of upwelling and downwelling confirm the experimental results of Leach et al. [44], who estimated gaining regions using distributed temperature sensing (DTS) fiber-optic cable. A comparison of the base- and high-flow discharges in Figure 4 reveals a higher contribution of stream water than groundwater in the hyporheic zone for high-flow discharge. Increasing the stream water flow discharge increases the variability of the hydrostatic pressure distribution on the streambed interface; hence, the hyporheic flow gradient is increased, which overwhelms the groundwater flow gradient. Consequently, the penetration depth increases, which increases the depth of the hyporheic zone.

4.2. The Role of the Heterogeneity of Sediment Permeability in Hyporheic Exchange

The complex pattern of temperature distribution observed in the field suggests an enhanced spatiotemporal variability of hyporheic flows. The implementation of the numerical model presented in Section 2.3 aims to capture the reach-scale flow patterns highlighted by the analyses of the temperature data collected in the field. The role of the heterogeneity of streambed sediment was investigated due to its strong impact on the distribution of hyporheic flow paths [52,53]. In particular, the effect of depth-decaying permeability on hyporheic fluxes was investigated in this study. As expected, the stream water penetrates into deeper sediment in the case of constant intrinsic permeability (Figure 5a), while the hyporheic zone is much shallower in the case of depth-decaying permeability (Figure 5b). This is basically due to the shielding effect of decreasing permeability with depth [8]. Low permeability sediment acts as a semi-impermeable surface that inhibits the penetration of stream water into deeper sediment. The high-flow discharge has a relatively shallower median hyporheic zone (Figure S3). The reason for this is the lower variability in the spatial gradient of the hydrostatic hydraulic head during high-flow discharges than during low- and base-flow discharges, at least for part of the study reach. The minimum and median values of hyporheic depth vary slightly between high flow 1 and 2 due to the slight change in the stream water flow discharge between these conditions (Table 2).

The results of this study indicate that increasing the flow discharge results in a larger distribution of hyporheic residence time. High-flow discharges cause a longer flow path, which consequently increases the residence time. On the other hand, the hyporheic fluxes with long residence time act like groundwater flow at upwelling reaches, so that stream water is prevented from penetrating into the sediment. This leads to a short flow path as well as a short residence time of hyporheic flow [38]. Therefore, upwelling and downwelling regions control the behavior of the residence time, with short residence times corresponding to upwelling zones and long residence times corresponding to the hyporheic fluxes starting their journey into the sediment from downwelling regions. The results of this study regarding hyporheic depth and residence time confirms the finding of Cardenas et al. [52] regarding the large dependency of the hyporheic flow process on the heterogeneity of streambed permeability.

Mojarrad et al. [15] observed variation in the fragmentation of coherent upwelling and downwelling zones for different stream orders (i.e., stream water depths). They found that the fragmentation of upwelling zones can vary depending on the large-scale groundwater contribution. However, the variation in the fragmentation of downwelling zones was negligible as there was no groundwater contribution in these areas. Mojarrad et al. [15] showed that increasing the stream order decreased the fragmentation of coherent upwelling zones but did not influence the fragmentation of coherent downwelling areas, which is consistent with our findings. In the present study, it was found that, in the case of a homogeneous subsurface, the stream water can penetrate the whole depth of the Quaternary deposits. Hence, the impact of a deeper flow streamline should be observed in the fragmentation length. As is shown in Figure 7a, high-flow discharges (i.e., a higher stream depth) are less fragmented compared to the low- and base-flow discharges. This is due to the effect of deeper flow on upwelling zones. On the other hand, Mojarrad et al. [15] showed that the fragmentation of downwelling zones varies slightly with stream order if the contribution of deep groundwater is negligible. Depth-decaying permeability prevents the deep penetration of stream water into the stream sediment, which is similar to the finding of Mojarrad et al. [15] that groundwater does not contribute to hyporheic fluxes in downwelling zones. Therefore, in this study, the fragmentation of upwelling areas differs slightly between flow discharges.

Field experiments consist of all the parameters that influence the hyporheic exchange. Heterogeneity in permeability of subsurface sediment is one of the factors that significantly controls the upwelling and downwelling regions, as well as the depth of hyporheic zone. Here, the result of field experiment showed that depth of hyporheic zone vary up to 50 cm in different flow discharges (Figure 4). This result confirms the numerical model results with the decaying intrinsic permeability. It was indicated that assuming homogeneous subsurface leads to hyporheic depth in the order of meters, whereas applying decaying in subsurface permeability results in much shallower hyporheic zone with the maximum depth of 40–60 cm (Figure 5). This result confirms the critical role of heterogeneity in subsurface flow, which should be considered in numerical modelling to improve the understanding of hyporheic flow processes.

5. Conclusions

This study investigated the impact of different stream discharges on the hyporheic zone using experimental and numerical approaches. We measured streambed temperature profiles at different locations along a boreal stream in northern Sweden. Different streamflow discharges were investigated by pumping water into the stream channel over a blocked weir. A numerical model was applied to investigate the impact of stream discharge on hyporheic exchange using geomorphological and hydrological data for the experimental reach. The gaining and losing conditions were evaluated using temperature data for the streambed sediment. In particular, this study evaluated locations of strong downwelling (at distances of 150, 200 m from the outflow of Lake Stortjärn), mild downwelling (at distances of 350 m from the outflow of Lake Stortjärn), mild upwelling (at distances of 550 and 975 m from the outflow of Lake Stortjärn), and strong upwelling (at distances of 620 and 1100 m from the outflow of Lake Stortjärn). The results show that increasing flow discharge increases hyporheic depth through the development of nested hyporheic flow paths. The contribution of groundwater to surface water is larger during low-flow discharge than during high-flow discharge, which is manifested as higher temperatures in the stream water. The modeling results agree with the experimentally measured temperature, showing a deeper penetration of stream water into sediment during high-flow discharges. Additionally, increasing streamflow discharge results in a larger distribution of the residence time of hyporheic fluxes. This is more evident in residence time distributions in which the high-flow discharges display larger interquartile ranges. Furthermore, the modeling results highlight the fact that heterogeneity in the permeability of the streambed sediment strongly controls the hyporheic flow exchange. The fragmentation of coherent upwelling length was defined as the shift of the CDF towards shorter lengths. The fragmentation of coherent upwelling length varied with streamflow discharge, with high-flow discharges being less fragmented. When permeability decayed with depth, flow paths were constrained to shallow regions; consequently, the upwelling fluxes were reduced and the fragmentation of the upwelling length was less variable between different flow discharges.

Supplementary Materials: The following are available online at http://www.mdpi.com/2073-4441/11/7/1436/s1, Figure S1: "Map showing the draining area every 50 m along the main river of the subcatchment, as well the main river (dark blue color), tributaries (light blue color), the lake Stortjärn (solid light blue region), and 50 m distance markers (red dots) from upstream", Figure S2: "Hydraulic conductivity decay function along the depth. Red color dashed lines represent the measured values from Morén [18], and blue color solid line is the mean exponential function for all the measured data", Figure S3: "Mean value of hyporheic zone depth under various flow condition assuming: (**a**) constant intrinsic permeability ($k = 10^{-9}$ [m^2]) for the entire subsurface region and (**b**) decaying intrinsic permeability (starting from $k(z = 0) = 10^{-9}$ [m^2] at the surface–subsurface water interface and decaying exponentially to $k(z = -1) = 10^{-12}$ [m^2] down to a depth of 1 m). In the case of vertically varying permeability, a constant permeability ($k = 10^{-12}$) was used for depths larger than 1 m.", Table S1: "Correlation between discharge and mean water depth using five flow data for every 50 m distance from upstream along the main river", Table S2: "Estimated mean water depth for different flow condition every 50 m from upstream".

Author Contributions: Conceptualization, B.B.M.; Investigation, B.B.M., A.B., and C.O.; Methodology, B.B.M.; Formal Analysis, A.B., and B.B.M.; Project Administration, B.B.M.; Software, B.B.M., and T.S.; Visualization, B.B.M., and A.B.; Writing—original draft preparation, B.B.M., A.B., and T.S.; Writing—review and editing B.B.M., A.B., C.O., and A.W., Supervision A.W.; Funding Acquisition A.W.

Funding: This project was supported by the EU's Horizon 2020 Research and Innovation Programme under Marie-Skłodowska-Curie grant agreement No. 641939 and the Swedish Radiation Safety Authority, SSM.

Acknowledgments: We would like to thank Anna Lupon (CEAB-CSIC); Nicolai Brekenfeld, University of Birmingham; and the crew of Krycklan Catchment Study (KCS) funded by SITES (VR), for their help in the field campaign.

Conflicts of Interest: The authors declare no conflict of interest.

References

1. Cardenas, M.B. Hyporheic zone hydrologic science: A historical account of its emergence and a prospectus. *Water Resour. Res.* **2015**, *51*, 3601–3616. [CrossRef]
2. Tonina, D.; Buffington, J.M. Effects of stream discharge, alluvial depth and bar amplitude on hyporheic flow in pool-riffle channels. *Water Resour. Res.* **2011**, *47*. [CrossRef]
3. Stanford, J.A.; Ward, J.V. The hyporheic habitat of river ecosystems. *Nature* **1988**, *335*, 64–66. [CrossRef]
4. Sawyer, A.H.; Bayani Cardenas, M.; Buttles, J. Hyporheic temperature 165 dynamics and heat exchange near channel-spanning logs. *Water Resour. Res.* **2012**, *48*. [CrossRef]

5. Harvey, J.W.; Böhlke, J.K.; Voytek, M.A.; Scott, D.; Tobias, C.R. Hyporheic zone denitrification: Controls on effective reaction depth and contribution to whole stream mass balance. *Water Resour. Res.* **2013**, *49*, 6298–6316. [CrossRef]

6. Boano, F.; Camporeale, C.; Revelli, R.; Ridolfi, L. Sinuosity-driven hyporheic exchange in meandering rivers. *Geophys. Res. Lett.* **2006**, *33*. [CrossRef]

7. Elliott, A.H.; Brooks, N.H. Transfer of nonsorbing solutes to a streambed with bed forms: Theory. *Water Resour. Res.* **1997**, *33*, 123–136. [CrossRef]

8. Gomez-Velez, J.D.; Krause, S.; Wilson, J.L. Effect of low-permeability layers on spatial patterns of hyporheic exchange and groundwater upwelling. *Water Resour. Res.* **2014**, *50*, 5196–5215. [CrossRef]

9. Stonedahl, S.H.; Harvey, J.W.; Packman, A.I. Interactions between hyporheic flow produced by stream meanders, bars, and dunes. *Water Resour. Res.* **2013**, *49*, 5450–5461. [CrossRef]

10. Sawyer, A.H.; Cardenas, M.B.; Bomar, A.; Mackey, M. Impact of dam operations on hyporheic exchange in the riparian zone of a regulated river. *Hydrol. Process.* **2009**, *23*, 2129–2137. [CrossRef]

11. McGlynn, B.L.; McDonnell, J.J.; Shanley, J.B.; Kendall, C. Riparian zone flowpath dynamics during snowmelt in a small headwater catchment. *J. Hydrol.* **1999**, *222*, 75–92. [CrossRef]

12. Malcolm, I.A.; Soulsby, C.; Youngson, A.F.; Petry, J. Heterogeneity in ground water–surface water interactions in the hyporheic zone of a salmonid spawning stream. *Hydrol. Process.* **2003**, *17*, 601–617. [CrossRef]

13. Schmidt, C.; Bayer-Raich, M.; Schirmer, M. Characterization of spatial heterogeneity of groundwater-stream water interactions using multiple depth streambed temperature measurements at the reach scale. *Hydrol. Earth Syst. Sci. Discuss.* **2006**, *3*, 1419–1446. [CrossRef]

14. Boano, F.; Revelli, R.; Ridolfi, L. Reduction of the hyporheic zone volume due to the stream-aquifer interaction. *Geophys. Res. Lett.* **2008**, *35*, L09401. [CrossRef]

15. Mojarrad, B.B.; Riml, J.; Wörman, A.; Laudon, H. Fragmentation of the hyporheic zone due to regional groundwater circulation. *Water Resour. Res.* **2019**, *55*, 1242–1262. [CrossRef]

16. Saar, M.O.; Manga, M. Depth dependence of permeability in the Oregon Cascades inferred from hydrogeologic, thermal, seismic, and magmatic modeling constraints. *J. Geophys. Res.* **2004**, *109*, B04204. [CrossRef]

17. Marklund, L.; Wörman, A. The use of spectral analysis-based exact solutions to characterize topography-controlled groundwater flow. *Hydrogeol. J.* **2011**, *19*, 1531–1543. [CrossRef]

18. Morén, I.; Wörman, A.; Riml, J. Design of remediation actions for nutrient mitigation in the hyporheic zone. *Water Resour. Res.* **2017**, *53*, 8872–8899. [CrossRef]

19. Brunke, M.; Gonser, T. The ecological significance of exchange processes between rivers and groundwater. *Freshwater Biol.* **1997**, *37*, 1–33. [CrossRef]

20. Conant, B.; Cherry, J.; Gillham, R. A PCE groundwater plume discharging to a river: Influence of the streambed and near-river zone on contaminant distributions. *J. Contam. Hydrol.* **2004**, *73*, 249–279. [CrossRef]

21. Kalbus, E.; Schmidt, C.; Bayer-Raich, M.; Leschik, S.; Reinstorf, F.; Balcke, G.; Schirmer, M. New methodology to investigate potential contaminant mass fluxes at the stream-aquifer interface by combining integral pumping tests and streambed temperatures. *Environ. Pollut.* **2007**, *148*, 808–816. [CrossRef] [PubMed]

22. Chapman, S.W.; Parker, B.L.; Cherry, J.A.; Aravena, R.; Hunkeler, D. Groundwater-surface water interaction and its role on TCE groundwater plume attenuation. *J. Contam. Hydrol.* **2007**, *91*, 203–232. [CrossRef] [PubMed]

23. Kalbus, E.; Reinstorf, F.; Schirmer, M. Measuring methods for groundwater? Surface water interactions: A review. *Hydrol. Earth Syst. Sci.* **2006**, *10*, 873–887. [CrossRef]

24. Anderson, M.P. Heat as a ground water tracer. *Groundwater* **2005**, *43*, 951–968. [CrossRef] [PubMed]

25. Conant, B. Delineating and quantifying ground water discharge zones using streambed temperatures. *Groundwater* **2004**, *42*, 243–257. [CrossRef]

26. Constantz, J. Heat as a tracer to determine streambed water exchanges. *Water Resour. Res.* **2008**, *44*. [CrossRef]

27. Rau, G.C.; Andersen, M.S.; McCallum, A.M.; Acworth, R.I. Analytical methods that use natural heat as a tracer to quantify surface water–groundwater exchange, evaluated using field temperature records. *Hydrogeol. J.* **2010**, *18*, 1093–1110. [CrossRef]

28. Krause, S.; Blume, T. Impact of seasonal variability and monitoring mode on the adequacy of fiber-optic distributed temperature sensing at aquifer-river interfaces. *Water Resour. Res.* **2013**, *49*, 2408–2423. [CrossRef]

29. Selker, J.; van de Giesen, N.; Westhoff, M.; Luxemburg, W.; Parlange, M.B. Fiber optics opens window on stream dynamics. *Geophys. Res. Lett.* **2006**, *33*. [CrossRef]

30. Selker, J.S.; Thévenaz, L.; Huwald, H.; Mallet, A.; Luxemburg, W.; Van De Giesen, N.; Stejskal, M.; Zeman, J.; Westhoff, M.; Parlange, M.B. Distributed fiber-optic temperature sensing for hydrologic systems. *Water Resour. Res.* **2006**, *42*. [CrossRef]

31. Krause, S.; Hannah, D.M.; Fleckenstein, J.H.; Heppell, C.M.; Kaeser, D.; Pickup, R.; Pinay, G.; Robertson, A.L.; Wood, P.J. Inter-disciplinary perspectives on processes in the hyporheic zone. *Ecohydrology* **2011**, *4*, 481–499. [CrossRef]

32. Boano, F.; Revelli, R.; Ridolfi, L. Modeling hyporheic exchange with unsteady stream discharge and bedform dynamics. *Water Resour. Res.* **2013**, *49*, 4089–4099. [CrossRef]

33. Gomez-Velez, J.D.; Wilson, J.L.; Cardenas, M.B.; Harvey, J.W. Flow and residence times of dynamic river bank storage and sinuosity-driven hyporheic exchange. *Water Resour. Res.* **2017**, *53*, 8572–8595. [CrossRef]

34. Malzone, J.M.; Anseeuw, S.K.; Lowry, C.S.; Allen-King, R. Temporal hyporheic zone response to water table fluctuations. *Groundwater* **2016**, *54*, 274–285. [CrossRef] [PubMed]

35. Malzone, J.M.; Lowry, C.S.; Ward, A.S. Response of the hyporheic zone to transient groundwater fluctuations on the annual and storm event time scales. *Water Resour. Res.* **2016**, *52*, 5301–5321. [CrossRef]

36. Schmadel, N.M.; Ward, A.S.; Lowry, C.S.; Malzone, J.M. Hyporheic exchange controlled by dynamic hydrologic boundary conditions. *Geophys. Res. Lett.* **2016**, *43*, 4408–4417. [CrossRef]

37. Trauth, N.; Fleckenstein, J.H. Single discharge events increase reactive efficiency of the hyporheic zone. *Water Resour. Res.* **2017**, *53*, 779–798. [CrossRef]

38. Singh, T.; Wu, L.; Gomez-Velez, J.D.; Lewandowski, J.; Hannah, D.M.; Krause, S. Dynamic Hyporheic Zones: Exploring the Role of Peak Flow Events on Bedform-Induced Hyporheic Exchange. *Water Resour. Res.* **2019**, *55*, 218–235. [CrossRef]

39. McCallum, J.L.; Shanafield, M. Residence times of stream-groundwater exchanges due to transient stream stage fluctuations. *Water Resour. Res.* **2016**, *52*, 2059–2073. [CrossRef]

40. Sickbert, T.; Peterson, E.W. The effects of surface water velocity on hyporheic interchange. *J. Water Resour. Prot.* **2014**, *6*, 327–336. [CrossRef]

41. Wroblicky, G.; Campana, M.; Valett, H.; Dahm, C. Seasonal variation in surface-subsurface water exchange and lateral hyporheic area of two stream-aquifer systems. *Water Resour. Res.* **1998**, *34*, 317–328 [CrossRef]

42. Laudon, H.; Taberman, I.; Agren, A.; Futter, M.; Ottosson-Löfvenius, M.; Bishop, K. The Krycklan Catchment Study: A flagship infrastructure for hydrology, biogeochemistry, and climate research in the boreal landscape. *Water Resour. Res.* **2013**, *49*, 7154–7158. [CrossRef]

43. Karlsen, R.H.; Grabs, T.; Bishop, K.; Buffam, I.; Laudon, H.; Seibert, J. Landscape controls on spatiotemporal discharge variability in a boreal catchment. *Water Resour. Res.* **2016**, *52*, 6541–6556. [CrossRef]

44. Leach, J.A.; Lidberg, W.; Kuglerová, L.; Peralta-Tapia, A.; Ågren, A.; Laudon, H. Evaluating topography-based predictions of shallow lateral groundwater discharge zones for a boreal lake-stream system. *Water Resour. Res.* **2017**, *53*, 5420–5437. [CrossRef]

45. Laudon, H.; Ottosson-Löfvenius, M. Adding snow to the picture: Providing complementary winter precipitation data to the Krycklan catchment study database. *Hydrol. Process.* **2016**, *30*, 2413–2416. [CrossRef]

46. Lupon, A.; Denfeld, B.A.; Laudon, H.; Leach, J.; Karlsson, J.; Sponseller, R.A. Groundwater inflows control patterns and sources of greenhouse gas emissions from streams. *Limnol. Oceanogr.* **2019**, *64*. [CrossRef]

47. Munz, M.; Oswald, S.E.; Schmidt, C. Sand box experiments to evaluate the influence of subsurface temperature probe design on temperature based water flux calculation. *Hydrol. Earth Syst. Sci.* **2011**, *15*, 3495–3510. [CrossRef]

48. Sterte, E.J.; Johansson, E.; Sjöberg, Y.; Karlsen, R.H.; Laudon, H. Groundwater-surface water interactions across scales in a boreal landscape investigated using a numerical modelling approach. *J. Hydrol.* **2018**, *560*, 184–201. [CrossRef]

49. Briggs, M.A.; Lautz, L.K.; Buckley, S.F.; Lane, J.W. Practical limitations on the use of diurnal temperature signals to quantify groundwater upwelling. *J. Hydrol.* **2014**, *519*, 1739–1751. [CrossRef]
50. Shanafield, M.; Hatch, C.; Pohll, G. Uncertainty in thermal time series analysis estimates of streambed water flux. *Water Resour. Res.* **2011**, *47*. [CrossRef]
51. Taniguchi, M. Evaluation of vertical groundwater fluxes and thermal properties of aquifers based on transient temperature-depth profiles. *Water Resour. Res.* **1993**, *29*, 2021–2026. [CrossRef]
52. Cardenas, M.B.; Wilson, J.L.; Zlotnik, V.A. Impact of heterogeneity, bed forms, and stream curvature on subchannel hyporheic exchange. *Water Resour. Res.* **2004**, *40*, W08307. [CrossRef]
53. Salehin, M.; Packman, A.I.; Paradis, M. Hyporheic exchange with heterogeneous streambeds: Laboratory experiments and modeling. *Water Resour. Res.* **2004**, *40*, W11504. [CrossRef]

Article

Characterization of Diffuse Groundwater Inflows into Streamwater (Part I: Spatial and Temporal Mapping Framework Based on Fiber Optic Distributed Temperature Sensing)

Hugo Le Lay [1,2], Zahra Thomas [1,*], François Rouault [1], Pascal Pichelin [1] and Florentina Moatar [2,3]

[1]   UMR SAS, Agrocampus Ouest, INRA, 35000 Rennes, France; lelayhugo@gmail.com (H.L.L.);
     francois.rouault@agrocampus-ouest.fr (F.R.); pascal.pichelin@agrocampus-ouest.fr (P.P.)

[2]   Laboratoire de GéoHydrosystèmes Continentaux (GéHCO), UPRES EA 2100, Université François-Rabelais,
     UFR des Sciences et Techniques, Parc de Grandmont, 37200 Tours, France; florentina.moatar@irstea.fr

[3]   Institut National de Recherche en Sciences et Technologies pour l'Environnement et l'Agriculture (Irstea),
     RiverLy, Centre de Lyon-Villeurbanne, 69625 Villeurbanne, France

*   Correspondence: zahra.thomas@agrocampus-ouest.fr or zthomas@agrocampus-ouest.fr;
     Tel.: +00-33-2-23-48-58-78

Received: 28 September 2019; Accepted: 6 November 2019; Published: 14 November 2019

Abstract: Although fiber optic distributed temperature sensing (FO-DTS) has been used in hydrology for the past 10 years to characterize groundwater–streamwater exchanges, it has not been widely applied since the entire annual hydrological cycle has rarely been considered. Properly distinguishing between diffuse and intermittent groundwater inflows requires longer periods (e.g., a few months, 1 year) since punctual changes can be lost over shorter periods. In this study, we collected a large amount of data over a one-year period using a 614 m long cable placed in a stream. We used a framework based on a set of hypotheses approach using thermal contrast between stream temperature and the atmosphere. For each subreach, thermal contrast was normalized using reference points assumed to lie outside of groundwater influence. The concepts and relations developed in this study provide a useful and simple methodology to analyze a large database of stream temperature at high spatial and temporal resolution over a one-year period using FO-DTS. Thus, the study highlighted the importance of streambed topography, since riffles and perched reaches had many fewer inflows than pools. Additionally, the spatial extent of groundwater inflows increased at some locations during high flow. The results were compared to the usual standard deviation of stream temperature calculated over an entire year. The two methods located the same inflows but differed in the mapping of their spatial extent. The temperatures obtained from FO-DTS open perspectives to understand spatial and temporal changes in interactions between groundwater and surface water.

Keywords: water temperature; groundwater–streamwater exchange; inflow mapping framework

---

1. Introduction

Groundwater inflows into streams play an important role in the ecological balance of streams, contributing chemicals that drive water quality [1–5], supporting baseflow [6–11] and providing refuge for fauna [12–21]. Locating and mapping inflows is thus of great interest for river management and ecological restoration [22–26]. However, groundwater inflows are usually driven by multiple factors, including hydraulic head gradients between streamwater and groundwater, stream geomorphology, and subsurface geology [27–29]. Multiscale processes involved in groundwater and streamwater exchange can be categorized as large- or small-scale [30]. Large-scale hydrological exchange is driven mainly by the spatial and temporal hydraulic head gradients between the stream and the surrounding groundwater.

By contrast, at a small-scale, water is pushed into the streambed due to interaction between the flow and morphological features of the streambed such as riffle–pool sequences. Variability in hydraulic conductivity influences both large- and small-scale processes [31–33]. The spatial distribution of geological heterogeneities influences stream–groundwater exchange [34,35]. Using a two-dimensional groundwater flow model to investigate the interface between the stream and groundwater hyporheic (below stream) zone, Wroblicky et al. [36] identified hydraulic conductivity of alluvial and streambed sediments as a major factor controlling stream–groundwater exchange. Mojarrad et al. [37] quantified effects of catchment-scale upwelling groundwater on hyporheic fluxes and tested a wide range of spatial scales. Their results identified streambed topographic structures as the predominant factor controlling the magnitude of hyporheic exchange fluxes. Predicting groundwater flow paths from groundwater–surface water analysis of easily available data, such as temperature, helps to identify catchment behaviors. Because groundwater inflows are heterogeneous in space and variable in time, locating and mapping them is challenging.

Many methods can be used to map groundwater inflow zones in rivers [38]. Some methods use a direct approach, such as measuring discharge in successive cross sections (i.e., differential gauging) to determine gains and losses along a reach. Others use natural markers of inflow such as specific biological communities [39–41]. Most, however, usually use tracers. For instance, many studies monitored radon concentrations or stable-isotope ratios to detect groundwater inflows to rivers [42–46]. Among other possible tracers (e.g., ions, contaminants), heat has been recognized as reliable for identifying exchanges between groundwater and surface water [47–49]. Most methods using heat as a tracer are based on punctual temperature measurements [50]. Such standard measurement techniques are usually based on placing temperature sensors directly into the water column [51], in a streambed piezometer or along thermal lances into shallow sediments [52]. However, all have a limited spatial range that does not allow the heterogeneity of groundwater–stream interactions, nor their possible temporal intermittency, to be mapped at sufficient resolution. Only the recent development of airborne thermal infrared imagery (TIR) [30,53] and fiber optic distributed temperature sensing (FO-DTS) has provided good spatial coverage of stream thermal heterogeneities. TIR is an indirect technique that measures the temperature of the stream surface only [54,55]. It has a larger spatial range than FO-DTS, but lower accuracy and temporal resolution [56].

FO-DTS sensors send a laser impulse down a fiber optic cable to infer temperature along the cable. The measured ratio of the temperature-dependent Raman backscattered signal (anti-Stokes) to the temperature-independent Raman signal (Stokes) provides the temperature, while the time required for the backscattered signal to return provides the temperature's location. This technique provides direct measurements every 0.25 m along a cable a few km long, with an accuracy as high as 0.05 °C depending on the brand, setup, and chosen configuration [57,58]. The spatial resolution and accuracy of FO-DTS can be used to map groundwater inflows, which are more thermally stable over time than streamwater. Indeed, stream temperature is usually influenced more by air temperature [59–61]; however where groundwater inflows, however, the stream temperature varies less over time. Calculating a simple standard deviation (SD) of temperature for a given period (from hours to years) and location is usually sufficient to identify stream points influenced by groundwater inflows (low SD) from points that are not (higher SD) [62]. However, this method can be problematic for intermittent inflows such as those during short high-water episodes, droughts, and floods. The time period over which thermal variability is calculated can obviously be adapted to address these episodes, but the approach still requires a long period of observations to properly distinguish groundwater from hyporheic recirculation [63]. Thus, potential changes in groundwater inflow dynamics could be hidden by transitional periods (e.g., beginning of high flows).

The present study applied a simple framework to a new study site to identify and interpret groundwater–surface water interactions. The framework maps groundwater inflows at individual timesteps using the thermal contrast between FO-DTS measurements in the hyporheic zone and the atmosphere. This framework thus attempts to characterize the spatiotemporal heterogeneity of groundwater inflows. After describing the testing of our framework in a second-order stream, we discuss its potentials and limits.

2. Materials and Methods

2.1. Study Site

We performed our study in an area in northeastern Brittany, France, called the Zone Atelier Armorique (ZAAr). The ZAAr is part of the International Long-Term Socio-Ecological Research Network LTSER (www.lter-europe.net). We collected a large amount of temperature data at high spatial and temporal resolutions. Measurements were made along a 614 m long reach of the Petit Hermitage, a second-order stream [64] that drains a 16 km^2 subwatershed and flows from south to north. The monitored reach represents the last several hundred meters of the Petit Hermitage were monitored until its confluence with its last tributary, the Vilqué (Figure 1c), which is a first-order stream draining an adjacent 2.34 km^2 watershed with a similar flow direction. The northern part of the ZAAr, where the monitoring occurred, lies on schist bedrock and loess. The southern part of the ZAAr, where the Petit Hermitage has its source, lies on granodiorite and altered hornfels. The bedrock of the entire area is overlain by a weathered zone considered to be the main unconfined aquifer of the area [65,66]. Soil and land-use characteristics reflect this north–south dichotomy: upstream soils (south) are characterized by a mix of silica sand and altered schist with forests. Downstream soils (north, monitored reach) are mainly aeolian and alluvial silts with wetlands, agricultural fields, and meadows. The streambed is thus a mix of silt and deposited organic matter, with sand coming from upstream. The climate in the region is oceanic, with mean annual precipitation of 965 mm and air temperatures generally ranging from 0 to 25 °C, with some exceptional events below 0 °C or above 30 °C. Rainfall and temperature data for the site were obtained from a weather station 1 km north (downstream) of the reach.

Previous studies offered good insight into the area [67,68] since results of long-term measurements were already available (e.g., soil moisture, piezometers, weather station, water quality, stream discharge, thermal dataloggers). In addition, the site was susceptible to dispersed groundwater inflows [69] and highly stratified reactivity [70], likely related to the high geological heterogeneities. The upstream monitored reach consists of a heavily monitored hillslope including a large wetland (Figure 1). The middle area is dominated by small woods followed by a meadow. The confluence is located in a woody wetland (hereafter, "swamp"). From this land-use typology, four subreaches were defined: wetland, woods, meadow, and swamp. Previous studies found the Vilqué tributary strongly influenced by groundwater throughout the year, although it was not accessible for direct measurements.

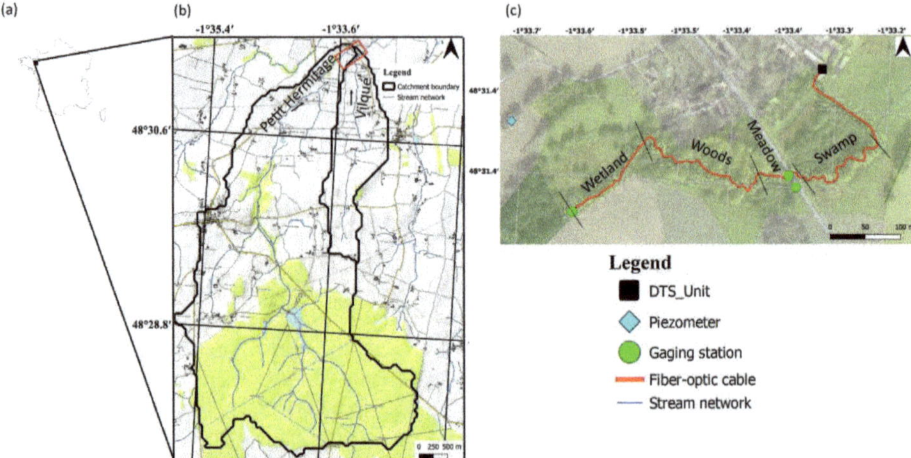

Figure 1. Location of the study site in the Zone Atelier Armorique in northeastern Brittany, France (**a**). The monitored stream, Petit Hermitage draining a 16 km^2 subwatershed whose small tributary drains a 2.34 km^2 area of Vilqué stream (**b**). A fiber optic cable 614 m long was placed in the main stream and connected to a distributed temperature sensing (DTS) unit in a nearby building (**c**). A piezometer was previously installed upstream of the site, as were gauging stations in the stream. Black lines indicate boundaries between subreaches with different geomorphologies. A weather station lies ca. 1 km to the north of the site (outside the map).

The longitudinal elevation of the Petit Hermitage streambed was measured with a theodolite (Leica) ca. every 4 m. The wetland, the most upstream subreach of the main stream, varies little in elevation, with small pools a few cm deep (Figure 2). However, the wetland has a steeper overall slope (ca. 2.4‰) than the rest of the stream (0.08‰) because a former deviation of the stream during the 19th century and the 1950s created a perched streambed in this reach (Figure 2). Despite high banks (ca. 50–80 cm above the streambed) and many bulrushes, it is sunnier than the following subreaches. The woods, with willows and bushes, are much more forested. It is also a wetland, but with more pronounced meandering and a clear succession of riffles and deep pools (every 40–50 cm). The following meadow subreach is straighter, with its left bank occupied by sheep. It is partly shaded because of bushes on the right bank, a narrower streambed and trees in its last third, but sunlight can still hit the stream surface at noon or in summer in some locations. A road bridge crosses over the stream in its last third. The swamp is the last subreach of the stream, which begins immediately after the confluence with the Vilqué tributary. It is similar to the woods but has smaller woody vegetation, more pronounced meanders and larger (although fewer) pools. Its banks (40 cm above the streambed) are also slightly lower than those of upstream subreaches (50–80 cm), so groundwater outcrops during high-water periods, and the soil remains wet most of the year. The most downstream part of this subreach was not accessible during the topography campaign due to fallen trees; thus, elevation data covered a shorter distance than the 614 m long thermal data from the FO-DTS cable.

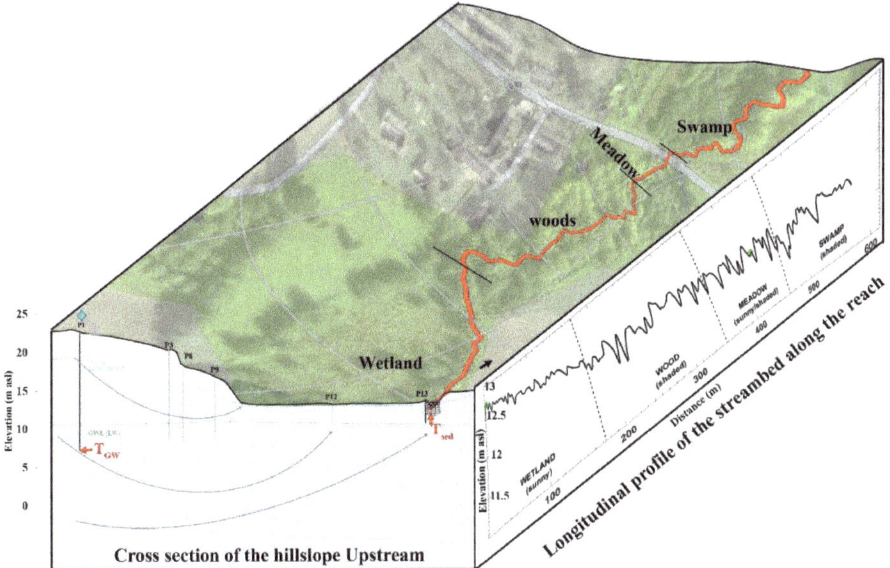

Figure 2. 3D view of the study area indicating the hillslope cross section of the upstream hillslope adjacent to the wetland sub-reach and longitudinal profile of the streambed of the Petit Hermitage. Hillslope cross section shows locations and depths of piezometers, including the one (blue diamond) used for groundwater temperature measurements (Tgw) and groundwater level for high water (HW) -dashed blue line- and low water (LW) -dashed green line period-. Tsed (FO-DTS) = sediment temperature from fiber optic distributed temperature sensing, m asl = m above sea level.

2.2. Piezometry and Differential Gauging

Large-scale groundwater movements and periods of potential stream loss and/or gain [71] were determined using the pre-existing network of piezometers and gauging stations (Figure 1). The depth of these piezometers varies from 15 m on the hillslope to 4.5 m near the stream in the wetland. Groundwater level was monitored by a minidiver submersible pressure transducer (HOBO Pro v2), with an accuracy of ±0.2 cm and a 5 min timestep. Only the 15 m deep piezometer, located on the crest of the hillslope, was used in this study (Figures 1 and 2a) to simplify interpretation by providing only the general hydraulic gradient with the stream. Streamwater stages were monitored using OTT Opheus sensors, with identical accuracy (±0.5 cm) and timestep (5 min), at two gauging stations (Figures 1 and 2). Discharge was measured at each gauging station ca. every two months using the salt dilution method [72]. The resulting discharges were used to refine pre-existing rating curves and thus infer stream discharge time series based on water level measurements at each gauging station. These discharges (L s^{-1}) were converted into specific flow rates Qs (L s^{-1} km^{-2}) by dividing them by their respective drained areas. These specific discharges were used to compare the relative groundwater contribution over time at these points. Groundwater and streamwater levels were expressed in m above sea level (m asl) for comparison purposes.

2.3. Temperature Measurements

Locations of groundwater inflows in the stream were collected using a distributed temperature sensor (Ultima XT-DTS, Silixa Ltd, London, UK) with a dual-ended duplexed configuration [73], 40 min time integration and 0.25 m spatial sampling. The fiber optic cable was a 4 mm wide Brussens cable (Brugg, Switzerland) protected by stainless steel armoring and polyamide. Short segments of the cable (20–30 m) were coiled in two water baths for calibration [57]. A cold bath inside a refrigerator and a heated and insulated warm bath were kept thermally homogeneous with bubblers and monitored with RBRsolo temperature loggers with an accuracy of 0.002 °C. The FO-DTS system was calibrated following van de Giesen et al. [73] using the reference temperatures in the baths. Due to a lack of electricity at the end of the cable, no validation bath could be set up. Temperature measurement of the FO-DTS system and a thermal lance (Umwelt-und Ingenieurtechnik GmbH, Germany) were compared to assess measurement accuracy broadly. The FO-DTS cable near the lance was found uncovered by sediment most of the year, so the instruments were compared with the lance set in the streamwater (z = +5 cm). Lance accuracy was estimated as 0.1 °C (manufacturer values). Measurement accuracy criteria indicated low values (0.016, 0.017, and 0.022 °C) for cold and warm baths compared to streamwater (0.015 and 0.018 °C) (Table 1). Le Lay et al. [74] highlight, in more detail, the concerns associated with DTS calibration and uncertainty analysis.

Table 1. Mean bias and root-mean-squared error (RMSE) of fiber optic temperature measurements following van de Giesen et al. [73]. Since no validation bath could be set up, three segments were used: cold calibration bath, warm calibration bath, and the farthest segment of the cable with a nearby independent temperature probe (UIT lance: depth = +5 cm; uncertainty = ±0.1 °C).

Location	Mean Bias (°C)	RMSE (°C)
Cold bath	0.016	0.022
Warm bath	0.016	0.017
Streamwater	0.150	0.180

Although the cable was 1000 m long, only 614 m of it were placed in the stream. We placed it in the thalweg and, when possible, in the middle of the stream, where hydraulic conductivity is likely to be highest [32]. The cable was buried in the streambed sediment, ca. 3 cm below the streambed surface, to prevent potential diffuse and intermittent groundwater inflows from being displaced by streamwater and thus going undetected [75,76]; this also kept the cable in place. Nonetheless, during the year-long monitoring, the cable was sometimes found uncovered by sediment in a few locations. Also, the streambed sometimes diverted from its original location, pushing the cable into sandbanks. Temperature data from the FO-DTS was considered to be that of the shallow sediment (T_{sed}) since the cable was, strictly speaking, set within the sediment. Air temperature (T_{air}) was obtained by averaging the temperature found along a 100 m long segment of fiber optic cable near the stream (straight section downstream of the swamp, Figure 1). Protected from direct solar radiation by trees and its northern orientation, the cable was also suspended on a fence to prevent contact with the ground.

2.4. Data Post-Processing: Framework for Spatiotemporal Mapping of Groundwater Inflows

The framework depends upon a series of three hypotheses.

The first, which is widely accepted, is that streamwater temperature (T_{sw}) varies in a range between T_{air} and groundwater temperature (T_{gw}) (Figure 3a). Evans et al. [61] stated that more than 80% of total thermal exchanges in streams occur at the air–water interface. By contrast, Caissie [60] demonstrated that groundwater, with approximately 15% of total thermal exchanges, tends to buffer the atmosphere's influence on stream temperature. Thus, T_{sw} should be closer to T_{air} than T_{gw}, regardless of the season or the hour, but it will be closer to T_{gw} when groundwater inflows are strong.

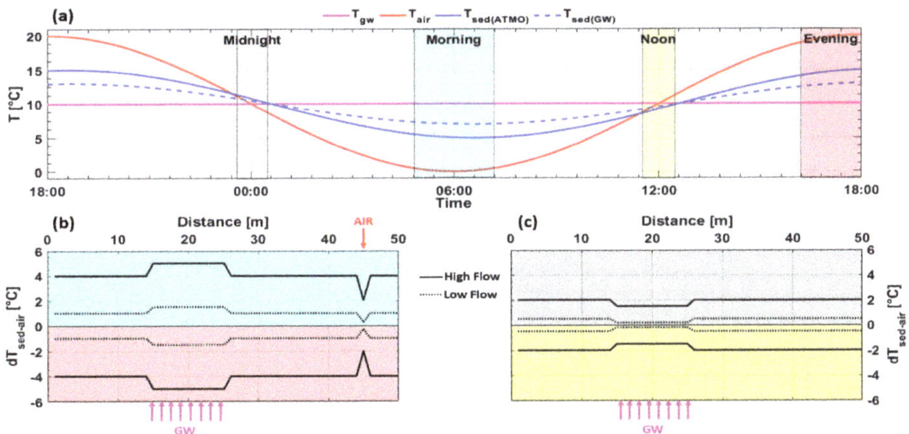

Figure 3. Conceptual diagram of the theory underlying the study's framework. (**a**) Theoretical diurnal temperatures of the atmosphere (T_{air}), groundwater (T_{gw}), and the sediment influenced mainly by each of them (T_{sed} (ATMO) and T_{sed} (GW), respectively). (Bottom) Theoretical differences between sediment and air temperatures ($dT_{sed-air}$) (Equation (1)) under high or low flow (**b**) in the morning and evening or (**c**) at midnight and midday along a given reach with groundwater (GW) inflow.

The second hypothesis, which is the first one in our framework, is to consider T_{sed} (obtained with FO-DTS) as similar to T_{sw} in order to have continuous information along the stream. Therefore, combining the two hypotheses, the first step to detect groundwater inflows is to calculate the thermal difference between T_{sed} and T_{air} at each point (i) and each timestep (t):

$$dT_{sed-air(t,i)} = T_{sed(t,i)} - T_{air(t)}. \tag{1}$$

This simple thermal contrast summarizes a conceptual presentation of temperature effects of the two main compartments (i.e., atmosphere (T_{air}) and groundwater (T_{gw})) on those of shallow sediment (T_{sed}) (Figure 3a).

The conceptual diagram of the theory underlying the study's framework is presented Figure 3. Based on data analysis, it is assumed that the sediment temperature of a stream point influenced by groundwater varies less diurnally and is closer to T_{gw} than that of a point influenced only by the atmosphere. T_{sed} lags slightly behind T_{air} because T_{sed} requires time to react to T_{air}. Depending on latitude, climate, geomorphology, and season, T_{air} differs most from T_{gw} in the morning (coldest) and evening (warmest) (Figure 3b), and differs least from T_{gw} around midnight and midday (transitional periods) (Figure 3c). When T_{air} and T_{gw} differ the most (morning and evening), the amplitude of $dT_{sed-air}$ always increases under groundwater influence (Figure 3b). When T_{air} and T_{gw} differ the least (midnight and midday), the relation between $dT_{sed-air}$ and groundwater inflow is inverted: the amplitude of $dT_{sed-air}$ always decreases under groundwater influence because T_{sed} requires time to react to T_{air} (Figure 3c). Regardless of the difference between T_{air} and T_{gw}, periods of high flow have much higher amplitude than those of low flow because the additional water volume of high water has more thermal inertia, which requires much more energy to increase in temperature (Figure 3b,c). Consequently, the atmosphere has less influence on stream temperature.

The third hypothesis, formulated from the previous concept (Figure 3), consists of normalizing thermal contrast using a reference point assumed to lie outside of groundwater influence. Additionally, since the information sought (groundwater influence) is related to the thermal contrast's amplitude and not its sign, the absolute value is calculated for clarity:

$$\text{diffT}_{\text{sed-air}(t,i)} = \left| \text{dT}_{\text{sed-air}(t,i)} - \min\left(\text{dT}_{\text{sed-air}(t,\text{ref})}\right) \right| \tag{2}$$

where $\text{dT}_{\text{sed-air}(t,\text{ref})}$ is the difference between T_{air} and T_{sed} at timestep t in a reference segment ref with no groundwater inflow. In Equation (2), the minimum value of $\text{dT}_{\text{sed-air}(t,\text{ref})}$ (i.e., that closest to T_{air}) in the reference segment is used to normalize data.

Like sunlight exposure, water stage and overall flow rate (no tributaries) can greatly change the influence of T_{air} [59,60,77,78]. These factors also vary along the stream, and special care should be taken for streams with clear and/or shallow water, such as those in our study area, since direct sunlight can influence FO-DTS measurements [79]. Ultimately, there were three requirements to for successful use of this method: (1) identify subreaches with similar geomorphology, (2) identify reference points for each reach, and (3) determine a threshold beyond which a $\text{diffT}_{\text{sed-air}}$ can be attributed to groundwater inflow. To this end, thermal contrast was normalized using a reference segment (Equation (2)) for each stream subreach.

The subreaches were determined empirically based on the overall sunlight exposure (shading and orientation) and whether a tributary was present or not. Thus, four geomorphological subreaches were defined, similar to those previously described (Figures 1 and 2):

(i)   The wetland, with a relatively straight channel, shallow flow and almost no high riparian vegetation;
(ii) The woods and upstream part of the meadow, with more pronounced meandering, alternating riffles and pools, and high trees or banks;
(iii) The downstream part of the meadow (including the bridge and the section after it), shaded with low trees and with a large streambed;
(iv) The swamp beyond the confluence with the Vilqué, always shaded with taller trees, larger meanders, and deeper flow, with alternating riffles and pools.

The reference points were determined based on $\text{diffT}_{\text{sed-air}}$ dynamics. According to our hypotheses, points influenced only by the atmosphere should have the lowest values, except for points outside of the stream or when $\text{dT}_{\text{sed-air}} \approx 0$ °C. First, for each subreach, short cable segments with low $\text{dT}_{\text{sed-air}}$—suggesting a strong atmospheric influence—were selected. These reference segments were preferentially chosen upstream of each subreach. Then, each potential reference segment was visually analyzed over time to exclude points accidentally exposed to the atmosphere. Finally, for each subreach and each timestep t, the minimum value of $\text{dT}_{\text{sed-air}}$ was selected and used to normalize data (Equation (2)).

Once normalized to their respective reference, the data from each subreach became more easily comparable. Small fluctuations in $\text{diffT}_{\text{sed-air}}$ remained, but they could be explained by a buffering effect due to different water stages (riffle, pool) or different depths of burial in the sediment. To automatically select locations most likely due to groundwater inflows and not only the buffering effect, a simple threshold was applied. The mean value of $\text{diffT}_{\text{sed-air}}$ along the entire monitored reach was calculated for each timestep: any points with values over it were considered potential anomalies due to groundwater inflows, while values below it were considered to be influenced by the atmosphere.

3. Results and Discussion

3.1. Flow Variability and Hydrological Behavior

From July to mid-November 2016, the stream had low specific discharge (1.2–2.8 L s^{-1} km^{-2}), both upstream of the site (Q_{sup}) and more downstream, in the site's meadow (Qs_{down}) (Figure 4a). Despite a few storm events from 16 July to 16 November, almost no floods were recorded in the stream, and both upstream and downstream discharges remained similar. A few episodes in which Qs_{down} decreased to Q_{sup} were observed in August and September 2016, indicating possible losses along the reach. However, the reality of losses is difficult to assess because of uncertainty in the gauging measurements. This period from July to mid-November 2016 was qualified as low flow.

By contrast, from late November 2016 to June 2017, the site had much larger specific discharge (up to 20 L s^{-1} km^{-2} in December), which would have been due to longer rainfall periods (Figure 4a). Unfortunately, no rainfall was recorded in autumn due to a malfunction in the weather station. During this period, Qs_{down} began to exceed Q_{sup} during storm events and even during periods with no rain, after the recession limb (Figure 4a). For instance, the difference increased from ca. 0.8 to 1.6 L s^{-1} km^{-2} during January, to 3.0 L s^{-1} km^{-2} in early March, although successive storm events increased uncertainty in baseflow estimates. Afterwards, the difference decreased until late May, when the locations had similar specific discharge (2.6 L s^{-1} km^{-2}). These observations of Qs_{down} higher than Q_{sup} when no runoff was recorded tended to indicate a gaining reach from at least December to March. Thus, the period from late November 2016 to May 2017, with successive floods and an overall increase in baseflow, was qualified as high flow. Flow variability provides a comprehensive means of assessing hydrological behavior [80]. Groundwater discharge seemed to be synchronized with the low-flow period (July to mid-November 2016), with the groundwater level in the upstream wetland subreach decreasing from ca. 13.0 to 11.3 m asl by mid-November (Figure 4b). During this period, the streamwater level remained at ca. 12.7 m asl. Recharge also began when high flow began in late November; however, groundwater remained lower than the stream (negative hydraulic gradient) until February (12.8 m asl). This behavior appeared unusual for the site, where the hydraulic gradient usually becomes positive as early as December. This extended recharge period was attributed to an exceptionally dry year for the area and perhaps the fact that the streambed had been raised (Figure 2). Ultimately, this upstream subreach had positive hydraulic gradient only from February to May 2017.

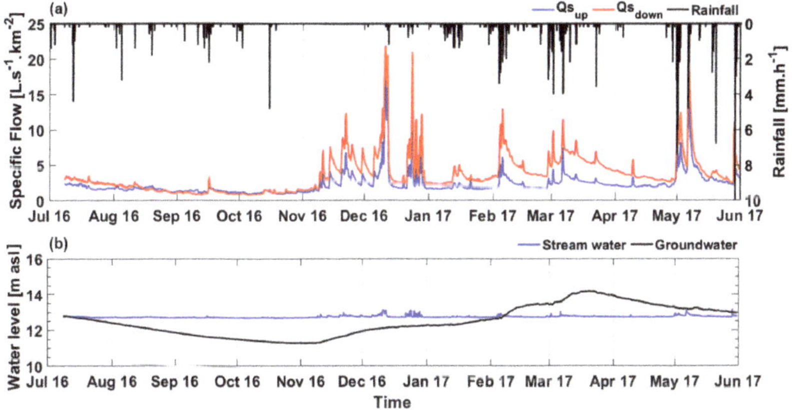

Figure 4. Hydrological processes in the study site over one year. (**a**) Stream discharge upstream and downstream of the reach and rainfall from a weather station 1 km downstream. (**b**) Groundwater and streamwater levels in the upstream wetland subreach.

3.2. Groundwater Inflow Mapping over Time and Space

Discharge and topography provided good insight into the hydrological processes involved. During low flow, the stream lost water, most likely in its upstream wetland subreach. During high flow and between storm events (e.g., early January 2017), discharge increased from upstream to downstream gauging stations, indicating groundwater inflows.

During the low-flow period (July to October 2016), T_{sed} was relatively high because T_{air} exceeded T_{gw} (Figure 5a). During the high-flow period (late November 2016 to May 2017), T_{gw} minus T_{air} (dT_{gw-air}) became increasingly positive, indicating the colder seasons of autumn and winter. T_{sed} followed this trend, decreasing to a minimum in late January before increasing again in spring. Within this overall synchronicity between T_{air} and Tsed, certain points in the stream reacted differently over time. For instance, several points in the sediment ca. 230 m (woods) and 475 m (swamp) from the beginning of the cable appeared cooler during warm periods and warmer during cold periods than the rest of the stream (Figure 5a).

Examination of $dT_{sed\text{-}air}$ showed not only these points of thermal anomaly but also periods during which T_{sed} of the entire stream deviated from T_{air} (Figure 5b). In addition, $T_{sed\text{-}air}$ of some of the points of thermal anomaly remained close to zero throughout the year (62, 350, 395, and 460 m), indicating artifacts due to non-submerged cable segments. Temporally, $T_{sed\text{-}air}$ varied more when T_{air} was close to T_{gw} (e.g., September to November 2016, March to May 2017), shifting between positive and negative values. Notably, during the transition from low flow to high flow (December 2016 to February 2017), $dT_{sed\text{-}air}$ and $dT_{gw\text{-}air}$ had high values at the same time. This high variability of $dT_{sed\text{-}air}$ illustrated the combined effect of a rapidly changing T_{air} and the lag time of T_{sed}. Because this behavior made interpretation more difficult, data were normalized to distinguish groundwater inflows from atmosphere-induced artifacts.

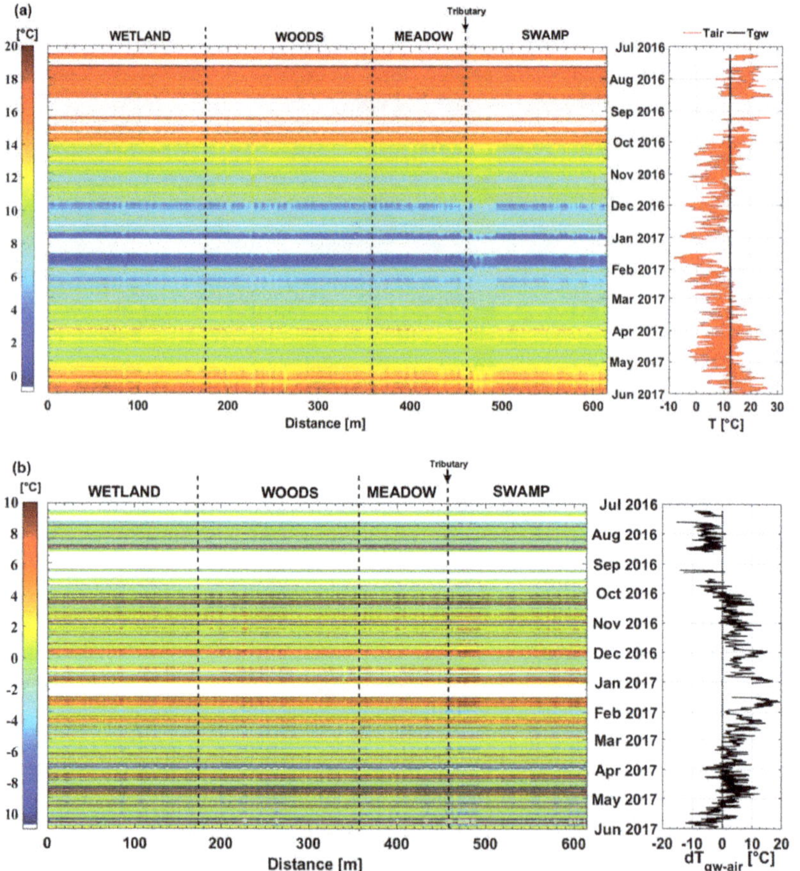

Figure 5. Spatiotemporal evolution of (**a**) streambed temperature (Tsed) in the Petit Hermitage compared to dynamics of air and groundwater temperature (T_{air} and T_{gw}, respectively) and (**b**) the difference between sediment and air temperature ($dT_{sed\text{-}air}$) compared to dynamics of T_{gw} minus T_{air} ($T_{gw\text{-}air}$). White bands indicate missing data.

After normalizing each subreach thermal anomaly, we observed large heterogeneities over time and space along the Petit Hermitage. Subreaches had many points with no groundwater inflows detected: the entire upstream wetland (0–180 m), the end of the woods (290–380 m), the entire meadow (380–450 m), and the middle of the swamp (500–540 m) (Figure 6b). The absence of thermal anomalies in the wetland subreach was most likely due to the perched streambed of the upstream zone, which likely made it lose water. The woods and swamp subreaches had several riffles (the latter from 500 to 550 m), and their higher elevations may have prevented groundwater inflows. By contrast, the meadow was more heterogeneous: although generally free of inflows during low flow (July to December), clear inflows appeared from January to June 2017. Thus, the swamp clearly experienced a seasonal effect in which groundwater flow paths were mobilized during high flow. Segments of the meadow always free of groundwater inflows (370–400 m) could also have been so because of several riffles.

The groundwater inflows themselves displayed different patterns throughout the year, such as intermittent and diffuse patterns in the meadow but constant patterns in the swamp. Constant inflows in the wetland (e.g., at 25 and 85 m) were apparently unrelated to streambed topography but could have been due to deep groundwater flow paths from the opposite, non-instrumented, side of the stream. Indeed, in the wetland, the stream's right side has a much steeper bank and is much closer to the hill; thus, the groundwater level in the right hillslope may have been higher than that measured in the left hillslope (Figure 4b). The potential of a distant groundwater flow path is also supported by the position of the cable in the thalweg, which has a greater probability of intercepting this kind of deep circulation [69,81]. Constant inflows were also detected in a segment of the wetland and woods subreaches (100–280 m), but they could not be related to any parameters besides proximity to the hillslope and the natural streambed (100–200 m) or the relatively low elevation (210–260 m) (Figure 6b). In addition, a seasonal effect starting in November 2016 may have increased inflows from 150–250 m. Finally, the segment of the swamp subreach after the confluence (460–614 m) was dominated mostly by inflows throughout the year. The especially clear inflow signal from 460 to 500 m may have been related to it occurring in the deepest pool recorded at the site.

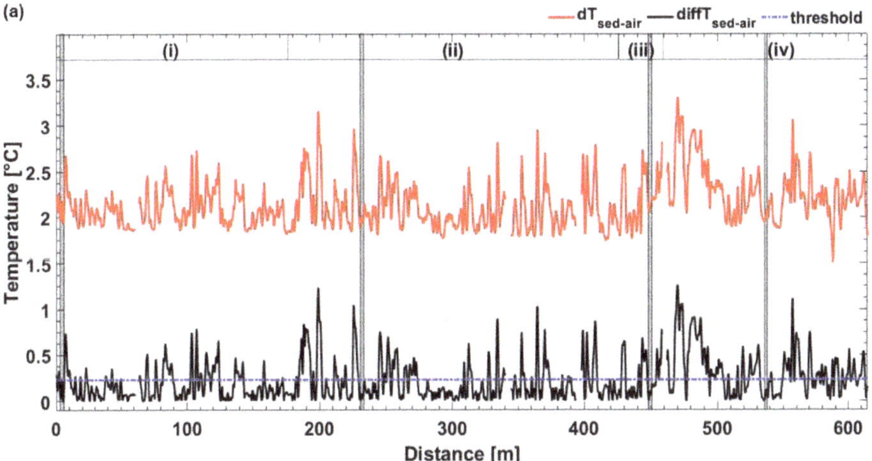

Figure 6. *Cont.*

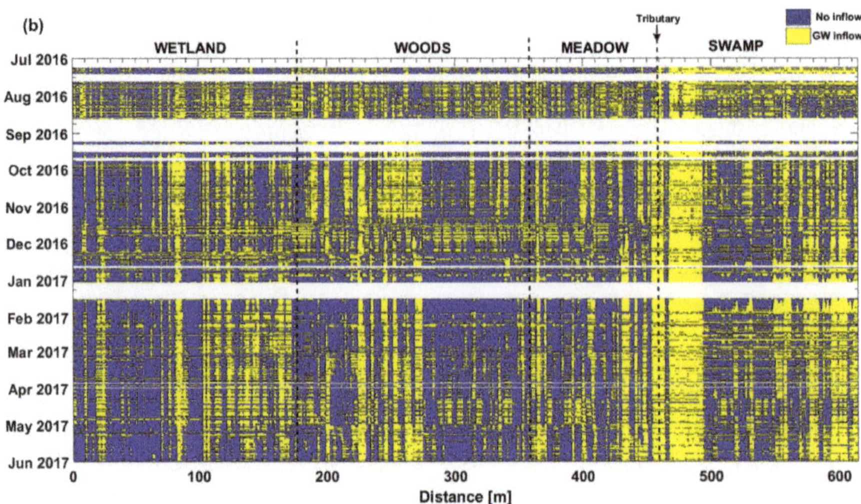

Figure 6. Application of the fiber optic distributed temperature sensing framework to map groundwater inflow. (**a**) Example of sediment temperature minus air temperature (d$T_{sed\text{-}air}$) at a given timestep (3 August 2016 at 14:04) and its corresponding normalized value (diff$T_{sed\text{-}air}$) using the d$T_{sed\text{-}air}$ of each geomorphological subreach's (i–iv) reference segment. Mean diff$T_{sed\text{-}air}$ at this timestep is used as a threshold above which a peak was considered to be groundwater inflow. (**b**) Spatiotemporal evolution of thermal anomalies indicating locations with or without groundwater (GW) inflows throughout the year. White columns indicate points where the sediment or fiber optic cable was frequently observed non-submerged. Note: Results of (**a**) generate data for one horizontal line of (**b**).

3.3. Comparing the SD Method and Framework Based on DTS

The thermal anomalies mapped using our framework were usually considered to be likely groundwater inflows. Comparing our results to the annual SD of T_{sed} shows that most low SDs were associated with diffuse and intermittent groundwater inflows detected more than half of the year, and vice versa (Figure 7). However, certain inflows detected more than half of the year (Figure 7) had a smaller spatial extent than those detected during shorter periods (e.g., 310–460 m during high flow) (Figure 6b). Other arguments support the groundwater inflow hypothesis. For example, some anomalies appeared at beginning of the high-flow period, especially in the meadow (310–460 m), as indicated by the net gain in discharge identified by differential gauging of Q_{sup} and Q_{sdown}. Moreover, most anomalies were located in pools, where intersection with groundwater was the most likely. Many studies have demonstrated that a hydraulic head gradient from upstream to downstream of a pool–riffle sequence is the cause of upwelling at the riffle tail [82–84]. In a study focused on water exchange between a stream and adjacent aquifer, Harvey et al. [85] showed that flow from the sediment to the stream occurred at transitions from steps to pools and vice versa. In our study, the generally continuous and turbulent stream flow throughout the year was considered sufficient to prevent thermal stratification, which also could have explained such differences in pools [86–88]. In addition, the downstream segment beyond 460 m lies in the swamp. Consequently, its banks and the surroundings were flooded or wet most of the year, indicating a likely positive hydraulic gradient from groundwater to the stream, which may explain the dense anomalies detected there (except for the shallowest sections).

Despite this indirect evidence, the nature of these thermal anomalies can be discussed further. For instance, Norman et al. [89] investigated, through flume experiments, the effects of bed topography on hyporheic flow. They identified that some anomalies could be caused by hyporheic flow recirculation instead of groundwater inflows. A cable buried deeper in the sediment may also experience greater lag time after the atmospheric signal. When our cable was removed, however, it had not been buried much deeper, except perhaps in the deepest pools, where estimating depth was difficult. In some locations (100–350 m), cable segments were uncovered by movement of the streambed [62] or even tangled up with branches and leaves. Nonetheless, the points directly in the streamwater, with low $diffT_{sed-air}$, did not behave so differently that they modified the detection threshold. Only segments of cable found non-submerged differed significantly ($diffT_{sed-air} \approx 0$) and were removed from the dataset. Even though the method still needs improvement, this study underscored the need to consider patterns and change in groundwater inflows over a long temporal scale completed by the Part II of this study focused on groundwater inflow quantification by coupling FO-DTS and vertical velocities inferred from thermal lances profiles in the hyporheic zone [90].

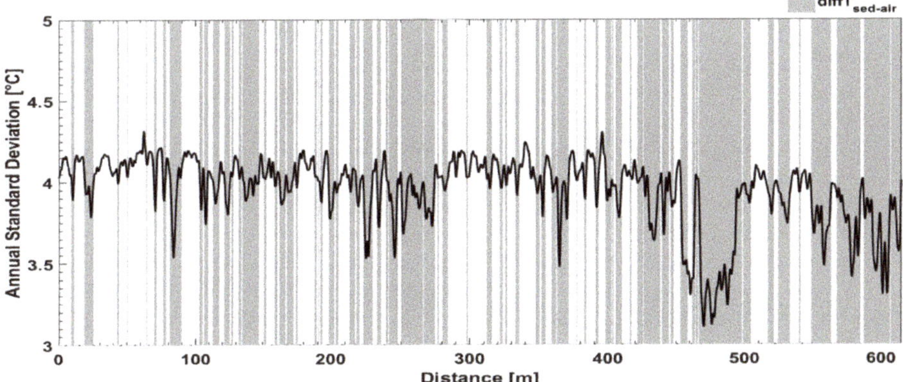

Figure 7. Locations of likely groundwater inflows detected more than half of the year by applying our framework (gray zones) compared to the annual standard deviation of sediment temperature (black line). Lower standard deviations also indicate likely groundwater inflows.

4. Conclusions

The concepts and relations developed in this study provide a useful and simple methodology to analyze a large database of stream temperature at high spatial and temporal resolution over a one-year period using FO-DTS. We used a simple approach based on thermal contrast normalized by a reference point assumed to lie outside of groundwater influence. The framework depends on a set of hypotheses for characterizing spatial heterogeneity and temporal intermittency and mapping groundwater inflow. Results highlighted the main temporal scale needed to identify effects from groundwater or the atmosphere. Diurnal temperature variation, as well as seasonal variability (high- and low-water periods), provided conceptual interpretation of thermal anomalies for determining patterns of groundwater–surface water exchange. Streambed topography was also important: riffle-and-pool sequences and perched reaches had many fewer inflows than pools. In addition, the spatial extent of groundwater inflows increased at some locations during the high-flow period. During high flow, the usual SD approach located the same inflows but underestimated their extent because the SD was integrated over the entire year. The inflow mapping method showed a clear distribution of groundwater inflows consistent with previous studies of the same stream. However, the results of thermal anomaly mapping could not distinguish groundwater inflows from hyporheic recirculation or overbuffering from sediment. Additional onsite piezometric measurements, vertical flow measurements in the hyporheic zone, and/or the use of isotopes and chemical tracers could help to determine the processes involved and identify sources and sinks. Also, mapping temporal intermittency can help improve the analysis and identification of biogeochemical and hydrological processes, especially in heterogeneous streams with geomorphological contrast or heavy human modification. This method may prove useful in domains such as hydrological modeling and environmental management, which require high spatial resolution and high-frequency data.

Author Contributions: Conceptualization, H.L.L. and Z.T.; Methodology, H.L.L., F.R., P.P. and Z.T.; Software, H.L.L.; Validation, H.L.L., Z.T. and F.M.; Formal Analysis, H.L.L. and Z.T.; Investigation, H.L.L., F.R., P.P. and Z.T.; Resources, H.L.L., F.R., P.P. and Z.T.; Data Curation, H.L.L. and Z.T.; Writing-Original Draft Preparation, H.L.L. and Z.T.; Writing-Review & Editing, Z.T.; Visualization, H.L.L..; Supervision, Z.T. and F.M.; Project Administration, Z.T. and F.M.; Funding Acquisition, Z.T. and F.M.

Funding: This research was funded by Agence de l'Eau Loire Bretagne, grant number [150417801].

Acknowledgments: Authors warmly thank Pitois for the shelter and electricity they kindly provided during the entire campaign. We also thank all technicians from UMR INRA AGROCAMPUS OUEST SAS who helped us place the fiber optic cable and perform the fieldwork. We also thank the ILSTER Zone Atelier Armorique.

Conflicts of Interest: The authors declare no conflict of interest.

References

1. Cox, M.H.; Su, G.W.; Constantz, J. Heat, chloride, and specific conductance as ground water tracers near streams. *Ground Water* **2007**, *45*, 187–195. [CrossRef]
2. Dybkjaer, J.B.; Baattrup-Pedersen, A.; Kronvang, B.; Thodsen, H. Diversity and Distribution of Riparian Plant Communities in Relation to Stream Size and Eutrophication. *J. Environ. Qual.* **2012**, *41*, 348–354. [CrossRef]
3. Ebersole, J.L.; Liss, W.J.; Frissell, C.A. Cold water patches in warm streams: Physicochemical characteristics and the influence of shading. *J. Am. Water Resour. Assoc.* **2003**, *39*, 355–368. [CrossRef]
4. Fernald, A.G.; Landers, D.H.; Wigington, P.J. Water quality changes in hyporheic flow paths between a large gravel bed river and off-channel alcoves in Oregon, USA. *River Res. Appl.* **2006**, *22*, 1111–1124. [CrossRef]
5. Vidon, P.; Allan, C.; Burns, D.; Duval, T.P.; Gurwick, N.; Inamdar, S.; Lowrance, R.; Okay, J.; Scott, D.; Sebestyen, S. Hot Spots and Hot Moments in Riparian Zones: Potential for Improved Water Quality Management1. *J. Am. Water Resour. Assoc.* **2010**, *46*, 278–298. [CrossRef]
6. Bucak, T.; Trolle, D.; Andersen, H.E.; Thodsen, H.; Erdogan, S.; Levi, E.E.; Filiz, N.; Jeppesen, E.; Beklioglu, M. Future water availability in the largest freshwater Mediterranean lake is at great risk as evidenced from simulations with the SWAT model. *Sci. Total Environ.* **2017**, *581*, 413–425. [CrossRef]
7. Freeze, R.A. Role of subsurface flow in generating surface runoff: 1. Base flow contributions to channel flow. *Water Resour. Res.* **1972**, *8*, 609–623. [CrossRef]
8. Hester, E.T.; Gooseff, M.N. Moving Beyond the Banks: Hyporheic Restoration Is Fundamental to Restoring Ecological Services and Functions of Streams. *Environ. Sci. Technol.* **2010**, *44*, 1521–1525. [CrossRef]
9. Kunkle, G.R. Computation of ground-water discharge to streams during floods, or to individual reaches during base flow, by use of specific conductance. In *US Geological Survey Professional Paper*; U.S. Geological Survey: Reston, VA, USA, 1965; pp. 207–210.
10. Nathan, R.; McMahon, T. Evaluation of automated techniques for base flow and recession analyses. *Water Resour. Res.* **1990**, *26*, 1465–1473. [CrossRef]
11. Wittenberg, H. Baseflow recession and recharge as nonlinear storage processes. *Hydrol. Process.* **1999**, *13*, 715–726. [CrossRef]
12. Alexander, M.D.; Caissie, D. Variability and comparison of hyporheic water temperatures and seepage fluxes in a small Atlantic salmon stream. *Ground Water* **2003**, *41*, 72–82. [CrossRef] [PubMed]
13. Dugdale, S.J.; Bergeron, N.E.; St-Hilaire, A. Temporal variability of thermal refuges and water temperature patterns in an Atlantic salmon river. *Remote Sens. Environ.* **2013**, *136*, 358–373. [CrossRef]
14. Dugdale, S.J.; Bergeron, N.E.; St-Hilaire, A. Spatial distribution of thermal refuges analysed in relation to riverscape hydromorphology using airborne thermal infrared imagery. *Remote Sens. Environ.* **2015**, *160*, 43–55. [CrossRef]
15. Ebersole, J.L.; Liss, W.J.; Frissell, C.A. Thermal heterogeneity, stream channel morphology, and salmonid abundance in northeastern Oregon streams. *Can. J. Fish. Aquat. Sci.* **2003**, *60*, 1266–1280. [CrossRef]
16. Ficke, A.D.; Myrick, C.A.; Hansen, L.J. Potential impacts of global climate change on freshwater fisheries. *Rev. Fish Biol. Fish.* **2007**, *17*, 581–613. [CrossRef]
17. Hester, E.T.; Doyle, M.W. Human Impacts to River Temperature and Their Effects on Biological Processes: A Quantitative Synthesis. *J. Am. Water Resour. Assoc.* **2011**, *47*, 571–587. [CrossRef]

18. Hillyard, R.W.; Keeley, E.R. Temperature-Related Changes in Habitat Quality and Use by Bonneville Cutthroat Trout in Regulated and Unregulated River Segments. *Trans. Am. Fish. Soc.* **2012**, *141*, 1649–1663. [CrossRef]

19. Lane, C.R.; Flotemersch, J.E.; Blocksom, K.A.; Decelles, S. Effect of sampling method on diatom composition for use in monitoring and assessing large river condition. *River Res. Appl.* **2007**, *23*, 1126–1146. [CrossRef]

20. Lisi, P.J.; Schindler, D.E.; Bentley, K.T.; Pess, G.R. Association between geomorphic attributes of watersheds, water temperature, and salmon spawn timing in Alaskan streams. *Geomorphology* **2013**, *185*, 78–86. [CrossRef]

21. Magoulick, D.D.; Kobza, R.M. The role of refugia for fishes during drought: A review and synthesis. *Freshw. Biol.* **2003**, *48*, 1186–1198. [CrossRef]

22. Baattrup-Pedersen, A.; Jensen, K.M.B.; Thodsen, H.; Andersen, H.E.; Andersen, P.M.; Larsen, S.E.; Riis, T.; Andersen, D.K.; Audet, J.; Kronvang, B. Effects of stream flooding on the distribution and diversity of groundwater-dependent vegetation in riparian areas. *Freshw. Biol.* **2013**, *58*, 817–827. [CrossRef]

23. Bayley, P.B. The flood pulse advantage and the restoration of river-floodplain systems. *Regul. Rivers Res. Manag.* **1991**, *6*, 75–86. [CrossRef]

24. Brown, A.G.; Lespez, L.; Sear, D.A.; Macaire, J.-J.; Houben, P.; Klimek, K.; Brazier, R.E.; Van Oost, K.; Pears, B. Natural vs. anthropogenic streams in Europe: History, ecology and implications for restoration, river-rewilding and riverine ecosystem services. *Earth-Sci. Rev.* **2018**, *180*, 185–205. [CrossRef]

25. Burkholder, B.K.; Grant, G.E.; Haggerty, R.; Khangaonkar, T.; Wampler, P.J. Influence of hyporheic flow and geomorphology on temperature of a large, gravel-bed river, Clackamas River, Oregon, USA. *Hydrol. Process.* **2008**, *22*, 941–953. [CrossRef]

26. Kurth, A.M.; Weber, C.; Schirmer, M. How effective is river restoration in re-establishing groundwater-surface water interactions?—A case study. *Hydrol. Earth Syst. Sci.* **2015**, *19*, 2663–2672. [CrossRef]

27. Hester, E.T.; Doyle, M.W.; Poole, G.C. The influence of in-stream structures on summer water temperatures via induced hyporheic exchange. *Limnol. Oceanogr.* **2009**, *54*, 355–367. [CrossRef]

28. Wawrzyniak, V.; Piégay, H.; Allemand, P.; Vaudor, L.; Goma, R.; Grandjean, P. Effects of geomorphology and groundwater level on the spatio-temporal variability of riverine cold water patches assessed using thermal infrared (TIR) remote sensing. *Remote Sens. Environ.* **2016**, *175*, 337–348. [CrossRef]

29. Winter, T.C. Landscape approach to identifying environments where ground water and surface water are closely interrelated. In Proceedings of the International Symposium on Groundwater Management Proceedings (ASCE, Ed.), San Antonio, TX, USA, 14–16 August 1995; pp. 139–155.

30. Varli, D.; Yilmaz, K. A Multi-Scale Approach for Improved Characterization of Surface Water—Groundwater Interactions: Integrating Thermal Remote Sensing and in-Stream Measurements. *Water* **2018**, *10*, 854. [CrossRef]

31. Fox, A.; Laube, G.; Schmidt, C.; Fleckenstein, J.H.; Arnon, S. The effect of losing and gaining flow conditions on hyporheic exchange in heterogeneous streambeds. *Water Resour. Res.* **2016**, *52*, 7460–7477. [CrossRef]

32. Genereux, D.P.; Leahy, S.; Mitasova, H.; Kennedy, C.D.; Corbett, D.R. Spatial and temporal variability of streambed hydraulic conductivity in West Bear Creek, North Carolina, USA. *J. Hydrol.* **2008**, *358*, 332–353. [CrossRef]

33. Hess, K.M.; Wolf, S.H.; Celia, M.A. Large-scale Natural gradient tracer test in sand and gravel, Cape Cod, Massachusetts 3. Hydraulic conductivity variability and calculated macrodispersivities. *Water Resour. Res.* **1992**, *28*, 2011–2027. [CrossRef]

34. Fleckenstein, J.H.; Niswonger, R.G.; Fogg, G.E. River-aquifer interactions, geologic heterogeneity, and low-flow management. *Ground Water* **2006**, *44*, 837–852. [CrossRef] [PubMed]

35. Schilling, O.S.; Irvine, D.J.; Hendricks Franssen, H.-J.; Brunner, P. Estimating the Spatial Extent of Unsaturated Zones in Heterogeneous River-Aquifer Systems. *Water Resour. Res.* **2017**, *53*, 10583–10602. [CrossRef]

36. Wroblicky, G.J.; Campana, M.E.; Valett, H.M.; Dahm, C.N. Seasonal variation in surface-subsurface water exchange and lateral hyporheic area of two stream-aquifer systems. *Water Resour. Res.* **1998**, *34*, 317–328. [CrossRef]

37. Mojarrad, B.B.; Riml, J.; Wörman, A.; Laudon, H. Fragmentation of the Hyporheic Zone Due to Regional Groundwater Circulation. *Water Resour. Res.* **2019**, *55*, 1242–1262. [CrossRef]

38. Kalbus, E.; Reinstorf, F.; Schirmer, M. Measuring methods for groundwater—surface water interactions: A review. *Hydrol. Earth Syst. Sci.* **2006**, *10*, 873–887. [CrossRef]

39. Griebler, C.; Lueders, T. Microbial biodiversity in groundwater ecosystems. *Freshw. Biol.* **2009**, *54*, 649–677. [CrossRef]

40. Iribar, A.; Sánchez-Pérez, J.M.; Lyautey, E.; Garabétian, F. Differentiated free-living and sediment-attached bacterial community structure inside and outside denitrification hotspots in the river–groundwater interface. *Hydrobiologia* **2008**, *598*, 109–121. [CrossRef]

41. Malard, F.; Hervant, F. Oxygen supply and the adaptations of animals in groundwater. *Freshw. Biol.* **1999**, *41*, 1–30. [CrossRef]

42. Frei, S.; Gilfedder, B.S. FINIFLUX: An implicit finite element model for quantification of groundwater fluxes and hyporheic exchange in streams and rivers using radon. *Water Resour. Res.* **2015**, *51*, 6776–6786. [CrossRef]

43. Frei, S.; Durejka, S.; Le Lay, H.; Thomas, Z.; Gilfedder, B.S. Hyporheic nitrate removal at the reach scale: Exposure times vs. residence times. *Water Resour. Res.* **2019**, in press. [CrossRef]

44. Genereux, D.P.; Hemond, H.F.; Mulholland, P.J. Use of radon-222 and calcium as tracers in a three-end-member mixing model for streamflow generation on the West Fork of Walker Branch Watershed. *J. Hydrol.* **1993**, *142*, 167–211. [CrossRef]

45. Kaandorp, V.P.; Doornenbal, P.J.; Kooi, H.; Peter Broers, H.; de Louw, P.G.B. Temperature buffering by groundwater in ecologically valuable lowland streams under current and future climate conditions. *J. Hydrol.* **2019**, *3*, 1–16. [CrossRef]

46. Mullinger, N.J.; Pates, J.M.; Binley, A.M.; Crook, N.P. Controls on the spatial and temporal variability of Rn-222 in riparian groundwater in a lowland Chalk catchment. *J. Hydrol.* **2009**, *376*, 58–69. [CrossRef]

47. Anderson, M.P. Heat as a ground water tracer. *Ground Water* **2005**, *43*, 951–968. [CrossRef]

48. Avery, E.; Bibby, R.; Visser, A.; Esser, B.; Moran, J. Quantification of Groundwater Discharge in a Subalpine Stream Using Radon-222. *Water* **2018**, *10*, 100. [CrossRef]

49. Constantz, J. Interaction between stream temperature, streamflow, and groundwater exchanges in Alpine streams. *Water Resour. Res.* **1998**, *34*, 1609–1615. [CrossRef]

50. Constantz, J. Heat as a tracer to determine streambed water exchanges. *Water Resour. Res.* **2008**, *44*, 1–20. [CrossRef]

51. Lee, J.-Y.; Lim, H.; Yoon, H.; Park, Y. Stream Water and Groundwater Interaction Revealed by Temperature Monitoring in Agricultural Areas. *Water* **2013**, *5*, 1677–1698. [CrossRef]

52. Schmidt, C.; Bayer-Raich, M.; Schirmer, M. Characterization of spatial heterogeneity of groundwater-stream water interactions using multiple depth streambed temperature measurements at the reach scale. *Hydrogeol. Earth Syst. Sci.* **2006**, *10*, 849–859. [CrossRef]

53. Harvey, M.C.; Hare, D.K.; Hackman, A.; Davenport, G.; Haynes, A.B.; Helton, A.; Lane, J.W.; Briggs, M.A. Evaluation of Stream and Wetland Restoration Using UAS-Based Thermal Infrared Mapping. *Water* **2019**, *11*, 1568. [CrossRef]

54. Lalot, E.; Curie, F.; Wawrzyniak, V.; Baratelli, F.; Schomburgk, S.; Flipo, N.; Piegay, H.; Moatar, F. Quantification of the contribution of the Beauce groundwater aquifer to the discharge of the Loire River using thermal infrared satellite imaging. *Hydrol. Earth Syst. Sci.* **2015**, *19*, 4479–4492. [CrossRef]

55. Wawrzyniak, V.; Piegay, H.; Poirel, A. Longitudinal and temporal thermal patterns of the French Rhne River using Landsat ETM plus thermal infrared images. *Aquat. Sci.* **2012**, *74*, 405–414. [CrossRef]

56. Hare, D.K.; Briggs, M.A.; Rosenberry, D.O.; Boutt, D.F.; Lane, J.W. A comparison of thermal infrared to fiber-optic distributed temperature sensing for evaluation of groundwater discharge to surface water. *J. Hydrol.* **2015**, *530*, 153–166. [CrossRef]

57. Hausner, M.B.; Suarez, F.; Glander, K.E.; van de Giesen, N.; Selker, J.S.; Tyler, S.W. Calibrating Single-Ended Fiber-Optic Raman Spectra Distributed Temperature Sensing Data. *Sensors* **2011**, *11*, 10859–10879. [CrossRef]

58. Tyler, S.W.; Selker, J.S.; Hausner, M.B.; Hatch, C.E.; Torgersen, T.; Thodal, C.E.; Schladow, S.G. Environmental temperature sensing using Raman spectra DTS fiber-optic methods. *Water Resour. Res.* **2009**, *45*, 11. [CrossRef]

59. Benyahya, L.; Caissie, D.; Satish, M.G.; El-Jabi, N. Long-wave radiation and heat flux estimates within a small tributary in Catamaran Brook (New Brunswick, Canada). *Hydrol. Process.* **2012**, *26*, 475–484. [CrossRef]

60. Caissie, D. The thermal regime of rivers: A review. *Freshw. Biol.* **2006**, *51*, 1389–1406. [CrossRef]

61. Evans, E.C.; McGregor, G.R.; Petts, G.E. River energy budgets with special reference to river bed processes. *Hydrol. Process.* **1998**, *12*, 575–595. [CrossRef]

62. Sebok, E.; Duque, C.; Engesgaard, P.; Boegh, E. Application of Distributed Temperature Sensing for coupled mapping of sedimentation processes and spatio-temporal variability of groundwater discharge in soft-bedded streams. *Hydrol. Process.* **2015**, *29*, 3408–3422. [CrossRef]

63. Collier, M.W. Demonstration of Fiber Optic Distributed Temperature Sensing to Differentiate Cold Water Refuge between Ground Water Inflows and Hyporheic Exchange. Master' Thesis, Oregon State University, Corvallis, OR, USA, 2008.

64. Strahler, A.N. Quantitative analysis of watershed geomorphology. *EosTrans. Am. Geophys. Union* **1957**, *38*, 913–920. [CrossRef]

65. Jaunat, J.; Huneau, F.; Dupuy, A.; Celle-Jeanton, H.; Vergnaud-Ayraud, V.; Aquilina, L.; Labasque, T.; Le Coustumer, P. Hydrochemical data and groundwater dating to infer differential flowpaths through weathered profiles of a fractured aquifer. *Appl. Geochem.* **2012**, *27*, 2053–2067. [CrossRef]

66. Lachassagne, P.; Wyns, R.; Dewandel, B. The fracture permeability of Hard Rock Aquifers is due neither to tectonics, nor to unloading, but to weathering processes. *Terra Nova* **2011**, *23*, 145–161. [CrossRef]

67. Thomas, Z.; Rousseau-Gueutin, P.; Kolbe, T.; Abbott, B.W.; Marçais, J.; Peiffer, S.; Frei, S.; Bishop, K.; Pichelin, P.; Pinay, G.; et al. Constitution of a catchment virtual observatory for sharing flow and transport models outputs. *J. Hydrol.* **2016**, *543*, 59–66. [CrossRef]

68. Thomas, Z.; Rousseau-Gueutin, P.; Abbott, B.W.; Kolbe, T.; Le Lay, H.; Marçais, J.; Rouault, F.; Petton, C.; Pichelin, P.; Le Hennaff, G.; et al. Long-term ecological observatories needed to understand ecohydrological systems in the Anthropocene: A catchment-scale case study in Brittany, France. *Reg. Environ. Chang.* **2019**, *19*, 363–377. [CrossRef]

69. Kolbe, T.; Marcais, J.; Thomas, Z.; Abbott, B.W.; de Dreuzy, J.R.; Rousseau-Gueutin, P.; Aquilina, L.; Labasque, T.; Pinay, G. Coupling 3D groundwater modeling with CFC-based age dating to classify local groundwater circulation in an unconfined crystalline aquifer. *J. Hydrol.* **2016**, *543*, 31–46. [CrossRef]

70. Kolbe, T.; de Dreuzy, J.-R.; Abbott, B.W.; Aquilina, L.; Babey, T.; Green, C.T.; Fleckenstein, J.H.; Labasque, T.; Laverman, A.M.; Marçais, J.; et al. Stratification of reactivity determines nitrate removal in groundwater. *Proc. Natl. Acad. Sci. USA* **2019**, *116*, 2494–2499. [CrossRef]

71. Tirado-Conde, J.; Engesgaard, P.; Karan, S.; Müller, S.; Duque, C. Evaluation of Temperature Profiling and Seepage Meter Methods for Quantifying Submarine Groundwater Discharge to Coastal Lagoons: Impacts of Saltwater Intrusion and the Associated Thermal Regime. *Water* **2019**, *11*, 1648. [CrossRef]

72. Calkins, D.; Dunne, T. A salt tracing method for measuring channel velocities in small mountain streams. *J. Hydrol.* **1970**, *11*, 379–392. [CrossRef]

73. van de Giesen, N.; Steele-Dunne, S.C.; Jansen, J.; Hoes, O.; Hausner, M.B.; Tyler, S.; Selker, J. Double-Ended Calibration of Fiber-Optic Raman Spectra Distributed Temperature Sensing Data. *Sensors* **2012**, *12*, 5471–5485. [CrossRef]

74. Le lay, H.; Thomas, Z.; Bour, O.; Rouault, F.; Pichelin, P.; Moatar, F. Experimental and model-based investigation of the effect of the free- surface flow regime on the detection threshold of warm water inflows. *Water Resour. Res.* **2019**, in press.

75. Krause, S.; Blume, T.; Cassidy, N. Investigating patterns and controls of groundwater up-welling in a lowland river by combining Fibre-optic Distributed Temperature Sensing with observations of vertical hydraulic gradients. *Hydrol. Earth Syst. Sci.* **2012**, *16*, 1775–1792. [CrossRef]

76. Lowry, C.S.; Walker, J.F.; Hunt, R.J.; Anderson, M.P. Identifying spatial variability of groundwater discharge in a wetland stream using a distributed temperature sensor. *Water Resour. Res.* **2007**, *43*, 1–9. [CrossRef]

77. Hebert, C.; Caissie, D.; Satish, M.G.; El-Jabi, N. Study of stream temperature dynamics and corresponding heat fluxes within Miramichi River catchments (New Brunswick, Canada). *Hydrol. Process.* **2011**, *25*, 2439–2455. [CrossRef]

78. Maheu, A.; Caissie, D.; St-Hilaire, A.; El-Jabi, N. River evaporation and corresponding heat fluxes in forested catchments. *Hydrol. Process.* **2014**, *28*, 5725–5738. [CrossRef]

79. Neilson, B.T.; Hatch, C.E.; Ban, H.; Tyler, S.W. Solar radiative heating of fiber optic cables used to monitor temperatures in water. *Water Resour. Res.* **2010**, *46*, 2–17. [CrossRef]

80. Westerberg, I.K.; Wagener, T.; Coxon, G.; McMillan, H.K.; Castellarin, A.; Montanari, A.; Freer, J. Uncertainty in hydrological signatures for gauged and ungauged catchments. *Water Resour. Res.* **2016**, *52*, 1847–1865. [CrossRef]

81. Toth, J. A theoretical analysis of groundwater flow in small drainage basins. *J. Geophys. Res.* **1963**, *68*, 4795–4812. [CrossRef]
82. Crispell, J.K.; Endreny, T.A. Hyporheic exchange flow around constructed in-channel structures and implications for restoration design. *Hydrol. Process.* **2009**, *23*, 1158–1168. [CrossRef]
83. Flipo, N.; Mouhri, A.; Labarthe, B.; Biancamaria, S.; Rivière, A.; Weill, P. Continental hydrosystem modelling: The concept of nested stream–aquifer interfaces. *Hydrol. Earth Syst. Sci.* **2014**, *18*, 3121–3149. [CrossRef]
84. Magliozzi, C.; Grabowski, R.C.; Packman, A.I.; Krause, S. Toward a conceptual framework of hyporheic exchange across spatial scales. *Hydrol. Earth Syst. Sci.* **2018**, *22*, 6163–6185. [CrossRef]
85. Harvey, J.W.; Bencala, K.E. The Effect of Streambed Topography on Surface-Subsurface Water Exchange in Mountain Catchments. *Water Resour. Res.* **1993**, *29*, 89–98. [CrossRef]
86. Matthews, K.R.; Berg, N.H.; Azuma, D.L.; Lambert, T.R. Cool water formation and trout habitat use in a deep pool in the sierra-nevada, california. *Trans. Am. Fish. Soc.* **1994**, *123*, 549–564. [CrossRef]
87. Nielsen, J.L.; Lisle, T.E.; Ozaki, V. Thermally stratified pools and their use by steelhead in northern california streams. *Trans. Am. Fish. Soc.* **1994**, *123*, 613–626. [CrossRef]
88. Webb, B.W.; Walling, D.E. Complex summer water temperature behaviour below a UK regulating reservoir. *Regul. Rivers-Res. Manag.* **1997**, *13*, 463–477. [CrossRef]
89. Norman, F.A.; Cardenas, M.B. Heat transport in hyporheic zones due to bedforms: An experimental study. *Water Resour. Res.* **2014**, *50*, 3568–3582. [CrossRef]
90. Le Lay, H.; Thomas, Z.; Rouault, F.; Pichelin, P.; Moatar, F. Characterization of diffuse groundwater inflows into stream water. Part II: Quantifying groundwater inflows by coupling FO-DTS and vertical flow velocities. *Water* **2019**, in press.

Article

Streambed Flux Measurement Informed by Distributed Temperature Sensing Leads to a Significantly Different Characterization of Groundwater Discharge

Troy E. Gilmore [1,2,*], Mason Johnson [1], Jesse Korus [1], Aaron Mittelstet [2], Marty A. Briggs [3], Vitaly Zlotnik [4] and Sydney Corcoran [1]

[1]   Conservation and Survey Division, School of Natural Resources, University of Nebraska-Lincoln, Lincoln, NE 68583, USA; mason.johnson@huskers.unl.edu (M.J.); jkorus3@unl.edu (J.K.); scorcoran2@huskers.unl.edu (S.C.)

[2]   Department of Biological Systems Engineering, University of Nebraska-Lincoln, Lincoln, NE 68583, USA; amittelstet2@unl.edu

[3]   U.S. Geological Survey, Hydrogeophysics Branch, 11 Sherman Place, Unit 5015, Storrs, CT 06269, USA; mbriggs@usgs.gov

[4]   Earth and Atmospheric Sciences, University of Nebraska-Lincoln, Lincoln, NE 68588, USA; vzlotnik1@unl.edu

*   Correspondence: gilmore@unl.edu; Tel.: +1-402-470-1741

Received: 26 September 2019; Accepted: 29 October 2019; Published: 5 November 2019

Abstract: Groundwater discharge though streambeds is often focused toward discrete zones, indicating that preliminary reconnaissance may be useful for capturing the full spectrum of groundwater discharge rates using point-scale quantitative methods. However, many direct-contact reconnaissance techniques can be time-consuming, and remote sensing (e.g., thermal infrared) typically does not penetrate the water column to locate submerged seepages. In this study, we tested whether dozens of groundwater discharge measurements made at "uninformed" (i.e., selected without knowledge on high-resolution temperature variations at the streambed) point locations along a reach would yield significantly different Darcy-based groundwater discharge rates when compared with "informed" measurements, focused at streambed thermal anomalies that were identified a priori using fiber-optic distributed temperature sensing (FO-DTS). A non-parametric U-test showed a significant difference between median discharge rates for uninformed (0.05 m·day^{-1}; $n = 30$) and informed (0.17 m·day^{-1}; $n = 20$) measurement locations. Mean values followed a similar pattern (0.12 versus 0.27 m·day^{-1}), and frequency distributions for uninformed and informed measurements were also significantly different based on a Kolmogorov–Smirnov test. Results suggest that even using a quick "snapshot-in-time" field analysis of FO-DTS data can be useful in streambeds with groundwater discharge rates <0.2 m·day^{-1}, a lower threshold than proposed in a previous study. Collectively, study results highlight that FO-DTS is a powerful technique for identifying higher-discharge zones in streambeds, but the pros and cons of informed and uninformed sampling depend in part on groundwater/surface water exchange study goals. For example, studies focused on measuring representative groundwater and solute fluxes may be biased if high-discharge locations are preferentially sampled. However, identification of high-discharge locations may complement more randomized sampling plans and lead to improvements in interpolating streambed fluxes and upscaling point measurements to the stream reach scale.

Keywords: groundwater discharge; distributed–temperature sensing; groundwater–surface water interaction; hyporheic zone

---

1. Introduction

Groundwater–surface water exchange can be highly variable in rate and direction across streambeds due to spatial heterogeneity in hydraulic gradient and sediment permeability [1–4]. As a result, a variety of methods have been employed to measure groundwater discharge at different spatial scales [5]. Point-scale vertical groundwater flux estimates are often based on hydraulic conductivity estimates from permeameter tests and vertical hydraulic gradient measured using piezomanometers or pressure sensors [6–9]. Other studies used point-scale measurements of streambed temperature (i.e., vertical profiles of streambed temperature) to estimate groundwater discharge [10–16]. Seepage meter footprints are typically larger than point scale [4,5,17] but, in the context of larger reach-scale measurements, are similar to point-scale measurements (a point-scale seepage device was recently developed [18]). Point-scale and similarly scaled measurements led to a greater understanding of spatial variability in groundwater discharge through streambeds [2,19,20], including stream reaches where reach mass balance and other larger-scale measurements are not possible. However, the necessary flux measurement density required for spatial interpolation and integration of results at the reach scale requires tradeoffs between feasible stream or river reach size (due to the time and/or equipment costs) and certainty in reach-scale estimates of groundwater discharge [19,21].

One reach-scale groundwater discharge "reconnaissance" method that is paired with point measurements of groundwater discharge is distributed temperature sensing with fiber-optic cables installed along the streambed (FO-DTS) [3,4,22,23]. In concept, fiber-optic cables (lengths on the order of 10^1 to 10^3 m) are placed on the streambed, giving a longitudinal "linear sample" of streambed interface temperature at typical resolution of 0.25 m to several meters. Streambed temperature anomalies, such as cool water patches during the summer, are often associated with focused groundwater discharge through the streambed [1,24,25]. Studies showed high rates of groundwater discharge associated with discrete temperature anomalies identified by the FO-DTS technique [4,17,26].

In a previous study where locations of focused groundwater discharge were identified using FO-DTS, temperature anomalies were described as "cool" zones (streambed temperature was comparatively low under summer conditions in the Quashnet River), where groundwater discharge was likely to be higher and more focused in space [4]. The term "ambient" was used to describe zones where the streambed interface was at a general average background temperature, and groundwater discharge in ambient zones was likely to be lower and more diffuse [4]. FO-DTS was used to select measurement locations in each zone, and groundwater discharge measurements from seepage meters at 29 locations (13 ambient, 16 cool) were reported. Median cool zone discharge (0.83 m·day^{-1}) was about five times the median discharge estimated from ambient temperature zone measurements (0.17 m·day^{-1}) even though the streambed was predominantly sandy and connected to a relatively homogeneous, permeable sand and gravel aquifer. Follow-up geophysical work found that the streambed was underlain by discontinuous peat lenses [27], driving spatially preferential groundwater discharge patterns, similar to that observed in other stream systems with low-permeability streambed lenses [3,20]. A threshold for FO-DTS detection of focused groundwater discharge in their stream system was hypothesized (0.4 m·day^{-1}), with the caveat that lower stream water velocity (compared to 0.5–1 m·s^{-1} in their study) and other thermal factors (e.g., shading) could lead to lower thresholds in other systems [4]. In a separate study, FO-DTS was used in a peat-dominated wetland stream to make direct comparisons between a "background" seepage meter measurement site and two other seepage meter sites located at thermal anomalies, referred to as "focused zones" [17] that showed enhanced discharge rates.

The ability to identify focused groundwater discharge using FO-DTS raises new questions about potential bias and uncertainty in point-scale sampling as a means to capture and integrate spatial variability in groundwater discharge through streambeds. In the absence of flux-reconnaissance temperature data, point measurements in streams were conducted at random or pre-determined locations in streambeds or conducted along lateral streambed transects distributed at regular along-profile intervals [28–34]. While at least some of these studies involved field reconnaissance of some kind (e.g., tests to ensure appropriate equipment for streambed conditions, and preliminary data on groundwater discharge), we refer to this streambed sampling approach as "uninformed" measurement, to reflect the lack of prior information on streambed temperature when measurement sites are selected. In contrast, the technique of using FO-DTS to identify and measure discharge at likely areas of focused discharge is referred to as "informed" measurement, to reflect that streambed interface temperature data anomalies inform the selection of measurement sites. In the following, terms "informed" and "uninformed" are applied to the different aspects of measurements (locations, data, etc.).

In our study, we established measurement locations along the length of a stream in the absence of any streambed temperature data (uninformed) and compared groundwater discharge rates to the rates measured at locations selected using FO-DTS temperature data (informed). Thus, this is the first study to directly compare informed and uninformed measurement approaches. The primary purpose of this study is (1) to determine whether informed measurements result in significantly higher estimates of groundwater discharge compared to uninformed measurements, (2) to determine whether informed measurements give a fundamentally different picture of variability in groundwater discharge within a stream reach when compared to uninformed measurement, and (3) to test previous estimates of the lower groundwater discharge threshold at which FO-DTS is still effective for distinguishing between areas of diffuse and focused discharge in streambeds.

2. Materials and Methods

Groundwater discharge measurements were made at points in a sandy streambed in central Nebraska using a Darcian flux approach. Point measurements were made at lateral three-point transects (left, center, and right locations across the streambed), which were distributed at regular intervals along the length of the study reach. A detailed arrangement of FO-DTS cable along the left, center, and right sides of the meandering stream was also used to locate apparent temperature anomalies on the streambed. Groundwater discharge measurements at the temperature anomaly sites (informed sampling) were then compared to measurements made at three-point lateral transects (uninformed sampling) to determine if and how the FO-DTS approach might influence the overall characterization of groundwater discharge into the reach.

2.1. Site Description

The study was conducted on a 700-m reach on the South Branch of the Middle Loup River (SBMLR) on the University of Nebraska Gudmundsen Sandhills Research Laboratory land located in the central Sand Hills of Nebraska (Figure 1). The site is located in the northern region of the High Plains aquifer, where saturated thickness is 300 m. The SBMLR, a tributary to the Middle Loup and Loup Rivers, is a sinuous, second-order stream that drains approximately 925 km^2 of pasture. The groundwater-fed stream has a stream width ranging from 2–5 m and water depth of 10–50 cm. Streamflow during the study period was about 250 L·s^{-1}.

The sand dunes that comprise the Sand Hills are vegetated and minimally grazed. The soils and shallow subsurface aquifer near the study site are primarily eolian sand and loamy and sandy alluvium. The streambed consists of sand with some organic matter. Data from a previous study conducted upstream of our study site, where the organic matter content is much higher, suggested mean groundwater discharge rates of 0.1 to 0.3 m·day^{-1} through the streambed (based on estimated 2-m ditch width) [35,36]. These estimates are at the lower end of the range of other studies conducted using FO-DTS [4,26,37] (flux ranged 0.05 to 7.2 m·day^{-1}) and similar to previous stream studies using physical measurement methods [2,18,21].

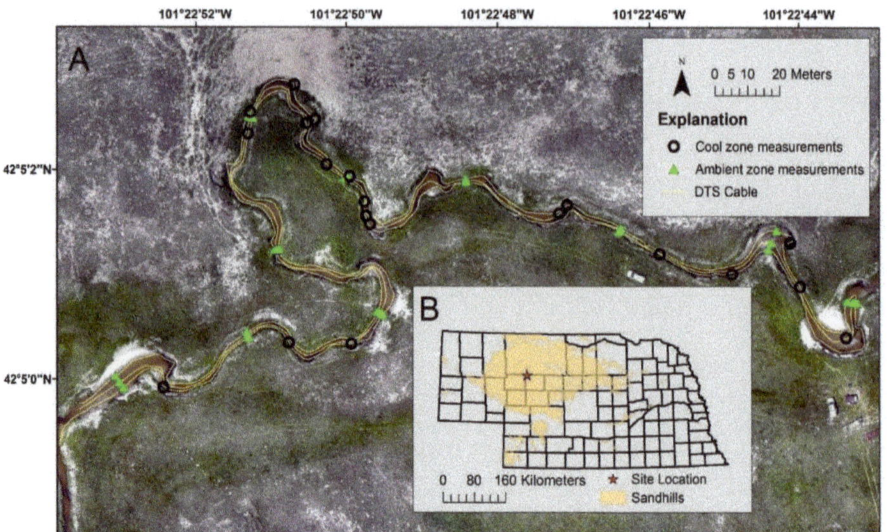

Figure 1. (**A**) Distributed temperature sensing (DTS) fiber-optic (FO) cable configuration on the streambed interface along ~700 m of stream channel. The cable was placed on the left bank, mid channel, and right bank at three distinct time periods throughout the study, as described in Section 2.2. Measurements were made at predetermined intervals, or "uninformed" locations along the fiber optic cable. Measurements were also made at "informed" locations of suspected high groundwater upwelling, which were indicated by cooler streambed temperatures sensed by the fiber optic cable. (**B**) Study site location in the central Sand Hills of Nebraska.

2.2. Identification of Uninformed and Informed Measurement Sites

Uninformed measurements were allocated at 16-m longitudinal intervals along the stream, for a total of 30 uninformed hydraulic measurements. The locations were predetermined without knowledge of streambed temperature dynamics. Beginning at the downstream end of the reach, intervals of 16-m longitudinal distance were measured, and a lateral three-point transect was marked across the stream (i.e., measurement locations at right bank, left bank, and center positions) at each interval.

To identify informed locations to measure streambed discharge, distributed temperature sensing with fiber-optic cables was used, and a total of 20 informed locations were identified. The FO-DTS unit was a Silixa-XT system provided by the Center for Transformative Environmental Monitoring Programs (CTEMPs). The FO-DTS control unit has a published linear spatial sampling resolution at 25 cm and an operating temperature from −40 to +65 °C at <0.01 °C resolution, although, in practice, the spatial resolution and temperature resolution are greater. FO-DTS theory applicable to our application can be found elsewhere [38–40].

The FO-DTS had 4000 m of Commscope flat-drop fiber-optic cable attached, although only 700 m of cable was deployed on the streambed at any given time during the study. The cable was attached to the control unit in double-ended mode, excess cable was left on a large spool, and temperature baths for DTS calibration were located between the spool and the stream. Temperature baths consisted of 10 m of cable placed in an ice bath in a cooler (0 to 3 °C) and an additional 10 m of cable placed in an ambient temperature water bath in a cooler. Each bath was equipped with a Silixa Pt100 temperature reference temperature probe, and an air bubbler circulator was used to avoid temperature stratification in the bath. Several air loops were also placed on the streambank or attached to stakes above the stream water level as spatial reference points.

The FO-DTS cable was secured on the streambed using clamps fastened onto vertical stakes (Figure 2). At the downstream end of the reach, approximately 100 m of stream length had FO-DTS cable deployed along the right bank, left bank, and center throughout the entire campaign. In the upstream part of the reach, the remaining 400 m of cable was sequentially placed on the right bank, then left bank, and then center of the stream during different time periods for groundwater discharge reconnaissance (starting 2 June 2017, 8 June 2017, and 9 June 2017, respectively). The movement of the cable to different streambed positions was done (1) to shorten the individual cable installation periods given the potential for precipitation events during the study period, and (2) to avoid temperature impacts of partial and/or temporary burial of large sections of the cable due to the mobile streambed. A third reason was that sequential installation also reduced the equipment (i.e., clamps and stakes) required for the detailed installation in the meandering channel.

Figure 2. (**A**) Fiber-optic cable deployed on the streambed, and (**B**) an example of areas where the cable was occasionally above the water line either due to difficulty in deploying the cable around bars in the channel, or due to a slight decrease in stream stage during the study period.

After cable deployment along the right, left, or center of the stream, the streambed and cable were left undisturbed for a 12-h period. The unit logged temperatures along the length of the cable every 10 s at 25-cm sampling resolution during the campaign period. When data collection was complete for the right bank (4 June 2017; Figure 3), the end section of the cable (roughly 400 m) was moved to the left bank and left undisturbed for another 12-h period. This cycle was completed for center stream as well. The locations of the FO-DTS cable and the uninformed and informed measurement locations were identified using a survey-grade Topcon Hiper V global positioning system (GPS).

For each of these cable deployment locations (right bank, left bank, and center), uncalibrated temperatures (Figure 4C) were used for a field analysis of anomalies (i.e., cool zones, since the study was conducted in the summer). These anomalies were determined based on visual analysis of a temperature versus optical distance plot downloaded from the FO-DTS unit in the field (i.e., based on a "snapshot" of 10-s-averaged temperature measurements). For the deployment along the right side of the stream, the 10 most pronounced cool zones were identified for an equal comparison to the 10 uninformed measurements conducted along the right side. For the center and left side of the streambed, there were fewer pronounced cool zones and, therefore, fewer informed measurement locations.

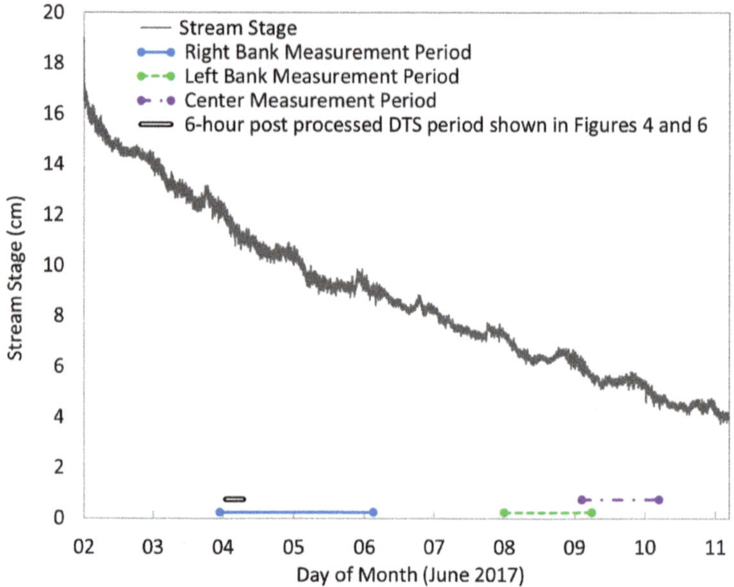

Figure 3. Field campaign timeline in the month of June 2017. Sampling time periods for vertical head gradient measurements along the right bank, left bank, and center channel are indicated by horizontal bars. Uninformed sampling occurred toward the beginning of each sampling period, while informed sampling occurred toward the end of each sampling period. The horizontal black and white bar indicates the time period over which data collected by the DTS was post-processed (to correct for temperature offsets) and as an example of calibrated and time-averaged FO-DTS temperature data.

After identification, the cooler sections of cable were located along the length of the physical cable based on cable length. Those cooler locations in the streambed were marked for informed measurements. The exact cable location where temperature anomalies existed were not known due to lack of precision in the cable sheathing. Even though the cable had numerical meter measurements, there were no tick marks on the sheath indicating where the meter marker started or ended, and there were also differences between the physical cable length and optical distance shown on the temperature plot. The long section of cable left on the spool may have also added some noise to the temperature data and contributed to this uncertainty. The general areas of cooler temperature zones were found first. A temperature probe (Fisherbrand™ Traceable™ Platinum Ultra-Accurate Digital Thermometer, seven-inch probe, ±0.05 °C accuracy) was then inserted into the streambed along the cable within these general areas. The point that had the lowest temperature based on the surrounding areas was identified as an informed location.

The DTS collected approximately 56,000 data points during the entire campaign. A subset of 2568 data points was post processed to represent the deployment period along the right bank and the 100 m of streambed where cable was left in the right bank, left bank, and center positions throughout the study for comparison to the field "snapshot" selection of informed sampling locations (e.g., Figures 3 and 4). Post-processed data represented a 6-h period between 1:00 and 7:00 p.m. on 4 June 2017. The data for this period were calibrated using a MATLAB® graphical user interface (GUI) created by CTEMPs. The GUI allows the user to determine the location of the calibration baths and splice box in the collected data and use that information to calibrate temperatures collected by the DTS unit. This double-ended calibration method allowed for accurate calibration along two different channels within the same fiber-optic cable.

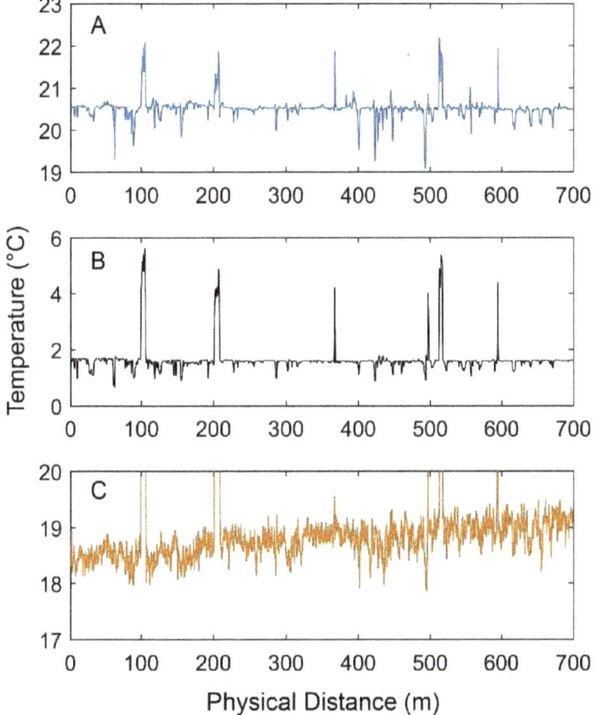

Figure 4. (**A**) Post-processed DTS mean streambed temperature data over a 6-h time period on 4 June 2017. Peaks in the data represent air loops while troughs represent areas of cooler temperature zones. (**B**) Standard deviation of temperature data for the same period. Large peaks in the data indicate high variability in air temperature at air loops. (**C**) Single snapshot of one temperature signal captured by the DTS on 4 June 2017. This raw (uncalibrated) dataset was used for selecting where informed physical measurements would be taken. Note that the full scale of temperature at air loops is not shown in (**C**).

2.3. Hydraulic Measurements

Vertical hydraulic gradient (VHG; i, m·m^{-1}) and conductivity (K_v) were measured at each of the informed and uninformed locations. The VHG was measured between groundwater and surface water using a light-oil piezomanometer [9]. Firstly, VHG was measured to avoid the disruption of the local vertical hydraulic gradient. The hand-built piezomanometer was equipped with a drive-point screen (5 cm), which was pushed into the streambed to a depth of 30 cm (top of screen). K_v was estimated using the falling head permeameter method [41]. A permeameter, or clear polycarbonate tube with incremental markings on the side, was inserted into the streambed at 30 cm and a falling head test [41] was conducted. K_v was then calculated for each sampling location as described in previous work [41].

With i and K_v known, groundwater flux (q) was calculated for each of the 50 informed and uninformed locations as $q = i \times K_v$.

A Sontek Flowtracker Acoustic Doppler Velocimeter was used for measuring stream discharge in the study reach and stream water velocity near the FO-DTS cable. Stream discharge measurements were made near the upstream and downstream ends of the study reach on 3 June 2017 and 5 June 2017 ($n = 3$). Stream velocity measurements were made at 45 locations with paired measurements made near the cable depth and higher in the water column ($n = 90$ stream velocity measurements).

2.4. Drone Imagery Acquisition

During the course of FO-DTS temperature data collection, a DJI Phantom 4© Quadcopter with a 12-megapixel camera was used to collect aerial imagery of the study site. The resulting high-resolution orthophoto aided in digitizing the fiber-optic cable on the streambed (Figure 1) where there was uncertainty in GPS locations due to the remote study site location. The aircraft was pre-programmed to run a gridded flight pattern with roughly 70% image overlap for photogrammetric processing. Over 500 images were captured during the flight and then processed with the photogrammetric mapping software Pix4D©. Ground control points (GCP) with known GPS coordinates were previously placed in the streambed and at locations along the stream bank. These GCPs were imported into the mapping software, tagged to targets in the imagery, and used to create a georeferenced, seamless orthophoto of the site with ~1.4-cm spatial resolution.

3. Results and Discussion

FO-DTS and vertical flux measurement methods were combined in this study to address three research questions, each of which is discussed in the sections below. The fourth and final section is focused on the FO-DTS application and study limitations.

3.1. Are Discharge Estimates at Informed and Uninformed Locations Significantly Different?

All VHGs were positive, which indicated a consistent upward vertical groundwater flux during the study period. Median groundwater discharge for informed measurements was 0.21 m·day^{-1}, which was significantly higher than the median of uninformed measurements (0.05 m·day^{-1}) based on the Wilcox rank sum non-parametric U-test that is most appropriate for non-normal distributions (Table 1). Mean groundwater discharge for informed measurements (0.27 m·day^{-1}) was substantially higher than for uninformed measurements (0.12 m·day^{-1}).

Table 1. Summary statistics for groundwater discharge (q) measurements in this study and in a previous study [4].

	This Study			Quashnet River [4]		
	All	**Uninformed** [a]	**Informed** [a]	**All**	**Ambient** [a]	**Cool** [a]
Median (m·day^{-1})	0.13	0.05 [b]	0.21 [b]	0.40	0.17	0.83
Mean (m·day^{-1})	0.18	0.12	0.27	0.58	0.19	1.07
Standard Deviation (m·day^{-1})	0.20	0.14	0.25	0.74	0.33	0.81
Coeffiecient of Variation (%)	110	113	91	126	174	76
Min (m·day^{-1})	<0.01	<0.01	<0.01	−0.55	−0.55	0.20
Max (m·day^{-1})	0.95	0.48	0.95	3.0	0.93	3.0
Measurements (-)	50	30	20	29	16	13

[a] A subtle difference in nomenclature, as described in the introduction: in previous work [4], groundwater discharge was measured at 29 locations, all selected based on fiber-optic distributed temperature sensing (FO-DTS) data, and classified results as ambient or cool zone measurements. In this study, we pre-selected measurement locations without any prior knowledge of streambed temperature (uninformed) and compare those discharge values with those measured at cool zones (informed). [b] Statistically different values ($\alpha = 0.05$, $p = 0.027$).

Both vertical hydraulic conductivity (K_v) and head gradient (i) were on average lower for uninformed measurement locations than for informed locations, but the greatest difference was in K_v (Table 2). Median K_v for uninformed measurement locations was about 32% of K_v from informed locations and statistically different (U-test; $\alpha = 0.05$, $p = 0.024$). The differences in K_v reflect how groundwater was preferentially discharging in areas with higher permeability. Uninformed i was about 82% of informed i. For informed measurement locations, variability (coefficient of variation) was greater for i than for K_v, and vice versa for uninformed measurement locations.

Table 2. Summary statistics for head gradient (i) and vertical hydraulic conductivity (K_v).

	All		Uninformed		Informed	
	i (m·m^{-1})	K_v (m·day^{-1})	i (m·m^{-1})	K_v (m·day^{-1})	i (m·m^{-1})	K_v (m·day^{-1})
Median	0.018	6.8	0.017	3.5 [a]	0.023	11.1 [a]
Mean	0.023	8.6	0.020	6.0	0.028	12.4
Standard Deviation	0.023	8.5	0.013	6.3	0.032	10.0
Coefficient of Variation	99%	99%	65%	105%	115%	80%

[a] Statistically different values ($\alpha = 0.05$, $p = 0.024$).

3.2. A Fundamentally Different View of Groundwater Discharge?

Although informed measurement locations yielded the highest groundwater discharge estimate (0.95 m·day^{-1}), uninformed measurements had greater variability in discharge than informed measurements (coefficient of variation of 113% versus 91%, Table 1). In both cases, variability in groundwater discharge was in the range (on a percentage basis) of variability commonly observed in other streams, although the differences between informed and uninformed measurements were less pronounced in this study compared to differences in variability found previously [4] (174% versus 76% for ambient versus cool zones, Table 1).

The relative frequency distribution of all groundwater discharge measurements in the SBMLR (Figure 5C) is typical of other distributions observed in streams with sandy streambeds and relatively homogeneous connected aquifers [2,21,28]. In general, these distributions indicate a high frequency of measurements at low-discharge locations, with a long tail toward higher groundwater discharge values. The frequency distribution for uninformed measurements in our study reach was also heavily weighted toward low discharge values (Figure 5A). In comparison, the informed measurement locations yielded a broader distribution, including higher discharge values, but with the bulk of discharge in the moderate range (Figure 5B). Overall, the frequency distributions for informed and uninformed measurements represent fundamentally different distributions, based on the Kolmogorov–Smirnov test (D = 0.45, p = 0.012), as did the frequency distributions for ambient and cool zones measurements in the Quashnet River [4] (p << 0.01, exact p-value could not be computed due to "ties" in the two datasets). Thus, the use of FO-DTS in these systems led to substantially different views of groundwater discharge, based on statistically significant differences in both the median discharge and the frequency distributions for informed and uninformed measurements.

The shapes of the relative frequency for informed measurements in the SBMLR (Figure 5B) was remarkably similar to that of the cool zone measurements in the Quashnet River (Figure 5E) [4], albeit with almost a factor of five difference in magnitude of discharge. The frequency of low discharge values is clearly lower compared to ambient or uninformed measurements in either the Quashnet River or SBMLR. In contrast, the uninformed measurements had a broader distribution than the ambient locations selected in the Quashnet River, likely because our uninformed measurements sampled the streambed more broadly than the ambient location measurements purposely at relatively warm streambed zones for that study. In other words, it was possible for our uninformed measurements to occur at locations with high-, intermediate-, or low-temperature streambed zones, whereas ambient sites in the Quashnet River [4] were purposefully located in warmer streambed zones for comparison with measurements selectively located at cool zones.

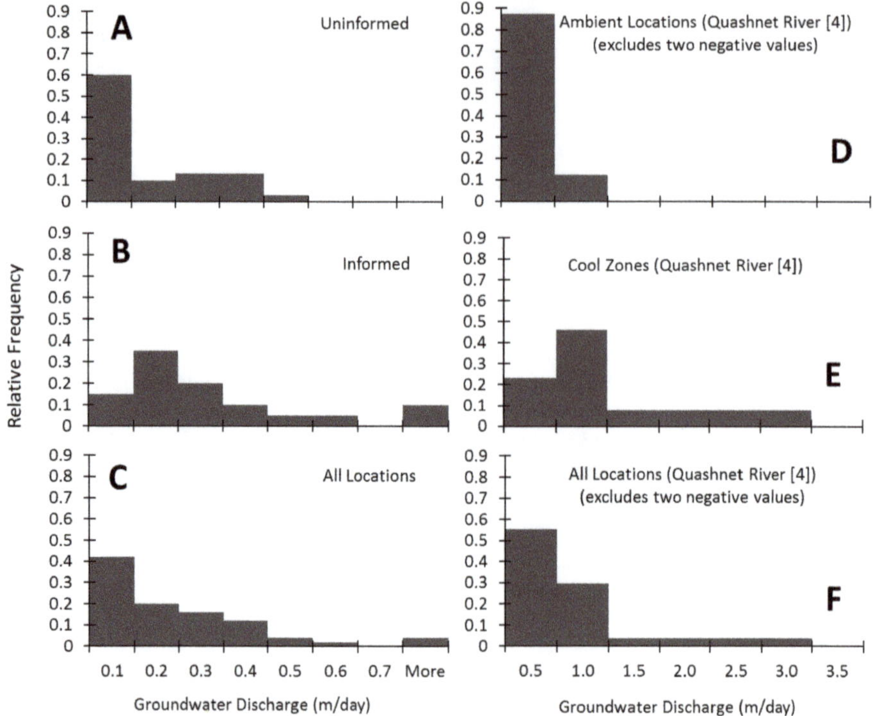

Figure 5. Relative frequency distributions of (**A**) informed, (**B**) uninformed, and (**C**) all groundwater discharge measurements in this study, alongside frequency distributions calculated from (**D**) ambient locations and (**E**) cool locations, and (**F**) calculated from all groundwater discharge measurement locations reported from the Quashnet River in a separate study [4]. Two negative (downward) flux estimates were excluded from the ambient and all measurement distributions from the Quashnet River. Labels on the horizontal axis indicate the upper limit for each bin.

The different observed frequency distributions are important to consider in future work, particularly for studies where the influence of groundwater chemistry on stream-water quality is a concern. There are potentially interesting tradeoffs between capturing groundwater chemistry at high-discharge points using FO-DTS, versus broader sampling that captures what appear to be high-frequency low-discharge points. For instance, studies of groundwater age and nitrate discharge from aquifers found highly variable groundwater age and nitrate concentrations in streambeds, and that high-versus low-discharge locations may be correlated with different nitrate fluxes [31,42]. Thus, it may seem intuitive to seek out and measure groundwater chemistry at high-discharge points using FO-DTS, but "missing" the high-frequency low-discharge points could potentially bias the overall estimates of chemical flux from groundwater and/or the distributions of groundwater age in aquifer discharge.

Furthermore, for studies where spatial interpolation is desired to upscale from streambed points to reach scale, it is important to capture both low and high discharge rates [20]. There may be opportunities in future work to exploit FO-DTS data in the field to better inform sampling strategy. For instance, streambed points may be selected at several temperature signatures (not just the warmest or coolest temperature anomalies) to capture the full spectrum of groundwater discharge rates. If temperature and groundwater discharge are reasonably correlated, FO-DTS may serve as a "line sample" along the streambed (e.g., Figure 6), and used to improve or evaluate spatial interpolation of streambed fluxes. On the other hand, a hybrid alternative is possible, where informed and uninformed sampling are combined to "randomly" sample the reach while also capturing potentially important high-discharge zones.

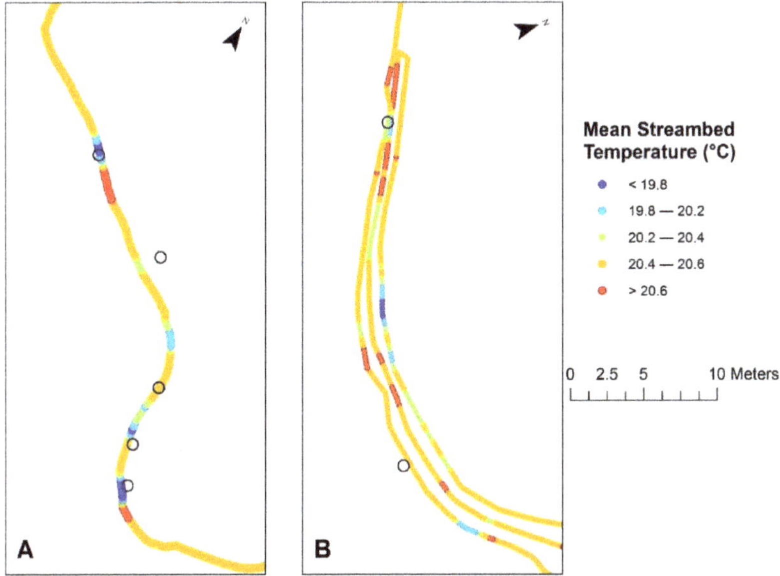

Figure 6. Calibrated mean streambed temperature for the time period shown in Figure 3. Informed sampling locations (shown as open circles) are along the right bank of the stream (**A**) in the upstream part of the reach where a single strand of cable was placed, and (**B**) in the downstream end of the reach where cable was left in the right bank, left bank, and center positions throughout the campaign.

3.3. A Lower Discharge Threshold for Effective Use of FO-DTS?

Median groundwater discharge was 0.13 m·day^{-1} in this study, relatively low compared to the recent similar study [4]. In that study, they conservatively estimated a threshold of 0.42 m·day^{-1} as a cutoff below which FO-DTS application might be less effective for identifying focused discharge locations in the coastal Quashnet River. However, they noted that an argument could also be made for a smaller threshold of 0.2 m·day^{-1} and that, in other stream systems with lower stream velocities (less than the typical 0.5–1 m·s^{-1} observed in the Quashnet River), an even lower threshold could be applicable. Our study reach in the SBMLR had an average stream velocity along the FO-DTS cable of about 0.3 m·s^{-1}. Thus, stream velocities and groundwater discharge rates at our site on the SBMLR are both lower than for the Quashnet River, suggesting the discharge rate threshold for successful application of FO-DTS for the geolocation of discharge points based on point-in-time temperature traces may indeed be <0.2 m·day^{-1}, especially where stream velocities and other variables are favorable for the creation of discharge-based thermal anomalies.

3.4. Raw vs. Calibrated FO-DTS Temperature Data and Study Limitations

One potential limitation to this study was the use of a less-precise "snapshot-in-time" approach to identify cool zones for informed measurement based on 10-s-averaged temperatures (Figure 4C; see also Section 2.2). A more robust in-field analysis of calibrated FO-DTS temperature data (Figure 4A,B), or even temperature averaged over a longer period (minutes, rather than seconds) may have led to more precise identification of cool zones. A subset of temperature data displayed in map view (Figure 6) suggests that the informed locations identified along the right bank in this study were reasonably consistent with post-processed and calibrated temperature data. Out of the seven measurement locations in Figure 6, five were located at or very near relatively cool locations as indicated by post-processed temperatures. The use of calibrated temperatures, mean streambed temperature, and temperature variance over time may eliminate some artefacts in the raw temperature data, and highlight areas that are consistently cooler, allowing for identification of subtle thermal anomalies corresponding to moderate flux zones. For example, parts of the cable in the direct sun could be affected more by solar radiation than parts of the cable that are in shaded areas. Integrating mean streambed temperature over a period of time may smooth out these direct solar warming-based variances [1].

However, more advanced approaches of post-processing FO-DTS temperature data in the field are time-consuming and could reduce the amount of physical measurements that can be made within a given study period. Lengthening the deployment measurement period increases potential for other artefacts as well. For instance, bedform migration in the stream channel can bury portions of the cable in the streambed, which would lead to a bias toward cooler streambed temperature, or possible misinterpretation of transience in groundwater discharge over time. Lengthened deployment and/or measurement periods also increase the risk of encountering significant transience in the system, including hydrological events that may create greater uncertainty in temperatures and even loss or damage of equipment (e.g., through deep burial of the cable, as we experienced at a larger river site). An approach that may help balance between lengthy post-processing and reserving sufficient time for physical measurements in the future is the use of temperature thresholds from FO-DTS temperature time series [4].

Due to topography, stream morphology, and the remote location of the research site, errors in GPS coordinates for measurement points and cable lengths mostly varied by roughly 0.10–1.0 m. Drone imagery was helpful for determining the cable location where GPS signal was poor. Other challenges included finding the exact locations of cool zones along the FO-DTS cable, as described previously in Section 2, and the time required to carefully deploy the cable along the right, left, and center of the meandering stream.

4. Conclusions

This field study quantitatively compared two different approaches for sampling streambed fluxes in space. The first approach used FO-DTS in a reconnaissance mode to identify likely zones of focused groundwater discharge in streambeds. We refer to this approach as informed measurement. These informed measurements were then compared to results from the more traditional approach of measuring streambed fluxes at evenly spaced transect locations selected without prior knowledge of streambed temperatures (uninformed measurements).

Streamflow in the Nebraska Sand Hills is dominated by groundwater discharge, which made our research site an ideal location for applying DTS methodology and for testing the hypothesis that informed groundwater discharge measurements would yield a significantly different view of groundwater discharge when compared with uninformed measurements. A non-parametric U-test showed a significant difference between median values for informed discharge measurements (0.17 m·day^{-1}) and uninformed measurement locations (0.05 m·day^{-1}). The frequency distributions for informed and uninformed measurements were also significantly different, based on a Kolmogorov–Smirnov test.

Areas of focused groundwater discharge were successfully identified in this study using a simple "snapshot-in-time" field analysis of FO-DTS data. Despite the less-precise raw data, results suggest a much lower threshold for the magnitude of groundwater discharge at which FO-DTS is applicable (approximately one-fifth of the value proposed previously [4]). Future studies that rely on more advanced post-processing of temperature data (to calibrate FO-DTS temperatures and remove artefacts in the data) may lead to even more precise identification of zones of focused groundwater discharge, although post-processing techniques need to be efficient enough to be conducted quickly in the field to guide sampling [43]. There has been a recent increase in the application of drone-based infrared thermal imaging for quick reconnaissance of groundwater discharge zones along streams at large scales [44];

however, infrared imaging does not penetrate the water column and low-to-moderate-discharge zones are likely to be missed.

Regardless of the approach to field analysis for FO-DTS, it is clear that temperature-based reconnaissance can lead to a substantially different view of groundwater discharge within a given stream reach. However, questions remain as to how application of FO-DTS may bias perceptions of overall average groundwater discharge into streams; it is important to carefully consider how FO-DTS fits into project objectives.

Author Contributions: Conceptualization, T.E.G. and M.A.B.; methodology, T.E.G., M.J. and J.K.; software, M.J.; validation, T.E.G., J.K. and A.M.; formal analysis, M.J., S.C. and T.E.G.; investigation, T.E.G., M.J., A.M., J.K., V.Z. and S.C.; resources, T.E.G., A.M. and J.K.; data curation, M.J., S.C. and T.E.G.; Writing—Original draft preparation, M.J.; Writing—Review and editing, T.E.G., A.M., J.K., V.Z. and M.A.B.; visualization, M.J. and T.E.G.; supervision, T.E.G.; project administration, T.E.G.; funding acquisition, T.E.G. and M.J.

Funding: This research was funded by the Daugherty Water for Food Global Institute, CTEMPS, and the US Department of Agriculture—National Institute of Food and Agriculture (Hatch project NEB-21-177). A portion of the analysis was funded by the USGS Toxic Substances Hydrology Program and by the National Science Foundation (EAR-1744719). The APC was funded by the Conservation and Survey Division—University of Nebraska.

Acknowledgments: The authors gratefully acknowledge Martin Wells for assistance with data collection. We are also grateful to Jacki Musgrave, John Nollette, and Andy Applegarth at the Gudmundsen Sandhills Research Laboratory for their insights and assistance in accessing the study site. Dave Gosselin also contributed to discussions about prior groundwater research at Gudmundsen. We gratefully acknowledge the CTEMPS support staff for their timely technical assistance during FO-DTS deployment and data processing. We also gratefully acknowledge three anonymous reviewers and one USGS reviewer who offered insightful suggestions that improved the manuscript.

Conflicts of Interest: The authors declare no conflicts of interest. The funders had no role in the design of the study; in the collection, analyses, or interpretation of data; in the writing of the manuscript, or in the decision to publish the results. Any use of trade, firm, or product names is for descriptive purposes only and does not imply endorsement by the US Government.

References

1. Briggs, M.A.; Lautz, L.K.; McKenzie, J.M. A comparison of fibre-optic distributed temperature sensing to traditional methods of evaluating groundwater inflow to streams. *Hydrol. Process.* **2012**, *26*, 1277–1290. [CrossRef]

2. Kennedy, C.D.; Genereux, D.P.; Corbett, D.R.; Mitasova, H. Spatial and temporal dynamics of coupled groundwater and nitrogen fluxes through a streambed in an agricultural watershed. *Water Resour. Res.* **2009**, *45*. [CrossRef]

3. Krause, S.; Blume, T.; Cassidy, N.J. Investigating patterns and controls of groundwater up-welling in a lowland river by combining Fibre-optic Distributed Temperature Sensing with observations of vertical hydraulic gradients. *Hydrol. Earth Syst. Sci.* **2012**, *16*, 1775–1792. [CrossRef]

4. Rosenberry, D.O.; Briggs, M.A.; Delin, G.; Hare, D.K. Combined use of thermal methods and seepage meters to efficiently locate, quantify, and monitor focused groundwater discharge to a sand-bed stream. *Water Resour. Res.* **2016**, *52*, 4486–4503. [CrossRef]

5. Kalbus, E.; Reinstorf, F.; Schirmer, M. Measuring methods for groundwater—surface water interactions: A review. *Hydrol. Earth Syst. Sci. Discuss.* **2006**, *10*, 873–887. [CrossRef]

6. Chen, X. Measurement of streambed hydraulic conductivity and its anisotropy. *Environ. Geol.* **2000**, *39*, 1317–1324. [CrossRef]

7. Chen, X. Streambed Hydraulic Conductivity for Rivers in South-Central Nebraska1. *Jawra J. Am. Water Resour. Assoc.* **2004**, *40*, 561–573. [CrossRef]

8. Chen, X.; Song, J.; Cheng, C.; Wang, D.; Lackey, S.O. A new method for mapping variability in vertical seepage flux in streambeds. *Hydrogeol. J.* **2009**, *17*, 519–525. [CrossRef]

9. Kennedy, C.D.; Genereux, D.P.; Corbett, D.R.; Mitasova, H. Design of a light-oil piezomanometer for measurement of hydraulic head differences and collection of groundwater samples. *Water Resour. Res.* **2007**, *43*. [CrossRef]

10. Briggs, M.A.; Lautz, L.K.; Buckley, S.F.; Lane, J.W. Practical limitations on the use of diurnal temperature signals to quantify groundwater upwelling. *J. Hydrol.* **2014**, *519*, 1739–1751. [CrossRef]

11. Constantz, J. Heat as a tracer to determine streambed water exchanges. *Water Resour. Res.* **2008**, *44*. [CrossRef]
12. Becker, M.W.; Georgian, T.; Ambrose, H.; Siniscalchi, J.; Fredrick, K. Estimating flow and flux of ground water discharge using water temperature and velocity. *J. Hydrol.* **2004**, *296*, 221–233. [CrossRef]
13. Fanelli, R.M.; Lautz, L.K. Patterns of Water, Heat, and Solute Flux through Streambeds around Small Dams. *Groundwater* **2008**, *46*, 671–687. [CrossRef] [PubMed]
14. Hatch, C.E.; Fisher, A.T.; Revenaugh, J.S.; Constantz, J.; Ruehl, C. Quantifying surface water–groundwater interactions using time series analysis of streambed thermal records: Method development. *Water Resour. Res.* **2006**, *42*. [CrossRef]
15. Keery, J.; Binley, A.; Crook, N.; Smith, J.W.N. Temporal and spatial variability of groundwater–surface water fluxes: Development and application of an analytical method using temperature time series. *J. Hydrol.* **2007**, *336*, 1–16. [CrossRef]
16. Schmidt, C.; Conant, B.; Bayer-Raich, M.; Schirmer, M. Evaluation and field-scale application of an analytical method to quantify groundwater discharge using mapped streambed temperatures. *J. Hydrol.* **2007**, *347*, 292–307. [CrossRef]
17. Lowry, C.S.; Walker, J.F.; Hunt, R.J.; Anderson, M.P. Identifying spatial variability of groundwater discharge in a wetland stream using a distributed temperature sensor. *Water Resour. Res.* **2007**, *43*. [CrossRef]
18. Solder, J.E.; Gilmore, T.E.; Genereux, D.P.; Solomon, D.K. A Tube Seepage Meter for In Situ Measurement of Seepage Rate and Groundwater Sampling. *Groundwater* **2016**, *54*, 588–595. [CrossRef]
19. Kennedy, C.D.; Genereux, D.P.; Mitasova, H.; Corbett, D.R.; Leahy, S. Effect of sampling density and design on estimation of streambed attributes. *J. Hydrol.* **2008**, *355*, 164–180. [CrossRef]
20. Conant, B. Delineating and Quantifying Ground Water Discharge Zones Using Streambed Temperatures. *Groundwater* **2004**, *42*, 243–257. [CrossRef]
21. Gilmore, T.E.; Genereux, D.P.; Solomon, D.K.; Solder, J.E.; Kimball, B.A.; Mitasova, H.; Birgand, F. Quantifying the fate of agricultural nitrogen in an unconfined aquifer: Stream-based observations at three measurement scales. *Water Resour. Res.* **2016**, *52*, 1961–1983. [CrossRef]
22. Matheswaran, K.; Blemmer, M.; Rosbjerg, D.; Boegh, E. Seasonal variations in groundwater upwelling zones in a Danish lowland stream analyzed using Distributed Temperature Sensing (DTS). *Hydrol. Process.* **2014**, *28*, 1422–1435. [CrossRef]
23. González-Pinzón, R.; Ward, A.S.; Hatch, C.E.; Wlostowski, A.N.; Singha, K.; Gooseff, M.N.; Haggerty, R.; Harvey, J.W.; Cirpka, O.A.; Brock, J.T. A field comparison of multiple techniques to quantify groundwater–surface-water interactions. *Freshw. Sci.* **2015**, *34*, 139–160. [CrossRef]
24. Sebok, E.; Duque, C.; Kazmierczak, J.; Engesgaard, P.; Nilsson, B.; Karan, S.; Frandsen, M. High-resolution distributed temperature sensing to detect seasonal groundwater discharge into Lake Væng, Denmark. *Water Resour. Res.* **2013**, *49*, 5355–5368. [CrossRef]
25. Hare, D.K.; Briggs, M.A.; Rosenberry, D.O.; Boutt, D.F.; Lane, J.W. A comparison of thermal infrared to fiber-optic distributed temperature sensing for evaluation of groundwater discharge to surface water. *J. Hydrol.* **2015**, *530*, 153–166. [CrossRef]
26. Poulsen, J.R.; Sebok, E.; Duque, C.; Tetzlaff, D.; Engesgaard, P.K. Detecting groundwater discharge dynamics from point-to-catchment scale in a lowland stream: Combining hydraulic and tracer methods. *Hydrol. Earth Syst. Sci.* **2015**, *19*, 1871–1886. [CrossRef]
27. Briggs, M.A.; Harvey, J.W.; Hurley, S.T.; Rosenberry, D.O.; McCobb, T.; Werkema, D.; Lane, J.W., Jr. Hydrogeochemical controls on brook trout spawning habitats in a coastal stream. *Hydrol. Earth Syst. Sci.* **2018**, *22*, 6383–6398. [CrossRef]
28. Browne, B.A.; Guldan, N.M. Understanding long-term baseflow water quality trends using a synoptic survey of the ground water-surface water interface, Central Wisconsin. *J. Environ. Qual.* **2005**, *34*, 825–835. [CrossRef]
29. Puckett, L.J.; Zamora, C.; Essaid, H.; Wilson, J.T.; Johnson, H.M.; Brayton, M.J.; Vogel, J.R. Transport and fate of nitrate at the ground-water/surface-water interface. *J. Environ. Qual.* **2008**, *37*, 1034–1050. [CrossRef]
30. Stelzer, R.; Drover, D.; Eggert, S.; Muldoon, M. Nitrate retention in a sand plains stream and the importance of groundwater discharge. *Biogeochemistry* **2011**, *103*, 91–107. [CrossRef]
31. Kennedy, C.D.; Genereux, D.P.; Corbett, D.R.; Mitasova, H. Relationships among groundwater age, denitrification, and the coupled groundwater and nitrogen fluxes through a streambed. *Water Resour. Res.* **2009**, *45*, W09402. [CrossRef]

32. Gilmore, T.E.; Genereux, D.P.; Solomon, D.K.; Solder, J.E. Groundwater transit time distribution and mean from streambed sampling in an agricultural coastal plain watershed, North Carolina, USA. *Water Resour. Res.* **2016**, *52*, 2025–2044. [CrossRef]

33. Tesoriero, A.J.; Duff, J.H.; Saad, D.A.; Spahr, N.E.; Wolock, D.M. Vulnerability of streams to legacy nitrate sources. *Environ. Sci. Technol.* **2013**, *47*, 3623–3629. [CrossRef] [PubMed]

34. Böhlke, J.K.; Denver, J.M. Combined use of groundwater dating, chemical and isotopic analyses to resolve the history and fate of nitrate contamination in two agricultural watersheds, atlantic coastal plain, Maryland. *Water Resour. Res.* **1995**, *31*, 2319. [CrossRef]

35. Gosselin, D.C.; Drda, S.; Harvey, F.E.; Goeke, J. Hydrologic Setting of Two Interdunal Valleys in the Central Sand Hills of Nebraska. *Groundwater* **1999**, *37*, 924–933. [CrossRef]

36. Gosselin, D.C.; (University of Nebraska, Lincoln, NE, USA). Personal communication, 4 October 2016.

37. Vogt, T.; Schneider, P.; Hahn-Woernle, L.; Cirpka, O.A. Estimation of seepage rates in a losing stream by means of fiber-optic high-resolution vertical temperature profiling. *J. Hydrol.* **2010**, *380*, 154–164. [CrossRef]

38. Selker, J.S.; Thévenaz, L.; Huwald, H.; Mallet, A.; Luxemburg, W.; van de Giesen, N.; Stejskal, M.; Zeman, J.; Westhoff, M.; Parlange, M.B. Distributed fiber-optic temperature sensing for hydrologic systems. *Water Resour. Res.* **2006**, *42*, W12202. [CrossRef]

39. Selker, J.; van de Giesen, N.; Westhoff, M.; Luxemburg, W.; Parlange, M.B. Fiber optics opens window on stream dynamics. *Geophys. Res. Lett.* **2006**, *33*. [CrossRef]

40. Tyler, S.W.; Selker, J.S.; Hausner, M.B.; Hatch, C.E.; Torgersen, T.; Thodal, C.E.; Schladow, S.G. Environmental temperature sensing using Raman spectra DTS fiber-optic methods. *Water Resour. Res.* **2009**, *45*, 11. [CrossRef]

41. Genereux, D.P.; Leahy, S.; Mitasova, H.; Kennedy, C.D.; Corbett, D.R. Spatial and temporal variability of streambed hydraulic conductivity in West Bear Creek, North Carolina, USA. *J. Hydrol.* **2008**, *358*, 332–353. [CrossRef]

42. Gilmore, T.E.; Genereux, D.P.; Solomon, D.K.; Farrell, K.M.; Mitasova, H. Quantifying an aquifer nitrate budget and future nitrate discharge using field data from streambeds and well nests. *Water Resour. Res.* **2016**, *52*, 9046–9065. [CrossRef]

43. Domanksi, M.; Quinn, D.; Day-Lewis, F.; Briggs, M.A.; Werkema, D.; Lane, J. DTSGUI: A Python program to process and visualize fiber-optic distributed temperature sensing data. *Groundwater*. (under review).

44. Harvey, M.C.; Hare, D.K.; Hackman, A.; Davenport, G.; Haynes, A.B.; Helton, A.; Lane, J.W.; Briggs, M.A. Evaluation of Stream and Wetland Restoration Using UAS-Based Thermal Infrared Mapping. *Water* **2019**, *11*, 1568. [CrossRef]

Article

Characterization of Diffuse Groundwater Inflows into Stream Water (Part II: Quantifying Groundwater Inflows by Coupling FO-DTS and Vertical Flow Velocities)

Hugo Le Lay [1,2], Zahra Thomas [1,*], François Rouault [1], Pascal Pichelin [1] and Florentina Moatar [2,3]

[1] UMR SAS, Agrocampus Ouest, INRA, 35000 Rennes, France; lelayhugo@gmail.com (H.L.L.); francois.rouault@agrocampus-ouest.fr (F.R.); pascal.pichelin@agrocampus-ouest.fr (P.P.)

[2] Laboratoire de GéoHydrosystèmes Continentaux (GéHCO), UPRES EA 2100, Université François-Rabelais, UFR des sciences et techniques, Parc de Grandmont, 37200 Tours, France; florentina.moatar@irstea.fr

[3] Institut National de Recherche en Sciences et Technologies pour l'Environnement et l'Agriculture (Irstea), RiverLy, Centre de Lyon-Villeurbanne, 69625 Villeurbanne, France

* Correspondence: zahra.thomas@agrocampus-ouest.fr or zthomas@agrocampus-ouest.fr; Tel.: +33-2-23-48-58-78

Received: 28 September 2019; Accepted: 16 November 2019; Published: 20 November 2019

Abstract: Temperature has been used to characterize groundwater and stream water exchanges for years. One of the many methods used analyzes propagation of the atmosphere-influenced diurnal signal in sediment to infer vertical velocities. However, despite having good accuracy, the method is usually limited by its small spatial coverage. The appearance of fiber optic distributed temperature sensing (FO-DTS) provided new possibilities due to its high spatial and temporal resolution. Methods based on the heat-balance equation, however, cannot quantify diffuse groundwater inflows that do not modify stream temperature. Our research approach consists of coupling groundwater inflow mapping from a previous article (Part I) and deconvolution of thermal profiles in the sediment to obtain vertical velocities along the entire reach. Vertical flows were calculated along a 400 m long reach, and a period of 9 months (October 2016 to June 2017), by coupling a fiber optic cable buried in thalweg sediment and a few thermal lances at the water–sediment interface. When compared to predictions of hyporheic discharge by traditional methods (differential discharge between upstream and downstream of the monitored reach and the mass-balance method), those of our method agreed only for the low-flow period and the end of the high-flow period. Our method underestimated hyporheic discharge during high flow. We hypothesized that the differential discharge and mass-balance methods included lateral inflows that were not detected by the fiber optic cable buried in thalweg sediment. Increasing spatial coverage of the cable as well as automatic and continuous calculation over the reach may improve predictions during the high-flow period. Coupling groundwater inflow mapping and vertical hyporheic flow allows flow to be quantified continuously, which is of great interest for characterizing and modeling fine hyporheic processes over long periods.

Keywords: water temperature; groundwater–stream; inflow quantification; vertical flow velocity

---

1. Introduction

Groundwater has great impact on stream ecology since it supports baseflow throughout the year [1–3] and stabilizes stream water quality [4,5], thus providing habitat for many species [6–8]. In their review of hydrologic exchange, Harvey and Goosef [9] outlined challenges in characterizing hydrologic connectivity, exchange flows, and related hydroecological processes. Tools from hydrological and ecological communities need to be combined to understand groundwater effects on stream flow and predict its future change. Many techniques such as seepage meters, mini-piezometric analysis [10,11], differential gauging [12,13], and chemical tracing [14,15] have been used to quantify groundwater inflow into streams. Each of these techniques has a specific spatial range, accuracy, and limitations [16–20]. Most have the disadvantage of being punctual, such as seepage meters giving information only for a given radius (<1 m). Moving from a fine scale to a coarser scale requires spatially integrative methods such as tracing in specific wells, which sample a large volume and obtain average concentrations, but do not allow for detailed characterization of spatial heterogeneities. Differential gauging allows for measurements at the reach level (a few m to km) but does not allow for detailed description along the longitudinal profile. Also, it requires additional measurements when considering the effects of tributaries. Among these techniques, temperature has been used to trace interactions between groundwater and surface water [21–23]. Since groundwater discharge affects the heat budget of the stream, its temperature can be modeled by solving a 1D transient advection–dispersion equation (Equation (1)) that requires stream temperature (T_w), flow velocity (V), dispersion coefficient (D_L), net heat flux (φ), specific heat of water (C_w), the density of water (ρ_w), and the average water column (d_w). Variable t is the time and z is the distance along the direction of flow.

$$\frac{\partial T_w}{\partial t} = -V \times \frac{\partial T_w}{\partial t} + D_L \times \frac{\partial^2 T_W}{\partial z^2} + \frac{\varphi}{C_w \times \rho_w \times d_w}. \tag{1}$$

The net heat flux term includes the processes related to stream discharge, atmosphere, groundwater inflow, and streambed conduction. Unidimensional analytical solutions of Equation (1) have been developed and used for decades to infer groundwater inflow from thermal profiles in the sediment [24]. These solutions use the attenuation and phase-shift of the atmosphere-influenced diurnal temperature signal in the sediment to infer vertical flow velocity in porous media, such as the hyporheic zone [25]. Benefiting from the development of affordable, yet precise, thermal sensors, the method has been used in various ways for many years [26,27]. Despite having good accuracy, the method suffers from a limited spatial range; since it is based on vertical thermal profiles in the sediment, it is punctual and usually unable to describe the spatial heterogeneity of inflows [28,29].

The recent use of fiber optic distributed temperature sensing (FO-DTS) has provided new opportunities [30]. Unlike other techniques, FO-DTS provides continuous measurements over space and time at high resolution. Depending on the brand, set-up, and configuration, this technique provides direct measurements with accuracy as high as 0.01 °C and sampling every 0.25 m along cables a few km long [31–33]. Taking advantage of the thermal contrast between groundwater and surface water during summer or winter, FO-DTS has been used mainly to characterize heterogeneous streams and locate focused groundwater inflows [34–36]. Many authors quantified in-stream measurements robustly at multiple times throughout the year [17,37,38]. However, the method is adapted for focused inflows that equal at least 2% of total stream discharge [39], as used in Part I of this study [40]. If groundwater inflows are weak and diffuse, FO-DTS cables must be buried in the sediment to prevent the signal from being displaced by stream flow [41].

To address the challenge of quantifying weak inflows at a high spatial resolution, Mamer and Lowry [42] used FO-DTS to calculate vertical flow velocities of groundwater inflows through the hyporheic zone. They installed two cables at two depths in a controlled flume and were able to estimate vertical flows with relatively good accuracy for uniform inflows. Their system slightly underestimated discharge for focused inflows, however, due to the low spatial resolution of their FO-DTS system (>1 m). Since FO-DTS measurements average the signal over a given distance, misalignment of the cable with the inflow can decrease flow estimates greatly (<25%). It was therefore hypothesized that higher spatial resolution could improve results. However, installing two fiber optic cables at two depths in a natural stream remains challenging, since fast surface flow can disturb the streambed over time [43].

Due to the spatial variability of groundwater inflows, information obtained from punctual (a few cm) or integrative (a few km) methods are difficult to interpret. Multi-scale approaches combining multiple measuring methods may considerably constrain estimates of fluxes between groundwater and surface water [44–46]. In part I of this two-part study [40], we developed a framework to locate and map groundwater inflow along a reach. Here, we focus on quantifying groundwater inflows, which is essential for investigating resilience of aquatic ecosystems to climate change [47–49]. In this research, vertical flows in the hyporheic zone along the stream were calculated to infer groundwater discharge into the stream. Vertical flow velocities were estimated by replacing the shallowest cable in Mamer and Lowry's study by a few punctual measurements at the stream-sediment interface [42]. Measurements were made at a high spatial resolution in a 400 m long reach in a second-order natural stream. Results from coupling temperature from DTS and from vertical velocities were compared to differential gauging performed over the entire reach and to discharge from heat-balance equations [50], which are more integrative over space. We discuss the potential and relevance of each method. The present article applied the framework presented in the previous article [40], which was dedicated to automatic mapping of such diffuse inflows with sub-hour resolution using FO-DTS. Our research framework consists of two parts to map and quantify diffuse and intermittent groundwater inflows into the stream.

2. Materials and Methods

2.1. Hydraulics of the Study Site

The geomorphology of the study site was presented in the previous article [40]. The reach actually monitored was 614 m long, however a quantification approach was applied between the two gauging stations because a tributary is located immediately downstream of the second gauging station (Figure 1). Thus, the following part of the study focused on the reach between the upstream and downstream gauging stations; it has no tributary and is 400 m long.

Stream water levels were measured at the gauging stations using OTT Orpheus-mini sensors (accuracy: ±0.5 cm, time-step: 5 min) and then converted into discharge using pre-existing rating curves. These rating curves were refined by punctual discharge measurements ca. every two months using salt dilution tests [50,51]. The upstream and downstream gauging stations in the monitored reach allowed for differential gauging to infer gains and losses of water between stations. To ensure that only exchanges along the hyporheic zone were considered, we extracted baseflow from hydrograms of discharge using the HYSEP program [52]. Subsurface and groundwater flows [53] were considered to be a single component of stream flow. Thus, upstream baseflow in the wetland sub-reach (Figure 1) was called Qb_{up}, while downstream baseflow in the meadow sub-reach was called Qb_{down}. The differential gauging $Q_{diff} = Qb_{down} - Qb_{up}$ was calculated to estimate the total losses or gains between the stations. Q_{diff} was then used as a reference for comparison with other quantification methods (described later).

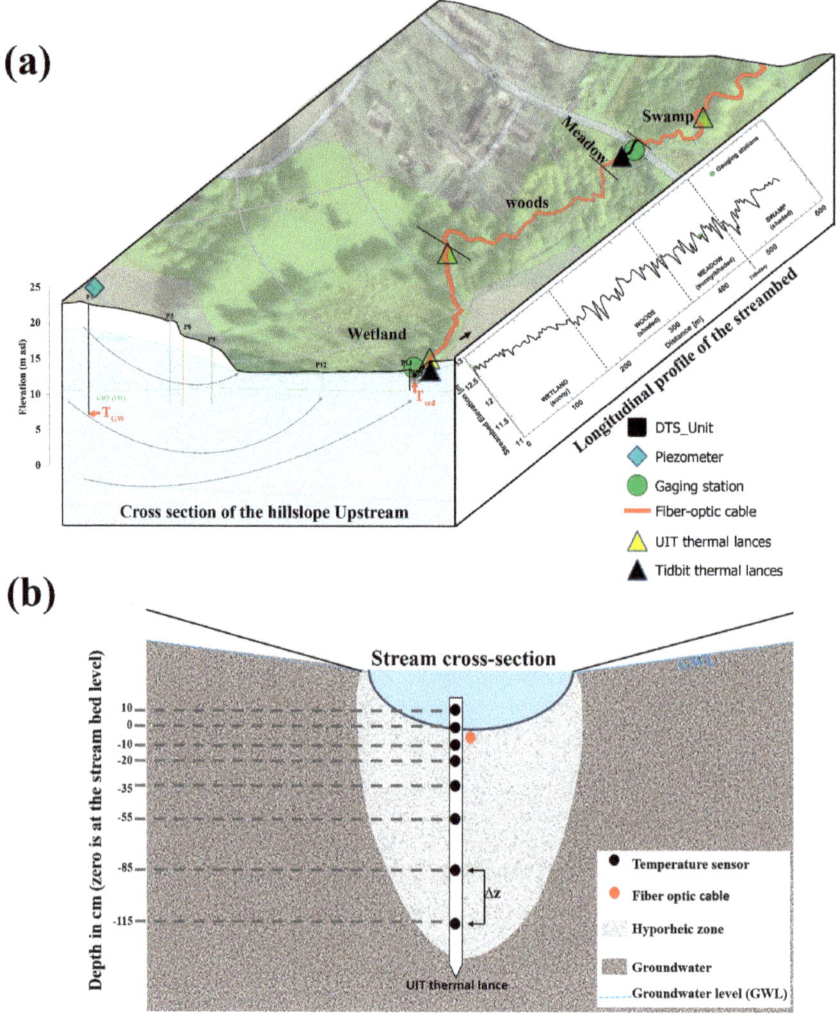

Figure 1. 3D view of the study area indicating the hillslope cross section and longitudinal profile.
(**a**) Black lines indicate boundaries between sub-reaches with different geomorphologies. The blue
diamond indicates the piezometer used to measure groundwater temperature (Tgw) and groundwater
level (GWL, dashed blue line). (**b**) Stream cross section showing Umwelt-und Ingenieurtechnik (UIT)
lance temperature sensors and depth in the hyporheic zone. Δz is the distance between two sensors
used to calculate flow at a groundwater inflow point and a neutral (non-inflow) point, respectively.

2.2. Temperature Measurements: DTS Set-up and Thermal Profiles

Thermal profiles in the hyporheic zone [54,55] were obtained using two types of temperature
lances. The first were traditional hand-made lances made from a double PVC (Polyvinyl Chloride) tube
and fine mesh with TidbiT v2 sensors at four depths: +10, −30, −65, and −100 cm. With an accuracy
of ±0.2 °C and a time-step of 5 min, they gave broad estimates of interactions between stream and
groundwater in the wetland and immediately before the confluence (Figure 1a). One of the sensors
(−30 cm) in the wetland broke down during the campaign, resulting in an incomplete profile. Because
of the scarcity of our vertical data and their low spatial resolution and accuracy, three thermal lances

of the second type (Umwelt-und Ingenieurtechnik (UIT) GmbH, Dresden, Germany) were installed. These lances provided thermal profiles with high spatial resolution (+5, 0, −10, −20, −35, −55, −85, and −115 cm) and good accuracy (±0.1 °C) and resolution (0.04 °C) (Figure 1b). These new lances were installed in October 2016, ca. four months after installation of the FO-DTS system. Consequently, the period considered for this study extended from October 2016 to July 2017.

Streambed temperature along the entire reach was measured with an Ultima XT-DTSTM (Silixa Ltd., UK). The cable was buried in the shallow sediment ($z \approx -3$ cm) to detect diffuse groundwater inflows despite the stream flow (Figure 1b). The configuration of the set-up, described thoroughly in Part I [40], had an accuracy of ca. 0.05 °C throughout the year with a 40 min time-step and 0.5 m spatial resolution. Groundwater temperature (T_{gw}) was measured [56] in the deepest and farthest piezometer on the hillside (Figure 1) using a minidiver submersible pressure transducer (HOBO Pro v2) (accuracy: ±0.2 °C, time-step: 5 min).

2.3. Framework for Quantifying Hyporheic Exchanges: Calculating Vertical Flows

All vertical flows were calculated using the VFLUX2 MATLAB toolbox [57]. The program deconvolutes temperature time series between two depths into velocity vectors by solving the 1-D heat transport equation derived from Equation (1) described by [24] for fluid-sediment systems:

$$\frac{\delta T}{\delta t} = K_e \frac{\delta^2 T}{\delta z^2} - q \frac{C_w}{C} \frac{\delta T}{\delta z} \tag{2}$$

where q is the vertical fluid flow in a downward direction (m s^{-1}), C is the volumetric heat capacity of the sediment (J m^{-3} °C^{-1}), and Cw is the volumetric heat capacity of the water (J m^{-3} °C^{-1}). K_e is the effective thermal diffusivity of the sediment, as described by the following equation:

$$K_e = \left(\frac{\lambda_0}{C}\right) + \beta |v_f| \tag{3}$$

where λ_0 is the thermal conductivity of the sediment (J s^{-1} m^{-1} °C^{-1}), β is the thermal dispersivity of the sediment, and v_f is the linear particle velocity (m s^{-1}).

To solve Equation (1), VFLUX2 uses, among others, analytical solutions of Hatch et al. [25]. These solutions use the phase-shift and difference in amplitude of the thermal signal between two sensors in the vertical direction to infer fluid velocity:

$$q = \frac{C}{C_w}\left(\frac{2K_e}{\Delta z} \ln A_r + \sqrt{\frac{\alpha + v_T^2}{2}}\right) \tag{4}$$

$$|q| = \frac{C}{C_w}\sqrt{\alpha - 2\left(\frac{4\pi \Delta t K_e}{P \Delta z}\right)^2} \tag{5}$$

where q is the volumetric flow (m^3 m^{-2} s^{-1}), Δz is the distance between two sensors set in the sediment (m), A_r is the ratio of amplitude between lower and upper sensors (dimensionless), v_T is the velocity of the thermal front (m s^{-1}), Δt is the time lag (speed of signal propagation) between temperature signals (s), and P is the period of the temperature signal (s). To obtain a linear velocity v, the following equation is applied:

$$v = \frac{q}{n_e} \tag{6}$$

where n_e is the effective porosity of the sediment (dimensionless). The sign of v determines the direction of the flow: Positive values indicate upward vertical flow (groundwater inflow), while negative values indicate downward flows (point of water loss).

2.4. Framework for Quantifying Hyporheic Exchanges: Coupling FO-DTS and Punctual Data

As mentioned, calculating vertical velocity requires at least two temperature measurements separated by a known distance Δz (Equations (3) and (4)). Mamer and Lowry [42] used two fiber optic cables positioned at two depths to obtain vertical flows with distance [48]. Since this configuration is difficult to control in natural streams, we replaced the shallowest cable—usually placed on the streambed—by punctual measurements of the UIT thermal lances at the water–sediment interface ($z = 0$ cm) (Figure 1b). This method assumes that (i) punctual temperature at the water–sediment interface is homogeneous in the reach and (ii) the distance between FO-DTS and this interface remains constant (here, $\Delta z = 3$ cm) over space and time.

Even though our method assumes homogeneous temperatures at the water–sediment interface, we could not ignore that external factors besides hyporheic exchange can modify it locally. For instance, solar radiation can cause local warming in a reach exposed to direct sunlight [58,59], and shading can influence stream temperature [60]. Since the monitored reach had different geomorphological traits (Figure 1a), we subdivided it into four sub-reaches, each of which was attributed a temperature at the water–sediment interface. This subdivision and the attribution of local temperatures was also expected to decrease errors due to the assumption of homogeneity. We also assumed that cable depth remained constant along the entire reach for the entire year of measurements. Obviously, streamflow uncovered it for short periods at some locations, and the streambed moved slightly during the experiment, but these factors were negligible at the reach scale. Uncovered segments of the cable were removed from our dataset when detected, and the cable was found to have remained at generally the same depth when removed in July 2017, after one year of measurements.

Given the thousands of measurements along the FO-DTS, we simplified estimation of vertical flows by selecting only one clear perennial inflow per sub-reach along the FO-DTS cable as a representative inflow point. The vertical velocity calculated for this representative inflow was then considered to represent the entire sub-reach and thus attributed to each groundwater inflow in this sub-reach. Therefore, we made a third assumption: Groundwater flow is homogeneous along a sub-reach. The same approach was applied to locations without inflow (i.e., "neutral"), assumed to be influenced by the atmosphere: One velocity at one representative location was applied to each neutral location along the sub-reach. Selection of these representative locations—inflows and neutral points—was based on groundwater inflow mapping (see Part I [40]). The number and location of inflows and neutral points to which these representative flows were applied were determined from the same map.

The UIT sensor set at the water–sediment interface ($z = 0$ cm) was used as the upper sensor for each calculation. At this depth, the sinusoidal diurnal signal has large amplitude. The FO-DTS point located above the representative groundwater inflow was used as the lower sensor for calculating flow (Figure 2). At this depth, and because of the groundwater influence, the diurnal signal is clearly dampened (ΔA) and greatly shifted in time (Δt) compared to that of the upper sensor (Equations (3) and (4)). The representative flow for neutral points was calculated directly using the UIT sensors at $z = -10$ cm. Its diurnal signal is not influenced by upward groundwater flow, so it is less dampened and phase-shifted. UIT lances were used for neutral points because the groundwater inflow mapping by FO-DTS (Part I [40]) could not completely distinguish true neutral points in the sediment from uncovered cable. In contrast, UIT lances were placed in zones without clear inflow, and the depths of their sensors were known with certainty throughout the year. Since no UIT lances were placed in the meadow (Figure 1), calculations in this sub-reach were based on the second UIT lance placed in the woods. UIT lances were preferred over the older TidbiT lances because of their proximity to the water–sediment interface (first 10 cm) and their greater accuracy (± 0.1 °C vs. ± 0.2 °C, respectively).

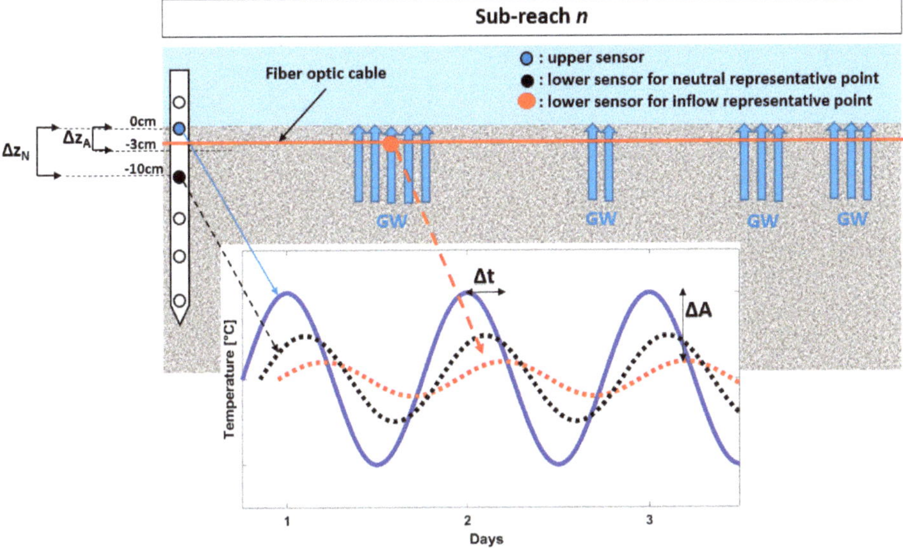

Figure 2. Conceptual representation of calculating vertical flow along a sub-reach. The upper sensor used for all velocity calculations is the thermal sensor at the water–sediment interface ($z = 0$ cm). A clear and perennial groundwater (GW) inflow is selected as a representative point and the fiber optic distributed temperature sensing (FO-DTS) point set above it is used as the lower sensor for calculating inflow velocity ($z = -3$ cm). A clear and perennial atmosphere-influenced stream water location was selected in each sub-reach as representative of "neutral" sections. In this study, all thermal lances were set in neutral locations without inflow: their stability in the sediment made them more accurate than the FO-DTS system in calculating the representative neutral flows ($z = -10$ cm).

Calculating vertical velocities based on analytical solution of 1D heat equation assumes that the fluid flow is vertical and one-dimensional. The temperature signal is assumed to be sinusoidal and there is no thermal gradient in sediment between the two sensors in the vertical direction [61]. As discussed in Irvine et al. [61], there is also a limitation regarding stream-bed heterogeneities which affect thermal properties of the sediments. Vandersteen et al. [62] developed a new method to calculate vertical flow called LPML (Local Polynomial Maximum Likelihood). This method was compared to VFLUX which was developed by Gordon et al. [57]. There were also major limitations with analytical solutions to the 1-D heat transport equation since streambed heterogeneity or non-vertical flow components are not considered. In our study, we parameterized input parameters of VFLUX2 (Table 1) based on sediment properties. Thus, a sampling campaign was performed in September 2017 to assess thermal properties of the sediment at multiple points along the stream. Each sub-reach was sampled due to geomorphological differences among sub-reaches. Sediments in the wetland and woods were similar to those in the confluence: dominated by silts and organic matter, with a sand fraction. In contrast, sediment in the meadow was exclusively sandy and relatively shallow. Dry bulk densities of these sediments were measured using the kerosene-displacement method [63]. Based on these densities, the thermal parameters required for calculations were estimated according to Lapham [64]. The standard deviation (SD) of each parameter was determined based on the uncertainty in bulk density measurements Lapham [64]. These SDs were used in 500 Monte Carlo iterations to estimate uncertainty in the vertical velocities calculated.

Table 1. Means (and standard deviations, if $\neq 0$) of input parameters of VFLUX2 [57] used in this study for each sub-reach.

Parameter	Symbol (Unit)	Sub-reach			
		Wetland	Woods	Meadow	Swamp
Fundamental period to filter	P (day)	1	1	1	1
Sediment porosity	n_e (-)	0.43 (0.01)	0.40 (0.01)	0.26 (0.07)	0.40 (0.01)
Sediment thermal dispersivity	b (m)	0.001	0.001	0.001	0.001
Sediment thermal conductivity	λ_0 (cal s^{-1} cm^{-1} °C^{-1})	0.0045 (0.0001)	0.0048 (0.0001)	0.0069 (0.0001)	0.0048 (0.0001)
Sediment volumetric heat capacity	C (cal m^{-3} °C^{-1})	0.643 (0.001)	0.630 (0.001)	0.560 (0.002)	0.630 (0.001)
Water volumetric heat capacity	C_w (cal m^{-3} °C^{-1})	1.00	1.00	1.00	1.00

2.5. Framework for Quantifying Hyporheic Exchanges: Comparing Vertical Velocities to Volumetric Discharge

In a punctual approach, Rosenberry et al. [35] compared their calculated vertical flows to seepage meters and found a strong correlation ($R^2 = 0.96$). Our method aimed to estimate groundwater inflows in a more continuous approach over an entire hydrological year. Since seepage meters were not available and not recommended for such long measurements, we compared our results not only to the differential gauging Q_{diff} but also to the heat-balance method of [37], adapted to the site. Since the FO-DTS cable was buried in the sediment to detect weak inflows (Figure 1b), stream water and groundwater did not mix much at the cable. Moreover, groundwater inflows at the site were expected to be too weak and diffuse to modify the stream temperature. Therefore, it was not possible to apply the heat-balance method to each point. Instead, we used it to estimate groundwater discharge between the 400 m separating the two gauging stations ($Q_{balance}$), in this form:

$$Q_{balance} = \frac{Qb_{down} \times T_{down} - Qb_{up} \times T_{up}}{T_{gw}} \tag{7}$$

where T_{up} and T_{down} are the temperatures at the TidbiT thermal lances at the upstream and downstream gauging stations, respectively.

Since the 400 m over which Q_{diff} and $Q_{balance}$ were estimated covered the first three sub-reaches upstream of the tributary, only vertical flows between the stations were considered for comparison. To compare volumetric discharge measurements along a reach (Q_{diff} and $Q_{balance}$) to linear discrete vertical velocities, we tested two approaches.

First, we converted the vertical flows (every 0.25 m) predicted by VFLUX2 into a volumetric discharge ($Q_{coupling}$) by considering a specific surface (S) for each punctual velocity (v), whether directed upward (groundwater inflow), downward (water loss) or null (neutral):

$$q_{coupling(t,i)} = v_{(t,i)} \times (l \times w) \tag{8}$$

where t is the time step, i is a point of the cable, l the length of the area and w the width of the area.

The length was set to 0.25 m (sampling distance of the DTS), but two widths were tested: 2 m (mean width of the streambed) and 0.5 m (within the 0.30–0.70 m range observed during low flow and equal to the spatial resolution of the DTS). Once $Q_{coupling}$ was calculated for each point, net discharge

($Q_{coupling}$) (gains − losses) along the reach due to interactions between the stream and the aquifer was calculated as follows:

$$Q_{coupling(t)} = \sum_{i}^{end} q_{coupling(t,i)}. \tag{9}$$

In the second approach, we converted the volumetric discharge into velocity (V_{diff}) using the following equation:

$$V_{diff} = \frac{Q_{diff}}{w \times L} \tag{10}$$

where L is the distance between the two gauging stations (400 m).

As for the previous approach, two widths were tested: 2 m and 0.50 m.

3. Results and Discussion

3.1. Groundwater Inflow Mapping

Normalized thermal anomalies mapped using the framework presented in Le Lay et al. [40] showed spatio-temporal variability over the study period (Figure 3). Neutral locations were assumed to be influenced only by the atmosphere (Figure 3b, in blue), unlike groundwater inflow locations (Figure 3b, in yellow). Some anomalies interpreted as inflows could have been due to hyporheic circulation or the cable becoming more deeply buried. Moreover, some inflows did not behave as expected given their location. For instance, strong and stable groundwater inflows were located upstream, in the wetland sub-reach, even though it was perched. The processes involved were suspected to be either groundwater inflows from the nearby hillslope or water drained from agricultural fields. Despite some uncertainty about certain inflows, we assumed here for convenience that all thermal anomalies (yellow zones in Figure 3) were indeed groundwater inflows. All representative groundwater inflows, chosen for their constancy over time, were located in pools.

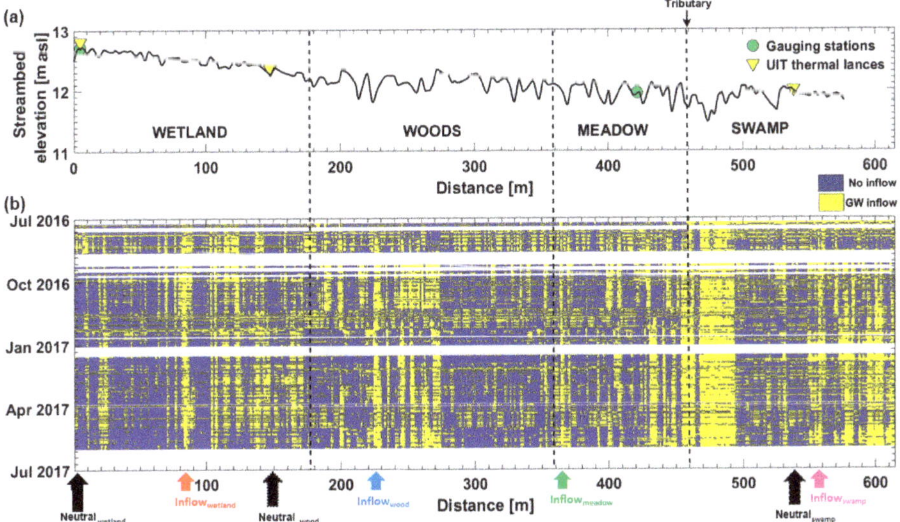

Figure 3. (a) Longitudinal profile of the streambed in the four sub-reaches of the monitored reach. UIT = Umwelt-und Ingenieurtechnik, m asl = m above sea level. (b) Spatio-temporal evolution of thermal anomalies [40]. Arrows indicate representative points of groundwater (GW) inflow and no inflow (Neutral). White strips show periods when the distributed temperature sensing system malfunctioned (horizontal) or points where the cable was often non-submerged (vertical).

3.2. Evolution of Thermal Profiles in the Hyporheic Zone

Overall, the temperature of the entire hyporheic zone varied over time and followed the seasonal cycle, with lower temperatures in winter (January 2017) and higher temperatures in summer (June 2017) (Figure 4). Nevertheless, temperature was vertically stratified throughout the study period, and that in the stream (z = +5 cm) varied much more than that at −115 cm (0–21 °C and 7.25–13.75 °C, respectively). The annual thermal cycle was thus clearly defined along the lances. During autumn and winter, shallower sensors recorded colder temperatures than deeper ones (up to 9 °C colder in January 2017) but with a consistent thermal stratification: all depths had distinctly different temperatures. Thermal stratification was also observed in late spring (May–June 2017), with higher temperatures at shallower depths (up to 8 °C warmer in June 2017). However, from February to early May 2017—the period defined as the peak of high flow in Part I [40]—temperature of the hyporheic zone was relatively homogeneous. Most thermal contrast (0.7–6.0 °C, depending on the day) was observed in the stream water (z = +5 cm) and at the water–sediment interface (z = 0 cm), while the remaining contrasts were small (0–3 °C). Moreover, short periods with little thermal contrast between deep and shallow depths were also observed in mid-November 2016, December 2016, mid-January 2017 and mid-May 2017. Similarly, some periods of even smaller contrast (as small as 0.7 °C) between February 2017 and May 2017 could be attributed to the same processes. This behavior could have been due to a surge of deeper groundwater in the sediment that decreased propagation of the diurnal signal in the sediment. Only thorough study of the phase-shift and difference in amplitude between two vertical locations (Equations (3) and (4)) could provide the amplitude and direction of flow over time in the hyporheic zone. We thus calculated vertical flow velocity from the data at −10 and 0 cm to determine whether this potential surge of groundwater truly discharged into the stream.

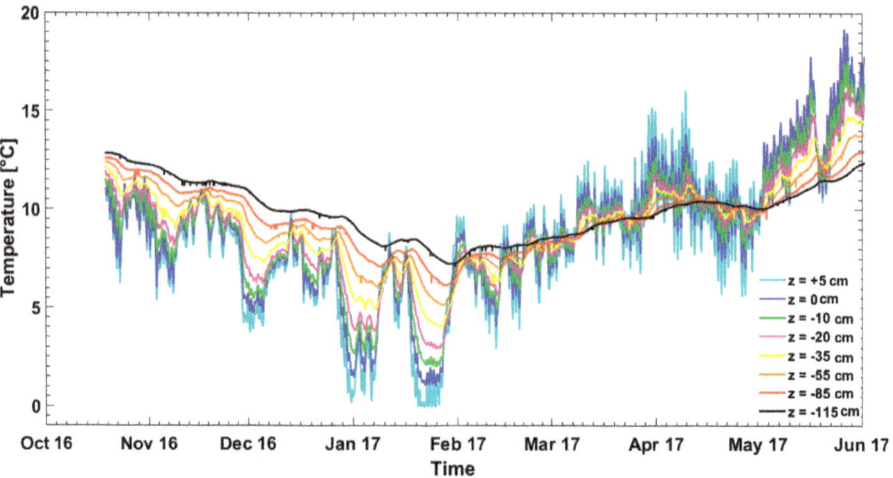

Figure 4. Example of temperature profile dynamics along the hyporheic zone from the thermal lance located in the wetland sub-reach.

Two-dimensional thermal profiles along the hyporheic zone at neutral locations revealed great similarity among the three points monitored (Figure 5). Colder surface temperatures propagated into the sediment from October 2016 to February 2017 along all three lances. In contrast, warmer surface temperatures propagated downward in late spring (May–June 2017). In between (February–May 2017), sediment temperatures remained relatively homogenous. Monthly mean profiles also showed quasi-identical thermal envelopes [65], with few differences above the streambed (z = +5 cm) during this period (Figure 5). The annual amplitude of temperature in the wetland, woods and swamp sub-reaches was 12.5 °C, 12.7 °C and 12.4 °C, respectively, at the water–sediment interface (z = 0 cm) and ca. 6.4 °C, 6.5 °C, and 6.2 °C, respectively, at the deepest point monitored in the hyporheic zone ($z = -115$ cm). Only the swamp (i.e., the most downstream sub-reach) had a slightly thinner thermal envelope (-55 to -115 cm), which indicates a stronger influence of groundwater in this part of the reach. Because the UIT lances were installed before mapping thermal anomalies, they were not placed in locations of clear groundwater inflow. This is consistent with field observations and the inflow mapping, which indicated that the swamp sub-reach was influenced greatly by groundwater. Overall, propagation of surface temperature through the hyporheic zone was clear, especially in winter from late December to February and in May–June. Temperature propagated faster above 55 cm than below it; however, in mid-December and mid-January, temperature homogenized quickly at 11–12 °C, close to that of groundwater. This thermal signature may suggest sudden groundwater inflow but could also represent downward flows with relatively warm temperature for the season. Whether these episodes were groundwater surges that discharged into the stream or warm downward flows was impossible to determine based solely on these qualitative data. We thus used the persistent difference in temperature between the water–sediment interface ($z = 0$ cm) and the first depth ($z = -10$ cm) to infer the vertical flow over time and its direction through the sediment.

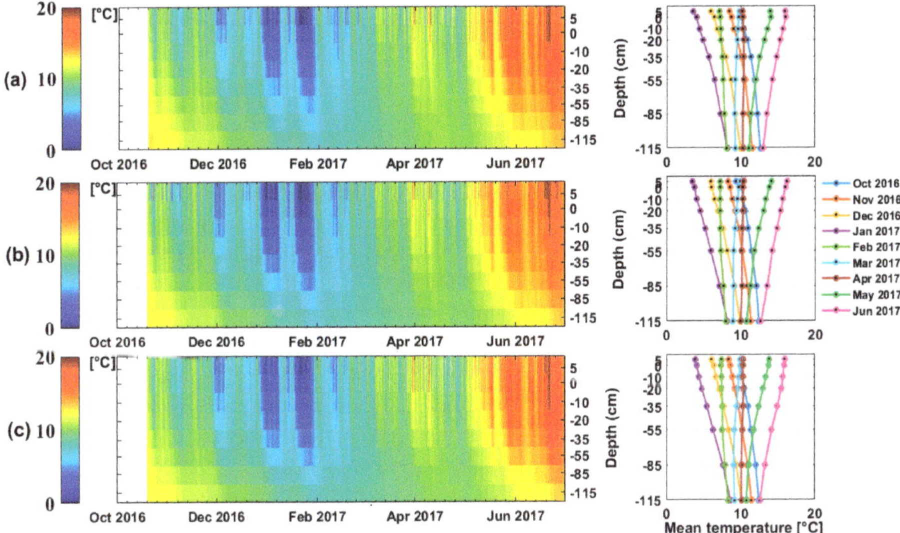

Figure 5. Thermal profiles along the hyporheic zone at locations with no groundwater inflows detected (**left**) at a time-step of 5 min or (**right**) as monthly means in the (**a**) wetland sub-reach (upstream), (**b**) interface between wetland and woods sub-reaches, and (**c**) swamp sub-reach (downstream).

3.3. Vertical Flow Velocities at Locations without Groundwater Inflows

At neutral locations (i.e., without inflow and assumed to be influenced only by the atmosphere), vertical flow velocities in the wetland, near the woods and in the swamp sub-reaches were generally low (0 to -1×10^{-5} m s^{-1}), directed downward (negative values), similar over space and constant over time (Figure 6). Nevertheless, larger downward velocities were observed in the swamp in mid-November 2016, December 2016 and late January 2017 (>ca. -2×10^{-5} m s^{-1}). Ultimately, no clear changes in direction (upward flows) or intensity (high velocities) could be related to the short periods when the temperature along the lance varied little. In fact, some of the downward velocity peaks in autumn 2016 occurred at the same time as these suspected groundwater surges. Part I of this study [40] observed flood events occurring at the same time as a weak negative hydraulic gradient at the site from November 2016 to January 2017. It is likely that these sudden floods temporarily inversed the hydraulic gradient and triggered downward flows (water loss).

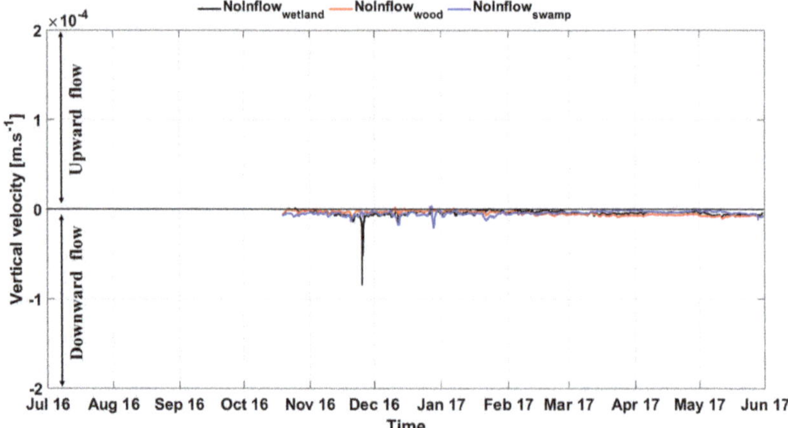

Figure 6. Vertical flow velocities along the hyporheic zone from locations without groundwater inflows (Neutral) estimated by VFLUX2 from $z2 = -10$ cm to $z1 = 0$ cm (i.e., negative values indicate downward flows).

Ultimately, the groundwater surges observed in autumn and during high flow (Figure 3) did not become groundwater discharges into the stream. With flows generally directed downward at a mean of -0.4×10^{-5} m s^{-1}, the neutral locations thus remained neutral most of the studied period. We therefore assumed that this was the case for all points considered neutral. These velocities were thus applied to all neutral points in each sub-reach to calculate global discharge ($Q_{coupling}$).

3.4. Vertical Flow Velocities in Groundwater–influenced Locations

Unlike velocities calculated at neutral points, most of those at groundwater–inflow points were directed upward (positive values), confirming groundwater inflows (Figure 7). Moreover, all were higher than the downward velocities at neutral points by one order of magnitude (up to 1.3×10^{-4} m s^{-1} in October and 1.8×10^{-4} m s^{-1} in April). Vertical velocities in all four sub-reaches showed similar patterns, ranging from 0.5–1.3×10^{-4} m s^{-1} at the end of the low-flow period (October 2016), decreasing from November 2016 to January 2017 (including even downward flows as high as -0.8×10^{-4} m s^{-1}) and then increasing again from February–June 2017 (0.3–1.8×10^{-4} m s^{-1}). At a shorter time scale, velocities in all sub-reaches increased and decreased simultaneously, differing only in the amplitude of these changes depending on location and time. The swamp always had higher velocities than the other three sub-reaches and never had downward flows, even during the overall decrease from November 2016 to January 2017. Vertical velocity in the meadow sub-reach varied the most over time, while that in the wetland and woods sub-reaches varied less. Velocity in the woods sub-reach decreased from 1.0 to 0.5×10^{-4} m s^{-1} in November 2016 and, except for common downward flow episodes in November and December 2016, remained approximately the same until June. Overall, vertical velocity in the wetland sub-reach was lower than those in the other sub-reaches from November-February.

This apparent difference between lower upstream (wetland and woods sub-reaches) vertical velocities and higher downstream (swamp sub-reach) vertical velocities was probably related to stream geomorphology. The upstream streambed's perched elevation probably constantly decreased its hydraulic gradient, explaining its lower velocities. In contrast, field observations had already determined that the swamp experienced regular groundwater surges: Its banks were often flooded or at least wet. Moreover, the swamp had a lower streambed with more pools than the rest of the reach. It thus seems likely that the swamp's higher vertical velocities were due to inflows caused by a higher hydraulic gradient. Velocities in the meadow sub-reach reflected the evolution of thermal anomalies (Figure 3b), being low from November 2016 to February 2017 but increasing greatly during the high-flow period, when the number of locations with inflows also increased. The meadow's greater variability in velocities was most likely due to its sandier sediment and the sensitivity of the VFLUX2 program to its higher thermal conductivity. Effects of sediment characteristics on hyporheic flow have been addressed by many authors [66–68]. The general decrease in vertical flows from November 2016 and February 2017 echoed the few similar episodes detected at neutral locations (Figure 5). Therefore, this decrease was probably caused by the same punctual gradient inversions following flood events in December.

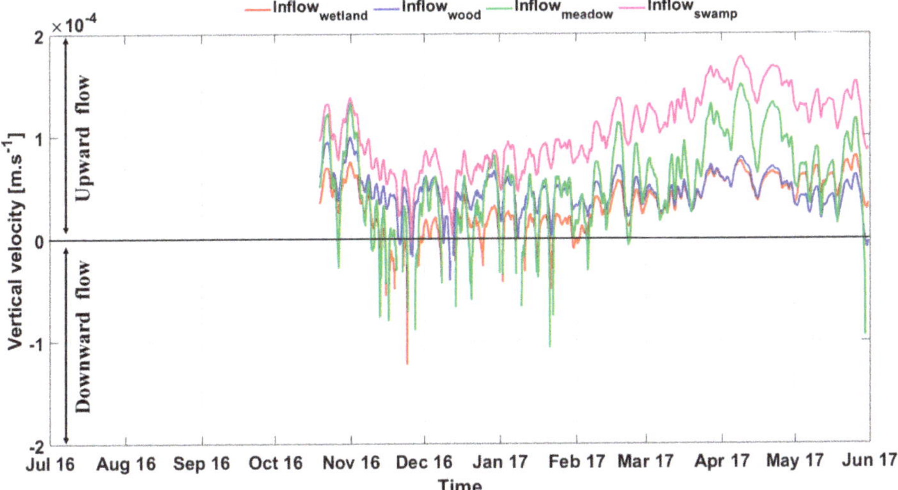

Figure 7. Vertical flow velocities calculated from temperatures at $z = 0$ cm and fiber optic distributed temperature sensing ($z = -3$ cm) for the wetland, woods, meadow, and swamp sub-reaches. Error bars are based on 500 Monte Carlo iterations with variable thermal parameters of the sediments.

Ultimately, the flows calculated across the entire reach clearly indicated positive groundwater discharges, showing the relevance of the inflow mapping method developed in Part I [40]. Moreover, despite having different amplitudes, these groundwater velocities were quite synchronized, supporting the assumption of homogeneous groundwater temperature. These velocities were thus applied to each point of their respective sub-reach that was considered to be an inflow in order to calculate global discharge ($Q_{coupling}$), which was then compared to the other quantification methods.

3.5. Comparison of Groundwater Discharge Estimation Methods

Discharge dynamics upstream and downstream clearly revealed the low- and high-flow periods previously identified. During the low-flow period (July to late October 2016), stream discharge was generally low upstream and downstream (mean = 22.3 and 24.6 L s⁻¹, respectively) with no large flood events (Figure 8). In contrast, discharge during the high-flow period from November 2016 to June 2017 varied more. The first large flood occurred in November 2016, which may explain the mostly downward flows calculated by VFLUX2 (Figures 6 and 7). During flood events, baseflow increased steadily upstream and downstream, reaching 49 and 109 L s⁻¹, respectively, in March 2017 (Figure 8). Discharge peaks in December 2016 and May 2017 were not considered as groundwater discharge since they were probably still a result of residual runoff.

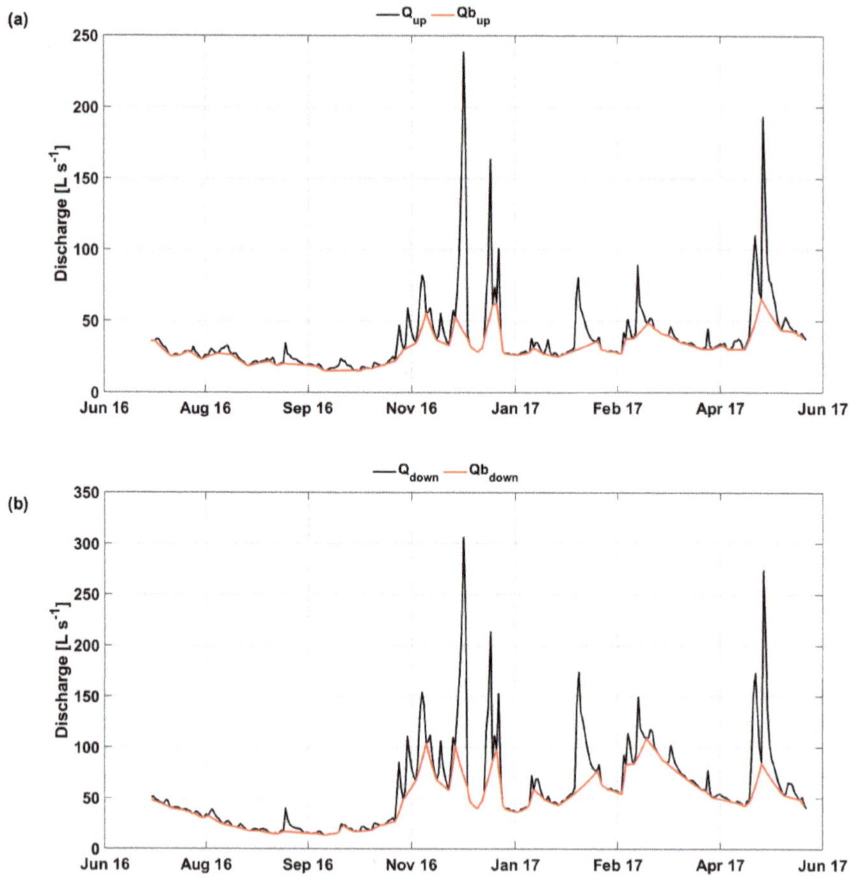

Figure 8. Stream discharge (Q) (including runoff) and baseflow (Qb) over time at the gauging station (**a**) upstream (wetland sub-reach) and (**b**) downstream (meadow sub-reach).

Upstream and downstream baseflows had similar dynamics, differing only in their amplitudes. Baseflow was clearly higher downstream than upstream during the high-flow period, which supported the idea that groundwater fed the reach during this period. Baseflows during the low-flow period were similar upstream and downstream and lay within the uncertainty in measurements, masking potential water gains or losses.

Figure 9 shows a comparison of the three methods for quantifying hyporheic discharge i.e., (i) $Q_{balance}$ from conservation of mass and energy, (ii) Q_{diff} from differential gauging, and (iii) $Q_{coupling}$ from coupling FO-DTS and vertical temperature profiles in hyporheic zone. Q_{diff} had weak, almost negative values from August–October 2016 (+8 to −6 L s^{-1}) (Figure 9), consistent with measurements during low flow, then increased until March 2017 (up to +62 L s^{-1}). Its net decrease after March was attributed to the beginning of root uptake by vegetation and evaporation in the riparian area [69]. $Q_{balance}$ was similar to Q_{diff} except for being higher than it in October 2016 and then lower than it during the high-flow period, especially in January and February 2017 (2–22 L s^{-1} lower). The larger differences were likely due to a lag time between hydraulics and stream temperature: During storms or heavy precipitation, stream water level increases quickly, but stream temperature does so more slowly because of its thermal inertia. This highlights limits of the $Q_{balance}$ approach over large distances: By assessing conservation of mass and energy, the longer the distance between measurements (here, nearly 400 m), the more mass and energy exchanges with the environment become likely, rendering the method less accurate. However, periods with low precipitation and moderate solar radiation (autumn and spring) yielded good results due to the short distance, high flow and shading at the site, which decreased energy inputs.

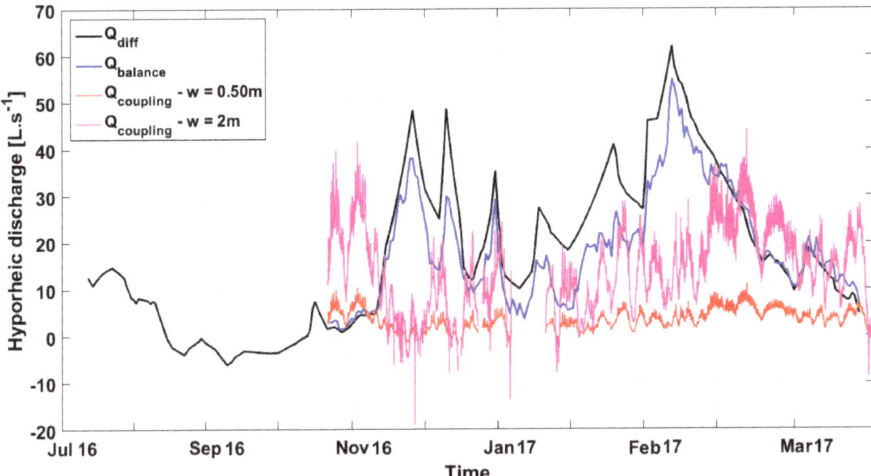

Figure 9. Comparison of three methods for quantifying hyporheic discharge along the 400 m long reach between gauging stations: differential gauging (Q_{diff}), heat-balance method ($Q_{balance}$), and the method coupling distributed temperature sensing mapping and vertical flow velocities ($Q_{coupling}$).

The hyporheic discharge estimated with our method ($Q_{coupling}$) (gains from groundwater inflows—potential losses from neutral points) depended on the width used to calculate it. For a width of 0.50 m, $Q_{coupling}$ was similar to Q_{diff} during low flow but much lower than Q_{diff} and $Q_{balance}$ during the high-flow period. For instance, during peak discharge in March 2017, $Q_{coupling}$ was nearly 55 L s⁻¹ (−90%) lower than Q_{diff}. This underestimation was lower in December 2016 and May 2017 but remained high (−65% to −85%). For a width of 2 m, $Q_{coupling}$ overestimated Q_{diff} by 15 L s⁻¹ (+700%) during low flow and generally underestimated it during high flow; however, underestimation was smaller, and even close to zero, in May 2017. Hyporheic discharge at the beginning of high flow (November–December 2016) was low or even negative (water loss) for both widths. This pattern was the expression of the downward flows observed at the site (Figures 6 and 7) but was in disagreement with Q_{diff}. In this context, the three methods showed consistent estimates of hyporheic discharge during low flow for the narrower width (0.5 m) and the end of the high-flow period. Hyporheic discharge occurring over the entire streambed width (2 m) was also underestimated during the high-flow period. Also, the order of the magnitude of discharge estimated by coupling DTS and the thermal gradient in the sediment is similar to those of groundwater inflows calculated using radon at the tributary outlet in the same study site [70]. In this study, groundwater inflow of ca. 0–5 L s⁻¹ was estimated along the Vilqué reach.

Since the hydrological year was relatively dry, the stream did not overflow its banks; consequently, its width never exceeded its minor bed. All three methods considered water gains and losses along the reach. Q_{diff} and $Q_{balance}$ spatially integrated discharge information between the stations, while $Q_{coupling}$ was based on velocities from groundwater inflows and neutral points.

Since the number of groundwater inflows and neutral points varied over time, $V_{coupling}$ was calculated as the mean of groundwater velocities in the wetland, woods, and meadow and velocities above neutral points in the wetland and woods. Comparisons of flow velocities through the streambed

(V_{diff} and $V_{coupling}$) yielded results similar to those of hyporheic discharge. During high flow, $V_{coupling}$ was similar to V_{diff} when a width of 2 m (i.e., the streambed) was assumed (Figure 10a). Some differences were observed in late November 2016 and March 2017 but the two velocities were particularly similar at the end of the high-flow period (late April and May). In contrast, during low flow, $V_{coupling}$ was closer to V_{diff} when a width of 0.5 m was assumed. Cumulative probability analysis revealed that the velocities in the hyporheic zone predicted by our method generally had the same mean and dynamics as those of V_{diff} when assuming a width of 2 m (Figure 10b). From these results, the few vertical velocities calculated with FO-DTS seemed representative of overall groundwater flow at the site. Based on estimated velocities, groundwater inflow rates seemed relatively uniform across the site.

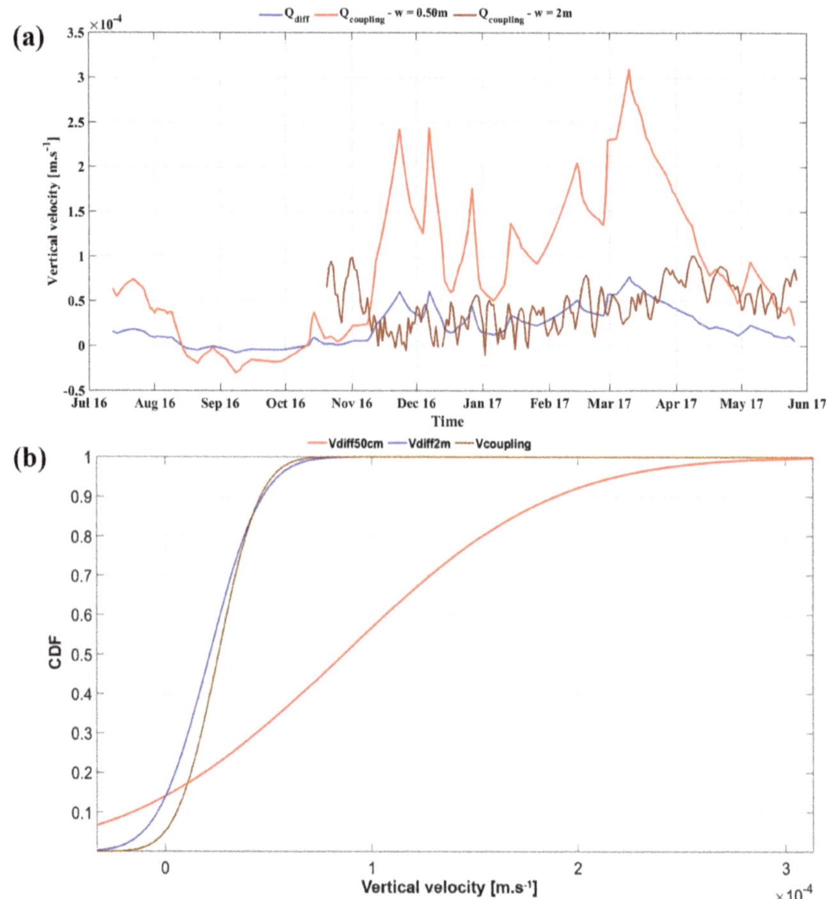

Figure 10. Linear velocities observed from differential gauging (V_{diff}) normalized using a stream 2 m or 0.5 m wide and mean vertical flow ($V_{coupling}$) calculated by coupling fiber optic distributed temperature sensing and thermal lances. (**a**) Dynamics of hyporheic flow over time. (**b**) Cumulative distribution function (CDF) of each hyporheic flow.

3.6. Limits of Our Method and Perspectives

In this study, we attempted to quantify exchanges between groundwater and surface water in a 400 m long reach by calculating vertical flow velocities in the sediment at every point of a FO-DTS cable. Given the large amount of data available, we relied on a few representative vertical flow velocities that we applied to the entire reach. Locations mapped as influenced (GW inflow) or not (neutral) by groundwater inflows (Figure 3b) were thus used to assess the spatial and temporal heterogeneity of groundwater–stream water exchange along the reach. Since only one fiber optic cable was buried in the sediment, but two temperatures at different depths are necessary to infer vertical flow, we coupled FO-DTS data with punctual data from thermal lances.

Generally, our coupling method provided mixed results that depended greatly on the area over which the aquifer and stream were assumed to exchange water. Since longitudinal heterogeneity was restricted to the FO-DTS sampling distance (0.25 m), width became the adjustable variable of this exchange area. Thus, during the low-flow period, our method provided acceptable results as long as the stream was assumed to be narrow (0.50 m). Assuming a wider stream (2 m) yielded results that disagreed with low-flow observations and led to overestimation of hyporheic discharge. In contrast, at the end of the high-flow period (May 2017), volumetric estimates were more accurate when we assumed a wide area (2 m), since the stream occupied the entire streambed during high flow. Nevertheless, during most of the high-flow period, our method underestimated net hyporheic discharge greatly. Although we cannot rule out the possibility that vertical flow velocities were underestimated or did not represent the actual groundwater discharge, some clues indicate other factors at play. First, flow calculated from differential gauging discharge was generally similar to that of our method during the high-flow period. In addition, vertical velocities calculated above groundwater inflows were relatively homogeneous across the site. Finally, the locations where the velocities were calculated were chosen for their constancy over time and their clearness on the map, which indicated probable strong inflows that were confirmed by velocity results. These results make it unlikely that we underestimated groundwater flow velocity.

Instead, we attribute the large underestimation of discharge during high flow mainly to the small spatial coverage of our FO-DTS system. Our only cable was located in the thalweg and buried in the sediment to detect even the weakest inflows. This strategy paid off during low flow: As long as the stream was narrow, the groundwater inflows detected by the cable were likely to be the main ones. By assuming an area with a width equal to the actual stream width, we estimated total discharge similar to that from differential gauging. During the high-flow period, however, the stream broadened, and new flow paths appeared in the sediment that we did not detect. To attempt to account for undetected inflows at the bottom of the streambed, we widened the area (2 m) of the previously detected inflows. Unfortunately, this approach assumed that the groundwater–stream exchange is uniform over the 2 m, which is false: Some points considered neutral could have been near a strong inflow that went undetected, and vice versa [71]. In addition, studies have shown that many inflows are located directly in stream banks [72–75]. Our experimental design could not identify such lateral inflows. Furthermore, field observations showed that the banks were often flooded by groundwater surges that occasionally flowed directly into the stream. In late spring, these many lateral inflows and bank overflows probably weakened due to evapotranspiration [69], and inflows at the bottom of the streambed dominated again, probably explaining why our method yielded better estimates during this period (Figure 9).

Ultimately, our method was able to estimate total hyporheic discharge of a small reach as long as the areas of groundwater inflows and neutral points are estimated accurately. In this context, installing additional cables in the streambed and along the banks appears necessary. In parallel, calculating vertical velocities for only a few locations and then applying them to all similar points along the reach yielded relatively good estimates when the spatial coverage was sufficient. However, this method also has limits. First, it requires previously mapping groundwater inflows (see Part I of this study [40]). Second, calculating total hyporheic discharge from so few velocities makes it sensitive to hydrological processes affecting them, such as punctual inversions of the hydraulic gradient [76]. One solution would be to combine an increase in spatial coverage with automatic estimation of vertical velocity at each location of the cable, which might improve estimates of total hyporheic discharge greatly. This solution could also help to distinguish inflows that are sensitive to the local hydraulic gradient from less sensitive inflows that follow deeper regional flow paths [77,78].

4. Conclusions

Coupling FO-DTS and temperature depth profiles to study groundwater–surface water interactions is quite challenging. This research outlines a framework to quantify groundwater inflows too weak or diffuse to be characterized by traditional FO-DTS methods. We coupled FO-DTS data in the sediment with punctual temperature data at the water–sediment interface to estimate vertical flow velocities along the entire cable. Velocities of representative points were calculated per sub-reach and applied to similar points along the cable. Comparison of these high spatial resolution velocities to those of spatially integrative methods revealed similar estimates of hyporheic discharge during the low-flow period but large underestimates during most of the high-flow period. This was attributed mainly to the limited coverage of the fiber optic cable, which could not detect lateral inflows during the high-flow period. Our method was also sensitive to punctual inversions of the hydraulic gradient that lead to slow vertical velocities.

Calculating vertical flows along the cable will improve the accuracy of the method and consequently discharge estimates during high flow. The high spatio-temporal resolution could help to distinguish groundwater inflows sensitive to local hydraulic changes from inflows that follow deeper regional flow paths or have different behaviors. Fine characterization of hyporheic exchanges (e.g., dynamics of deep inflows vs. lateral inflows, detection of downward flows) might be greatly useful for managing streams and studying hydro-biochemical processes.

Author Contributions: Conceptualization, H.L.L. and Z.T.; Methodology, H.L.L., F.R., P.P. and Z.T.; Software, H.L.L.; Validation, H.L.L., Z.T. and F.M.; Formal Analysis, H.L.L. and Z.T.; Investigation, H.L.L., F.R., P.P. and Z.T.; Resources, H.L.L., F.R., P.P. and Z.T.; Data Curation, H.L.L. and Z.T.; Writing—original Draft Preparation, H.L.L. and Z.T.; Writing—review & Editing, Z.T.; Visualization, H.L.L.; Supervision, Z.T. and F.M.; Project Administration, Z.T. and F.M.; Funding Acquisition, Z.T. and F.M.

Funding: This research was funded by Agence de l'Eau Loire Bretagne, grant number [150417801].

Acknowledgments: Authors warmly thank Pitois for the shelter and electricity they kindly provided during the entire campaign. We also thank all technicians from UMR INRA AGROCAMPUS OUEST SAS who helped us place the fiber optic cable and perform the field work. We also thank the ILSTER Zone Atelier Armorique.

Conflicts of Interest: The authors declare no conflict of interest.

References

1. Freeze, R.A. Role of subsurface flow in generating surface runoff: 1. Base flow contributions to channel flow. *Water Resour. Res.* **1972**, *8*, 609–623. [CrossRef]
2. Hester, E.T.; Gooseff, M.N. Moving Beyond the Banks: Hyporheic Restoration Is Fundamental to Restoring Ecological Services and Functions of Streams. *Environ. Sci. Technol.* **2010**, *44*, 1521–1525. [CrossRef]
3. Hester, E.T.; Brooks, K.E.; Scott, D.T. Comparing reach scale hyporheic exchange and denitrification induced by instream restoration structures and natural streambed morphology. *Ecol. Eng.* **2018**, *115*, 105–121. [CrossRef]
4. Alexander, M.D.; Caissie, D. Variability and comparison of hyporheic water temperatures and seepage fluxes in a small Atlantic salmon stream. *Groundwater* **2003**, *41*, 72–82. [CrossRef]
5. Barlow, J.R.; Coupe, R.H. Groundwater and surface-water exchange and resulting nitrate dynamics in the Bogue Phalia basin in northwestern Mississippi. *J. Environ. Qual.* **2012**, *41*, 155–169. [CrossRef]
6. Baxter, C.V.; Hauer, F.R. Geomorphology, hyporheic exchange, and selection of spawning habitat by bull trout (*Salvelinus confluentus*). *Can. J. Fish. Aquat. Sci.* **2000**, *57*, 1470–1481. [CrossRef]
7. Hillyard, R.W.; Keeley, E.R. Temperature-Related Changes in Habitat Quality and Use by Bonneville Cutthroat Trout in Regulated and Unregulated River Segments. *Trans. Am. Fish. Soc.* **2012**, *141*, 1649–1663. [CrossRef]
8. Vidon, P.; Allan, C.; Burns, D.; Duval, T.P.; Gurwick, N.; Inamdar, S.; Lowrance, R.; Okay, J.; Scott, D.; Sebestyen, S. Hot Spots and Hot Moments in Riparian Zones: Potential for Improved Water Quality Management1. *J. Am. Water Resour. Assoc.* **2010**, *46*, 278–298. [CrossRef]

9. Harvey, J.; Gooseff, M. River corridor science: Hydrologic exchange and ecological consequences from bedforms to basins. *Water Resour. Res.* **2015**, *51*, 6893–6922. [CrossRef]

10. Lee, D.R.; Hynes, H. Identification of groundwater discharge zones in a reach of Hillman Creek in southern Ontario. *Water Qual. Res. J.* **1978**, *13*, 121–134. [CrossRef]

11. Libelo, E.L.; MacIntyre, W.G. Effects of surface-water movement on seepage-meter measurements of flow through the sediment-water interface. *Appl. Hydrogeol.* **1994**, *2*, 49–54. [CrossRef]

12. McCallum, J.L.; Cook, P.G.; Berhane, D.; Rumpf, C.; McMahon, G.A. Quantifying groundwater flows to streams using differential flow gaugings and water chemistry. *J. Hydrol.* **2012**, *416*, 118–132. [CrossRef]

13. Ruehl, C.; Fisher, A.; Hatch, C.; Los Huertos, M.; Stemler, G.; Shennan, C. Differential gauging and tracer tests resolve seepage fluxes in a strongly-losing stream. *J. Hydrol.* **2006**, *330*, 235–248. [CrossRef]

14. Frei, S.; Gilfedder, B. FINIFLUX: An implicit finite element model for quantification of groundwater fluxes and hyporheic exchange in streams and rivers using radon. *Water Resour. Res.* **2015**, *51*, 6776–6786. [CrossRef]

15. Pittroff, M.; Frei, S.; Gilfedder, B. Quantifying nitrate and oxygen reduction rates in the hyporheic zone using 222Rn to upscale biogeochemical turnover in rivers. *Water Resour. Res.* **2017**, *53*, 563–579. [CrossRef]

16. Westhoff, M.C.; Savenije, H.H.G.; Luxemburg, W.M.J.; Stelling, G.S.; van de Giesen, N.C.; Selker, J.S.; Pfister, L.; Uhlenbrook, S. A distributed stream temperature model using high resolution temperature observations. *Hydrol. Earth Syst. Sci.* **2007**, *11*, 1469–1480. [CrossRef]

17. Briggs, M.A.; Lautz, L.K.; McKenzie, J.M. A comparison of fibre-optic distributed temperature sensing to traditional methods of evaluating groundwater inflow to streams. *Hydrol. Process.* **2012**, *26*, 1277–1290. [CrossRef]

18. Gonzalez-Pinzon, R.; Ward, A.S.; Hatch, C.E.; Wlostowski, A.N.; Singha, K.; Gooseff, M.N.; Haggerty, R.; Harvey, J.W.; Cirpka, O.A.; Brock, J.T. A field comparison of multiple techniques to quantify groundwater–surface- water interactions. *Freshw. Sci.* **2015**, *34*, 139–160. [CrossRef]

19. Kalbus, E.; Reinstorf, F.; Schirmer, M. Measuring methods for groundwater surface water interactions: A review. *Hydrol. Earth Syst. Sci.* **2006**, *10*, 873–887. [CrossRef]

20. Poulsen, J.R.; Sebok, E.; Duque, C.; Tetzlaff, D.; Engesgaard, P.K. Detecting groundwater discharge dynamics from point-to-catchment scale in a lowland stream: Combining hydraulic and tracer methods. *Hydrol. Earth Syst. Sci.* **2015**, *19*, 1871–1886. [CrossRef]

21. Anderson, M.P. Heat as a ground water tracer. *Groundwater* **2005**, *43*, 951–968. [CrossRef]

22. Constantz, J. Interaction between stream temperature, streamflow, and groundwater exchanges in Alpine streams. *Water Resour. Res.* **1998**, *34*, 1609–1615. [CrossRef]

23. Tristram, D.A.; Krause, S.; Levy, A.; Robinson, Z.P.; Waller, R.I.; Weatherill, J.J. Identifying spatial and temporal dynamics of proglacial groundwater–surface-water exchange using combined temperature-tracing methods. *Freshw. Sci.* **2015**, *34*, 99–110. [CrossRef]

24. Stallman, W.R. Steady One-Dimensional Fluid Flow in a Semi-Infinite Porous Medium with Sinusoidal Surface Temperature. *J. Geophys. Res.* **1965**, *70*, 2821–2827. [CrossRef]

25. Hatch, C.E.; Fisher, A.T.; Revenaugh, J.S.; Constantz, J.; Ruehl, C. Quantifying surface water–groundwater interactions using time series analysis of streambed thermal records: Method development. *Water Resour. Res.* **2006**, *42*, 14. [CrossRef]

26. Briggs, M.A.; Lautz, L.K.; McKenzie, J.M.; Gordon, R.P.; Hare, D.K. Using high-resolution distributed temperature sensing to quantify spatial and temporal variability in vertical hyporheic flux. *Water Resour. Res.* **2012**, *48*, 1–16. [CrossRef]

27. Briggs, M.A.; Lautz, L.K.; Buckley, S.F.; Lane, J.W. Practical limitations on the use of diurnal temperature signals to quantify groundwater upwelling. *J. Hydrol.* **2014**, *519*, 1739–1751. [CrossRef]

28. Käser, D.H.; Binley, A.; Heathwaite, A.L.; Krause, S. Spatio-temporal variations of hyporheic flow in a riffle-step-pool sequence. *Hydrol. Process. Int. J.* **2009**, *23*, 2138–2149. [CrossRef]

29. Krause, S.; Blume, T.; Cassidy, N. Investigating patterns and controls of groundwater up-welling in a lowland river by combining Fibre-optic Distributed Temperature Sensing with observations of vertical hydraulic gradients. *Hydrol. Earth Syst. Sci.* **2012**, *16*, 1775–1792. [CrossRef]

30. Hurtig, E.; Grosswig, S.; Kuhn, K. Distributed fibre optic temperature sensing: A new tool for long-term and short-term temperature monitoring in boreholes. *Energy Sources* **1997**, *19*, 55–62. [CrossRef]

31. Hausner, M.B.; Suarez, F.; Glander, K.E.; van de Giesen, N.; Selker, J.S.; Tyler, S.W. Calibrating Single-Ended Fiber-Optic Raman Spectra Distributed Temperature Sensing Data. *Sensors* **2011**, *11*, 10859–10879. [CrossRef]

32. Tyler, S.W.; Selker, J.S.; Hausner, M.B.; Hatch, C.E.; Torgersen, T.; Thodal, C.E.; Schladow, S.G. Environmental temperature sensing using Raman spectra DTS fiber-optic methods. *Water Resour. Res.* **2009**, *45*, 11. [CrossRef]

33. Van de Giesen, N.; Steele-Dunne, S.C.; Jansen, J.; Hoes, O.; Hausner, M.B.; Tyler, S.; Selker, J. Double-Ended Calibration of Fiber-Optic Raman Spectra Distributed Temperature Sensing Data. *Sensors* **2012**, *12*, 5471–5485. [CrossRef]

34. Moffett, K.B.; Tyler, S.W.; Torgersen, T.; Menon, M.; Selker, J.S.; Gorelick, S.M. Processes controlling the thermal regime of saltmarsh channel beds. *Environ. Sci. Technol.* **2008**, *42*, 671–676. [CrossRef]

35. Rosenberry, D.O.; Briggs, M.A.; Delin, G.; Hare, D.K. Combined use of thermal methods and seepage meters to efficiently locate, quantify, and monitor focused groundwater discharge to a sand-bed stream. *Water Resour. Res.* **2016**, *52*, 4486–4503. [CrossRef]

36. Selker, J.S.; Thevenaz, L.; Huwald, H.; Mallet, A.; Luxemburg, W.; de Giesen, N.V.; Stejskal, M.; Zeman, J.; Westhoff, M.; Parlange, M.B. Distributed fiber-optic temperature sensing for hydrologic systems. *Water Resour. Res.* **2006**, *42*, 1–8. [CrossRef]

37. Selker, J.S.; van de Giesen, N.; Westhoff, M.; Luxemburg, W.; Parlange, M.B. Fiber optics opens window on stream dynamics. *Geophys. Res. Lett.* **2006**, *33*, 1–4. [CrossRef]

38. Westhoff, M.C.; Gooseff, M.N.; Bogaard, T.A.; Savenije, H.H.G. Quantifying hyporheic exchange at high spatial resolution using natural temperature variations along a first-order stream. *Water Resour. Res.* **2011**, *47*, 1–13. [CrossRef]

39. Lauer, F.; Frede, H.G.; Breuer, L. Uncertainty assessment of quantifying spatially concentrated groundwater discharge to small streams by distributed temperature sensing. *Water Resour. Res.* **2013**, *49*, 400–407. [CrossRef]

40. Le Lay, H.; Thomas, Z.; Rouault, F.; Pichelin, P.; Moatar, F. Characterization of diffuse groundwater inflows into stream water. Part I: Spatial and temporal mapping framework based on fiber optic distributed temperature sensing. *Water* **2019**, *11*, 2389. [CrossRef]

41. Lowry, C.S.; Walker, J.F.; Hunt, R.J.; Anderson, M.P. Identifying spatial variability of groundwater discharge in a wetland stream using a distributed temperature sensor. *Water Resour. Res.* **2007**, *43*, 1–9. [CrossRef]

42. Mamer, E.A.; Lowry, C.S. Locating and quantifying spatially distributed groundwater/surface water interactions using temperature signals with paired fiber-optic cables. *Water Resour. Res.* **2013**, *49*, 7670–7680. [CrossRef]

43. Sebok, E.; Duque, C.; Engesgaard, P.; Boegh, E. Application of Distributed Temperature Sensing for coupled mapping of sedimentation processes and spatio-temporal variability of groundwater discharge in soft-bedded streams. *Hydrol. Process.* **2015**, *29*, 3408–3422. [CrossRef]

44. Unland, N.P.; Cartwright, I.; Andersen, M.S.; Rau, G.C.; Reed, J.; Gilfedder, B.S.; Atkinson, A.P.; Hofmann, H. Investigating the spatio-temporal variability in groundwater and surface water interactions: A multi-technique approach. *Hydrol. Earth Syst. Sci.* **2013**, *17*, 3437–3453. [CrossRef]

45. Tirado-Conde, J.; Engesgaard, P.; Karan, S.; Müller, S.; Duque, C. Evaluation of Temperature Profiling and Seepage Meter Methods for Quantifying Submarine Groundwater Discharge to Coastal Lagoons: Impacts of Saltwater Intrusion and the Associated Thermal Regime. *Water* **2019**, *11*, 1648. [CrossRef]

46. Varli, D.; Yilmaz, K. A Multi-Scale Approach for Improved Characterization of Surface Water—Groundwater Interactions: Integrating Thermal Remote Sensing and in-Stream Measurements. *Water* **2018**, *10*, 854. [CrossRef]

47. Carlson, A.K.; Taylor, W.W.; Schlee, K.M.; Zorn, T.G.; Infante, D.M. Projected impacts of climate change on stream salmonids with implications for resilience-based management. *Ecol. Freshw. Fish* **2017**, *26*, 190–204. [CrossRef]

48. Hendriks, D.M.D.; Kuijper, M.J.M.; van Ek, R. Groundwater impact on environmental flow needs of streams in sandy catchments in the Netherlands. *Hydrol. Sci. J.* **2014**, *59*, 562–577. [CrossRef]

49. Laize, C.L.R.; Meredith, C.B.; Dunbar, M.J.; Hannah, D.M. Climate and basin drivers of seasonal river water temperature dynamics. *Hydrol. Earth Syst. Sci.* **2017**, *21*, 3231–3247. [CrossRef]

50. Hongve, D. A revised procedure for discharge measurement by means of the salt dilution method. *Hydrol. Process.* **1987**, *1*, 267–270. [CrossRef]

51. Calkins, D.; Dunne, T. A salt tracing method for measuring channel velocities in small mountain streams. *J. Hydrol.* **1970**, *11*, 379–392. [CrossRef]

52. Sloto, R.A.; Crouse, M.Y. HYSEP: A Computer Program for Streamflow Hydrograph Separation and Analysis. In *Water–Resources Investigations Report*; USGS Numbered Series; U.S. Geological Survey: Reston, VA, USA, 1996; p. 46. [CrossRef]
53. Freeze, R.A. Streamflow generation. *Rev. Geophys.* **1974**, *12*, 627–647. [CrossRef]
54. Boulton, A.J.; Findlay, S.; Marmonier, P.; Stanley, E.H.; Valett, H.M. The functional significance of the hyporheic zone in streams and rivers. *Annu. Rev. Ecol. Syst.* **1998**, *29*, 59–81. [CrossRef]
55. Cardenas, M.B. Hyporheic zone hydrologic science: A historical account of its emergence and a prospectus. *Water Resour. Res.* **2015**, *51*, 3601–3616. [CrossRef]
56. Thomas, Z.; Rousseau-Gueutin, P.; Kolbe, T.; Abbott, B.W.; Marçais, J.; Peiffer, S.; Frei, S.; Bishop, K.; Pichelin, P.; Pinay, G.; et al. Constitution of a catchment virtual observatory for sharing flow and transport models outputs. *J. Hydrol.* **2016**, *543*, 59–66. [CrossRef]
57. Gordon, R.P.; Lautz, L.K.; Briggs, M.A.; McKenzie, J.M. Automated calculation of vertical pore-water flux from field temperature time series using the VFLUX method and computer program. *J. Hydrol.* **2012**, *420*, 142–158. [CrossRef]
58. Benyahya, L.; Caissie, D.; Satish, M.G.; El-Jabi, N. Long-wave radiation and heat flux estimates within a small tributary in Catamaran Brook (New Brunswick, Canada). *Hydrol. Process.* **2012**, *26*, 475–484. [CrossRef]
59. Caissie, D. The thermal regime of rivers: A review. *Freshw. Biol.* **2006**, *51*, 1389–1406. [CrossRef]
60. Petrides, A.C.; Huff, J.; Arik, A.; van de Giesen, N.; Kennedy, A.M.; Thomas, C.K.; Selker, J.S. Shade estimation over streams using distributed temperature sensing. *Water Resour. Res.* **2011**, *47*, 1–4. [CrossRef]
61. Irvine, D.J.; Cranswick, R.H.; Simmons, C.T.; Shanafield, M.A.; Lautz, L.K. The effect of streambed heterogeneity on groundwater–surface water exchange fluxes inferred from temperature time series. *Water Resour. Res.* **2015**, *51*, 198–212. [CrossRef]
62. Vandersteen, G.; Schneidewind, U.; Anibas, C.; Schmidt, C.; Seuntjens, P.; Batelaan, O. Determining groundwater–surface water exchange from temperature-time series: Combining a local polynomial method with a maximum likelihood estimator. *Water Resour. Res.* **2015**, *51*, 922–939. [CrossRef]
63. Abrol, I.; Palta, J. Bulk density determination of soil clods using rubber solution as a coating material. *Soil Sci.* **1968**, *106*, 465–468. [CrossRef]
64. Lapham, W.W. Use of temperature profiles beneath streams to determine rates of vertical ground-water flow and vertical hydraulic conductivity. In *Water Supply Paper*; U.S. Geological Survey: Reston, VA, USA, 1989. [CrossRef]
65. Stonestrom, D.A.; Constantz, J. *Using Temperature to Study Stream-Ground Water Exchanges*; U.S. Geological Survey Fact Sheet: Menlo Park, CA, USA, 2004; pp. 1–4.
66. Broecker, T.; Teuber, K.; Sobhi Gollo, V.; Nützmann, G.; Lewandowski, J.; Hinkelmann, R. Integral Flow Modelling Approach for Surface Water–Groundwater Interactions along a Rippled Streambed. *Water* **2019**, *11*, 1517. [CrossRef]
67. Gooseff, M.N.; Anderson, J.K.; Wondzell, S.M.; LaNier, J.; Haggerty, R. A modelling study of hyporheic exchange pattern and the sequence, size, and spacing of stream bedforms in mountain stream networks, Oregon, USA. *Hydrol. Process.* **2005**, *19*, 2915–2929. [CrossRef]
68. Yao, C.; Lu, C.; Qin, W.; Lu, J. Field Experiments of Hyporheic Flow Affected by a Clay Lens. *Water* **2019**, *11*, 1613. [CrossRef]
69. Poblador, S.; Thomas, Z.; Rousseau-Gueutin, P.; Sabaté, S.; Sabater, F. Riparian forest transpiration under the current and projected Mediterranean climate: Effects on soil water and nitrate uptake. *Ecohydrology* **2019**, *12*, e2043. [CrossRef]
70. Frei, S.; Durejka, S.; Le Lay, H.; Thomas, Z.; Gilfedder, B.S. Hyporheic nitrate removal at the reach scale: Exposure times vs. residence times. *Water Resour. Res.* **2019**. [CrossRef]
71. Duff, J.H.; Toner, B.; Jackman, A.P.; Avanzino, R.J.; Triska, F.J. Determination of groundwater discharge into a sand and gravel bottom river: A comparison of chloride dilution and seepage meter techniques. *Int. Ver. Theor. Angew. Limnol. Verh.* **2000**, *27*, 406–411. [CrossRef]
72. Briggs, M.A.; Voytek, E.B.; Day-Lewis, F.D.; Rosenberry, D.O.; Lane, J.W. Understanding Water Column and Streambed Thermal Refugia for Endangered Mussels in the Delaware River. *Environ. Sci. Technol.* **2013**, *47*, 11423–11431. [CrossRef]

73. Hare, D.K.; Briggs, M.A.; Rosenberry, D.O.; Boutt, D.F.; Lane, J.W. A comparison of thermal infrared to fiber-optic distributed temperature sensing for evaluation of groundwater discharge to surface water. *J. Hydrol.* **2015**, *530*, 153–166. [CrossRef]

74. Shope, C.L.; Constantz, J.E.; Cooper, C.A.; Reeves, D.M.; Pohll, G.; McKay, W.A. Influence of a large fluvial island, streambed, and stream bank on surface water–groundwater fluxes and water table dynamics. *Water Resour. Res.* **2012**, *48*, 1–18. [CrossRef]

75. Van Balen, R.T.; Kasse, C.; De Moor, J. Impact of groundwater flow on meandering; example from the Geul River, The Netherlands. *Earth Surf. Process. Landf.* **2008**, *33*, 2010–2028. [CrossRef]

76. Winter, T.C.; Harvey, J.W.; Franke, O.L.; Alley, W.M. *Ground Water and Surface Water: A Single Resource*; United States Geological Survey Circular; DIANE Publishing Inc.: Darby, PA, USA, 1998; Volume 1139. [CrossRef]

77. Kolbe, T.; Marcais, J.; Thomas, Z.; Abbott, B.W.; de Dreuzy, J.R.; Rousseau-Gueutin, P.; Aquilina, L.; Labasque, T.; Pinay, G. Coupling 3D groundwater modeling with CFC-based age dating to classify local groundwater circulation in an unconfined crystalline aquifer. *J. Hydrol.* **2016**, *543*, 31–46. [CrossRef]

78. Kolbe, T.; de Dreuzy, J.-R.; Abbott, B.W.; Aquilina, L.; Babey, T.; Green, C.T.; Fleckenstein, J.H.; Labasque, T.; Laverman, A.M.; Marçais, J.; et al. Stratification of reactivity determines nitrate removal in groundwater. *Proc. Natl. Acad. Sci. USA* **2019**, *116*, 2494–2499. [CrossRef]

Article

Field Experiments of Hyporheic Flow Affected by a Clay Lens

Congcong Yao, Chengpeng Lu *, Wei Qin and Jiayun Lu

College of Hydrology and Water Resources, Hohai University, Nanjing 210098, China
* Correspondence: luchengpeng@hhu.edu.cn; Tel.: +86-258-378-7683

Received: 29 June 2019; Accepted: 1 August 2019; Published: 3 August 2019

Abstract: As a typical water exchange of surface water and groundwater, hyporheic flow widely exists in streambeds and is significantly affected by the characteristics of sediment and surface water. In this study, a low-permeability clay lens was chosen to investigate the influence of the streambed heterogeneity on the hyporheic flow at a river section of the Xin'an River in Anhui Province, China. A 2D sand tank was constructed to simulate the natural streambed including a clay lens under different velocity of surface water velocity. Heat tracing was used in this study. In particular, six analytical solutions based on the amplitude ratio and phase shift of temperatures were applied to calculate the vertical hyporheic flux. The results of the six methods ranged from −102.4 to 137.5 m/day and showed significant spatial differences. In view of the robustness of the calculations and the rationality of the results, the amplitude ratio method was much better than the phase shift method. The existence of the clay lens had a significant influence on the hyporheic flow. Results shows that the vertical hyporheic flux in the model containing a clay lens was lower than that for the blank control, and the discrepancy of the hyporheic flow field on both sides of the lens was obvious. Several abnormal flow velocity zones appeared around the clay lens where the local hyporheic flow was suppressed or generally enhanced. The hyporheic flow fields at three test points had mild changes when the lens was placed in a shallow layer of the model, indicating that the surface water velocity only affect the hyporheic flow slightly. With the increasing depth of the clay lens, the patterns of the hyporheic flow fields at all test points were very close to those of the hyporheic flow field without a clay lens, indicating that the influence of surface water velocity on hyporheic flow appeared gradually. A probable maximum depth of the clay lens was 30 to 40 cm, which approached the bottom of the model and a clay lens buried lower than this maximum would not affect the hyporheic flow any more. Influenced by the clay lens, hyporheic flow was hindered or enhanced in different regions of streambed, which was also depended on the depth of lens and surface water velocity. Introducing a two-dimensional sand tank model in a field test is an attempt to simulate a natural streambed and may positively influence research on hyporheic flow.

Keywords: clay lens; heat tracing; field experiments; hyporheic flow

---

1. Introduction

As typical linear surface water bodies, rivers play an important role in many geological and ecological processes [1]. The hyporheic zone, which is the saturated zone alongside and beneath the streambed, is an active zone of surface water and groundwater interaction [2]. Driven by hydraulic head gradients, the stream water flows into and out of the hyporheic zone, which induces the exchange of mass and energy [3–6]. Owing to the unique characteristics of the hyporheic flow and biochemical environment, the hyporheic zone has significant effects on river–aquifer systems, including the solute transport [7–9], oxygen and nitrogen circulations [10–13], particle movement (e.g., colloid and heavy-metal ion) [14–17], and biological processes [18].

Hyporheic flow is mainly dominated by the patterns of the streambed, including its topography and structure [2,19,20]. The streambed topography, which changes the distribution of the hydraulic head, is one of the driving factors of the hyporheic flow [3,21–23]. The structure, which is the distribution of hydraulic conductivity (K), influences the hyporheic flow field and residence time [24–26]. Scouring and deposition lead to the redistribution of streambed sediment, resulting in the stratification which is ubiquitous in a natural streambed.

Fox et al. [27] constructed a 2D heterogeneous structure to evaluate the effects of different flow conditions on the hyporheic flow. Pryshlak et al. [28] chose a low-gradient stream with 300 different bimodal K fields to analyze the effects of heterogeneity in K on the hyporheic flux, residence time, and spatial pattern in a hyporheic zone. A variety of studies indicated that a heterogeneous streambed produces temporally and spatially variable hyporheic flows [29–32]. However, simulations of a heterogeneous streambed were random and unrealistic, and the condition of surface water were lack of consideration.

In this study, we examine the effects of a clay lens on the hyporheic flow. The clay lens is a typical low-permeable media and generally composed of peat and silt deposits. This impervious layer is widely distributed in streambeds, aquifers, and other strata, and mainly affects stream-groundwater interactions, mass migrations, and the exploitation of ground resources [33–35]. Studies about the relations between a clay lens and the hyporheic flow are rare, and there remain many challenges in determining the mechanism of hyporheic exchange in a heterogeneous streambed.

The application of heat as a tracer to determine hyporheic flow has become a general method in studies of hyporheic flows [36–39], groundwater recharge and discharge [40–44], and stream ecology [45,46]. As a novel and natural tracer, heat is pollution-free on the environment and can be monitored with low cost. Recently, the development of distributed temperature sensing technology has provided a means to obtain high-resolution temperature data, which is vital in delineating the hyporheic flow [47].

In this experimental study, we investigated the influence of a clay lens on the hyporheic flow in a natural stream. A 2D sand tank made of aluminum alloys and gauze was designed to simulate a heterogenous streambed for better quantitative and spatial control of the clay lens. This "artificial streambed" was placed to several sections of a river to experience different surface water velocities. As far as we know, there has been no research on hyporheic flow in a natural stream by using a sand tank. Comparing with the laboratory tests [48–50] and numerical simulations [51], this study reflects the characteristic of the mountain river in summer, making the hydraulic conditions of hyporheic flow more realistic. The vertical hyporheic flux (VHF) were calculated from raw temperature time series, which were monitored by temperature sensors buried in the sand tank. The specific objectives were (a) to analyze the characteristics of hyporheic flow in a streambed under the effect of a clay lens; and (b) demonstrate the relationship between the clay lens and hyporheic flow.

2. Site Description and Method

2.1. Site Description

The study sites were located in the Jinyuan River, a second-order river within the Xin'an River of Anhui Province, China. The length of the river was 373 km, and the area of catchment was 12,100 km^2. This catchment was bounded by Mount Huang and Yangtze River to the west and north, respectively. To the east and south, the catchment was bounded by Mount Tianmu and Mount Baiji.

The runoff of this catchment was influenced by the characteristics of the soil and the distribution of vegetation. The former was mainly related to the soil bulk density and porosity, and the latter depended on the land cover and land use. The soils in the study area included loam and clay. The vegetation coverage in high-altitude areas was larger than that in low-altitude areas, where part of the native vegetation had been destroyed owing to cultivation.

The lithology in the catchment was dominated by Proterozoic group strata covered with Quaternary sediment. Regional groundwater included pore water and fracture water in loose rock mass, and the pumpage capacity in a single well was up to 120–720 m^3/day. Owing to the large fluctuation of the surface landform and the poor storage ability of rock, groundwater recharged from precipitation was rapidly lateral discharged into the nearby valley in the form of a depression spring.

The distribution of the annual and monthly average precipitation was uneven and ranged from 913.9 mm to 2708.4 mm. Sixty percent of the annual precipitation occurred from April to July. The annual average evaporation was 851 mm, and the evaporation from May to August accounted for 50% of one year. Considering the spatial distribution, the runoff decreased from southwest to northeast, which was similar to the trend of precipitation. Most of the time during our field experiments the weather was sunny, and there was only a 7 mm rainfall event on 10 July.

The study river reach was nearly 100 m long, encompassing a section of tributary (Figure 1). The streambed consisted of fluvial sediments covering the bedrock where the average thickness of the sediment was 0.15 m. In this study, continuous observation of the water level and temperature in the streambed, river, and air were conducted over a ten-day period from 10 July 2018 to 20 July 2018.

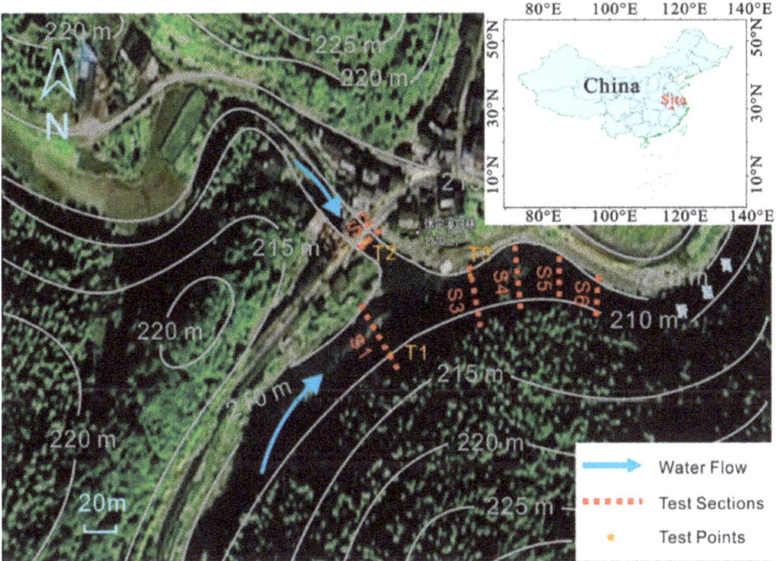

Figure 1. Study area of Jinyuan River and the hypsometric map of study area. Measurements of hydrologic factors, e.g., flow velocity, water depth, and section width, were taken at test sections S1, S2, S3, S4, S5, and S6. Field tests were conducted at test points T1, T2, and T3 located at S1, S2, and S3, respectively.

2.2. Hydraulic Characterization and Water Quality

Measurements of the hydraulic characterization of the stream and water quality in the study reach were conducted before the field test. The widths of six sections were determined with a digital level (DL201, South). A propeller-type current meter (LS10, Hydro-Bios) was used to measure the flow velocity of the surface water. The river discharge was calculated based on the weighted average velocity and water depth.

Sections S1 and S3 were located upstream and downstream of the main channel, respectively. The flow velocity, and water depth at S1 was much higher than those at S3, which meant that the upper reaches featured a higher drop with torrents and the flow in the downstream area become much slowly. Section S2 was located at tributary. Details of the hydraulic characterization are listed in Table 1.

Table 1. Hydraulic characterization of six sections and three test points.

Section	Width, m	Maximum Depth, m	Average Flux, m^3/s	Test Points	Distance to River Bank, m	Flow Velocity, m/s
S1	24.0	0.95	6.46	T1	5.8	0.46
S2	9.7	0.62	1.47	T2	3.9	0.37
S3	27.6	0.55	1.93	T3	5.0	0.15
S4	27.0	0.72	3.82			
S5	24.6	0.82	4.38			
S6	22.4	0.92	4.52			

In addition to the hydraulic characterization, several kinds of water samples from the stream, well, and soil were collected and measured in situ (Table 2). T1, T2, and T3 were selected as three sampling sites for surface water. Additionally, soil water at three different depths was collected at a natural overland flow site. In comparing the water quality, the stream water sample was clear and had lower solute concentrations, while the groundwater and soil water were muddy and alkaline, which meant that quality of the groundwater and soil water was relatively worse than surface water.

Table 2. Water quality of stream, groundwater, and soil water. A, B, and C were stream water samples from T1, T2, and T3. D was groundwater sample from well. E, F, and G were soil water samples at depths of 10 cm, 20 cm, and 30 cm, respectively.

Sample	Temperature (°C)	EC (µs)	Salinity (ppt)	TDS (mg/L)	pH
A	25.3	43.3	0.02	30.8	7.46
B	23.9	30.6	0.01	21.8	7.48
C	24.5	30.3	0.01	21.8	7.49
D	26.1	933	0.46	662	9.43
E	28.1	148.5	0.07	105	8.55
F	28.1	192	0.09	137	8.22
G	27.3	364	0.18	258	9.28

2.3. Analysis Methods of Vertical Hyporheic Flux

Time series of temperature in stream, air, and groundwater have significantly different trends. This makes it possible to reflect the exchange of water and mass between the stream and groundwater. Hatch et al. [52] proposed an analytical model for determining the streambed seepage using time series thermal data. The analytical model defined the relationships between the seepage and the amplitude ratio (A_r) or the lag time ($\Delta\phi$) of the temperature signal. On this basis, Luce et al. [53], Keery et al. [54], and McCallum et al. [55] modified and proposed new solution methods. All of these heat tracing models are based on a one-dimensional conduction–advection–dispersion equation [56–58]:

$$\frac{\partial T}{\partial t} = \kappa_e \frac{\partial^2 T}{\partial z^2} - \frac{n v_f}{\gamma} \frac{\partial T}{\partial z} \tag{1}$$

where T is the temperature (varies with time t and depth z); κ_e is the effective thermal diffusivity; $\gamma = \rho c / \rho_f c_f$, the ratio of the heat capacity of the streambed to the fluid ($\rho_f c_f$ is the heat capacity of the fluid, and ρc is the heat capacity of the saturated sediment–fluid system); n is the porosity, and v_f is the vertical fluid velocity (positive = upwards flow) [52].

The solution to equation (1), given periodic variations in temperature at the top of the half-space, is modified from Goto et al. [59] and Stallman [57], and is rearranged to solve for the velocity of a thermal front as a function of amplitude and phase relations (Equations (2) and (3), or v_{Ar} and $v_{\Delta\Phi}$, respectively):

$$v_{Ar} = \frac{2\kappa_e}{\Delta z} \ln A_r + \sqrt{\frac{\alpha + v^2}{2}} \tag{2}$$

$$v_{\Delta\Phi} = \sqrt{\alpha - 2\left(\frac{\Delta\Phi 4\pi\kappa_e}{P\Delta z}\right)^2} \tag{3}$$

where A is the amplitude of the temperature variations at the upper boundary, P is the period of temperature variations ($P = 1/f$, where f is the frequency), and $\alpha = \sqrt{v^4 + (8\pi \times \kappa_e/P)^2}$. Here, A_r is the ratio of the amplitude ($A_r = A_d/A_s$), and $\Delta\Phi$ is the phase-shift variation between measurement points at different depths. The fluid velocities can be calculated from the thermal front velocities based on the relationships $v_{f,Ar} = v_{Ar}\gamma$ and $v_{f,\Delta\Phi} = v_{\Delta\Phi}\gamma$. In addition, the effective thermal diffusivity κ_e is defined as:

$$\kappa_e = \frac{\lambda_e}{\rho c} = \frac{\lambda_0}{\rho c} + \beta|v_f| \tag{4}$$

where λ_e is the effective thermal conductivity, λ_0 is the baseline thermal conductivity (in the absence of fluid flow), and β is the thermal dispersivity.

Gordon et al. [60] developed a computer program named VFLUX to calculate the vertical water flux (in short, VHF). The analytical solutions were provided by Hatch et al. [52] to calculate the time series of the flux using either the amplitude ratio (the method HatchA) or phase shift (the method HatchP) of diurnal signals between two temperature time series.

On this basis, Keery et al. [54] improved and proposed a new analytical solution that was based on the amplitude ratio (the method KeeryA) or phase shift (the method KeeryP) without thermal dispersivity. In addition, Luce et al. [53] and McCallum et al. [55] presented novel analytical methods utilizing the amplitude and phase information, respectively (herein referred to as the method of McCallum and Luce). As the influence of nonideal field conditions on the amplitude ratio and phase-shift models has not been investigated thoroughly, Irvine et al. [61] included and evaluated many factors (e.g., non-sinusoidal temperature signals, unsteady flows, and multidimensional flows) within an updated version of VFLUX2. This model is composed of all the above methods. The sediment and thermal properties should be input before the program is run.

Raw temperature time series are recorded by sensors buried in the streambed. Dynamic Harmonic Regression (DHR) is used to isolate the diurnal signals in order to extract the amplitude and phase angle information. DHR is a method for nonstationary time series analysis that is particularly useful for extracting harmonic signals from dynamic environmental systems [62]. Here, DHR can produce time-varying apparent amplitude and phase coefficients for a time series at a user-specified frequency (10 min in this study).

Thermal parameters are considered constant in spatial and temporal distribution because of the relatively thermal homogeneity of streambed sediment. The values are determined through field observation, referring to several published studies [38,63,64].

2.4. Design of Temperature Sensors and Field Test

To simulate a natural streambed, we constructed a 2D sand-channel model in which the frame was welded together by aluminum alloy and wire netting wrapped in gauze, and was paved internally in order to fill the experimental material (sand, clay lens, and temperature sensors) and allow for water flow. The size of the model was 50 cm in length, 10 cm in width, and 50 cm in height (Figure 2a). The size of the clay lens was 10 cm in length, 5 cm in width, and 10 cm in height. Since the width of model and the height of lens were equal, when the lens was placed horizontally it attached the two sides of model (Figure 2b). Sensors were buried in the model to record the temperature distribution of this "artificial streambed". The interval between sensors in the horizontal and vertical directions was 10 cm and refined to 5 cm around the clay lens (from x = 10 cm and z = 10 cm to x = 40 cm and z = 40 cm, Figure 2b).

Based on the orthogonal test method, three test points (T1, T2, and T3) were chosen to test the influences of the surface water velocity. At each point, the depth of the clay lens was changed from 20 cm to 40 cm (Figure 2b). Twelve test scenarios are listed in Table 3. In addition to the "artificial streambed", five temperature sensors were fixed on PVC tubes at 2-cm intervals. The tubes were then inserted into the natural streambed to record the temperatures at different depths from 2 cm to 10 cm (Figure 2c). The water levels, air pressure, and temperature were monitored by a level meter (Solinst 3001 Levelogger).

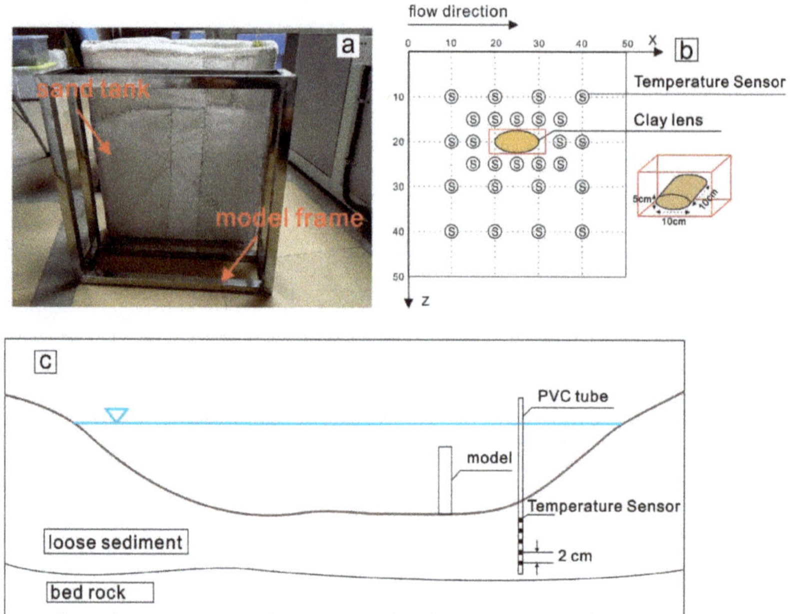

Figure 2. (**a**) Photo of model frame and sand tank before completion of construction. (**b**) Temperature sensors and clay lens distribution when depth of lens is 20 cm. Relative position when depth is 30 cm and 40 cm is consistent. (**c**) Sketch map of a cross-section of the study site.

Table 3. Summary of test scenarios, depths, and test points. N1–N12 are test numbers.

Test Points	No Lens	Depth (m)		
		0.2	0.3	0.4
T1	N1	N2	N3	N4
T2	N5	N6	N7	N8
T3	N9	N10	N11	N12

In the test period, the model was placed at three test points (T1, T2, and T3), and the depth of the lens changed from 0.2 m to 0.4 m (Table 3). The temperature sensor was an iButton Thermochrons (model DS1922L) manufactured by Dallas Semiconductor. It had a 17.35-mm diameter and 6-mm thickness and was quite wearproof, making it suitable for fieldwork. Users can set different parameters according to their work needs. In this study, the accuracy, resolution, and response time of the sensors were 0.5 °C, 0.0625 °C, and 5 min.

3. Results

3.1. Temperature and Water Level of Natural Streambed

Figure 3 shows the temperature fluctuation in the air and natural streambed at different depths for T1, T2, and T3 from July 11 to July 20. Temperature changes with time formed a quasi-sine wave and change smoothly. Some small irregular fluctuations appeared during the test period, possibly because of changes in the solar radiation and wind. At the end of each test, the temperature exhibited a rapid increase because the sensors were moved out from the stream.

Compared with the streambed temperature, the amplitude of the air temperature was significantly larger. The maximum and minimum of the air temperature were 30.9–35.7 °C and 21.8–22.9 °C, respectively, while in the streambed they were 26.4–30.7 °C and 22.7–25.2 °C. The air-temperature curve fluctuated more sharply, which were distributed in a serrated pattern owing to the great difference in heat capacity between air and water (the air temperature is more easily varied by solar radiation).

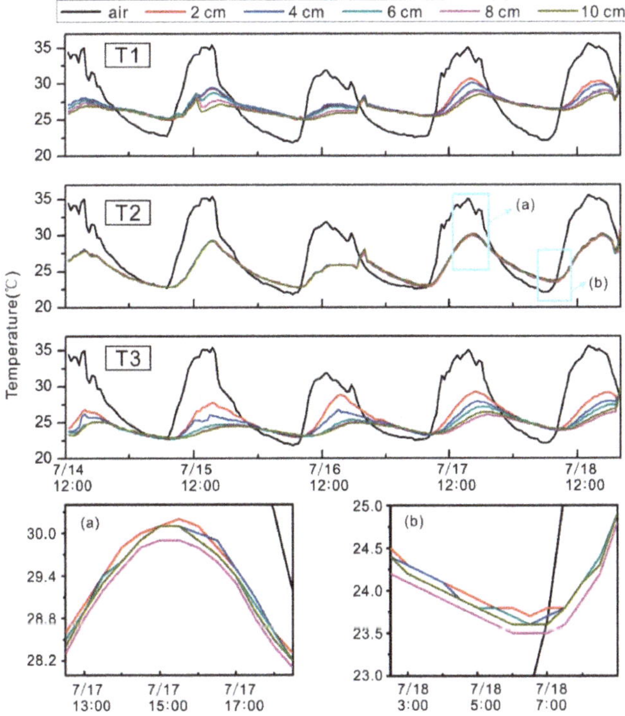

Figure 3. Variations of temperature and water level with time. Two periods of temperature were chosen to partially enlarge the curve (**a**,**b**).

3.2. Comparison of Six Methods in VFLUX2

Table 4 lists the values of the parameters used in VFLUX2. The data was selected from several similar field tests [43,65,66]. Figure 4 showed the mean and coefficient of variation (Cv) of the VHF calculated by six analytic methods. The six methods had significant differences that the order of VHF ranged from 10^{-2} to 10^{2} m/day. Hence, the results of HatchA and KeeryA were enlarged and shown in Figure 4.

Table 4. Physical properties and thermal parameters used in VFLUX2.

	Porosity	Density (kg m^{-3})	Volumetric Heat Capacity (cm^3 C)	Thermal Conductivity (s cm °C)	Dispersivity
Water		1000	0.5		
Sand	0.372	2580	1		
Saturated Sediment				0.0045	0.001

Specifically, the VHF as calculated by the two amplitude methods (HatchA and KeeryA) were extremely close. The means of HatchA and KeeryA at T1, T2, and T3 were 0.142, 2.169, and −0.190 m/day, and 0.142, 2.123, and −0.184 m/day, respectively. The variation in the VHF was apparent only at T2, where Cv of HatchA and KeeryA were 0.023 and 0.222, respectively. Since the variation was too small, it was difficult to find the fine distinctions in Figure 4. In addition, calculations of HatchP and KeeryP at T2 were missing, i.e., only included the results at T1 and T3; this was mainly caused by the limitation of the phase shift methods.

Based on the raw temperature data and hydrogeological investigations, the streambed in the study reach was bedrock covered with loose sediment whose average thickness was around 15 to 20 cm. Thus, the interval between the two sensors was only 2 cm (as mentioned in Section 2.4). Hence, the heat transport was sometimes abnormal at T3. The temperatures at deeper points reached the maximum earlier than those at shallower points. That is, the phase shift was negative, and thus the VHF could not be calculated by the phase shift methods (HatchP and KeeryP).

The methods proposed by Hatch and Keery used a single information (amplitude ratio or phase shift), while the methods proposed by Luce and McCallum combined the amplitude ratio with the phase shift. However, calculations of these "synthesis" methods at three test points were quite different. The means of Luce and McCallum at T1, T2, and T3 were −0.853, −38.892, and −0.433 m/day, and −0.432, −38.892, and −0.403 m/day, respectively. In addition to different points, it was apparent that the variation in the VHF at the same point at different depths indicated that Cv from Luce and McCallum were much higher among the six methods.

From a comparison of the six methods, the methods based on the amplitude ratio and phase shift have a broader scope and higher robustness, which can be used in most cases. The phenomenon of missing calculations is insignificant. However, the results from Luce and McCallum are more easily affected by environmental factors and exhibit great fluctuations, which may lead to lower reliability and even errors. Additionally, the phase shift method has a higher requirement for raw temperature data. In this study, the changes in the amplitude ratio met the test demands. Therefore, the method of HatchA was used to calculate the VHF.

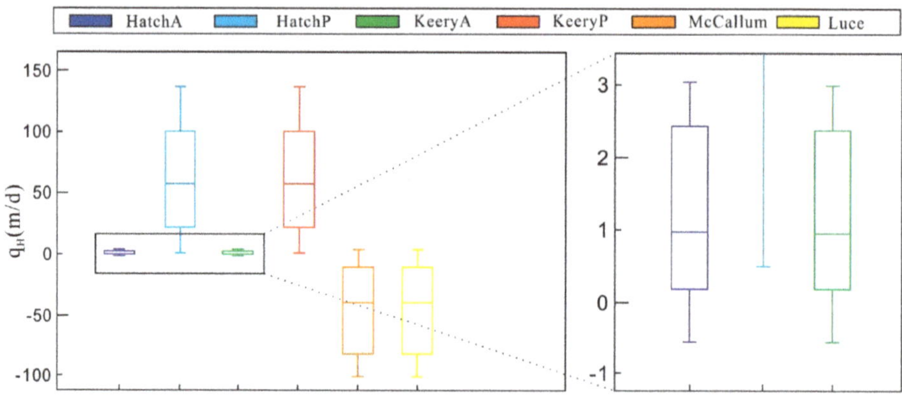

Figure 4. Box plots of hyporheic fluxes by six analytic methods.

3.3. Range and Spatial Distribution of Vertical Hyporheic Flux

Figure 5 shows the ranges of the VHF calculated by HatchA at depths from 0.03 m to 0.09 m, which were determined by the depths of the sensors and the rule of slide windows. The absolute value of the VHF at T1 was lowest at the three test points where the mean ranged from −0.178 to 0.247 m/day and the median ranged from −0.126 to 0.227 m/day. The VHF was close to 0, indicating that the hyporheic flow was relatively weak. It is noted that the positive and negative signs represent the direction of the VHF, i.e., a positive value indicates water flowing from the stream to the streambed.

The VHF at T2 ranged from 1.401 to 2.952 m/day, and decreased with depth except at 6 cm. This was an abnormal value of up to 6.790 m/day. T2 was located at a tributary where the particles of streambed sediment were coarser, including a few types of gravel. Thus, the hydraulic conductivity of the streambed at T2 was higher. Hyporheic flow was greater in the shallow layer and weakened in the deep layer. The direction of hyporheic flow at T3 included upwelling and downwelling flows, and this situation was similar as that of T1. However, the VHF in the shallow layer exhibited upwelling, which is different with T1. With an increase in depth, the VHF weakened gradually and transformed to a downwelling at the depth of 7 cm. The closer to the bedrock, the higher the VHF.

The VHF in a natural streambed is intimately linked to the sediments of the streambed and the surface water velocity. In general, at a site with deep water depth and high surface water velocity, the streambed has always been scoured and moved, and thus the VHF is suppressed significantly by the horizontal water flow. By contrast, in a low-velocity zone, the condition of surface water velocity is more beneficial to hyporheic flow in the vertical direction. In addition, as a special kind of porous flow, the lithology and construction of sediment obviously modifies the spatial distribution of hydraulic conductivity and then influences the hyporheic flux and residence time of the hyporheic flow.

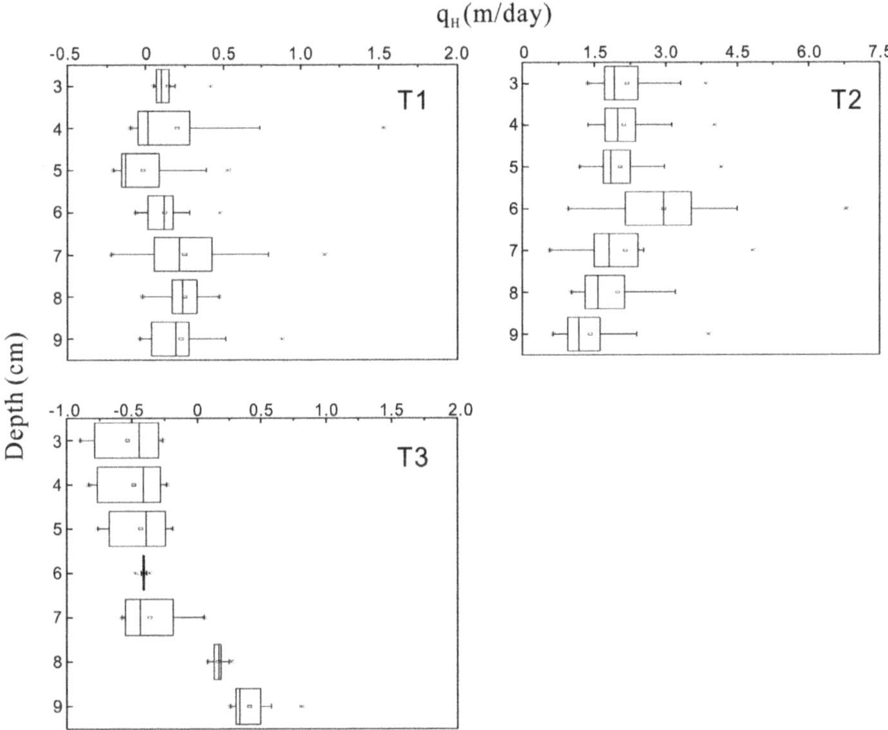

Figure 5. Statistics of vertical hyporheic flux (VHF) at different depths for T1, T2, and T3. Number of VHFs depended on frequency of temperature monitoring. Here, VHF was calculated at the interval of 30 min.

The field experiments were embedded in the natural environment in order to analyze the influence of the clay lens and surface water velocity on the VHF. Each experiment lasted more than 6 h to ensure the reliability of the raw temperature data and that the amplitude and phase signal could be extracted correctly. Figure 6 shows the spatial distribution of VHF at the artificial streambed without clay lens at

T1, T2, and T3, indicating changes in the VHF with time. The pattern of the surface water velocity at each point probably does not change a lot owing to the short test period for each experiment.

The VHF field varied slightly during the test period. The area of the high-velocity zone increased slightly and was mainly distributed around the deep zone (>25 cm). Focusing on the characteristics of the three test points, the VHF was lower in the shallow zone (<15 cm) and had small changes spatially. Several maximums ranging from 2.24 to 3.14 m/day appeared in the deep zone (>25 cm). In general, the VHF fields at T1 and T3 were similar probably because these two points were located in the main stream. Specifically, the VHF at T1 ranged from 0.73 to 3.14 m/day and from 0.97 to 2.57 m/day at T3. At T2, the maximum VHF occurred at around 35 cm in depth, which was deeper than those at T1 and T3. The topography of the tributary streambed had an obvious undulation, producing a multistage terrace, hydraulic drop, and hydraulic jump. This may be why the VHF field at T2 was different with those of the other two test points.

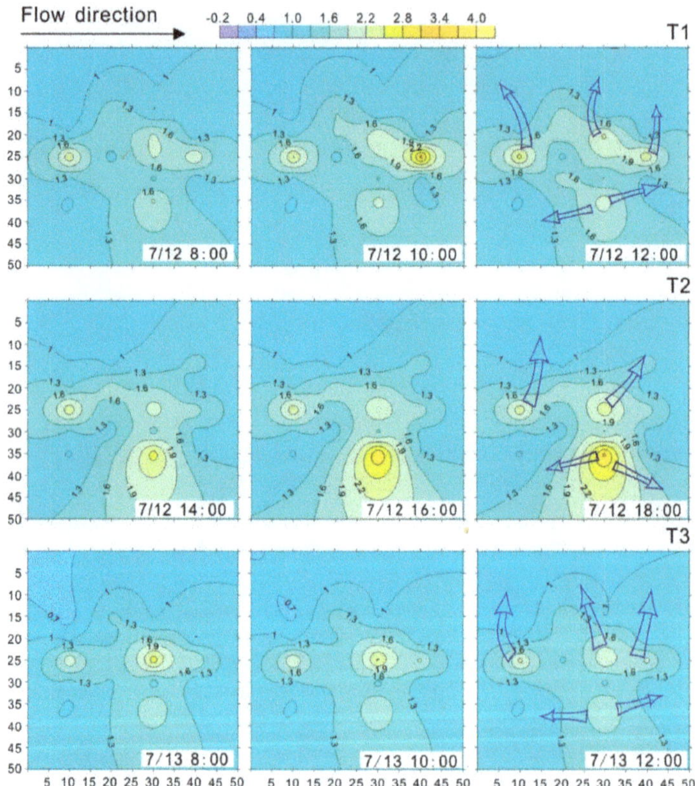

Figure 6. VHF field at T1, T2, and T3, calculated based on spatial temperature series. Contours were generated by the kriging method. The field experiments was carried out from 6:00 to 18:00 every day, and the interval between two VHF fields was 2 h. Surface water flowed from left to right.

4. Discussion

4.1. Influencing Factors on Hyporheic Flux

VHF fields influenced by clay lens depths and surface water velocity. Different from the blank control, the existence of the clay lens divided the artificial streambed into two parts, including the upper area and the downside area of the clay lens. Several pairs of sensors were separated at the two sides of the clay lens. As a kind of low-permeable sediment, the VHF located at the point covered with a clay lens was blocked by the clay and the VHF value should be extremely low, which was not consistent with the theoretical value calculated by these "separated" sensor pairs. Therefore, the VHF in the covered zone should be excluded in the spatial analyses.

Figure 7 shows the influences of the clay lens and surface water velocity on the VHF. Jin et al. [17] found that the clogged layer with fine particles could reduce the hyporheic flow. In most scenarios with a clay lens, the VHF was generally less than the value for the blank control. The existence of a clay lens suppressed the vertical hyporheic flow to some extent. It was clear that the clay lens affected the local VHF, such as at 20 cm in depth, where the contours were distorted and moved toward the lens. In the region of the clay lens, the contours became more concentrated, meaning that the VHF dropped rapidly. The VHF often appeared below the lens (30 cm in depth) and above the lens (40 cm in depth at T1 and T3). Gomez-Velez et al. [34] posited that low-permeability layers induce hydrodynamic sequestration owing to the relocation and emergence of new stagnation zones.

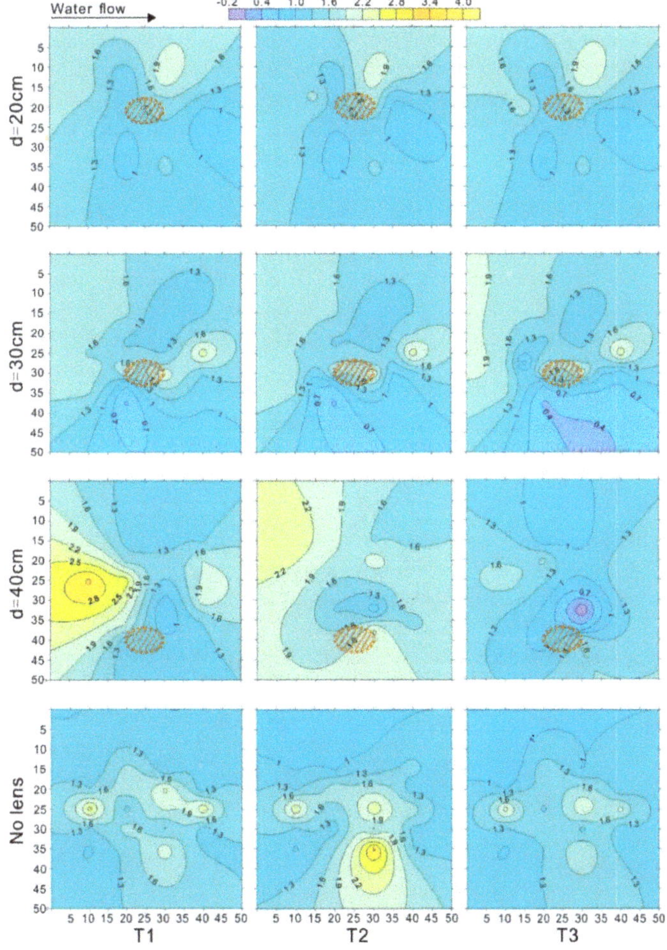

Figure 7. VHF fields of 12 scenarios. Brown ellipse represents clay lens, which is located at 20 cm, 30 cm, and 40 cm in depth. Columns from left to right are fields for T1, T2, and T3, respectively.

At T3, when the clay lens was located at 40 cm in depth, VHF contours were shown as concentric circles, which was similar as the stagnation zones mentioned in Gomez-Velez et al. [34]. Influenced by the surface water flow, the spatial distribution of the VHF produced an apparent distinction along the flow direction. When the clay lens was at 40 cm in depth at T1, a high-velocity zone of water flow was distributed around the area of 25 cm in depth near the upstream boundary. At T2 at 40 cm in depth, the hyporheic flow field did not form a closed high-velocity zone, but the VHF upstream was higher than that downstream.

Stonedahl et al. [67] found a strong correlation between the flow rate and the placement of high-permeability grid cells in regions of high hydraulic head gradients. In this study, the area near the upstream was prone to suffer scouring. At T1, the surface flow velocity and water depth were highest among the three points. Interestingly, the distinctions between upstream and downstream only existed when the lens was at 40 cm in depth. That is, when the lens was at 20 cm or 30 cm, this distinction could not be observed. Except for the distinction between upstream and downstream at the same points, the distinction between different points when the lens was at the same depth was also subtle.

The main differences are reflected in the depth of the clay lens, which might cause the changes in the VHF field. When the lens is in the shallow or middle layer (<30 cm), the hyporheic flow is dominated by the location of clay lens. In this situation of shallow clay lens, the influence of the surface water velocity is not significant. Thus, at the same depth, the VHF fields at each point are similar. However, while the clay lens was placed at the depth of 40 cm, the influence of the surface water velocity is obvious. It seems that the clay lens affects the hyporheic flow in local areas among the different surface water conditions.

4.2. Enhancement and Hindering Effects on Hyporheic Flow

For the effects of clay lens, hindering effect was more widely observed. Recently, enhancement effects were found through physical experiments and numerical simulations. Moreover, the hindering effects and enhancement effects could coexist in some cases. Su et al. [51] developed a two-dimensional dune-generated hyporheic flow model using the VS2DH model and indicated that a clay lens in streambed can hinder or enhance hyporheic flow, depending on its relative spatial location to dunes. Lu et al. [49] found the similar rule in physical model that when the clay lens was located on middle of sand dune, the effect transformed from hindering to enhancement.

In this study, hyporheic flow was embodied by the pattern of flow field, which has difficulties on distinguishing the effect of clay lens. Hence, each VHF field at different test points with a clay lens was subtracted by the blank control for corresponding sites. The new VHF fields reflected the influence of the clay lens and surface water velocity more clearly (Figure 8). Two effects of enhancement and hindering were distributed in different area and the domain of each area was no significant difference. Focus on the hindering effect, the existence of a clay lens suppressed the hyporheic flow in the surrounding area of the lens. The shape of the suppressed area also changed with the depth of clay lens. For the three test points, the VHF field at T2 was different from those of T1 and T3 to some extent, especially at a depth of 20 cm. This was indirectly affected by the surface water velocity. Our experiments graphically demonstrated that the enhancement and hindering effects of clay lens on hyporheic flow was also coexist in natural river, however it still remained challenges to quantify the range and transformation of these two effects.

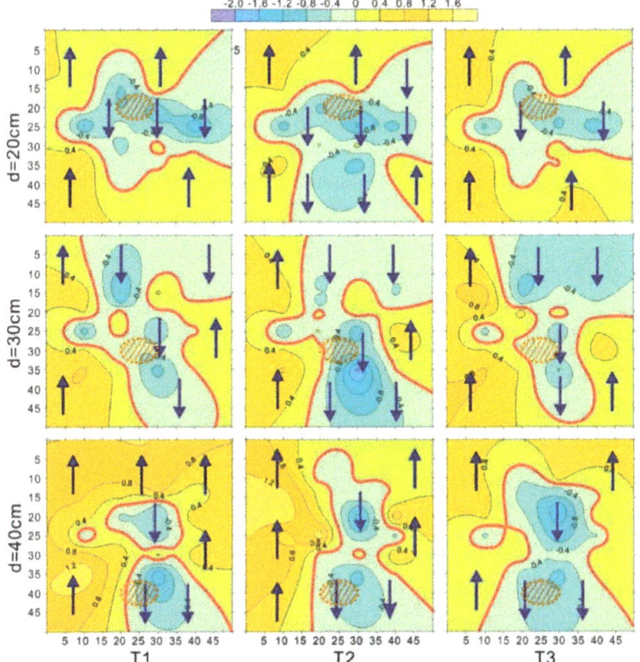

Figure 8. VHF field of scenarios deducting value of blank control. Directions of arrows represent signs of end results, i.e., up arrow means results are greater than 0 and the hyporheic flow was enhanced. On the other hand, down arrow means results are smaller than 0 and the hyporheic flow was hindered. Red lines are zero contours with no significant change.

5. Conclusions

This study used spatial temperature series data observed in a natural streambed and an artificial streambed model to study the behavior of the vertical hyporheic flux in a field test. The test in a natural streambed showed that temperature fluctuations in the streambed and air had significant differences. As the depth increased, the amplitude of the temperature curve decreased, and the temperature changes lagged in time. In general, the phase shift was less than 0, and the VHF could not be calculated by the methods of phase shift. VFLUX2 was used to calculate the VHF via six methods, which should be compared and chosen for application at specific sites. In this study, the method of amplitude ratio proposed by Hatch et al. [52] was satisfied either in robustness or in authenticity.

Field experiments in an artificial streambed showed that the vertical hyporheic flux remained stable in whole. The appearances of several extreme points resulted in significant changes to the VHF. In general, these abnormal areas of vertical hyporheic flux located in deep layers (between 25 cm and 40 cm) and the vertical flux was almost invariable.

Focusing on the heterogeneity of the sediment, the influence of a clay lens was explored by changing the depth of the clay lens. The surface water velocity was another factor that changed the characteristics of the hyporheic flow. Based on the field experiments, the relationship between the two factors was dependent with the depth of clay lens.

Comparing different depths, the influence in depth has a threshold, which was 30 to 40 cm in this study. At a value less than the threshold, the vertical hyporheic exchange is suppressed by the clay lens. When the depth exceeds the threshold, the hyporheic flux increased and appeared a high-velocity zone in the upstream bed, indicating that the hyporheic flow pattern is dominated by the surface water velocity, and the influence of the clay lens is weakened. The influence of the clay lens existed in the local area that it decreases the local VHF and produces a "stagnation zone". This kind of abnormal velocity zone, including high-velocity zones and stagnation zones, is probably produced by the clay lens owing to rearrangement of the hydraulic conductivity distribution.

Enhancement and hindering effects were verified and distributed in different regions through the field experiments. In natural streambed, surface water velocity and depth of clay lens were the two main factors on the transformation of these two effects. However, limited by the conditions, revealing the range and transformation of two effects was still a challenge, and also the future research direction.

Author Contributions: All authors contributed extensively to the work presented in this paper. C.Y., C.L., W.Q. and J.L. performed the experiments and data curation; C.L. provided funding acquisition; C.Y. and C.L. wrote the paper. C.Y. and C.L. finished the review and editing.

Funding: This work was supported by the National Key R&D Program of China (2018YFC0407701), Natural Science Foundation of Jiangsu (BK20181035), and Fundamental Research Funds for the Central Universities (2019B10514). This study does not necessarily reflect the views of the funding agencies.

Acknowledgments: Many thanks to Lei Wei, Lei Qiu, Jiajie Huang and Dan Lou for their assistant in field work. We would like to thank Editage (www.editage.cn) for English language editing. This study does not necessarily reflect the views of the funding agencies.

Conflicts of Interest: The authors declare no conflict of interest.

References

1. Boano, F.; Revelli, R.; Ridolfi, L. Reduction of the hyporheic zone volume due to the stream-aquifer interaction. *Geophys. Res. Lett.* **2008**, *35*, L09401. [CrossRef]

2. Song, J.; Cheng, D.; Zhang, J.; Zhang, Y.; Long, Y.; Zhang, Y.; Shen, W. Estimating spatial pattern of hyporheic water exchange in slack water pool. *J. Geogr. Sci.* **2019**, *29*, 377–388. [CrossRef]

3. Boano, F.; Revelli, R.; Ridolfi, L. Bedform-induced hyporheic exchange with unsteady flows. *Adv. Water Resour.* **2007**, *30*, 148–156. [CrossRef]

4. Cardenas, M.B.; Wilson, J.L. Dunes, turbulent eddies, and interfacial exchange with permeable sediments. *Water Resour. Res.* **2007**, *43*, 199–212. [CrossRef]

5. Cardenas, M.B.; Wilson, J.L. Exchange across a sediment–water interface with ambient groundwater discharge. *J. Hydrol.* **2007**, *346*, 69–80. [CrossRef]

6. Marzadri, A.; Tonina, D.; Bellin, A. A semianalytical three-dimensional process-based model for hyporheic nitrogen dynamics in gravel bed rivers. *Water Resour. Res.* **2011**, *47*, W11518. [CrossRef]

7. Fernald, A.G.; Wigington, P.J., Jr.; Landers, D.H. Water quality changes in hyporheic flow paths between a large gravel bed river and off-channel alcoves in Oregon, USA. *River Res. Appl.* **2006**, *22*, 1111–1124. [CrossRef]

8. Kibichii, S.; Feeley, H.B.; Baars, J.R.; Kelly-Quinn, M. The influence of water quality on hyporheic invertebrate communities in agricultural catchments. *Mar. Freshw. Res.* **2015**, *66*, 805–814. [CrossRef]

9. Nyberg, L.; Calles, O.; Greenberg, L. Impact of short-term regulation on hyporheic water quality in a boreal river. *River Res. Appl.* **2010**, *24*, 407–419. [CrossRef]

10. Cardenas, M.B.; Ford, A.E.; Kaufman, M.H.; Kessler, A.J.; Cook, P.L.M. Hyporheic flow and dissolved oxygen distribution in fish nests: The effects of open channel velocity, permeability patterns, and groundwater upwelling. *J. Geophys. Res. Biogeosci.* **2016**, *121*, 3113–3130. [CrossRef]

11. Kaufman, M.H.; Cardenas, M.B.; Buttles, J.; Kessler, A.J.; Cook, P.L.M. Hyporheic hot moments: Dissolved oxygen dynamics in the hyporheic zone in response to surface flow perturbations. *Water Resour. Res.* **2017**, *53*, 6642–6662. [CrossRef]

12. Storey, R.G.; Williams, D.D.; Fulthorpe, R.R. Nitrogen processing in the hyporheic zone of a pastoral stream. *Biogeochemistry* **2004**, *69*, 285–313. [CrossRef]

13. Strayer, D.; May, S.; Nielsen, P.; Wollheim, W.; Hausam, S. Oxygen, organic matter, and sediment granulometry as controls on hyporheic animal communities. *Arch. Hydrobiol.* **1997**, *140*, 131–144. [CrossRef]

14. Drummond, J.D.; Larsen, L.G.; Gonzalez-Pinzon, R.; Packman, A.I.; Harvey, J.W. Less fine particle retention in a restored versus unrestored urban stream: Balance between hyporheic exchange, resuspension, and immobilization. *J. Geophys. Res. Biogeosci.* **2018**, *123*, 1425–1439. [CrossRef]

15. Harvey, J.W.; Drummond, J.D.; Martin, R.L.; Mcphillips, L.E.; Packman, A.I.; Jerolmack, D.J.; Stonedahl, S.H.; Aubeneau, A.F.; Sawyer, A.H.; Larsen, L.G. Hydrogeomorphology of the hyporheic zone: Stream solute and fine particle interactions with a dynamic streambed. *J. Geophys. Res. Biogeosci.* **2012**, *117*, 2740–2742. [CrossRef]

16. Jin, G.; Jiang, Q.; Tang, H.; Li, L.; Barry, D.A. Density effects on nanoparticle transport in the hyporheic zone. *Adv. Water Resour.* **2018**, *121*, 406–418. [CrossRef]

17. Jin, G.; Zhang, Z.; Tang, H.; Yang, X.; Li, L.; Barry, D.A. Colloid transport and distribution in the hyporheic zone. *Hydrol. Process.* **2019**, *33*, 932–944. [CrossRef]

18. Battin, T.J. Hydrodynamics is a major determinant of streambed biofilm activity: From the sediment to the reach scale. *Limnol. Oceanogr.* **2000**, *45*, 1308–1319. [CrossRef]

19. Gooseff, M.N.; Anderson, J.K.; Wondzell, S.M.; Lanier, J.; Haggerty, R. A modeling study of hyporheic exchange pattern and the sequence, size and spacing of stream bedforms in mountain stream networks, Oregon, USA. *Hydrol. Process.* **2006**, *20*, 2443–2457. [CrossRef]

20. Jonsson, K.; Johansson, H.; Wörman, A. Hyporheic exchange of reactive and conservative solutes in streams: Tracer methodology and model interpretation. *J. Hydrol.* **2003**, *278*, 153–171. [CrossRef]

21. Elliott, A.H.; Brooks, N.H. Transfer of nonsorbing solutes to a streambed with bed forms: Laboratory experiment. *Water Resour. Res.* **1997**, *33*, 137–151. [CrossRef]

22. Elliott, A.H.; Brooks, N.H. Transfer of nonsorbing solutes to a streambed with bed forms: Theory. *Water Resour. Res.* **1997**, *33*, 123–136. [CrossRef]

23. Harvey, J.W.; Bencala, K.E. The effect of streambed topography on surface-subsurface water exchange in mountain catchments. *Water Resour. Res.* **1993**, *29*, 89–98. [CrossRef]

24. Marion, A.; Packman, A.I.; Zaramella, M.; Bottacin-Busolin, A. Hyporheic flows in stratified beds. *Water Resour. Res.* **2008**, *44*, 542–547. [CrossRef]

25. Packman, A.I.; Marion, A.; Zaramella, M.; Cheng, C.; Gaillard, J.F.; Keane, D.T. Development of layered sediment structure and its effects on pore water transport and hyporheic exchange. *Water Air Soil Pollut. Focus* **2006**, *6*, 433–442. [CrossRef]

26. Salehin, M.; Packman, A.I.; Paradis, M. Hyporheic exchange with heterogeneous streambeds: Laboratory experiments and modeling. *Water Resour. Res.* **2004**, *40*, 309–316. [CrossRef]

27. Fox, A.; Laube, G.; Schmidt, C.; Fleckenstein, J.H.; Arnon, S. The effect of losing and gaining flow conditions on hyporheic exchange in heterogeneous streambeds. *Water Resour. Res.* **2016**, *52*, 7460–7477. [CrossRef]

28. Pryshlak, T.T.; Sawyer, A.H.; Stonedahl, S.H.; Soltanian, M.R. Multiscale hyporheic exchange through strongly heterogeneous sediments. *Water Resour. Res.* **2016**, *51*, 9127–9140. [CrossRef]

29. Genereux, D.P.; Leahy, S.; Mitasova, H.; Kennedy, C.D.; Corbett, D.R. Spatial and temporal variability of streambed hydraulic conductivity in West Bear Creek, North Carolina, USA. *J. Hydrol.* **2008**, *358*, 332–353. [CrossRef]

30. Koltermann, C.E.; Gorelick, S.M. Fractional packing model for hydraulic conductivity derived from sediment mixtures. *Water Resour. Res.* **1995**, *31*, 3283–3297. [CrossRef]

31. Leek, R.; Wu, J.; Wang, L.; Hanrahan, T.P.; Barber, M.E.; Qiu, H.X. Heterogeneous characteristics of streambed saturated hydraulic conductivity of the Touchet River, south eastern Washington, USA. *Hydrol. Process.* **2010**, *23*, 1236–1246. [CrossRef]

32. Sawyer, A.H.; Cardenas, M.B. Hyporheic flow and residence time distributions in heterogeneous cross-bedded sediment. *Water Resour. Res.* **2009**, *45*, 2263–2289. [CrossRef]

33. Arnon, S.; Krause, S.; Gomez-Velez, J.D.; De Falco, N. Effects of low-permeability layers in the hyporheic zone on oxygen consumption under losing and gaining groundwater flow conditions. In Proceedings of the 2017 AGU Fall Meeting Abstracts, New Orleans, LA, USA, 11–25 December 2017.

34. Gomez-Velez, J.D.; Krause, S.; Wilson, J.L. Effect of low-permeability layers on spatial patterns of hyporheic exchange and groundwater upwelling. *Water Resour. Res.* **2014**, *50*, 5196–5215. [CrossRef]

35. Huo, S.; Jin, M.; Liang, X. Impacts of low-permeability clay lens in vadose zone onto rainfall infiltration and groundwater recharge using numerical simulation of variably saturated flow. *J. Jilin Univ.* **2013**, *43*, 1579–1587. (In Chinese)

36. Bhaskar, A.S.; Harvey, J.W.; Henry, E.J. Resolving hyporheic and groundwater components of streambed water flux using heat as a tracer. *Water Resour. Res.* **2012**, *48*, 1076. [CrossRef]

37. Kim, H.; Lee, K.-K.; Lee, J.-Y. Numerical verification of hyporheic zone depth estimation using streambed temperature. *J. Hydrol.* **2014**, *511*, 861–869. [CrossRef]

38. Naranjo, R.C.; Pohll, G.; Niswonger, R.G.; Stone, M.; Mckay, A. Using heat as a tracer to estimate spatially distributed mean residence times in the hyporheic zone of a riffle-pool sequence. *Water Resour. Res.* **2013**, *49*, 3697–3711. [CrossRef]

39. Zlotnik, V.; Tartakovsky, D.M. Interpretation of heat-pulse tracer tests for characterization of three-dimensional velocity fields in hyporheic zone. *Water Resour. Res.* **2018**, *54*, 4028–4039. [CrossRef]

40. Bakker, M.; Calije, R.; Schaars, F.; van der Made, K.; de Haas, S. An active heat tracer experiment to determine groundwater velocities using fiber optic cables installed with direct push equipment. *Water Resour. Res.* **2015**, *51*, 2760–2772. [CrossRef]

41. Briggs, M.A.; Lane, J.W.; Snyder, C.D.; White, E.A.; Johnson, Z.C.; Nelms, D.L.; Hitt, N.P. Shallow bedrock limits groundwater seepage-based headwater climate refugia. *Limnologica* **2018**, *68*, 142–156. [CrossRef]

42. Hare, D.K.; Boutt, D.F.; Clement, W.P.; Hatch, C.E.; Davenport, G.; Hackman, A. Hydrogeological controls on spatial patterns of groundwater discharge in peatlands. *Hydrol. Earth Syst. Sci.* **2017**, *21*, 1–39. [CrossRef]

43. Rau, G.C.; Andersen, M.S.; McCallum, A.M.; Acworth, R.I. Analytical methods that use natural heat as a tracer to quantify surface water–groundwater exchange, evaluated using field temperature records. *Hydrogeol. J.* **2010**, *18*, 1093–1110. [CrossRef]

44. Reeves, J.; Hatch, C.E. Impacts of three-dimensional nonuniform flow on quantification of groundwater-surface water interactions using heat as a tracer. *Water Resour. Res.* **2016**, *52*, 6851–6866. [CrossRef]

45. Briggs, M.A.; Johnson, Z.C.; Snyder, C.D.; Hitt, N.P.; Kurylyk, B.L.; Lautz, L.; Irvine, D.J.; Hurley, S.T. and Lane, J.W. Inferring watershed hydraulics; cold-water habitat persistence using multi-year air and stream temperature signals. *Sci. Total Environ.* **2018**, *636*, 1117–1127. [CrossRef]

46. Burkholder, B.K.; Grant, G.E.; Haggerty, R.; Khangaonkar, T.; Wampler, P.J. Influence of hyporheic flow and geomorphology on temperature of a large, gravel-bed river, Clackamas River, Oregon, USA. *Hydrol. Process.* **2008**, *22*, 941–953. [CrossRef]

47. Selker, J.S.; Thévenaz, L.; Huwald, H.; Mallet, A.; Luxemburg, W.; van de Giesen, N.; Stejskal, M.; Zeman, J.; Westhoff, M.; Parlange, M.B. Distributed fiber-optic temperature sensing for hydrologic systems. *Water Resour. Res.* **2006**, *42*, W12202. [CrossRef]

48. Lu, C.; Zhuang, W.; Wang, S.; Zhu, X.; Li, H. Experimental study on hyporheic flow varied by the clay lens and stream flow. *Environ. Earth Sci.* **2018**, *77*, 482. [CrossRef]

49. Lu, C.; Yao, C.; Su, X.; Jiang, Y.; Yuan, F.; Wang, M. The Influences of a Clay Lens on the Hyporheic Exchange in a Sand Dune. *Water* **2018**, *10*, 826. [CrossRef]

50. Fox, A.; Boano, F.; Arnon, S. Impact of losing and gaining streamflow conditions on hyporheic exchange fluxes induced by dune-shaped bed forms. *Water Resour. Res.* **2014**, *50*, 1895–1907. [CrossRef]

51. Su, X.; Shu, L.; Lu, C. Impact of a low-permeability lens on dune-induced hyporheic exchange. *Hydrol. Sci. J.* **2018**, *63*, 818–835. [CrossRef]

52. Hatch, C.E.; Fisher, A.T.; Revenaugh, J.S.; Constantz, J.; Ruehl, C. Quantifying surface water-groundwater interactions using time series analysis of streambed thermal records: Method development. *Water Resour. Res.* **2006**, *42*, W10410. [CrossRef]

53. Luce, C.H.; Tonina, D.; Gariglio, F.; Applebee, R. Solutions for the diurnally forced advection-diffusion equation to estimate bulk fluid velocity and diffusivity in streambeds from temperature time series. *Water Resour. Res.* **2013**, *49*, 488–506. [CrossRef]

54. Keery, J.; Binley, A.; Crook, N.; Smith, J.W.N. Temporal and spatial variability of groundwater–surface water fluxes: Development and application of an analytical method using temperature time series. *J. Hydrol.* **2007**, *336*, 1–16. [CrossRef]

55. McCallum, A.M.; Andersen, M.S.; Rau, G.C.; Acworth, R.I. A 1-D analytical method for estimating surface water-groundwater interactions and effective thermal diffusivity using temperature time series. *Water Resour. Res.* **2012**, *48*, 76–78. [CrossRef]

56. Lapham, W.W. Use of Temperature Profiles Beneath Streams to Determine Rates of Vertical Ground-Water Flow and Vertical Hydraulic Conductivity. In *U.S. Geological Survey Water-Supply Paper (USA)*; US Geological Survey: Reston, VA, USA, 1989.

57. Stallman, R.W. Steady one-dimensional fluid flow in a semi-infinite porous medium with sinusoidal surface temperature. *J. Geophys. Res.* **1965**, *70*, 2821–2827. [CrossRef]
58. Suzuki, S. Percolation measurements based on heat flow through soil with special reference to paddy fields. *J. Geophys. Res.* **1960**, *65*, 2883–2885. [CrossRef]
59. Goto, S.; Yamano, M.; Kinoshita, M. Thermal response of sediment with vertical fluid flow to periodic temperature variation at the surface. *J. Geophys. Res. Solid Earth* **2005**, *110*, B01106. [CrossRef]
60. Gordon, R.P.; Lautz, L.K.; Briggs, M.A.; McKenzie, J.M. Automated calculation of vertical pore-water flux from field temperature time series using the VFLUX method and computer program. *J. Hydrol.* **2012**, *420*, 142–158. [CrossRef]
61. Irvine, D.J.; Lautz, L.K.; Briggs, M.A.; Gordon, R.P.; McKenzie, J.M. Experimental evaluation of the applicability of phase, amplitude, and combined methods to determine water flux and thermal diffusivity from temperature time series using VFLUX 2. *J. Hydrol.* **2015**, *531*, 728–737. [CrossRef]
62. Young, P.C.; Pedregal, D.J.; Tych, W. Dynamic harmonic regression. *Forecast* **1999**, *18*, 369–394. [CrossRef]
63. Constantz, J. Heat as a tracer to determine streambed water exchanges. *Water Resour. Res.* **2008**, *44*, W00D10. [CrossRef]
64. Lu, C.; Chen, S.; Zhang, Y.; Su, X.; Chen, G. Heat tracing to determine spatial patterns of hyporheic exchange across a river transect. *Hydrogeol. J.* **2017**, *25*, 1633–1646. [CrossRef]
65. Lautz, L.K. Impacts of nonideal field conditions on vertical water velocity estimates from streambed temperature time series. *Water Resour. Res.* **2010**, *46*, W01509. [CrossRef]
66. Schmidt, C.; Bayer-Raich, M.; Schirmer, M. Characterization of spatial heterogeneity of groundwater-stream water interactions using multiple depth streambed temperature measurements at the reach scale. *Hydrol. Earth Syst. Sci.* **2006**, *10*, 849–859. [CrossRef]
67. Stonedahl, S.H.; Sawyer, A.H.; Stonedahl, F.; Reiter, C.; Gibson, C. Effect of heterogeneous sediment distributions on hyporheic flow in physical and numerical models. *Groundwater* **2018**, *56*, 934–946. [CrossRef]

Article

Integral Flow Modelling Approach for Surface Water-Groundwater Interactions along a Rippled Streambed

Tabea Broecker [1,*], Katharina Teuber [1], Vahid Sobhi Gollo [1], Gunnar Nützmann [2,3], Jörg Lewandowski [2,3] and Reinhard Hinkelmann [1]

[1] Chair of Water Resources Management and Modeling of Hydrosystems, Technische Universität Berlin, 13355 Berlin, Germany
[2] Ecohydrology Department, Leibniz-Institute of Freshwater Ecology and Inland Fisheries, 12587 Berlin, Germany
[3] Geography Department, Humboldt-University Berlin, 10099 Berlin, Germany
* Correspondence: tabea.broecker@uwi.tu-berlin.de; Tel.: +30-314-72-444

Received: 18 June 2019; Accepted: 16 July 2019; Published: 22 July 2019

Abstract: Exchange processes of surface and groundwater are important for the management of water quantity and quality as well as for the ecological functioning. In contrast to most numerical simulations using coupled models to investigate these processes, we present a novel integral formulation for the sediment-water-interface. The computational fluid dynamics (CFD) model OpenFOAM was used to solve an extended version of the three-dimensional Navier–Stokes equations which is also applicable in non-Darcy-flow layers. Simulations were conducted to determine the influence of ripple morphologies and surface hydraulics on the flow processes within the hyporheic zone for a sandy and for a gravel sediment. In- and outflowing exchange fluxes along a ripple were determined for each case. The results indicate that larger grain size diameters, as well as ripple distances, increased hyporheic exchange fluxes significantly. For higher ripple dimensions, no clear relationship to hyporheic exchange was found. Larger ripple lengths decreased the hyporheic exchange fluxes due to less turbulence between the ripples. For all cases with sand, non-Darcy-flow was observed at an upper layer of the ripple, whereas for gravel non-Darcy-flow was recognized nearly down to the bottom boundary. Moreover, the sediment grain sizes influenced also the surface water flow significantly.

Keywords: groundwater-surface water interactions; integral model; computational fluid dynamics; hyporheic zone; OpenFOAM; ripples

---

1. Introduction

Hyporheic exchange—the exchange of stream and shallow subsurface water—is controlled by pressure gradients along the streambed surface and subsurface groundwater gradients. Over multiple scales, the bedform induced hyporheic exchange was identified as a crucial process for the biogeochemistry and ecology of rivers [1–10]. On large and intermediate scales, stream stage differences, meander loops or bars can generate hyporheic exchange. Accordingly, it is possible to control surface water-groundwater exchange by river stage manipulation e.g., to manage the inflow of saline groundwater into a river [11]. A decrease of the groundwater level, in turn, impacts surface water infiltration up to a maximum where groundwater and surface water are disconnected. This condition is achieved when the clogging layer does not cross the top of the capillary zone above the water table [12]. On small scales, river sediments usually form topographic features such as dunes or ripples. The flowing fluid encounters an uneven surface on the permeable streambed, which results in an irregular pattern in the pressure along that surface and induces hyporheic exchange [11–13].

Within theoretical, experimental, and computational studies the general mechanics of the bedform induced hyporheic exchange were examined over the past decades. By manipulating streambed morphology, stream discharge, and groundwater flow, experiments have been used to study driving forces for the hyporheic exchange intensively [14–17]. At submerged structures such as pool-riffle sequences or ripples, turbulences, eddies or hydraulic jumps may occur. Packman et al. [15], Tonina and Buffington [18], Voermans et al. [19] and other studies showed, that turbulence influences hyporheic exchange and should not be ignored. Facing these complex three-dimensional flow dynamics at the sediment-water interface, it can be challenging to establish suitable flume experiments or field studies. Computational fluid dynamics has proven to be a viable alternative. The majority of these studies have focused on surface-subsurface coupled models. Reasons for the application of different models for the surface and the subsurface are for example the strong temporal variability in streams including relatively high velocities, whereas the velocities and temporal variabilities in the groundwater are usually several orders of magnitude smaller, leading to different applied equations for the stream and the subsurface. Often, the two computational domains are linked by pressure. Pressure distributions from a surface water model are consequently used for a coupled groundwater model [20–26]. However, also fully coupled models such as the Integrated Hydrology Model [27] or HydroGeoSphere have already been successfully applied [28–30]. Within these models, open channel flow is described by the two-dimensional diffusion-wave approximation of the St. Venant equations, whereas the three-dimensional Richards equation is used for the subsurface. Water and solute exchange flux terms enable to simultaneously solve one system of equations for both flow regimes.

For many coupled surface-subsurface models, the Darcy law is applied within the sediment. However, especially for coarse bed rivers, this law may cause errors in the presence of non-Darcy hyporheic flow [15]. Following Bear [31], the linear assumption of the Darcy law is only valid if the Reynolds number does not exceed a value between 1 and 10. Applying Darcy's law in non-Darcy-flow areas leads to an overestimating of groundwater flow rates [32]. Packman et al. [15] investigated hyporheic exchange through gravel beds with dune-like morphologies and applied the modified Elliot and Brooks model [33]. They realized that the model did not perform well—among other reasons—due to non-Darcy flow in the near-surface sediment which was not considered in the model. One possible solution to model groundwater in non-Darcy-flow areas is e.g., to use the Darcy-Brinkmann equation instead of the Darcy law. However, there is an additional parameter—the effective viscosity—which has to be determined.

In the present study, an extended version of the three-dimensional Navier–Stokes equations after Oxtoby et al. [34] is used for the whole system comprising the stream as well as the subsurface. For the application in the groundwater, sediment porosity, as well as an additional drag term, are included into the Navier–Stokes equations. The model is consequently also applicable for high Reynolds numbers within the subsurface where the Darcy law cannot be applied. To our knowledge, this solver was never used for the hyporheic zone before. We apply the new integral solver to evaluate the effect of ripple geometries and surface hydraulics on hyporheic exchange processes, based on the study by Broecker et al. [35] who investigated free surface flow and tracer retention over streambeds and ripples without considering the subsurface. In Broecker et al. [35] the three-dimensional Navier–Stokes equations were solved in combination with an implemented transport equation. In that study, ripple sizes, spacing as well as flow velocities affected pressure gradients and tracer retention considerably. Seven simulation cases were examined varying ripple height, length, distance, and flow rate. The investigated ripple geometries and flow rates are mainly transferred to the present study. Only case 6 is not used for the present study, as the irregular distance between the ripples gave no significant new findings compared to equal distances [35]. In contrast to Broecker et al. [35], the present study examines both free surface flow and subsurface flow. The aim of the present study is to evaluate the impact of ripple dimensions, lengths, spacing and surface velocity on flow dynamics within the hyporheic zone using a new integral model.

2. Materials and Methods

2.1. Geometry and Mesh

The analyzed geometry consists of a prismatic domain with a length of 15 m, a width of 1 m and a height of 1.5 m at the inlet and 1 m at the outlet. A weir structure is included in front of the outlet to fix the water level. A rippled area of approximately 3 m is introduced 6 m downstream of the inlet. The model geometry of the reference case with the corresponding initial water depth can be seen in Figure 1.

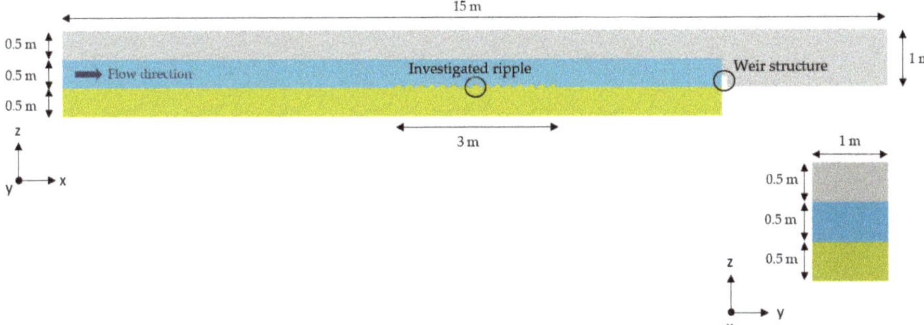

Figure 1. Model geometry and initial condition for the water level (sediment: yellow, water: blue, air: gray); top: front view, bottom right: cross-section.

The mesh has been discretized using the three-dimensional finite element mesh generator gmsh. The ripple parameters are based on the approach of Broecker et al. [35]. Table 1 summarizes the most important ripple parameter values. Unstructured elements were chosen in the x-z-plane to depict the curved profile of the ripples. Thereafter the elements have been extruded with 10 layers in the y-direction to produce a three-dimensional mesh. Different meshes with similar mesh conditions have been created for the varying ripple geometries. Smaller element sizes were chosen for surface water and the subsurface, as detailed processes in the air phase are not of interest for the present study. The air phase was only included to account for water level fluctuations which showed to have a significant effect on the pressure distribution at the streambed [31]. The mesh resolution has been examined by calculating the fraction of the turbulent kinetic energy in the resolved motions after Pope [36], who suggest to resolve 80% of the turbulent kinetic energy for a well resolved large eddy simulation (LES). In our simulation we are around this range with 75–83% resolution for the water phase with the applied meshes using the LES turbulence model (see Section 2.3). Lower resolutions were observed for the air-phase, which is not of interest for our simulations.

Table 1. Simulation cases including ripple geometries and flow rates.

Case	1 (Reference Case)	2	3	4	5	6
ripple height (cm)	5.6	1.4	11.2	5.6	5.6	5.6
ripple length (cm)	20	5	40	40	20	20
ripple distance (cm)	0	0	0	0	20	0
flow rate (m^3/s)	0.5	0.5	0.5	0.5	0.5	0.25

2.2. Numerical Model

To simulate exchange processes of surface water and groundwater, the open source software Open Source Field Operation and Manipulation (OpenFOAM) version 2.4.0 has been used. A solver called "porousInter" has been applied. This solver was developed by Oxtoby et al. [34] and is based on the interFoam solver by OpenFOAM. PorousInter is a multiphase solver for immiscible fluids and extends the three-dimensional Navier–Stokes equations by the consideration of soil porosity and effective grain size diameter. For our simulations two phases—water and air—are considered to allow water level fluctuations. Since the porousInter–solver does not account for the solid fraction of the soil, values that are represented by $[\ ]^f$ are averaged only over the pore space volume. The conservation of mass and momentum are defined after Oxtoby et al. [34] as:

Mass conservation equation

$$\varphi \nabla \cdot [\vec{U}]^f = 0 \tag{1}$$

Momentum conservation equation

$$\varphi \left(\frac{\partial [\rho]^f [\vec{U}]^f}{\partial t} + \cdot [\vec{U}]^f \nabla([\rho]^f [\vec{U}]^f) \right) = -\varphi \nabla [p]^f + \varphi [\mu]^f \nabla^2 [\vec{U}]^f + \varphi [\rho]^f \vec{g} + D \tag{2}$$

where φ is the soil porosity (-); $\vec{U}$ is the velocity (m/s); ρ is the density (kg/m^3); t is time (s); p is pressure (Pa); μ is the dynamic viscosity (Ns/m^2), g is the gravitational acceleration (m/s^2) and D an additional drag term (kg/(m^{2}s^2)). The drag term was developed by Ergun [37] and accounts for momentum loss by means of fluid friction with the porous medium and flow recirculation within the sediment. To consider flow recirculation, an effective added mass coefficient is included after van Gent [38]. The porous drag term is defined as:

$$D = -\left(150 \frac{1-\varphi}{d_p \varphi} [\mu]^f + 1.75 [\rho]^f [\vec{U}]^f \right) \frac{1-\varphi}{d_p} [\vec{U}]^f - 0.34 \frac{1-\varphi}{\varphi} \frac{[\rho]^f \partial [\vec{U}]^f}{\partial t} \tag{3}$$

with d_p (m) as effective grain size diameter.

PorousInter uses the volume of fluid (VOF) approach. Consequently, multiple phases are treated as one fluid with changing properties [39]. The indicator fraction α (-) varies between zero for the air phase and one for the water phase. The water-air interface is captured by a convective transport equation:

$$\varphi \frac{\partial [\alpha]^f}{\partial t} + \varphi \nabla \cdot ([\alpha]^f [\vec{U}]^f) = 0 \tag{4}$$

The dynamic viscosity and the density of each fluid are calculated according to their fraction as:

$$\mu = \alpha \mu_w + \mu_a (1 - \alpha) \tag{5}$$

$$\rho = \alpha \rho_w + \rho_a (1 - \alpha) \tag{6}$$

The subscripts w and a denote the fluids water and air.

2.3. Turbulence

Turbulent properties have been captured by a large eddy simulation (LES) turbulence model (see also Section 3.1). Eddies up to a certain size were consequently directly resolved, whereas for small eddies a subgrid model is used. For the present study, the Smagorinski subgrid scale model [40] has been applied.

A measure $M(\vec{x},t)$ for the turbulence resolution was calculated after Pope [36]:

$$M(\vec{x},t) = \frac{k_r(\vec{x},t)}{K(\vec{x},t) + k_r(\vec{x},t)} \tag{7}$$

where $K(\vec{x},t)$ defines the turbulent kinetic energy of the resolved motions by:

$$K(\vec{x},t) = \frac{1}{2}(\vec{U} - \vec{U}_{mean})(\vec{U} - \vec{U}_{mean}) \tag{8}$$

and $k_r(\vec{x}, t)$ defines the turbulent kinetic energy of the residual motions. The solver by Oxtoby et al. [34] had to be adjusted to write $k_r(\vec{x}, t)$ automatically. $K(\vec{x}, t)$ and $k_r(\vec{x}, t)$ were calculated and averaged for the whole running time. A measure $M(\vec{x}, t) = 0.25$ corresponds to a resolution of 75% of the turbulent kinetic energy.

2.4. Boundary and Initial Conditions

Figure 2 shows the most important boundary conditions. The inlet of the boundary is divided into two fractions: for the air and for the water phase. The parameter α is fixed accordingly at the inlet. For the water phase the discharge is set to 0.5 m^3/s for case 1–5 and to 0.25 m^3/s for case 6, whereas for the air phase a total pressure condition is defined with a total pressure of 0 Pa. The total pressure condition specifies the given total pressure for outflow and the dynamic pressure subtracted from the total pressure for inflow. Next to the air phase at the inlet, the total pressure definition is applied for the upper boundary and at the outlet. The streambed is surrounded by walls. Consequently, the velocity is set to 0 m/s with a no flow condition.

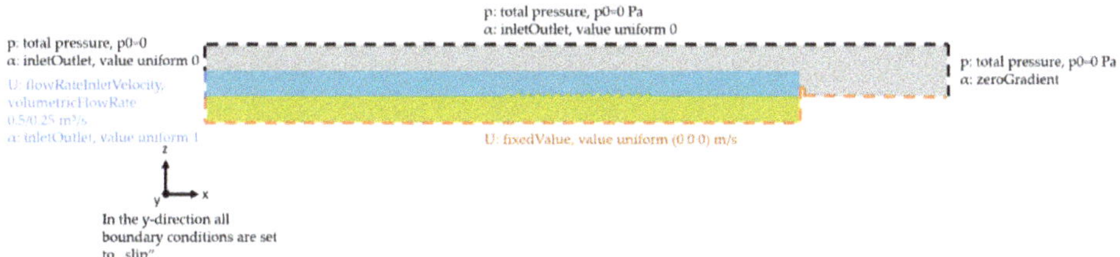

Figure 2. Boundary conditions.

In OpenFOAM a definition of a constant water level at the outlet is challenging [41]. Therefore, a weir structure is established as a barrier to keep a constant water level for our model. The water flows freely over the weir top. Behind the weir, the water level decreases before it flows out of the model. This method is described e.g., in Bayon-Barrachina and Lopez-Jimenez [42]. For case 1–5 and case 6 different heights for the weir structure were chosen, since the water level is affected by the flow rate of the surface water. For the weir structure and at the whole bottom of the model, an impervious no slip condition is used. All boundary conditions in the third dimension contain slip conditions.

For the sediment, two materials are chosen: coarse sand with a grain size diameter of 2 mm and medium gravel with a grain size diameter of 1.5 cm. Both sediments have an effective porosity of 0.25 and are considered to be homogeneous. According to Freeze and Cherry [32] coarse sand has a hydraulic conductivity of about 10^{-2} m/s and medium gravel of about 10^{-1} m/s. These values are only estimations and are not considered in our simulations. An initial water level of 1 m is set from the inlet up to the weir structure (see Figure 1).

2.5. Validation

To ensure reliable behavior of the integral model concerning the hydraulics for the interaction of groundwater and surface water, the solver was tested based on two applications. The seepages through dams with different water levels and dam geometries were compared with numerical and analytical solutions.

First, flow through a rectangular dam with a constant water level at both sides was investigated. The dam width amounts to 16 m and the dam height to 24 m. The dam height is equal to the water level at the left side of the dam. A median grain size diameter of 2 mm and a porosity of 0.25 were defined which correspond to a sandy dam filling. At the right hand, the water level is fixed to 4 m. The seepage through the dam was compared with two numerical solutions after Westbrook [43] and Aitchison and Coulson [44] and with a one-dimensional [45] as well as with a two-dimensional analytical solution after Di Nucci [46] (see Figure 3). The seepage calculated with the integral solver was in between the two-dimensional analytical solution after Di Nucci and the numerical solutions after Westbrook and Aitchison and Coulson. A large deviation was recognized for the one-dimensional solution. Based on the two-dimensional velocities observed in the numerical simulation, an analytical one-dimensional solution is obviously not adequate.

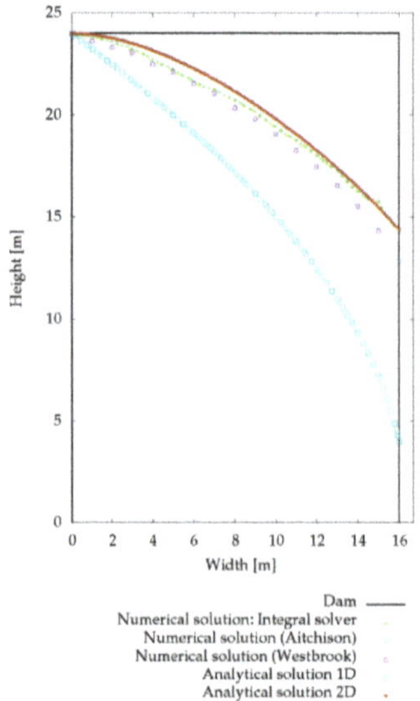

Figure 3. Seepage calculated with 1D and 2D analytical and numerical solutions for a rectangular dam.

For the second validation case, the seepage through a homogeneous dam with a constant water level of 1.9 m on the left side was compared with an analytical solution by Kozeny [47] and an analytical solution by Casagrande [48]. The latter is an improvement of the solution by Kozeny. The dam has a height of 2.2 m and a width of 8.7 m. The two-dimensional mesh consists of 750,000 rectangular elements with a width of 0.02 m in x-and y-direction. The dam material properties were the same as in the first validation case. A good agreement can be recognized for the simulated water levels with the calculated analytical data (see Figure 4). At the entrance and at the outlet, the results gained with the integral model were closer to the solution after Casagrande compared to the solution after Kozeny.

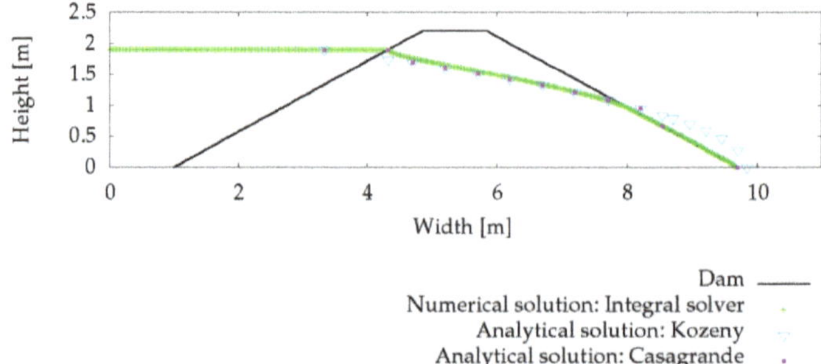

Figure 4. Seepage through a homogeneous dam after Kozeny [45], Casagrande [48] and calculated with the integral solver.

Our test simulations showed that the integral flow model can predict the interaction of surface and groundwater with reasonable accuracy compared to analytical and numerical solutions.

3. Results and Discussion

In the following section, results for the reference case (see Table 1, case 1) will be presented for both sandy and gravel sediments. Based on these results, the influence of the different ripple parameters and surface water discharge on the flow field will be analyzed, including pressure and velocity distributions as well as hyporheic exchange fluxes after 5 min simulation time. For all cases we focused on a single ripple in the center of a series of ripples. For the quantification of the fluxes, fluxes through the cell faces at the investigated ripple are calculated at the intersections of surface water and sediment. The ripple is divided into an area left and right from the ripple crest (see Figure 5). In- as well as outflowing fluxes for both sides as well as the sum of these fluxes divided by the face area—defined as "total flux"—are determined. The fluxes are averaged for the time frame of 60–300 s due to non-steady flow conditions.

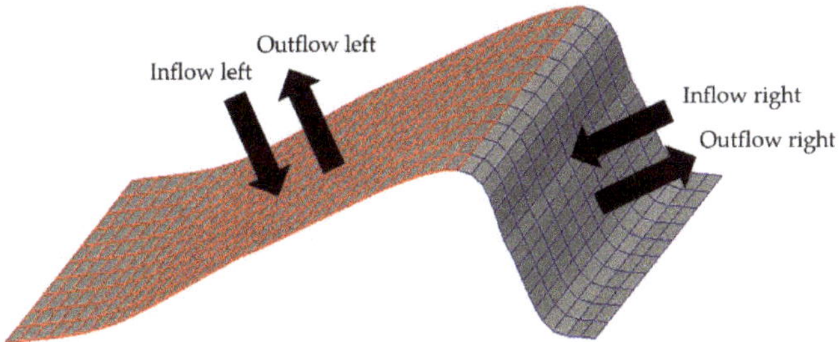

Figure 5. In- and outflowing fluxes at the left and right side of the ripple crest.

3.1. Reference Case

For the reference case (see Table 1, case 1), the discharge amounts to 0.5 m^3/s, the ripple length to 20 cm and the height to 5.6 cm. Figure 6 shows the pressure distribution and velocity vectors at the investigated ripple (see Figure 1) for case 1 with a sandy and a gravel sediment. The solver solves the pressure term p_rgh as the static pressure minus the hydrostatic pressure (ρgz with z as coordinate vector). The highest pressure is observed at the last third of the upstream face of the ripple. Low pressure is present at the ripple crest and the first two-thirds of the upstream face as well as downstream the crest. As these pressure differences lead to hyporheic exchange, flow occurs in downstream and upstream directions from high to low pressure. The described flow paths fit well to the results by Fox et al. [49], where the exchange of water between surface and subsurface was illustrated based on tracer experiments in the laboratory at a rippled sandy streambed. Also Thibodeaux and Boyle [50], Elliott and Brooks [14] and Janssen et al. [51] came to similar results from laboratory experiments with triangular bedforms. Fehlman [52] and Shen et al. [53] presented non-hydrostatic pressure distributions at triangular bed forms which were also similar to our results with pressure peaks at the middle of the stoss face, pressure minimum at the crest with low pressure remaining at the lee face until the pressure increases again at the stoss face of the following ripple. The description of the principal pressure pattern at the observed ripple in our simulations is valid for the sand as well as for the gravel, though the pressure values differ. Due to the higher resistance of the sand compared to gravel, higher pressure gradients are observed. Conversely, it behaves in terms of subsurface velocities: higher velocities are determined in the gravel sediment compared to the less permeable sand.

The applied LES turbulence model allows to resolve large parts of the turbulence at the streambed directly. Hence, between each ripple pair, eddies are identified. Comparing Figure 6 left and Figure 6 right, it is obvious, that the flow field in the surface water depends on the properties of the sediment: While in the sand, two eddies (clockwise as well as counterclockwise) can be recognized between the ripples, for the gravel only one eddy is apparent which flows in clockwise direction. This indicates, that a simple successively coupling via pressure distributions from surface water as a boundary condition for a groundwater model as for the example in [23,24] is not adequate, since not only the surface water influences the subsurface, but also the subsurface affects the surface flow conditions. The feedback from the subsurface to the surface is consequently also important. The clockwise eddies are located in the area where surface water introduces into the ripple, while the counterclockwise ripple is located in the area where the water within the ripple flows back to the surface water.

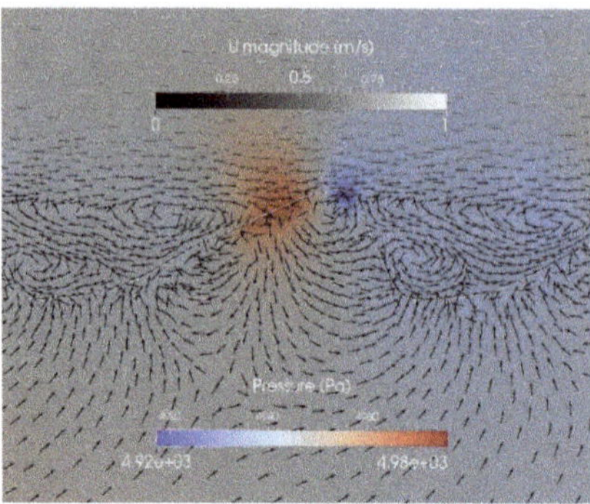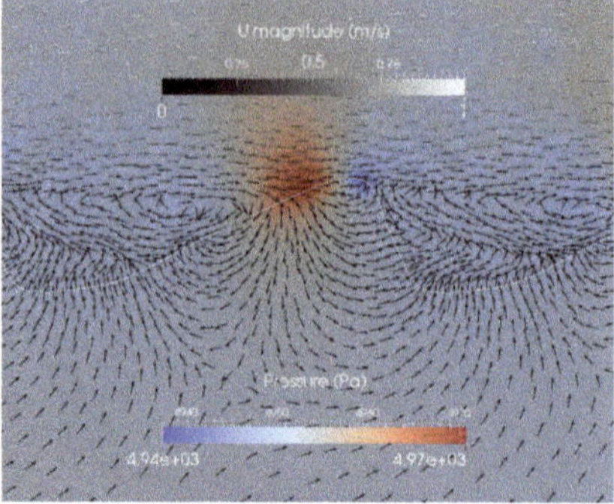

Figure 6. Pressure distribution and velocity vectors at a sandy (**left**) and gravel (**right**) ripple for case 1 (Table 1). The white line indicates the sediment-water interface. The colors indicate the pressure distribution. Please note that the scaling is different in the right and the left panel. The arrows indicate flow directions of the surface and the subsurface flow. To visualize the intensity of the flow U a grey is used.

Tables 2 and 3 show the in- and outflowing fluxes for all cases as well as the total fluxes per ripple area for gravel and sand according to Figure 5 As described already above, in- as well as outflow are observed at the ripple lee while at the ripple stoss the outflow is dominant. The total flux is much higher for the gravel with 1.8×10^{-2} m^3/s/m^2 than for the sand with 2.7×10^{-3} m^3/s/m^2. This fits to the flume experiment results by Tonina and Buffington [17], where larger hyporheic exchange was claimed for gravel compared to sandy sediments.

Table 2. Hyporheic fluxes of a single ripple in the center of a series of ripples for case 1–6 (sand). Right and left indicate the part of the ripple right and left of the ripple crest (compare Figure 4).

Case	Inflow Left (m^3/s)	Inflow Right (m^3/s)	Inflow Sum (m^3/s)	Outflow Left (m^3/s)	Outflow Right (m^3/s)	Outflow Sum (m^3/s)	Total Flux [1] (m^3/s/m^2)
1	2.9×10^{-4}	3.8×10^{-5}	3.3×10^{-4}	2.1×10^{-4}	1.2×10^{-4}	3.3×10^{-4}	2.7×10^{-3}
2	1.4×10^{-4}	6.3×10^{-6}	1.4×10^{-4}	3.7×10^{-5}	1.3×10^{-4}	1.6×10^{-4}	5.1×10^{-3}
3	6.6×10^{-4}	5.3×10^{-5}	7.2×10^{-4}	4.9×10^{-4}	2.4×10^{-4}	7.3×10^{-4}	3.0×10^{-3}
4	4.0×10^{-4}	6.1×10^{-5}	4.6×10^{-4}	3.3×10^{-4}	1.6×10^{-4}	4.9×10^{-4}	2.2×10^{-3}
5	4.2×10^{-4}	6.0×10^{-5}	4.8×10^{-4}	1.7×10^{-4}	2.9×10^{-4}	4.6×10^{-4}	3.9×10^{-3}
6	1.2×10^{-4}	2.0×10^{-5}	1.4×10^{-4}	9.6×10^{-5}	4.8×10^{-5}	1.4×10^{-4}	2.9×10^{-4}

[1] Total flux = (mag (inflow left) + mag (inflow right) + mag (outflow left) + mag (outflow right))/area.

Based on the overall high velocities within the sediment our simulations indicate, that non-Darcy-flow is present in the whole ripple nearly down to the bottom boundary for the gravel bed and to a part of the sandy bed (see Figure 7). At the near-surface area at the crest of the gravel ripple, Reynolds numbers up to 1770 were recognized, while for the sandy bed Reynolds numbers up to 330 were determined. For a better illustration of the non-Darcy-flow areas, Reynolds numbers up to 10 are illustrated in Figure 7. Consequently, dark red areas have a Reynolds number that equals or is higher than 10. Due to lower permeability, the flow velocities of the surface water influenced the sandy sediment less than the gravel bed with high permeability. The explicit modeling of the hyporheic zone with Darcy's law is not possible in river beds with such coarse grain sizes since groundwater flow rates would be overestimated. Facing e.g., contaminant transport depending on residence time serious misperceptions could appear. The Reynolds number distribution of the following cases were similar to the reference case: for the whole gravel ripple down to the bottom non-Darcy-flow is apparent, while for the sand a small layer at the interface as well as the crest shows non-Darcy-flow areas. Only for case 5 with a distance of 20 cm between each ripple, there is even more non-Darcy-flow within the sandy ripple.

Table 3. Hyporheic fluxes of a single ripple in the center of a series of ripples for case 1–6 (gravel). Right and left indicate the part of the ripple right and left of the ripple crest (compare Figure 4).

Case	Inflow Left (m^3/s)	Inflow Right (m^3/s)	Inflow Sum (m^3/s)	Outflow Left (m^3/s)	Outflow Right (m^3/s)	Outflow Sum (m^3/s)	Total Flux [1] (m^3/s/m^2)
1	2.2×10^{-3}	2.5×10^{-5}	2.2×10^{-3}	1.0×10^{-3}	1.2×10^{-3}	2.2×10^{-3}	1.8×10^{-2}
2	5.6×10^{-4}	2.9×10^{-5}	5.9×10^{-4}	1.6×10^{-4}	3.7×10^{-4}	5.2×10^{-4}	1.8×10^{-2}
3	4.5×10^{-3}	3.8×10^{-5}	4.6×10^{-3}	1.5×10^{-3}	2.1×10^{-3}	3.6×10^{-3}	1.7×10^{-2}
4	3.5×10^{-3}	0	3.5×10^{-3}	2.0×10^{-3}	1.9×10^{-3}	3.9×10^{-3}	1.7×10^{-2}
5	3.6×10^{-3}	0	3.6×10^{-3}	8.4×10^{-4}	2.2×10^{-3}	3.1×10^{-3}	2.7×10^{-2}
6	9.3×10^{-4}	1.4×10^{-5}	9.4×10^{-4}	4.6×10^{-4}	5.2×10^{-4}	9.8×10^{-4}	1.9×10^{-3}

[1] Total flux = (mag (inflow left) + mag (inflow right) + mag (outflow left) + mag (outflow right))/area.

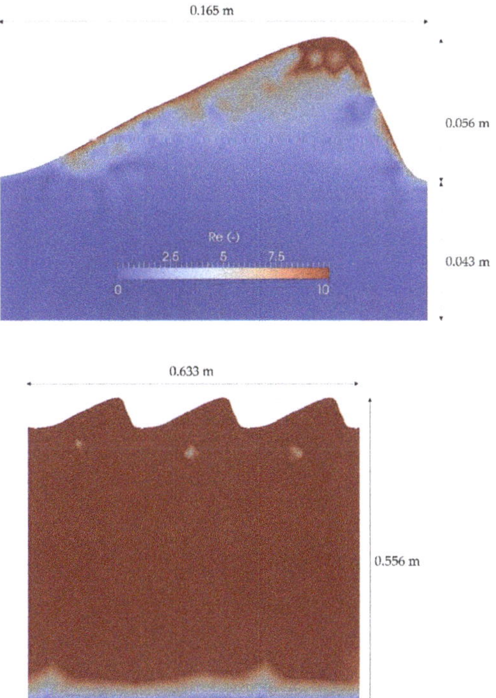

Figure 7. Reynolds numbers at a sandy (**top**) and gravel (**bottom**) ripple for case 1 (Table 1).

Janssen et al. [51] stated that the largest discrepancies of most CFD simulations of flow over ripples and dunes occur in the eddy zone. Especially for Reynolds-averaged Navier–Stokes turbulence models this is a known weakness. Therefore, we have chosen a LES turbulence model. At the same time, we are aware of the computational limitation, which is additionally increased by the calculation of the three-dimensional Navier–Stokes equations in the sediment in contrast to the commonly applied Darcy law. However, facing the growing availability of computational sources and the observed non-Darcy-flow areas in the investigated cases, we apply a promising tool for analyzing integral surface-subsurface flow processes with high resolution.

3.2. Ripple Dimension

For cases 2 and 3 the ripple length to height ratio is the same as for the reference case (see Table 1), but the ripple height and length are quartered for case 2 and doubled for case 3. Figure 8 shows the velocity and pressure distributions for the investigated ripples in the middle for case 2 for sand and gravel. The general pressure pattern for case 2 for sand and gravel as well as for the reference case are similar: the lowest pressure occurs at the crest and the highest pressure upstream of the crest. But the high-pressure area related to the ripple size is much higher for case 2 than for the reference case. Related to the ripple face area at the interface, we consequently expect higher inflow rates compared to the reference case, which can be seen in Tables 2 and 3. The total flux per area is higher for case 2 with 5.1×10^{-3} m^3/s/m^2 and 1.81×10^{-2} m^3/s/m^2 than for the reference case with 2.7×10^{-3} m^3/s/m^2 and 1.84×10^{-2} m^3/s/m^2.

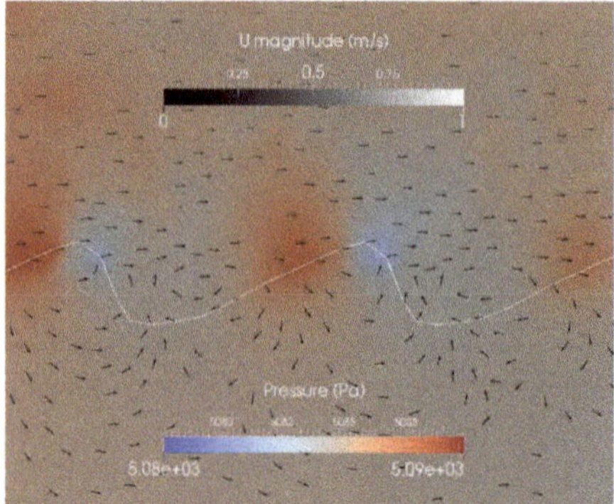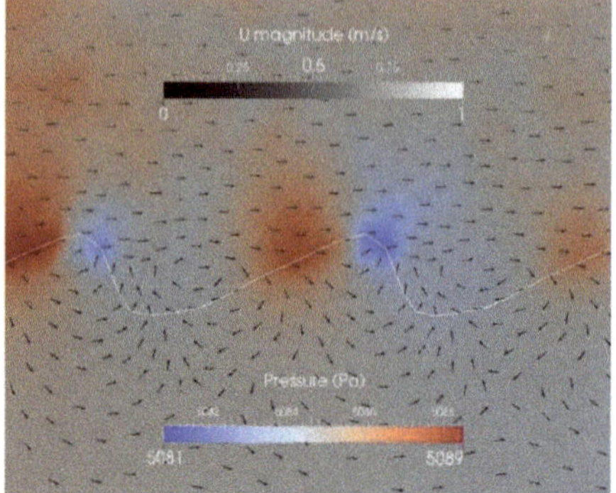

Figure 8. Pressure distribution and velocity vectors at a sandy (left) and gravel (right) ripple for case 2 (Table 1). The white line indicates the sediment-water interface. The colors indicate the pressure distribution. Please note that the scaling is different in the right and the left panel. The arrows indicate flow directions of the surface and the subsurface flow. To visualize the intensity of the flow U a grey is used.

The flow field within the ripple is again directed upstream and downstream from the point of highest pressure. For both simulations (case 2 sand and gravel) only one eddy was recognized in a clockwise direction. For case 3 high turbulence was recognized between the ripples with up to five eddies (see Figure 9). Significantly high- and low-pressure areas are recognized in the turbulent phase of the surface area especially for the simulation with sandy sediment. For this simulation two instead of one inflow area can be recognized at the upstream face of the ripple. Between these inflow areas, there is an outflow area. Another outflow area is located upstream of the lower inflow area, but the main outflow occurs downstream of the ripple crest. In the simulation of the gravel ripple, less eddies are observed than for the simulation with the sand. For the gravel ripple only one inflow area is present. The outflow is located similar to case 1 and 2: upstream from the inflow area and downstream from the crest.

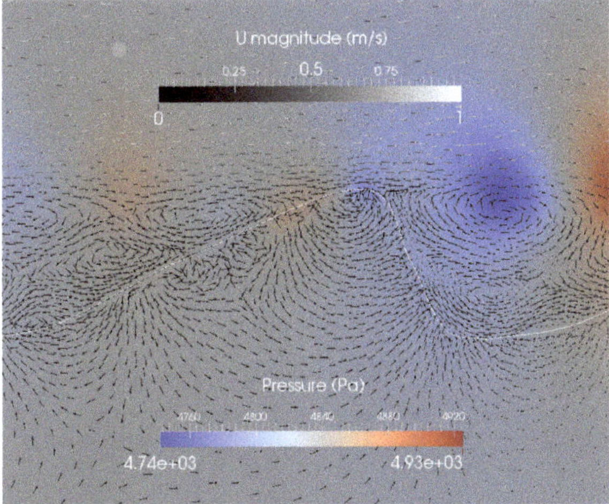

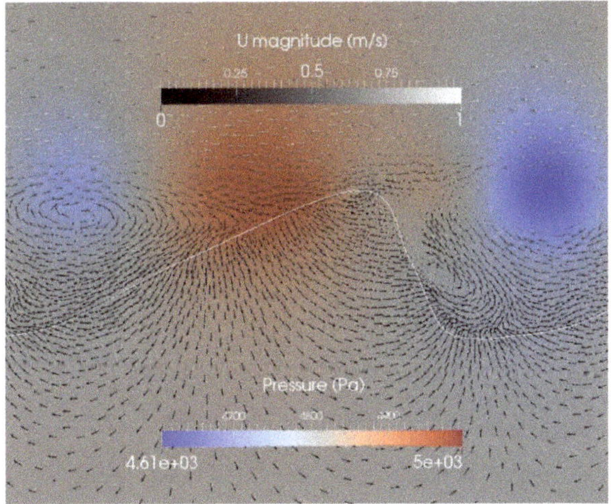

Figure 9. Pressure distribution and velocity vectors at a sandy (top) and gravel (bottom) ripple for case 3 (Table 1). The white line indicates the sediment-water interface. The colors indicate the pressure distribution. Please note that the scaling is different in the right and the left panel. The arrows indicate flow directions of the surface and the subsurface flow. To visualize the intensity of the flow U a grey is used.

The total fluxes per area are bigger for case 3 with sand (3.0×10^{-3} m³/s/m²) compared to the reference case (case 1) with sand (2.7×10^{-3} m³/s/m²). For the gravel the opposite is true (case 3). 1.7×10^{-2} m³/s/m² and case 1: 1.8×10^{-2} m³/s/m²). The extremely high turbulence between the ripples for the sand could be an explanation for that. The results for cases 2 and 3 with gravel and sand show, that a general statement about the influence of the ripple size is not possible, as there is a complex relation between the size and the material leading to different turbulence and pressure distributions, where also a threshold can be conceivable. Tonina and Buffington [16] presented results from a laboratory experiment with a pool-riffle channel and came to the same conclusion that hyporheic exchange does not necessarily decrease with lower bed form amplitudes. Closer investigations with more simulations including additional ripple size variations would be necessary for a more profound interpretation.

3.3. Ripple Length

For case 4 the ripple height equals the reference case, but the ripple length is doubled with 40 cm. This leads to less turbulence between the ripples (compare Figures 6 and 10). The bigger area leads to higher fluxes per ripple, but the total flux per area is smaller compared to the reference case for sand as well as for gravel (case 4: 2.2×10^{-3} m^3/s/m^2 and 1.7×10^{-2} m^3/s/m^2, case 1: 2.7×10^{-3} m^3/s/m^2 and 1.8×10^{-2} m^3/s/m^2). The pressure distribution is very similar to the reference case (case 1). The decisive difference is probably again the turbulence, which is higher for large height-to-length-ratios as already described by Broecker et al. [31].

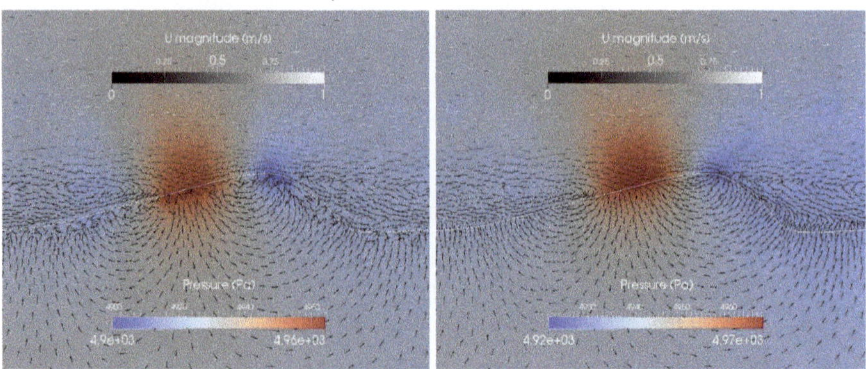

Figure 10. Pressure distribution and velocity vectors at a sandy (**left**) and gravel (**right**) ripple for case 4 (Table 1). The white line indicates the sediment-water interface. The colors indicate the pressure distribution. Please note that the scaling is different in the right and the left panel. The arrows indicate flow directions of the surface and the subsurface flow. To visualize the intensity of the flow U a grey is used.

3.4. Ripple Distance

Figure 11 shows the velocity and pressure distributions for case 5 with the same ripple geometry as for the reference case, but with a distance between the ripples of 20 cm. This distance leads to higher pressure gradients for gravel and sand compared to the reference case. The flow fields within the ripples are similar to the reference case. But for this case there are also in- and outflow areas at the flat streambed between the ripples for both simulations. Eddies occur between the investigated ripples, but due to the distance, they are more elongated than for the reference case (case 1). Since the pressure gradients are higher for case 5 compared to the reference case and the area is the same for both cases, (the area is only the ripple area, not the flat part between the ripples) the total flux is higher for gravel as well as for sand compared to the reference case with both sediments (case 5: 3.9×10^{-3} m^3/s/m^2 and 2.7×10^{-2} m^3/s/m^2, case 1: 2.7×10^{-3} m^3/s/m^2 and 1.8×10^{-2} m^3/s/m^2). Broecker et al. [31] already presumed higher hyporheic exchange for this case compared to the reference case, based on the higher pressure gradients. To our knowledge, distances between the ripples were never investigated so far for hyporheic zone processes, apart from Broecker et al. [31] where only a surface water model was used.

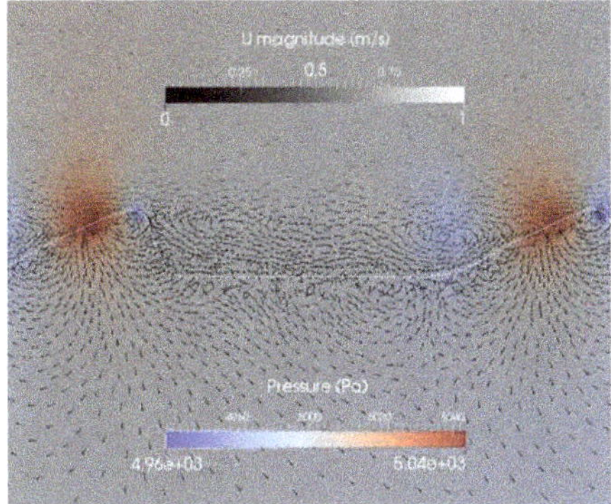 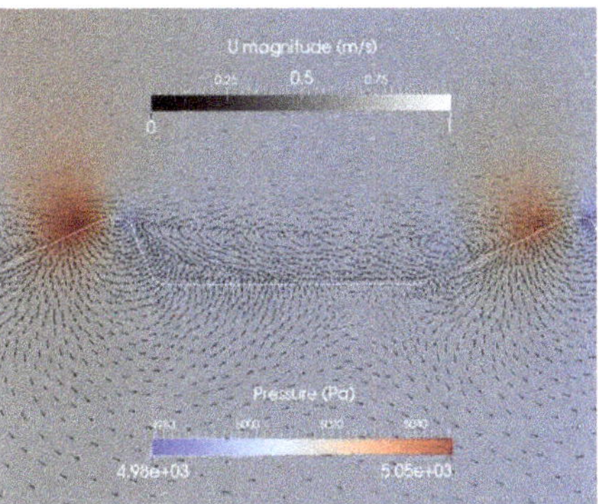

Figure 11. Pressure distribution and velocity vectors at a sandy (**left**) and gravel (**right**) ripple for case 5 (Table 1). The white line indicates the sediment-water interface. The colors indicate the pressure distribution. Please note that the scaling is different in the right and the left panel. The arrows indicate flow directions of the surface and the subsurface flow. To visualize the intensity of the flow U a grey is used.

Compared to the investigated sandy ripples described above, the non-Darcy-flow areas of case 5 are significantly larger (compare Figures 7 and 12). Due to the distance between the ripples, higher velocities reach the ripple stoss which influence the velocities within the ripple.

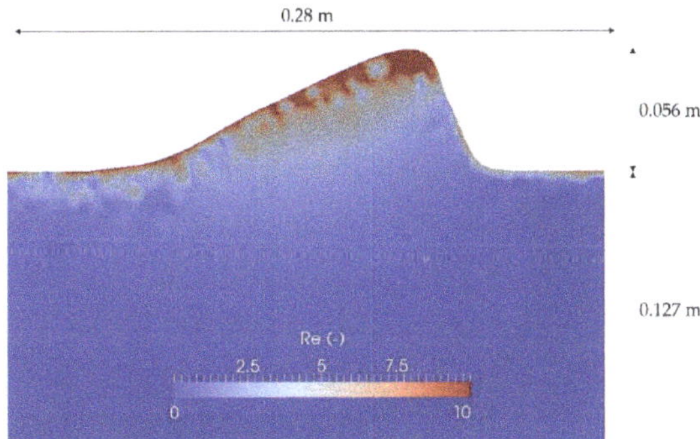

Figure 12. Reynolds numbers at a sandy ripple for case 5

3.5. Flow Rate

For case 6 the discharge was set to 0.25 m^3/s (for case 1–5 the discharge was 0.5 m^3/s). The ripple geometry is the same as for the reference case (case 1). Comparing the reference case with case 6, it is obvious that both flow discharges show qualitatively similar flow fields. The flow velocities within the ripples decrease due to lower surface water velocities. Nevertheless, there is still a layer with Reynolds numbers higher than 10, which is slightly smaller than for the reference case (see Figure 13). The hyporheic fluxes are decreased compared to the reference case (case 6: 2.9×10^{-4} m^3/s/m^2 and 1.9×10^{-3} m^3/s/m^2, case 1: 2.7×10^{-3} m^3/s/m^2 and 1.8×10^{-2} m^3/s/m^2). This fits with laboratory observations e.g., by Marion et al. [54] and Elliott and Brooks [14].

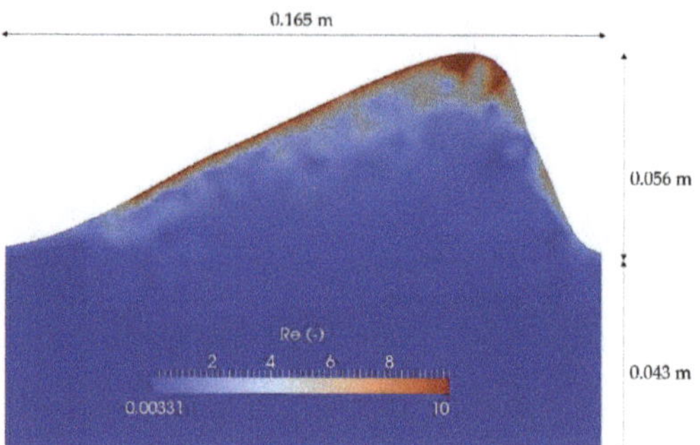

Figure 13. Reynolds numbers at a sandy ripple for case 6.

4. Conclusions

CFD simulations were designed to simultaneously examine both surface and subsurface flow processes with an extended version of the three-dimensional Navier–Stokes equations. Based on two simulations for seepages through dams, it was shown that the applied model can describe the interaction of groundwater and surface water. The validated CFD model was applied to investigate the impact of bed form structures, grain sizes and surface flow discharges on hyporheic exchange processes. The examined ripple structures changed the streambed pressure and created in- and outflowing fluxes at the interface which were calculated for each case study for a representative ripple in the middle and play a significant role in biogeochemical processes within the hyporheic zone. It was shown that not only the surface water influences the flow within the sediment, but also the sediment properties lead to a change of the flow field within the surface water. Consequently, we claim, that a simple coupling of surface water with a closed boundary at the streambed, which is commonly used, is not appropriate at least for coarse sediments. Moreover, non-Darcy-flow areas were observed for all cases within the sediment. For the sandy sediment, the non-Darcy-flow areas are restricted to the upper layer and the crest of the investigated ripples. For the gravel non-Darcy-flow was observed almost down to the bottom of the model. The application of the Darcy law in these areas would lead to an overestimation of flow rates. For equations that can be applied in non-Darcy-flow areas in the subsurface such as the Darcy-Brinkmann-equation, additional parameters such as the effective viscosity have to be determined.

Comparing the extended Navier–Stokes equations with the commonly used coupling of surface water with a Darcy-law-model, the integral model is definitely more time consuming than the coupled models. The model shows direct feedbacks from surface to subsurface and vice versa, is applicable also in non-Darcy-flow areas and provides high resolution results. The applied LES turbulence model gives additional insights about the turbulence at the interface which has a high impact on hyporheic exchange.

The general flow paths were the same for almost all simulations. Upstream of the crest, in high pressure areas, surface water flows into the ripple. Subsurface water flows out of the ripple towards the crest as well as upstream of the inflow area. Only for the ripple with the highest dimension multiple inflow areas were recognized upstream of the crest. Higher turbulences were generally observed for sandy ripples compared to gravel ripples. Gravel ripples always showed higher hyporheic exchange fluxes compared to sand ripples. Three ripple dimensions with the same height to length ratios were examined, but no clear relationship to exchange fluxes was found. For longer ripples, the exchange was slightly smaller due to less turbulence, while distances between the ripples increased the hyporheic exchange fluxes drastically. Also, in the flat streambed sections exchange was observed. Decreasing flow rates lead to decreasing exchange fluxes.

Numerous of the observations of our simulations were already seen in laboratory experiments. Our simulations allowed to get a deeper understanding of the present velocity and pressure distribution at the interface and to determine in- and outflowing fluxes, which can be important for the understanding and prediction of hydrological, chemical, and biological processes. In contrast to other coupled models, it is applicable in non-Darcy-flow areas and allows to simultaneously simulate the surface and subsurface with one system of equation for surface and groundwater. We can develop upscaling approaches where we quantify the exchange rates depending on the ripple geometry and other variables with the high resolution three-dimensional integral model to serve as sink/source terms in one- or two-dimensional shallow water flow models. The shallow water equations are based on vertical averaged velocities (not discretizing the vertical dimension) and are generally applied on coarser scales. In a next step, also transport equations will be included in the presented integral model.

Author Contributions: Methodology, T.B. and K.T.; Software, T.B. and K.T.; Validation, T.B.; Investigation, T.B. and V.S.G.; Writing—original draft, T.B.; Writing—review and editing, T.B., R.H., G.N. and J.L.; Visualization, T.B.; Supervision, R.H. G.N. and J.L.

Funding: The funding provided by the German Research Foundation (DFG) within the Research Training Group 'Urban Water Interfaces' (GRK 2032) is gratefully acknowledged.

Acknowledgments: Parts of the simulations were computed on the supercomputers of Norddeutscher Verbund für Hoch- und Höchstleistungsrechnen in Berlin.

Conflicts of Interest: The authors declare no conflict of interest.

References

1. Boulton, A.J.; Findlay, S.; Marmonier, P.; Stanley, E.H.; Valett, H.M. The functional significance of the hyporheic zone in streams and rivers. *Annu. Rev. Ecol. Syst.* **1998**, *29*, 59–81. [CrossRef]
2. Brunke, M.; Gonser, T. The ecological significance of exchange processes between rivers and groundwater. *Freshw. Biol.* **1997**, *37*, 1–33. [CrossRef]
3. Cardenas, M.B. Hyporheic zone hydrologic science: A historical account of its emergence and a prospectus. *Water Resour. Res.* **2015**, *51*, 3601–3616. [CrossRef]
4. Dahm, C.N.; Grimm, N.B.; Marmonier, P.; Valett, H.M.; Vervier, P. Nutrient dynamics at the interface between surface waters and groundwaters. *Freshw. Biol.* **1998**, *40*, 427–451. [CrossRef]
5. Findlay, S. Importance of surface-subsurface exchange in stream ecosystems: The hyporheic zone. *Limnol. Oceanogr.* **1995**, *40*, 159–164. [CrossRef]
6. Gomez-Velez, J.D.; Harvey, J.W.; Cardenas, M.B.; Kiel, B. Denitrification in the Mississippi River network controlled by flow through river bedforms. *Nat. Geosci.* **2015**, *8*, 941. [CrossRef]
7. Harvey, J.; Gooseff, M. River corridor science: Hydrologic exchange and ecological consequences from bedforms to basins. *Water Resour. Res.* **2015**, *51*, 6893–6922. [CrossRef]
8. Stonedahl, S.H.; Harvey, J.W.; Packman, A.I. Interactions between hyporheic flow produced by stream meanders, bars, and dunes. *Water Resour. Res.* **2013**, *49*, 5450–5461. [CrossRef]
9. Stonedahl, S.H.; Harvey, J.W.; Wörman, A.; Salehin, M.; Packman, A.I. A multiscale model for integrating hyporheic exchange from ripples to meanders. *Water Resour. Res.* **2010**, *46*. [CrossRef]
10. Schaper, J.L.; Posselt, M.; McCallum, J.L.; Banks, E.W.; Hoehne, A.; Meinikmann, K.; Shanafield, M.A.; Batelaan, O.; Lewandowski, J. Hyporheic Exchange Controls Fate of Trace Organic Compounds in an Urban Stream. *Environ. Sci. Technol.* **2018**, *52*, 12285–12294. [CrossRef]
11. Harvey, J.W.; Bencala, K.E. The Effect of streambed topography on surface-subsurface water exchange in mountain catchments. *Water Resour. Res.* **1993**, *29*, 89–98. [CrossRef]
12. Winter, T.C.; Harvey, J.W.; Franke, O.L.; Alley, W.M. *Ground Water and Surface Water; a Single Resource*; Diane Publishing Inc.: Darby, PA, USA, 1998; Volume 1139.
13. Wondzell, S.M.; Gooseff, M. Geomorphic controls on hyporheic exchange across scales: Watersheds to particles. *Treatise Geomorphol.* **2013**, *9*, 203–218.
14. Elliott, A.H.; Brooks, N.H. Transfer of nonsorbing solutes to a streambed with bed forms: Laboratory experiments. *Water Resour. Res.* **1997**, *33*, 137–151. [CrossRef]
15. Packman, A.I.; Salehin, M.; Zaramella, M. Hyporheic Exchange with Gravel Beds: Basic Hydrodynamic Interactions and Bedform-Induced Advective Flows. *J. Hydraul. Eng.* **2004**, *130*, 647–656. [CrossRef]

16. Mutz, M.; Kalbus, E.; Meinecke, S. Effect of instream wood on vertical water flux in low-energy sand bed flume experiments. *Water Resour. Res.* **2007**, *43*. [CrossRef]

17. Tonina, D.; Buffington, J.M. Hyporheic exchange in gravel bed rivers with pool-riffle morphology: Laboratory experiments and three-dimensional modeling. *Water Resour. Res.* **2007**, *43*. [CrossRef]

18. Tonina, D.; Buffington, J.M. A three-dimensional model for analyzing the effects of salmon redds on hyporheic exchange and egg pocket habitat. *Can. J. Fish. Aquat. Sci.* **2009**, *66*, 2157–2173. [CrossRef]

19. Voermans, J.; Ghisalberti, M.; Ivey, G. The variation of flow and turbulence across the sediment–water interface. *J. Fluid Mech.* **2017**, *824*, 413–437. [CrossRef]

20. Saenger, N.; Kitanidis, K.P.; Street, R. A numerical study of surface-subsurface exchange processes at a riffle-pool pair in the Lahn River, Germany. *Water Resour. Res.* **2005**, *41*, 12424. [CrossRef]

21. Cardenas, M.B.; Wilson, J.L. Dunes, turbulent eddies, and interfacial exchange with permeable sediments. *Water Resour. Res.* **2007**, *43*. [CrossRef]

22. Bardini, L.; Boano, F.; Cardenas, M.B.; Revelli, R.; Ridolfi, L. Nutrient cycling in bedform induced hyporheic zones. *Geochim. Cosmochim. Acta* **2012**, *84*, 47–61. [CrossRef]

23. Trauth, N.; Schmidt, C.; Maier, U.; Vieweg, M.; Fleckenstein, J.H. Coupled 3-D stream flow and hyporheic flow model under varying stream and ambient groundwater flow conditions in a pool-riffle system. *Water Resour. Res.* **2013**, *49*, 5834–5850. [CrossRef]

24. Trauth, N.; Schmidt, C.; Vieweg, M.; Maier, U.; Fleckenstein, J.H. Hyporheic transport and biogeochemical reactions in pool-riffle systems under varying ambient groundwater flow conditions. *J. Geophys. Res. Biogeosci.* **2014**, *119*, 910–928. [CrossRef]

25. Chen, X.; Cardenas, M.B.; Chen, L. Hyporheic Exchange Driven by Three-Dimensional Sandy Bed Forms: Sensitivity to and Prediction from Bed Form Geometry. *Water Resour. Res.* **2018**, *54*, 4131–4149. [CrossRef]

26. Chen, X.; Cardenas, M.B.; Chen, L. Three-dimensional versus two-dimensional bed form-induced hyporheic exchange. *Water Resour. Res.* **2015**, *51*, 2923–2936. [CrossRef]

27. VanderKwaak, J.E. Numerical Simulation of Flow and Chemical Transport in Integrated Surface-Subsurface Hydrologic Systems. Ph.D. Thesis, University of Waterloo, Waterloo, ON, Canada, 1999.

28. Brunner, P.; Simmons, C. HydroGeoSphere: A Fully Integrated, Physically Based Hydrological Model. *Groundwater* **2011**, *50*, 170–176. [CrossRef]

29. Brunner, P.; Cook, P.G.; Simmons, C.T. Hydrogeologic controls on disconnection between surface water and groundwater. *Water Resour. Res.* **2009**, *45*. [CrossRef]

30. Alaghmand, S.; Beecham, S.; Jolly, I.D.; Holland, K.L.; Woods, J.A.; Hassanli, A. Modelling the impacts of river stage manipulation on a complex river-floodplain system in a semi-arid region. *Environ. Model. Softw.* **2014**, *59*, 109–126. [CrossRef]

31. Bear, J. *Dynamics of Fluids in Porous Media*; American Elsevier Publishing Company: New York, NY, USA, 1972.

32. Freeze, R.A.; Cherry, J.A. *Groundwater*; Prentice-Hall: Upper Saddle River, NJ, USA, 1979.

33. Packman, A.I.; Brooks, N.H.; Morgan, J.J. A physicochemical model for colloid exchange between a stream and a sand streambed with bed forms. *Water Resour. Res.* **2000**, *36*, 2351–2361. [CrossRef]

34. Oxtoby, O.; Heyns, J.; Suliman, R. A finite-volume solver for two-fluid flow in heterogeneous porous media based on OpenFOAM. In Proceedings of the 7th Open Source CFD International Conference, Hamburg, Germany, 24–25 October 2013. [CrossRef]

35. Broecker, T.; Elsesser, W.; Teuber, K.; Özgen, I.; Nützmann, G.; Hinkelmann, R. High-resolution simulation of free-surface flow and tracer retention over streambeds with ripples. *Limnologica* **2018**, *68*, 46–58. [CrossRef]

36. Pope, S.B. Ten questions concerning the large-eddy simulation of turbulent flows. *New J. Phys.* **2004**, *6*, 35. [CrossRef]

37. Ergun, S. Fluid Flow Through Packed *Columns. Chem. Eng. Prog.* **1952**, *48*, 89–94.

38. van Gent, M. *Wave Interaction with Permeable Coastal Structures*; Elsevier Science: Amsterdam, The Netherlands, 1995; Volume 95.

39. Hirt, C.W.; Nichols, B.D. Volume of fluid (VOF) method for the dynamics of free boundaries. *J. Comput. Phys.* **1981**, *39*, 201–225. [CrossRef]

40. Smagorinsky, J. General circulation experiments with the primitive equations. *Mon. Weather Rev.* **1963**, *91*, 99–164. [CrossRef]

41. Thorenz, C.; Strybny, J. On the numerical modelling of filling-emptying system for locks. In Proceedings of the 10th International Conference on Hydroinformatics, Hamburg, Germany, 14–18 July 2012.

42. Bayon-Barrachina, A.; Lopez-Jimenez, P.A. Numerical analysis of hydraulic jumps using OpenFOAM. *J. Hydroinform.* **2015**, *17*, 662–678. [CrossRef]

43. Westbrook, D.R. Analysis of inequality and residual flow procedures and an iterative scheme for free surface seepage. *Int. J. Numer. Methods Eng.* **1985**, *21*, 1791–1802. [CrossRef]

44. Aitchison, J.; Coulson, C.A. Numerical treatment of a singularity in a free boundary problem. *Proc. R. Soc. Lond. A Math. Phys. Sci.* **1972**, *330*, 573–580. [CrossRef]

45. Kobus, H.; Keim, B. *Grundwasser. Taschenbuch der Wasserwirtschaft*; Blackwell Wissenschaftsverlag: Hoboken, NJ, USA, 2001; pp. 277–313.

46. Di Nucci, C. A free boundary problem for fluid flow through porous media. *arXiv* **2015**, arXiv:1507.05547. Available online: https://arxiv.org/abs/1507.05547 (accessed on 14 July 2015).

47. Lattermann, E. *Wasserbau-Praxis: Mit Berechnungsbeispielen Bauwerk-Basis-Bibliothek*; Beuth Verlag GmbH: Berlin, Germany, 2010.

48. Casagrande, A. *Seepage Through Dams*; Harvard University Graduate School of Engineering: Cambridge, MA, USA, 1937.

49. Fox, A.; Boano, F.; Arnon, S. Impact of losing and gaining streamflow conditions on hyporheic exchange fluxes induced by dune-shaped bed forms. *Water Resour. Res.* **2014**, *50*, 1895–1907. [CrossRef]

50. Thibodeaux, L.J.; Boyle, J.D. Bedform-generated convective transport in bottom sediment. *Nature* **1987**, *325*, 341–343. [CrossRef]

51. Janssen, F.; Cardenas, M.B.; Sawyer, A.H.; Dammrich, T.; Krietsch, J.; de Beer, D. A comparative experimental and multiphysics computational fluid dynamics study of coupled surface–subsurface flow in bed forms. *Water Resour. Res.* **2012**, *48*. [CrossRef]

52. Fehlman, H.M. *Resistance Components and Velocity Distributions of Open Channel Flows Over Bedforms*; Colorado State University: Fort Collins, CO, USA, 1985.

53. Shen, H.W.; Fehlman, H.M.; Mendoza, C. Bed Form Resistances in Open Channel Flows. *J. Hydraul. Eng.* **1990**, *116*, 799–815. [CrossRef]

54. Marion, A.; Bellinello, M.; Guymer, I.; Packman, A. Effect of bed form geometry on the penetration of nonreactive solutes into a streambed. *Water Resour. Res.* **2002**, *38*, 27-1–27-12. [CrossRef]

Article

Solute Transport and Transformation in an Intermittent, Headwater Mountain Stream with Diurnal Discharge Fluctuations

Adam S. Ward [1,*], Marie J. Kurz [2,3], Noah M. Schmadel [4], Julia L.A. Knapp [5,6], Phillip J. Blaen [7,8], Ciaran J. Harman [9], Jennifer D. Drummond [7], David M. Hannah [7], Stefan Krause [7,10], Angang Li [11], Eugenia Marti [12], Alexander Milner [7], Melinda Miller [1], Kerry Neil [1], Stephen Plont [13,14], Aaron I. Packman [11], Nathan I. Wisnoski [15], Steven M. Wondzell [16] and Jay P. Zarnetske [14]

[1] O'Neill School of Public and Environmental Affairs, Indiana University, Bloomington, IN 47401, USA; mgerhart@indiana.edu (M.M.); kerryneil17@gmail.com (K.N.)
[2] The Academy of Natural Sciences of Drexel University, Philadelphia, PA 19104, USA; mk3483@drexel.edu
[3] Department of Hydrogeology, Helmholtz Centre for Environmental Research-UFZ, 04318 Leipzig, Germany
[4] U.S. Geological Survey Earth System Processes Division, Reston, VA, 20192, USA; nschmadel@usgs.gov
[5] Department of Environmental Systems Science, ETH Zurich, 8092 Zurich, Switzerland; julia.knapp@usys.ethz.ch
[6] Center for Applied Geoscience, University of Tübingen, 72074 Tübingen, Germany
[7] School of Geography, Earth & Environmental Sciences, University of Birmingham, Edgbaston, Birmingham B15 2TT, UK; pblaen@gmail.com (P.J.B.); J.Drummond@bham.ac.uk (J.D.D.); D.M.HANNAH@bham.ac.uk (D.M.H.); S.Krause@bham.ac.uk (S.K.); a.m.milner@bham.ac.uk (A.M.)
[8] Yorkshire Water, Bradford BD6 2SZ, UK
[9] Department of Environmental Health and Engineering, Johns Hopkins University, Baltimore, MD 21218, USA; charman1@jhu.edu
[10] University Claude Bernard Lyon 1, LEHNA - Laboratory of Ecology of Natural and Man-Impacted Hydrosystems, 69622 Lyon, France
[11] Department of Civil and Environmental Engineering, Northwestern University, Evanston, IL 60208, USA; angang-li@u.northwestern.edu (A.L.); a-packman@northwestern.edu (A.I.P.)
[12] Integrative Freshwater Ecology Group, Center for Advanced Studies of Blanes (CEAB-CSIC), 17300 Blanes, Spain; eugenia@ceab.csic.es
[13] Department of Biological Sciences, Virginia Polytechnic Institute and State University, Blacksburg, VA 24061, USA; plontste@vt.edu
[14] Department of Earth and Environmental Sciences, Michigan State University, East Lansing, MI 48824, USA; jpz@msu.edu
[15] Department of Biology, Indiana University, Bloomington, IN 47405, USA; wisnoski@indiana.edu
[16] US Forest Service, Pacific Northwest Research Station, 3200 SW Jefferson Way, Corvallis, OR 97331, USA; steve.wondzell@usda.gov
* Correspondence: adamward@indiana.edu

Received: 9 September 2019; Accepted: 10 October 2019; Published: 23 October 2019

Abstract: Time-variable discharge is known to control both transport and transformation of solutes in the river corridor. Still, few studies consider the interactions of transport and transformation together. Here, we consider how diurnal discharge fluctuations in an intermittent, headwater stream control reach-scale solute transport and transformation as measured with conservative and reactive tracers during a period of no precipitation. One common conceptual model is that extended contact times with hyporheic zones during low discharge conditions allows for increased transformation of reactive solutes. Instead, we found tracer timescales within the reach were related to discharge, described by a single discharge-variable StorAge Selection function. We found that Resazurin to Resorufin (Raz-to-Rru) transformation is static in time, and apparent differences in reactive tracer were due to interactions with different ages of storage, not with time-variable reactivity. Overall we found reactivity was highest in youngest storage locations, with minimal Raz-to-Rru conversion in waters older than about 20 h of storage in our study reach. Therefore, not all storage in the study reach has the same potential biogeochemical function and increasing residence time of solute storage does not necessarily increase reaction potential of that solute, contrary to prevailing expectations.

Keywords: unsteady-state discharge; solute transport; intermittent stream; diurnal discharge fluctuations; reactive tracers; headwaters; river corridor; hyporheic; resazurin

1. Introduction

Discharge dynamics are known to control both transport and chemical transformations in the river corridor. Discharge is often considered a master variable in river corridors because it directly controls residence times of river water, and thus, many chemical reaction processes [1,2]. However, there are a wide range of interactions between transport and transformation, including changes in the relative importance of downstream transport and hydrologic exchange flows [3,4], changes in timescales of contact with reactive storage zones [5,6], and changes in reactivity as streams wet and dry [7]. Despite the recognized role of discharge as a primary control on both transport and transformation processes individually, few studies have considered the joint responses of transport and transformation to dynamic hydrologic forcing. Thus, our overarching goal in this study is to explain how daily fluctuations in discharge control both transport and transformation of reactive tracers in an intermittent stream reach.

Discharge-related hydrological forcing influences river corridor exchange fluxes between river water and reactive sediments and related subsurface transit time distributions, such as in response to storm events [8–12], tides [13–16], baseflow recession [5,17–22], diurnal fluctuations due to evapotranspiration in the catchment and river corridor [22–24], glacial melt or snowmelt [25–28], dam releases [29–32], or wastewater treatment plant operations [33]. Even in intermittent streams where all discharge can become subsurface, discharge dynamics may lead to variation in travel times through the study reach. Dynamics in sediment porewater flow arise due to variation in hydraulic gradients that can cause compression or expansion of reactive storage zones as discharge rises and falls [34–45]. Changes in discharge also affect mixing with in-stream storage [46–49], initiate bank storage [50,51], and possibly change the geometry of subsurface flow paths themselves [52,53]. Overall, we expect the relative role of storage to be maximized during periods of lowest discharge, resulting in extended transit times through the study reach. For example, during low flow conditions, we expect the contact between stream water and the streambed to be maximized due to: (1) the largest bed area per unit discharge [54–56]; (2) exchange flux representing the largest fraction of streamflow [4,36,57]; and (3) less light attenuation with depth in the water column, as would be expected during higher flows [58–60]. Intermittent flow ensures that neither water nor solutes are able to bypass the reactive streambed and hyporheic zone, eliminating the potential shunting of unreacted solutes through the study reach [61].

Reaction rates in the river corridor vary directly in response to dynamics in discharge because they scale with the ratio of river water volume to the biogeochemically active river bed [62]. Stream hydraulics, therefore, may control which microbial or algal communities thrive and thus contribute to high reaction rates [63–65]. High flows may scour streambeds and catalyze either a net reduction in reach-scale reactivity as the microbial community recolonizes or stimulate productivity due to delivery of limiting reagents [66–69]. Reaction rates are known to vary in response to wetting and drying of streambeds [7], where oxidizing and reducing conditions are varied in response to water levels in intermittent streams. The major mechanisms of disturbance-recovery (e.g., scour, desiccation) that affect metabolic activity after large perturbations may not be dominant during regular wetting and drying of diurnal fluctuations in discharge, because diurnal wetting and drying is more frequent and less extreme than large disturbances. However, the changes in contact times with reactive surface area and fluctuations in solute availability may still drive diurnal changes in metabolic activity.

Solute tracers are commonly used to measure reach-scale transport and transformation processes in river corridors, but usually during low flow conditions with steady discharge. One tool that has proven effective in assessment of combined transport and transformation is the Resazurin–Resorufin (hereafter Raz–Rru, respectively) tracer system, in which Raz undergoes an irreversible transformation to Rru in the presence of metabolically active microbial communities [70–73]. The Raz–Rru system has broadly been used to identify the portion of transient storage that is metabolically active, and has been used to assess reactivity in biofilms [74], the benthic zone [75,76], vegetation beds [77], stream- and lakebed sediments [78–81], and whole stream reaches [82–85]. Interpretation of Raz–Rru information commonly relies upon inverse modeling that enforces a conceptual model with binary divisions of the system into zones such as channel or storage (based on the importance of advection and relative timescales of storage) or metabolically active or inactive storage (based on transformation rates) [83,86].

One challenge with modeling solute tracer transport and transformation is parameter uncertainty and equifinality [18,86–90]. Further, in their standard form, these models have not been formulated to address time-variable transport or transformation, and do not account for the possibility that the system is more complex than the binary differentiation of stream and storage zone. Moreover, common model formulations are not appropriate for interpreting experimental data in intermittent streams, as they require the presence of a continuously flowing stream [91–93], though exceptions do exist [23]. To overcome these limitations, new approaches, such as the StorAge Selection (SAS) model, take transport as a continuum and can simulate continuously variable discharge [94–96].

The overall objective of this study is to assess how reach-scale transport and transformation of reactive tracers vary in response to diurnal discharge fluctuations in an intermittent, headwater stream. We expect that transit times will be longest and reach-scale transformation highest during periods of low discharge because water and solute tracers will experience extended contact with highly reactive streambeds and hyporheic zones. Conversely, during higher-discharge periods, we expect water and solutes to have less contact with hyporheic zones, because previously stored water will be upwelling to the stream, compressing hyporheic flowpaths and forcing tracer to remain in and near the stream channel in shortened flow paths. To test these hypotheses, we conducted a series of four solute tracer experiments, including both conservative and reactive solute tracers (Uranine and Resazurin, respectively) during a period when evapotranspiration was causing diurnal fluctuations in discharge. We also measured in-stream anion and cation concentrations during a 34 h period to assess changes in transport and transformation that may be apparent in fluctuations of the ambient water chemistry.

2. Methods

2.1. Site Description

Experiments were conducted in Watershed 01 (WS01) in the H.J. Andrews Experimental Forest (Western Cascade Mountains, Oregon, USA) during July and August 2016. The study stream is steep and characterized by a pool–step morphology. The valley is bedrock-constrained both vertically and laterally and overlain by colluvium deposited from hillslope failures. The catchment and study reach have been investigated extensively and well characterized in past field and modeling studies [5,23,53,84,97–102].

For this study, we established an experimental reach within WS01 of about 60 m in length. Pressure transducers (U20L-04 model, Onset) measured stream stage at three locations within the study reach at 15-min intervals (see additional details in related publications [5,53]). The study reach was selected to include a reach that would be spatially intermittent during the entire study period, ensuring that all stream water and solutes would turnover through the hyporheic zone. The site includes a large gravel wedge and step through which 100% of streamflow is transported through the subsurface. Vegetation in the basin is primarily red alder in the valley bottom, with stands of Douglas fir of varied ages and shrubs on the hillslopes [24]. Stream discharge varied due to evapotranspiration in the basin [5,22–24,95,103–108], but contiguous surface flow along the reach was never observed (i.e., the reach was spatially intermittent for the entire study period). We monitored the maximum and minimum extent of surface water during the study, flagging the most expansive and contracted conditions within the study reach and mapping their positions with a topographic survey (Figure 1).

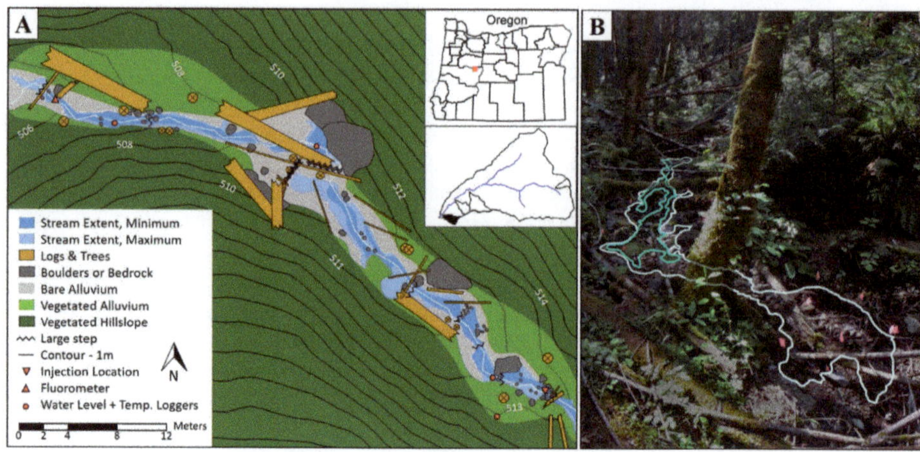

Figure 1. (**A**) Plan-view map of the study reach, including surface water extent at the diurnal minimum and maximum, channel morphology, reach topography, and logger locations. (**B**) Image looking upstream at the large step in the middle of the reach. Annotation added to show the wetted perimeter at the diurnal maximum (light blue) and minimum (darker blue) extent.

2.2. Stream Solute Tracer Experiments

To measure reach-scale transport and transformation processes, we conducted a series of four replicate solute tracer (both conservative and reactive) injections between 29 July and 7 Aug-2016, a baseflow period free of precipitation. During the study period, diurnal peak and minimum discharges occurred at approximately 13:00 and 04:00, respectively (Figure 2G–I). These instantaneous injections were timed to start at minimum, rising, peak, and falling discharge conditions, with each tracer test measured for more than 24 h, and thus each tracer test underwent the full range of diurnal discharge fluctuations. For each injection, we released known masses of NaCl, Resazurin (Raz), and Uranine (Ura) at a location about 550 m upstream from the weir in WS01 (Figure 1A; Table 1). We monitored in-stream concentrations of Uranine, Resazurin, and Resorufin (Rru; daughter product of Raz) at 10-sec frequency at a location about 55 m downstream of the injection point (GGUN-FL30 Fluorometer, Albilla, Inc.). Sensors for Ura, Raz, and Rru were calibrated using a laboratory fluorometer and standards made with stream water from the site. Grab samples were collected regularly throughout the injections and run on the laboratory fluorometer to correct for drift.

We measured in-stream fluid conductivity about 5 m downstream of the injection location and converted measured specific conductance to concentration of NaCl using a calibration curve developed by adding known masses of NaCl to water from the study site. These observations were used to calculate stream discharge via dilution gauging for the minimum, rising, and peak discharge conditions prevailing during the different tracer injections (logger failure on falling limb injection). The stream channel above and within the 5-m dilution gaging reach is primarily bedrock, giving us confidence that these measurements captured the majority of the water flowing through the valley. We used stage measurements from an in-stream water level logger co-located with the fluid conductivity monitoring location to construct a linear stage–discharge curve from these three discharge values. Using the regression and the 15 min stage record, we estimated continuous discharge throughout the study period.

We interpreted Ura time series to describe conservative transport through the study reach. While Ura is known to photodecay, the study reach is well-shaded and has a heavy tree canopy. As a proxy for reactive transport and metabolic activity through our reach, we assessed Raz and Rru time series. Raz-to-Rru transformation is a single reactive pathway commonly associated with aerobic respiration and is a reliable proxy for respiration in our study system [83,84,109].

Table 1. Summary of tracer injections.

Injection Code	Date and Time	NaCl (g)	Ura (g)	Raz (g)	Q at Release (L s^{-1})
Falling	17:15 29-July-2016	899.8	2.504	8.003	0.70
Rising	05:02 1-August-2016	599.0	2.504	8.005	0.59
Peak	10:39 3-August-2016	601.2	2.495	8.002	0.76
Minimum	23:52 5-August-2016	698.8	2.501	7.999	0.24

To characterize differences between the observed Ura breakthrough curves, we calculated a series of summary metrics to describe the transport. Notably, we used mass flux rather than concentration as the basis for these calculations to account for the time-variable discharge in the system. First, we calculated the time at which 99% of the recovered tracer exited the reach (t_{99}) and the travel time for the peak through the study reach (t_{peak}) based on the observed solute tracer timeseries. Next, observed concentrations ($C(t)$) and discharges ($Q(t)$) were converted to normalized mass flux ($m(t)$) as:

$$m(t) = \frac{Q(t)C(t)}{\int_{t=0}^{t=t_{99}} Q(t)C(t)dt} \tag{1}$$

The first temporal moment (M_1) is the median travel time through the study reach, calculated as:

$$M_1 = \int_{t=0}^{t=t_{99}} tm(t)dt \tag{2}$$

Next, 2nd and 3rd order central temporal moments (μ_2 and μ_3) were calculated as:

$$\mu_n = \int_{t=0}^{t=t_{99}} (t - M_1)^n m(t)dt \tag{3}$$

where n is the moment order. These moments can be combined to describe the symmetrical spreading normalized to travel time (coefficient of variation, CV) and asymmetrical late-time tailing (skewness, γ) as:

$$CV = \frac{\mu_2^{1/2}}{M_1} \tag{4}$$

and

$$\gamma = \frac{\mu_3}{\mu_2^{3/2}} \tag{5}$$

Additionally, we calculated holdback (H), which places the tracer in a range of plug flow ($H = 0$) to only dispersive transport ($H = 1$) [109], where higher values of H indicate increasing importance of non-advective transport processes:

$$H = \frac{1}{M_{1,norm}} \int_{t=0}^{M_1} F(t)dt \tag{6}$$

$$F(t) = \int_{t=0}^{t} m(\tau)d\tau \tag{7}$$

where F is a dummy variable for integration. Finally, we calculated apparent dispersion (D_{app}) as:

$$D_{app} = \frac{\mu_2 L^2}{2M_1} \tag{8}$$

where L is the length of the study segment.

2.3. StorAge Selection Modeling

We used the StorAge Selection (SAS) approach to interpret conservative solute transport through the study reach [94–96]. Briefly, the SAS approach considers outflow as a combination of waters sampled from different ages within the total storage volume between the injection and observation locations. The approach does not require an arbitrary division of surface from subsurface water, nor does it force observations to be fit by various transport mechanisms (e.g., advection, dispersion). Furthermore, this approach allows us to examine the potential importance of down-valley underflow (i.e., down-valley flow in the subsurface [23,110–112]) by incorporating this as an additional flux through the stream reach whose rate we hope to infer from the data. Importantly, the SAS approach is time-variable, whereas the summary metrics we calculated above (e.g., temporal moments) are integrative through time. Here, we closely follow the approach of Harman et al. [95] in describing the SAS function for the channel (Ω_C) as a shifted gamma distribution with offset S_{min}, shape parameter α, and scale parameter β (which is related to the mean, μ, as $\beta = \mu/\alpha$):

$$\Omega_C(S_T, t) = \frac{\gamma\left(\alpha, \frac{S_T - S_{min}}{\beta}\right)}{\Gamma(\alpha)} \quad (S_T > S_{min}) \tag{9}$$

where $\Gamma(\alpha)$ and $\gamma(\alpha, x)$ are the complete and incomplete gamma functions. In keeping with the conceptual model of Ward et al. [23], we represented the down-valley underflow with a separate SAS function (Ω_U), assumed to be an exponential distribution with scale parameter (S_U):

$$\Omega_U(S_T, t) = \frac{1}{S_U} exp\left(-\frac{S_T}{S_U}\right) \tag{10}$$

The SAS parameters and underflow discharge (Q_U) are each linked to the observed channel discharge by the relationships:

$$S_{min}(t) = S_{minref} exp\left(k_{S_{min}}(Q - Q_{ref})\right) \tag{11}$$

$$\alpha(t) = \alpha_{ref} exp\left(k_\alpha(Q - Q_{ref})\right) \tag{12}$$

$$\mu(t) = \mu_{ref} exp\left(k_\mu(Q - Q_{ref})\right) \tag{13}$$

$$S_U(t) = S_{Uref} exp\left(k_{SU}(Q - Q_{ref})\right) \tag{14}$$

$$Q_U(t) = Q_{Uref} exp\left(k_{QU}(Q - Q_{ref})\right) \tag{15}$$

where the reference discharge, Q_{ref}, is 0.55 L s^{-1}, (the approximate mean discharge during the period) at which the parameters take on values $S_{min,ref}$, α_{ref}, μ_{ref}, S_{Uref}, and Q_{Uref}. The sensitivity parameters k_{Smin}, α_{ref}, and μ_{ref} determine the direction and strength of the relationship between the parameters and discharge and can be positive or negative.

Parameters were inferred by Bayesian parameter estimation using DREAM [113,114]. Model likelihoods were estimated by comparing the log likelihood of each parameter set to observed conservative transport of Ura through the study reach, assuming a Gaussian error (σ_{error}). This error was inferred alongside the model parameters. DREAM was run with 12 chains for 100,000 generations each. After 10,000 generations, convergence was observed and prior generations were discarded. For the majority of the analysis below, the maximum likelihood parameter set was used. We report results as the maximum likelihood parameter set, and the 10th through 90th percentile range of the parameters in the retained generations was used to evaluate parameter identifiability.

2.4. In-Stream Biogeochemistry

To test for diurnal fluctuations in solute concentrations that might be associated with fluctuations in discharge, we collected samples from the stream channel just above the tracer injection site every 2 h for 34 h, from 10:00, 3 Aug 2016 to 14:00, 4 Aug 2016, contemporaneously with the third solute tracer injection. All sample collection, preservation, and analyses follow the methods described by Ward et al. [114]. Briefly, aliquots were filtered on-site for nutrient and major ion analysis using a 0.45 μm PES filter and for carbon analysis using a 0.2 μm cellulose acetate filter into 60 mL HDPE bottles. All bottles and syringes were triple rinsed with sample water. All samples were chilled immediately after collection, and nutrient/major ion samples were frozen within 24 h of collection until analysis. Samples were analyzed for anions (Cl^-, SO_4^-; Dionex Ion Chromatograph, University of Birmingham), nutrients (NO_3^-, NH_3, PO_4^{3-}; Skalar Nutrient Analyzer, University of Birmingham), and dissolved organic carbon (DOC; Shimadzu TOC-Analyser, Michigan State University). Finally, we used Mann–Kendall tests to evaluate the likelihood of monotonically increasing or decreasing trends relating discharge to in-stream concentration for each analyte, interpreting $p > 0.10$ as indicating that a monotonic trend between discharge and concentration is unlikely.

3. Results and Discussion

3.1. Conservative Solute Tracer Transport Fluctuates Consistently with Unsteady Discharge

Observations of conservative tracer (Ura) isolate the role of time-variable discharge on residence times within the study reach. In-stream time series of concentrations (Figure 2A) and mass flux (Figure 2D) are visually consistent between injections, indicating that diurnal discharge fluctuations are controlling the timing of conservative mass transport and storage. In all four injections (falling, rising, peak, minimum), mass flux is reduced at the daily minimum discharge conditions, and rebounds rapidly when discharge and surface water extent increase. For the injections conducted on the rising and peak discharge conditions, the highest concentrations are advected through the study reach so rapidly that they do not experience minimum discharge conditions (i.e., peaks prior to minimum discharge; Figure 2A). As a result, the highest concentration water was preferentially exported from the study reach under higher discharges rather than being stored. Consistent with this preferential export, the late-time tail is no longer visible as the second rising limb of discharge occurs (X = 60 h since 00:00 on day of injection; Figure 2A). In contrast, the peak concentration water from the tracer test remained in the study reach during the minimum discharge conditions (Figure 2A). Consequently, peak concentration water was stored in the reach rather than rapidly exported during minimum discharge conditions.

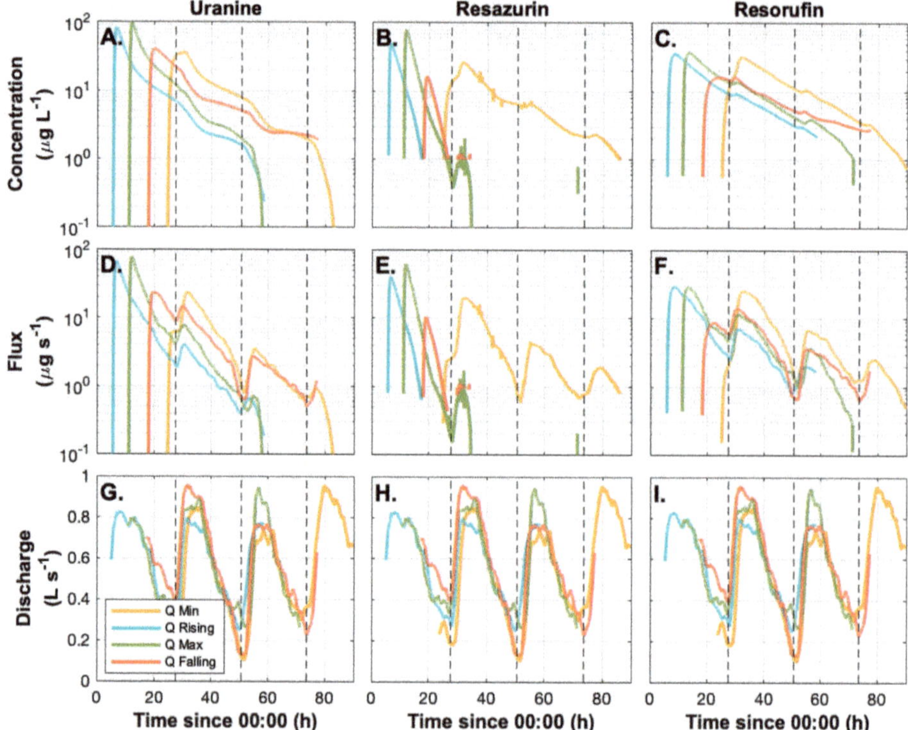

Figure 2. Time series of concentration (panels **A–C**), mass flux (panels **D–F**), and discharge (panels **G–I**) for all experiments, and all tracers including Uranine (conservative; panels A, D, E), Resazurin (parent compound; panel B, E, H), and Resorufin (daughter product; panels C, F, I). The dashed vertical lines indicate the time of lowest discharge.

Despite the differences in timing of advection of the highest concentration water relative to the time of minimum discharge, we found few differences between the injections with respect to calculated solute transport metrics. For example, we found no significant trends between discharge and any of the Ura breakthrough curve metrics (Table 2). Similarly, a single SAS function produced reasonable representations of the observed Ura time series for all injections ($\sigma_{error} = 3.2$ µg L^{-1}). Moreover, all SAS model parameters were well constrained by the data. The S_{min} was the parameter least sensitive to discharge. Both Q_U and S_U were well constrained and sensitive to discharge. The best-fit parameters vary by a factor of 4–6-fold (Table 3). However, the underflow time scale estimated as S_U/Q_U varied from about 21 h at low flow to about 31 h at high flow. The SAS function varied systematically with discharge (Figure 3I–L and Figure 4). The long-term mean SAS function has an offset of about 2.5 m^3, representing the approximate plug flow volume in the study reach. The SAS function peaks around 15 m^3 of storage (the most frequently exported age-ranked storage), and then decays toward the oldest storage.

Table 2. Summary metrics for Uranine mass flux time series.

Discharge at Time of Injection	t_{peak} (h)	t_{99} (h)	M_1 (h)	CV	Skewness	Apparent Dispersion ($\times 10^4$ m^2 h^{-1})	Holdback
Falling	2.3	55.3	14.3	0.85	1.35	1.55	0.61
Rising	1.7	45.0	8.5	1.04	1.80	1.39	0.68
Peak	1.4	39.3	8.2	1.01	1.48	1.26	0.68
Minimum	7.9	53.5	15.3	0.70	1.46	1.15	0.67

Table 3. Maximum likelihood parameter values from DREAM parameter estimation. Values in parentheses show the 10–90th percentile ranges. Columns show the value at median discharge, sensitivity (1/k), and parameter value at Q_{10} (about 0.28 L s^{-1}) and Q_{90} (about 0.83 L s^{-1}).

	Parameter Values			Parameter Sensitivity
Parameter	Reference Value	Value at Q_{10}	Value at Q_{90}	$1/k_X$ *
Storage offset, S_{min} (L)	3336 (*3288, 3617*)	3728	2981	−1.25 (*−1.61, −1.33*)
Mean of gamma distribution, μ (L)	29,063 (*28,981, 29,713*)	37,406	22,513	−1.10 (*−1.11, −1.02*)
Gamma distribution shape parameter, α (unitless)	1.63 (*1.57, 1.64*)	2.50	1.06	−0.645 (*−0.717, −0.632*)
Underflow discharge, Q_U (L s^{-1})	0.390 (*0.387, 0.405*)	0.954	0.158	−1.03E-3 (*−1.06E-3, −1.00E-3*)
Underflow scale parameter, S_U (L)	36,305 (*35,945, 38,837*)	73,414	17,802	−0.393 (*−0.403, −0.358*)

* Values in each row are $1/k_X$, where X is the parameter being described in each row. For example, the values in the top row are for $1/k_{S_{min}}$.

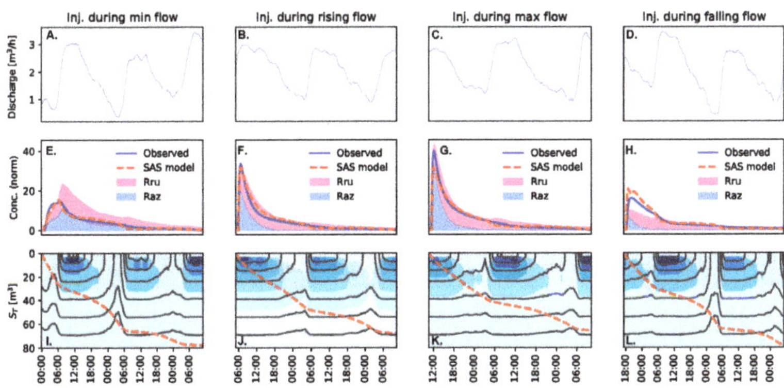

Figure 3. Time-series of discharge (panels **A–D**), observed and simulated concentration normalized by input mass (panels **E–H**), and density of the SAS function (panels **I–L**; darker color indicates higher probability). A vertical section along the bottom row can be interpreted as a probability density function of the age-ranked storage being exported from the study reach at any given time step.

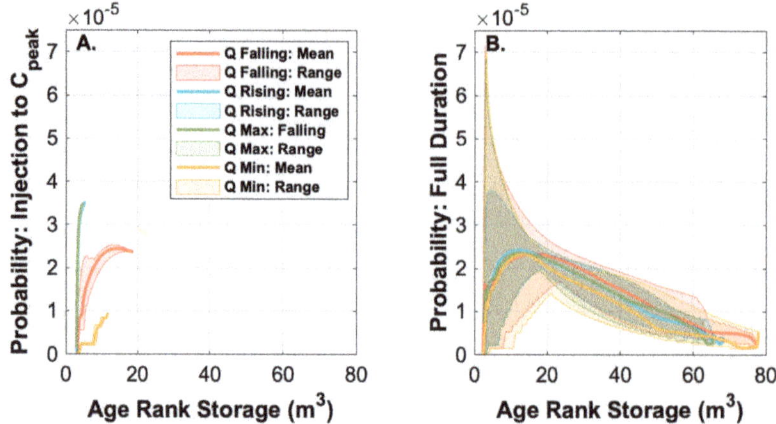

Figure 4. (**A**) Mean and range of SAS functions between tracer injection and the first diurnal peak in discharge after each injection. Note that the solute tracer injected at minimum discharge is too young to be exported prior to peak discharge, compared to all other injections which are biased toward export of the youngest age-ranked storage. (**B**) The full range of and overall mean SAS function for each injection. Time-variable SAS functions are displayed in the supplemental animation, showing how early-time disparities converge over the course of the entire time series.

During the rising, peak, and falling discharge injections, the tracer age was almost perfectly aligned with the ages being predominantly exported rather than stored. In contrast, tracer was too young to be exported rapidly during the minimum discharge injection. Thus, tracer injected at minimum discharge was predominantly stored in the reach, ultimately labeling older age-ranked storage that is more slowly and consistently exported from the study reach. These differences are attributable to the timing of the study release, as all four injections were well-described by a single SAS function and did not exhibit wide variation in summary metrics such as temporal moments. Indeed, the evolution of SAS functions, and particularly the conditions experienced shortly after each injection, may dominate overall transport and transformation dynamics in the study reach.

3.2. Observed Variation in Solute Transport is Primarily Due to Storage Release Timing Relative to an Oscillating Diurnal Cycle in the Underlying Transport Processes

The best-fit SAS function reveals that high-discharge periods are dominated by the release of younger water, and lower-discharge periods by the release of older water (Figure 3). Conceptually, this means periods of higher discharge preferentially shunt younger water through and out of the system, bypassing older storage. In contrast, during periods of low discharge, the water entering the upstream end of the study segment cannot be rapidly transported through and out of the reach. Instead, water entering during low-discharge conditions is almost entirely stored, and older water is displaced and released from the study reach. This means that the general principles of the pulse-shunt model [61] may be operating on a daily basis for a small headwater stream experiencing diurnal fluctuations in discharge. Overall, we found a single parameterization of the SAS function provided a strong fit to observed data, which indicates that the same consistent diurnal cycles of transport processes were present during all four injections and varied systematically with diurnal fluctuations in discharge. However, the observed in-stream time series demonstrate that the timing of the solute tracer release relative to the diurnal cycle is important in determining which age-ranked storage is sampled by the solute tracer, and how the tracer is ultimately stored in and released from the study reach.

We generally found that the occurrence of minimum discharge conditions serves as a diurnal reset on transport in the system. The observed Ura time series are systematically varied before reaching the first diurnal minimum discharge condition after injection (X = 24:00 in Figure 2). After this time of minimum discharge, the slopes of the falling limbs are similar and responses of concentration and flux are synchronized with minimum discharge conditions. As such, we focus on the time between the injection and first minimum discharge occurrence as critical in defining differences between injections. Transport between the injection and minimum discharge condition, when youngest storage is mostly being exported from the system, are important determinants of the observed in-stream tracer time series in the first 24 h after 00:00 on the day of release.

3.3. Spatial and Temporal Variation in Raz-to-Rru Transformation

3.3.1. Rru Production Does Not Appear to Change with Discharge

How the coupled transport and transformation of Raz and the resultant Rru varies between injections is inconsistent with Ura. This inconsistency indicates that fluctuations in residence time are driving reaction progress rather than changes in reaction rates. For the rising, peak, and falling discharge conditions, in-stream Raz is predominantly transported through the reach prior to the first rising limb after injection (Figure 2B). In contrast, Raz injected during minimum discharge conditions is stored in the reach and slowly exported during the 2.5 days after its initial release. This is notable for two reasons. First, Raz is stored and released in substantially higher amounts than in other injections, with a time series more comparable to the conservative Ura tracer than the other Raz time series. Second, the late-time Raz is being exported in its unreacted form for the minimum discharge release, whereas all other injections have little-to-no measurable unreacted Raz after minimum discharge.

Reaction potential occurred throughout the study period, as evidenced by rapid Rru production in the study reach for all injections (Figure 2). The slopes of the rising limb of Rru production are similar, which we interpret as a comparable transformation rate across all injections occurring along the shortest and fastest flowpaths through the reach. Notably, peak Rru concentrations occurred with the rising limb for the injections at falling and minimum discharge (compared to pre-rising limb peaks for the rising and peak discharge injections). For the injection at minimum discharge, Rru continued to be exported from the study reach in approximately equal concentration to Raz for 60 h after the injection. In all other cases, Rru exceeded Raz (i.e., reacted tracer exceeds unreacted tracer) after the first rising limb of the hydrograph.

As a second line of evidence, ambient in stream water chemistry exhibits no discernable trend with in-stream discharge (Figure 5; $p > 0.10$ for all solutes). We note here that in addition to discharge, other biogeochemically important factors likely varied on a diurnal basis, including dissolved oxygen, pH, solar radiation, and water temperature. We interpret the nearly steady-state concentrations for DOC, phosphate, and nitrate as indicative that no systematic shift in metabolism occurs in response to diurnal fluctuations in discharge or other factors (e.g., water temperature). Instead, we argue the long-term integration of the nitrogen, carbon, and phosphorous cycles in the system appears to be at approximately steady-state under the diel hydrologic forcing of flows in the stream. This interpretation is consistent with past studies in the same watershed, which concluded that biological limitations are the primary control on nutrient cycling during low flows [115].

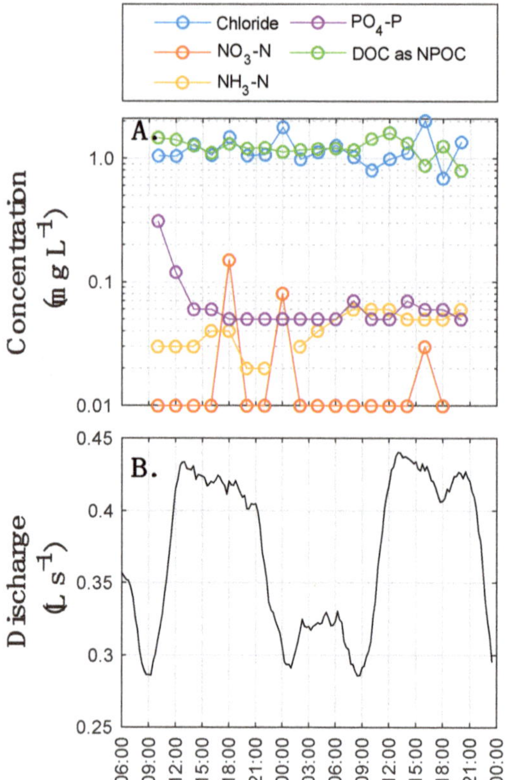

Figure 5. (**A**) Time series of nutrients and anions in the stream during diurnal fluctuations; (**B**) discharge during data collection.

3.3.2. Older Storage is Not as Metabolically Active as Younger Storage

While the transformation rates did not appear to be time-variable in our system, it does not mean that rates were spatially homogeneous. Instead, we found that both transport and transformation were controlled by injection timing, and ultimately the timescales of storage that were sampled by the tracers. During all injections, we observed rapid transformation of Raz-to-Rru associated with the youngest 20 h of age-ranked storage in the system (Figure 6A). However, during the three highest discharge injections, Raz was primarily transported through the study reach in less than 20 h, meaning that unreacted Raz largely bypassed storage within the reach rather than interacting with and documenting those domains. In contrast, the lowest discharge injection occurred at a time when water entering the study reach was stored rather than exported, allowing Raz to interact with and sample older age-ranked storage. Thus, the transformation of Raz-to-Rru is primarily associated with water spending less than 20 h in the study reach (Figure 6A), which corresponds to the youngest 40 m^3 of age-ranked storage in the reach (Figure 6B). Our observed higher transformation in the younger, presumably surficial, flowpaths can be explained by existing studies showing that the shallowest layer at the top of the streambed is the most metabolically active [75,116–118]. Further, this study demonstrates that the timing of tracer injection and concurrent flow conditions can influence the "window of detection" [118] of particular smart tracers for quantifying geochemical and ecological processes in the river corridor.

The persistence of unreacted Raz and the lack of Raz-to-Rru transformation in waters older than about 20 h (i.e., age-ranked storage 2–40 m^3) indicates that older storage is not metabolically active on the timescales relevant to the Raz additions (Figure 6). There are three possible explanations for this behavior. First, it is possible that the redox conditions along these flowpaths do not favor the Raz-to-Rru conversion, which is primarily sensitive to aerobic and facultative anaerobic conditions [73]. However, we found bulk oxic conditions in the hyporheic water [114]. While anoxic microsites may exist, they are not likely to be abundant in the coarse streambed substrate of the site [119]. Overall, it is not expected that these bulk oxic conditions of the site would cause the expected inhibition of Raz transformation. Second, and more likely, the metabolic activity in the older age-ranked storage is low because of stoichiometric limitations of metabolic pathways affecting Raz. Similar to other detrital-based, heterotrophic streams, low nitrogen and phosphorus concentrations at our study site could limit metabolic activity [120]. The bioavailability of these nutrients likely limit microbial metabolism in our system under all flow conditions [116,121,122]. However, we would expect these limitations to be minimized along shallow, intermittently inundated, and hydrologically youngest flowpaths that experience warmer temperatures, are nearer to light exposure, increased reactive surface area, and availability of fresh organic matter sources. Lastly, metabolic activity may be energetically constrained on timescales relevant to the water residence time along these longer and hydrologically older flowpaths. Previous work has suggested that buried particulate organic matter from debris flows promotes hyporheic microbial respiration at the study site [122]. While this buried organic matter, likely western hemlock (*Tsuga heterophylla*) and Douglas fir (*Pseudotsuga menziesii*) wood and needles, may persistently increase metabolic activity of the oldest water at the site, it is also likely a low-quality organic matter source with less reaction potential, relative to the younger organic matter along the shallower hyporheic flowpaths, given its composition and old age within the stream corridor [123]. Together, there are multiple potential limitations on the Raz-to-Rru transformation that may exist in the older water at the site. Further work is needed to identify the location of the older hydrologic storage at the site, which will allow for the in situ sampling of the pore water chemistry needed to directly test these mechanistic constraints on Raz-to-Rru transformation.

While the exact location and distribution of the metabolically active storage within the study reach is unknown, we can ground our SAS-based interpretation with calculations of the known extent of hyporheic exchange in our study system. We estimate a surface water volume of 4.4 m^3 (typical width and depth of 0.8 and 0.1 m along the 55 m length) in the study reach, leaving approximately an additional 35.6 m^3 of metabolically active storage in the subsurface of the study reach (estimated as transformations in the youngest 40 m^3 of age-ranked storage minus the in-stream volume based on visual inspection of minimal transformation of Raz-to-Rru in older storage; Figure 6B). Notably, the SAS interpretation indicated about 2.5 m^3 as the plug flow volume, suggesting 1.9 m^3 of in-stream transient storage. On average, this would represent a metabolically active storage zone with a cross-sectional area of 0.65 m^2 along the study reach, which is about 8 times larger than the average cross-sectional area of the stream. This cross-sectional area is smaller than the broad penetration of tracer into the riparian zone observed during longer-duration injections monitored by electrical resistivity and in a network of monitoring wells [19–22]. However, this is consistent with a persistent, near-stream hyporheic zone [21] that would have been sampled by an instantaneous injection of tracer due to window of detection limitations [124,125]. Similarly, past studies found a metabolically active storage zone of about 2.45 times the area of the stream in a nearby headwater catchment [83].

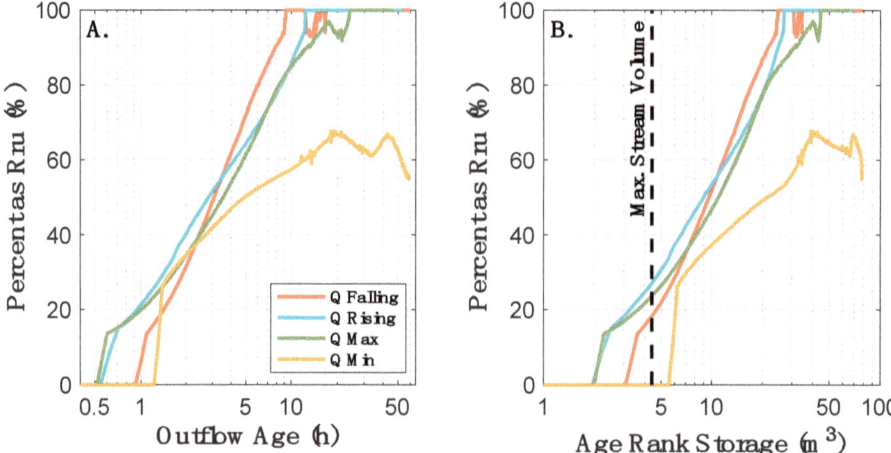

Figure 6. Fraction of outflowing Raz that has been transformed to Rru as a function of (**A**) outflow age, and (**B**) age-ranked storage in the study reach.

4. Conclusions

The objective of this study was to assess how reach-scale transport and transformation of reactive tracers vary in response to diurnal discharge fluctuations in an intermittent, headwater stream. Based on visual inspection, tracer transport varies depending upon discharge dynamics at the time of release into the system. However, a single time-variable SAS function was able to represent the variation with discharge in the system, and no summary metrics exhibited a significant trend with discharge. Thus, we conclude the transport patterns describe a single StorAge Selection function that varies with discharge, at least across the range of discharge observed in our study. Variation with time of release is visually apparent (Figure 2), but these differences are not sufficient to alter the overall storage and release of tracer at the timescales associated with the entire reach.

With respect to solute transformation potential of the stream, we expected diurnal fluctuations in discharge would alter contact times and overall observed Raz-to-Rru conversion in the system. Instead, we found little variation in Raz-to-Rru reaction rate through time. Our results suggest that the biological processes controlling the Raz-to-Rru transformation are static, and variations in the reactive transport of Raz are primarily due to variations in residence time and locations of water and solute storage, not transformation processes. This is in opposition to the simplest interpretation of the observed time series, which appear to indicate different transformation rates as a function of discharge (e.g., less transformation for the minimum discharge injection). However, with the aided insight of the SAS analysis, we see that, across all injections, Raz-to-Rru primarily occurs in the water with an outflow age of less than about 20 h, representing the youngest 40 m^3 of age-ranked storage in the reach. By injecting at different discharge conditions and exploring how age of stored water varied across those flow conditions, we were able to sample the metabolic activity of different storage locations within the system and show that the youngest storage has higher metabolic activity than the oldest storage. Therefore, not all storage in the stream has the same potential biogeochemical function, and increasing residence time of solute storage does not necessarily increase reaction potential of that solute.

Supplementary Materials: The following are available online at http://www.mdpi.com/2073-4441/11/11/2208/s1: Video S1: Evolution of SAS functions.

Author Contributions: Conceptualization, A.S.W., M.J.K., N.M.S., J.L.A.K., P.J.B., D.M.H., S.K., E.M., and J.P.Z.; data curation, A.S.W. and S.P.; formal analysis, M.J.K., N.M.S., J.L.A.K., C.J.H., J.D.D., M.M., and S.M.W.; funding acquisition, A.S.W., D.M.H., S.K., A.M., A.I.P., S.M.W., and J.P.Z.; investigation, M.J.K., N.M.S., J.L.A.K., P.J.B., J.D.D., A.L., K.N., M.M., S.P., and N.I.M.; methodology, M.J.K. and J.L.A.K.; project administration, A.S.W. and J.P.Z.; resources, A.S.W., D.M.H., S.K., E.M., A.I.P., and J.P.Z.; visualization, C.J.H.; writing—original draft, A.S.W.

and M.J.K.; writing—review and editing, M.J.K., N.M.S., J.L.A.K., P.J.B., C.J.H., J.D.D., D.M.H., S.K., A.L., E.M., A.M., K.N., M.M., S.P., A.I.P., N.I.W., S.M.W., and J.P.Z.

Funding: This research was funded by the Leverhulme Trust (Where rivers, groundwater and disciplines meet: A hyporheic research network), the UK Natural Environment Research Council (Large woody debris—A river restoration panacea for streambed nitrate attenuation? NERC NE/L003872/1), and the European Commission supported HiFreq: Smart high-frequency environmental sensor networks for quantifying nonlinear hydrological process dynamics across spatial scales (project ID 734317). Data and facilities were provided by the HJ Andrews Experimental Forest and Long Term Ecological Research program, administered cooperatively by the U.S. Department of Agriculture Forest Service Pacific Northwest Research Station, Oregon State University, and the Willamette National Forest and funded, in part, by the National Science Foundation under Grant No. DEB-1440409. Additional support to individual authors is acknowledged from National Science Foundation (NSF) awards EAR 1652293, EAR 1417603, EAR 1846855, and EAR 1446328, Department of Energy (DOE) award DE-SC0019377, and DOE's Office of Biological and Environmental Research via the Mercury Scientific Focus Area at Oak Ridge National Laboratory.

Acknowledgments: Any use of trade, firm, or product names is for descriptive purposes only and does not imply endorsement by the U.S. Government nor any author. Underlying data for this study are hosted by CUAHSI's HydroShare at: http://www.hydroshare.org/resource/d353ce4a7dc643c8a566504bfd578a57.

Conflicts of Interest: The authors declare no conflict of interest.

References

1. Boano, F.; Harvey, J.W.; Marion, A.; Packman, A.I.; Revelli, R.; Ridolfi, L.; Worman, A. Hyporheic flow and transport processes: Mechanisms, models, and biogeochemical implications. *Rev. Geophys.* **2014**, *52*, 2012RG000417. [CrossRef]

2. Maher, K. The role of fluid residence time and topographic scales in determining chemical fluxes from landscapes. *Earth Planet. Sci. Lett.* **2011**, *312*, 48–58. [CrossRef]

3. Bardini, L.; Boano, F.; Cardenas, M.B.; Revelli, R.; Ridolfi, L. Nutrient cycling in bedform induced hyporheic zones. *Geochim. Cosmochim. Acta* **2012**, *84*, 47–61. [CrossRef]

4. Harvey, J.; Gomez-Velez, J.; Schmadel, N.M.; Scott, D.; Boyer, E.; Alexander, R.; Eng, K.; Golden, H.; Kettner, A.; Konrad, C.; et al. How Hydrologic Connectivity Regulates Water Quality in River Corridors. *J. Am. Water Resour. Assoc.* **2019**, *55*, 369–381. [CrossRef]

5. Ward, A.S.; Schmadel, N.M.; Wondzell, S.M. Time-Variable Transit Time Distributions in the Hyporheic Zone of a Headwater Mountain Stream. *Water Resour. Res.* **2018**. [CrossRef]

6. Van Meerveld, H.J.I.; Kirchner, J.W.; Vis, M.J.P.; Assendelft, R.S.; Seibert, J. Expansion and contraction of the flowing stream network changes hillslope flowpath lengths and the shape of the travel time distribution. *Hydrol. Earth Syst. Sci. Discuss.* **2019**, *2006*, 1–18. [CrossRef]

7. Pinay, G.; Clément, J.C.; Naiman, R.J. Basic Principles and Ecological Consequences of Changing Water Regimes on Nitrogen Cycling in Fluvial Systems. *Environ. Manage.* **2002**, *30*, 481–491. [CrossRef] [PubMed]

8. Ward, A.S.; Gooseff, M.N.; Voltz, T.J.; Fitzgerald, M.; Singha, K.; Zarnetske, J.P. How does rapidly changing discharge during storm events affect transient storage and channel water balance in a headwater mountain stream? *Water Resour. Res.* **2013**, *49*, 5473–5486. [CrossRef]

9. Dudley-Southern, M.; Binley, A. Temporal responses of groundwater-surface water exchange to successive storm events. *Water Resour. Res.* **2015**, *51*, 1112–1126. [CrossRef]

10. Schmadel, N.M.; Ward, A.S.; Lowry, C.S.; Malzone, J.M. Hyporheic exchange controlled by dynamic hydrologic boundary conditions. *Geophys. Res. Lett.* **2016**, *43*, 4408–4417. [CrossRef]

11. Malzone, J.M.; Anseeuw, S.K.; Lowry, C.S.; Allen-King, R. Temporal Hyporheic Zone Response to Water Table Fluctuations. *Groundwater* **2015**, *54*, 274–285. [CrossRef] [PubMed]

12. Malzone, J.M.; Lowry, C.S.; Ward, A.S. Response of the hyporheic zone to transient groundwater fluctuations on the annual and storm event time scales. *Water Resour. Res.* **2016**, *52*, 1–20. [CrossRef]

13. Sawyer, A.H.; Shi, F.; Kirby, J.T.; Michael, H.A. Dynamic response of surface water-groundwater exchange to currents, tides, and waves in a shallow estuary. *J. Geophys. Res. Ocean.* **2013**, *118*, 1749–1758. [CrossRef]

14. Musial, C.; Sawyer, A.H.; Barnes, R.; Bray, S.; Knights, D. Surface water–groundwater exchange dynamics in a tidal freshwater zone. *Hydrol. Process.* **2016**, *30*, 739–750. [CrossRef]

15. Bianchin, M.S.; Smith, L.; Beckie, R.D. Defining the hyporheic zone in a large tidally influenced river. *J. Hydrol.* **2011**, *406*, 16–29. [CrossRef]

16. Knights, D.; Sawyer, A.H.; Barnes, R.T.; Musial, C.T.; Bray, S. Tidal controls on riverbed denitrification along a tidal freshwater zone. *Water Resour. Res.* **2017**, *53*, 799–816. [CrossRef]

17. Payn, R.A.; Gooseff, M.N.; McGlynn, B.L.; Bencala, K.E.; Wondzell, S.M. Channel water balance and exchange with subsurface flow along a mountain headwater stream in Montana, United States. *Water Resour. Res.* **2009**, *45*. [CrossRef]

18. Ward, A.S.; Payn, R.A.; Gooseff, M.N.; McGlynn, B.L.; Bencala, K.E.; Kelleher, C.A.; Wondzell, S.M.; Wagener, T. Variations in surface water-ground water interactions along a headwater mountain stream: Comparisons between transient storage and water balance analyses. *Water Resour. Res.* **2013**, *49*, 3359–3374. [CrossRef]

19. Ward, A.S.; Fitzgerald, M.; Gooseff, M.N.; Voltz, T.J.; Binley, A.M.; Singha, K. Hydrologic and geomorphic controls on hyporheic exchange during base flow recession in a headwater mountain stream. *Water Resour. Res.* **2012**, *48*, W04513. [CrossRef]

20. Ward, A.S.; Gooseff, M.N.; Fitzgerald, M.; Voltz, T.J.; Singha, K. Spatially distributed characterization of hyporheic solute transport during baseflow recession in a headwater mountain stream using electrical geophysical imaging. *J. Hydrol.* **2014**, *517*, 362–377. [CrossRef]

21. Wondzell, S.M.; Swanson, F.J. Seasonal and Storm Dynamics of the Hyporheic Zone of a 4th-Order Mountain Stream. I: Hydrologic Processes. *J. North Am. Beth. Soc.* **1996**, *15*, 3–19.

22. Voltz, T.J.; Gooseff, M.N.; Ward, A.S.; Singha, K.; Fitzgerald, M.; Wagener, T. Riparian hydraulic gradient and stream-groundwater exchange dynamics in steep headwater valleys. *J. Geophys. Res. Earth Surf.* **2013**, *118*, 953–969.

23. Ward, A.S.; Schmadel, N.M.; Wondzell, S.M. Simulation of dynamic expansion, contraction, and connectivity in a mountain stream network. *Adv. Water Resour.* **2018**, *114*, 64–82.

24. Bond, B.J.; Jones, J.A.; Moore, G.; Phillips, N.; Post, D.; McDonnell, J.J. The zone of vegetation influence on baseflow revealed by diel patterns of streamflow and vegetation water use in a headwater basin. *Hydrol. Process.* **2002**, *16*, 1671–1677.

25. Koch, J.C.; McKnight, D.M.; Neupauer, R.M. Simulating unsteady flow, anabranching, and hyporheic dynamics in a glacial meltwater stream using a coupled surface water routing and groundwater flow model. *Water Resour. Res.* **2011**, *47*, 1–15. [CrossRef]

26. Wlostowski, A.N.; Gooseff, M.N.; McKnight, D.M.; Lyons, W.B. Transit Times and Rapid Chemical Equilibrium Explain Chemostasis in Glacial Meltwater Streams in the McMurdo Dry Valleys, Antarctica. *Geophys. Res. Lett.* **2018**, *45*, 13322–13331. [CrossRef]

27. Lowry, C.S.; Deems, J.S.; Loheide, S.P.; Lundquist, J.D. Linking snowmelt-derived fluxes and groundwater flow in a high elevation meadow system, Sierra Nevada Mountains, California. *Hydrol. Process.* **2010**, *24*, 2821–2833.

28. Ii, S.P.L.; Lundquist, J.D. Snowmelt-induced diel fluxes through the hyporheic zone. *Water Resour. Res.* **2009**, *45*, 1–9.

29. Sawyer, A.H.; Cardenas, M.B.; Bomar, A.; Mackey, M. Impact of dam operations on hyporheic exchange in the riparian zone of a regulated river. *Hydrol. Process.* **2009**, *23*, 2129–2137.

30. Gerecht, K.E.; Cardenas, M.B.; Guswa, A.J.; Sawyer, A.H.; Nowinski, J.D.; Swanson, T.E. Dynamics of hyporheic flow and heat transport across a bed-to-bank continuum in a large regulated river. *Water Resour. Res.* **2011**, *47*, W03524. [CrossRef]

31. Hancock, P.J.; Boulton, A.J. The effects of an environmental flow release on water quality in the hyporheic zone of the Hunter River, Australia. *Hydrobiologia* **2005**, *552*, 75–85. [CrossRef]

32. Liu, D.; Zhao, J.; Chen, X.; Li, Y.; Weiyan, S.; Feng, M. Dynamic processes of hyporheic exchange and temperature distribution in the riparian zone in response to dam-induced water fluctuations. *Geosci. J.* **2018**, *22*, 465–475. [CrossRef]

33. Paul, M.T.; Meyer, J.L. Streams in the Urban Landscape. *Annu. Rev. Ecol. Evol. Syst.* **2001**, *32*, 333–365. [CrossRef]

34. Butturini, A.; Sabater, F. Importance of transient storage zones for ammonium and phosphate retention in a sandy-bottom Mediterranean stream. *Freshw. Biol.* **1999**, *41*, 593–603. [CrossRef]

35. Fabian, M.W.; Endreny, T.A.; Bottacin-Busolin, A.; Lautz, L.K. Seasonal variation in cascade-driven hyporheic exchange, northern Honduras. *Hydrol. Process.* **2011**, *25*, 1630–1646. [CrossRef]

36. Wondzell, S.M. The role of the hyporheic zone across stream networks. *Hydrol. Process.* **2011**, *25*, 3525–3532. [CrossRef]

37. Zarnetske, J.P.; Gooseff, M.N.; Brosten, T.R.; Bradford, J.H.; McNamara, J.P.; Bowden, W.B. Transient storage as a function of geomorphology, discharge, and permafrost active layer conditions in Arctic tundra streams. *Water Resour. Res.* **2007**, *43*, W07410. [CrossRef]

38. Hart, D.R.; Mulholland, P.J.; Marzolf, E.R.; DeAngelis, D.L.; Hendricks, S.P. Relationships between hydraulic parameters in a small stream under varying flow and seasonal conditions. *Hydrol. Process.* **1999**, *13*, 1497–1510. [CrossRef]

39. Jin, H.S.; Ward, G.M. Hydraulic characteristics of a small Coastal Plain stream of the southeastern United States: Effects of hydrology and season. *Hydrol. Process.* **2005**, *19*, 4147–4160. [CrossRef]

40. Karwan, D.L.; Saiers, J.E. Influences of seasonal flow regime on the fate and transport of fine particles and a dissolved solute in a New England stream. *Water Resour. Res.* **2009**, *45*, W11423. [CrossRef]

41. Morrice, J.A.; Valett, H.M.; Dahm, C.N.; Campana, M.E. Alluvial characteristics, groundwater-surface water exchange and hydrological retention in headwater streams. *Hydrol. Process.* **1997**, *11*, 253–267.

42. Schmid, B.H. Can Longitudinal Solute Transport Parameters Be Transferred to Different Flow Rates? *J. Hydrol. Eng.* **2008**, *13*, 505–509.

43. Schmid, B.H.; Innocenti, I.; San, U.; Sanfilippo, U. Characterizing solute transport with transient storage across a range of flow rates: The evidence of repeated tracer experiments in Austrian and Italian streams. *Adv. Water Resour.* **2010**, *33*, 1340–1346.

44. Ward, A.S.; Gooseff, M.N.; Singha, K. How does subsurface characterization affect simulations of hyporheic exchange? *GroundWater* **2013**, *51*, 14–28.

45. Wondzell, S.M. Effect of morphology and discharge on hyporheic exchange flows in two small streams in the Cascade Mountains of Oregon, USA. *Hydrol. Process.* **2006**, *20*, 267–287.

46. Jackson, T.R.; Haggerty, R.; Apte, S.V.; O'Connor, B.L. A mean residence time relationship for lateral cavities in gravel-bed rivers and streams: Incorporating streambed roughness and cavity shape. *Water Resour. Res.* **2013**, *49*, 3642–3650.

47. Jackson, T.R.; Haggerty, R.; Apte, S.V. A fluid-mechanics based classification scheme for surface transient storage in riverine environments: Quantitatively separating surface from hyporheic transient storage. *Hydrol. Earth Syst. Sci.* **2013**, *17*, 2747–2779.

48. Jackson, T.R.; Haggerty, R.; Apte, S.V.; Coleman, A.; Drost, K.J. Defining and measuring the mean residence time of lateral surface transient storage zones in small streams. *Water Resour. Res.* **2012**, *48*, 1–20.

49. Briggs, M.A.; Gooseff, M.N.; Arp, C.D.; Baker, M.A. A method for estimating surface transient storage parameters for streams with concurrent hyporheic storage. *Water Resour. Res.* **2009**, *45*. [CrossRef]

50. Pinder, G.F.; Sauer, S.P. Numerical simulation of flood wave modification due to bank storage effects. *Water Resour. Res.* **1971**, *7*, 63–70.

51. Cooper, H.; Rorabaugh, M. *Groundwater Movments and Bank Storage Due to Flood Stages in Surface Streams*; US Government Printing Office: Washington, DC, USA, 1963.

52. Ward, A.S.; Schmadel, N.M.; Wondzell, S.M.; Gooseff, M.N.; Singha, K. Dynamic hyporheic and riparian flow path geometry through base flow recession in two headwater mountain stream corridors. *Water Resour. Res.* **2017**, *53*, 3988–4003.

53. Schmadel, N.M.; Ward, A.S.; Wondzell, S.M. Hydrologic controls on hyporheic exchange in headwater mountain streams. *Water Resour. Res.* **2017**, *53*, 6260–6278.

54. Seitzinger, S.P.; Styles, R.V.; Boyer, E.W.; Alexander, R.B.; Billen, G.; Howarth, R.W.; Mayer, B.; Van Breemen, N. Nitrogen retention in rivers: Model development and application to watersheds in the northeastern USA. *Biogeochemistry* **2002**, *57*, 199–237.

55. Wollheim, W.M.; Vo, C.J.; Vorosmarty, C.J.; Peterson, B.J.; Seitzinger, S.P.; Hopkinson, C.S.; Vörösmarty, C.J. Relationship between river size and nutrient removal. *Geophys. Res. Lett.* **2006**, *33*, 4.

56. Basu, N.B.; Rao, P.S.C.; Thompson, S.E.; Loukinova, N.V.; Donner, S.D.; Ye, S.; Sivapalan, M. Spatiotemporal averaging of in-stream solute removal dynamics. *Water Resour. Res.* **2011**, *47*. [CrossRef]

57. Harvey, J.W.; Gooseff, M.N. River corridor science: Hydrologic exchange and ecological consequences from bedforms to basins. *Water Resour. Res.* **2015**, *51*, 6893–6922. [CrossRef]

58. McConville, M.B.; Cohen, N.M.; Nowicki, S.M.; Lantz, S.R.; Hixson, J.L.; Ward, A.S.; Remucal, C.K. A field analysis of lampricide photodegradation in Great Lakes tributaries. *Environ. Sci. Process. Impacts* **2017**, *19*. [CrossRef]

59. Young, R.G.; Huryn, A.D. Comment: Improvements to the diurnal upstream-downstream dissolved. *Can. J. Fish. Aquat. Sci.* **1998**, *55*, 1784–1785.

60. Hall, R.O.; Yackulic, C.B.; Kennedy, T.A.; Yard, M.D.; Rosi-Marshall, E.J.; Voichick, N.; Behn, K.E. Turbidity, light, temperature, and hydropeaking control primary productivity in the Colorado River, Grand Canyon. *Limnol. Oceanogr.* **2015**, *60*, 512–526.

61. Raymond, P.A.; Saiers, J.E.; Sobczak, W.V. Hydrological and biogeochemical controls on watershed dissolved organic matter transport: Pulse- shunt concept. *Ecology* **2016**, *97*, 5–16.

62. Alexander, R.B.; Smith, R.A.; Schwarz, G.E. Effect of stream channel size on the delivery of nitrogen to the Gulf of Mexico. *Nature* **2000**, *403*, 758–761. [PubMed]

63. Hart, D.D.; Finelli, C.M. Physical-biological coupling in streams: The pervasive effects of flow on benthic organisms. *Am. Rev. Ecol. Syst.* **1999**, *30*, 363–395.

64. Rauter, A.; Weigelhofer, G.; Waringer, J.; Battin, T.J. Transport and metabolic fate of sewage particles in a recipient stream. *J. Environ. Qual.* **2005**, *34*, 1591–1599. [CrossRef] [PubMed]

65. Aubeneau, A.F.; Hanrahan, B.; Bolster, D.; Tank, J.L. Biofilm growth in gravel bed streams controls solute residence time distributions. *J. Geophys. Res. Biogeosciences* **2016**, *121*, 1840–1850.

66. Marti, E.; Grimm, N.B.; Fisher, S.G. Pre-and post-flood retention efficiency of nitrogen in a Sonoran Desert stream. *J. North Am. Benthol. Soc.* **1997**, *16*, 805–819. [CrossRef]

67. Uehlinger, U. Resistance and resilience of ecosystem metabolism in a flood-prone river system. *Freshw. Biol.* **2000**, *45*, 319–332.

68. Uehlinger, U.; Kawecka, B.; Robinson, C.T. Effects of experimental floods on periphyton and stream metabolism below a high dam in the Swiss Alps (River Spöl). *Aquat. Sci.* **2003**, *65*, 199–209.

69. Bernal, S.; von Schiller, D.; Sabater, F.; Martí, E. Hydrological extremes modulate nutrient dynamics in mediterranean climate streams across different spatial scales. *Hydrobiologia* **2013**, *719*, 31–42.

70. Guerin, T.F.; Mondido, M.; McClenn, B.; Peasley, B. Application of resazurin for estimating abundance of contaminant-degrading micro-organisms. *Lett. Appl. Microbiol.* **2001**, *32*, 340–345. [CrossRef]

71. O'Brien, J.M.; Wilson, I.; Orton, T.; Pognan, F. Investigation of the Alamar Blue (resazurin) fluorescent dye for the assessment of mammalian cell cytotoxicity. *Eur. J. Biochem.* **2000**, *267*, 5421–5426. [CrossRef]

72. Knapp, J.L.A.; González-Pinzón, R.; Haggerty, R. The Resazurin-Resorufin System: Insights From a Decade of "Smart" Tracer Development for Hydrologic Applications. *Water Resour. Res.* **2018**, *54*, 6877–6889. [CrossRef]

73. Haggerty, R.; Argerich, A.; Martí, E.; Marti, E. Development of a "smart" tracer for the assessment of microbiological activity and sediment-water interaction in natural waters: The resazurin-resorufin system. *Water Resour. Res.* **2008**, *44*, W00D01. [CrossRef]

74. Haggerty, R.; Ribot, M.; Singer, G.A.; Marti, E.; Argerich, A.; Agell, G.; Battin, T.J. Ecosystemrespiration increases with biofilm growth and bed forms: Flume measurements with resazurin. *J. Geophys. Res. Biogeosciences* **2014**, *119*, 2220–2230. [CrossRef]

75. Knapp, J.L.A.; González-Pinzón, R.; Drummond, J.D.; Larsen, L.G.; Cirpka, O.A.; Harvey, J.W. Tracer-based characterization of hyporheic exchange and benthic biolayers in streams. *Water Resour. Res.* **2017**, *53*, 1575–1594. [CrossRef]

76. Folegot, S.; Krause, S.; Mons, R.; Hannah, D.M.; Datry, T. Mesocosm experiments reveal the direction of groundwater–surface water exchange alters the hyporheic refuge capacity under warming scenarios. *Freshw. Biol.* **2018**, *63*, 165–177. [CrossRef]

77. Kurz, M.J.; Drummond, J.D.; Mart??, E.; Zarnetske, J.P.; Lee-Cullin, J.; Klaar, M.J.; Folegot, S.; Keller, T.; Ward, A.S.; Fleckenstein, J.H.; et al. Impacts of water level on metabolism and transient storage in vegetated lowland rivers: Insights from a mesocosm study. *J. Geophys. Res. Biogeosciences* **2017**, *122*, 628–644. [CrossRef]

78. Baranov, V.; Lewandowski, J.; Romeijn, P.; Singer, G.; Krause, S. Effects of bioirrigation of non-biting midges (Diptera: Chironomidae) on lake sediment respiration. *Sci. Rep.* **2016**, *6*, 1–10. [CrossRef]

79. Baranov, V.; Lewandowski, J.; Krause, S. Bioturbation enhances the aerobic respiration of lake sediments in warming lakes. *Biol. Lett.* **2016**, *12*, 20160448. [CrossRef]

80. Comer-Warner, S.A.; Romeijn, P.; Gooddy, D.C.; Ullah, S.; Kettridge, N.; Marchant, B.; Hannah, D.M.; Krause, S. Thermal sensitivity of CO_2 and CH_4 emissions varies with streambed sediment properties. *Nat. Commun.* **2018**, *9*. [CrossRef]

81. Romeijn, P.; Comer-Warner, S.A.; Ullah, S.; Hannah, D.M.; Krause, S. Streambed Organic Matter Controls on Carbon Dioxide and Methane Emissions from Streams. *Environ. Sci. Technol.* **2019**, *53*, 2364–2374. [CrossRef]

82. Blaen, P.J.; Kurz, M.J.; Drummond, J.D.; Knapp, J.L.A.; Mendoza-Lera, C.; Schmadel, N.M.; Klaar, M.J.; Jäger, A.; Folegot, S.; Lee-Cullin, J.; et al. Woody debris is related to reach-scale hotspots of lowland stream ecosystem respiration under baseflow conditions. *Ecohydrology* **2018**, *11*, 1–9. [CrossRef]

83. Argerich, A.; Haggerty, R.; Martí, E.; Sabater, F.; Zarnetske, J.P.; De Barcelona, U.; Haggerty, R.; Sabater, F.; Zarnetske, J.P.; Marti, E.; et al. Quantification of metabolically active transient storage (MATS) in two reaches with contrasting transient storage and ecosystem respiration. *J. Geophys. Res.* **2011**, *116*, G03034. [CrossRef]

84. González-Pinzón, R.; Haggerty, R.; Myrold, D.D. Measuring aerobic respiration in stream ecosystems using the resazurin-resorufin system. *J. Geophys. Res. Biogeosciences* **2012**, *117*, 1–10. [CrossRef]

85. Lemke, D.; Liao, Z.; Wöhling, T.; Osenbrück, K.; Cirpka, O.A. Concurrent conservative and reactive tracer tests in a stream undergoing hyporheic exchange. *Water Resour. Res.* **2013**, *49*, 3024–3037. [CrossRef]

86. Kelleher, C.; Ward, A.S.; Knapp, J.L.A.; Blaen, P.J.; Kurz, M.J.; Drummond, J.D.; Zarnetske, J.P.; Hannah, D.M.; Mendoza-Lera, C.; Schmadel, N.M.; et al. Exploring Tracer Information and Model Framework Trade-Offs to Improve Estimation of Stream Transient Storage Processes. *Water Resour. Res.* **2019**, *55*, 3481–3501. [CrossRef]

87. Kelleher, C.A.; Wagener, T.; McGlynn, B.L.; Ward, A.S.; Gooseff, M.N.; Payn, R.A. Identifiability of transient storage model parameters along a mountain stream. *Water Resour. Res.* **2013**, *49*, 5290–5306. [CrossRef]

88. Ward, A.S.; Kelleher, C.A.; Mason, S.J.K.; Wagener, T. A software tool to assess uncertainty in transient-storage model parameters using Monte Carlo simulations. *Freshw. Sci.* **2016**, *36*. [CrossRef]

89. Wagner, B.J.; Harvey, J.W. Experimental design for estimating parameters of rate-limited mass transfer: Analysis of stream tracer studies. *Water Resour. Res.* **1997**, *33*, 1731–1741. [CrossRef]

90. Wagener, T.; Kollat, J. Numerical and visual evaluation of hydrological and environmental models using the Monte Carlo analysis toolbox. *Environ. Model. Softw.* **2007**, *22*, 1021–1033. [CrossRef]

91. Bencala, K.E.; Walters, R.A. Simulation of solute transport in a mountain pool-and-riffle stream: A transient storage model. *Water Resour. Res.* **1983**, *19*, 718–724. [CrossRef]

92. Haggerty, R.; Reeves, P. *STAMMT-L Version 1.0 User's Manual*; Sandia National Laboratories: Albuquerque, NM, USA, 2002; pp. 1–76.

93. Runkel, R.L. *One-dimensional Transport with Inflow and Storage (OTIS): A Solute Transport Model for Streams and Rivers*; Information Services [distributor]; ISBN Water-Resources Investigations Report 98–4018; US Dept. of the Interior, US Geological Survey: Washington, DC, USA, 1998.

94. Harman, C.J. Time-variable transit time distributions and transport: Theory and application to storage-dependent transport of chloride in awatershed. *Water Resour. Res.* **2015**, *51*, 1–30. [CrossRef]

95. Harman, C.J.; Ward, A.S.; Ball, A. How does reach-scale stream-hyporheic transport vary with discharge? Insights fromrSAS analysis of sequential tracer injections in a headwater mountain stream. *Water Resour. Res.* **2016**, *52*, 7130–7150. [CrossRef]

96. Ward, A.S.; Wondzell, S.M.; Schmadel, N.M.; Herzog, S.; Zarnetske, J.P.; Baranov, V.; Blaen, P.J.; Brekenfeld, N.; Chu, R.; Derelle, R.; et al. Spatial and temporal variation in river corridor exchange across a 5th order mountain stream network. *Hydrol. Earth Syst. Sci. Discuss.* **2019**, 1–39. [CrossRef]

97. Dyrness, C.T. Hydrologic properties of soils on three small watersheds in the western Cascades of Oregon. *USDA For. Ser. Res. Note PNW* **1969**, *111*, 17.

98. Swanson, F.J.; Jones, J.A. Geomorphology and hydrology of the HJ Andrews experimental forest, Blue River, Oregon. *F. Guid. to Geol. Process. Cascadia* **2002**, *36*, 289–314.

99. Swanson, F.J.; James, M.E. *Geology and Geomorphology of the HJ Andrews Experimental Forest, Western Cascades, Oregon*; U.S. Department of Agriculture: Portland, OR, USA, 1975.

100. Jefferson, A.; Grant, G.E.; Lewis, S.L. A River Runs Underneath It: Geological Control of Spring and Channel Systems and Management Implications, Cascade Range, Oregon. In *Advancing the Fundamental Sciences: Proceedings of the Forest Service National Earth Sciences Conference, San Diego, CA, USA, 18–22 October 2004*; U.S. Forest Service, Pacific Northwest Research: Portland, OR, USA, 2004; Volume 1, pp. 18–22.

101. Deligne, N.I.; Mckay, D.; Conrey, R.M.; Grant, G.E.; Johnson, E.R.; O'Connor, J.; Sweeney, K. Field-trip guide to mafic volcanism of the Cascade Range in Central Oregon—A volcanic, tectonic, hydrologic, and geomorphic journey. *Sci. Investig. Rep.* **2017**, *110*. [CrossRef]

102. Cashman, K.V.; Deligne, N.I.; Gannett, M.W.; Grant, G.E.; Jefferson, A. Fire and water: Volcanology, geomorphology, and hydrogeology of the Cascade Range, central Oregon. *F. Guid.* **2009**, *15*, 539–582.

103. Gooseff, M.N.; Wondzell, S.M.; McGlynn, B.L.; Science, I.G.; Well-being, H.; Northwest, P.; Forestry, O.; Resources, L. On the relationships among temporal patterns of evapo-transpiration, stream flow and riparian water levels in headwater catchments during baseflow. In Proceeding of the 36th IAH Congress: Integrating Groundwater Science and Human Well-Being, Toyama, Japan, 27–31 October 2008; pp. 1–10.

104. Wondzell, S.M.; Gooseff, M.N.; McGlynn, B.L. Flow velocity and the hydrologic behavior of streams during baseflow. *Geophys. Res. Lett.* **2007**, *34*, L24404. [CrossRef]

105. Cain, M.R.; Ward, A.S.; Hrachowitz, M. Ecohydrologic separation alters interpreted hydrologic stores and fluxes in a headwater mountain catchment. *Hydrol. Process.* **2019**, *33*, 2658–2675. [CrossRef]

106. Barnard, H.R.; Graham, C.B.; Van Verseveld, W.; Brooks, J.; Bond, B.J.; McDonnell, J.J. Mechanistic assessment of hillslope transpiration controls of diel subsurface flow: A steady-state irrigation approach. *Ecohydrology* **2010**, *3*, 133–142. [CrossRef]

107. Perry, T.D.; Jones, J.A. Summer streamflow deficits from regenerating Douglas-fir forest in the Pacific Northwest, USA. *Ecohydrology* **2017**, *10*, 1–13. [CrossRef]

108. Knapp, J.L.A.; Cirpka, O.A. A Critical Assessment of Relating Resazurin–Resorufin Experiments to Reach-Scale Metabolism in Lowland Streams. *J. Geophys. Res. Biogeosciences* **2018**, *123*, 3538–3555. [CrossRef]

109. Danckwerts, P. Continuous flow systems. Distribution of residence times. *Chem. Eng. Sci.* **1953**, *2*, 1–13. [CrossRef]

110. Castro, N.M.; Hornberger, G.M. Surface-subsurface water interactions in an alluviated mountain stream channel. *Water Resour. Res.* **1991**, *27*, 1613–1621. [CrossRef]

111. Kennedy, V.C.; Jackman, A.P.; Zand, S.M.; Zellweger, G.W.; Avanzino, R.J.; Walters, R.A. Transport and concentration controls for chloride, strontium, potassium and lead in Uvas Creek, a small cobble-bed stream in Santa Clara County, California, USA: 2. Mathematical modeling. *J. Hydrol.* **1984**, *75*, 67–110. [CrossRef]

112. Laloy, E.; Vrugt, J.A. High-dimensional posterior exploration of hydrologic models using multiple-try DREAM (ZS) and high-performance computing. *Water Resour. Res.* **2012**, *48*, 1–18. [CrossRef]

113. Vrugt, J.; ter Braak, C.; Diks, C.; Hyman, J.; Robinson, B.; Higdon, D. Accelerating Markov-Chain Monte Carlo simulation by differential evolution with self-adaptive randomized subspace sampling. *Int. J. Nonlinear Sci. Numer. Simulatoin* **2009**, *10*, 273–290. [CrossRef]

114. Ward, A.S.; Zarnetske, J.P.; Baranov, V.; Blaen, P.J.; Brekenfeld, N.; Chu, R.; Derelle, R.; Drummond, J.D.; Fleckenstein, J.; Garayburu-Caruso, V.; et al. Co-located contemporaneous mapping of morphological, hydrological, chemical, and biological conditions in a 5th order mountain stream network, Oregon, USA. *Earth Syst. Sci. Data Discuss.* **2019**, 1–27. [CrossRef]

115. Argerich, A.; Haggerty, R.; Johnson, S.L.; Wondzell, S.M.; Dosch, N.; Corson-rikert, H.; Ashkenas, L.R.; Pennington, R.; Thomas, C.K. Comprehensive multiyear carbon budget of a temperate headwater stream. *J. Geophys. Res. - Biogeosciences* **2016**, *121*, 1306–1315. [CrossRef]

116. Roche, K.R.; Shogren, A.J.; Aubeneau, A.; Tank, J.L.; Bolster, D. Modeling Benthic Versus Hyporheic Nutrient Uptake in Unshaded Streams With Varying Substrates. *J. Geophys. Res. Biogeosciences* **2019**, *124*, 367–383. [CrossRef]

117. Schaper, J.L.; Posselt, M.; Bouchez, C.; Jaeger, A.; Nuetzmann, G.; Putschew, A.; Singer, G.; Lewandowski, J. Fate of Trace Organic Compounds in the Hyporheic Zone: Influence of Retardation, the Benthic Biolayer, and Organic Carbon. *Environ. Sci. Technol.* **2019**, *53*, 4224–4234. [CrossRef] [PubMed]

118. Harvey, J.W.; Wagner, B.J. Quantifying hydrologic interactions between streams and their subsurface hyporheic zones. In *Streams and Ground Waters*; Jones, J.B., Mulholland, P.J., Eds.; Elsevier: Amsterdam, The Netherlands, 2000; pp. 3–44.

119. Briggs, M.A.; Day-Lewis, F.D.; Zarnetske, J.P.; Harvey, J.W. A physical explanation for the development of redox microzones in hyporheic flow. *Geophys. Res. Lett.* **2015**, *42*, 4402–4410. [CrossRef]

120. Rosemond, A.D.; Benstead, J.P.; Bumpers, P.M.; Gulis, V.; Kominoski, J.S.; Manning, D.W.P.; Suberkropp, K.; Wallace, J.B. Experimental nutrient additions accelerate terrestrial carbon loss from stream ecosystems. *Science (80-.).* **2015**, *347*, 1142–1144. [CrossRef] [PubMed]

121. Vanderbilt, K.L.; Lajtha, K.; Swanson, F.J. Biogeochemistry of unpolluted forested watersheds in the Oregon Cascades: Temporal patterns of precipitation and stream nitrogen fluxe. *Biogeochemistry* **2003**, *62*, 87–118. [CrossRef]
122. Corson-Rikert, H.A.; Wondzell, S.M.; Haggerty, R.; Santelmann, M. V Carbon dynamics in the hyporheic zone of a headwater mountain streamin the Cascade Mountains, Oregon. *Water Resour. Res.* **2016**, *52*, 7556–7576. [CrossRef]
123. Webster, J.R.; Benfield, E.F.; Ehrman, T.P.; Schaeffer, M.A.; Tank, J.E.; Hutchens, J.J.; D'Angelo, D.J. What happens to allochthonous material that falls into streams? A synthesis of new and published information from Coweeta. *Freshw. Biol.* **1999**, *41*, 687–705. [CrossRef]
124. Harvey, J.W.; Wagner, B.J.; Bencala, K.E. By Was S); Hyporheic Exchange Persisted When Base Flow Was Decreasing By. *Water Resour.* **1996**, *32*, 2441–2451. [CrossRef]
125. Reynolds, K.N.; Loecke, T.D.; Burgin, A.J.; Davis, C.A.; Riveros-Iregui, D.; Thomas, S.A.; St. Clair, M.A.; Ward, A.S. Optimizing Sampling Strategies for Riverine Nitrate Using High-Frequency Data in Agricultural Watersheds. *Environ. Sci. Technol.* **2016**, acs.est.5b05423.

 *water*

Article

Impact of Bed Form Celerity on Oxygen Dynamics in the Hyporheic Zone

Philipp Wolke [1,2], **Yoni Teitelbaum** [3], **Chao Deng** [3], **Jörg Lewandowski** [1,4] **and Shai Arnon** [3,*]

[1] Department Ecohydrology, Leibniz-Institute of Freshwater Ecology and Inland Fisheries, Müggelseedamm 310, 12587 Berlin, Germany; philipp.wolke@web.de (P.W.); lewe@igb-berlin.de (J.L.)

[2] Institute of Geological Sciences, Department of Earth Sciences, Freie Universität Berlin, Malteserstr. 74-100, 12249 Berlin, Germany

[3] Zuckerberg Institute for Water Research, The Jacob Blaustein Institutes for Desert Research, Ben-Gurion University of the Negev, 84990 Midreshet Ben-Gurion, Israel; ytbaum@gmail.com (Y.T.); marschao@163.com (C.D.)

[4] Geography Department, Humboldt University Berlin, Rudower Chaussee 16, 12489 Berlin, Germany

* Correspondence: sarnon@bgu.ac.il; Tel.: +972-8-6563-510

Received: 26 September 2019; Accepted: 6 December 2019; Published: 22 December 2019

Abstract: Oxygen distribution and uptake in the hyporheic zone regulate various redox-sensitive reactions and influence habitat conditions. Despite the fact that fine-grain sediments in streams and rivers are commonly in motion, most studies on biogeochemistry have focused on stagnant sediments. In order to evaluate the effect of bed form celerity on oxygen dynamics and uptake in sandy beds, we conducted experiments in a recirculating indoor flume. Oxygen distribution in the bed was measured under various celerities using 2D planar optodes. Bed morphodynamics were measured by a surface elevation sensor and time-lapse photography. Oxygenated zones in stationary beds had a conchoidal shape due to influx through the stoss side of the bed form, and upwelling anoxic water at the lee side. Increasing bed celerity resulted in the gradual disappearance of the upwelling anoxic zone and flattening of the interface between the oxic (moving fraction of the bed) and the anoxic zone (stationary fraction of the bed), as well as in a reduction of the volumetric oxygen uptake rates due shortened residence times in the hyporheic zone. These results suggest that including processes related to bed form migration are important for understanding the biogeochemistry of hyporheic zones.

Keywords: hyporheic exchange; bed form migration; moving streambed; ripples; planar optodes

---

1. Introduction

The hyporheic zone (HZ) plays a major role in various physical, chemical, and biological processes in streams and rivers [1–3]. The HZ is often characterized as the zone in which flow paths of stream water enter the sediment, and re-emerge back to the stream after spending some time in the subsurface [4]. Water flux through the HZ is termed hyporheic exchange flux (HEF). The mixing of stream water and groundwater occurs in the HZ, and this may change the physical properties (e.g., temperature) and chemical composition of the water. Such mixing leads to steep chemical gradients, and therefore, the HZ is a highly-reactive environment with intense chemical turnover rates, which are catalyzed by microorganisms [2,5–7]. This includes nutrient cycling, metal transformation, and contaminant degradation [8–10]. Understanding the processes in the HZ, therefore, has implications for water resources management and stream restoration [11,12].

Dissolved oxygen is the key solute driving the metabolism of organisms living in rivers and within the HZ. The consumption rate of oxygen by microorganisms is an indicator of biochemical turnover rates, and dictates the redox zonation in the streambed.

The flux of oxygen across the sediment–water interface and the size of the oxygenated zone is strongly regulated by bed topography, stream water velocity, and oxygen consumption rate [13,14]. Therefore, variations of the flow regime, river morphology, and the regional groundwater flow pattern alter the exchange across the sediment–water interface, and thus, also the extent of the HZ [15,16]. Because HEF in sandy rivers is primarily induced by bed forms, their ubiquitous occurrence makes them the most important contributor to HEF and associated biogeochemical processes [17]. The driving forces of flow through the HZ are the hydraulic head variations at the immediate sediment–water interface, which are developed by water flowing over bed forms. Solute transport by advective flow due to the head difference along the bed forms is often called "advective pumping" [18], and is often used to model transport processes in sandy sediments.

Sandy and silty streambeds are highly sensitive to shear forces and are mobile under flow conditions that are commonly found in streams and rivers. The motion of the streambed can be characterized by the speed of its movement (i.e., celerity). During the movement of the bed, sediment particles roll from the stoss side and avalanche down the lee side of the bed form. During such sand movement, the trapping and release of solute occur as bed forms propagate, and contribute to the exchange by advection. This physical exchange mechanism is termed "turnover" [18]. During bed form motion, advective pumping and turnover can occur simultaneously, and their relative contributions result in altered flow paths and residence time distribution of pore water as compared to pure advection under stationary bed conditions. With increasing bed form celerity, turnover becomes more dominant as compared to advective pumping. Ultimately, when the bed form celerity is greater than the pore water velocity, the penetration of solutes is mostly restricted to the extent of the moving bed forms, and the exchange between the sediment layer affected by bed form celerity and the immobile sediment layer underneath is primarily limited to dispersion and diffusion. Thus, bed form celerity can have a major influence on the distribution and residence time of water entering the sediment, as well on oxygen availability for microorganism and redox processes [19–21]. However, despite the potential impact of moving streambeds on ecological and biogeochemical processes, there is a scarcity of such studies, because studying moving streambeds is much more challenging than studying stagnant streambeds.

Oxygen dynamics under moving bed form conditions have so far only been considered in a few modelling studies [19,21,22], and in two experimental studies in marine systems [23,24]. The only experimental study on a freshwater system focused on mobile dunes, several meters in size [25]. To the best of our knowledge, controlled experiments have never been conducted in a freshwater system. The main objective of the present study is to quantify the effect of bed form celerity on oxygen dynamics and oxygen consumption in moving bed forms.

2. Materials and Methods

2.1. Experimental Setup, Conditions, and Approach

The experiments were conducted in a recirculating indoor flume with a working section dimension of 260 cm in length and 29 cm in width, as described in detail by De Falco et al. [26]. A variable-speed pump controlled the water flow in the flume channel, while the discharge was measured by a magnetic flow meter (SITRANS F M, MAG 5100 W Siemens, Nordborg, Denmark) integrated in the return pipe between pump and flume inlet. Water in the flume was kept between 24 and 25 °C by a chiller (TR/TC 10, TECO Refrigeration Technologies, Ravenna, Italy). Natural sandy sediment from the Yarqon River in central Israel was used in the experiments [27]. The sediment was excavated from the upper 10 cm of the streambed, and was taken directly to the laboratory. The sediment was kept moist in ambient air for three days, during which it was mixed several times. Afterwards, the sediment was wet-sieved to remove all particles larger than 2 mm and was added to the flume channel, which was filled with deionized water. Flow was initiated soon after the sediment was added to the channel. The recirculating system carried all the sediment washed from the outlet of the channel, and reintroduced it through the inlet so that the net amount of sediment in the flume would remain unchanged.

A total of eleven experiments were performed with different flow conditions (Table 1). Each flow condition was duplicated, except for the streamflow velocity of 0.32 m·s^{-1}, which was repeated three times. The duplicates were labeled Set 1 and Set 2, respectively, to elucidate possible temporal changes, and because true replication is not possible when using such large experimental systems. Each experiment was centered around the measurement of oxygen distribution in the sediment, which was conducted by planar optodes. Each batch of oxygen measurements took a few hours, with images captured every 5–15 min. Bed form morphodynamics were quantified using time-lapse digital photographs to quantify the bed form length and celerity, and depth tracking with an acoustic profiler to quantify the bed height. Temperature, pH, oxygen saturation, and electrical conductivity (EC) were measured in the stream water with a handheld meter (WTW, Multi 3320, Weilheim, Germany) before and after oxygen distribution measurements. A waiting time of approximately 24 h was implemented before the next run started, letting the system adapt to the new flow conditions, which was defined as a stable distribution of oxygen within the sediments. HEF was measured under all stream water velocities after finishing all oxygen measurements. Afterwards, sediment samples were collected from the mobile and immobile section of the bed. Porosity, organic matter content, and grain size distribution were determined based on these sediment samples. Hydraulic conductivity was calculated from the particle size distribution using HydrogeoSieveXL and the Hazen equation [28].

2.2. Bed Form Morphodynamics and Sediment Characterization

Bed form height was evaluated with a Doppler velocimeter (Nortek Vectrino II Profiler, Rud, Norway), which was fixed in a single known position, and by using the depth tracking mode to measure the distance to the bed as the bed moved. The bed topography dynamics were logged for up to six days under the slowest celerity, and two days under all other celerities, to ensure that the number of moving bed forms was sufficient for statistical analysis (Table 1). The time series were evaluated with the "find peaks" method, in which bed form troughs are represented by local minima and crests by local maxima. A minimum required prominence level of 5 mm was set in order to exclude small height variations on larger bed forms, which would otherwise be identified as individual bed forms themselves. The mean bed height was calculated using the arithmetic mean of the recorded bed form troughs and crests. The final bed form height of each condition was determined as the average of the data of the two or three experimental runs, respectively.

The 1-dimensional bottom tracking measures could not be used to calculate the bed celerity. Therefore, bed form celerity was measured with a digital camera (Nikon D5300, with AF-P DX NIKKOR 18–55 mm VR Lens, Ayuthaya, Thailand) by taking images at the same place as the optode, and simultaneously with oxygen imaging in order to ensure correct coupling between the oxygen distribution and the bed form shape (Appendix A, Figure A1). The bed form celerity during the experiments was determined by using an ad-hoc Python code. This software made it possible to track the pixels of bed form crests and troughs on the recorded images over several images. The average position change over time of these pixels was translated into the celerity.

Porosity was measured by filling a container with a known volume with saturated sediment (100 mL) and measuring the loss of weight after drying at 105 °C for two days. Organic matter content was measured by loss through ignition after burning a 5 g sediment sample at 450 °C for 4 h. The methods followed the protocols shown in Klute et al. [29].

2.3. Hyporheic Exchange Flux

A total of 19 tracer experiments were carried out to determine HEF under different flow conditions. In these experiment, 100 g NaCl was added to the surface water, which increased the EC of the overlying water from ~400 to ~900 µS·cm^{-1}. The EC in the overlying water was recorded every 5 s, and HEF was calculated from the initial decline of the EC due to hyporheic exchange, as described by Packman et al. [30] and Fox et al. [31]. Briefly, the interactions between the streamflow and the sediment produced pressure head variations over bed forms, with high pressure zones on the stoss side and low pressure zones on the lee side of the bed forms.

The outcome of this is a two-dimensional advective flow field within the sediments, with water flowing into (on the stoss side), through, and then out of the bed (on the lee side). Shortly after the addition of the tracer NaCl to the stream water, the EC of the stream water was higher than that of the pore water. Thus, HEF delivered high-EC water into the bed while porewater with lower EC returned to the stream and diluteed the stream water, i.e. reduceed the EC. Solving mass balance equations allowed us to extract the exchange flux from the initial slope of the EC decline, as explained in detail by Packman et al. [30] and Fox et al. [31].

2.4. Oxygen Imaging with Planar Optodes

The distribution of oxygen in the sediment was measured with a planar optode system (VisiSensTD, Presens GmbH, Regensburg, Germany). An optode with a size of 10 × 15 cm was attached to the inner glass side wall of the flume such that it captured both the streambed and its interface with the stream water (Appendix A, Figure A1). The basic physical principle of oxygen sensing with planar optodes is the dynamic quenching of a luminescence indicator substance in the presence of oxygen. The method has great advantages over conventional oxygen measurements, since it does not consume oxygen, and allows non-invasive 2D imaging of oxygen in the bed to be undertaken with high precision [32]. The flume was covered with black cloth during the entire duration of the experiment, since the fluorescent substances in the optode are photosensitive and degrade with increased exposure to light. Light was used only for short periods when photographs were taken.

2.5. Data Analysis

Image processing techniques were used to combine the topographic information from the digital image of the sediment–water boundary with the spatial oxygen saturation information of the calibrated optode images. This allowed us to exclude areas of oxygenated surface water from the optode images. In order to use the optode images, it was assumed that they were representative of the porous medium, i.e., that no significant wall effects occurred. The images were batch processed using an ad-hoc Python code with the open source computer vision library OpenCV [33]. A flow diagram of the processing steps is given in Appendix A (Figure A2).

The preprocessed images allowed us to determine the size and mean oxygen saturation of the oxygenated area at high resolution (pixel size was 15.625 μm^2). The oxygenated zone was defined here using a threshold value of 15% oxygen saturation. The size of oxic zone A_{ox} was calculated by Equation (1):

$$A_{ox} = A_{pix} \sum_{i=1}^{n} P_i \tag{1}$$

where pixel P_i is 1 if it is located in the streambed and the oxygen saturation is above the threshold (otherwise, P_i is 0), A_{pix} is the area of one pixel (15.625 μm^2), and n is the total number of pixels of the planar optode. The mean oxygen concentration of the oxic zone C_{oxz} is calculated by Equation (2):

$$C_{oxz} = \frac{1}{n} \sum_{i=1}^{n} P_i \cdot C_i \tag{2}$$

where C_i is the oxygen concentration of the individual pixel.

The volumetric oxygen consumption rates R (mg $l^{-1} \cdot h^{-1}$) of the pore water were calculated with two different methods: the "maximum uptake method", and the "delta method". Both methods are based on averaging the flux and oxygen concentrations within the two-dimensional porous domain, and thus, represent more reliable results than previously used 1-dimensional approaches [34,35].

We refer to this process in the text as oxygen uptake. The first method, suggested by Ahmerkamp et al. [21], is referred to as the maximum uptake method, R_{max}, represented with Equation (3):

$$R_{max} = \frac{F_{\downarrow O_2}}{\delta \times \theta \times 24\,h} \tag{3}$$

where θ is the porosity, $F_{\downarrow O_2}$ (Equation (4)) represents the flux of oxygen into the bed:

$$F_{\downarrow O_2} = C_{sw} \times q_h \tag{4}$$

and the mean oxygen penetration depth δ is calculated with Equation (5):

$$\delta = \frac{A_{ox}}{W_{opt}} \tag{5}$$

where C_{sw} is the oxygen concentrations in the stream water, q_h is the HEF, and W_{opt} is the optode width. In this case, it was assumed that all oxygen transported from the stream water into the sediment by HEF was consumed within the sediment, and that no oxygen had returned to the stream water by upwelling flow paths.

The second method is referred to as the delta method (Equation (6)).

$$R_{delta} = \frac{F_{\downarrow O_2} - F_{\uparrow O_2}}{\delta \times \theta \times 24\,h} \tag{6}$$

where $F_{\uparrow O_2}$ (Equation (7)) represents the flux of oxygen out of the bed:

$$F_{\uparrow O_2} = C_{oxz} \times q_h \tag{7}$$

The delta method is based on the assumption that the HEF that enters the subsurface with flux q_h and oxygen concentration C_{sw} reemerges to the surface water with the same q_h but with an altered oxygen concentration due to the microbial consumption. For brevity, the mean oxygen concentration in the oxic zone (C_{oxz}) is used.

In addition to the experimentally-determined HEF, the HEF due to pumping was calculated following Elliott and Brooks [18] theory by using Equation (8).

$$\bar{q} = \frac{K \times k \times h_m}{\pi} \tag{8}$$

where $\bar{q}$ refers to spatially averaged flux into the bed, K is hydraulic conductivity, k is the wavenumber (given by the equation $k = 2\pi/\lambda$, where λ is the bed form wavelength), and h_m is the head variation over the bed form (see more details in Elliott and Brooks [18]).

Finally, the flushing time of the bed was calculated following the definition of Monsen et al. [36], which is an integrative approach consisting of taking the oxygenated pore water volume of the hyporheic zone (A_{ox} multiplied by the flume width and θ) and dividing by the volumetric flux through the oxygenated volume (i.e., HEF multiplied by the flume width and optode width, W_{opt}). This only serves as an approximation of time it takes to flush the oxygenated volume despite the fact that flow paths in the streambed has variable lengths and that water velocities are lognormally distributed [14]. The concept of flushing time is widely used in much of the hyporheic zone literature, although in many cases it is referred to as residence time [37,38].

3. Results

3.1. Water and Sediment Characteristics

The turbidity of the surface water increased with stream water velocity, from 62 NTU under stationary conditions up to 478 NTU under the maximum stream velocity of 0.37 m·s^{-1} (Table 1). EC was also not constant, and increased gradually from 291 μS·cm^{-1} to 362 μS·cm^{-1} during the two weeks of the experiments. pH and oxygen were relatively stable, ranging between 8.1–8.2 and 92.8% to 98.7% for pH and oxygen, respectively.

Comparisons between the sediment in the moving fraction (upper few cm) and the immobile fraction below revealed that the mobile fraction had a slightly higher porosity (36% vs. 33.6%), but that these differences were not significant (t-test, $p > 0.05$). This is also true for the median grain diameters, which were 0.264 mm and 0.298 in the moving fraction and immobile fraction, respectively. The hydraulic conductivity was calculated based on the grain size distribution, and was found to be almost the same for the moving and immobile fractions (4.62×10^{-4} m·s^{-1}). Significantly higher organic matter content was observed in the immobile fraction as compared to the mobile sediment layer (1.38% vs. 1.05%, t-test $p < 0.05$).

The calculated bed form celerities and mean heights of bed forms under each experimental run are given in the Table 1. It was observed that the bed form celerity under the same flow conditions could be temporally variable, while some bed forms tended to accelerate or decelerate at certain time points without clear explanation. This is expressed in the increased standard deviation under higher celerities (e.g., Table 1 and Supplementary Material, Figures S1–S9 and Movies S1–S9).

It was also observed that suspended fine material deposits resulted in a very thin layer at the interface between the bed and the stream water when the bed forms were stationary (slowest flow conditions). The mean bed form length was 13.3 ($\pm$ 3.8) cm, but there was no clear trend with celerity. Finally, the mean height of the bed forms increased from 1.13 cm to 1.37 cm between the slowest and intermediate celerity (0.04 m·h^{-1} and 0.14 m·h^{-1}). However, this is increase in mean height was somewhat smaller than the variation in height within each flow condition, which can be exemplified with the observed standard deviations that ranged between 0.42 and 0.60 cm.

Table 1. Physical and chemical properties of the water and sediment during the experiments.

Run No./Set No.	Stream Water Velocity (m·s^{-1})	Bed Form Height (cm)[1]	Bed Form Celerity (m·h^{-1})[1]	Water Depth (cm)	Temp. (°C)	EC (μS·cm^{-1})	Turbidity (NTU)
1/1	0.16	1.50 (N/A)	0.000 (N/A)	14.2	24.9	291	62
7/2	0.16	1.94 (N/A)	0.000 (N/A)	14.2	24.3	339	189
3/1	0.25	1.13 (0.42)	0.035 (0.000)	13.8	24.9	322	126
8/2	0.25	1.13 (0.42)	0.049 (0.000)	13.7	24.1	345	137
4/1	0.28	1.37 (0.51)	0.140 (0.004)	14.2	24.4	326	141
9/2	0.29	1.37 (0.51)	0.135 (0.006)	13.9	24.8	355	308
2/1	0.32	1.38 (0.46)	0.394 (0.110)	14.2	24.5	299	267
5/1	0.32	1.38 (0.46)	0.275 (0.045)	14.2	24.4	331	230
10/2	0.33	1.38 (0.46)	0.375 (0.105)	13.9	24.2	359	375
6/1	0.36	1.41 (0.60)	0.699 (0.287)	14.3	24.4	335	311
11/2	0.37	1.41 (0.60)	0.644 (0.080)	13.8	24.6	362	478

[1] standard deviations shown in parenthesis.

3.2. Dynamics of HEF and Oxygen Distribution

HEF increased monotonically from the slowest stream water velocity (stationary streambed) until it reached a maximum at stream water velocity of about 0.35 m·s^{-1} (celerity of 0.35 m·h^{-1}). Further increase in stream water velocity resulted in a decrease of HEF (Figure 1). The variability among the different experiments with the same flow conditions was attributed mainly to experimental sensitivity, which was larger at faster stream water velocities. Exchange fluxes were modeled to calculate predictions based on advective pumping [18]. Modelling was conducted using a hydraulic conductivity of 4.62×10^{-4} m·s^{-1}. For stationary beds, the modelled results were similar to the measurements, but deviations increased as celerity increased (Figure 1).

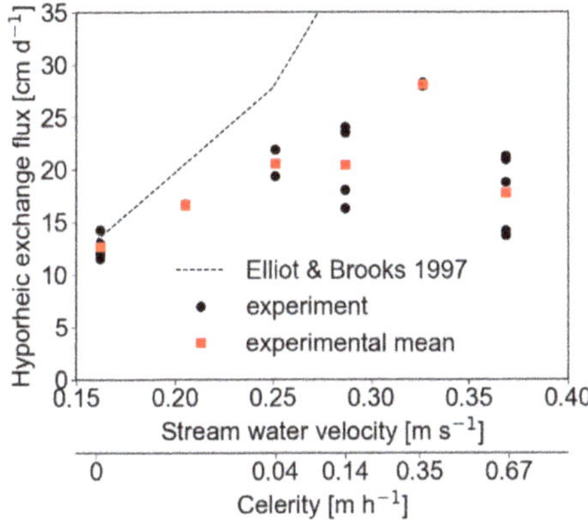

Figure 1. Relationship between HEF and stream water velocity and bed form celerity (bed form celerity axis is not scaled). The dashed line shows modelled results for stationary bed forms by including only the effect of pumping. The predicted exchange continuously increases in a linear manner until it reaches 74 cm·s^{-1} under the stream water velocity of 0.37 m·s^{-1} (data not shown).

The oxygenated zone under stationary bed conditions had a typical conchoidal shape (Figure 2A). A distinctly developed upwelling zone of anoxic water at the bed form trough was observed under stationary conditions, but gradually diminished until it disappeared under faster velocities (Figure 2). These upwelling zones of deeper pore water with low oxygen concentrations are often termed "chimneys" [21]. When these chimneys disappeared at higher celerities, the moving fraction of the bed forms were usually well oxygenated, and a thin transition zone with a steep oxygen gradient towards the deeper parts of the bed were very typical (Figure 2 and Supplementary Material, Figures and Movies S5–S9).

It was also observed that as the bed forms moved, the chimneys disappeared and reappeared when the next bed form passed through the optode (Supplementary Material, Figures S2: image 3–8 and S4: image 1–8, and Movies S2, S4). However, above a celerity of 0.35 m·h^{-1}, the chimneys were no longer observed, and the interface between the oxic and anaerobic zone became flatter, as compared to slower celerities. It is important to mention that the oxygen scale in Figure 2 slightly exceeds 100% oxygen saturation (bright color). This stems from the technical inability to provide the same lighting to the whole optode. Because light is focused more in the center, deviations are usually observed in the edges of the planar optodes. Additionally, planar optodes are less precise at high oxygen saturation levels compared to lower ones. Corrections of the surface water oxygen saturation could be done easily, since we measured oxygen saturation in the stream water using an electrode. It was found that the mean oxygen saturation of the pixels above 100% of all experimental runs was 107.3% ± 1.1%, and the area they occupied was 8.9% ± 7.4%. The high standard deviation was caused by larger proportions of these pixels at celerities above 0.35 m·h^{-1}.

The size of the oxygenated zone increased dramatically right after the bed forms started to move, but a more modest increase in its size was observed as velocity increased further (Figure 3). For example, during stationary conditions, the oxygenated areas in the two runs were 23.39 and 31.23 cm^2, as compared to 53.73 and 41.79 cm^2 under slow celerity of 0.04 m·h^{-1}. Interestingly, the first runs of all experimental conditions show less oxygenated sediment in the subsurface than second runs, with the slow celerity experiment being an exception (Figure 3B). Moreover, the size of the oxygenated area was positively linked to turbidity (Figure 4).

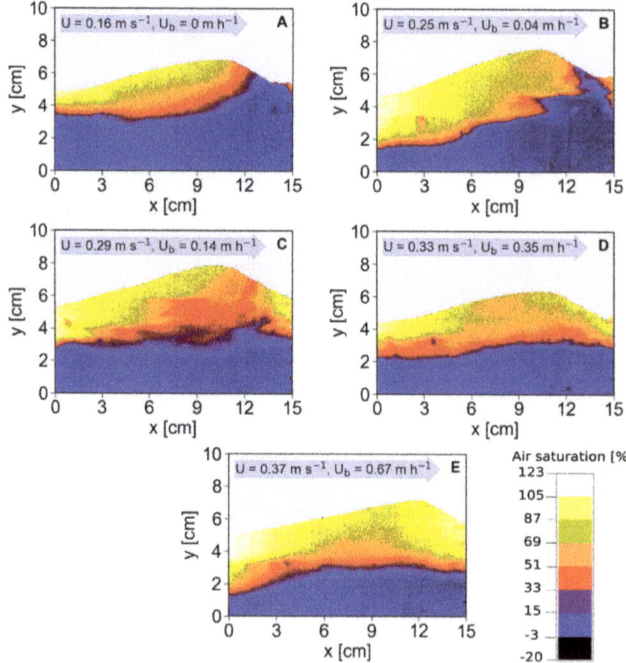

Figure 2. Optode images showing spatial oxygen distributions in the streambed under increasing streamflow velocities (U) and bed form celerities (U_B), from stationary conditions to the fastest bed form celerity. Streamflow velocities (U) and bed form celerities (U_B) are shown on Panels (**A–E**). Images and videos of the other flow conditions are given in the Supplementary Material.

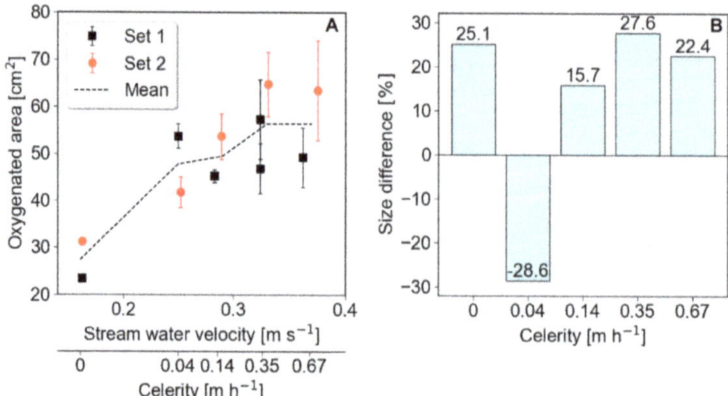

Figure 3. The spatial and temporal mean of the total oxygenated area under different stream water velocities and celerities (**A**). Each condition is displayed chronologically as Set 1 and 2, and the dotted line is the arithmetic mean. Error bars denote standard deviations of each measurement. In general, the areas in Set 2 were greater than in Set 1, except for bed form celerity of 0.14 m·h^{-1}, as shown by the percentage difference in area size between Set 1 and 2 (**B**).

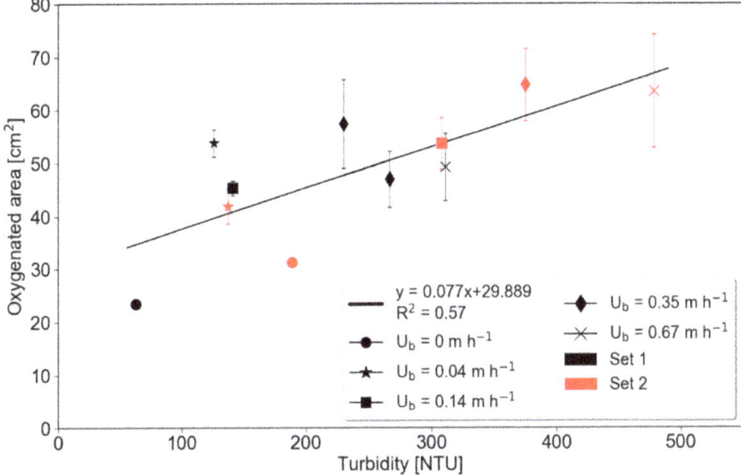

Figure 4. Relationship between oxygenated area and surface water turbidity. The experimental conditions, labelled by their celerities, are additionally separated into Set 1 and 2 (first and second run of the same celerity). Whiskers denote standard deviations.

3.3. Oxygen Fluxes and Uptake Rates

The oxygen flux into the sediment was derived with Equation (4) using the mean of the HEF, as shown in Figure 1, and the oxygen concentrations in the water (Table 1). The calculated oxygen fluxes following Equation (4) trailed the pattern of the HEF (Table 2). Table 2 also displays modeled fluxes following the model presented by Ahmerkamp et al. [20]. Nevertheless, although the prediction following Ahmerkamp et al. were of the same order of magnitude (and only slightly lower for stationary bed forms), increasing deviations were observed as celerity increased. Oxygen uptake rates that were calculated with the delta method covered a range from 22 to 75.1 μmol·L^{-1}·h^{-1}, and were considerably lower than rates of the maximum uptake method with 125.1 to 252.4 μmol·L^{-1}·h^{-1} (Figure 5). The oxygen uptake rates showed a decreasing trend with increasing stream water velocity and associated bed form migration celerity in both model approaches. The aforementioned shift of the size of the total oxygenated area between experimental runs (Figure 3) translates into higher oxygen uptake rates of the experimental results from Set 1. An exception is again the first run under the slow celerity (0.04 m·h^{-1}), Run 2, which was the first measurement with a moving bed. It exhibits an unusually high value, but is not included in calculating the trends in Figure 5A,B (marked as an outlier in the legend). Because HEF, oxygen fluxes, and the oxygenated zone were all influenced by bed form celerity in a complex manner, the resulting oxygen consumption rates varied as well, and were negatively correlated with increasing flushing times (Figure 5C,D).

Table 2. Calculated oxygen flux into the sediment (Equation (4)), modelled oxygen influx based on the equations provided by Ahmerkamp et al. [20], and calculated oxygen flux out of the sediment based on the assumption underlying the delta method (Equation (7)).

Run No./Set No.	Oxygen Influx (mmol·m^2·d^{-1}) [1]	Oxygen Outflux (mmol·m^2·d^{-1}) [1]	Modelled Oxygen Influx (mmol·m^2·d^{-1})
1/1	28.48 (2.32)	20.07 (1.64)	14.67
7/2	29.59 (2.41)	18.98 (1.55)	10.55
3/1	47.48 (2.96)	34.69 (2.16)	11.09
8/2	48.11 (2.99)	34.92 (2.18)	15.66
4/1	47.91 (7.92)	34.74 (5.75)	14.43

Table 2. *Cont.*

Run No./Set No.	Oxygen Influx (mmol·m^2·d^{-1}) [1]	Oxygen Outflux (mmol·m^2·d^{-1}) [1]	Modelled Oxygen Influx (mmol·m^2·d^{-1})
9/2	47.50 (7.85)	33.02 (5.46)	11.69
2/1	65.01 (0.38)	45.84 (0.27)	21.32
5/1	65.80 (0.39)	49.13 (0.29)	17.00
10/2	65.73 (0.39)	50.69 (0.30)	14.48
6/1	42.48 (7.78)	31.43 (5.75)	12.16
11/2	42.91 (7.86)	35.52 (6.50)	9.11

[1] standard deviations shown in parenthesis.

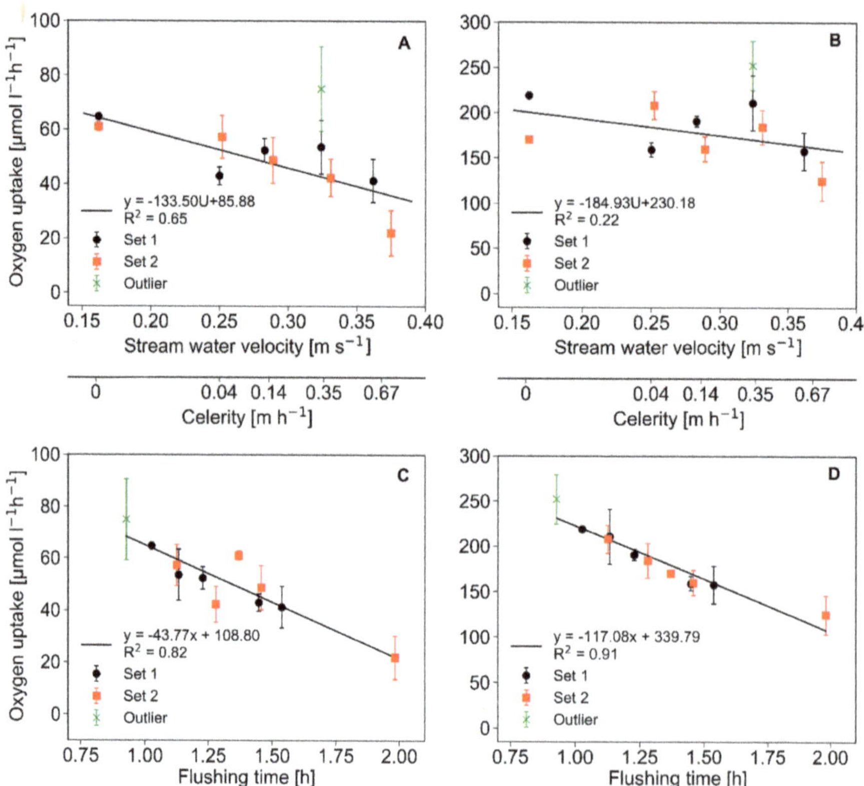

Figure 5. Oxygen uptake rates under different stream water velocities, bed form celerities, and flushing times of the oxygenated zone. The delta method (**A**) and (**C**) is based on the assumption that a proportion of oxygen in the hyporheic zone is transported back to the surface water. The maximum uptake method (**B**) and (**D**) is based on the assumption that all oxygen transported into the subsurface is consumed. The standard deviations are shown as error bars.

4. Discussion

4.1. Bed Form Morphodynamics and Flow

The sizes and shapes of the bed forms that were captured during the oxygen measurements were relatively similar under the different flow conditions (Table 1 and Supplementary Material, Figures and Movies S1–S9), i.e., similar to bed forms reported in other studies [39,40], and similar to bed forms predicted by models [22]. However, the amount of captured bed forms could not cover the whole range of variability that was expected, since previous studies have suggested that tens of bed forms are required to statistically characterize their variability. Although bed form morphodynamics were extensively studied in the past, efforts were made to collapse their statistical behavior into simple scaling laws, rather than to discuss their variability [41]. Consequently, it has been assumed in many biogeochemical studies that bed form shapes are unchanging, even though they are constantly in motion. Thus, they were modelled using a frame that moves at the bed form celerity [21,22]. Using this approach did not include temporal dynamics such as those shown here.

In addition, some bed parameters changed due to grain sorting processes, which slightly changed the grain size and porosity, and significantly reduced the organic matter content in the mobile fraction, as compared to the non-mobile fraction. It was also observed that the mean height of the bed forms slightly increased with increasing stream water velocity, which is common behavior in sandy streambeds [42,43]. The most prominent difference that was observed under different velocities was the change of the bed form celerity, which occurred due to the change in the shear forces over the bed. Despite the fact that bed form celerity was not calculated over prolonged periods that are sufficient to establish statistical distributions of the temporal movement dynamics, the observed celerities were paired to the measurements of oxygen uptake (Figures 3 and 5), thus making it possible to link celerity to oxygen uptake. In addition, the trend of celerity increase with increasing velocities was similar to the trends observed in previous experimental studies with sand [20,39].

HEF increased nearly linearly with velocity until reaching a maximum value at a celerity of $0.35\,\mathrm{m\cdot h^{-1}}$. This trend is different from what was seen under slower streamflow velocities with stationary bed forms, where HEF increases proportionally to the square of the stream water velocity [44,45]. This change in the trend towards a linear relationship was not directly discussed in earlier studies on solute exchange, but can probably be attributed to the increase of turnover as stream water velocity increases, which reduces the overall HEF [18]. When turnover becomes more dominant, HEF actually declines, as shown in Figure 1 for celerities above $0.35\,\mathrm{m\cdot h^{-1}}$, and exemplified by Ahmerkamp et al. [21] and Bottacin-Busolin and Marion [46].

4.2. Dynamics of Oxygen Patterns

The oxygen distribution pattern in the experiment with stationary bed forms matches those shown by Kaufman et al. [13] and by Precht et al. [23]. Under stationary conditions, the upwelling water through the troughs resulted in sharp transition between the anaerobic water and the stream water at the water-sediment interface (Figure 2A). Under slow and intermediate celerities (0.04–$0.14\,\mathrm{m\cdot h^{-1}}$), advective pumping was still the dominant transport process for pore water, and the characteristic pattern of stationary bed forms, with low oxygen concentrations in chimney-like structures, are still present (Figure 2B–C and Supplementary material, Figures S1–S2: image 5–12, S3: images 14, 18–20, S4: images 3–9, Movies S1, S2, S3 and S4). The dominance of exchange due to turnover started at higher celerity ($> 0.35\,\mathrm{m\cdot h^{-1}}$) where the 2D conchoidal shapes disappeared, and the oxic-anoxic interface occurred along a straight horizontal interface. This pattern was even more pronounced at the fastest celerity used in the present study ($0.67\,\mathrm{m\cdot h^{-1}}$). Under these conditions, the downward transport of oxygen and other solutes is mostly limited by dispersion and diffusion processes [21]. The patterns of oxygen distribution (e.g., conchoidal shape and chimneys) are commonly visible in studies using optodes [13], and reflect well the patterns that were observed in mathematical models of transport in streambeds [21,22]. Therefore, despite the fact that we cannot strictly show that wall-effects do not exist, it may imply that the bias in advective transport is relatively small, as shown also in previous studies that used dye to trace flow patterns in bed forms [20,31,47].

The present study reveals that there is also dynamic behavior of the oxygen distribution in the sediment. For example, the offset of the redox chimneys and its relicts in the direction of the bed form stoss side during intermediate celerity of $0.14\,\mathrm{m\cdot h^{-1}}$ are indicators of longer residence times in the vicinity of the points where the seepage velocity is small (Supplementary Material, Figures S3, S4 and

Movies S3, S4). The fraying of the redox chimneys was described in detail neither by the modelling studies nor by the two experimental studies of Precht et al. [23] and Ahmerkamp et al. [24]. This might be due to a lower spatial sampling and imaging resolution of the aforementioned experimental studies compared to our study. The fraying of the oxygen near the upwelling zone resulted in an irregular interface that can be explained by dispersion [47,48]. Fraying of the redox chimneys could occur also due to the dynamics between advancing or receding advective oxygen fronts and local consumption rates. In order to evaluate whether transport or reaction rates are dominating those patterns, a more detailed analysis is required. This can be done using the Damköhler number, which compares the relative rates of advective transport to reaction rates [21,24]. This can be done locally, but requires a detailed numerical modeling analysis of the flow field, which we did not conduct in this study.

Hysteresis effects on oxygenated areas upon increasing and decreasing velocity were shown by Kaufman et al. [13]. The size of the oxygenated area depends on whether the flow is accelerating or decelerating. In our case, the dynamics of the bed form movement led to different distributions of oxygen in the bed (Supplementary Material, comparison between Figures S1 and S2, S4, S5: image 1–12 and Movie S1, S2, S4, S5). Such hysteresis occurred when larger bed forms induced an advective pulse, pushing the oxic–anoxic boundary deeper into the bed, and leaving deeply-oxygenated sediment in its tail. A following, smaller bed form, having smaller advective pumping, cannot generate deep flow, and consequently cuts off deeper bed regions from oxygen resupply. The ascending oxycline can enable bed forms to reinstall temporal reconnections of deeper anoxic water with the surface (Supplementary Material, Figures and Movies S1 and S2; S4).

The complex temporal dynamics of oxygen that were observed here, as well as by Kaufman et al. [13], are expected to play an important role in redox-dependent reactions. Current modeling studies do not take into account such stochastic behavior of bed form movement, despite the fact that such statistics are well studied among geomorphologists [48]. Analyzing reaction rates in bed forms may result in underestimates when bed form dynamics are not included. For example, Kessler et al. [19] concluded in their modeling study that bed form celerity has only small effect on coupled nitrification–denitrification reactions. However, a more recent modeling study showed that both nitrification and net denitrification increase with celerity, and that both are higher under migrating bed forms. However, coupled nitrification–denitrification rates are higher under stationary bed forms. Overall, bed form migration results in a reduction of the nitrogen removal efficiency (i.e., the amount of the nitrogen that enters the hyporheic zone relative to the amount that is removed) [22]. The latter is more related to our experimental results, since it used solute levels that are more relevant to streams; in contrast, Kessler et al. [19] used conditions that were more relevant to coastal sediments. Recently, Zheng et al. [22] came up with a more detailed explanation of why different modeling studies show different patterns, and why at the moment this topic is still not fully understood because of the relatively small numbers of modeled scenarios.

4.3. Oxygen Consumption

The calculated volumetric uptake rates (delta method: 22–75.1 μmol L^{-1}·h^{-1}, maximum uptake method: 125.1–252.4 μmol L^{-1}·h^{-1}) show the efficiency of the microbial community at recycling nutrients. The rates calculated with the maximum uptake method were approximately the same as in De Falco et al. (2018), who used sediments from the same location, but used microelectrodes to measure oxygen concentrations. These rates are on the upper end of the range of respiration rates measured in marine environments (10–144 μmol L^{-1}·h^{-1}), which are usually more oligotrophic than streams [24]. The oxygen uptake rates decreased monotonically with increasing celerity, and were highly correlated with flushing time. Similar trends of declining consumption rates with increasing celerity and flushing time were seen, regardless of the method that we used to calculate the absolute rates. The assumptions that all oxygen is consumed may not be valid in all systems. Indeed, marine environments with smaller reaction rates may lead to significant masses of oxygen flowing out of the bed. For example, Kessler et al. [19] showed that only ~25% of the oxygen is consumed during flow in the subsurface. The delta method takes such behavior into account. However, the delta method assumes that all water in the hyporheic zone is transported back to the surface water with the mean concentration of the oxygenated zone. This holds true for regions with high turnover and for shallow flow paths, but not for deeper-penetrating, slower flow paths, where most or all of the oxygen is consumed. Therefore, the real respiration rates will be in the range between the values of both calculation methods. It can be concluded that the maximum uptake method is probably more correct under conditions with long residence times and high organic matter contents. The delta method more effectively represents conditions with high pore-water exchange rates.

The results of the present study suggest that it is important to consider the effects of particle mobilization and particle deposition, especially in the case of moving bed forms. For example, the respiration rates in the North Sea shelf were lower at sites affected by bed form celerity due to the wash out of organic matter from the moving fraction of the bed [24]. We observed similar patterns in the upper section of the flume sediment, which had a lower organic matter content compared to the deeper immobile sediment (Table 1; organic matter was 1.05% vs. 1.38% in the mobile and immobile sediment, respectively). The reduced amount of organic carbon may be caused by the degradation and resuspension of fine material during bed form movement. Indeed, increasing streamflow velocity and celerity led to the transfer of organic and clay particles to the surface water column, and resulted in higher turbidity and larger oxygenated zones in the bed (Table 1, Figure 4). When flow and celerity are decreased again, the organic and clay particles are deposited, and may shift the flow paths in the bed due to clogging, or may create active zones due to the concentration of organic matter. The deposition varies with velocity and celerity, and can be located near the sediment–water interface near the inflow zone [49], or at the interface between the mobile-immobile fractions of the bed. Such sorting and deposition processes may greatly influence the fluxes due to clogging or reaction rates because of the flushing of organic matter from the reactive zones [50].

The majority of hyporheic studies that focus on biogeochemistry and ecology were conducted under stagnant bed conditions. In contrast, studies on the morphology of streambeds included the motion of the sediments. However, the connection between these topics remains loose, despite the fact that bed migration dramatically changes the physical environment in the bed. The results presented in this study imply that it is essential to incorporate the processes affected by bed form migration into future modeling of stream bed processes, with implications for river network modeling [51,52]. All biogeochemical processes are affected by the flux of nutrients from the water into the sediments, where most of the microbial activity occurs. The zone that is affected by HEF is also critical for process quantification, since the volume of sediment that is involved in a reaction dictates also the amount of biomass that is active (e.g., for oxidation). We have shown that within the hyporheic zone, different flow conditions can result in complex interfaces between oxic and anaerobic zonation, which can lead to variable volumetric reaction rates due to the changing fluxes [22]. While modeling of the aforementioned processes is essential for understanding process coupling, special efforts are needed to incorporate other processes that are known to affect biogeochemical processes but which are currently neglected. These include, for example, fine particle transport and deposition, the physical impacts of sediment motion on biofilms, etc.

5. Conclusions

This study has clearly shown, in a series of highly-controlled flume experiments, that bed form celerity has a significant impact on the dynamics of oxygen distribution and uptake rates in the HZ. To the best of our knowledge, this is the first study that shows high spatial and temporal resolution of two-dimensional oxygen distribution during bed form migration in a freshwater system. The transport of oxygen into the bed was purely advection-dominated in a stationary bed, but gradually changed into a turnover-dominated system as celerity increases. This resulted in an increase in the streambed volume that was exposed to oxygen, but not necessarily in higher oxygen fluxes into the sediment. However, because oxygen uptake depends on the combination between the volume of the oxygenated bed, biomass, and oxygen fluxes, increasing celerity resulted in a reduction in the average volumetric oxygen uptake rate. Part of this was due to the washout and deposition of fine particles, including organic material, and due to a reduction in the flushing times of the hyporheic zone.

Supplementary Materials: The following are available online at http://www.mdpi.com/2073-4441/12/1/62/s1, Figure S1: Oxygen distribution during bed form celerity of 0.04 m·h^{-1} (run 3). Images from left to right, Figure S2: Oxygen distribution during bed form celerity of 0.04 m·h^{-1} (run 8). Images from left to right, Figure S3: Oxygen distribution during bed form celerity of 0.14 m·h^{-1} (run 4). Images from left to right, Figure S4: Oxygen distribution during bed form celerity of 0.14 m·h^{-1} (run 9). Images from left to right, Figure S5: Oxygen distribution during bed form celerity of 0.35 m·h^{-1} (run 2). Images from left to right, Figure S6: Oxygen distribution during bed form celerity 0.35 m·h^{-1} (run 5). Images from left to right, Figure S7: Oxygen distribution during bed form celerity of 0.35 m·h^{-1} (run 10). Images from left to right, Figure S8: Oxygen distribution during bed form celerity of 0.67 m·h^{-1} (run 6). Images from left to right. Figure S9: Oxygen distribution during bed form celerity of 0.67 m·h^{-1} (run 11). Images from left to right. Movie S1: Run 3, Movie S2: Run 8, Movie S3: Run 4, Movie S4: Run 9, Movie S5: Run 2, Movie S6: Run 5, Movie S7: Run 10, Movie S8: Run 6, Movie S9: Run 11

Author Contributions: Conceptualization, P.W., S.A. and J.L.; methodology, P.W., and C.D.; software, P.W. and Y.T.; formal analysis, P.W., C.D., and Y.T.; data curation, P.W; writing—original draft preparation, P.W., and S.A.; writing—review and editing, all authors; funding acquisition, S.A, P.W., and J.L. All authors have read and agreed to the published version of the manuscript.

Funding: This research was supported by the Israel Science Foundation (grant 682/17) and by the Young Scientist Exchange Program of the German Federal Ministry of Education and Research (BMBF) and the Israeli Ministry of Science, Technology and Space (MOST) in the framework of the German-Israeli Cooperation in Water Technology Research (grant number YSEP-122).

Acknowledgments: We thank Soeren Ahmerkamp for assistance with modeling calculations.

Conflicts of Interest: The authors declare no conflict of interest.

Appendix A

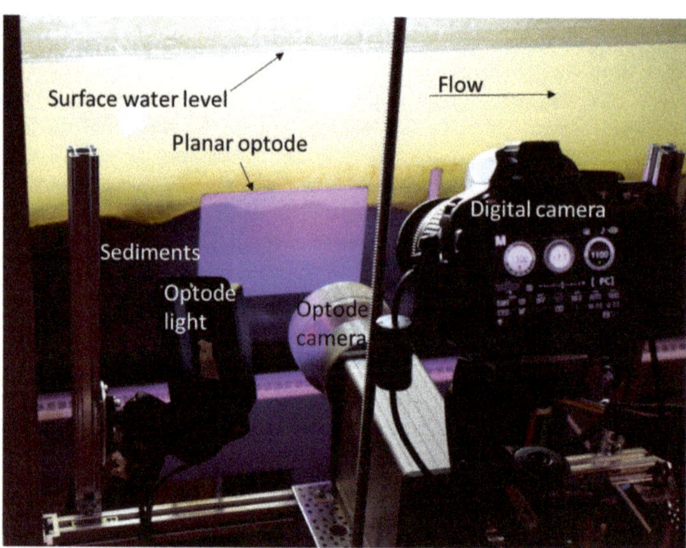

Figure A1. Experimental set-up of the planar optodes.

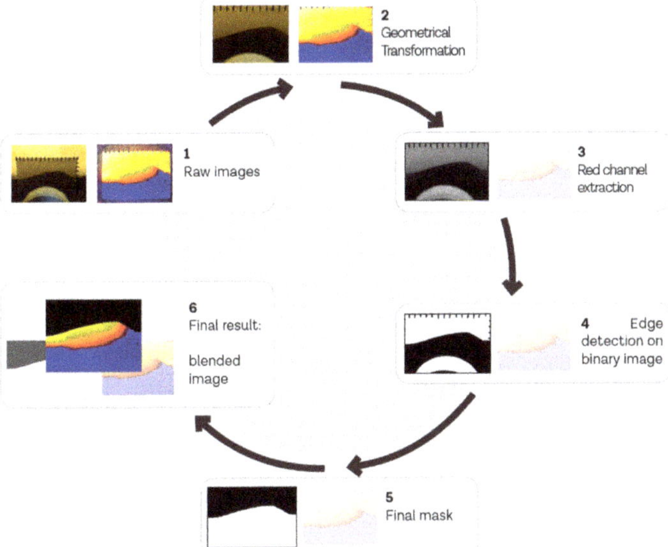

Figure A2. Flow diagram showing the stepwise image processing procedure to merge the topography information from a digital image with the spatial oxygen saturation information from the optode into one image.

References

1. Boulton, A.J.; Findlay, S.; Marmonier, P.; Stanley, E.H.; Valett, H.M. The functional significance of the hyporheic zone in streams and rivers. *Annu. Rev. Ecol. Syst.* **1998**, *29*, 59–81. [CrossRef]
2. Boano, F.; Harvey, J.W.; Marion, A.; Packman, A.I.; Revelli, R.; Ridolfi, L.; Wörman, A. Hyporheic flow and transport processes: Mechanisms, models, and biogeochemical implications. *Rev. Geophys.* **2014**, *52*, 603–679. [CrossRef]
3. Lewandowski, J.; Arnon, S.; Banks, E.; Batelaan, O.; Betterle, A. Is the Hyporheic Zone Relevant beyond the Scientific Community? *Water* **2019**, *11*, 2230. [CrossRef]
4. Harvey, J.W.; Bencala, K.E. The effect of streambed topography on surface-subsurface water exchange in mountain catchment. *Water Resour. Res.* **1993**, *29*, 89–98. [CrossRef]
5. Cardenas, M.B. Hyporheic zone hydrologic science: A historical account of its emergence and a prospectus. *Water Resour. Res.* **2015**, *51*, 3601–3616. [CrossRef]
6. Battin, T.J.; Besemer, K.; Bengtsson, M.M.; Romani, A.M.; Packmann, A.I.; Packman, A.I.; Packmann, A.I. The ecology and biogeochemistry of stream biofilms. *Nat. Rev. Microbiol.* **2016**, *14*, 251–263. [CrossRef]
7. Gandy, C.J.; Smith, J.W.N.; Jarvis, A.P. Attenuation of mining-derived pollutants in the hyporheic zone: A review. *Sci. Total Environ.* **2007**, *373*, 435–446. [CrossRef]
8. Jaeger, A.; Posselt, M.; Betterle, A.; Schaper, J.; Mechelke, J.; Coll, C.; Lewandowski, J. Spatial and temporal variability in attenuation of polar organic micropollutants in an urban lowland stream. *Environ. Sci. Technol.* **2019**, *53*, 2383–2395. [CrossRef]
9. Schaper, J.L.; Posselt, M.; Mccallum, J.L.; Banks, E.W.; Hoehne, A.; Meinikmann, K.; Shanafield, M.A.; Batelaan, O.; Lewandowski, J.J. Hyporheic exchange controls fate of trace organic compounds in an urban stream. *Environ. Sci. Technol.* **2018**, *52*, 12285–12294. [CrossRef]
10. Nagorski, S.A.; Moore, J.N. Arsenic mobilization in the hyporheic zone of a contaminated stream. *Water Resour. Res.* **1999**, *35*, 3441. [CrossRef]
11. Boulton, A.J.; Datry, T.; Kasahara, T.; Mutz, M.; Stanford, J.A. Ecology and management of the hyporheic zone: Stream-groundwater interactions of running waters and their floodplains. *J. N. Am. Benthol. Soc.* **2010**, *29*, 26–40. [CrossRef]

12. Hester, E.T.; Gooseff, M.N. Moving beyond the banks: Hyporheic restoration is fundamental to restoring ecological services and functions of streams. *Environ. Sci. Technol.* **2010**, *44*, 1521–1525. [CrossRef] [PubMed]

13. Kaufman, M.H.; Cardenas, M.B.; Buttles, J.; Kessler, A.J.; Cook, P.L.M. Hyporheic hotmoments: Dissolved oxygen dynamics in the hyporheic zone in response to surface flow perturbations. *Water Resour. Res.* **2017**, *53*, 1–21. [CrossRef]

14. Reeder, W.J.; Quick, A.M.; Farrell, T.B.; Benner, S.G.; Feris, K.P.; Tonina, D. Spatial and temporal dynamics of dissolved oxygen concentrations and bioactivity in the hyporheic zone. *Water Resour. Res.* **2018**, 2112–2128. [CrossRef]

15. Fox, A.; Laube, G.; Schmidt, C.; Fleckenstein, J.H.; Arnon, S. The effect of losing and gaining flow conditions on hyporheic exchange in heterogeneous streambeds. *Water Resour. Res.* **2016**, *52*, 613–615. [CrossRef]

16. Cardenas, M.B. Surface water-groundwater interface geomorphology leads to scaling of residence times. *Geophys. Res. Lett.* **2008**, *35*, L08402. [CrossRef]

17. Gomez-Velez, J.D.; Harvey, J.W.; Cardenas, M.B.; Kiel, B. Denitrification in the Mississippi River network controlled by flow through river bedforms. *Nat. Geosci.* **2015**, *8*, 941–945. [CrossRef]

18. Elliott, H.; Brooks, N.H. Transfer of nonsorbing solutes to a streambed with bed forms: Theory. *Water Resour. Res.* **1997**, *33*, 123–136. [CrossRef]

19. Kessler, A.J.; Cardenas, M.B.; Cook, P.L.M. The negligible effect of bed form migration on denitrification in hyporheic zones of permeable sediments. *J. Geophys. Res.* **2015**, *120*, 1–11. [CrossRef]

20. Elliott, A.H.; Brooks, N.H.; Elliott, H.; Brooks, N.H. Transfer of nonsorbing solutes to a streambed with bed forms: Laboratory experiments. *Water Resour. Res.* **1997**, *33*, 137–151. [CrossRef]

21. Ahmerkamp, S.; Winter, C.; Janssen, F.; Kuypers, M.M.M.; Holtappels, M. The impact of bedform migration on benthic oxygen fluxes. *J. Geophys. Res. Biogeosciences* **2015**, *120*, 1–14. [CrossRef]

22. Zheng, L.; Cardenas, M.B.; Wang, L.; Mohrig, D. Ripple effects: Bedform morphodynamics cascading into hyporheic zone biogeochemistry. *Water Resour. Res.* **2019**. [CrossRef]

23. Precht, E.; Franke, U.; Polerecky, L.; Huettel, M. Oxygen dynamics in permeable sediments with wave-driven pore water exchange. *Limnol. Oceanogr.* **2004**, *49*, 693–705. [CrossRef]

24. Ahmerkamp, S.; Winter, C.; Krämer, K.; Beer, D.D.; Janssen, F.; Friedrich, J.; Kuypers, M.M.M.; Holtappels, M. Regulation of benthic oxygen fluxes in permeable sediments of the coastal ocean. *Limnol. Oceanogr.* **2017**, *62*, 1935–1954. [CrossRef]

25. Rutherford, J.C.; Latimer, G.J.; Smith, R.K. Bedform mobility and benthic oxygen uptake. *Water Res.* **1993**, *27*, 1545–1558. [CrossRef]

26. De Falco, N.; Boano, F.; Bogler, A.; Bar-Zeev, E.; Arnon, S. Influence of stream-subsurface exchange flux and bacterial biofilms on oxygen consumption under nutrient-rich conditions. *J. Geophys. Res. Biogeosci.* **2018**, *123*, 1–14. [CrossRef]

27. Arnon, S.; Avni, N.; Gafny, S. Nutrient uptake and macroinvertebrate community structure in a highly regulated Mediterranean stream receiving treated wastewater. *Aquat. Sci.* **2015**, *77*, 623–637. [CrossRef]

28. Devlin, J.F. HydrogeoSieveXL: An Excel-based tool to estimate hydraulic conductivity from grain-size analysis. *Hydrogeol. J.* **2015**, *23*, 837–844. [CrossRef]

29. Klute, A.A. *Methods of Soil Analysis. Part 1*, 2nd ed.; Inc. Publishers: Madison, WI, USA, 1986.

30. Salehin, M.; Packman, A.I.; Paradis, M. Hyporheic exchange with heterogeneous streambeds: Laboratory experiments and modeling. *Water Resour. Res.* **2004**, *40*, W11504. [CrossRef]

31. Fox, A.; Boano, F.; Arnon, S. Impact of losing and gaining streamflow conditions on hyporheic exchange fluxes induced by dune-shaped bed forms. *Water Resour. Res.* **2014**, *50*, 1–13. [CrossRef]

32. Klein, S.; Worch, E.; Knepper, T.P. Occurrence and spatial distribution of microplastics in river shore sediments of the rhine-main area in Germany. *Environ. Sci. Technol.* **2015**, *49*, 6070–6076. [CrossRef] [PubMed]

33. Bradski, G. The OpenCV Library. *Dr. Dobb's J. Softw. Tools* **2000**, *120*, 122–125.

34. Rao, A.M.F.; McCarthy, M.J.; Gardner, W.S.; Jahnke, R.A. Respiration and denitrification in permeable continental shelf deposits on the South Atlantic Bight: Rates of carbon and nitrogen cycling from sediment column experiments. *Cont. Shelf Res.* **2007**, *27*, 1801–1819. [CrossRef]

35. Polerecky, L.; Franke, U.; Werner, U.; Grunwald, B.; de Beer, D. High spatial resolution measurement of oxygen consumption rates in permeable sediments. *Limnol. Oceanogr.* **2005**, *3*, 75–85. [CrossRef]

36. Monsen, N.E.; Cloern, J.E.; Lucas, L.V.; Monismith, S.G. A comment on the use of flushing time, residence time, and age as transport time scales. *Limnol. Oceanogr.* **2002**, *47*, 1545–1553. [CrossRef]

37. Gomez, J.D.; Wilson, J.L.; Cardenas, M.B. Residence time distributions in sinuosity-driven hyporheic zones and their biogeochemical effects. *Water Resour. Res.* **2012**, *48*, W09533. [CrossRef]
38. Cardenas, M.B.; Wilson, J.L.; Haggerty, R. Residence time of bedform-driven hyporheic exchange. *Adv. Water Resour.* **2008**, *31*, 1382–1386. [CrossRef]
39. Packman, A.I.; Brooks, N.H. Hyporheic exchange of solutes and colloids with moving bed forms. *Water Resour. Res.* **2001**, *37*, 2591–2605. [CrossRef]
40. Van Den Berg, J. Bedform migration and bed-load transport in some rivers and tidal environments. *Sedimentology* **1987**, *34*, 681–698. [CrossRef]
41. Bradley, R.W.; Venditti, J.G. Reevaluating dune scaling relations. *Earth Sci. Rev.* **2017**, *165*, 356–376. [CrossRef]
42. Ernstsen, V.; Winter, C.; Becker, M.; Bartholdy, J. Tide-controlled variations of primary- and secondary-bedform height: Innenjade tidal channel (Jade Bay, German Bight). In *River, Coastal and Estuarine Morphodynamics*; Vionnet, C., Perillo, G., Latrubesse, E., Garcia, M., Eds.; Taylor & Francis Group: London, UK, 2009; pp. 779–786.
43. Mehmet, Y. On the determination of ripple geometry. *J. Hydraul. Eng.* **1985**, *111*, 1148–1155.
44. Packman, A.; Salehin, M.; Zaramella, M. Hyporheic exchange with gravel beds: Basic hydrodynamic interactions and bedform-induced advective flows. *J. Hydraul. Eng.* **2004**, *130*, 647–656. [CrossRef]
45. Arnon, S.; Gray, K.A.; Packman, A.I. Biophysicochemical process coupling controls nitrate use by benthic biofilms. *Limnol. Oceanogr.* **2007**, *52*, 1665–1671. [CrossRef]
46. Bottacin-Busolin, A.; Marion, A. Combined role of advective pumping and mechanical dispersion on time scales of bed form-induced hyporheic exchange. *Water Resour. Res.* **2010**, *46*, 1–12. [CrossRef]
47. Ren, J.H.; Packman, A.I. Coupled stream-subsurface exchange of colloidal hematite and dissolved zinc, copper, and phosphate. *Environ. Sci. Technol.* **2005**, *39*, 6387–6394. [CrossRef]
48. Jerolmack, D.; Mohrig, D. Interactions between bed forms: Topography, turbulence, and transport. *J. Geophys. Res. Earth Surf.* **2005**, *110*, 1–13. [CrossRef]
49. Fox, A.; Packman, A.I.; Boano, F.; Phillips, C.B.; Arnon, S. Interactions between suspended kaolinite Deposition and hyporheic exchange flux under losing and gaining flow conditions. *Geophys. Res. Lett.* **2018**, *45*, 4077–4085. [CrossRef]
50. Hartwig, M.; Borchardt, D. Alteration of key hyporheic functions through biological and physical clogging along a nutrient and fine-sediment gradient. *Ecohydrology* **2014**, *8*, 961–975. [CrossRef]
51. Gomez-velez, J.D.; Harvey, J.W. A hydrogeomorphic river network model predicts where and why hyporheic exchange is important in large basins. *Geophys. Res. Lett* **2014**, *41*, 1–10. [CrossRef]
52. Singh, T.; Wu, L.; Gomez-Velez, J.D.; Lewandowski, J.; Hannah, D.M.; Krause, S. Dynamic hyporheic zones: Exploring the role of peak flow events on bedform-induced hyporheic exchange. *Water Resour. Res.* **2019**, *55*, 218–235. [CrossRef]

Article

Determining the Discharge and Recharge Relationships between Lake and Groundwater in Lake Hulun Using Hydrogen and Oxygen Isotopes and Chloride Ions

Zhiming Han, Xiaohong Shi *, Keli Jia, Biao Sun, Shengnan Zhao and Chenxing Fu

College of Water Conservancy and Civil Engineering, Inner Mongolia Agricultural University, Hohhot 010018, China; m15029258074@163.com (Z.H.); kelijia58@126.com (K.J.); sunbiao@imau.edu.cn (B.S.); zhaoshengnan2005@163.com (S.Z.); fcx7@sina.com (C.F.)

* Correspondence: imaushixiaohong@163.com

Received: 15 November 2018; Accepted: 30 January 2019; Published: 3 February 2019

Abstract: This study examined the discharge and recharge relationships between lake and groundwater in Lake Hulun using a novel tracer method that tracks hydrogen and oxygen isotopes and chloride ions. The hydrogen and oxygen isotopes in precipitation falling in the Lake Hulun Basin were compared with those in water samples from the lake and from the local river, well and spring water during both freezing and non-freezing periods in 2017. The results showed that the local meteoric water line equation in the Lake Hulun area is $\delta D = 6.68\ \delta^{18}O - 5.89‰$ ($R^2 = 0.96$) and the main source of water supply in the study area is precipitation. Long-term groundwater monitoring data revealed that the groundwater is effectively recharged by precipitation through the aeration zone. Exchanges between the various compounds during the strong evaporative fractionation process in groundwater are responsible for the gradual depletion of $\delta^{18}O$. The lake is recharged by groundwater during the non-freezing period, as shown in the map constructed to show the recharge and discharge relationships between the lake and groundwater. The steadily rising lake water levels in the summer mean that the water level before the freeze is high and consequently the water in the lake drains into the surrounding groundwater via faults along both sides of the lake during the frozen period. The groundwater is discharged into the lake in the west and into the Urson River in the east due to the Cuogang uplift.

Keywords: Lake Hulun; groundwater; frozen and non-frozen; hydrogen and oxygen isotopes; chloride

---

1. Introduction

The five major types of lake regions in China are categorized by the natural environment, regional differences and characteristics of the water resources. The lakes in the Mongolian Plateau are typical of those in the cold and arid regions in the northern part of China. The distinct regional features and water cycle of these lakes serve as significant indicators of historical climate change and thus provide a good starting point for studies on environmental change. Investigations of the lakes in these cold and dry areas supplement the existing literature and help build our understanding of the processes involved [1,2]. Thanks to recent advances in element geochemistry, stable isotopes of liquid water have become an effective way to study the transformations between atmospheric precipitation, surface water and groundwater, especially in arid and semi-arid regions where such transformations are complex [3]. The ionic composition of natural water is influenced by a number of factors, including the chemical composition of the water, lithological formations, climate and human activities in the area. The transfer features of chloride ions are similar to those of liquid water and chlorine has a diffusion coefficient that is closer to water than any other ion. In addition, chlorine is not involved in the earth chemical reactions commonly found in aquifers. Taken together, these three factors make chlorine a useful measure of conservativeness [4]. The environmental isotopes and chloride ions in the water can thus be used to document a great deal of information on the processes involved in the water cycle, and therefore constitute an ideal method for investigating the regional hydrological cycle [5].

When analyzing the relationship between δD and $\delta^{18}O$ in different water bodies, most researchers have utilized either the Global Meteoric Water Line (GMWL) [6] or the Local Meteoric Water Line (LMWL), depending on the data collected by the Global Network of Isotopes in Precipitation (GNIP) sites that are nearest to their study areas. However, the atmospheric water lines in different regions may vary due to differences in the many factors that affect the stable isotope fractionation from the water vapor source to raindrops [7]. The nearest GNIP site to Lake Hulun is 700 km to the southeast in the city of Qigihar in the Greater Khingan Range; the readings taken there are unlikely to accurately represent the distinct characteristics of precipitation in the study area. A new sampling point in the vicinity of Xin Barag Right Banner, 30 km to the south of Lake Hulun, was thus established for this study in June 2016 to collect atmospheric precipitation including both rainfall and snow in order to characterize the precipitation in the Hulun area.

Previous research on Hulun's lake water has primarily utilized the balance model for meteorological and hydrological data [8,9]. The main problem with such model is that it is not straightforward to determine the equilibrium items. Also, the Lake Hulun Basin, which shares borders with Russia and Mongolia, has a complex water system. For instance, the basin connects the Krulun River, which originates in Mongolia to the south, and the Erguna River which demarcates the China-Russia border to the north. The lake's unique geographical location and natural environment make it an important place for ecotourism, fisheries and scientific research. The combination of hydrogen and oxygen isotopes and chloride ions can effectively reduce the uncertainty caused by a single research method. Hence for this study, the water from Lake Hulun and its surrounding groundwater, springs and major inflowing rivers were sampled in April 2017 (frozen period) and July 2017 (non-frozen period). A comparative analysis was conducted to investigate the variations in the hydrogen and oxygen isotope and chloride ion content in each type of water during the two periods (frozen and non-frozen). Revealing the hydrogen-oxygen isotope and chloride ion cycle between the water cycles in each water body and its influencing factors, along with the recharge relationship between the lake water and groundwater. The study's findings have important implications for the sustainable exploitation and utilization of water resources and policymaking for water resource conservation in the Lake Hulun Basin.

2. Study Area

Lake Hulun, which is also known as Lake Dalai, is located between Xin Barag Left Banner and Xin Barag Right Banner in the western Hulunbuir steppe, at a longitude of 117°00′10″ E–117°41′40″ E, and a latitude of 48°30′40″ N–49°20′40″ N. The lake is 93 km long, reaching a maximum width of 41 km, and has a total area of 2339 square kilometers. The average depth of the lake is 5 m, with a maximum depth of 8 m. Lake Hulun is the fourth largest freshwater lake in China and the largest in northern China. Lake Hulun's water system is an integral part of the Erguna water system. The lake's main tributaries include Lake Bel, and the Hailar, Urson, Kherlan and Xinkaihe Rivers. The Lake Hulun Basin is located in the high-latitude west Hulun Buir Plateau, which consists of semi-arid grassland and is in the temperate continental climate zone, with a prevailing northwest wind all year and an annual average temperature of −0.5 degrees Celsius. The annual precipitation is 200–319 mm, about 80% of which falls in June, July and August. The amount of evaporation is 6–7 times that of precipitation. Winter begins in early October and ends in early May. The lake is frozen for up to 180 days each year, with a maximum ice thickness of up to 1.2 m. The summer begins in late June and ends in early August. The climate at Lake Hulun can thus be characterized as a long, cold winter, a windy, dry spring, a short, cool summer, and rapid cooling in autumn with an early frost [10].

The lake lies at the junction of the Erguna Xingkai geosynclinal system and the Inner Mongolia-Greater Khingan Range Lixi geosynclinal fold system and was formed in the late glacial period about 13 thousand years ago. It is a tectonic lake formed by the long-term accumulation of river water in a fault basin [11]. The lake basin takes the shape of an irregular oblique rectangle lying in a generally northeast-southwest direction, which was formed by the Cuogang Fault in the east and the Xishan Fault in the west. In the northwest of the basin, low mountain ranges composed of Mesozoic intermediate-acid volcanic rocks stretch from the northeast to the southwest. The strata of the wider lakeshore plains in the northern, southern, and eastern sides of the lake are mainly covered with quaternary sediments and the low hills along the southern side of the lake are composed of basalt.

3. Method

3.1. Sampling and Analysis

Water samples were collected in April and July 2017 respectively, including 13 samples from the lake, 25 from wells, 7 from rivers and 2 from springs. Precipitation (including 31 rain samples and 7 snow samples) was collected by precipitation collector. If there were several precipitation processes in a day, they were merged into one sample for determination. In order to avoid the influence of dry deposition, the precipitation is retrieved immediately after the end of the day. The sampling sites are shown in Figure 1. The samples were field filtered through 0.45-μm filters and stored in tightly capped plastic bottles for transport. Chloride analyses were conducted using an ICS-90 Ion Chromatograph (DIONEX Corp, Sunnyvale, CA, USA). The hydrogen and oxygen isotope analyses were conducted using an LGR DT100 Liquid Isotope Analyzer (LGR, San Jose, CA, USA), and the test accuracy was ±0.2‰. The isotope values are expressed as deviations in per mill (‰) relative to standard values of Vienna Standard Mean Ocean Water (VSMOW), shown as below:

$$\delta(‰) = \left[\frac{R_{sample}}{R_{standard}} - 1 \right] \times 1000. \tag{1}$$

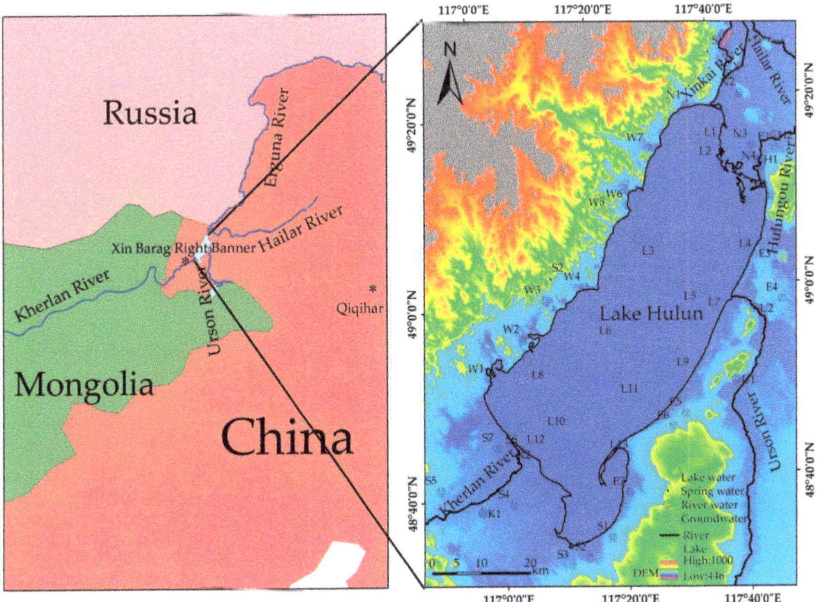

Figure 1. Map of sampling sites

3.2. Statistical Tests

In order to better verify the difference between hydrogen and oxygen isotopes in lake water, groundwater and river water during the frozen and non-frozen periods, we conducted statistical tests on the results of hydrogen and oxygen isotopes. In addition, according to the quantity and distribution characteristics of various water samples, Rank sum test for groundwater and Student's t-test for lake and river water samples were used respectively.

4. Results and Discussion

4.1. Local Meteoric Water Line in the Lake Hulun Area

Due to the different origins, types and degrees of isotopic fractionation in the precipitation, there are inevitably discrepancies in terms of the slopes and intercepts between the GMWL and LMWL for a specific region. It is therefore important to establish the LMWL before any investigation of the isotopic characteristics in a region is conducted. Based on an analysis of the hydrogen and oxygen isotopes in the local precipitation from June 2016 to September 2017, the snow water's δD ranges from $-253‰$ to $-184‰$, the δ^{18}O value ranges from $-33.4‰$ to $-23.6‰$, the δD of the rainwater ranges from $-137‰$ to $-16‰$ and the δ^{18}O value ranges from $-19.1‰$ to $-1.3‰$. Based on this data, a weighted average of the hydrogen and oxygen isotope in rainwater can now be calculated and the LMWL in the Lake Hulun area determined (Figure 2). The equation for LMWL in the study area is therefore defined as follows:

$$\delta D = 6.68\,\delta^{18}O - 5.98‰ \qquad \left(R^2 = 0.96\right). \tag{2}$$

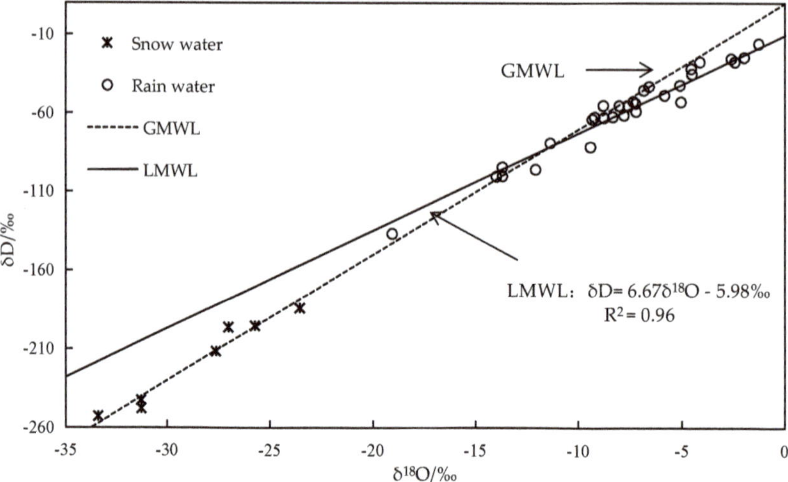

Figure 2. Meteoric lines in the Lake Hulun area.

As Figure 2 shows, the slope and intercept of the LMWL in the Lake Hulun area deviate somewhat from the global average levels. The slope of less than 8 indicates that the water vapor in the precipitation comes from a number of disparate origins that contain different ratios of these stable hydrogen and oxygen isotopes. This also suggests that some evaporation occurs during precipitation [12]. It is interesting to note that samples from snow water are uniformly distributed along the GMWL. This can be explained by the fact that the cold winter climate in the Lake Hulun area promotes the gathering of heavy isotopes in the remaining water after the snow begins to form in the clouds. As the amount of fractionation is far less than that seen for precipitation evaporation, the isotope levels in snow water are lower than those in rainwater. Comparisons of the precipitation equations for northeast China ($\delta D = 7.24\ \delta^{18}O + 1.96‰$) [13], Lake Dali ($\delta D = 6.686\ \delta^{18}O - 8.3124‰$) [14], and Lake Wuliangsuhai ($\delta D = 6.3\ \delta^{18}O - 5.6‰$) [15] show that the slopes and intercepts of the equations for the three main lakes in Inner Mongolia plateau are very similar. This suggests that the precipitation in all the lake areas in Inner Mongolia plateau will share broadly similar characteristics, with the minor differences observed indicating that in the area around each lake evaporation influences the precipitation slightly differently. The slope of the LMWL in the Lake Hulun area is less steep than that generally seen in northeast China, likely because the Greater Khingan Mountain Range blocks the moist air from the summer monsoon, which is the main source of precipitation in the Hulun Buir Grassland area, arriving from the Pacific. In addition, the lake water raises the temperature in the area, causing strong evaporation and resulting in a lack of $\delta^{18}O$ in the precipitation. The precipitation features on the eastern and western sides of Greater Khingan Mountain Range are thus significantly different. The marked differences in the climatic characteristics in the eastern and western parts of China also affect the composition of the isotopes in the atmospheric precipitation [16], making it necessary to gather data to establish the local LMWL in the Lake Hulun area.

4.2. Changes in the Water Isotopes in the Lake Hulun Basin

The results of the stable isotope analysis are shown in Table 1 and indicate that the δD values recorded in the lake water ranged from $-84.5‰$ to $-63.2‰$ and the $\delta^{18}O$ values ranged from $-9.9‰$ to $-6.1‰$, while for spring water these ranged from $-103.3‰$ to $-89.1‰$ and from $-12.9‰$ to $-10.9‰$, respectively. In the river water, the δD values ranged from $-110‰$ to $-62‰$ and the $\delta^{18}O$ values from $-13.9‰$ to $-5.2‰$; the δD values in the well water ranged from $-211‰$ to $-100‰$ and the $\delta^{18}O$ values ranged from $-27.6‰$ to $-8.4‰$. Table 1 shows that the evaporation intensity for each type of water during the frozen period is lower than that during the non-frozen period. As the variation of hydrogen and oxygen isotopes in water bodies is influenced by evaporative fractionation [17]. Among them, the δD values and the $\delta^{18}O$ values of lake water passed the Rank sum test of $p < 0.01$ (n = 13), and the groundwater (including two springs water) passed the Student's t-test of $p < 0.01$ (n = 27). The Student's *t*-test of the δD values and the $\delta^{18}O$ values of river water were $p = 0.252$ and $p = 0.249$ (n = 7), respectively. This was attributed to the small number of river water samples which came from four different. In April, when the lake is frozen, the temperature falls below zero degree Celsius in the Lake Hulun Basin and the surface water is covered with ice, which greatly reduces the water evaporation and interferes with the normal evaporative fractionation of the water body. In contrast, in July the water evaporates under the influence of the relatively high temperatures in the region, resulting in a significantly higher accumulation of heavy isotopes in the lake, spring and river water than in the groundwater.

The variations in the levels of the hydrogen and oxygen isotopes in the lake, river and groundwater during frozen and non-frozen periods (Figure 3) indicate that the accumulation of these heavier isotopes in the lake and river water is higher in the non-frozen period than when the lake is frozen. The degree of this fractionation is significant and stable and conforms to the thermodynamic isotopic effect [18,19]. It is interesting to note that the levels of the hydrogen and oxygen isotopes were lower in July than in April in the groundwater collected from wells along the western side of the lake that lie along the fault zone to the west of the lake basin (Figure 3I). A tentative explanation for this phenomenon is that it arises due to the recharge from lake water to groundwater along the western lakeshore, although this can only be confirmed by further research. In addition, intense evaporative fractionation of the hydrogen and oxygen isotopes was observed in two wells S1 and S3, both of which are located along the southeastern shore of the lake. The δD levels in these two wells during the frozen period were found to be $-211‰$ and $-210‰$, respectively, and the $\delta^{18}O$ levels in the wells were $-27.6‰$ and $-27.8‰$, respectively, which are close to the levels of hydrogen and oxygen isotopes found in snow (Figure 3II). Interestingly, during the non-frozen period, the δD levels measured in the two wells were $-145‰$ and $-164‰$, respectively, and the $\delta^{18}O$ levels were $-19.3‰$ and $-21.1‰$, respectively, slightly lower than the levels of these isotopes in the local rainwater. This suggests that the main source of groundwater in this area is the recharge from the rain. The melting and infiltration of snow and the exchange intensity of groundwater are relatively weak. Field research in this part of the study area, which is located to the southeast of the lake area, found the soil to consist of Quaternary Holocene Aeolian dune rock, sand hills and ridges with coarse-fine and silt grains. The wells are generally 1–2 m below the surface in this area, with no nearby rivers or springs. This observation further supports our conclusion that the primary recharge source for shallow groundwater is atmospheric precipitation. The temporal variations in the isotope levels in the four wells in the third box (N1–N4; groundwater) and in the two wells close to the Urson River (U1 and U2; river water) are relatively small. The evaporative fractionation is weak, and is not seasonal, which indicates that both the groundwater to the north of Lake Hulun and the Urson River itself have their own stable recharge sources.

Table 1. Levels of hydrogen and oxygen isotopes and the results of the hydrochemical analysis of the water in the Lake Hulun Basin during the frozen and non-frozen study periods.

Type of Water Body	Descriptive Statistics	April, 2017			July, 2017		
		δD/‰	$\delta^{18}O$/‰	Cl^-/(mg·L^{-1})	δD/‰	$\delta^{18}O$/‰	Cl^-/(mg·L^{-1})
Lake water (n = 13)	Min.	−84.5	−9.9	97	−66.7	−7.4	177
	Max.	−65.6	−6.9	201	−63.2	−6.1	191
	Mean	−71.9	−7.9	165	−65.2	−6.6	185
	SD	4.6	0.7	31	0.9	0.4	5
River water (n = 7)	Min.	−106.3	−13.9	2	−110.2	−14.1	1
	Max.	−74.9	−8.4	38	−61.5	−5.2	20
	Mean	−94.5	−11.8	15	−82.8	−9.0	11
	SD	13.0	2.2	14	18.9	3.3	8
Well water (n = 25)	Min.	−210.9	−27.6	5	−164.2	−21.1	18
	Max.	−81.8	−9.6	1600	−79.0	−11.3	2592
	Mean	−105.7	−13.7	269	−99.3	−13.5	389
	SD	32.7	4.5	406	18.7	2.5	627
Spring water (n = 2)	Min.	−99.8	−12.6	8	−93.9	−12.2	13
	Max.	−97.4	−12.0	38	−90.7	−11.0	62
	Mean	−98.6	−12.3	23	−92.3	−11.6	37
	SD	1.21	0.34	15	1.62	0.61	24

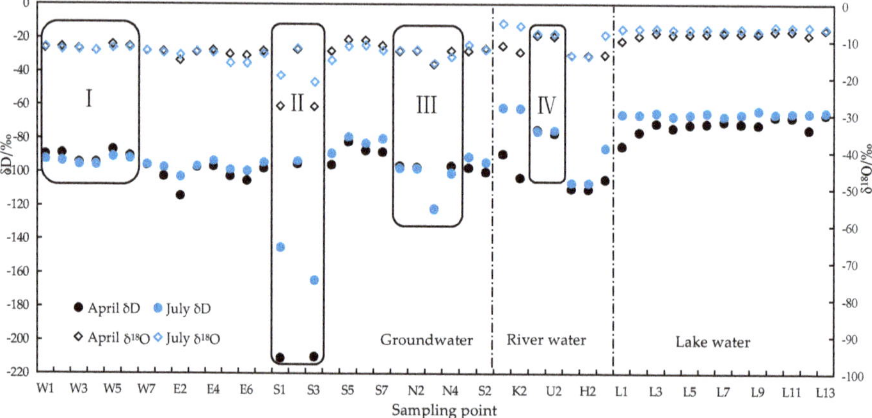

Figure 3. Levels of hydrogen and oxygen isotopes in different types of water body in the Lake Hulun area during frozen and non-frozen periods.

4.3. Analysis of the Isotopes in Water Bodies in the Lake Hulun Area

The relationships between the hydrogen and oxygen isotopes in precipitation, water, river, well and spring in the Lake Hulun area during the non-frozen period are shown in Figure 4a. Although the values of δD and δ^{18}O in the lake and river water both fall below the LMWL, the values of δD and δ^{18}O for the lake water are more stable and intense than that of those for the river water. This suggests that the wide lake surface of Lake Hulun is subject to strong evaporation, leading to excessive fractionation of the isotopes. It accords with the fact that Lake Hulun is located in an arid/ semi-arid area where evaporation is greater than precipitation. The following EL_1 equation for the evaporation trend line for the Lake Hulun area is thus obtained through a regression analysis of the hydrogen and oxygen isotope levels in the lake and its inflowing rivers.

$$\delta D = 4.99\ \delta^{18}O - 32.89\text{‰} \qquad \left(R^2 = 0.89\right). \tag{3}$$

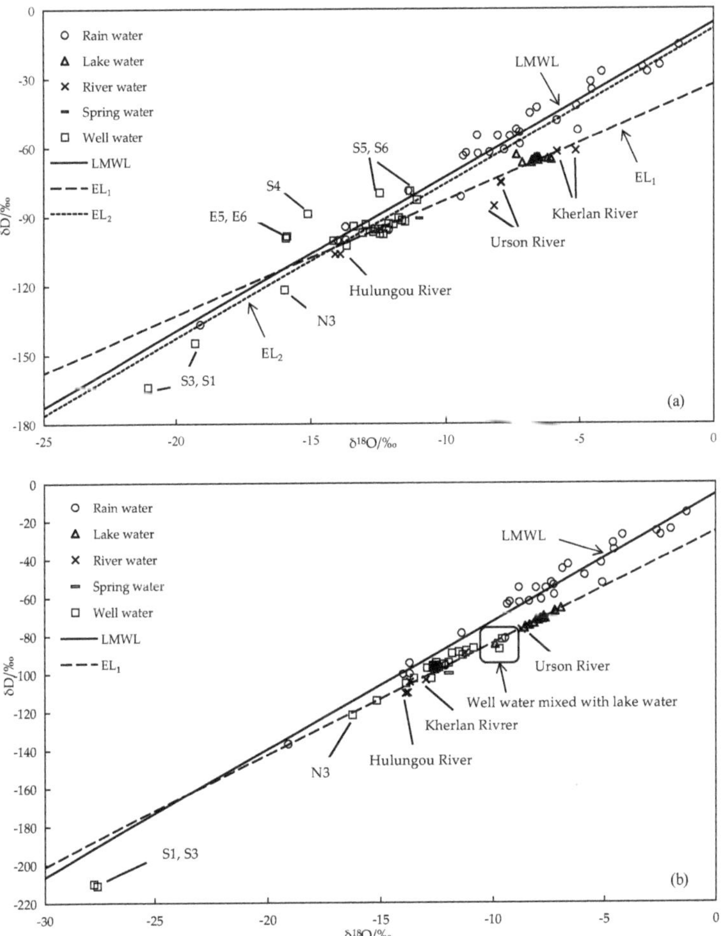

Figure 4. Relationship between the levels of hydrogen and oxygen isotopes in lake, river, groundwater and precipitation in the Lake Hulun area during non-frozen (**a**) and frozen (**b**) periods. EL_1: the evaporation trend line for the Lake Hulun area. EL_2: the evaporation trend line of the groundwater.

The evaporation line EL_1 for the lake and its surrounding rivers falls below the LWML as the isotope levels increase. This phenomenon often occurs in a dry climate or closed basin. The fact that the values of water samples fall at different positions on the evaporation trend line indicates the different degrees of evaporation experienced by each of the different water bodies in the study area [20]. Compared with the slopes of 4.9678 [21] obtained for the Lake Daihai region and 4.66 [22] for the area around Lake Wuliangsuhai in the Inner Mongolia plateau, the degree of evaporation in the Lake Hulun area is closer to that for Lake Daihai than Lake Wuliangsuhai. The isotope levels at the intersection between the evaporation line and the global meteoric water line determine the isotopic composition of the initial recharge source of the lake [23,24]. In this case, the δD and $\delta^{18}O$ values at the intersection of the weighted averages of the initial precipitation that can effectively recharge Lake Hulun are approximately $-100\permil$ and $-15.9\permil$, indicating a combination of precipitation, groundwater and river water. Field research also found that the elevation of the groundwater in the study area is generally higher than that of the lake water level, which is the hydraulic condition required for recharge from groundwater to the lake. The data presented in Table 1 also show that the average Cl^- concentration in the lake water is lower in April than it is in July. With the advent of the rainy season, the volume of the runoff carried by the major inflowing rivers increases considerably, but given that the mean concentration of Cl^- in these rivers is relatively low, at only around $11-15$ mg/L, this suggests that the significant increase in the Cl^- concentration in the lake water is actually due to recharge by groundwater. This provides further evidence that groundwater is the main source of recharge for Lake Hulun, supplemented by direct recharge from the rivers and precipitation. The following EL_2 equation for the evaporation trend line of the groundwater is thus obtained through a regression analysis of the hydrogen and oxygen isotopes in the groundwater in the Lake Hulun area:

$$\delta D = 6.69\,\delta^{18}O - 8.82\permil \qquad \left(R^2 = 0.79\right). \tag{4}$$

As Figure 4a shows, when the lake is frozen the trend lines for EL_2 and LWML are close to each other and both the groundwater and precipitation samples are distributed in close proximity to these two lines. This further corroborates our conclusion that the main recharge source of groundwater is precipitation, and the evaporative fractionation of the groundwater is stronger than that for precipitation. Meanwhile, it was found that the water samples of E5, E6, S4, S5 and S6, are all located above LWML, a significant deviation. Similar to the case of the Heihe River, which originates from the Qilian Mountains [25]. This phenomenon is likely due to a change in the isotope balance, which dilutes the $\delta^{18}O$ thus causing the $\delta^{18}O$ levels measured in the samples to drift downwards.

As to whether the precipitation can be effectively recharged into the groundwater through the aeration zone, it is first necessary to study the transport of rainwater in the soil. Many research methods focus on indoor rainfall infiltration experiments to test soil profiles at different depths in the study area [26,27]. However, since the area covered in the present study is large, and the local geology complicated due to the many different types of geological landforms surrounding the lake, indoor precipitation infiltration experiments cannot be effectively applied in Lake Hulun area. Our research team from Inner Mongolia Agricultural University has conducted longitudinal studies in the Lake Hulun area to monitor the groundwater levels before and after the rainy season (mainly from June to August) around the lake for four consecutive years. The results of the monitoring, which are shown in Table 2, indicate that the groundwater level around Lake Hulun after the rainy season is generally higher than before the rainy season, confirming that precipitation does indeed recharge the groundwater. According to the regional hydrogeological data, bedrock fissure water is widely distributed to the west of Lake Hulun, which is mainly composed of Mesozoic volcano rocks and Hercynian and Yanshan-period granite, with pore water in the loose rock formed during the Quaternary period being distributed on the top. On the eastern side of the lake, fissure phreatic water in volcanic rocks is widely distributed and is mainly found in the small basins between hills, at weathering fractures, and in vein-shaped structural fissures of volcanic rocks in the Upper Khingan

Range and granite formed during the early Yanshan Period. Pore-water in a strip of Holocene lacustrine sand is distributed along the Lake Hulun Beach area. Hence, even in parts of the Lake Hulun area where the annual evaporation is up to 7 times the precipitation that falls, precipitation infiltration can achieve the maximum possible soil water holding capacity and thus recharge the shallow groundwater through infiltration into the saturated zone. These findings are in sharp contrast to the situation in Lakes Wuliangsuhai and Daihai, where precipitation in the surrounding area cannot effectively recharge the groundwater [21,22].

Table 2. Groundwater levels around Lake Hulun (2012 to 2015).

NO.	Longitude	Latitude	Elevation (m)	Buried Depth (m)							
				2012.7.13	2012.8.15	2013.6.15	2013.8.11	2014.6.26	2014.8.4	2015.6.8	2015.8.12
1	116°59′13.6″	48°51′52.8″	615.8	14.2	14.1	13.69	12.42	11.50	11.48	10.64	10.59
2	117°05′43.1″	48°56′18.7″	546.5	0.92	0.87	1.05	0.72	0.75	0.74	0.87	0.85
3	117°09′46.2″	49°00′17.1″	554.5	1.76	1.76	2	1.48	1.44	1.38	1.55	1.32
4	117°10′52.8″	49°00′29.4″	550.7	5.1	5.1	5.54	4.8	5.00	4.95	5.4	5.1
5	117°09′53.0″	49°01′14.4″	570.3	4.02	3.93	5.32	3.65	3.80	3.30	3.69	3.42
6	117°12′59.0″	49°02′50.1″	578.2	14.77	14.65	14.73	14.46	14.30	14.34	14.63	14.48
7	117°14′49.8″	49°02′51.1″	567.4	7.6	7.56	7.75	7.23	6.91	6.50	6	5.54
8	117°15′10.1″	49°03′15.0″	572.6	6.5	4.67	4.6	3.08	3.86	3.87	3.82	3.82
9	117°17′28.2″	49°04′47.5″	567.2	3.1	3.11	3.32	2.4	3.27	3.16	3.22	3.11
10	117°19′13.6″	49°04′36.6″	559.3	2.9	2.71	2.12	1.5	2.28	2.23	2.36	2.87
11	117°20′07.5″	49°08′20.0″	570.2	4.23	4.2	4.19	4	4.20	4.12	4.18	4.15

The analysis of the isotope levels in each of the different water bodies in the Lake Hulun area during the frozen period, shown in Figure 4b, reveals that the water samples are evenly distributed along the evaporation line. Some samples from wells are mixed with those from the lake, leading to homogenization of the mixed isotopes. A comparison with Figure 4a reveals that there are traces of isotopes in snow water at two sites, N1 and N3, during the summer. Among the three major inflowing rivers, the isotopes in the Hulungou River and the Urson River remain stable across the seasons, while those in the Kherlon River demonstrate significant changes. The values for the hydrogen and oxygen isotopes in July are higher in the Kherlon River than in the Lake Hulun, which indicates that in the summer the Kherlon River is strongly influenced by evaporative fractionation. This may be explained by the slow flow velocity of the river, its wide riverbed and even the"urban heat island effect" to some extent since the Kherlon River flows through Xin Barag Right Banner into the lake. The Hulungou River, which flows into Lake Hulun from the north, has significantly lower levels of the hydrogen and oxygen isotopes than the surrounding groundwater and the relatively lower levels seen in lake water near the river mouth are thus likely due to its recharge into the lake water diluting the effect of the nearby groundwater. The reverse is the case for the Urson River, which has accumulated more heavy hydrogen and oxygen isotope and hence has higher levels of both isotopes than either the surrounding groundwater or the lake water. The Urson River flows into Lake Hulun to the right of high plains located along the Cuogang Uplift Belt, where rocks are exposed to open air and joints and fractures in the strata are formed, which facilitates the direct infiltration of atmospheric precipitation. Therefore, it can be surmised that the reason for the relatively dense accumulation and consistently high levels of hydrogen and oxygen isotopes in the Urson River are due to the reception of recharge from groundwater from the high plains that has travelled through rock fractures along the river's left bank.

The values found for the hydrogen and oxygen isotopes in well water at the N3 site, shown in the third box in Figure 3, are quite consistent. Comparing Figure 4a with Figure 4b, the water samples from N3 are below LWML, deviating significantly from the evaporation line for groundwater and surface water, indicating that the water samples collected at this site originate from a different source or combination of sources than the other samples. The δD and δ^{18}O of the well water in April and July are −121.61‰ and −16.24‰, and −121.75‰ and −15.95‰, respectively. The levels of both isotopes are lower than those in the nearby Hulungou River, which suggests that the river is not the main recharge source of the groundwater in this area. Field research found the pH value of the groundwater near site

N3 to be weakly alkaline, with the only exception being the well water sample collected at the site itself, which was below 7. Interviews with local herdsmen revealed that the amount of fluorine ions in the well water exceeds the amount permitted in drinking water and that the alkalinity or acidity and degree of mineralization in two wells that are only a few hundred meters apart can be dramatically different. The hydrogeological map shows that this area is located in the transitional fracture zone between Mount Huoerpo, Mount Huoerwula and the Zhalainuoer depression belt northeast of Lake Hulun. This fracture zone stretches across the entire area. The Derbugan deep fault also crosses the area along the eastern margin of Lake Hulun. The concentration of Cl^- in Well N3 was 72 mg/L in April and 134.7 mg/L in July, which suggests that the lake cannot be recharged by either infiltration or fracture. The presence of fracture recharge from water bodies containing lower levels of the hydrogen and oxygen isotopes cannot be ruled out, however, and this remains to be tested in future research.

4.4. Distribution Features and Tracer Analysis of Cl^- in Water Bodies in the Hulun Lake Area

In terms of tracer analysis, the most important advantage of Cl^- is its conservativeness after the atmospheric sediment into the soil layer. Therefore, the conservativeness property of Cl^- in soil water and groundwater is better than that of tritium in the water bodies during convection, dispersion and diffusion. Some amount of tritium will be lost in its evaporative and diffusive gaseous form, while Cl^- is completely nonvolatile under normal conditions. Therefore, Cl^- will stay in the aqueous water without entering the atmosphere and will not be removed by normal geochemical processes. Therefore, Cl^- serves as an excellent tracer to study the recharge of groundwater [28]. The analysis of the spatial distribution of Cl^- in water bodies in the basin can help further understand the recharge and discharge relationships between surface water and groundwater in the basin.

As shown in Figure 5, the Cl^- concentration in the Hulun Lake lake Hulun Lake in different periods of remained relatively constant. The Cl^- concentration in the inflowing rivers into the lake during the frozen and non-frozen periods is far less than that in the lake and surrounding groundwater. In addition, the concentrations in the surrounding wells near the lake vary greatly and change significantly in different periods. In the water samples from Well E5 and E6 in the right of the lake basin, concentrations of Cl^- were significantly higher than those from other wells. It can be explained by the influence of the landform, i.e., the Cuogang uplift zone. This is consistent with the recharge relationship from groundwater to Urson River in the above analysis of hydrogen and oxygen isotope. Meanwhile, the area is mainly composed of the high plains with flat terrain, strong evaporation, and weak exchange between groundwater, which accelerates mineralization and concentration of Cl^-. In the water samples from Well E3 and E4 in the north of the lake where Urson River flows into the lake, the concentration of Cl^- in groundwater during the frozen and non-frozen periods is significantly higher than that in water. The hydrogen and oxygen isotope in the two wells remain constant across the seasons. Besides, the two newly formed lakes caused by the crevasse due to the rising water level in the Hulun Lake are located here [29]. It indicates that this area is lower than the Hulun Lake. Regional hydrogeological data and field research shows that Cuogang uplift belt has been developed in the east of Hulun Lake, and the trans-tensional fault is well developed on the belt. This suggests that the effect of tensile stress lasts rather long, which subsequently cuts Mesozoic tectonic layers into pieces. The tensional fault between Huoerpo Mountain and Huoerwula Mountain is formed in the direction of N50° and west in this area. The southern slope of Huoerpo Mountain is steep, while Huoerwula Mountain is isolated with a steep slope on the northern and eastern sides, demonstrating the characteristics of tensional fault and connecting the Hulun Lake with newly formed lakes. According to the indications of hydrogen and oxygen isotope and Cl^-, in addition to the outflow from the Hulun Lake into newly formed lakes, water in the Hulun Lake can also recharge the groundwater in this area through the underground fault.

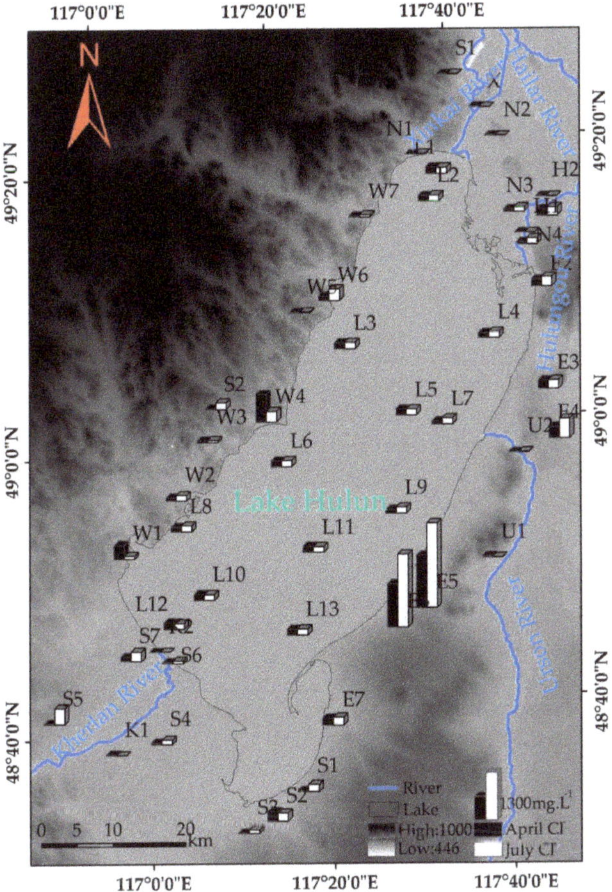

Figure 5. Distribution of Cl⁻ concentrations in water bodies in the Lake Hulun area during frozen and non-frozen periods.

The concentration of Cl⁻ in Wells W1 and W4 on the western side of the lake basin were significantly higher when the lake was frozen than when it was not, with Cl⁻ concentrations of 421 mg/L and 854 mg/L, respectively, in April, and 88 mg/L and 325 mg/L in July, markedly different from the levels measured in other wells along the western side. This suggests that during the frozen season these two wells receive recharges from water bodies with higher Cl⁻ concentrations compared to those tapped in the non-frozen period. According to the data shown in Figure 3, the amounts of the hydrogen and oxygen isotopes present in the groundwater during the frozen period are abnormally high compared to those during the non-frozen period. Figure 4b includes data points indicating that some well water becomes mixed with lake water during the frozen period; the field researchers noted that these two wells are both less than 500 m away from the lake shore. Wells W1 and W4 are 19 m and 60 m deep, respectively. Being so close to the lake, the elevation of the water level at these locations is approximately the same as that of the lake water level, leading to poor water quality and salinity exceeding the level permitted by the water quality criteria for drinking water; the water in Well W4 is yellow and turbid. The Xishan fault is located along the western lakeshore in the lake basin. The specific geographic location and climatic characteristics of Lake Hulun mean that every year the winter starts from at the end of October, when the lake starts to freeze and ends in early May when the lake ice starts to melt. The rate of evaporation reaches its maximum in May and June, with monthly evaporation of up to 270–370 mm, leading to the volume of lake water decreasing and a noticeable drop in the lake's water level. In July and August, the amount of precipitation in the area reaches its peak, and this precipitation recharges the lake via infiltration, increasing the amount of runoff in the inflowing rivers and a rise in the lake's water level. From October onwards the temperature in the Lake Hulun Basin drops sharply and lake ice starts to form. As the winter progresses the lake ice gradually expands downward, further elevating the lake surface and putting pressure on the lakebed. Although the 0.5 m layer of sludge at the bottom of the lake is very effective at sealing leaks when the lake is not frozen, after a cycle in which the water level of the lake falls and rises again lake water is discharged through the faults on both sides of the lake basin during the frozen period.

4.5. The Recharge and Discharge Relationship between the Lake and Groundwater in Lake Hulun

The local climate, geology, geomorphology and geological structure determine the hydrogeological conditions in the study area. Lake Hulun is located within the third Neocathaysian subsidence belt in the western Hailar basin. The geological structure formed by in-situ stress is highly complex in terms of its nature, order and distribution; the situation is further complicated by the fact that much of it is covered and hidden underground by the Cenozoic strata. The emergence of new satellite lakes formed by two crevasses suggests that as an occlusive tectonic lake, Lake Hulun has no other discharge outlets than those provided by the fault zones that lie along both sides of the lake and that the velocity and volume of the discharge into these fault zones are both limited. In terms of the relationship between geological structure and hydrogeology, it seems likely that the main reason why the occlusive Lake Hulun, which is located in a basin and has no effluent water outlet, can survive salinization is that the lake water recharges the underground water along both sides of the tensional fracture.

Based on the field research and the prior research in this area reported in the literature, and taking into account the evolutionary characteristics of the hydrogen and oxygen isotopes and the Cl^- levels in the lake and the nearby rivers and springs measured for two typical periods, a diagram depicting the recharge and discharge relationship of the lake and groundwater in Lake Hulun can now be constructed (Figure 6). The directions of the arrows in the figure indicate the recharge and discharge flows in or out of the lake and/or groundwater: ①⑤ indicate the recharge from the lake water into the groundwater through the deep fracture on both sides of the lake; ②④ represents the recharge from groundwater flowing into the lake by means of shallow runoffs; ③ indicates the recharge passing through precipitation fissures on the high plains due to the Cuogang Uplift Belt into the groundwater and eventually into lake on the western side, and into groundwater and eventually the Urson River on the eastern side; ⑥ shows how the runoff from the Hulungou River recharges the lake as well as the groundwater; and ⑦ indicates the possible recharge from exogenous deep water or initial water into the groundwater in this area through the fault channels evidenced by the far lower Cl^- concentrations and levels of hydrogen and oxygen isotopes measured in Well B3, which is located at the intersection of the Zhalainuoer Depression Belt and the fracture belt.

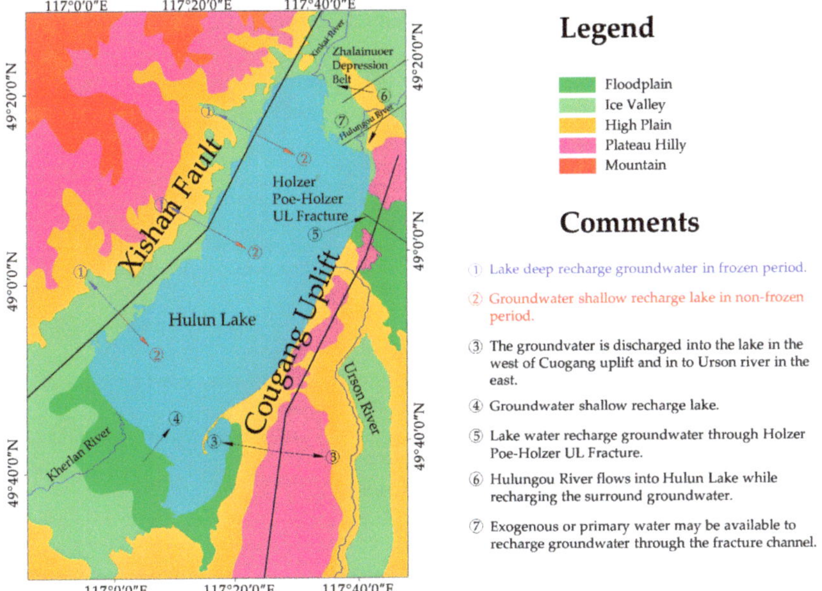

Figure 6. The recharge/discharge relationship of the lake and groundwater in Lake Hulun.

5. Conclusions

1. The analysis of the hydrogen and oxygen isotope levels in precipitation at the observation sites presented in this paper has established the LMWL for Lake Hulun to be $\delta D = 6.68\ \delta^{18}O - 5.89$ ‰ ($R^2 = 0.96$), which reflects the unique local climate in the Lake Hulun area. Comparing the slopes and intercepts of the LMWL for Lake Hulun and two other major lakes in the Inner Mongolia plateau, we found the precipitation for these three typical lakes in Inner Mongolia plateau to be quite similar, with the slight differences observed suggesting different degrees of evaporation of the precipitation that fell for each. The amounts of heavy hydrogen and oxygen isotopes measured in rivers, wells and springs near Lake Hulun all fall along the local meteoric water line, indicating that their main recharge source is precipitation.

2. Analyzing the data from April and July 2017, revealed that the intensity of evaporation from the study area's water bodies during the non-frozen period is generally higher than that during the frozen period, with the only exceptions being several of the wells. The area's groundwater is heavily evaporated and fractionated during the non-frozen period, leading to multiple isotopic exchanges among the various compounds as the groundwater undergoes strong evaporative fractionation, gradually depleting the levels of $\delta^{18}O$. Precipitation in the Lake Hulun area infiltrates and recharges the shallow groundwater in the lake basin very effectively. During the non-frozen period, precipitation recharges the lake in the form of subsurface flow.

3. The comprehensive analysis conducted for this study has demonstrated that Lake Hulun is recharged by groundwater during the non-frozen period and when the lakes freeze up discharges lake water into the surrounding groundwater through the fault zones that lie along both sides of the lake. The water cycles in these different periods also help guard against over-salinization. The hydrogeological conditions of the surrounding groundwater are influenced by the local climate, geology, geomorphology, geological structure, resulting in complicated and changeable recharge and discharge relationships between the lake's surface water and the groundwater in the Lake Hulun Basin.

Author Contributions: Investigation, Z.H. and B.S.; Data curation, Z.H., X.S., and C.F.; Formal analysis, Z.H. and K.J.; Methodology, Z.H., S.Z. and X.S.; Writing-original draft, Z.H.; Writing-review & editing, Z.H. and X.S.

Funding: This work was supported by the Chinese National Natural Science Foundation (Nos. 51339002, 51669022, 51509133 and 51779118), and Natural Science Foundation of Inner Mongolia (No. 2012MS0406).

Acknowledgments: We would like to thank Qihui Wu, Jingjing Liu and Ziyang Guo from the Water Environment Protection and Restoration Group of Inner Mongolia Agriculture University for sampling assistance.

Conflicts of Interest: The authors declare no conflicts of interest.

References

1. Shi, C.X.; Wang, X.J. *An Introduction to Chinese Lake*; Nanjing Institute of Geography and Limnology, Chinese Academy of Sciences: Beijing, China, 1989; pp. 1–3. (In Chinese)
2. Hu, R.J.; Jiang, F.Q.; Wang, Y.J. On the Importance of Research on the Lakes in Arid Land of China. *Arid Zone Res.* **2007**, *24*, 137–140. (In Chinese)
3. Zhang, Y.H.; Wu, Y.Q.; Wen, X.H. Application of environmental isotope in water cycle. *J. Adv. Water Sci.* **2006**, *17*, 738–745.
4. Herczeg, A.L.; Edmunds, W.M. Inorganic Ions as Tracers. In *Environmental Tracers in Subsurface Hydrology*; Springer: New York, NY, USA, 2000; pp. 31–77.
5. Dansgaard, W. Stable isotopes in precipitation. *Tellus* **1964**, *16*, 436–468. [CrossRef]
6. Vouillamoz, J.M.; Valois, R.; Lun, S.; Caron, D.; Arnout, L. Can groundwater secure drinking-water supply and supplementary irrigation in new settlements of North-West Cambodia? *Hydrogeol. J.* **2016**, *24*, 1–15. [CrossRef]
7. Zhu, G.F.; Li, J.F.; Shi, P.J. Relationship between sub-cloud secondary evaporation and stable isotope in precipitation in different regions of China. *Environ. Earth Sci.* **2016**, *75*, 876. [CrossRef]
8. Xiao, J.; Chang, Z.; Wen, R.; Zhai, D.; Itoh, S.; Lomtatidze, Z. Holocene weak monsoon intervals indicated by low lake levels at hulun lake in the monsoonal margin region of Northeastern Inner Mongolia, China. *Holocene* **2009**, *19*, 899–908. [CrossRef]
9. Cai, Z.; Jin, T.; Li, C. Is China's fifth-largest inland lake to dry-up? Incorporated hydrological and satellite-based methods for forecasting Hulun lake water levels. *J. Adv. Water Resour.* **2016**, *94*, 185–199. [CrossRef]
10. Chen, X.; Chuai, X.; Yang, L. Climatic warming and overgrazing induced the high concentration of organic matter in Lake Hulun, a large shallow eutrophic steppe lake in northern China. *J. Sci. Total Environ.* **2012**, *431*, 332–338. [CrossRef]
11. Xue, B.; Qu, W.; Wang, S. Lake level changes documented by sediment properties and diatom of Hulun Lake, China since the late Glacial. *J. Hydrobiol.* **2003**, *498*, 133–141. [CrossRef]
12. Tsujimura, M.; Abe, Y.; Tanaka, T. Stable isotopic and geochemical characteristics of groundwater in Kherlen River basin, a semi-arid region in eastern Mongolia. *J. Hydrol.* **2007**, *333*, 47–57. [CrossRef]
13. Li, X.F.; Zhang, M.J.; Ma, Q. Characteristics of stable isotopes in precipitation over Northeast China and its water vapor sources. *J. Environ. Sci.* **2012**, *33*, 2924–2931. (In Chinese)
14. Liu, X.X. Study on the relationship between hydrogen and oxygen isotope and water quality in Dalinuoer Lake. *J. Lake Sci.* **2015**, *27*, 1159–1167. (In Chinese)
15. Zhu, D.N. Research on Non-Natural Lakes of Chilly and Arid Region Based on Isotope Technology and Hydrogeochemistry. Ph.D. Thesis, Inner Mongolia Agricultural University, Inner Mongolia, China, 2014. (In Chinese)
16. Horita, J. *Isotopes in the Water Cycle*; Saline Waters; Springer: Dordrecht, The Netherlands, 2005; pp. 271–287.
17. Biggs, T.W.; Lai, C.T.; Chandan, P.; Lee, R.M.; Messina, A.; Lesher, R.S.; Khatoon, N. Evaporative fractions and elevation effects on stable isotopes of high elevation lakes and streams in arid western Himalaya. *J. Hydrol.* **2015**, *522*, 239–249. [CrossRef]
18. Jerzy, N.; Jerzy, G.; Lesả, B.; Jerzy, R. The altitude effect on the isotopic composition of snow in high mountains. *J. Glaciol.* **2017**, *27*, 99–111.
19. Chao, L.; Yang, S.; Lian, E.; Yang, C.; Kai, D.; Liu, Z. Damming effect on the Changjiang (Yangtze River) river water cycle based on stable hydrogen and oxygen isotopic records. *J. Geochem. Explor.* **2016**, *165*, 125–133.

20. Reid, R.; Gibson, J. Water balance along a chain of tundra lakes: A 20-year isotopic perspective. *J. Hydrol.* **2014**, *519*, 2148–2164.
21. Chen, J.S.; Zhang, Z.W.; Liu, Z. Isotope study of recharge relationships of water sources in Wuliangsuhai Lake and its surrounding areas. *J. Water Resour. Prot.* **2013**, *29*, 12–18. (In Chinese)
22. Chen, J.S.; Ji, B.C.; Liu, Z. Isotopic and hydro-chemical evidence on the origin of groundwater through deep-circulation ways in Lake Daihai region, Inner Mongolia plateau. *J. Lake Sci.* **2013**, *25*, 521–530. (In Chinese)
23. Yang, X.; Liu, T.; Xiao, H. Evolution of megadunes and lakes in the Badain Jaran Desert, Inner Mongolia, China during the last 31,000 years. *Quat. Int.* **2003**, *104*, 99–112. [CrossRef]
24. Yang, X.P.; Ma, N.; Dong, J.F. Recharge to the inter-dune lakes and Holocene climatic changes in the Badain Jaran Desert, western China. *Quat. Res.* **2010**, *73*, 10–19. [CrossRef]
25. Xue, Q.; Zhang, M.; Wang, S. Preliminary research on hydrogen and oxygen stable isotope characteristics of different water bodies in the Qilian Mountains, northwestern Tibetan Plateau. *Environ. Earth Sci.* **2016**, *75*, 1491.
26. Zagana, E.; Obeidat, M.; Kuells, C.; Udluft, P. Chloride, hydrochemical and isotope methods of groundwater recharge estimation in eastern Mediterranean areas: A case study in Jordan. *Hydrol. Process.* **2010**, *21*, 2112–2123. [CrossRef]
27. Song, X.; Wang, S.; Xiao, G.; Wang, Z.; Xin, L.; Peng, W. A study of soil water movement combining soil water potential with stable isotopes at two sites of shallow groundwater areas in the North China Plain. *Hydrol. Process.* **2010**, *23*, 1376–1388. [CrossRef]
28. Edmunds, W.; Fellman, E.; Goni, I. Spatial and temporal distribution of groundwater recharge in northern Nigeria. *Hydrogeol. J.* **2002**, *10*, 205–215. [CrossRef]
29. Zhang, Z.B. *Hulun Lake Zhi*; Inner Mongolia Culture Press: Inner Mongolia, China, 1998. (In Chinese)

Article

Dissolved Inorganic Geogenic Phosphorus Load to a Groundwater-Fed Lake: Implications of Terrestrial Phosphorus Cycling by Groundwater

Catharina Simone Nisbeth [1], Jacob Kidmose [2], Kaarina Weckström [2,3], Kasper Reitzel [4], Bent Vad Odgaard [5], Ole Bennike [2], Lærke Thorling [2], Suzanne McGowan [6], Anders Schomacker [7,8], David Lajer Juul Kristensen [1] and Søren Jessen [1,*]

[1] Department of Geosciences and Natural Resource Management (IGN), University of Copenhagen, Øster Voldgade 10, 1350 Copenhagen K, Denmark; csnm@ign.ku.dk (C.S.N.); dlajer@gmail.com (D.L.J.K.)

[2] Geological Survey of Denmark and Greenland (GEUS), Øster Voldgade 10, 1350 Copenhagen K, Denmark; jbki@geus.dk (J.K.); kaarina.weckstrom@helsinki.fi (K.W.); obe@geus.dk (O.B.); lts@geus.dk (L.T.)

[3] Ecosystems and Environment Research Programme (ECRU) / Helsinki Institute of Sustainability Science, P.O. Box 65 (Viikinkaari 1), 00014, University of Helsinki, Yliopistonkatu 4, 00100 Helsinki, Finland

[4] Institute of Biology, University of Southern Denmark, Campusvej 55, DK-5230 Odense M, Denmark; reitzel@biology.sdu.dk

[5] Institute of Geoscience, University of Aarhus, Høegh-Guldbergs Gade 2, 8000 Aarhus C, Denmark; bvo@geo.au.dk

[6] School of Geography, University of Nottingham, University Park Nottingham NG7 2RD, UK; suzanne.mcgowan@nottingham.ac.uk

[7] Natural History Museum of Denmark (SNM), Øster Voldgade 5-7, 1350 Copenhagen K, Denmark; anders.schomacker@bio.ku.dk

[8] Department of Geosciences, UiT The Arctic University of Norway, Postboks 6050 Langnes, N-9037 Tromsø, Norway

*   Correspondence: sj@ign.ku.dk

Received: 28 June 2019; Accepted: 16 October 2019; Published: 24 October 2019

Abstract: The general perception has long been that lake eutrophication is driven by anthropogenic sources of phosphorus (P) and that P is immobile in the subsurface and in aquifers. Combined investigation of the current water and P budgets of a 70 ha lake (Nørresø, Fyn, Denmark) in a clayey till-dominated landscape and of the lake's Holocene trophic history demonstrates a potential significance of geogenic (natural) groundwater-borne P. Nørresø receives water from nine streams, a groundwater-fed spring located on a small island, and precipitation. The lake loses water by evaporation and via a single outlet. Monthly measurements of stream, spring, and outlet discharge, and of tracers in the form of temperature, $\delta^{18}O$ and $\delta^{2}H$ of water, and water chemistry were conducted. The tracers indicated that the lake receives groundwater from an underlying regional confined glaciofluvial sand aquifer via the spring and one of the streams. In addition, the lake receives a direct groundwater input (estimated as the water balance residual) via the lake bed, as supported by the artesian conditions of underlying strata observed in piezometers installed along the lake shore and in wells tapping the regional confined aquifer. The groundwater in the regional confined aquifer was anoxic, ferrous, and contained 4–5 µmol/L dissolved inorganic orthophosphate (DIP). Altogether, the data indicated that groundwater contributes from 64% of the water-borne external DIP loading to the lake, and up to 90% if the DIP concentration of the spring, as representative for the average DIP of the regional confined aquifer, is assigned to the estimated groundwater input. In support, paleolimnological data retrieved from sediment cores indicated that Nørresø was never P-poor, even before the introduction of agriculture at 6000 years before present. Accordingly, groundwater-borne geogenic phosphorus can have an important influence on the trophic state of recipient surface water ecosystems, and groundwater-borne P can be a potentially important component of the terrestrial P cycle.

Keywords: geogenic phosphorus; dissolved inorganic orthophosphate; transport; eutrophication; groundwater-surface water interaction

---

1. Introduction

Eutrophication poses a major threat to freshwater lake ecosystems worldwide [1–3] and is assumed to be particularly controlled by input of phosphorus (P) [4,5]. Identifying the different pathways of P transport within catchments and determining their relative importance is crucial for optimal freshwater ecosystems management [6–9]. The general perception is that surface and near-surface hydrological pathways are the main transport routes for external P. This is due to the common assumption that P is retained readily in the aquifer sediments by adsorption and metal complex precipitation, predominantly in the oxic zone [10–12].

Over the last few decades, however, evidence of groundwater-transported P has increased [7–9,13–16], indicating that P is not necessarily immobilized in aquifers. Furthermore, there has been increasing awareness that P-rich groundwater can play an important role regarding the trophic state of lakes [14,17–19] especially for groundwater-fed aquatic systems [8,9,17,18]. Even in cases where groundwater only makes up a minor part of the lake's annual water budget, it can still control the trophic state of the lake, if the discharging groundwater contains a high concentration of P [6,20,21].

The geochemical environment in the aquifer may play an important role regarding the mobilization of P within the aquifer. Walter [13] found that P was mobile in groundwater in two distinct geochemical environments: (1) In anoxic environments with a high concentration of dissolved ferrous iron, and (2) in suboxic environments where a continued phosphorus loading has filled all available sorption sites. Phosphorus may also be mobile in oxic environments with low iron concentrations or if the iron is immobilized as FeS. This is consistent with the recorded correlation between sediment redox conditions and mobilization of inorganic P, where the main mechanisms responsible for sediment P release is reductive dissolution of Fe-oxides [11,12,22].

In many studies [7–9,13–15,17–19] the source of P in the groundwater, discharging to a lake, is either shown or expected to be anthropogenic, or it remains undetermined whether the P source is natural or anthropogenic. However, total phosphorus (TP) concentrations above the 0.3 μM (10 μg P/L) threshold for oligotrophic freshwater environments [23] have been recorded in deeper, reduced aquifers [8], where P is highly unlikely to be derived from anthropogenic activities. If groundwater from these aquifers discharges into a lake, naturally released (i.e., geogenic) groundwater-borne P may contribute significantly to the trophic state of the lake [24].

The objective of the current study was to investigate whether geogenic groundwater-borne P can be the dominant external contributor of P. This was done by (1) determining the water budget of a Danish lake, Nørresø, (2) investigating the different water-borne external sources of P, and calculating their relative contributions. In order to put the year-long study into a long-term context the results were (3) combined with results of a parallel paleolimnological study.

The internal release of P, which may play an important factor controlling the concentrations of soluble P in the water column [23,25,26], and thus biotic populations during summer blooms, was not included in the present study. Rather, this paper focuses on the identification of the original, external source of the phosphorus now accumulated in the lake and its sediment.

2. Materials and Methods

2.1. Study Site

Brahetrolleborg Nørresø (55°08′50″ N, 10°22′40″ E, hereafter for simplicity termed Nørresø) is a lake located in a clay till moraine landscape. The lake formed during the final stage of the Weichselian glaciation [27], in the southwestern part of Fyn, Denmark (Figure 1). Historical maps show that the topographical catchment of Nørresø was dominated by forest through at least the past two centuries [28].

Nørresø is Fyn's third largest lake, with a geographical surface area of 69.3 ha, including a small island, Lucieø (0.14 ha), located in the eastern end of the lake. The lake has a mean and maximum water depth of 2.3 m and 5.7 m [29], respectively, and the current annual mean elevation of the lake water table is 40.76 ± 0.1 m (seasonal variation) above sea level (ASL).

Visual inflow of water to the lake can be observed from a natural spring elevated 1 m above the lake's water level, located on the island, Lucieø (Figure 1d). The Lucieø spring is simply referred to as "the spring" in this paper. In addition, nine small streams discharge into the lake. They are, for the purpose of this paper, numbered according to their importance in relation to the determined dissolved inorganic orthophosphate (DIP) loading (kg P/yr), with stream 1 and stream 9 representing, respectively, the highest and the lowest external DIP load. Seven of the streams are ephemeral (Nos. 2 and 4–9), whereas the last two streams (Nos. 1 and 3) and the spring feed water to the lake all year round. The perennial stream 1 receives drainage from a meadow, while the ephemeral stream 2 drains a small lake, Hjertesø, north of Nørresø. Hence the P-loading from stream 2 probably reflects the P-cycling processes within Hjertesø. Just one stream (No. 3) receives water from drainage pipes connected to a large agricultural field located west of Nørresø (Figure 1b). Water is only discharging from the lake via one surface water outlet located at the western end of the lake.

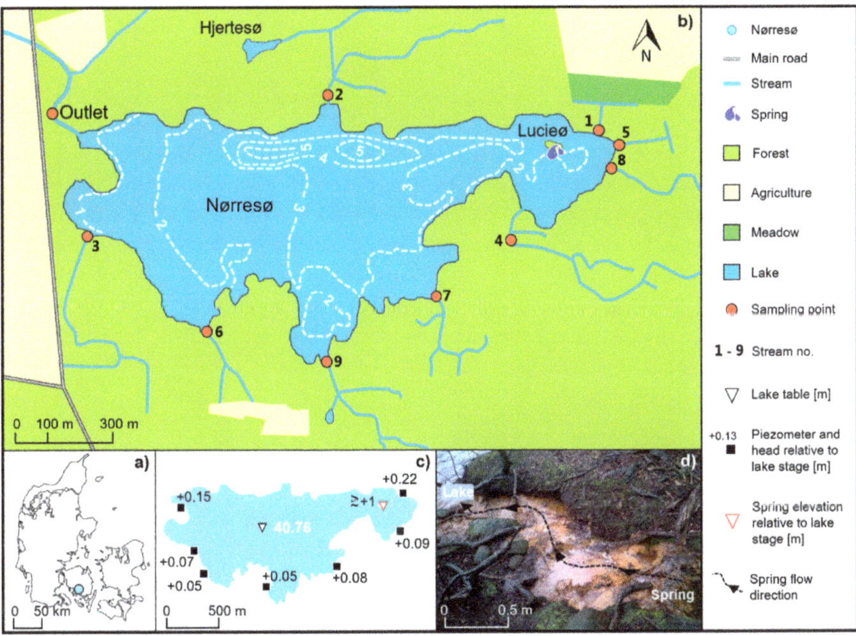

Figure 1. (**a**) Location of the Lake Nørresø in the southern part of Fyn, Denmark. (**b**) Overview of the numbering of streams, and the locations of stream sampling points and the outlet. The locations of the small island, Lucieø, hosting a spring and the small lake, Hjertesø, are also indicated. The dashed white lines indicate lake bathymetry (in meters; [29]). (**c**) The relative hydraulic head in seven piezometric wells placed at ~3 m depth. (**d**) Photo of the spring on the island Lucieø (black arrows indicate flow direction). Iron-oxides stain the bed of the 4-m-long stream which connects the spring to the lake.

Lake stratification has not been observed in Nørresø [27,29], most likely due to the shallow depth relative to the surface area combined with exposure to westerly winds; the lake is therefore assumed to be a well-mixed lake.

2.2. Data Collection

During a year-long study starting in June 2016 and ending in June 2017, a variety of different methods were applied to investigate the lake, according to the objective of this study. In order to understand the hydrogeology and determine the water budget, four different techniques were used: (1) Observations of lithology (including hand drillings and hydrogeophysics), (2) hydraulic head, (3) discharge measurement (from the outlet, spring, and streams), and (4) a water budget equation (using data obtained through the current study and precipitation and evaporation data extracted from a database). Distinction between streams fed with groundwater from a deep regional confined aquifer as opposed to surface runoff water or groundwater from shallower circulation systems was based on values of $\delta^{18}O$ and δ^2H of water combined with the chemical speciation and temperature of the water matrix.

The only bioavailable form of P is orthophosphate (PO_4^{3-}), which is the dominant P species in DIP [30]. The focus was, therefore, on the external water-borne DIP loadings. This is further justified as the majority of P entering Nørresø is dissolved inorganic P (cf. [31]). It is, therefore, assumed that TP mainly consisted of DIP and thus the measured DIP concentrations were representative of P entering Nørresø (i.e., particulate P and organic P is of secondary importance; further discussion in Section 4). Quantification of the external DIP loads from the different external water-borne sources were calculated by multiplication of each source's monthly discharge and associated measured DIP concentration summed over a year. The methods of investigation are further described below.

2.2.1. Lithology

The surficial geology close to the lake shore and on Lucieø was mapped from 13 hand drillings using an Eijkelkamp hand auger and one Geoprobe® (Salina, Kansas) drilling (see Figure 2 for locations and lithological logs). Eleven of the drillings penetrated to 2 m depth, two went down to 5 m depth, and the Geoprobe® drilling was 20 m deep. A broader understanding of the geological setting was obtained from lithological logs from three nearby water supply wells (Figure 2) for which data were extracted from the Danish borehole JUPITER-database [32]. The three water supply wells all are less than 3 km from the spring and penetrate down to a confined glaciofluvial sand aquifer (screen depths within 42–58 m below terrain).

Figure 2. Overview of Nørresø and the surrounding area, and locations of piezometers, electrical resistivity tomography (ERT) profiles and the lakebed sediment core site. Lithological logs from the drillings (simplified to combine the shallow peat and the underlying clay till) and water supply wells are also included.

In addition, two electrical resistivity tomography (ERT) profiles northeast (NE ERT) and southeast (SE ERT) of the lake, i.e., close to Lucieø (Figure 2) were conducted (Supplementary Materials Section S1 and Figure S1). Interpretation of the inverted data was based on literature resistivity values [33].

2.2.2. Hydraulic Heads

Local interaction between surface water and groundwater can be detected by observations of vertical piezometeric head gradients [34]. Thus, seven piezometers were installed at the lakeshore (Figure 1c) using a pneumatic hammer. They were all placed in clay till and their 10-cm-long screen was installed typically 3 m but up to 12 m below surface. The hydraulic head was measured using a Solinst 101 water level meter or a Solinst 102 coaxial cable fitted with a P4 probe. Elevation of reference points was recorded using a Trimble differential GPS.

The lake water table was measured every 4 hours using two Onset$^{®}$ pressure transducers with an auto logger function (HOBO Water Level 30′ U20L-01). The pressure transducers were installed at the eastern and western lakeshore. A third similar pressure transducer, installed on land, was used to correct for the atmospheric pressure.

The direction of the underground flow in the confined sand aquifer was estimated based on hydraulic head measurements recorded in the JUPITER-database [32].

2.2.3. Water Budget Equation

A general lake water budget equation can be written in the form [35]:

$$0 = N + S_{in} + GW_{in} - E - S_{out} - GW_{out} + \Delta St \tag{1}$$

where N is precipitation, S_{in} is surface water input (stream discharge in streams 1–9), GW_{in} is groundwater input, E is evaporation, S_{out} is surface water output, GW_{out} is groundwater output, and ΔSt denotes storage change.

Groundwater discharge to the lake (GW_{in}) was divided into two components: Measured groundwater flow (GW_{meas}) and an estimated groundwater input (GW_{est}) calculated as the residual of the water balance. GW_{meas} includes the spring and stream 1. Discussion of this, and other assumptions regarding the water budget components, is presented in Section 4. Precipitation (N) and evaporation (E) values for the catchment were extracted from a database compiled by the Danish National Monitoring Program for Aquatic Environment and Nature [36], where the potential evaporation are quantified based on a modified Makkink equation. The used data are from a 20 × 20 km gridded dataset interpolated from the Danish Metrological Institute's automated climate stations. The interpolations of potential evaporation for the grid-cell of Nørresø are based on the stations Assens (55°15′ N, 9°53′ E) 32 km WNW of Nørresø, Årslev (55°19′ N, 10°26′ E) 18 km NNE of Nørresø, and Sydfyns Flyveplads (55°01′ N, 10°34′ E) 18 km SE of Nørresø. The climate stations measure hydrological variables every 10 min. The gridded values used for the study were daily averages for the investigated period. Storage change (ΔSt) in the lake was calculated from the average monthly lake water table variation.

2.2.4. Frequency of Collection of Hydrological and Hydrochemical Data

Stream discharge (Q) measurements and water sampling were conducted concurrently on a monthly basis throughout one hydrological year (6 June 2016–21 June 2017) from all streams with a discharge above quantification limit (1 L/min), as well as from the spring and the outlet. Attempts to increase stream discharge measurement frequency via high-frequency monitoring of stream water level (h) using data logger-equipped pressure transducers were unsuccessful. For a successful Q/h-relation to be established, h needs to be controlled by Q. However, water levels in the streams were less affected by stream discharge than by accumulation of leaves and other organic debris in the streams, and by its high-precipitation-event-based removal by flushing or by humans as part of routine stream management.

2.2.5. Outlet, Spring, and Stream Discharge

A cutthroat flume from Baski, Inc. (Englewood, CO, USA) [37,38] equipped with 6.5″ entrance wing walls was used to measure stream discharge of the outlet (S_{out}; 8″ throat width), and the spring and streams 1–9 (S_{in}; 1″ throat width).

Triplicate measurements for determination of uncertainty of individual 1″ throat width Baski flume measurements were conducted twice in stream 3, once in stream 5 and once in the spring, at conditions representing nearly the full range of stream flow measurements (not including the outlet), from ~6 L/min (stream 3, June 2017) to 64 L/min (stream 9, December 2016). For each uncertainty determination, measurements were carried out within 30 min at three locations a few meters apart each other along each stream.

The uncertainty expressed as one standard deviation (σ) as a function of the discharge rate (Q) determined by the cutthroat flume was found by linear regression which resulted in the equation $\sigma = 0.157 \times Q$ ($r^2 = 0.99$). This equation was then used to determine σ for each specific discharge rate measurement (Figure 6).

2.2.6. Water Sampling and Analysis

All water samples were filtered in the field with 0.22 µm Minisart Sartorius cellulose acetate membrane filters directly into polyethylene vials. Samples for DIP and Fe^{2+} analysis were preserved with 0.1 mL 2 M H_2SO_4 per 5 mL water and ferrozine per 3 mL water, respectively.

Determination of Fe^{2+} and DIP were done spectrophotometrically in the laboratory according to the methods of Stookey [39] and Murphy and Riley [40], respectively. Initially, we also measured the S^{-2} concentration. However, none of the samples contained any S^{-2}, hence sampling for S^{-2} was discontinued.

Major cations (Ca^{2+}, Mg^{2+}, Na^+, K^+, and NH_4^+) and anions (F^-, Cl^-, Br^-, NO_3^-, and SO_4^{2-}) were analyzed using ion chromatography (IC) and alkalinity was measured by Gran-titration. These analyses were conducted at the hydrogeochemical laboratory at the Section of Geology, University of Copenhagen, Denmark.

The stable isotopes $\delta^{18}O$ and δ^2H were analyzed on a Picarro Cavity Ring-Down Spectrometer (CRDS) L2120-i at the Geological Survey of Denmark and Greenland, Copenhagen. Standard deviations were equal to or lower than 0.18‰ for $\delta^{18}O$ and 0.45‰ for δ^2H.

DIP, Fe^{2+}, O_2, and temperature data of the water matrix in the water supply wells were extracted from the JUPITER-database [32] and assumed representative of the confined aquifer (i.e., representing deep groundwater). Water samples collected from the east and west end of the lake were combined with outlet samples under the term 'lake' in the remainder of the paper as the lake is assumed to be well mixed.

Aqueous speciation and saturation index calculations were done using the free software PHREEQC 3.0 developed by David L. Parkhurst and C.A.J. Appelo (United States Geological Survey, Denver, Colorado, USA).

2.2.7. Field Measurements

Measurements of dissolved O_2 were conducted in the field using a WTW oxi-3310 IDS Portable Dissolved Oxygen Meter (accuracy: ±0.5% of value). Values were recorded upon complete stabilization. Temperature was read from the probe. Calibration of the probe was performed in the field in the morning before use. pH and electrical conductivity were measured but are not reported in this paper since there were no significant differences internally between the streams and the spring.

2.2.8. Paleolimnological Analyses

The parallel paleolimnological study conducted within the overall project presents results based on diatom, plant pigment, and pollen analyses of a 6-m-long sediment core section sampled in the center of the lake (Figure 2). These 6 m of the core consist of gyttja accumulated since about 7500 cal yr BP.

The core chronology is based on accelerator mass spectrometry (AMS) radiocarbon dating of terrestrial plant remains (to avoid hard-water effects) found in the Nørresø sediment core. The obtained radiocarbon ages were calibrated to calendar ages (cal yr BP) using IntCal13 [41]. The first obtained ages are presented in this study, while more sediment samples, to allow for a more detailed chronology, are currently being dated.

Elements were analyzed via X-ray fluorescence (XRF) using an ITRAX core scanner (COX Analytical Systems, Mölndal, Sweden). The scanning of the core was performed at the Natural History Museum of Denmark. To normalize the XRF data in order to compensate for the effects of water content, organic matter, and compaction of the sediment through depth, all data were divided with the total scatter (Rhodium incoherent + Rayleigh). A moving average of five neighboring points was calculated for all elements.

Diatom samples were prepared from ca. 0.2 g of freeze-dried sediment, which was prepared following standard methodologies [42]. Briefly, the sediment was oxidized at 90 °C for 6 hours using hydrogen peroxide (H_2O_2, 30%) in order to remove organic material. Carbonates were then dissolved by addition of a few drops of HCl (35%). The test tubes in which samples were treated were subsequently filled with distilled water and left to settle for 12 h. Residues were then washed several times using demineralized water. A few drops of the final suspension were dried on a cover slip and subsequently mounted in Naphrax® for identification, which was conducted using an optical microscope (Leica DMLB), equipped with phase contrast, at a magnification of ×1000.

Pigments were quantitatively extracted in an acetone:methanol:water (80:15:5) mixture. Extracts were left overnight at −10 °C, filtered with a polytetraflourethylen (PTFE) 0.2 µm filter and dried under nitrogen gas. A known quantity was re-dissolved into an injection solution of a 70:25:5 mixture of (1) acetone, (2) ion-pairing reagent (IPR; 0.75 g of tetra butyl ammonium acetate and 7.7 g of ammonium acetate in 100 mL water), and (3) methanol and injected into a high-performance liquid chromatography (HPLC) unit. Pigment extracts were separated using an Agilent 1200 series separation module with quaternary pump. The mobile phase consisted of solvent A (80:20 methanol: 0.5 M ammonium acetate), solvent B (9:1 acetonitrile: water), and solvent C (ethyl acetate) with the stationary phase consisting of a Thermo Scientific ODS Hypersil (Thermo Fisher Scientific, Waltham, MA, USA) column (205 × 4.6 mm; 5 µm particle size). Eluted pigments passed through a photo-diode array detector and UV-visible spectral characteristics were scanned at between 350 and 750 nm. Quantification was based on scanning peak areas at 435 nm and calibrating to a set of commercial standards (DHI LAB Products, Hørsholm, Denmark). Pigment concentrations are reported as molecular weights of pigments in nanomoles per unit dry weight of sediments.

Samples for pollen analysis were prepared from 1 cm^3 of sediment following Faegri & Iversen [43]. Briefly, 37% HCl was used to remove calcium carbonate and samples rinsed 3 times before boiling in a water bath in centrifugal tubes in 10 mL 10% KOH for about 10 min. After rinsing twice with water, 10 mL concentrated HF (40%) was added and the samples boiled in a water bath for 20–25 min. The samples were then centrifuged and rinsed once with 10 mL 10% HCl and once with water, before 10 mL of concentrated acetic acid was added and the samples were rinsed. This was followed by acetolysis. Finally, the samples were suspended in tert-butanol and transferred to a preparation glass. Drops of silicon oil (AK 2000) were added, before placing the samples for evaporation in a heating cabinet at 50 °C for 2 days. Pollen was counted using a light microscope at a magnification of ×400, higher magnification (×1000) was used for identifying taxonomically difficult pollen types. At least 300 terrestrial pollen grains were counted per sample. The pollen sum, which the percentages are based on, includes all terrestrial pollen taxa.

3. Results

3.1. Lithology

In general, the 13 hand drillings all showed two types of sediment: A 0.2–0.7 m thick top layer of peat (not shown in Figure 2) followed by grey clay-rich till (seen in Figure 2). The lower boundary of the till was not detected. The grey color of the till indicated reduced conditions. The same clay till was also found in the 20 m deep Geoprobe® drilling and in the lithological logs of the three water supply wells. The water supply wells all penetrate a ≥35 m thick layer of clay till which confines a deep anoxic glaciofluvial sand aquifer. Furthermore, clay till was indicated by slow (few months) equilibration of hydraulic pressures in the piezometers screened at up to 12 m below ground. The two ERT profiles (located close to the spring) underlined the occurrence of a thick clay layer (>15 m) confining a sand layer. However, the ERT profiles also indicated that the deep-lying sand layer appears to be connected to the terrain through a thin sloping sand layer (Figure S1 in Supplementary Materials). Hence, the spring could possibly be connected to the deep confined regional aquifer via such sand layers.

3.2. Hydraulic Heads

Groundwater flow in the underlying confined aquifer is toward the southwest. Hydraulic heads in the regional aquifer (42–49 m ASL) were 1–8 m higher than the lake level (40.76 m ASL), resulting in upward hydraulic gradients. The upward gradients were confirmed in all seven of the lakeshore piezometers which had water levels ranging from 0.05–0.22 m above the lake level (Figure 1c).

3.3. Water Samples

3.3.1. Stable Isotopes of Water

The isotopic composition of the lake differed significantly from that of the streams, the spring, and the confined aquifer. Lake water samples had the most enriched values of δ^2H (−38.1‰ to −0.9‰) and $\delta^{18}O$ (−5.0‰ to −3.2‰) as well as the largest seasonal variation (Figure 3a). However, during no season did the lake water isotopic signature overlap with that of the other groups. The isotope data plotted along a line with a distinctly lower slope (~5) than that of the local meteoric water line (LMWL, $\delta^2H = 7.48 \times \delta^{18}O + 5.36$; [44]). This can be explained by the lake's relatively long hydraulic retention time of 1.6 years, which enabled evaporation to significantly enrich the isotopic composition.

The nine streams, the spring, and the confined aquifer were subdivided into three groups according to their isotopic composition; the subdivision is shown in Figure 3b,c, which are subsets of Figure 3a. Isotopic group 1 represents the samples where the isotope compositions were almost stable throughout the year. Group 1 consists of stream 1, the spring, and the confined aquifer where the δ^2H and $\delta^{18}O$ (mean δ^2H = −58.2‰, −58.3‰, and −57.9‰, respectively, mean $\delta^{18}O$ = −8.6‰ for all three) are significantly depleted relative to rest of the streams and clusters within a relatively narrow range (1 std. dev. (σ) for δ^2H is 0.3–0.4‰, and 0.2‰ for $\delta^{18}O$), thus indicating a shared source of water between these three sources. Isotopic group 2 consists of streams 3, 5, 6, 8, and 9. Their isotopic composition lies likewise within a relatively narrow range (σ for δ^2H is 0.8–0.9‰, and 0.1–0.3‰ for $\delta^{18}O$), though isotopically enriched (mean δ^2H are −52.7‰, −53.0‰, −52.4‰, −52.9‰, and −52.6‰, respectively, and mean $\delta^{18}O$ is −7.9‰ for all five sources) relative to group 1 (Figure 3b). Isotopic group 3 consists of streams 2, 4, and 7 and represents the three streams with the largest seasonal variation (σ for δ^2H is 1.5–2.8‰, and 0.3–0.6‰ for $\delta^{18}O$).

Stream 2 stands out in particular with signatures ranging from −57.7‰ to −47.2‰ for δ^2H and −8.4‰ to −6.3‰ for $\delta^{18}O$, thus containing the isotopically heaviest water of the streams and spring (Figure 3b,c). The stream 2 data, moreover, plot along a line with a lower slope than the LMWL, indicating an evaporation trend similar to that of the lake water in Nørresø (Figure 3a). This is consistent with stream 2 seasonally getting water from the small lake, Hjertesø (cf. Figure 1b).

The δ-values of isotopic group 1 were significantly different from the δ-values of both groups 2 and 3 with a level of significance of 5%. In contrast, there was no significant difference between the δ-values of isotopic groups 2 and 3.

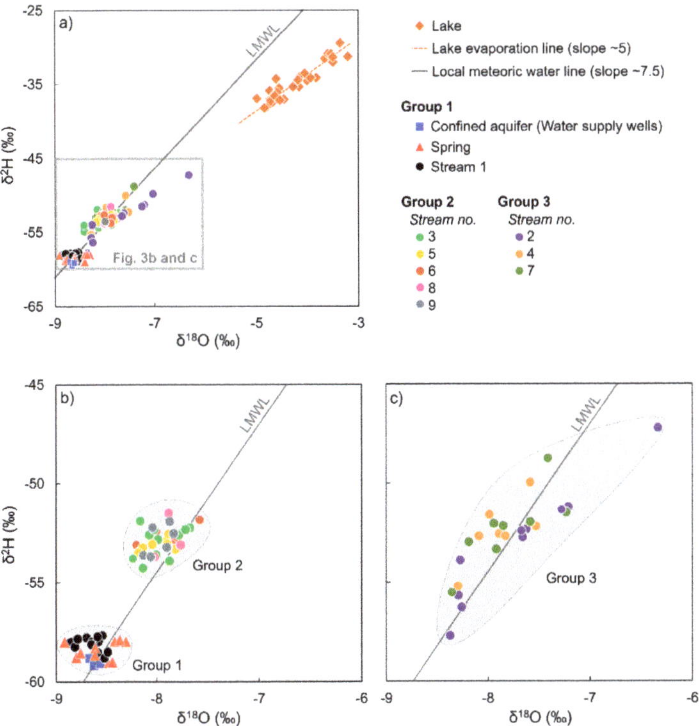

Figure 3. $\delta^{18}O$ and δ^2H values for (**a**) all streams, the spring, the water supply wells and the lake. Panels (**b**,**c**) are subsets of panel (**a**); note the different range of the *y*- and *x*-axes. Group 1 and 2 in (**b**) represent clustered isotope data, while group 3 in (**c**) represents fluctuating isotope data. Local meteoric water line (LMWL) is from [44].

3.3.2. Ca^{2+} and Alkalinity

Ca^{2+} and alkalinity (dominantly bicarbonate, HCO_3^-) made up >80% of the total charge equivalents in the water compositions in all the streams, the spring, the confined aquifer, and the lake with a molar Ca:HCO$_3$ stoichiometry for all data close to 1:2 (Figure 4). Accordingly, the process of (calcium) carbonate dissolution by carbonic acid [45] can generally be assumed to control the water chemistry in this freshwater ecosystem.

When measured Ca^{2+} and alkalinity data for the different streams and the spring were grouped with respect to their isotopic grouping (cf. Figure 3), one can see the same trends as for the isotopic composition (Figure 4). Group 1 clustered within a relatively small range, group 2 clustered likewise, however within a larger interval. Group 3 showed a fairly large scatter but the data generally still fall close to the 1:2 line.

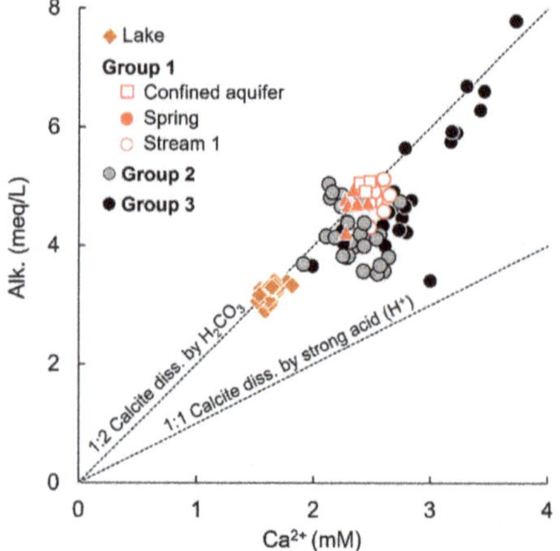

Figure 4. Concentrations of Ca^{2+} and alkalinity of the streams, the spring, and the lake (Nørresø). The group number refers to the same stream grouping as presented for the isotope data (Figure 3). Lines indicate molar calcium:alkalinity ratios corresponding to calcite dissolution by carbonic acid (1:2) and strong acids (1:1).

3.3.3. Dissolved Inorganic Phosphorus

As compared to the other streams, DIP concentrations measured in the spring and stream 1 were very stable and relatively high (Figure 5a), from 4 to 5 μM (σ of 0.2) in the spring and from 2 to 4 μM (σ of 0.4) in stream 1. The same high DIP concentrations were found in the confined aquifer (mean of 5 μM, n = 4) tapped by the water supply wells. This corresponded to the grouping by similar trends in isotope data where group 1 deviated from the other two groups.

In contrast, there was no correlation between the isotopically determined groups 2 and 3, and the respective streams' DIP concentrations. Among groups 2 and 3, stream 2 showed the highest and most fluctuating DIP concentrations (σ of 3.4), with values from 2 μM up to as high as 16 μM in December 2016. DIP concentrations in stream 3 were somewhat lower, from 1.3 to 3.6 μM (σ of 0.6). The rest of the streams (Nos. 4–9) all had mean DIP concentrations less the 1.5 μM. The lowest DIP concentrations were measured in streams 5, 7, and 9, with DIP mean concentrations of 0.3 μM.

DIP concentrations in the lake varied throughout the year, from 2 to 12 μM (σ of 2.9). Thus in most months the concentration in the lake was significantly higher than in any of the streams.

3.3.4. Fe^{2+}, O_2 and NO_3^- Concentrations

The highest concentrations of Fe^{2+} were found in the spring and stream 1 with a mean of 36.5 μM and 23.8 μM, respectively (Figure 5b). Concurrently, low concentrations of O_2 were measured with means of 0.01 mM and 0.09 mM for the spring and stream 1, respectively (Figure 5c). Similar Fe^{2+} (31 μM) and O_2 (<0.008 mM) concentrations were found in the regional confined aquifer. The high concentration of Fe^{2+} and DIP in the aquifer could potentially favor the formation of vivianite [46]. However, according to a PHREEQC speciation (Section 2.2.6), all samples collected in this study were subsaturated with respect to vivianite.

The Fe^{2+} and O_2 concentrations in the spring, stream 1, and the confined aquifer (i.e., group 1) differed significantly from all the other streams (i.e., groups 2 and 3), which showed concentrations of Fe^{2+} generally below 3 μM and concentrations of O_2 much closer to saturation, which was 0.2–0.4 mM, depending on temperature (Figure 5b,c); with the exception of stream 9, in which low concentrations of O_2 were measured (mean of 0.04 mM). Low Fe^{2+} and high O_2 concentrations were likewise measured in the lake water column with means of 0.5 μM Fe^{2+} and 0.3 mM O_2.

The relatively high nitrate concentration in stream 3 (mean: 0.63 mM) differed from all the other streams and the spring (Figure 5d). The lowest nitrate concentrations were recorded in the spring, stream 1, and the confined aquifer (i.e., group 1) with a mean of 0.02 mM. The other streams and the lake all had a mean nitrate concentration in the interval 0.03 to 0.06 mM with the exception of stream 4 (mean: 0.20 mM) and 5 (mean: 0.15 mM).

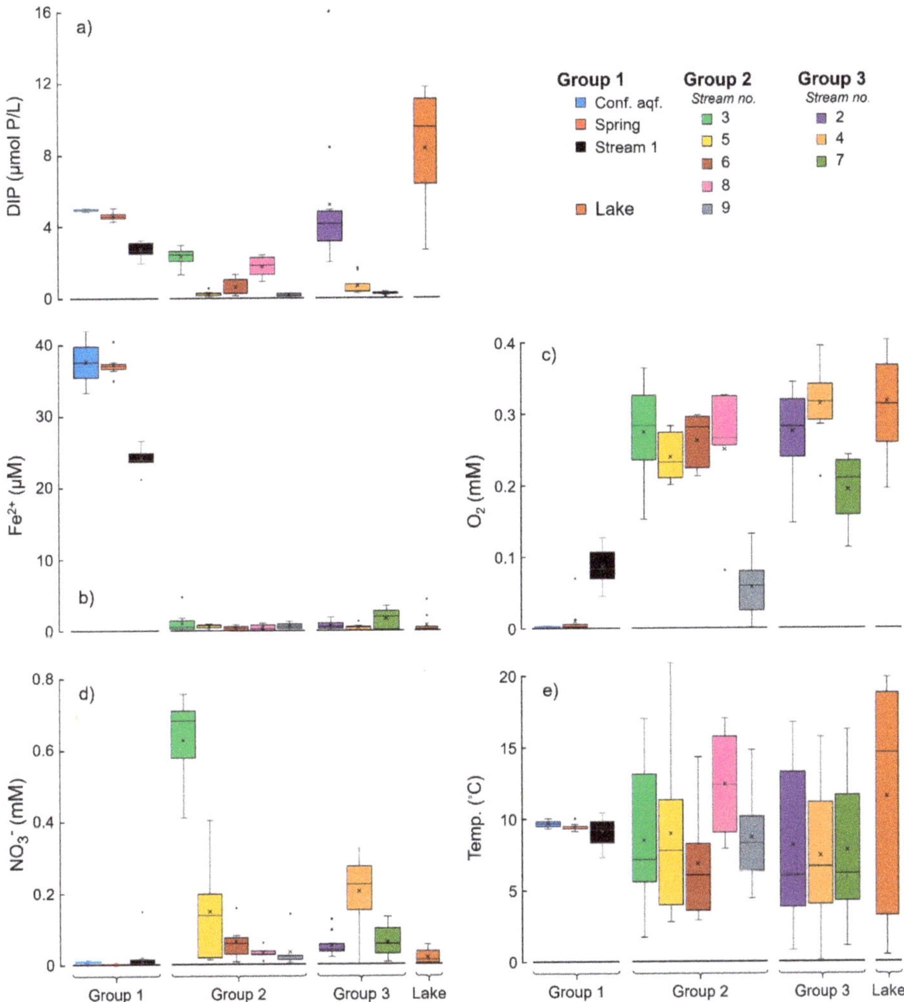

Figure 5. Boxplot showing concentrations of dissolved inorganic orthophosphate (DIP) (**a**), Fe^{2+} (**b**), O_2 (**c**), and NO_3^- (**d**) and temperature (**e**) in the streams, the spring, Lake Nørresø, and in the regional confined aquifer. Note that along the *x*-axis the data for the different streams, the spring, and confined aquifer are grouped with respect to their isotopic grouping (cf. Figure 3). Lower and upper whisker indicates the minimum and maximum values, respectively, excluding statistical outliers. Median and mean value is depicted in each box by a straight horizontal line and a cross, respectively.

3.3.5. Temperature

The temperature in the spring and stream 1 was relatively constant at ~9 °C throughout the sampling period, corresponding well to the mean temperature measured in the regional confined aquifer (~9.5 °C). In contrast, temporal variability in temperature from 0 to 20 °C was recorded in all other streams (i.e., groups 2 and 3) and in the lake (Figure 5e).

3.4. Stream and Spring Discharge and DIP Load to the Lake

Figure 6 shows the rates of stream and spring discharge (L/s) and the corresponding DIP load (g P/day) to the lake (note that streams 5–9 are summed). The depicted uncertainty bands (shaded areas) for the discharge in Figure 6a are based on the conducted discharge measurement uncertainty determination and represent ±1 standard deviation. The uncertainty bands for the DIP load in Figure 6b only reflect the discharge uncertainty as the analysis uncertainty for the DIP concentration is assumed relative small.

Throughout the year stream 1 accounted for the largest water inflow, with an annual mean discharge of 2 L/s (Figure 6a). The discharge in stream 1 started to decrease at the end of August and then increased again in October, likely caused by a gradual blockage and later cleaning of an under-road connecting tube upstream of the measurement point, which is typically carried out in the late autumn. In contrast, the spring showed a very stable discharge throughout the sampling period, though with an annual mean discharge of 0.8 L/s, which was less than half of the value for stream 1. Nevertheless, stream 1 and the spring were both important contributors of external DIP due to their high DIP concentrations (cf. Figure 5a), which corresponded to average DIP loads of 15.4 g P/day and 11.3 g P/day, respectively (Figure 6b).

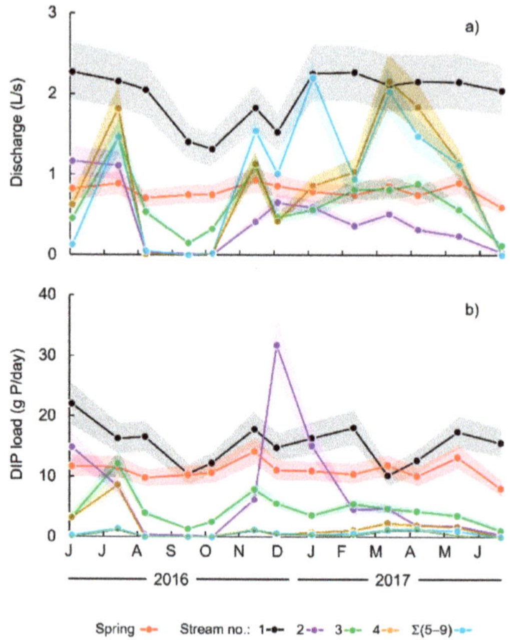

Figure 6. (**a**) Discharge rates and (**b**) DIP load to Lake Nørresø from the spring (red) and streams 1 (black), 2 (purple), 3 (green), and 4 (brown), and the sum of streams 5 to 9 (blue). Shaded area behind the lines represents ±1 standard deviation of the cutthroat flume stream discharge measurement.

In mid-June to mid-August 2016 and at the beginning of December 2016 to mid-February 2017, stream 2 also contributed with a substantial amount of DIP. The average load of DIP in these two periods were 11.6 g P/day and 23.5 g P/day, respectively. However, during the rest of the year, stream 2 had a low to non-existing DIP load, which can be attributed to low to non-existing discharge in stream 2 during these months (Figure 6a). Accordingly, stream 2 only had an annual mean DIP load of 7.5 mg P/L and thus stream 1 and the spring were the two main DIP contributors.

Streams 3 and 4 both had approximately the same annual mean discharge rate as the spring (stream 3: 0.7 L/s; stream 4: 0.9 L/s). However, as their DIP concentrations were low (cf. Figure 5a), their DIP loads were likewise significantly less than that of the spring (stream 3: 4.9 g P/day; stream 4: 1.8 g P/day).

Streams 5 to 9 contributed with the lowest DIP loads ranging from 0.0 to 0.3 g P/day as a result of low DIP concentrations and low discharge rates, all less than 0.5 L/s.

3.5. Annual Water Budget and DIP Load to Nørresø

The total water outflow from Nørresø was 10.2×10^5 m^3/year (Figure 7a), as calculated for the hydrological year 6 June 2016 to 21 June 2017. Taking into account a lake water volume of approximately 16×10^8 m^3, this corresponded to a lake hydraulic retention time of 1.6 years.

Precipitation and evaporation represented the major water input and output, respectively, to Nørresø, accounting for 43% and 54%.

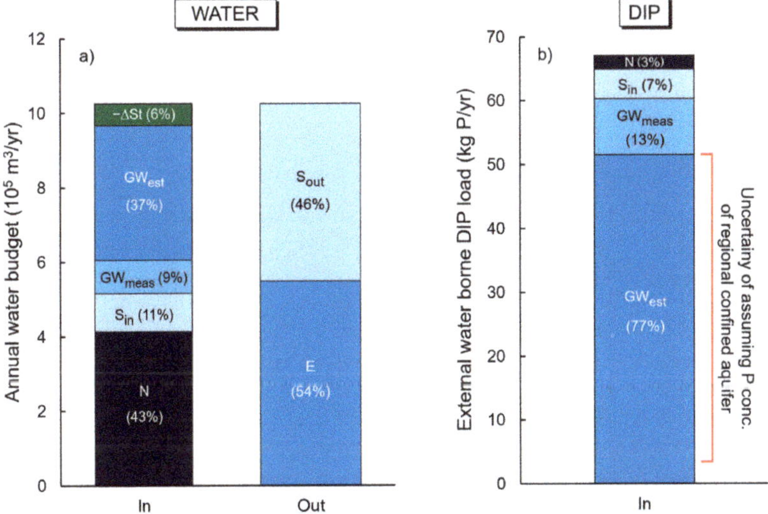

Figure 7. The annual water budget (**a**) and the annual external water-borne DIP loading (**b**) to Lake Nørresø from June 2016 to June 2017. In (**b**), the column is for the assumption that the groundwater input, GW$_{est}$, derived from the water balance carried a DIP concentration representing the regional confined aquifer, i.e., the spring's 4.6 µM. Percentages in brackets indicate the relative water (**a**) or phosphorus (P) (**b**) contribution of precipitation (N), surface water input (S$_{in}$, sum of streams 2–9), measured groundwater input (GW$_{meas}$, sum of the spring and stream 1), estimated groundwater input (GW$_{est}$, water budget residual), evaporation (E), surface water output (S$_{out}$, via the outlet), and storage change (ΔSt, the lake water level decreased during the investigated hydrological year).

The isotopic and hydrochemical composition of the groundwater, the spring, and water in the nine streams were used as tracers to differentiate streams discharging 'deep' groundwater derived from the regional confined aquifer from streams discharging 'shallow' groundwater, as derived from shallower circulation systems, or surface runoff. Based on this, the spring and only one of the streams (No. 1; located near the spring) was assumed to discharge deep groundwater from the regional confined aquifer to the lake (the derivation of this is presented in Section 4). Conditional on this differentiation, the deep groundwater discharge was partly measured (GW$_{meas}$) and accounted for 9% of the total water input. Shallow groundwater or surface runoff discharging from the remaining eight streams

(Nos. 2–9) only accounted for 11% of the water input. Surface water output via the outlet accounted for 46% of the water loss.

By combining the water outputs from the lake and taking into account the decline in the water level during the investigated hydrological year (ΔSt), it is evident that the water output exceeded the detected total water input (cf. Equation 1). The missing water in the water budget is believed to have stemmed from undiscovered groundwater seepage areas at the bottom of the lake. This estimated groundwater (GW_{est}) input accounts of 37% the water input to the lake, so that the total groundwater input adds up to 46% ($GW_{meas} + GW_{est}$). Bearing in mind the continuous upwards hydraulic gradients to the lake (cf. Section 3.2), a negligible downwards groundwater recharge from the lake was assumed.

For the DIP budget an assumption has to be made regarding the DIP load carried by the estimated groundwater input. Assuming the estimated groundwater input came from the regional confined aquifer and carried an average DIP concentration as represented by the spring (4.6 µM), the estimated and measured groundwater input represented by far the largest share of the external water-borne DIP to the lake, contributing 77% and 13% of the DIP load, respectively (Figure 7b). Also, with this assumption, the annual water-borne external DIP load to Nørresø amounted to 67 kg P/year (Figure 7b). Stream discharge composed of surface runoff or shallow groundwater (i.e., streams 2–9) accounted for just 7% of the DIP load of which almost half of the DIP load is from stream 2 alone. Precipitation contributed with only ~3% of the annual water-borne DIP load to the lake, assuming that the precipitation DIP load equaled the atmospheric deposition of 1.3 mol TP/ha/year [47]. The uncertainty related to the assumption of the DIP concentration of the GW_{est} component will be discussed in Section 4.

3.6. Paleolimnological Analyses

Figure 8 presents percentage abundances of dominant diatom species in the sediment stratigraphy, aphanizophyll, which is a pigment specific to filamentous cyanobacteria, normalized titanium data, and the percentage of tree/shrub and herb pollen indicative of land use changes in the catchment area of Nørresø. Dominant diatom species throughout the core include *Cyclotella comensis*, *Cyclotella comta*, *Stephanodiscus medius*, *Stephanodiscus neoastraea*, and *Stephanodiscus parvus*. The lower part of the core, from ca. 7500–5000 cal yr BP, is characterized by the highest abundances of *Cyclotella comensis* and *Stephanodiscus medius*, after which especially *Cyclotella comta* becomes more abundant. *Stephanodiscus parvus*, which is a common constituent of hypertrophic lakes, was relatively abundant throughout the core (ca. 10–20%), reaching ca. 40% in the most recent parts of the sediment stratigraphy. Aphanizophyll concentrations were ca. 0.5–1.5 nmol/g dry weight (DW) throughout most of the core, apart from two periods of elevated concentrations between ca. 5000 and 3300 cal yr BP and around 1500 cal yr BP. Tree and shrub pollen dominated throughout the sediment stratigraphy. Their percentage abundance was ≥90% until ca. 3000 cal yr BP, after which forest clearance (for pastures and fields) was evident as a decrease in tree/shrub pollen abundance and an increase in herb pollen during two distinctive periods. However, even here the total tree/shrub percentage did not decrease below ca. 60%, suggesting a mainly forested lake catchment throughout Holocene. This was supported by titanium (Ti), an unambiguous indicator of allochthonous inputs from the catchment, which showed low amounts and little variability unrelated to the forest clearance suggested by pollen data, and to the algal indicators of lake trophy.

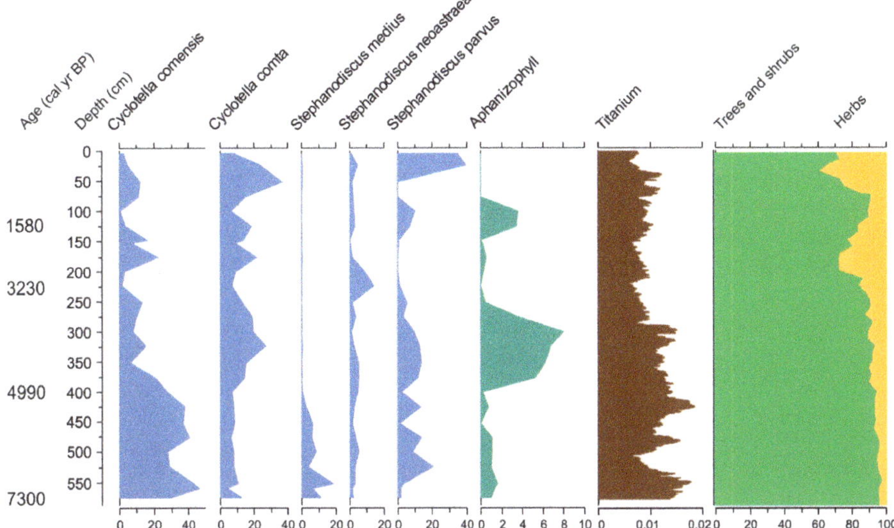

Figure 8. Percentage abundance of dominant diatom species, concentrations (nmol/g DW) of the filamentous cyanobacteria pigment aphanizophyll, normalized titanium data, and the percentage abundance of tree/shrub and herb pollen in Lake Nørresø sediment core. Sediment depth and calibrated ages are given on the left.

4. Discussion

Present day DIP concentrations (2–20 µM) in the lake exceeded by far the threshold for the trophic class eutrophic (TP > 1 µM; [23]) and in some months even the threshold for the hypereutrophic class (TP > 3 µM; [23]). Hence, under the current condition the lake did not achieve "good ecological status". In order to restore and improve eutrophic freshwater ecosystems it is necessary to identify and determine the relative P load of the different external sources of DIP to the lake [6–9].

Using geochemical attributes as a tracer [45,48], we discuss in the following subchapters which streams were mostly fed by water from the regional confined aquifer and which were mostly fed from other sources, such as groundwater with a shallower circulation system or surface runoff. Subsequently the main external sources of water-borne DIP was identified.

4.1. Identifying Deep Groundwater Discharge

The 14 drillings conducted near the lakeshore and the lithological logs from the three nearby water supply wells, together with the slow hydraulic response of the seven piezometers, indicated clayey till from within the upper 1–2 m below terrain to more than 35 m depth, which showed that the catchment area of Nørresø and the lake itself is located in a low permeable lithology.

The presence of the (artesian) spring on Lucieø demonstrated, however, that a confined aquifer must exist below the till bed, which was also evident from the lithological logs of the water supply wells.

In humid climate zones such as Denmark, it is generally considered that the δ^2H and $\delta^{18}O$ isotope composition in groundwater is a representative mixture of weighted average annual rainfall [49], and that the groundwater temperature corresponds closely to the average annual air temperature. Accordingly, both parameters were assumed constant in the regional confined aquifer throughout the year. The constant stable isotopic composition of the spring water matrix as well as its constant temperature of ~9 °C (Danish groundwater temperature is 8–9 °C; [44]) therefore substantiated that the spring discharged groundwater from the regional confined aquifer, rather than shallowly circulated groundwater or surface runoff, to the lake. This was further consistent with the identical isotopic composition of the spring and the confined aquifer (i.e., both below the isotopic group 1). Using the same line of reasoning of identical isotopic composition and constant temperature, water in stream 1 most likely also originated directly from the confined aquifer. Accordingly, stream 1 likely drain the meadow from a constant supply of confined aquifer groundwater, rather than from surface runoff or precipitation.

The assumption that the spring and stream 1 are fed by groundwater from the regional confined aquifer is further substantiated by the highly comparable chemical compositions (high concentrations of $Fe^{2+} > 20$ μM and low concentrations of $O_2 < 0.004$ mM) of the confined aquifer water matrices, with the spring and stream 1 water matrices. Concurrently, the elevated concentrations of DIP in the spring (mean 4.6 μM) and stream 1 (mean ~3 μM) were consistent with the high TP concentrations measured in the confined aquifer (mean ~5 μM).

Indications that the regional confined aquifer approached the terrain close to the area where the spring and stream 1 are located are found in the ERT data, acquired between the meadow and the east end of Nørresø (NE ERT). The ERT data profile (not shown) reflected a sloping layer of sand, connecting a deep-lying sand layer (i.e., the confined aquifer) and the terrain. This observation supported the suggestion that the spring on Lucieø and stream 1 tap the regional aquifer.

According to the annual water budget, groundwater discharge to the lake accounted for 46% of the total water input. One quarter of this groundwater input was directly measured as the summed discharge of stream 1 and the spring, while three quarters were deducted from the water budget.

From a stratigraphic viewpoint, it seemed highly unlikely that a sand seam through which the groundwater was discharging occurred only at one location. Thus, it is likely that that the confined aquifer also approached the surface at other locations under the lake. Accordingly, this would explain the relatively large estimated groundwater component. The significant elevation of the spring, relative to the lake surface (>+1 m), indicated further that the estimated groundwater input must enter the lake via small-scale discrete discharge zones in the lakebed, rather than through diffuse inflow over one or more laterally extensive connection(s) with the sandy aquifer, which would equalize the hydraulic pressure of the aquifer (spring) and the lake much more.

4.2. Identifying Discharge of Shallow Groundwater or Surface Runoff

The above water budget considerations showed that isotopic group 1 (i.e., the spring, stream 1, and the regional confined aquifer) represented groundwater with more or less constant δ-values. The isotopic composition of group 1 was significantly different from the δ-values in group 2 (i.e., streams 3, 5, 6, 8, and 9) as well as in group 3 (i.e., streams 2, 4, and 7) (Figure 3). This strongly suggested that the discharging water from the streams in these two groups had different origins and different flow paths. The temporal variation in the isotopic δ-values was more pronounced in isotopic group 3 (for stream 2 in particular), as compared to group 2, which again was more variable than group 1 (i.e., groundwater from the regional confined aquifer).

Sub-annual fluctuations in water isotopic composition, when they occur close to the LMWL, most likely reflected seasonal isotopic variations in the precipitation feeding the streams and, consequently, sub-annual residence times for at least a part of the stream discharge. This short-term variability may be associated to surface runoff or groundwater in shallow circulation systems, or a mixture of these.

The ephemeral (streams 2 and 4–9) or variable-discharge (stream 3) characteristics of these streams (i.e., 2–9) further supported a larger fraction of water with sub-annual residence times in contrast to continual base flow dominance. In addition, streams 2–9 all showed large temperature fluctuations and oxidized conditions ($Fe^{2+} < 1.4$ μM and close-to-saturation O_2) which did not reflect the anoxic groundwater. However, these parameter characteristics may have resulted from exposure to ambient conditions along longer stream stretches rather than from the source of the water. Nevertheless, streams 2–9 are likely to primarily represent discharging shallow groundwater or surface runoff, rather than water from the deep regional confined aquifer.

In contrast to the groundwater from the confined aquifer, the DIP concentrations were relatively low in the majority of the streams discharging shallow groundwater or surface runoff, with the exception of stream 2. Stream 2, as mentioned earlier, seasonally discharged water from a small lake, Hjertesø, which explains its elevated DIP concentrations and its more extreme temporal δ-values variation caused by evaporation from this lake's water surface.

4.3. Water Budget and External Water-Borne DIP Load to Nørresø

According to the water budget, precipitation accounted for 43% of the total annual water input to the lake, whereas it only contributed ~3% of the annual water-borne DIP load to the lake. This demonstrated that there was no direct relationship between water input and P loading between different sources. Shallow groundwater water or surface runoff (i.e., discharge of streams 2–9) accounted for 11% of the water input and just 7% of annual water-borne DIP load, of which stream 2 (feeding water from Hjertesø) accounted for almost half. The measured groundwater input (sum of the spring and stream 1, GW_{meas}) accounted for just 9% of the annual water input to the lake, but was still the second largest DIP contributor, accounting for 13% of the total water-borne DIP load to lake.

When GW_{meas} was combined with the estimated groundwater (GW_{est}) component and if the latter was assumed to carry an average DIP concentration representative for the regional confined aquifer (the spring's 4.6 µM), then groundwater discharge contributed to half of the total annual water input to the lake and with 90% of the annual water-borne external DIP load to the lake. However, this assumption (i.e., of the (unmeasured) DIP concentration of GW_{est}) introduced uncertainty in the DIP budget. If the estimated groundwater input was assumed to carry a much smaller DIP concentration, for example the 0.3 µM DIP concentration of streams 5, 7, and 9 (Section 3.3.3), then the DIP load related to GW_{est} would have been reduced from 52 kg P/year to just 3.4 kg P/year. This offset is indicated as uncertainty to the right of the DIP budget column in Figure 7b. A low DIP concentration could reflect that the estimated groundwater input was actually shallow (not deep) groundwater or even unmeasured surface runoff. While the uncertainty range allows a reduction in the total DIP load by almost 50 kg P/year, to just 19 kg P/year, still the actually measured groundwater-borne DIP load of ~9 kg P/year would account for nearly half of the total DIP load, i.e., 9 out of 19 kg P/year. Furthermore, neglecting any unmeasured surface runoff so that GW_{est} is assumed to be some source of groundwater with at least a 0.3 µM DIP concentration, then groundwater-borne DIP will amount to at least 64% of the annual DIP load to the lake.

This serves as a clear example that groundwater-borne DIP can be a significant source of P entering freshwater ecosystems. The estimated groundwater P contribution was at a significantly higher level than previously assumed for this lake. Fyn's County [27] estimated that groundwater only accounted for 16% of the annual water-borne TP load to Nørresø. Although that study dealt with TP and not DIP, the different results underline the importance of a thorough study of water and P budget components for understanding P budgets.

One of the main challenges in identifying hydrological budgets for a lake lies in quantifying the groundwater fluxes [16]. As in the present study, this flux has often been determined as the residual of the water budget [34]. In cases where both groundwater discharge and groundwater recharge are unknown only the net groundwater component (i.e., the difference of discharge and recharge) is identified. Considering different P concentrations of groundwater and lake water, this results in erroneous P budgets based on the "residual" approach [16]. The present approach is, thus, only applicable for estimations of P loadings if groundwater-lake gradients consistently show groundwater discharge to the lake (like in the present study).

Bearing in mind that the estimated groundwater component in the present study is quantified as the residual in the water budget, all measurement uncertainties are consequently accumulated in the estimated groundwater component. The studied hydrological year was unusual in the sense that evaporation exceeded precipitation; usually for Danish catchments the opposite is the case. Nevertheless, the evaporation component was not measured directly and if its estimation in the present year led to an overestimation, then this would result also in an overestimation of the groundwater DIP loading.

In contrast, the measured water output via the outlet represented a minimum water discharge. Installation of the flume introduced resistance to the stream flow, which required equilibration to a new higher stationary water level in the stream reach upstream the flume. The conditions in all streams (1–9) allowed such equilibration within a few minutes. In the outlet, however, equilibration would require the level of the whole lake to increase to the new stationary water level which, of course, was practically infeasible. Therefore, the flume readings (which reflected the upstream water level) must have been too low, resulting in too low discharge determinations. The underestimation of the outlet discharge implies that the water balance residual, and thereby the estimated groundwater input (GW_{est}) and associated DIP load, were also underestimated.

Accordingly, the interpretation of the water budget and associated DIP loadings is somewhat uncertain. Nevertheless, a large certainty persists regarding the amounts and relative importance of the actual measured groundwater-borne DIP load vs. the DIP load fed by streams 2–9. Therefore, the groundwater-borne DIP load to the lake constituted minimum ca. half, and likely at least 64% of the total water-borne DIP load. Uncertainties are mainly related to the total DIP load (amount) and to which percentage in between 64% and 90% of the total DIP load is correct for the groundwater-borne DIP contribution.

The findings in the present study underline the benefit of conducting a combined investigation of both water chemistry, isotopic tracers, temperature, and water volume fluxes, with a dense temporal resolution. Nørresø has many small inlets and the frequent measurements of fluxes and geochemistry of these streams has revealed that the lake is dominated by groundwater-borne DIP input to the lake, although groundwater input is of only secondary importance in the lake's water budget.

4.4. Origin of the Dissolved Inorganic Phosphorus

The sources of DIP in the water discharging in streams 2–9 should be found at or near the surface, rather than in the regional confined aquifer. The DIP concentration of stream 2 is obviously controlled by internal P cycling in the small lake (Hjertesø) drained by this stream as concluded above. Stream 3 received water from drainage pipes connected to a large agricultural field located west of Nørresø (Figure 2) and contained elevated concentrations of NO_3^- (average 0.63 mM; Figure 5d). Thus, the DIP in stream 3, which is the third most important stream in terms of annual DIP loading to the lake, properly originated from fertilizers.

NO_3^- was also detected in streams 4 and 5 with average concentrations of 0.2 and 0.15 mM, respectively. Stream 4 is located close to a conifer plantation and stream 5 is close to an agricultural field (Figure 2). The (small; Figure 6) DIP loading associated with streams 4 and 5 may accordingly also be influenced by agriculture.

The observed elevated DIP concentrations in the spring and stream 1 is attributed to the discharge of P-rich groundwater from the regional confined aquifer. The probability that anthropogenic-induced P-enriched water has percolated down through the thick confining clay layer and subsequently reached the underling sand-aquifer resulting in the elevated groundwater TP concentrations is highly unlikely [35,50]. The mobile P in the confined aquifer, hence, more likely resulted from mobilization of geogenic phosphorus within the aquifer.

Numerous studies show a correlation between the mobilization of DIP and sediment redox conditions [11,12,22]. Particularly, the reduction of Fe-oxides [11,15], and in some cases also manganese oxides [13], will lead to a mobilization of DIP. The high concentrations of Fe^{2+} and the reducing conditions in the regional confined aquifer are thus consistent with the theory of natural release of DIP from reductive dissolution of Fe-oxides.

A shift in the redox condition might explain why the DIP concentration in stream 1 was lower than the one detected in the spring, even though the water is evaluated to have the same origin. The discharging water in stream 1 had a longer flow path through oxidizing conditions from the groundwater seepage area (the meadow, Figure 1) to the lake, giving time for Fe-oxide precipitation and thus depletion of DIP via co-precipitation and adsorption.

Spontaneous precipitation of Fe-oxides will most likely also occur when anoxic, ferrous P-rich groundwater enters the oxic lake water column. This process would affect lake DIP concentrations through adsorption or co-precipitation. However, the Fe-oxides with adsorbed P will be accumulated on the lakebed and become a part of the internal mobile P pool, which potentially is released to the water column during summer anoxia at the lakebed. The impact of internally released P and the role of accumulated Fe-oxides with adsorbed P in Nørresø were investigated and discussed by Nisbeth et al. [31]. Their results strongly suggests that groundwater-transported geogenic P was the original source of the now internally cycled P and that this mechanism was mainly driven by the reductive dissolution of Fe-oxides.

The elevated P concentration in the groundwater could also be attributed to degradation of buried organic matter. However, the observed mol:mol ratios of dissolved NH_4^+-to-DIP in groundwater of the confined aquifer (~4; [32]) were somewhat lower than the N:P ratio of organic matter (~16), indicating that organic matter-bound P is not the main component of the P in the groundwater of the confined aquifer.

4.5. Paleolimnological Indicators for P Sources to the Lake

To further substantiate the hypothesis that groundwater-transported geogenic P has been the dominant external source of P entering the lake, we applied diatom, pigment, XRF, and pollen analyses on a ca. 6-m-long sediment core section collected from the central part of the lake (Figure 2).

Diatom analyses of core samples, dating from before the introduction of agriculture (i.e., older than 6000 cal yr BP), showed dominance of *Cyclotella comensis*, *C. comta*, and *Stephanodiscus medius*, with smaller contributions (up to ca. 10% of the assemblage) by *S. neoastraea* and *S. parvus*, indicating that even in the pristine state the lake was already meso-eutrophic [51]. This strongly suggested that natural nutrient enrichment of Nørresø has occurred. The diatom assemblages of the whole sediment record suggested, in addition, that the lake water was never P-poor, which is consistent with current conditions (Figure 8).

Carotenoid pigments from filamentous bloom-forming cyanobacteria (aphanizophyll and/or myxoxanthophyll) are almost consistently present in the sediment record of Nørresø for the last 7500 years (Figure 2). Such cyanobacteria are commonly associated with low N:P ratios because of their ability to fix atmospheric nitrogen. Their persistence in Nørresø is probably connected with the high P supply. Naturally high abundances of cyanobacteria are observed in other lakes on glacial moraine landscapes where groundwater supply dominates [52]. Debate still exists, however, about the relative roles of groundwater supply and internal recycling of P in such lakes with very low flushing rates [53]. Nevertheless, the strong indication of filamentous cyanobacteria persistence throughout the last 7500 years of Nørresø's history supports the assumption of groundwater being an important source of DIP to the lake.

For some lakes particulate P accounts for a substantial proportion of the total external P loading [54], especially during high rainfall were soil erosion is accelerated. The monthly measurements of discharge and water chemical parameters may not have captured high flow events. Hence, particulate P could constitute an impotent external P source entering Nørresø. Pollen analysis indicates, however, that the lake catchment was predominantly forested throughout its Holocene history (Figure 8). This is consistent with historical and present day conditions, where 88% of the topographical catchment is covered by forest and 12% by agriculture [27,28]. A dominantly forested catchment makes particulate P transport to the lake by soil erosion less likely. Furthermore, low and relatively invariable concentrations of the soil erosion indicator Ti [55] (Figure 8) indicates that little erosion from the catchment occurred throughout the history of Nørresø. Also, the variability of Ti is unrelated to changes in the algal trophic indicators. This strongly suggest that other main transporters of P, such as DIP with groundwater, are more important for the trophic state of Nørresø.

A possible additional source of P could be leaves and other organic detritus from the catchment causing gradual and progressing eutrophication through the lake's ontogeny. There has, however, never been documented long-term natural eutrophication due to catchment organic detritus. In contrast, ample documentation of the opposite process (i.e., natural long-term oligotrophication) occurring in ecosystems without additional anthropogenic or geogenic influence such as during interglacials (terrestrial [56], lacustrine [57]) as well as during the Holocene in lakes in subtropical [58], temperate [59], and boreal [60] zones is published. Accordingly, it is highly unlikely that the natural eutrophication of Nørresø through the Holocene was driven by P input from catchment organic detritus. Combining the paleolimnological results with the modern measurements strongly suggests groundwater to be a highly important external source of DIP in the case of Nørresø and potentially in the case of other surface water bodies. The first important implication of this conclusion is that very significant releases of geogenic DIP to groundwater may occur naturally within aquifers or near the surface in the recharge area.

In the present study area, the elevated DIP concentrations in the deep confined aquifer signify that the P release occurred long before the groundwater reached the near-lake area (water supply wells with elevated groundwater TP concentrations are more the 3 km away from the spring). The second implication therefore is that phosphorus can be transported with groundwater over at least distances of a couple of km.

In the case of Nørresø, some deforestations during the periods 3000–2000 ca yr BP and again 800–0 cal yr BP seems to have occurred, caused by domestic grazing and arable farming. These anthropogenic disturbances could have caused some cultural eutrophication, as, for example, seen in the last ca. 800 years as peak abundances of *Cyclotella comta* and *Stephanodiscus parvus*. However, the dominance of meso-eutrophic diatom assemblages together with the occurrence of filamentous cyanobacteria long before increased anthropogenic pressure on the catchment area of Nørresø are consistent with the current situation where just a fraction (~10%) of the external DIP load is believed to originate from fertilizers and nutrient enrichment from non-cultural sources is of major importance for the trophic status of the lake.

5. Conclusions

This cross-disciplinary study found the presence of a regional artesian aquifer with high concentrations of geogenic P, which ultimately end up in the lake via groundwater discharge. This was deduced from combining the water budget-derived groundwater contribution with observations of lithology, hydraulic heads, and temperature, isotopic, and hydrochemical tracers. In addition, results from paleolimnological studies indicated that the lake was never P-poor due to dissolved P inputs, even before agricultural disturbances. In combination, these results are of relevance for understanding global terrestrial P cycling in that they provide evidence for:

- Phosphorus transport with groundwater over long (km) distances;
- Natural significant releases of geogenic phosphorus to groundwater;
- Groundwater-borne dissolved inorganic phosphorus can potentially be the main contributor of P to surface freshwater ecosystems. In the studied lake, Nørresø, groundwater inputs accounted for 90% of the total annual external DIP load, even though groundwater constituted only one third of water input to the lake.

Finally, the study demonstrates the clear benefits in the field of water resources research of combining lithological observations with dense temporal observations of multiple tracers, hydraulic heads and water volume fluxes, and paleolimnological observations.

Supplementary Materials: The following are available online at http://www.mdpi.com/2073-4441/11/11/2213/s1, Figure S1: Electrical Resistivity Tomography profiles.

Author Contributions: Conceptualization, C.S.N., S.J., and J.K.; Methodology, C.S.N., S.J., and J.K.; Formal analysis, C.S.N. and K.R.; Investigation, C.S.N. and D.L.J.K., Paleolimnological part done by K.W., O.B., B.V.O., L.T., S.M., and A.S.; Resources, GeoCenter Denmark; Data curation, C.S.N.; Writing—original draft preparation, C.S.N.; Writing—review and editing, All; Visualization, C.S.N. and S.J., except Figure 8, K.W.; Supervision, S.J. and J.K.; Project administration, S.J., O.B., and J.K.; Funding acquisition, K.W., S.J., and J.K.

Funding: This research was funded by GeoCenter Denmark, grant number 6-2015.

Acknowledgments: The authors are thankful to the estate of Brahetrolleborg and Catharina Reventlow-Mourier for kindly providing access to Nørresø. Thanks are also given to Jörg Lewandowski (IGB, Berlin) and three anonymous reviewers for providing thorough and critical comments to the manuscript. We further thank Marie-Louise Siggaard-Andersen (Globe Institute, University of Copenhagen) for assistance in the paleolimnological analysis, Anne Thoisen (IGN) for her assistance in the hydrogeochemical laboratory, Ingelise Møller Balling (GEUS) for sharing her expertise regarding the interpretation of the ERT profiles, Aia A. Eriksen and Hans Henrik Havn for their supporting works, Tai Lund Jørgensen for his proof reading, and Peter Engesgaard (IGN) for his good advice and assistance during especially the early part of the study.

Conflicts of Interest: The authors declare no conflict of interest.

References

1. Søndergaard, M.; Jeppesen, E. Anthropogenic impacts on lake and stream ecosystems, and approaches to restoration. *J. Appl. Ecol.* **2007**, *44*, 1089–1094. [CrossRef]

2. UNEP. *Environment for the Future We Want*; Progress Press Ltd: Valletta, Malta, 2012; ISBN 9789280731774.

3. EEA. *Assessment of Global Megatrends—An Update. Global Megatrend 1: Diverging Global Population Trends*; European Environment Agency: Copenhagen, Denmark, 2014.

4. Hecky, R.E.; Kilham, P. Nutrient limitation of phytoplankton in freshwater and marine environments: A review of recent evidence on the effects of enrichment1. *Limnol. Oceanogr.* **1988**, *33*, 796–822. [CrossRef]

5. Blake, R.E.; O'Neil, J.R.; Surkov, A.V. Biogeochemical cycling of phosphorus: Insights from oxygen isotope effects of phosphoenzymes. *Am. J. Sci.* **2005**, *305*, 596–620. [CrossRef]

6. Vanek, V. The interactions between lake and groundwater and their ecological significance. *Stygologia* **1987**, *3*, 1–23.

7. Kilroy, G.; Coxon, C. Temporal variability of phosphorus fractions in Irish karst springs. *Environ. Geol.* **2005**, *47*, 421–430. [CrossRef]

8. Holman, I.P.; Whelan, M.J.; Howden, N.J.K.; Bellamy, P.H.; Willby, N.J.; Rivas-Casado, M.; McConvey, P. Phosphorus in groundwater—An overlooked contributor to eutrophication? *Hydrol. Process.* **2008**, *22*, 5121–5127. [CrossRef]

9. Holman, I.P.; Howden, N.J.K.; Bellamy, P.; Willby, N.; Whelan, M.J.; Rivas-Casado, M. An assessment of the risk to surface water ecosystems of groundwater P in the UK and Ireland. *Sci. Total Environ.* **2009**, *408*, 1847–1857. [CrossRef] [PubMed]

10. Heiberg, L.; Pedersen, T.V.; Jensen, H.S.; Kjaergaard, C.; Hansen, H.C.B. A comparative study of phosphate sorption in lowland soils under oxic and anoxic conditions. *J. Environ. Qual.* **2010**, *39*, 734–743. [CrossRef] [PubMed]

11. Kjaergaard, C.; Heiberg, L.; Jensen, H.S.; Hansen, H.C.B. Phosphorus mobilization in rewetted peat and sand at variable flow rate and redox regimes. *Geoderma* **2012**, *173–174*, 311–321. [CrossRef]

12. Prem, M.; Hansen, H.C.B.; Wenzel, W.; Heiberg, L.; Sørensen, H.; Borggaard, O.K. High Spatial and Fast Changes of Iron Redox State and Phosphorus Solubility in a Seasonally Flooded Temperate Wetland Soil. *Wetlands* **2015**, *35*, 237–246. [CrossRef]

13. Walter, D.A.; Rea, B.A.; Sollenwerk, K.G.; Savoie, J. *Geochemical and Hydrologic Controls on Phosphorus Transport in a Sewage Contaminated Sand and Gravel Aquifer Near Ashumet Pond, Cape Cod, Massachusetts*; Technical Report; United States Geological Survey: Reston, VA, USA, 1996.

14. Burkart, M.R.; Simpkins, W.W.; Morrow, A.J.; Gannon, J.M. Occurrence of total dissolved phosphorus in unconsolidated aquifers and aquitards in Iowa. *J. Am. Water Resour. Assoc.* **2004**, *40*, 827–834. [CrossRef]

15. Griffioen, J. Extent of immobilisation of phosphate during aeration of nutrient-rich, anoxic groundwater. *J. Hydrol.* **2006**, *320*, 359–369. [CrossRef]

16. Lewandowski, J.; Meinikmann, K.; Nützmann, G.; Rosenberry, D.O. Groundwater—The disregarded component in lake water and nutrient budgets. Part 2: Effects of groundwater on nutrients. *Hydrol. Process.* **2015**, *29*, 2922–2955. [CrossRef]

17. Meinikmann, K.; Lewandowski, J.; Hupfer, M. Phosphorus in groundwater discharge—A potential source for lake eutrophication. *J. Hydrol.* **2015**, *524*, 214–226. [CrossRef]

18. Kidmose, J.; Nilsson, B.; Engesgaard, P.; Frandsen, M.; Karan, S.; Landkildehus, F.; Søndergaard, M.; Jeppesen, E. Focused groundwater discharge of phosphorus to a eutrophic seepage lake (Lake Væng, Denmark): Implications for lake ecological state and restoration. *Hydrogeol. J.* **2013**, *21*, 1787–1802. [CrossRef]

19. Kazmierczak, J.; Müller, S.; Nilsson, B.; Postma, D.; Czekaj, J.; Sebok1, E.; Jessen, S.; Karan, S.; Stenvig Jensen, C.; Edelvang, K.; et al. Groundwater flow and heterogeneous discharge into a seepage lake: Combined use of physical methods and hydrochemical tracers. *Water Resour. Res.* **2016**, *52*, 9109–9130.

20. Shaw, R.D.; Shaw, J.F.H.; Fricker, H.; Prepas, E.E. An integrated approach to quantify groundwater transport of phosphorus to Narrow Lake, Alberta. *Limnol. Oceanogr.* **1990**, *35*, 870–886. [CrossRef]

21. Kenoyer, G.; Anderson, M.P. Groundwater's dynamic role in regulating acidity and chemistry in a precipitatuon-dominated lake. *J. Hydrol.* **1989**, *109*, 287–306. [CrossRef]

22. Szilas, C.P.; Borggaard, O.K.; Hansen, H.C.B. Potential iron and phosphate mobilization during flooding of soil material. *Water. Air. Soil Pollut.* **1998**, *106*, 97–109. [CrossRef]

23. Wetzel, R.G. The Phosphorus Cycle. In *Limnology: Lake and River Ecosystems*; Academic Press: San Diego, CA, USA, 2001; pp. 242–250, ISBN 9780127447605.

24. Anderson, N.J. Naturally eutrophic lakes: Reality, myth or myopia? *Trends Ecol. Evol.* **1995**, *10*, 137–138.

25. Søndergaard, M.; Jensen, J.P.; Jeppesen, E. Internal phosphorus loading in shallow Danish lakes. *Hydrobiologia* **1999**, *408–409*, 145–152. [CrossRef]

26. House, W.A. Geochemical cycling of phosphorous in rivers. *Appl. Geochem.* **2003**, *18*, 739–748. [CrossRef]

27. Fyn's County. *Nørresø 1989–1993, Lake Monitoring in Fyn's County*; Technical Report No. 1; Fyn's County: Odense, Denmark, May 1994; 70p.

28. Odgaard, B.; Møller, P.F.; Wolin, J.A.; Rasmussen, P.; Anderson, N.J. *Brahetrolleborg Nørresø. Palæolimnologisk Undersøgelse og Oplandsanalyse*; Company Report (48); Denmark's Geological Surveys: Copenhagen, Denmark, 1995; pp. 1–56.

29. Høy, T. *Danmarks Søer—Søerne i Fyns Amt*; Strandbergs Forlag: Charlottenlund, Denmark, 2000.

30. Moorleghem, V.C.; De Schutter, N.; Smolders, E.; Merckx, R. The bioavailability of colloidal and dissolved organic phosphorus to the alga Pseudokirchneriella subcapitata in relation to analytical phosphorus measurements. *Hydrobiologia* **2013**, *709*, 41–53. [CrossRef]

31. Nisbeth, C.S.; Jessen, S.; Bennike, O.; Kidmose, J.; Reitzel, K. Role of Groundwater-Borne Geogenic Phosphorus for the Internal P Release in Shallow Lakes. *Water* **2019**, *11*, 1783. [CrossRef]

32. Geological Survey of Denmark and Greenland (GEUS). Danish National Borehole Achieve (Jupiter). 2018. Available online: www.geus.dk (accessed on 21 September 2018).

33. Jørgensen, F.; Sandersen, P.B.E.; Auken, E. Imaging buried Quaternary valleys using the transient electromagnetic method. *J. Appl. Geophys.* **2003**, *53*, 199–213. [CrossRef]

34. Rosenberry, D.O.; LaBaugh, J.W. *Field Techniques for Estimating Water Fluxes Between Surface Water and Ground Water Techniques and Methods*; United States Geological Survey: Reston, VA, USA, 2008.

35. Fitts, C.R. *Groundwater Science*, 2nd ed.; Academic Press: Waltham, MA, USA, 2002; ISBN 9780123847058.

36. Det Nationale Overvågningsprogram for Vandmiljø og Natur (NOVANA). Available online: http://novana.dmi.dk/ (accessed on 3 April 2018).

37. Skogerboe, G.V.; Bennett, R.S.; Walker, W.R. *Selection and Installation of Cutthroat Flumes for Measuring Irrigation and Drainage Water*; Technical Bulletin 120; Colorado State University: Denver, CO, USA, December 1973.

38. Skogerboe, G.V.; Bennett, R.S.; Walker, W.R. Generalized Discharge Relations for Cutthroat Flumes. *J. Irrig. Drain. Divis.* **1972**, *98*, 569–583.

39. Stookey, L.L. Ferrozine—A New Spectrophotometric Reagent for Iron. *Anal. Chem.* **1970**, *42*, 779–781. [CrossRef]

40. Murphy, J.; Riley, J.P. A modified single solution method for the determination of phosphate in natural waters. *Anal. Chim. Acta* **1986**, *27*, 31–36. [CrossRef]

41. Reimer, P.J.; Bard, E.; Bayliss, A.; Beck, J.W.; Blackwell, P.G.; Ramsey, C.B.; Buck, C.E.; Cheng, H.; Edwards, R.L.; Friedrich, M.; et al. IntCal13 and Marine13 Radiocarbon Age Calibration Curves 0–50,000 Years cal BP. *Radiocarbon* **2013**, *55*, 1869–1887. [CrossRef]

42. Battarbee, R.W.; Jones, V.J.; Flower, R.J.; Cameron, N.G.; Bennion, H.; Carvalho, L.; Juggins, S. Diatoms. In *Tracking Environmental Change Using Lake Sediments. Terrestrial, Algal, and Siliceous Indicators*; Smol, J.P., Birks, H.J.P., Last, W.M., Eds.; Kluwer Academic Publishers: Dordrecht, The Netherlands, 2001; Volume 3, pp. 155–202, ISBN 9780306476686.

43. Faegri, K.; Iversen, J. *Textbook of Pollen Analysis*, 3rd ed.; Munksgaard: Copenhagen, Denmark, 1975.

44. Müller, S.; Stumpp, C.; Sørensen, J.H.; Jessen, S. Spatiotemporal variation of stable isotopic composition in precipitation: Post-condensational effects in a humid area. *Hydrol. Process.* **2017**, *31*, 3146–3159. [CrossRef]

45. Cirkel, D.G.; Van Beek, C.G.E.M.; Witte, J.P.M.; Van der Zee, S.E.A.T.M. Sulphate reduction and calcite precipitation in relation to internal eutrophication of groundwater fed alkaline fens. *Biogeochemistry* **2014**, *117*, 375–393. [CrossRef]

46. O'Connell, D.W.; Mark Jensen, M.; Jakobsen, R.; Thamdrup, B.; Joest Andersen, T.; Kovacs, A.; Bruun Hansen, H.C. Vivianite formation and its role in phosphorus retention in Lake Ørn, Denmark. *Chem. Geol.* **2015**, *409*, 42–53. [CrossRef]

47. Ellermann, T.; Bossi, R.; Christensen, J.; Løfstrøm, P.; Monies, C.; Grundahl, L.; Geels, C. *Atmosfærisk Deposition 2014*; Technical Report 163; Department of Environmental Sciences (DCE): Aarhus, Denmark, 2015.

48. Appelo, C.A.J.; Postma, D. *Geochemistry, Groundwater and Pollution*, 2nd ed.; Balkema: Rotterdam, The Netherlands, 2005; ISBN 9781439833544.

49. Clark, I.D.; Fritz, P. *Environmental Isotopes in Hydrogeology*; Lewis: Boca Raton, FL, USA; New York, NY, USA, 1997; ISBN 1566702496.

50. Cucarella, V.; Renman, G. Phosphorus sorption capacity of filter materials used for on-site wastewater treatment determined in batch experiments—A comparative study. *J. Environ. Qual.* **2009**, *38*, 381–392. [CrossRef] [PubMed]

51. Battarbee, R.W.; Juggins, S.; Gasse, F.; Anderson, N.J.; Bennion, H.; Cameron, N.G.; Ryves, D.B.; Pailles, C.; Chalie, F.; Telford, R. *European Diatom Database (EDDI): An Information System for Palaeoenvironmental Reconstruction*; Environmental Change Research Centre (ECRC) Report 81; University College London: London, UK, 2001.

52. McGowan, S.; Britton, G.; Haworth, E.; Moss, B. Ancient blue-green blooms. *Limnol. Oceanogr.* **1999**, *44*, 436–439. [CrossRef]

53. Kilinc, S.; Moss, B. Whitemere, a lake that defies some conventions about nutrients. *Freshw. Biol.* **2002**, *47*, 207–218. [CrossRef]

54. Carpenter, S.R.; Booth, E.G.; Kucharik, C.J.; Lathrop, R.C. Extreme daily loads: Role in annual phosphorus input to a north temperate lake. *Aquat. Sci.* **2014**, *77*, 71–79. [CrossRef]

55. Cohen, A.S. *Paleolimnology: The History and Evolution of Lake Systems*; Oxford University Press: New york, NY, USA, 2003.

56. Kuneš, P.; Odgaard, B.V.; Gaillard, M.J. Soil phosphorus as a control of productivity and openness in temperate interglacial forest ecosystems. *J. Biogeogr.* **2011**, *38*, 2150–2164. [CrossRef]

57. Kupryjanowicz, M.; Fiłoc, M.; Czerniawska, D. Occurrence of slender naiad (Najas flexilis (Willd.) Rostk. & W. L. E. Schmidt) during the Eemian Interglacial—An example of a palaeolake from the Hieronimowo site, NE Poland. *Quat. Int.* **2018**, *467*, 117–130.

58. Kenney, W.F.; Brenner, M.; Curtis, J.H.; Arnold, T.E.; Schelske, C.L. A holocene sediment record of phosphorus accumulation in shallow Lake Harris, Florida (USA) offers new perspectives on recent cultural eutrophication. *PLoS ONE* **2016**, *11*, e0147331. [CrossRef]

59. Odgaard, B.V. The Holocene vegetation history in northern West Jutland, Denmark. *Opera Bot.* **1994**, *123*, 1–171.

60. Norton, S.A.; Perry, R.H.; Saros, J.E.; Jacobson, G.L.; Fernandez, I.J.; Kopáček, J.; Wilson, T.A.; SanClements, M.D. The controls on phosphorus availability in a Boreal lake ecosystem since deglaciation. *J. Paleolimnol.* **2011**, *46*, 107–122. [CrossRef]

 *water*

Article

Role of Groundwater-Borne Geogenic Phosphorus for the Internal P Release in Shallow Lakes

Catharina S. Nisbeth [1,*], Søren Jessen [1], Ole Bennike [2], Jacob Kidmose [2] and Kasper Reitzel [3]

[1]   Department of Geosciences and Natural Resource Management (IGN), University of Copenhagen, Øster Voldgade 10, 1350 Copenhagen K, Denmark
[2]   Geological Survey of Denmark and Greenland (GEUS), Øster Voldgade 10, 1350 Copenhagen K, Denmark
[3]   Institute of Biology, University of southern Denmark (SDU), Campusvej 55, 5230 Odense M, Denmark
*   Correspondence: csnm@ign.ku.dk

Received: 28 June 2019; Accepted: 18 August 2019; Published: 27 August 2019

Abstract: This study explores the under-investigated issue of groundwater-borne geogenic phosphorus (P) as the potential driving factor behind accumulation of P in lake sediment. The annual internally released P load from the sediment of the shallow, hypereutrophic and groundwater-fed lake, Nørresø, Denmark, was quantified based on total P (TP) depth profiles. By comparing this load with previously determined external P loadings entering the lake throughout the year 2016–2017, it was evident that internal P release was the immediate controller of the trophic state of the lake. Nevertheless, by extrapolating back through the Holocene, assuming a groundwater P load corresponding to the one found at present time, the total groundwater P input to the lake was found to be in the same order of magnitude as the total deposit P in the lake sediment. This suggests that groundwater-transported P was the original source of the now internally cycled P. For many lakes, internal P cycling is the immediate controller of their trophic state. Yet, this does not take away the importance of the external and possibly geogenic origin of the P accumulating in lake sediments, and subsequently being released to the water column.

Keywords: internal phosphorus release; geogenic phosphorus; groundwater-borne phosphorus; groundwater-surface water interaction; phosphorus cycling

---

1. Introduction

It is well established that eutrophication of freshwater lake ecosystems is controlled by nutrient input, and in particular the availability of phosphorus (P) [1–4]. In many eutrophic lakes, internal P release, derived from P accumulated in the sediment, is the most important component of the annual P cycle [5–9]. Various processes affect the exchange of P between the water column and the sediment. These include chemical (e.g., redox conditions, nitrate and sulphate availability, pH, alkalinity), biological (e.g., biomineralization), and physical processes (e.g., bioturbation, resuspension, discharging groundwater) [10–16].

The redox conditions in the surface sediment strongly affect the sediment's capacity to bind P [10], especially for sediments with a high iron content [15,17–19]. Phosphorus is sorbed to Fe-oxides under oxidized conditions, while under reducing conditions, microbial respiration uses the ferric iron as electron acceptors, subsequently remobilizing some of the sorbed P to the water column [14,17,19]. Accordingly, higher sediment temperatures during summer lead to accelerated biomineralization [20], which is accompanied by an increase in P release due to development of anoxia in the surface sediment [5,6,20]. In winter, lower temperatures inhibit biomineralization and enable a sufficient supply of oxygen to the sediments to establish a high redox potential, which leads to P retention by Fe-oxides. In addition, a downward flux of P bound in organic matter occurs as primary producers sink to the bottom. Thus, P is particularly re-accumulated in the sediment during winter. This seasonally-dependent process results in higher lake-water P concentrations during the summer than in winter [9,21–23]. Throughout the year, though mainly during the summer, a continuous long-term P mobilization driven by mineralization of slowly biodegradable organic compounds [23] and reducible Fe-oxides buried deeper in the sediment in preceding years, may also occur. It should be emphasized, however, that only a proportion of the Fe-oxides buried in the sediments are re-mobilized and returned to the water column. The rest is permanently buried in the lakes sediments [23].

Phosphorus accumulated in the lake sediment originates from external P sources entering the lake. The magnitude of the P pool accumulated in the sediment is lake-specific, depending on loading history, flushing rate, and chemical characteristics of the sediment [4,5,22,24]. Particulate P (PP) entering the lake most often settles directly by gravity on the lake-bed [25]. In contrast, dissolved P (DP) remains in the water column until incorporated in organic matter, co-precipitated with calcium carbonate or adsorbed by Fe-hydroxides, which all eventually sink to the bottom. When looking for the root cause of the trophic state of a lake, internal P release is therefore only indirectly relevant.

External P loads are generally associated with surficial hydrological pathways [26–29], due to the common assumption that P is retained readily in the vadose zone, during percolation, before it reaches the aquifer [30,31].

Awareness of groundwater-transported P has increased in recent years and been proven as a more important contributor of external P than previously assumed [32–38]. Substantiating evidence for this hypothesis was reported by Kenoyer and Anderson [39], Vanek [40], Shaw et al. [41], Kidmose et al. [36], and Meinikmann et al. [42]. These studies found that groundwater-borne P can account for more than half of the external total P (TP) load to a freshwater eco-system. So far, however, studies of external groundwater-P loading to eutrophic lakes have not considered the importance of internal P release. Moreover, research focusing on internal P release does not, to our knowledge, include cases where groundwater-P loading is the main external contributor of P.

Furthermore, enriched P concentrations in eutrophic freshwater ecosystems are generally considered to originate from anthropogenic surficial-deposited P sources such as agricultural fertilizers, sewer overflows, wastewater treatment plants, and septic tanks [28,43–46]. The impact of long-term loadings by geogenic P, released by natural processes, is only rarely considered and documented, especially in relation to groundwater-transported P to freshwater ecosystems.

The objectives of this study were therefore to: (1) investigate whether groundwater-borne geogenic P can potentially be the driving factor behind accumulation of P in a lake's sediment, (2) investigate whether this could also apply for lakes where internal P cycling is the immediate controller of the trophic state of a surface freshwater ecosystem, and (3) assess the possibility that the P accumulated in the sediment, where a fraction is being internally cycled, may have entered the lake with groundwater throughout the Holocene.

2. Materials and Methods

2.1. Study Site

Nørresø is a shallow (mean depth 2.3 m, maximum depth 5.7 m [47]), high alkalinity (water column alkalinity ~3 meq/L), and hypereutrophic (annual mean: 330 µg P/L) lake located in a glacial moraine landscape in southwestern Fyn, Denmark (55°08′50″ N, 10°22′40″ E) (Figure 1a). The lake has a surface area of 69.3 ha, including three small islands: Lucieø, in the eastern part of the lake, with a prominent perennial spring, and two unnamed islets. Nine small streams discharge into the lake (Figure 1b), numbered 1 to 9 in order of decreasing annual P loading to the lake. This numbering was based on the relative size of the annual dissolved inorganic P (DIP)-flux quantified by Nisbeth et al. [48]. Stream 1 and the spring on Lucieø both represent anoxic groundwater discharging from a regional confined aquifer, with geogenic P and ferrous iron concentrations of ~150 µg P/L and 1700 µg Fe/L, respectively [49]. Stream 2 drains a small lake, Hjertesø, north of Nørresø. The other streams represent surface water of which just one, stream 3, receives water from drainage pipes connected to a large agricultural field located to the west of Nørresø (Figure 1b). Nørresø has one surface water outlet at the western end of the lake.

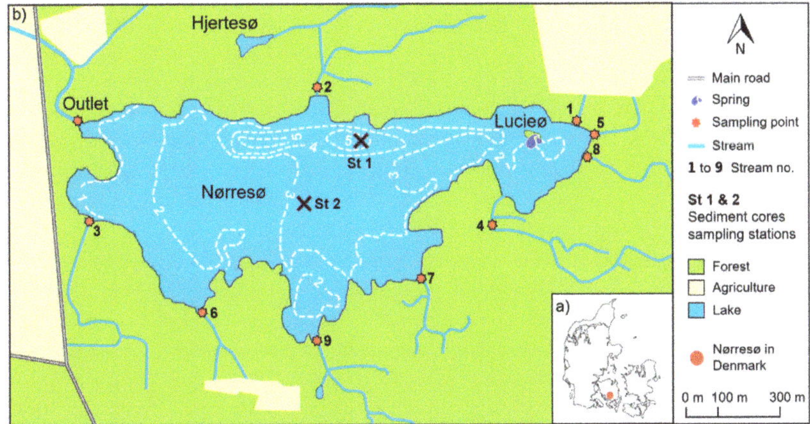

Figure 1. (**a**) Location of Nørresø in the southern part of Fyn, Denmark. (**b**) Locations of the outlet, the nine inlet streams and their respective numbering, and of station 1 and 2 from where sediment were cored in the present study. The small island, Luciø, hosts a spring. A small lake, Hjertesø, north of Nørresø, is indicated. Dashed white lines indicate lake bathymetry in meters [47].

Table 1 shows the water balance of Nørresø compiled by Nisbeth et al. [48]. Quantification of the different numbers in the water balance was inferred from a year-long record of the lake's water levels, stream and spring discharges to the lake, and the water output via the outlet. Precipitation and evaporation data were extracted from a database compiled by the Danish National Monitoring Program for Aquatic Environment and Nature [50]. Groundwater was assumed to enter the lake via the spring and stream 1, combined with hot spots in the lake-bed. The latter groundwater contribution was calculated as the residual in the water balance. During the hydrological year from the 6th of June 2016 to the 21st of June 2017, Nørresø received 450,000 m^3 groundwater compared with 110,000 m^3 surface of water. Precipitation and evaporation amounted to 410,000 m^3 and 550,000 m^3 respectively, and the outlet discharge was 480,000 m^3. The lake's hydraulic retention time is 1.6 years.

Table 1. Water balance components of Nørresø (6 June 2016 to 21 June 2017, [48]).

In (m^3/yr)	
Groundwater input	450,000
Precipitation	410,000
Surface water input	110,000
Out (m^3/yr)	
Lake evaporation	550,000
Outlet	480,000
Storage change (m^3/yr)	−60,000
Lake hydraulic retention time (yr)	1.6

The current mean lake-water table is 40.76 m above sea level and the seasonal variation is 0.1 m. The elevation of the water table has varied over millennia. From 13,000 cal. yr BP to about 7500 cal. yr BP, the lake was a closed basin with a low water level. After 7500 cal. yr BP, the lake had an outlet [51]. In the early 1900s, according to historical maps, the lake was somewhat more extensive with a water level of 41.2 m above sea level [48].

Besides water-borne contributions of P to the lake, other potential sources of P include bird droppings from a colony of cormorants (*Phalacrocorax carbo*) found at the margin of Nørresø, and leaf fall. Historically, the area around Nørresø has seen major changes in cultivation. For instance, from 1785 to 1885, half of the area was forest, with the other half used for agriculture. Today, 90% of the catchment is forested. Nonetheless, pollen analyses indicate that the lake was surrounded by forest or at least a rim of forest along its perimeter during the Holocene [52]. Lake stratification was not observed in Nørresø [47,53], most likely due to the shallow depths relative to the surface area, combined with exposure to westerly winds. The lake is therefore assumed to be well-mixed. Paleolimnological studies indicate that Nørresø is, and probably was throughout most of the Holocene, a eutrophic lake [51].

2.2. Data Collection

In order to meet the objectives of the current study, the overall research approach was to: (*i*) Establish the P contribution to the lake via discharging groundwater, on the basis of previously determined external P loads entering the lake throughout the year 2016–2017 [48]. (*ii*) Use Vollenweiders P mass loading model [54] to estimate the amount of accumulated total dissolved P (TDP) in the lake-water column driven by external P loadings. (*iii*) Estimate the annual internal-P release contribution to the lake-water column based on the summer vs. winter-deficit in the total P content in the lake-bed sediments. (*iv*) Evaluate the possible role of geogenic groundwater-borne P as a controlling factor for P accumulated during the Holocene. The evaluation was based on an estimate of the total amount of P accumulated in the sediment, which was compared with an estimate of the total groundwater P input to the lake, from the beginning of the sediment-P accumulation and until present.

2.2.1. External Phosphorus (P) Loadings

Investigation of the different external waterborne sources of P entering Nørresø was conducted by Nisbeth et al. [48] during 2016 to 2017. Their quantification of the external P loadings was based on the water balance of the lake (cf. Table 1) and the associated measured P concentrations of the different water components. The groundwater contribution was assumed to come from the Lucieø spring and stream 1, combined with a calculated residual groundwater input, based on the water balance.

Observed TDP concentrations for the lake, the outlet, as well as streams 1, 2, and 3, were similar in values (within ~10%) to TP monitoring concentrations published by the Danish Ministry of Environment and Food [55]. It is therefore assumed that TP mainly consists of TDP and thus the measured TDP concentrations are representative of the P balance at Nørresø (i.e., particulate P is of secondary importance).

2.2.2. Water Sampling and Dissolved P Analysis

Water sampling from the spring, the nine inlet streams, and the outlet were conducted, on average, once every month from June 2016 to December 2017 (see Figure 1 for sampling points). The samples were vacuum filtered in the field with 0.22 μm Minisart Sartorius cellulose acetate membrane filters, directly into polyethylene vials, and then preserved with 2 vol% 2 M H_2SO_4. See Nisbeth et al. [48] for a more detailed sampling description. Dissolved inorganic P concentrations in the samples were determined spectrophotometrically by the molybdenum-reactive P method [56]. TDP was measured as DIP after wet oxidation by potassium peroxydisulphate. Dissolved organic P (DOP) was calculated as the difference between total TDP and DIP. Due to the fully mixed conditions in Nørresø, TDP, and DIP concentrations for the outlet are assumed to be representative of the entire lake.

2.2.3. Mass-Loading Model of lake Water Column P

It is possible to evaluate whether the accumulated TP concentration in a lake water column is predominantly controlled by external P loadings, by using the P mass loading model of Vollenweider [54], usually expressed as:

$$TP_{lake} = \frac{TP_{in}}{1 + \sigma\tau_w} \tag{1}$$

where, TP_{lake} is the TP concentration in the lake (mg/m^3 or μg/L), TP_{in} is the flow-weighted TP input concentration (mg/m^3 or μg/L), τ_w is the lake hydraulic retention time (yr), and σ is a first-order rate constant (yr^{-1}) for P loss.

Two important assumptions apply in the derivation of Equation (1). (*i*) The lake is well mixed and at a steady state, thus the TP concentration in the outlet represents the TP concentration in the fully mixed lake. (*ii*) The TP can be removed from the water column by either advection (i.e., via the outlet) or by one or more first-order processes occurring within the lake. The latter is assumed to represent net loss due to sedimentation, since this is the dominant internal process driving TP concentration in the water column [57]. Note that the model does not consider P input from internal P release, hence it expresses an estimation of the accumulated TP in the water column if external loads are the only input of P.

Brett and Benjamin [58] conducted a statistical assessment of various mathematical expressions for σ. They found that σ was the best predictor of TP_{lake} when expressed as the inverse function of the lake's hydraulic retention time:

$$\sigma = k\tau_w^x \tag{2}$$

where, $k = 1.12 \pm 0.08$ yr$^{-0.47}$ and $x = 0.53 \pm 0.03$ are constants. By combining Equation (1) and Equation (2), we obtain:

$$TP_{lake} = \frac{TP_{in}}{1 + k\tau_w^x \times \tau_w} \tag{3}$$

Equation (3) was applied to the data from Nørresø assuming well-mixed conditions and that TDP corresponds to TP concentrations.

2.2.4. Estimation of the Internal P Release Load

In the present study, the internal P release load, which is released to the water column during a year, is assumed to be constituted by: (*i*) A rapid seasonal mobilized P pool ($P_{seasonal\_released}$) released from the upper sediment, combined with (*ii*) a slowly long-term mobilized P pool ($P_{longterm\_released}$), mobilized by mineralization of slowly biodegradable organic compounds and Fe-oxides under low redox potential.

Phosphorus in freshly settled material, containing easily degradable organic matter and redox-sensitive P, accumulated during the winter, is rapidly mobilized and released during the summer. The depth of the sediment layer in which the rapid mobilization of P occurs is hence reflected by a seasonal variation in the sediment TP depth profile. This layer is referred to as the active mobilization layer. In Nørresø the lower boundary of the active mobilization were found at 7 cm depth.

The $P_{seasonal\_released}$ pool was estimated as the difference in TP content (in mg P/g DW/cm) between the winter and summer cores in the active mobilization layer (i.e., upper 7 cm), multiplied by the depth-specific dry-weight density (g DW/cm^3). This quantity difference was then multiplied by the lake surface area, to obtain the total amount of $P_{seasonal\_released}$ in the lake.

Assuming that the TP content in the sediment during summer represents the P remaining in the sediment after the release of the easily mobilized P, a decrease in the summer TP content with increasing depth in the active mobilization, may reflect the mobilization of the long-term mobilizable P pool.

The annual released P from the $P_{longterm\_released}$ pool was hence calculated as the difference in TP (in mg P/g DW/cm) between the upper sediment sample and the TP content at 7 cm depth, multiplied by the mean sediment dry-weight density (0.071 g DW/cm^3) and subsequently divided by the age of the sediment at 7 cm. According to a previously established age-depth model for Nørresø, the upper 7 cm of the sediment represents 50 years of deposition [51]. The $P_{longterm\_released}$ pool is further based on assumptions of constant P-sedimentation rates during this 50-year period.

The mineralization rate presumably decreases with increasing depth once these easily degradable compounds are gone. Hence, the TP content decreases exponentially, reaching a steady state at the lower boundary of the active mobilization layer. Accordingly, a significant decline in the TP content beneath the active mobilization is believed more likely to reflect a change in the sedimentation rate than an additional mobilization of P.

2.2.5. Total P Sediment Depth Profiles

Lake sediment cores ($n = 12$) were collected on two dates, on 24 July 2017 (summer) and on 9 April 2018 (winter). The winter coring was conducted relatively late due to an unusually cold and long winter period, during which the lake was covered by ice [59]. The sampling was conducted with a Kajak gravity corer [60] using a 40 cm long core with an inner diameter of 5.2 cm. At each of the dates, three cores were collected at each of the two sampling stations (i.e., $n = 2$ stations $\times$ 3 cores $\times$ 2 dates = 12 cores in total). Station 1 (Figure 1b), represented the deepest part of the lake (water depth ~5 m), and station 2, represented the mean water depth of 2.3 m.

In the laboratory, each sediment core was sectioned in 1 cm intervals from 0 to 10 cm depth. Thereafter, three additional 1 cm samples were collected at 15, 20, and 25 cm.

The total P content of the individual sediment samples was determined after drying the sediment for 24 h at 105 °C followed by combustion at 520 °C for 8 h. The combusted sample was boiled for one hour in 120 °C 1 M HCl. Total P was then measured as DIP in the extract by the molybdenum-reactive P method [56]. Each sediment sample was analyzed individually, after which mean values for each sectioned depth of the summer and winter cores respectively, were calculated by combining the six sediment samples from the two stations ($n = 6$ for each depth interval).

The dry weight (DW) content of the sediment was determined gravimetrically, before and after drying the samples at 105 °C.

2.2.6. Sediment-P Composition

A general idea of the P composition of the sediment was obtained by determining the depth profiles of eight different P fractions in the sediment. In accordance with the P sequential extraction technique presented by Paludan and Jensen [61], quantification was conducted by extracting: (*i*) Loosely sorbed water-soluble P (P_{Water}), (*ii*) redox-sensitive bicarbonate-dithionite (BD; 0.11 M NaHCO$_3$ + 0.11 M Na$_2$S$_2$O$_4$) extractable P (P_{BD}), (*iii*) 0.1 M NaOH extractable P (P_{NaOH}), (*iv*) 0.5 M extractable HCl-soluble P (P_{HCl}), and finally, (*v*) residual P not extracted (P_{Res}).

The humic-bound organic P fraction (OP$_{Humic}$) was determined by acidification of the NaOH extractant to pH ~1 with 1.5 mL 2 M H$_2$SO$_4$, which led to the flocculation of humic acids after two days [61]. The humic acids was separated by filtration with Whatman GF/C filters, combusted (520 °C, 8 h), and extracted by 1 M 120 °C warm HCl for 1 h before analysis as DIP. Non-reactive P (nrP), which is assumed to represent organic labile P compounds [62,63], was calculated as the sum of the differences between TDP and DIP in the P_{Water}, P_{BD}, and P_{NaOH} fractions. P_{Res} was determined on the remaining sediment after the sequential extraction technique. The sediment was dried and combusted at 520 °C for 8 h, followed by one hour in 120 °C 1 M HCl, and P_{Res} was then measured as DIP in the extract.

Inorganic P in the different extracts (DIP) was determined by the molybdenum-reactive P method [56], and TP in the extracts (TDP) was detected by inductively coupled plasma-optical emission spectrometry (ICP-OES; detection limit 4 µg/L), with the exception of P_{HCl}, since this fraction only represents inorganic P. In the result section the different P fractions are named according to the form in which they occurred in the sediment (e.g., as either inorganic P (IP) or organic P (OP)).

As for the TP depth profiles, mean values for each sectioned depth of the summer and winter cores were calculated respectively, by combining the six sediment samples from the two stations.

3. Results

3.1. External Waterborne Dissolved P Loadings

The total annual external dissolved P load to Nørresø amounted to 97 kg P/yr (Table 2). The external TDP load included the supply of P via groundwater (84 kg P/yr), surface water (9 kg P/yr), and dry and wet deposition (4 kg P/yr), see Table 2. Applying the P mass loading model of Vollenweider [54] (Equation (3)), and assuming a lake hydraulic retention time of 1.6 year (Table 1), the external P load entering Nørresø would result in an accumulation of 40 kg P in the water column.

The annual quantity of TDP leaving the lake via the one outlet amounted to 126 kg P/yr.

Table 2. Annual P loading to and from Lake Nørresø during 2016 to 2017.

	kg P/yr
External TDP loads	
Groundwater [a]	84
Surface water [a]	9
Atmospheric deposition [b]	4
Total external TDP load	97
Internal P release load	
$P_{seasonal\_released}$ [c]	680
$P_{longterm\_released}$ [d]	40
Total internal P load	720
TDP Output	126
Net P increase in the water column during summer	650

[a] Ref. [48]. [b] Estimate based on national average of total wet and dry deposition of organic and inorganic P (both 0.02–0.04 kg P/ha) [64]. [c] Rapid seasonal mobilized P pool, release from the upper 7 cm of the sediment. [d] The annual released P from the long-term mobilized P pool, driven by mineralization of slowly biodegradable organic compounds and slowly reducible Fe-oxides buried in the sediment.

3.2. Internal P Release Load

A higher TP content was measured in the winter sediment, relative to the summer sediment, in the upper 7 cm of the lake-bed (Figure 2). The TP content in both profiles decreased exponentially with increasing depth, going from 2.7 and 2.3 mg P/g DW in the uppermost sediment respectively, reaching a steady state with similar winter and summer concentrations at a depth of 7 cm (~1.7 mg P/g DW). From 7 to 10 cm depth the TP content in both profiles were almost constant at 1.6 mg P/g DW.

This indicates that the lower boundary of the active mobilization layer is located at a depth of 7 cm (Figure 2). Below 10 cm depth (i.e., beneath the active mobilization layer) the winter and summer TP sediment content remained similar, though with a significant decline in the TP content. This suggest that a change in the sedimentation condition has occurred.

Assuming a lake-bed area of 69.3 ha, an active mobilization layer of 7 cm, and that the sediment cores are representative for the whole lake, the $P_{seasonal\_released}$ pool mobilized rapidly during the summer amounted to 680 kg P (i.e., the difference between TP content during winter and summer).

By further assuming that the summer TP depth profile in Figure 2 represents the P sediment content after the release of the easily mobilized P, the total $P_{longterm\_released}$ pool which can be mobilized over many years amounted to 1828 kg P. Assuming that the mineralization rate of the slowly biodegradable organic compounds and reducible Fe-oxides have remained constant over the last 50 years, the $P_{longterm\_released}$ pool mobilized per year amounted to 40 kg P/yr.

The total internal P release load is, hence, estimated to be 720 kg P/yr (Table 2).

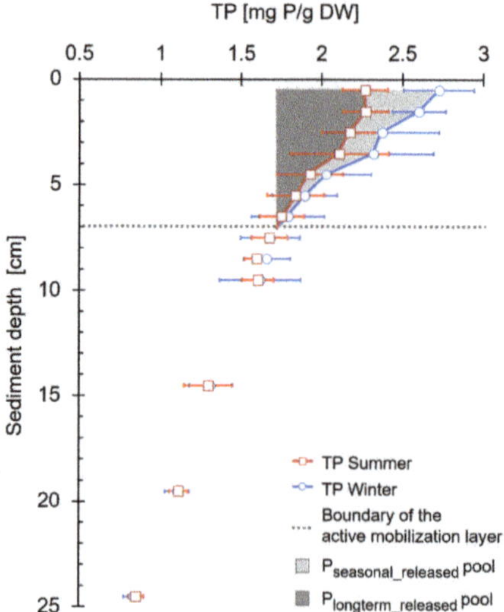

Figure 2. Total phosphorus (TP) content in the sediment in the summer (June 2017) and winter (April 2018). Data presented are mean concentrations of the sediment samples of the six cores collected in each season, the error bars shows the standard deviation. The two pools that make up the internal released P are highlighted in gray. For location of the sediment core sampling stations, see Figure 1.

3.3. Sediment-P Composition

The seasonal variation in the TP content in the upper 7 cm of the lake-bed can mainly be attributed to a seasonal difference in the redox-sensitive IP_{BD} fraction (Figure 3). The exponential decrease in the sediment's TP content down through the upper 7 cm was also primarily due to a decrease in the IP_{BD}, together with a minor decrease in the nrP (i.e., organic labile P) and the IP_{Water} fraction. At 7 to 10 cm depth, both IP_{BD}, nrP and the IP_{Water} fraction had reached a steady state.

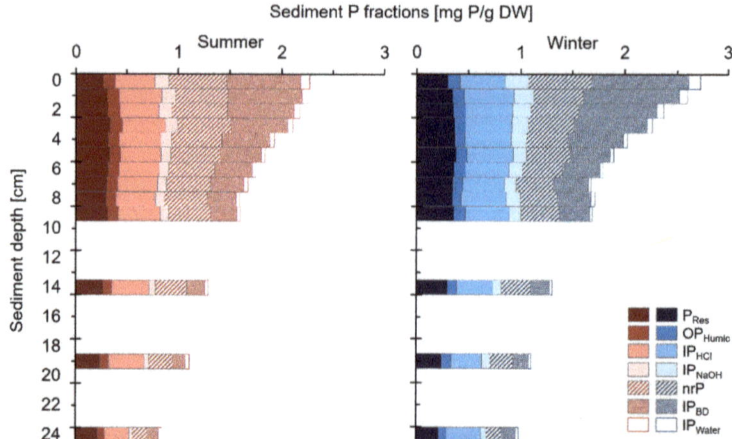

Figure 3. Sediment-P fractions from Nørresø in summer (June 2017; left) and winter (April 2018; right). Data presented are mean concentrations of the sediment samples of the six cores collected in each season. For location of the sediment core sampling stations, see Figure 1.

The four fractions P_{Res}, OP_{Humic}, IP_{HCl}, and IP_{NaOH} were relatively constant with depth and displayed similar concentrations in summer and winter in the upper 10 cm of the lake-bed, indicating steady state conditions during the accumulation of the upper 10 cm.

At 15, 20, and 25 cm sediment depth, the amounts of the otherwise stable P_{Res}, OP_{Humic}, IP_{HCl}, and IP_{NaOH} factions had decreased relative to the upper 10 cm. A decrease in the IP_{BD}, nrP, and the IP_{Water} fractions could also be observed at 15, 20, and 25 cm sediment depth even though these fractions were at a steady state at 7 to 10 cm.

3.4. Dissolved P Concentrations in the Lake, Spring, and Streams

In the lake water column, DIP accounted for the largest share of TDP throughout the year (Figure 4. TDP and DIP concentrations followed the same annual pattern, peaking in August and September, and reaching a minimum concentration in April. If both 2016 and 2017 are taken into account, the mean concentration of TDP and DIP during the peak corresponds to 507 and 336 µg P/L, respectively). During the period with minimum concentrations in April, the mean TDP and DIP concentrations were 98 and 96 µg P/L, respectively. Accordingly, DOP was detected mainly from the end of July to the beginning of November.

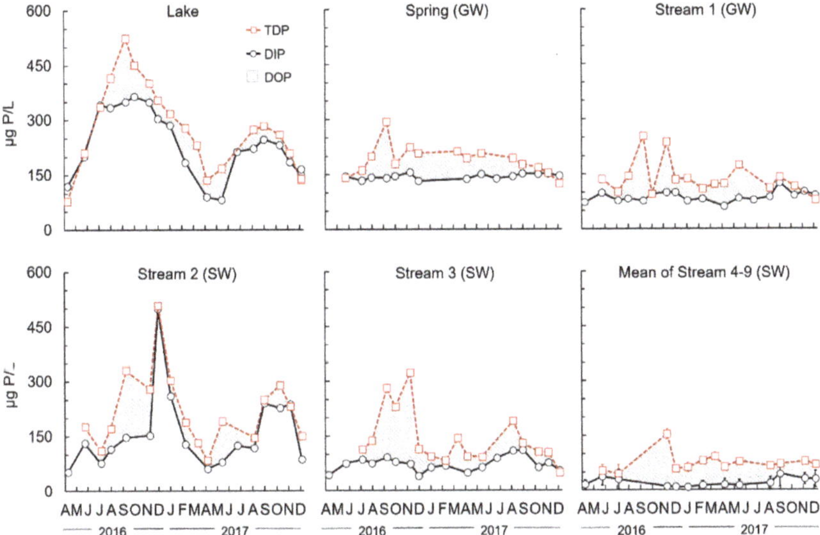

Figure 4. Concentrations of total dissolved phosphorus (TDP), dissolved inorganic phosphorus (DIP), and dissolved organic phosphorus (DOP) in the lake, spring and streams 1–9 at Nørresø during 2016–2017. Lower left panel shows the mean of streams 4–9 the error bars shows the standard deviation. Spring and stream 1 represent groundwater (GW), and streams 2–9 represent surface water (SW).

The TDP increase in the water column during summer relatively to the lake's background TDP concentration before the onset of internal P loading, as represented by the winter lake-water concentrations [4,65,66] amounted to 650 kg P (Table 2), assuming a lake water volume of 1.6 mill. m^3 (2.3 m × 69.3 ha).

Figure 4 shows the TDP and DIP concentrations, for the four main sources of external P loading from measured inflows. The spring and stream 1 represented groundwater and had fairly stable DIP concentrations throughout the year (142 µg P/L and 84 µg P/L, respectively) whereas the DOP concentration varied. DIP and DOP in stream 2 had a pronounced seasonality, both peaking (500 µg P/L) in December. Stream 2, which represents the main contributor of surface water P, discharges water from a small lake/wetland area called Hjertesø. Thus, the annual internal P cycling occurring in Hjertesø propagates to the annual DIP variation in stream 2. Stream 3 had a relatively high DOP concentration from August to November whereas the DIP concentration was relatively stable at 70 µg P/L. The rest of the streams (numbers 4 to 9) all had low DIP and DOP concentrations.

4. Discussion

By estimating the external and internal P loadings to a naturally eutrophic, groundwater-fed lake, it became clear that under the current conditions, internally released P was by far the dominant P loading to the lake. Thus, internally released P was the immediate controller of the trophic state of the freshwater ecosystem. The accumulated P in the sediment, and hence the formation of the mobile P pool, was likely conditional upon groundwater-borne geogenic P entering the lake throughout the Holocene, which will be discussed in more detail below. Our findings demonstrate that groundwater-borne geogenic P should be seen as a potential contributor to eutrophication of a freshwater ecosystem.

4.1. Current Lake P Budget

4.1.1. External P Loadings

Groundwater-borne P constituted the primary external P load to the lake, accounting for 88% of the total external P load. The remaining 12% of P was entering the lake via surface water discharging from stream 2 to 9, and atmospheric deposition.

In general, when high P concentrations are observed in discharging groundwater to a lake, it is often shown to be, or expected to be, from an anthropogenic source [32–35,38,42,67]. Nisbeth et al.'s [48] investigation of the external DIP sources indicated, however, that the presence of P in the groundwater discharging into Nørresø was most likely driven by a natural release of geogenic P within a regional artesian, confined aquifer. Hence, causing a natural eutrophication of the lake.

Nisbeth et al. [48] measured a constant groundwater discharge from the spring and stream 1. Thus, the fairly stable and dominant DIP concentration in the spring and stream 1 indicates that groundwater P input, and hence the external P input, remains stable throughout the year. Stream 1 runs under a dirt road via a pipe, which occasionally clogs. This leads to large variations in the water table upstream of the pipe. When clogged, parts of the surrounding wetland are flooded. Phosphorus mobilization processes during flooding may explain the occasionally high DOP concentration.

Under current conditions, the contribution of external P can only explain 8% of the measured accumulated TDP within the lake water column, according to Vollenweider's P mass-loading model. Van der Molen and Boers [9] and Søndergaard et al. [65] emphasize that Vollenweider's model underestimates the TP concentration for lakes in which internal P release was the dominant load. An underestimate of TP (TDP in this study) in Nørresø would substantiate internal P release as playing a dominant role. Nonetheless, the corresponding equilibrium-TDP concentration in the water column, driven merely by the external TDP input (26 µg P/L; based on Equation (3)), would still nearly classify as eutrophic (TP > 30 µg P/L; [4]).

Droppings from the cormorant colony found at Nørresø may contribute an additional external P load to the water column [68]. Previous studies indicate that bird droppings only represent a small fraction [68–70] of the total contribution to TP loads in eutrophic lakes. However, the impact that birds may have on the P budgets of lakes is so far not well documented. Thus, an estimate of the P contribution from the cormorant colony remains uncertain. A preliminary investigation of the DIP concentrations in the lake water column near the colony of cormorants did not indicate elevated DIP concentrations near the colony. This could suggest that the soil has the capacity to retain the P from the bird droppings, and hence indicates a low impact from the cormorant colony [69].

Leaf fall from the forests covering the main part of the area around the lake may also end up in the lake, representing yet another external P load. The leaves probably settle by gravity on the lake-bed, followed by biodegradation, potentially contributing P over time as a part of the internal release.

4.1.2. Internal P Loading

The total annual internally mobilized and released P from the sediment to the water column was estimated to amount to 720 kg P/yr. Comparing this load with the external dissolved P entering the lake during a year (97 kg), it is evident the internal P load is by far the dominant P contributor to the lake under the current conditions. The significant increases in the TPD concentration in the water column during the summer months of August and September can further be seen as an indication of internal, rather than external, P loading dominance [5,9,65,71,72]. Quantitatively, the TDP increase in the water column (650 kg P) more or less equaled the P released from the sediment. The internal P load might even have been much higher than estimated because a strong downward direct P-flux (sedimentation) can be expected in such a productive lake. In parallel, the estimated TDP increase in the water column only represents the net P accumulation, and thus does not directly reflect the gross amount of P released from the sediment.

The net retention of P in the sediment may change over time as the relative ratio of the downward and upward flux of P changes. The stable concentration of the P_{Res}, OP_{Humic}, IP_{HCl}, and IP_{NaOH} fractions within the upper 10 cm of the sediment indicate stable P input within this period, since these fractions are generally considered to be unaffected by diagenesis in the sediment [66].

The significant decline in the TP content and change in the immobile P composition in the sediment at 15 cm depth indicate that the steady state of the lake changed around 50 to 75 years ago. Accordingly, the decline in the TP below the 15 cm is therefore not considered to be an expression of mobilization processes but rather an indication of a lower sedimentation rate. This is supported by findings from another study, where it was shown that the diatom-inferred TP concentrations increased from 30 µg P/L in the mid-1800s to 120 µg P/L at around 1970 [73].

Of the total annual internal P load, consisting of the seasonally and long-term released P, P rapidly released from the sediment during the summer (i.e., the $P_{seasonal\_released}$ pool) constitutes the majority (95%). The remaining 5% of P originates from the long-term released P pool (i.e., the $P_{longterm\_released}$ pool).

Based on the sediment-P composition, the $P_{seasonal\_released}$ pool primarily consists of the IP_{BD} fraction, i.e., Fe-oxide-bound IP. This strongly suggests that the rapid seasonal P release is mainly controlled by reductive dissolution of the iron oxides in the sediment [6,15,74]. In contrast, $P_{longterm\_released}$ partly consists of nrP. In general, nrP from the P_{NaOH} fraction represents by far the largest share of the nrP [61] and consists of organic P such as fulvic acids [75], monoester phosphate and condensed inorganic polyphosphates [76]. The mobilization of P driven by degradation of organic labile P compounds can, however, be estimated to account for just 21% of the total annually released $P_{seasonal\_released}$ pool. Mobilization of buried Fe-oxide-bound IP (relatively resilient towards reductive dissolution) is, as for the $P_{seasonal\_released}$ pool, the dominant mobilization process of the $P_{seasonal\_released}$ pool, accounting for 67%.

The accumulation of Fe-bound IP in the sediment can be attributed to a mechanism, often referred to as the 'iron curtain' [77]. In this process, soluble ferrous compounds in anoxic water precipitate as amorphous iron oxides and hydroxides when entering an oxidized environment, which subsequently removes dissolved P from the water column. The same conditions occur at Nørresø. Anoxic groundwater, accounting for 46% of the water input to the lake, continuously transport dissolved ferrous iron and dissolved geogenic P to the oxygenated lake (cf. [48]). Accordingly, Fe-oxide-bound IP continuously precipitates and settles by gravity on the lake-bed where it accumulates over time. Ferrous iron was not detected in the discharging surface water [48], thus indicating that groundwater is the main driving factor of the accumulated IP_{BD} fraction. Correlation between a pronounced IP_{BD} fraction and discharging Fe-rich, anoxic groundwater was also found by Chambers and Odum [77] and Pacini and Gächter [78].

4.2. Long-Term Influence of External TDP

Although the internal TDP load supersedes the external load, it is important to emphasize that internal P load is only indirectly relevant for the trophic state of a lake. The accumulated P in the sediment originates from external P sources entering and subsequently sedimented on the lake-bed.

Evidence that these processes have occurred in Nørresø at longer timescales is presented in Nisbeth et al. [48], where a 7.2 m long sediment core, sampled near station 2 (Figure 1), is described. The upper 6 m of the core consists of gyttja which began to accumulate around 7500 cal. yr BP.

Preliminary diatom and pigment analyses indicate that Nørresø has never been P-poor, but rather has been meso- to eutrophic, even in its pristine state. Pigment data further indicates a persistent presence of cyanobacteria and algae blooms during the past 7500 years. High abundances of cyanobacteria have been recorded at other lakes in glacial moraine landscapes, where groundwater P supply dominates [79]. This situation is similar to the present conditions at Nørresø, where groundwater-transported TDP represents the largest contributor of external TDP (88%). This substantiates the assumption that the deposited P is, and probably always has been, conditional on a continuous TDP transport to the lake via groundwater.

Assuming that the data in Figure 3 are representative of the entire lake area, and the TP concentration in the sediment stagnates at 25 cm depth, up to about 1 mg TP/g DW is permanently deposited in the sediment. Assuming a 6 m gyttja layer thickness with a density of 0.10 g DW/cm^3 (ϱ from 25 cm depth) in the lake, the total deposited P in Nørresø would amount to 374 tons. This is likely an overestimate as the thickness of the gyttja probably decreases towards the bank. Nevertheless, if the groundwater P input is of the same order of magnitude as the total deposited P in the sediment, it is likely to be the driving factor behind accumulation of P in the lake. The calculation of total deposited P only accounts for the net-accumulated P in the sediment, and not the total P input to the lake. Since the amount of P leaving the lake via the outlet over the 7500 years is not included, the groundwater input should in fact exceed the total deposited P (i.e., 374 tons).

The P loss recorded in our own short study period is certainly not representative of the conditions occurring through the Holocene, as it would imply that no P was stored in the sediment.

Assuming a continuous groundwater TDP load equal to the current (84 kg P/yr), from the beginning of the gyttja accumulation until the present (~7500 yrs), the total groundwater P input adds up to 630 tons. If the lake had an outlet in its earlier state, the external P load would also have been higher than today in order to generate the accumulated gyttja. It can be argued that the P concentration in the groundwater was probably higher in the past, as the geogenic P pool in the aquifer is slowly being depleted over time, thus leading to a decrease in groundwater-P concentration. A higher surface-water P contribution in the past could, likewise, be the underlying reason for the accumulated P in the sediment. However, there is no reason to expect that past surface water P load was higher than today, since (*i*) P started to accumulate long before anthropogenic influences, and (*ii*) the lake has been surrounded by forest, or at least a rim of forest along its perimeter, throughout the Holocene.

Despite the uncertainties in extrapolating back through the Holocene based on limited modern-day data, the estimated amounts of groundwater-borne P and P contained in the lake sediment are of the same magnitude. The groundwater TDP input even exceeds the estimated amount of permanently deposited P in the sediment, suggesting that groundwater-borne geogenic P was the controlling mechanism for the accumulation of P in the sediment during the Holocene.

5. Conclusions

This study demonstrates and expands upon how groundwater-borne geogenic P can play an important role in the eutrophic state of a freshwater ecosystem. Our findings demonstrate that even in lakes where internal P cycling is the immediate controller of their trophic state, groundwater-transported geogenic P can potentially be the original source of the internally cycled P.

For naturally eutrophic lakes, this indicates that:

- Geogenic groundwater-borne P can be the dominant cause of P accumulation: the amount of geogenic groundwater-borne P brought to the lake during Holocene is in the same order of magnitude as the total accumulated P in the sediment.
- Even though the seasonal release of internal P controls seasonal fluctuations in water column P, groundwater-borne geogenic P may be the actual eutrophicating factor.

When investigating internally released P, one should not ignore the importance of the original, and possibly natural and geogenic, source of this P.

Author Contributions: Conceptualization, C.S.N., K.R., J.K. and S.J.; Methodology, C.S.N. and K.R.; Validation, C.S.N. and K.R.; Formal analysis, C.S.N.; Investigation, C.S.N., K.R., O.B., J.K. and S.J.; Resources, C.S.N. and K.R.; Writing—original draft preparation, C.S.N.; Writing—review and editing, C.S.N., K.R., O.B., J.K. and S.J.; Visualization, C.S.N.; Supervision, K.R., J.K. and S.J.; project administration, S.J, O.B. and J.K.; funding acquisition, S.J. and J.K.

Funding: This research was funded by Geocenter Denmark, grant number 6-2015.

Acknowledgments: The authors are grateful to the estate of Brahetrolleborg and Catharina Reventlow-Mourier for kindly providing access to Nørresø. Thanks are also given to Rikke Orloff Holm (SDU) and Carina Kronborg Lohmann (SDU) for assistance with the sequential extraction of the sediment and analysis of TDP.

Conflicts of Interest: The authors declare no conflict of interest.

References

1. Blake, R.E.; O'Neil, J.R.; Surkov, A.V. Biogeochemical cycling of phosphorus: Insights from oxygen isotope effects of phosphoenzymes. *Am. J. Sci.* **2005**, *305*, 596–620. [CrossRef]
2. Hecky, R.E.; Kilham, P. Nutrient limitation of phytoplankton in freshwater and marine environments: A review of recent evidence on the effects of enrichment1. *Limnol. Oceanogr.* **1988**, *33*, 796–822. [CrossRef]
3. Schindler, D. Eutrophication and recovery in experimental lakes—Implications for lake management. *Science.* **1974**, *184*, 897–899. [CrossRef] [PubMed]
4. Wetzel, R.G. The phosphorus cycle. In *Limnology: Lake and River Ecosystems*; Academic Press: San Diego, CA, USA, 2001; pp. 242–250. ISBN 9780127447605.
5. Søndergaard, M.; Jensen, J.P.; Jeppesen, E. Role of sediment and internal loading of phosphorus in shallow lakes. *Hydrobiologia* **2003**, *506–509*, 135–145. [CrossRef]
6. Smolders, A.J.P.; Lamers, L.P.M.; Lucassen, E.C.H.E.T.; Van Der Velde, G., Roelofs, J.G.M. Internal eutrophication: How it works and what to do about it—A review. *Chem. Ecol.* **2006**, *22*, 93–111. [CrossRef]
7. Boström, B.; Andersen, J.M.; Fleischer, S.; Jansson, M. Exchange of phosphorus across the sediment-water interface. *Hydrobiologia* **1988**, *170*, 229–244. [CrossRef]
8. Paytan, A.; Roberts, K.; Watson, S.; Peek, S.; Chuang, P.C.; Defforey, D.; Kendall, C. Internal loading of phosphate in lake erie central basin. *Sci. Total Environ.* **2017**, *579*, 1356–1365. [CrossRef] [PubMed]
9. Van der Molen, D.T.; Boers, P.C.M. Influence of internal loading on phosphorus concentration in shallow lakes before and after reduction of the external loading. *Hydrobiologia* **1994**, *275*, 379–389. [CrossRef]
10. Murray, G.C.; Hesterberg, D. Iron and phosphate dissolution during abiotic reduction of ferrihydrite-boehmite mixtures. *Soil Sci. Soc. Am. J.* **2006**, *70*, 1318. [CrossRef]
11. Hupfer, M.; Gloess, S.; Grossart, H.P. Polyphosphate-accumulating microorganisms in aquatic sediments. *Aquat. Microb. Ecol.* **2007**, *47*, 299–311. [CrossRef]
12. Hupfer, M.; Lewandowski, J. Oxygen controls the phosphorus release from lake sediments—A long-lasting paradigm in limnology. *Int. Rev. Hydrobiol.* **2008**, *93*, 415–432. [CrossRef]
13. Hansen, K.; Mouridsen, S.; Kristensen, E. The impact of Chironomus plumosus larvae on organic matter decay and nutrient (N, P) exchange in a shallow eutrophic lake sediment following a phytoplankton sedimentation. *Hydrobiologia* **1998**, *364*, 65–74. [CrossRef]
14. Golterman, H.L. The calcium-and iron bound phosphate phase diagram. *Hydrobiologia* **1988**, *159*, 149–151. [CrossRef]
15. Penn, M.R.; Auer, M.T.; Doerr, S.M.; Driscoll, C.T.; Brooks, C.M.; Effler, S.W. Seasonality in phosphorus release rates from the sediments of a hypereutrophic lake under a matrix of pH and redox conditions. *Can. J. Fish. Aquat. Sci.* **2000**, *57*, 1033–1041 [CrossRef]

16. House, W.A. Geochemical cycling of phosphorous in rivers. *Appl. Geochem.* **2003**, *18*, 739–748. [CrossRef]

17. Heiberg, L.; Pedersen, T.V.; Jensen, H.S.; Kjaergaard, C.; Hansen, H.C.B. A comparative study of phosphate sorption in lowland soils under oxic and anoxic conditions. *J. Environ. Qual.* **2010**, *39*, 734–743. [CrossRef] [PubMed]

18. Kjaergaard, C.; Heiberg, L.; Jensen, H.S.; Hansen, H.C.B. Phosphorus mobilization in rewetted peat and sand at variable flow rate and redox regimes. *Geoderma* **2012**, *173–174*, 311–321. [CrossRef]

19. Prem, M.; Hansen, H.C.B.; Wenzel, W.; Heiberg, L.; Sørensen, H.; Borggaard, O.K. High spatial and fast changes of iron redox state and phosphorus solubility in a seasonally flooded temperate wetland soil. *Wetlands* **2015**, *35*, 237–246. [CrossRef]

20. Jensen, H.S.; Andersen, F.O. Importance of temperature, nitrate, and pH for phosphate release from aerobic sediments of four shallow, eutrophic lakes. *Limnol. Oceanogr.* **1992**, *37*, 577–589. [CrossRef]

21. Bennion, H.; Smith, M.A. Variability in the water chemistry of shallow ponds in southeast England, with special reference to the seasonality of nutrients and implications for modelling trophic status. *Hydrobiologia* **2000**, *436*, 145–158. [CrossRef]

22. Ekholm, P.; Malve, O.; Kirkkala, T. Internal and external loading as regulators of nutrient concentrations in the agriculturally loaded lake Pyhajarvi (southwest Finland). *Hydrobiologia* **1997**, *345*, 3–14. [CrossRef]

23. Søndergaard, M.; Jensen, P.J.; Jeppesen, E. Retention and internal loading of phosphorus in shallow, eutrophic lakes. *Sci. World* **2001**, *1*, 427–442. [CrossRef] [PubMed]

24. Jeppesen, E.; Kristensen, P.; Jensen, J.P.; Søndergaard, M.; Mortensen, E.; Lauridsen, T. Recovery resilience following a reduction in external phosphorus loading of shallow, eutrophic Danish lakes: Duration, regulating factors and methods for over-coming resilience. *Mem. Ist. Ital. Idrobiol.* **1991**, *48*, 127–148.

25. Van der Grift, B.; Osté, L.; Schot, P.; Kratz, A.; Van Popta, E.; Wassen, M.; Griffioen, J. Forms of phosphorus in suspended particulate matter in agriculture-dominated lowland catchments: Iron as phosphorus carrier. *Sci. Total Environ.* **2018**, *631*, 115–129. [CrossRef] [PubMed]

26. Heathwaite, A.L.; Burke, S.P.; Bolton, L. Field drains as a route of rapid nutrient export from agricultural land receiving biosolids. *Sci. Total Environ.* **2006**, *365*, 33–46. [CrossRef] [PubMed]

27. Heathwaite, A.L.; Dils, R.M. Characterising phosphorus loss in surface and subsurface hydrological pathways. *Sci. Total Environ.* **2000**, *251*, 523–538. [CrossRef]

28. Withers, P.J.A.; Haygarth, P.M. Agriculture, phosphorus and eutrophication: A European perspective. *Soil Use Manag.* **2007**, *23*, 1–4. [CrossRef]

29. Hodgkinson, R.A.; Chambers, B.J.; Withers, P.J.A.; Cross, R. Phosphorus losses to surface waters following organic manure applications to a drained clay soil. *Agric. Water Manag.* **2002**, *57*, 155–173. [CrossRef]

30. Gilliom, R.J.; Patmont, C.R. Lake phosphorus loading from septic systems by seasonally perched groundwater. *Water Pollut. Control. Fed.* **1983**, *55*, 1297–1305.

31. Zanini, L.; Robertson, W.D.; Ptacek, C.J.; Schiff, S.L.; Mayer, T. Phosphorus characterization in sediments impacted by septic effluent at four sites in central Canada. *J. Contam. Hydrol.* **1998**, *33*, 405–429. [CrossRef]

32. Burkart, M.R.; Simpkins, W.W.; Morrow, A.J.; Gannon, J.M. Occurrence of total dissolved phosphorus in unconsolidated aquifers and aquitards in Iowa. *J. Am. Water Resour. Assoc.* **2004**, *40*, 827–834. [CrossRef]

33. Holman, I.P.; Whelan, M.J.; Howden, N.J.K.; Bellamy, P.H.; Willby, N.J.; Rivas-Casado, M.; McConvey, P. Phosphorus in groundwater—An overlooked contributor to eutrophication? *Hydrol. Process.* **2008**, *22*, 5121–5127. [CrossRef]

34. Holman, I.P.; Howden, N.J.K.; Bellamy, P.; Willby, N.; Whelan, M.J.; Rivas-Casado, M. An assessment of the risk to surface water ecosystems of groundwater P in the UK and Ireland. *Sci. Total Environ.* **2009**, *408*, 1847–1857. [CrossRef] [PubMed]

35. Kilroy, G.; Coxon, C. Temporal variability of phosphorus fractions in Irish karst springs. *Environ. Geol.* **2005**, *47*, 421–430. [CrossRef]

36. Kidmose, J.; Nilsson, B.; Engesgaard, P.; Frandsen, M.; Karan, S.; Landkildehus, F.; Søndergaard, M.; Jeppesen, E. Focused groundwater discharge of phosphorus to a eutrophic seepage lake (Lake Væng, Denmark): Implications for lake ecological state and restoration. *Hydrogeol. J.* **2013**, *21*, 1787–1802. [CrossRef]

37. Lewandowski, J.; Meinikmann, K.; Nützmann, G.; Rosenberry, D.O. Groundwater—The disregarded component in lake water and nutrient budgets. Part 2: Effects of groundwater on nutrients. *Hydrol. Process.* **2015**, *29*, 2922–2955. [CrossRef]

38. Walter, D.A.; Rea, B.A.; Sollenwerk, K.G.; Savoie, J. *Geochemical and Hydrologic Controls on Phosphorus Transport. in a Sewage Contaminated Sand and Gravel Aquifer Near Ashumet Pond, Cape Cod, Massachusetts*; Technical report No. 95-381; United States Geological Survey Water-Supply Paper 2463: Washington, DC, USA, 1996.

39. Kenoyer, G.; Anderson, M.P. Groundwater's dynamiv role in regulating acidity and chemistry in a precipitatuon-dominated lake. *J. Hydrol.* **1989**, *109*, 287–306. [CrossRef]

40. Vanek, V. The interactions between lake and groundwater and their ecological significance. *Stygologia* **1987**, *3*, 1–23.

41. Shaw, R.D.; Shaw, J.F.H.; Fricker, H.; Prepas, E.E. An integrated approach to quantify groundwater transport of phosphorus to Narrow Lake, Alberta. *Limnol. Oceanogr.* **1990**, *35*, 870–886. [CrossRef]

42. Meinikmann, K.; Lewandowski, J.; Hupfer, M. Phosphorus in groundwater discharge - A potential source for lake eutrophication. *J. Hydrol.* **2015**, *524*, 214–226. [CrossRef]

43. Heathwaite, A.L.; Dils, R.M.; Liu, S.; Carvalho, L.; Brazier, R.E.; Pope, L.; Hughes, M.; Phillips, G.; May, L. A tiered risk-based approach for predicting diffuse and point source phosphorus losses in agricultural areas. *Sci. Total Environ.* **2005**, *344*, 225–239. [CrossRef] [PubMed]

44. Orderud, G.I.; Vogt, R.D. Trans-disciplinarity required in understanding, predicting and dealing with water eutrophication. *Int. J. Sustain. Dev. World Ecol.* **2013**, *20*, 404–415. [CrossRef]

45. Sharpley, A.N.; Daniel, T.; Sims, T.; Lemunyon, J.; Stevens, R.; Parry, R. *Agricultural Phosphorus and Eutrophication*, 2nd ed.; U.S. Department of Agriculture, Agricultural Research Service: Lincoln, NE, USA, 2003.

46. Smith, V.H.; Schindler, D.W. Eutrophication science: Where do we go from here? *Trends Ecol. Evol.* **2009**, *24*, 201–207. [CrossRef] [PubMed]

47. Høy, T. *Danmarks Søer—Søerne i Fyns Amt*; Strandbergs Forlag: Charlottenlund, Denmark, 2000.

48. Nisbeth, C.S.; Kidmose, J.; Weckström, K.; Reitzel, K.; Odgaard, B.V.; Thorling, L.; McGowan, S.; Schomacker, A.; Kristensen, D.L.J.; Jessen, S. Dissolved inorganic geogenic phosphorus load to a groundwater-fed lake: Implications of terrestrial phosphorus cycling by groundwater. *Spec. Issues Water* **2019**, in press.

49. GEUS. (Geological Survey of Denmark and Greenland) 2018 Danish National Borehole Achieve (Jupiter). Available online: www.geus.dk (accessed on 21 September 2018).

50. NOVANA. (Det Nationale Overvågningsprogram for Vandmiljø og Natur). Available online: http://novana.dmi.dk/ (accessed on 3 April 2018).

51. Bennike, O.; Odgaard, B.V.; Eriksen, A.A.; McGowan, S.; Siggaard-Andersen, M.L.; Schomacker, A.; Jessen, S.; Kazmierczak, J.; Olsen, J.; Rasmussen, P.; et al. Multi-proxy records of late Quaterny environmental changes based on lake sediment records from Vængsø and Nørresø, Denmark. Unpublished work.

52. Eriksen, A.A. Holocene Catchment Changes at and Brahetrolleborg Nørresø and Lake Eutrophication Response. Master's Thesis, Aarhus University, Aarhus, Denmark, 2017.

53. Fyn's County. *Nørresø 1989–1993, Lake Monitoring in Fyn's County*; Technical report No.1; Fyn's County: Odense, Denmark, 1994.

54. Vollenweider, R.A. Advances in defining critical loading levels for phosphorus in lake eutrophication. *Mem. dell'Istituto Ital. Idrobiol. Doll. Marco Marchi Verbania Pallanza* **1976**, *33*, 53–83.

55. ODA. Surface Water Monitoring, Aarhus University. Available online: https://odaforalle.au.dk/main.aspx (accessed on 25 February 2019).

56. Murphy, J.; Riley, J.P. A modified single solution method for the determination of phosphate in natural waters. *Anal. Chim. Acta* **1986**, *27*, 31–36. [CrossRef]

57. Vollenweider, R.A. Input-output models with special reference to the phosphorus loading concept in limnology. *Schweiz. Z. Hydrol.* **1975**, *37*, 53–84.

58. Brett, M.T.; Benjamin, M.M. A review and reassessment of lake phosphorus retention and the nutrient loading concept. *Freshw. Biol.* **2008**, *53*, 194–211. [CrossRef]

59. DMI. The Danish Meteorological Institute. Available online: https://www.dmi.dk/vejrarkiv/ (accessed on 25 February 2019).

60. Renberg, I.; Hansson, H. The HTH sediment corer. *J. Paleolimnol.* **2008**, *40*, 655–659. [CrossRef]

61. Paludan, C.; Jensen, H.S. Sequential extractiom of phosphorus in freshwater wetland and lake sediment: Significance of humic acids. *Wetlands* **1995**, *15*, 365–373. [CrossRef]

62. Rydin, E. Potentially mobile phosphorus in Lake Erken sediment. *Water Res.* **2000**, *34*, 2037–2042. [CrossRef]

63. Reitzel, K. Separation of aluminum bound phosphate from iron bound phosphate in freshwater sediment by sequential extraction procedure. *Phosphates Sediments* **2005**, 109–117.

64. Ellermann, T.; Bossi, R.; Christensen, J.; Løfstrøm, P.; Monies, C.; Grundahl, L.; Geels, C. *Atmosfærisk Deposition 2014*; Technical report No. 163; Department of Environmental Sciences (DCE): Aarhus, Denmark, 2015.

65. Søndergaard, M.; Jensen, J.P.; Jeppesen, E. Internal phosphorus loading in shallow Danish lakes. *Hydrobiologia* **1999**, *408/409*, 145–152. [CrossRef]

66. Reitzel, K.; Hansen, J.; Andersen, F.; Hansen, K.S.; Jensen, H.S. Lake restoration by dosing aluminum relative to mobile phosphorus in the sediment. *Environ. Sci. Technol.* **2005**, *39*, 4134–4140. [CrossRef] [PubMed]

67. Griffioen, J. Extent of immobilisation of phosphate during aeration of nutrient-rich, anoxic groundwater. *J. Hydrol.* **2006**, *320*, 359–369. [CrossRef]

68. Marion, L.; Clergeau, P.; Brient, L.; Bertru, G. The importance of avian-contributed nitrogen (N) and phosphorus (P) to Lake Grand-Lieu, France. *Hydrobiologia* **1994**, *279–280*, 133–147. [CrossRef]

69. Gwiazda, R.; Jarocha, K.; Szarek-Gwiazda, E. Impact of a small cormorant (Phalacrocorax carbo sinensis) roost on nutrients and phytoplankton assemblages in the littoral regions of a submontane reservoir. *Biologia* **2010**, *65*, 742–748. [CrossRef]

70. Søndergaard, M.; Lauridsen, T.L. *Fugle og Karpers Påvirkning af Søer*; Technical report No. 84; Department of Environmental Sciences (DCE): Aarhus, Denmark, 2014.

71. Phillips, G.; Jackson, R.; Bennett, C.; Chilvers, A. The importance of sediment phosphorus release in the restoration of very shallow lakes (The Norfolk Broads, England) and implications for biomanipulation. *Hydrobiologia* **1994**, *275–276*, 445–456. [CrossRef]

72. Welch, E.B.; Cooke, G.D. Internal phosphorus loading in shallow lakes: Importance and control. *Lake Reserv. Manag.* **2005**, *21*, 209–217. [CrossRef]

73. Odgaard, B.; Møller, P.F.; Wolin, J.A.; Rasmussen, P.; Anderson, N.J. *Brahetrolleborg Nørresø. Palæolimnologisk Undersøgelse og Oplandsanalyse*; Company Report (48); Denmark's Geological Surveys: Copenhagen, Denmark, 1995; pp. 1–56.

74. Ding, S.; Wang, Y.; Wang, D.; Li, Y.Y.; Gong, M.; Zhang, C. In situ, high-resolution evidence for iron-coupled mobilization of phosphorus in sediments. *Nat. Sci. Rep.* **2016**, *6*, 1–11. [CrossRef]

75. Malcolm, R.L. The uniqueness of humic substances in each of soil, stream and marine environments. *Anal. Chim. Acta* **1990**, *232*, 19–30. [CrossRef]

76. Reitzel, K.; Ahlgren, J.; Gogoll, A.; Jensen, H.S.; Rydin, E. Characterization of phosphorus in sequential extracts from lake sediments using 31 P nuclear magnetic resonance spectroscopy. *Can. J. Fish. Aquat. Sci.* **2006**, *63*, 1686–1699. [CrossRef]

77. Chambers, R.; Odum, W. Porewater oxidation, dissolved phosphate and the iron curtain. *Biogeochemistry* **1990**, *10*, 37–52. [CrossRef]

78. Pacini, N.; Gächter, R. Speciation of riverine particulate phosphorus during rain events. *Biogeochemistry* **1999**, *47*, 87–109. [CrossRef]

79. McGowan, S.; Britton, G.; Haworth, E.; Moss, B. Ancient blue-green blooms. *Limnol. Oceanogr.* **1999**, *44*, 436–439. [CrossRef]

Article

Evaluation of Temperature Profiling and Seepage Meter Methods for Quantifying Submarine Groundwater Discharge to Coastal Lagoons: Impacts of Saltwater Intrusion and the Associated Thermal Regime

Joel Tirado-Conde [1,*], Peter Engesgaard [1], Sachin Karan [2], Sascha Müller [1] and Carlos Duque [3]

[1] Department of Geosciences and Natural Resource Management, University of Copenhagen, 1350 Copenhagen, Denmark
[2] Department of Geochemistry, Geological Survey of Denmark and Greenland (GEUS), 1350 Copenhagen, Denmark
[3] WATEC, Department of Geoscience, Aarhus University, 8000 Aarhus, Denmark
* Correspondence: jtc@ign.ku.dk

Received: 2 July 2019; Accepted: 7 August 2019; Published: 9 August 2019

Abstract: Surface water-groundwater interactions were studied in a coastal lagoon performing 180 seepage meter measurements and using heat as a tracer in 30 locations along a lagoon inlet. The direct seepage meter measurements were compared with the results from analytical solutions for the 1D heat transport equation in three different scenarios: (1) Homogeneous bulk thermal conductivity (K_e); (2) horizontal heterogeneity in K_e; and (3) horizontal and vertical heterogeneity in K_e. The proportion of fresh groundwater and saline recirculated lagoon water collected from the seepage experiment was used to infer the location of the saline wedge and its effect on both the seepage meter results and the thermal regime in the lagoon bed, conditioning the use of the thermal methods. The different scenarios provided the basis for a better understanding of the underlying processes in a coastal groundwater-discharging area, a key factor to apply the best-suited method to characterize such processes. The thermal methods were more reliable in areas with high fresh groundwater discharge than in areas with high recirculation of saline lagoon water. The seepage meter experiments highlighted the importance of geochemical water sampling to estimate the origin of the exchanged water through the lagoon bed.

Keywords: heat as a tracer; temperature profiles; seepage meter; lagoon; coastal areas; seawater intrusion; groundwater discharge; seawater-groundwater interactions

---

1. Introduction

Exchange fluxes between groundwater (GW) and surface water (SW) bodies are important due to the hydrological [1] and ecological [2,3] consequences for aquatic plants and animals. Understanding the temporal and spatial heterogeneity of exchange fluxes is important as it can affect water resource management plans. Although GW-SW interactions have been extensively studied [4], it remains a challenge to accurately measure the exchange fluxes in-situ and also decide on where and how often [5].

Lewandowski et al. [6] discussed the historical evolution of studies investigating Lacustrine groundwater discharge (LGD), i.e., exchange fluxes between lakes and groundwater versus that of submarine groundwater discharge (SGD), i.e., exchange fluxes between sea/lagoons and groundwater. The current study is interested in evaluating the methods for quantifying SGD and its inherent spatial heterogeneity. Duque et al. [7] recently showed that spatial heterogeneity in exchange fluxes was more important than temporal heterogeneity and highlighted the impact of a correct characterization of the spatial heterogeneity on the discharge estimations, so this study focuses on capturing SGD as a snapshot in time.

Seepage meters, such as the ones designed by Lee [8], are the only tool to directly measure LGD and SGD [9]. The use of seepage meters to quantify SGD is often reported [7,10,11], but it is a labor intense methodology and it has certain limitations [9,12]. Other techniques (see review by Rosenberry et al. [5]) that have been extensively applied are the use of groundwater head measurements from piezometers [13] and the use of heat as a tracer [14–16], among others [17], that indirectly derive the horizontal/vertical flux by the application of Darcy's law or by the use of analytical solutions of the 1D heat transport equation. Most of these methods rely on prior information of hydraulic and thermal characteristics of the sediment, which may not be as well-known and predictable in coastal environments as they are in other hydrological systems [18]. In other cases (e.g., [19]), the time series of temperature data at different depths have been used to estimate exchange vertical fluxes. The analytical models for heat transport also require the specification of upper and lower temperature boundary conditions. The upper boundary condition is often set to what was measured at the sediment-water interface, while the lower boundary condition is assumed to be equal to the average groundwater temperature (equal to the average seasonal air temperature) at some depth below the measurement point. While the thermal method has found extensive use for LGD, it has not been used that much to study SGD [18,20,21].

In coastal areas, the GW-SW exchange, or SGD, is influenced by the density-driven flow produced by the differences between the low-density fresh groundwater (FGW) and the higher density saline surface water (SSW). These differences generate a mixing zone with a dynamic behavior that depends on the salinity of both FGW and SSW as well as on the magnitude of the groundwater discharge flux to the coast. Befus et al. [18] summarized a field study using heat as a tracer in a conceptual model of a groundwater-bay system that showed the well-known complex flow with a saltwater interface, a salt water plume at the high tide mark, and a freshwater tube with fresh groundwater discharge. However, they also superimposed a model of the thermal regime, which due to the ~3 °C difference in the average annual temperature of groundwater and water in the bay developed into distinct thermal regimes, where the subsurface near the bay was warmer than discharging groundwater. The complexity of the flow field and thermal regime therefore will affect the choice of how to specify boundary conditions for heat transport modeling and possibly also how well two different methods like the seepage meter and thermal methods compare.

In this study, 180 seepage meter measurements in 30 different locations along a coastal lagoon inlet were combined with indirect estimations of the vertical flux using heat as a tracer in those same 30 locations. The proportion of discharging fresh groundwater and recirculated saline lagoon water was used to infer the location of the saline wedge and the consequences that it had on the seepage meter results. The time series of the temperature in deeper wells were used to assess the boundary conditions for the thermal regime of the lagoon sediments that affected the applicability of the thermal methods.

The objectives of this work are therefore to: (1) Test and compare the in-situ direct measurements of groundwater discharge to a coastal lagoon using seepage meters with indirect estimates using heat as a tracer; (2) determine the effect of vertical and horizontal sediment heterogeneity in bulk thermal conductivity on the results obtained by the use of analytical solutions of the heat transport equation; (3) address the impact of lagoon water recirculation on the subsurface thermal regime and on the results of the GW-SW exchange for both the direct measurements and the estimation using heat as a tracer

2. Materials and Methods

2.1. Study Area

Ringkøbing Fjord is a coastal lagoon in Western Jutland, Denmark, bordering the North Sea (latitude 55°58′40″ N, longitude 8°14′21″ E), covering an area of 300 km^2 (Figure 1). The mean water depth is 1.9 m and although the deepest points are up to 5 m, more than 25% of the lagoon has water depths of less than 0.5 m. The shallow depths also dominate the study site from the eastern shore line to the elongated island offshore (Figure 1b). The lagoon contains brackish water (5–15‰, salinity of 6–15 g L^{-1}) due to the connection to the sea to the west (controlled by a sluice) and the seasonal variation of the fresh water discharge mainly from the Skjern River (annual average of 50 m^3 s^{-1}). The tides are not an issue [18], although the operation of the sluice and storms can generate small flooding of the shore line. The annual mean precipitation and evaporation are 1050 and 630 mm, respectively [22]. More details of the hydrology can be found in Haider et al. [23], Duque et al. [7,20], and Müller et al. [24].

The study area is one of several human-made inlets on the eastern side of the lagoon (Figure 1c). The geology is predominantly Pleistocene fluvio-glacial sand, including relatively thin layers of silt or lower permeable materials, as well as paleo-channels with higher hydraulic conductivity. At the eastern border, close to the shore, the dense coastal vegetation covers the sandy sediments, creating a layer of organic material that can extend into the fjord for tens of meters [20]. The lagoon bed is composed of homogeneous sand with small wind- and wave-controlled ripples. The orography both in the lagoon and the surrounding areas is mostly flat, with small variations (in the order of 10–20 cm) of the sediment bed topography offshore. The boreholes indicate the existence of an unconfined aquifer of at least 12 m thickness. The hydraulic gradient, obtained from the water level measurements in the boreholes on land and lagoon stage, is 0.005–0.010 (east to west direction) [24].

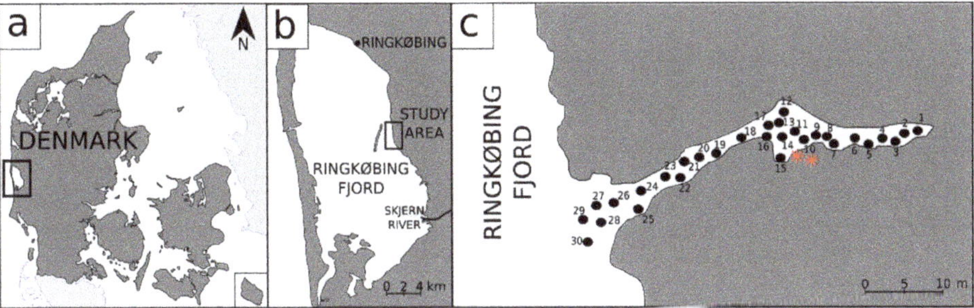

Figure 1. (**a**) The location of Ringkøbing Fjord, (**b**) the study area, the water depth in the lagoon is ~30–50 cm from the shore line and 2 km offshore to the elongated island, (**c**) The lagoon inlet. The black numbered dots indicate the location of the measuring points. The red stars indicate the location of the deep wells used to obtain the average annual groundwater temperature.

2.2. Fluxes from Seepage Meters

Thirty Lee-type seepage meters [8] were installed on 4 July, 2017 along a 43 m long transect, covering most of the inlet to the lagoon and the first few meters offshore (Figure 1c). Each seepage meter covered an area of 0.25–0.29 m^2, and were installed by inserting them around 20 cm into the lagoon bed. The space between the lagoon bed and the top part of the seepage meter was between one and two centimeters to assure that the flow of water to or from the surface was possible. A valve located on one side of the seepage meter allowed connecting a plastic bag to measure the exchange fluxes with the lagoon. The seepage meters were left for several days without attached bags before sampling to allow the brackish lagoon water inside the seepage meters to be displaced by discharging water. The seepage meter experiments were performed six times over a period of two consecutive days, 9–10 July, 2017. These 180 seepage measurements had an average duration of 140 min, with the shortest and the longest experiments of 118 and 175 min, respectively. The plastic bags were pre-filled with 1 L of tap water and weighed before being attached to the seepage meters. After the experiment, the plastic bags were weighed again. The electrical conductivity (EC) of the water in the bags attached to the seepage meters was measured before and after the experiments. The EC was used to approximate the total dissolved solids (TDS) and the density (ρ) [25] of the water in the bag in order to convert from mass to volume. The exchange flux was then calculated based on the difference in the volume divided by the experimental duration. The exchange fluxes are here reported as a Darcy flux, q = Q/A, where Q is the measured volumetric flux (as described above) and A is the area of the seepage meter.

The EC of the tap (initial water in the bags) and the lagoon water (end members) were used to discern between the fresh groundwater component of the exchange flux and the component corresponding to the brackish lagoon water recirculation in the exchange fluxes. This was based on a mass balance:

$$V_o EC_o + V_{ex} EC_{ex} = V_f EC_f \tag{1}$$

$$V_{ex} = V_f - V_o \tag{2}$$

where V_o and EC_o are the initial water volume (L) with its corresponding EC (mS cm^{-1}) in the plastic bag prior to the experiment, V_{ex} and EC_{ex} are the volume and the EC of water exchanged during the duration of the experiment, and V_f and EC_f are the volume and the EC of water in the bag after running the experiment. The only unknown is EC_{ex}:

$$EC_{ex} = \frac{V_f EC_f - V_o EC_o}{V_{ex}} \tag{3}$$

Then, the mass balance of the exchanged water was calculated:

$$V_{ex} EC_{ex} = V_{gw} EC_{gw} + V_{lw} EC_{lw} \tag{4}$$

$$V_{lw} = V_{ex} - V_{gw} \tag{5}$$

$$V_{gw} = \frac{V_{ex}(EC_{ex} - EC_{lw})}{EC_{gw} - EC_{lw}} \tag{6}$$

where V_{gw} and EC_{gw} are the volume and the EC of the groundwater exchanged during the experiment and V_{lw} and EC_{lw} are the volume and the EC of recirculated lagoon water exchanged during the experiment. The relation between V_{gw} and V_{ex} and between V_{lw} and V_{ex} yield the groundwater component of the exchanged volume of water (%GW) and the recirculated brackish lagoon water component of the exchanged volume of water (%LW), respectively:

$$\%GW = \frac{V_{gw}}{V_{ex}} \times 100 \tag{7}$$

$$\%LW = \frac{V_{lw}}{V_{ex}} \times 100 \tag{8}$$

Or, in terms of EC:

$$\%GW = \left(1 - \frac{EC_{lw} - EC_{ex}}{EC_{lw}}\right) \times 100 \tag{9}$$

$$\%LW = \frac{EC_{lw} - EC_{ex}}{EC_{lw}} \times 100 \tag{10}$$

2.3. Fluxes from Temperature Profiling

The vertical temperature distribution in the lagoon bed next to each seepage meter was measured by inserting a probe 0.5 m vertically into the lagoon bed. The probe had 10 temperature sensors (thermocouples with an accuracy of ±0.2 °C) located at different depths (0, 2.5, 5, 7.5, 10, 15, 20, 30, 40 and 50 cm from the sediment-water interface). The temperature was recorded after an equilibration period of 10 min based on previous field experiments [20]. Two analytical solutions were fitted to the vertical sediment temperature profiles. This process was performed manually to account for possible measurement errors or inconsistencies. The fitting of the analytical solutions to the temperatures obtained from the deepest temperature sensors was prioritized as they are located further from the lagoon water. The effects of the diurnal temperature variations near the lagoon bed were thus minimal.

Bredehoeft and Papadopulos [26] proposed a steady-state analytical solution (BP) for estimating vertical fluxes from the temperature-depth profiles:

$$T(z) = T_s + \left(T_g - T_s\right) \frac{\exp\left[N_{pe}\left(\frac{z}{L}\right)\right] - 1}{\exp\left[N_{pe}\right] - 1} \tag{11}$$

where $T(z)$ is the (steady-state) temperature at a given depth z(m) in degrees Celsius, T_S is the (steady-state) lagoon water temperature in degrees Celsius, T_g is the temperature of the groundwater at a given depth L(m) in degrees Celsius, and N_{pe} is the Peclet number representing the ratio of convection to conduction:

$$N_{pe} = \frac{q_z \rho_f c_f L}{K_e} \tag{12}$$

where q_z is the vertical fluid flux or exchange flux (m s^{-1}), ρ_f is the density of the fluid (kg m^{-3}), c_f is the specific heat capacity of the fluid (J kg^{-1} °C^{-1}) and K_e is the bulk thermal conductivity (W m^{-1} °C^{-1}).

The BP was applied considering two different scenarios: (1) Homogeneous sediment composition with respect to K_e in the study area (Figure 2a); (2) heterogeneous K_e (Figure 2b) due to the different organic matter fractions in the lagoon bed, based on the observation of decreasing organic matter in the top sediment offshore. To reproduce these two scenarios, two different parameterizations were used. The values of the parameters considered for the homogeneous scenario were chosen assuming a sandy sediment (Table 1).

Table 1. The parameters used for the application of a steady-state analytical solution (BP).

Parameter	Value	Unit	Symbol
Density of the fluid (assuming discharging fresh water dominated)	1000.8	kg m^{-3}	ρ
Specific heat capacity of the fluid	4192	J kg^{-1} °C^{-1}	c_f
Distance to stable groundwater temperature	5	m	L
Effective thermal conductivity (homogeneous scenario)	1.84	W m^{-1} °C^{-1}	K_e
Effective thermal conductivity (heterogeneous scenario):			
Zone 1 (points 1 to 11)	0.72	W m^{-1} °C^{-1}	
Zone 2 (points 12 to 17)	1.46	W m^{-1} °C^{-1}	K_e
Zone 3 (points 18 to 30)	1.82	W m^{-1} °C^{-1}	
Groundwater temperature (annual mean)	9.8	°C	T_g
Lagoon temperature	Top sensor	°C	T_s

The parameters used for the heterogeneous scenario were based on the work of Duque et al. [20] and are also listed in Table 1. Duque et al. [20] observed that the thermal conductivity in the top part of the lagoon bed was not homogeneous. It is distinguished between the different types of sediments along the fjord inlets related to the amount of vegetation present. Close to the shore, the vegetation was abundant and the thermal conductivity of the top part of the lagoon bed was low (0.72 W m^{-1} °C^{-1}). Further offshore, the amount of vegetation decreased until eventually disappearing and the thermal conductivity of the top part of the lagoon bed increased until reaching a value typical for sandy sediments (1.82 W m^{-1} °C^{-1}). The lagoon water temperature (T_s) was obtained from the top temperature sensor in the vertical probe, located at the lagoon water-sediment interface. Unlike the remaining parameters, the groundwater temperature (T_g) was based on the mean annual groundwater temperature collected in previous studies from the different wells within the study area (Figure 3). This value was chosen as it represents the overall thermal behavior of deep groundwater in the study area. As the BP solution assumes a steady state, it is necessary to choose a groundwater temperature with minor variability during the measurement period, a condition that could not be met if using groundwater temperature at shallow depths. Furthermore, as temporal variation of exchange fluxes in the study area are small [7], the use of the mean annual groundwater temperature provides a robust lower boundary condition for this method, representing the average thermal characteristics of the system. The density of the fluid was assumed to be dominated by discharging fresh water as the area is close to the shore and the measured EC values were relatively small.

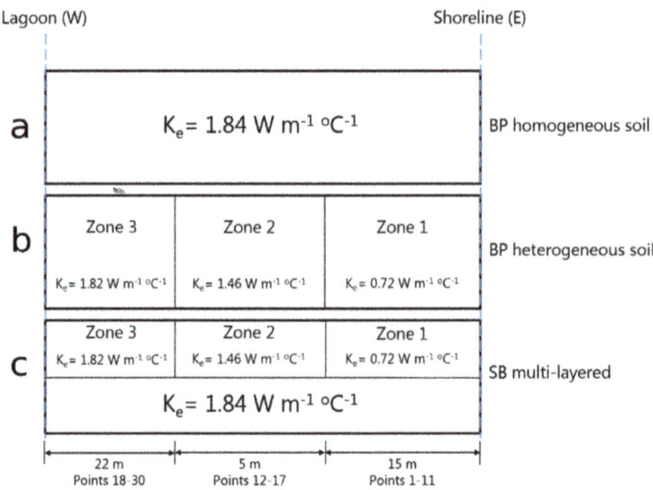

Figure 2. A schematic cross section of the study area with different zonation and parametrization for the lagoon bed temperature analysis. (**a**) BP homogeneous soils scenario with a single thermal conductivity, (**b**) BP heterogeneous soil scenario. Three zones are defined, each of them with a different thermal conductivity, (**c**) SB multi-layered. See text for the definition of the two BP and SB analytical solutions. Three different zones and two layers per zone are defined. Each zone has a different thermal conductivity for the upper layer. All zones have the same thermal conductivity for the lower layer.

The vertical heterogeneity in thermal conductivity (layering) was likely to be present on site [27] and its effect on the estimation of fluxes were also assessed. Previous thermal conductivity measurements mapped the top 10–15 cm of the lagoon bed [20]. It is the hypothesis that the thermal conductivity at greater depths is more homogeneous as sediments are less affected by organic matter depositions which originated from the root activity and plant debris, and the consolidation increases with depth, decreasing porosity [27]. Accounting for the layering of the thermal conductivity, the analytical solution by Shan and Bodvarsson [28] was applied.

Shan and Bodvarsson [28] modified the Bredehoeft and Papadopulos [26] method to allow for the introduction of different thermal conductivity values representing a multi-layer system (SB):

$$T_i(z) = C_{i,1}e^{q_z z/\alpha_i} + C_{i,2} \ (i = 1, 2, \ldots, n) \tag{13}$$

where $T_i(z)$ is the temperature at a given depth z in layer i, $C_{i,1}$ and $C_{i,2}$ are two integral constants, q_z is the vertical fluid flux across all the layers and α_i is the thermal diffusivity of the ith-layer, defined as the ratio between the effective thermal conductivity (K_e) and the heat capacity ($\rho c = 4.18 \times 10^6$ J/m^3/°K). It is important to note that, as described in Kurylyk et al. [29], Shan and Bodvarsson [28] incorrectly defined thermal diffusivity. Here, α_i is obtained from the volumetric heat capacity of the water (ρc) and should not be confused by the volumetric heat capacity of the medium. By setting the origin at the surface of the top layer, the depth z increases downward and the base of each layer at a depth d_i, the thickness of each layer is the difference of its two boundary coordinates $b_i = d_i - d_{i-1}$, where $d_0 = 0$.

Assuming constant temperatures at the top and the bottom of the system (T_s and T_g respectively), the boundary conditions are:

$$T_1(0) = T_s; \; T_n(d_n) = T_g \tag{14}$$

$$T_i(d_i) = T_{i+1}(d_i) \; (i = 1, 2, \ldots, n-1) \tag{15}$$

$$\alpha_i \left(\frac{dT_i}{dz}\right)_{z=d_i} = \alpha_{i+1} \left(\frac{dT_{i+1}}{dz}\right)_{z=d_i} \; (i = 1, 2, \ldots, n-1) \tag{16}$$

Equation (14) are the top and the bottom temperature boundary conditions, (15) is the temperature continuity condition at the layer interfaces, and (16) is the thermal flux continuity condition.

The two integral constants are:

$$C_{i,2} = \frac{aT_s - T_g}{a - 1} \; (i = 1, 2, \ldots, n) \tag{17}$$

$$C_{1,1} = \frac{T_g - T_s}{a - 1} \tag{18}$$

$$C_{(i+1),1} = e^{q_z d_i \left(\frac{1}{\alpha_i} - \frac{1}{\alpha_{i+1}}\right)} C_{i,1} \; (i = 1, 2, \ldots, n-1) \tag{19}$$

where the parameter a is defined as:

$$a = e^{q_z d_n / \alpha_{eff}} \tag{20}$$

where d_n is the total thickness of the n layers and α_{eff} is the effective thermal diffusivity of the n layers:

$$\alpha_{eff} = d_n / \sum_{i=1}^{n} \left(\frac{b_i}{\alpha_i}\right) \tag{21}$$

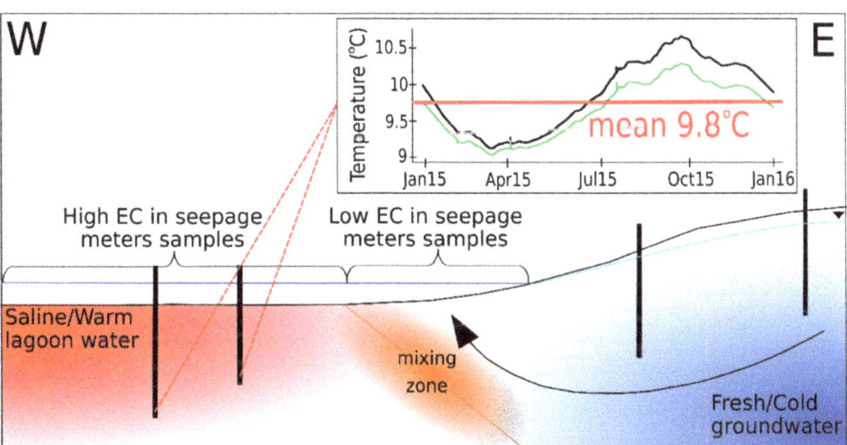

Figure 3. A schematic (not to scale) cross-section of the study area. In summer, fresh cold groundwater discharges into the lagoon from the east, creating a mixing zone when it encounters the saline slightly warmer lagoon water (a saline wedge is generated due to the density difference). The EC measurements east of the mixing zone showed lower EC values than the measurements west of the mixing zone. The mean annual temperature (9.8 °C) measured in two deep wells (4 m depth, black vertical lines, see Figure 1c) was used as a lower boundary condition for the BP method.

Following Duque et al. [20], three different zones were defined for the application of SB, each representing a 2-layer system with a lower thermal conductivity top layer and a higher thermal conductivity bottom layer (Figure 2c). The first zone, with a top layer thermal conductivity of 0.72 W m^{-1} °C^{-1}, corresponds to the initial 15 m from the shore (points 1 to 11, Figure 1), the second zone, with a top layer thermal conductivity of 1.46 W m^{-1} °C^{-1}, corresponds to the next 5 m (points 12 to 17) and the third zone, with a top layer thermal conductivity of 1.82 W m^{-1} °C^{-1}, corresponds to the rest of the points located more offshore (points 18 to 30). The thermal conductivity value chosen for the bottom layer was 1.84 W m^{-1} °C^{-1}, representing sand. The last zone therefore has a nearly uniform thermal conductivity. The temperatures measured by the top and bottom sensors of the vertical probe were used to fix T_s and T_g respectively. The location of the interface between the two layers (the base of the top layer) was based on the best possible fit of the SB solution.

3. Results

3.1. Seepage Meter Fluxes

The differences between estimated fluxes using the density of each sample compared to using a constant density (1000 kg m^{-3}) were negligible (lower than ± 0.1 cm day^{-1} for 156 of the 180 samples). Thus, only the results obtained using the same density for all the samples are shown and discussed in the following. The fluxes measured during the second day were generally higher (25 of the 30 points for rounds 4 to 6). A trend in the variability of the measured fluxes at each point indicated an increase in the standard deviation when the distance to the shore increased (Figure 4). The maximum and minimum standard deviations (2.65 cm day^{-1} and 0.11 cm day^{-1}) correspond to locations 24 and 22 respectively, with a separation of only 5 m. The arithmetic mean seepage fluxes (red diamonds, Figure 4) are used to study the spatial variability of the exchange fluxes, Figure 5a.

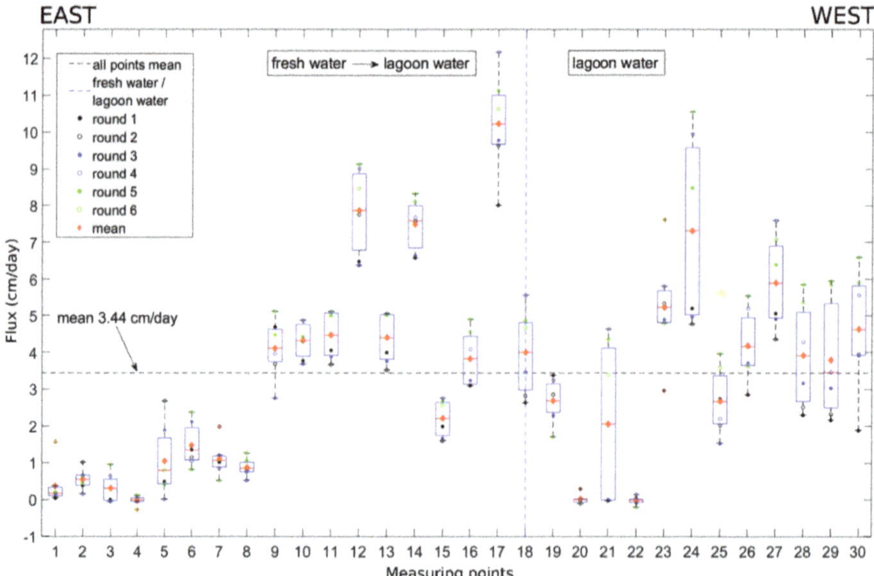

Figure 4. The exchange fluxes corresponding to the seepage meter experiments for each round of measurements at each measuring point. The red diamond corresponds to the mean value. The black dashed line shows the mean flux for all locations and all rounds. The blue dashed line points to the location from which no fresh water was found in the water samples to the left of this line (see Section 3.4).

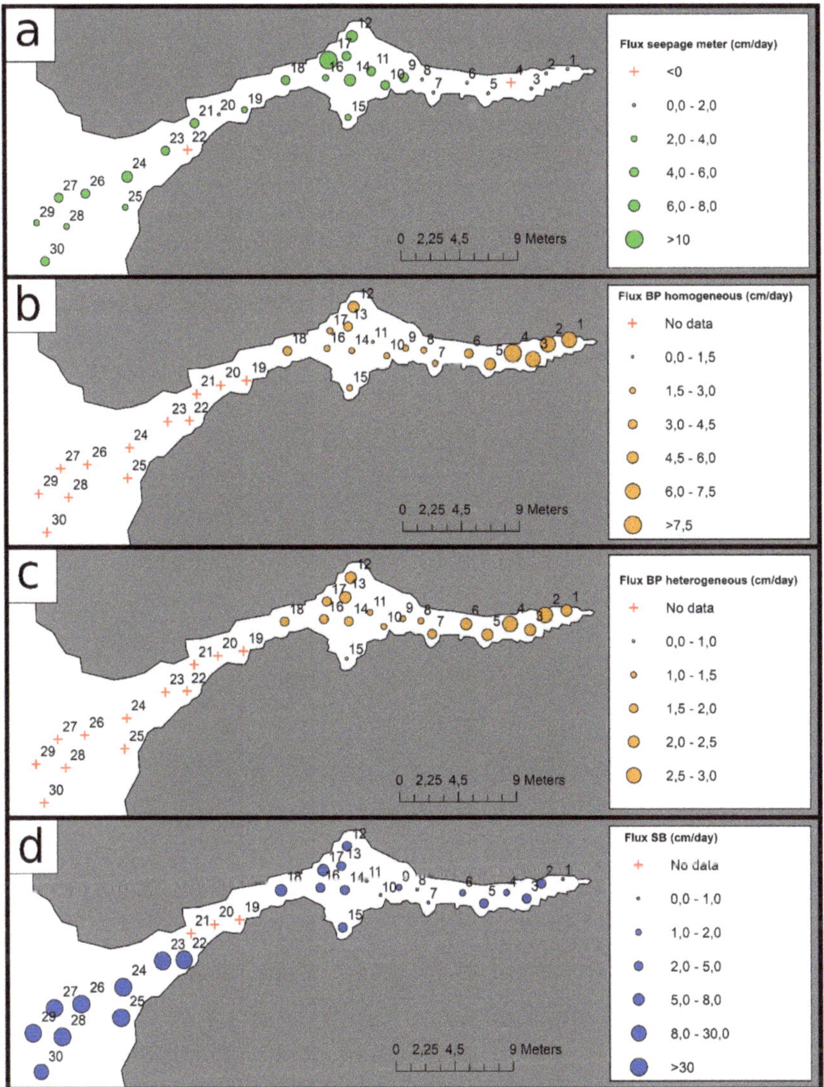

Figure 5. The spatial distribution of the measured and calculated exchange fluxes: (**a**) The mean from seepage meters, (**b**) calculated with the Bredehoeft and Papadopulos (1965) method for the homogeneous scenario, (**c**) calculated with the Bredehoeft and Papadopulos (1965) method for the heterogeneous scenario and (**d**) calculated with the Shan and Bodvarsson (2004) method.

The exchange fluxes ranged from -0.02 to 10.22 cm day^{-1} with an overall mean of 3.44 cm day^{-1} (Figure 4) and a standard deviation of 2.63 cm day^{-1}. Further, 28 of the 30 points showed a positive groundwater–lagoon water exchange flux, indicating a predominant process of groundwater discharge to the lagoon. The two negative exchange fluxes, indicating groundwater recharge, were very small (-0.02 cm day^{-1}) compared with the rest of the measured fluxes. There is not an obvious pattern in the distribution of the discharge fluxes (Figure 5a), but two different higher discharge zones are distinguished. The first is located 18 m offshore (points 12 to 14 and 17) and the second is located from 30 to 43 m offshore (points 21, 23, 24 and 26 to 30).

3.2. Fluxes Estimated with BP Solution to Heat Transport Equation

The calculation of fluxes using the BP solution was merely possible for 18 of the 30 points (points 1 to 18). In some locations (points 19 and 21), the temperature of the lagoon bed showed a small decrease for the first 15 cm and then dropped more than 4°C over 5 cm (from the sensor located at 15 cm depth to the sensor located at 20 cm depth) (Figure 6). In other locations (points 20 and 22 to 30), the temperature stabilized at 16–17 °C at 35-40 cm depth (Figure 6), far from the 9.8 °C expected for groundwater in the study area (Figure 5b,c). In all these locations (19–30), it was not possible to satisfactorily fit the BP to the data. It was noted that the vertical temperature data not fitted by the BP was located farthest offshore.

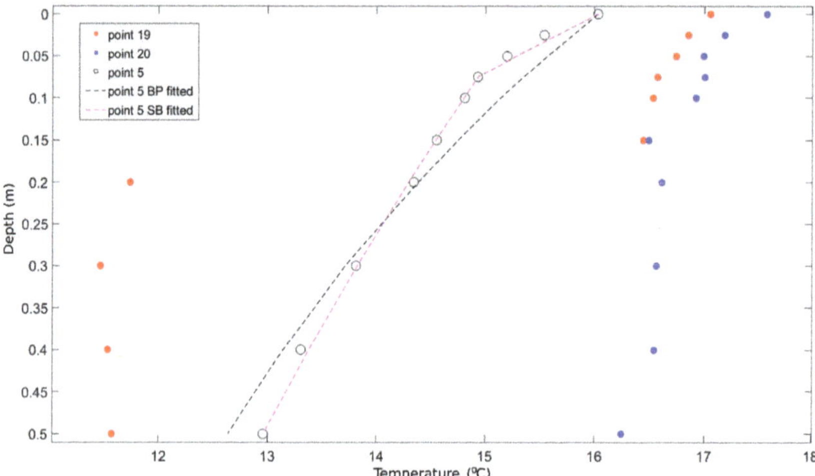

Figure 6. The temperature-depth profiles for points 5, 19 and 20. An example of reasonably fitted data is point 5 (empty black dots), fitted with both BP (black dashed line) and SB (magenta dashed line). An anomalous change in the temperature can be observed for point 19 (red dots) from depth 0.15 m to depth 0.2 m. The temperature at point 20 (blue dots) remained above 16 °C along the entire 0.5 m profile.

For the scenario of homogeneous sediment, the exchange fluxes ranged from 0.86 to 6.05 cm day^{-1} with a mean of 3 cm day^{-1} and a standard deviation of 1.56 cm day^{-1}. A general trend was observed of higher exchange fluxes close to the shore and decreased with the distance offshore (Figure 5b). Thus, these data indicate that most of the groundwater discharge took place within the first meters of the lagoon.

For the scenario of heterogeneous sediment, the exchange fluxes ranged from 0.69 to 2.25 cm day^{-1} with a mean of 1.44 cm day^{-1} and a standard deviation of 0.47 cm day^{-1}. In this case, the higher discharge values were also located close to the shore, but a second zone with high values can be observed at points 12 to 17 (Figure 5c).

3.3. Fluxes Estimated with SB Solution to Heat Transport Equation

The calculation of the fluxes using the SB solution was possible in 27 of the 30 points (19–21 showed anomalous temperatures, see explanation in Section 3.2). The main difference was that the multi-layer approach allowed the authors to fit the data even if the temperature of the lowest sensor for each measurement was far from the expected groundwater temperature.

The exchange fluxes ranged from 0.09 to 34.6 cm day^{-1} with a mean of 12.9 cm day^{-1} and a standard deviation of 14.8 cm day^{-1}. The high standard deviation indicates that there were extreme values. The fluxes above 8 cm day^{-1} were obtained only for offshore points (22 to 30). At points closer to the shore (1 to 18), the maximum calculated flux was 7.78 cm day^{-1} with a mean of 2.77 cm day^{-1}. There were no obvious patterns in the distribution of the fluxes (Figure 5d), but two different high-discharge zones were distinguished: The first zone located 18 m offshore (points 12 to 18) and the second zone located from 30 to 43 m offshore (points 22 to 30).

3.4. Fresh Groundwater and Lagoon Water Proportion in Exchange Fluxes

The proportion of fresh groundwater in the seepage meter samples showed that after point 18, there was little to no fresh water in the exchange flux (Figure 7). Further, the highest percentage of groundwater collected was approximately 78%, indicating that even the points located close to the shore were influenced by brackish lagoon water.

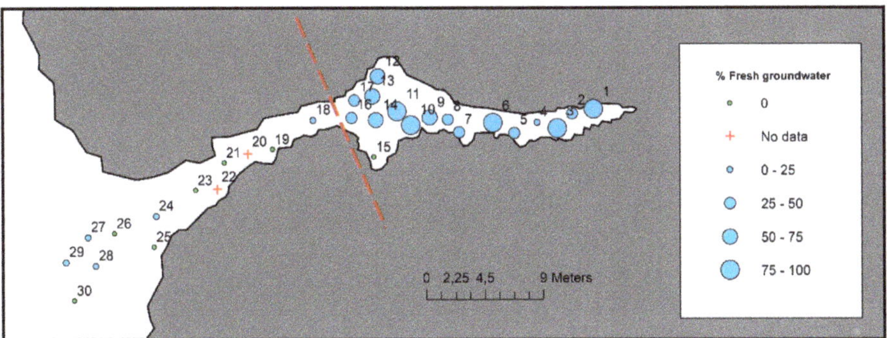

Figure 7. The mean percentage of fresh groundwater calculated from the seepage meters experiment. The red dashed line indicates the approximate location of the brackish lagoon water–fresh groundwater interface.

3.5. Method Comparison

The comparison between the two analytical solutions and seepage meter results in points 1 to 18 are analyzed due to lack of data for all methods at the remaining measuring points. This showed that fluxes obtained with both BP and SB for points closer to shore (1 to 8) were generally much higher than those obtained with the seepage meters (Figure 8). In the remaining points, the fluxes obtained with the seepage meters were generally higher than those obtained with the temperature approach.

The mean difference between the fluxes calculated with BP and the fluxes measured with seepage meters was 3.55 cm day^{-1} with a standard deviation of 1.96 cm day^{-1} for the scenario with homogeneous soil and 2.66 cm day^{-1} with a standard deviation of 2.33 cm day^{-1} for the scenario with heterogeneous soil. The mean difference between the fluxes calculated with SB and the fluxes measured with seepage meters was 2.14 cm day^{-1} with a standard deviation of 1.56 cm day^{-1}.

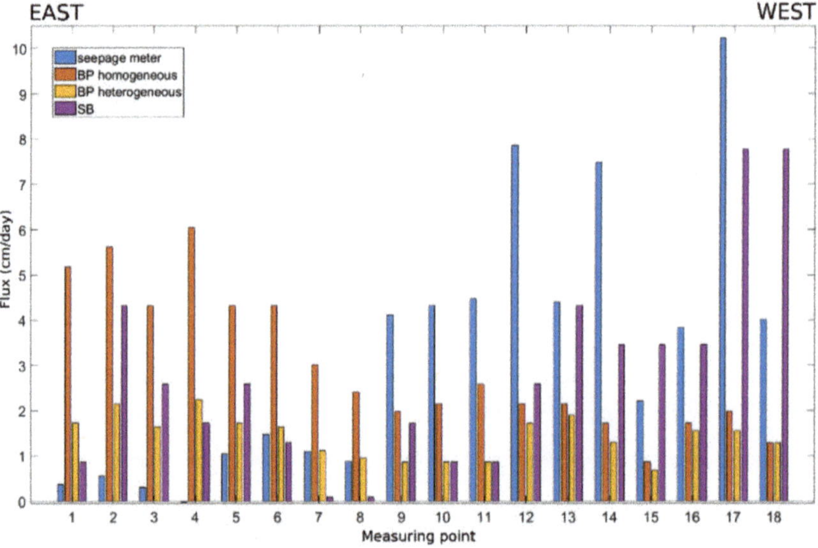

Figure 8. The mean of exchange fluxes measured with seepage meters and calculated using analytical solutions for temperature profiles.

4. Discussion

4.1. Comparison of Seepage Meter and Thermal Methods

The results show the existence of an exchange of water between the subsurface and the lagoon. The measurements with seepage meters indicated that this exchange was generally from the subsurface to the lagoon (groundwater discharge) with only two locations showing groundwater recharge from the lagoon. However, seepage at those locations were negligible as the measured fluxes (-0.02 cm day^{-1}) fell below the detection limit [11]. Only the discharge fluxes were inferred using the two analytical solutions for the heat transport in the lagoon bed. It is important to note that in some cases, it was not possible to reproduce the temperature-depth data with the analytical solutions.

The seepage meter results show a spatial distribution with two main zones of groundwater discharge. A similar behaviour was observed by Duque et al. [20] and corresponded to the position of the saline wedge showing a first peak in the discharge 18 m offshore and to a wave-induced pumping flux yielding a second peak, 30–43 m offshore [24]. A similar phenomenon was also observed by Shinn et al. [30]. The first main discharge zone matched the end of significant fresh groundwater component in the exchanged water from the seepage meters experiment (around point 18).

The BP method was able to reproduce the temperature profiles for the first 18 points. Those 18 locations corresponded to the area in which a portion of fresh groundwater was measured in the seepage meter samples. The fluxes at these locations decreased as the distance to the shore increased (Figure 5c). This agrees with the commonly assumed decreasing trend of groundwater discharge with the distance offshore in the surface water bodies such as lakes [31]. This indicates that the BP method is not applicable if the origin of the discharged groundwater is lagoon water.

The SB method was able to reproduce most of the temperature profiles and the calculated fluxes showed a pattern similar to the ones obtained from the seepage meters (Figure 5d). In this case, the method was applicable also when there was no fresh water in the exchanged water. Despite this, the fluxes calculated for the offshore points (22 to 30) were very high in comparison with the results obtained by other methods used at other points, and also in comparison with previous studies in the study area [7,20]. This may indicate that, although the method can be applied and the temperature-depth profiles are correctly fitted with the SB analytical solution, the results obtained are not realistic and should be considered with care.

The analytical solutions to reproduce the temperature-depth data suggested a higher discharging flux than the seepage meters for the first 6–8 measuring points. For the remaining points, the fluxes measured with the seepage meters were higher. This may be caused by the proximity of the saline wedge around points 10–12. On the one hand, in the fresher part of the saline wedge, the main component of the exchanged flux is fresh groundwater and it can be well captured by the thermal methods. On the other hand, closer to the saline wedge the proportion of the recirculated lagoon water increased (Figure 3), altering the thermal signal in the groundwater. Thus, the reliability of the thermal methods decreases, because it is difficult to precisely define an appropriate lower boundary condition. A common temperature of 9.8 °C was used at all points at a depth of 5 m, but this may not necessarily be representative. Although the groundwater discharge fluxes obtained from the different methods were in the same order of magnitude, it is clear that both the location and the range of fluxes differ considerably depending on the methodology. This has been reported in other studies (e.g., [32]) and highlights the importance of complementing methods for obtaining groundwater exchange fluxes with other hydrological measurements.

When analysing the proportion of fresh groundwater and lagoon water in the seepage meter samples, it was observed that many samples did not have any fresh groundwater. This indicates that the water flowing to the lagoon at these points is recirculated brackish lagoon water (Figure 3).

4.2. Direct Measurements (Seepage Meters): Pros and Cons

The use of seepage meters for obtaining the groundwater discharge fluxes to Ringkøbing fjord is the most straightforward methodology used in this study. The fact that the flux is obtained by a simple mass balance makes it a robust and reliable method. Furthermore, as the exchanged water is stored in the collecting plastic bags, a geochemical analysis can be performed, complementing the dataset with insights regarding the origin of the discharged groundwater. Care must be taken, however, when interpreting the water quality of seepage water as the seepage meter can by itself change the oxidation state of the lagoon bed [33]. Performing several rounds of measurements, as described in this work, allows for a better estimation of the robustness of the method. The measured fluxes had an average standard deviation of less than 1 cm day^{-1} for all six rounds, which indicates that the method is reliable.

However, this method has some noticeable drawbacks that may make it not applicable in some situations. An extended description of such situations can be found in [34]. In this particular case, the most important ones were: First, the installation of the seepage meters has to be performed in advance before the experiment takes place [8]. In the case of geochemical sampling of the exchanged groundwater, the amount of time needed for the seepage meters to replace the trapped surface water with discharged groundwater (if any) is inversely proportional to the exchange flux, which is unknown before the experiment takes place. This implies that the discharge flux needs to be assessed prior to the experiment based on previous studies or by the use of other methods (for example, the thermal method), thus increasing the uncertainty of the results. In this case, the equipment was installed approximately one week before the data collection. Once the measurements were performed and the exchange fluxes known, this study could confirm that the time from the installation until the experiment was long enough, but it could not have been the case. Second, the amount of human resources needed for an intense sampling campaign, as the one described here, is extensive [35]. Decreasing the number of measuring points is not always a good option, as it has been observed that to obtain reliable results, many seepage meters are needed [36,37]. Third, the methodology has some associated uncertainty. Shinn et al. [30] observed that the volumetric data obtained from seepage meters in areas exposed to currents, waves and swells, like the studied area and almost any costal area, must be used with caution as the induced flux from wave-induced pumping affects the measurements. Further, Lee [8] highlighted the potential issues in the measurements due to the improper placement of the cylinder.

The seepage meter installation is a manual activity and it may not be possible to ensure that all requirements for reliable data acquisition are met. Finally, the use of correction coefficients to better estimate the exchange flux has been studied for many years (e.g., [8,38]) and it may be needed. This study did not apply any of the correction coefficients from the literature. This parameter seems to be rather equipment dependent and for the comparison of methods carried out in this work, the authors considered it appropriate to use the original value obtained from the mass balance.

4.3. Analytical Solutions for Temperature-Depth Profiles (BP and SB): Pros and Cons

The temperature-depth data collection was performed by one person during one of the field campaign days. The two temperature probes and two logger boxes were used to obtain the streambed temperature profiles. Each measurement takes ten minutes which implies that in approximately three hours, all measurements were done. Both the temperature probes and the logger boxes can be brought to the field in a small van and can be carried by a single person. This is a noticeably smaller amount of time and human resources needed in comparison with the seepage meter measurements. Therefore, the field campaigns are cheaper and more measurements can be done in the same amount of time.

The use of heat as a tracer, and specifically analytical solutions to study temperature profiles, has several limitations that need to be taken into account.

Both the BP and SB rely on the assumption of vertical and steady water and heat flow [26,28]. These conditions are rarely found outside laboratory experiments and are unknown a priori, unless there is some knowledge of the specific site. In this case, a location near the saline wedge is expected to have exchange fluxes (if any) with a strong vertical component due to the density driven flow. However, the locations influenced by a recirculating flux may be biased by a horizontal or transverse flux, which may be responsible for a weaker performance of the analytical methods within our study. Furthermore, the data collection took place in mid-July assuring that the temperature fluctuation of the surface is as small as possible (higher fluctuations with changing weather conditions are expected in spring and autumn). Thus, the assumption of steady-state is more likely to be appropriate.

The BP assumes that the media in which the temperature is measured is homogeneous, but this is the exception rather than the rule in natural systems and previous studies indicated that this was not the case in this study site. Furthermore, several parameters need to be known a priori: The average steady groundwater temperature; the depth at which this temperature is reached; the thermal conductivity of the medium. Here, several temperature-depth profiles were measured in boreholes around the study area, thus allowing the assessment of the groundwater temperature regime (9.8 °C at 5 m). The subsurface is on average 1.8 °C warmer in the saline sector compared with the fresh groundwater, similarly to what was observed by Befus et al. [18].

The SB allows for the definition of a multi-layered system. This study differentiated three different zones in this study area, providing the basis for studying the effect of the horizontal and vertical heterogeneity in the effective bulk thermal conductivity. The bulk thermal conductivity values [20] were used as a framework to consider a more complex hydrogeological system. Fitting the data with the SB solution was considerably easier than with the BP due to the flexibility on the lower boundary condition (lower sensor temperature versus steady groundwater temperature). This decreases the amount of uncertainty derived from the assumption of a steady groundwater temperature at unknown depth. At the same time, it also forces the solution to satisfy this condition, which does not allow taking into account the bigger scale thermal behaviour of the system (i.e., the thermal regime at greater depths). With this solution, the estimated fluxes deviated substantially from the results of previous studies [7,24]. Fixing the lower boundary condition to the temperature of the sensor located at a depth of 50 cm (instead of using a deep groundwater steady temperature as in the BP method) focuses the calculation of the flux to only the upper part of the lagoon bed, disregarding the rest of the subsurface. This may lead to under/overestimations of the flux if the processes occurring in that upper section are not representative of the overall behaviour of the system. Also, the fact that these fluxes were obtained for the more offshore points indicates that the thermal methods usually applied to study GW-SW interactions in fresh water environments seem less reliable, when there is a strong component of saline water recirculation in the flux exchange. Therefore, obtaining unrealistic results in some locations and the use of more parameters that may be unknown (number of layers and thickness, thermal conductivity for each layer) are important limitations to take into account before using this method.

Some of the temperature profiles were very uniform (temperature not decreasing much with depth). This could actually indicate that there was a dominant vertical downward flow in the top 50 cm (contrary to other measurements) or that the flow was not strictly vertical upwards having a significant horizontal or transversal component. The complex flow field near a salt water wedge may be the explanation, so the use of 1D solutions is then not warranted or, at least, it is only the vertical component that is estimated in many cases. The comparison of these data and the seepage meter results helps to distinguish in which cases this may occur. That is, the differences between the fluxes estimated by the thermal and seepage meter methods could be explained by the fact that the seepage meter, located right at the lagoon bed, is a composite measure of the discharging flux, whereas, it is only the average vertical flux component that is measured over 50 cm by the thermal method. However, the fact that this phenomenon (nearly uniform temperature profiles) is observed at the more offshore points seems to confirm that the applicability of the thermal methods is not well suited for these environments.

5. Conclusions

The use of different methodologies to study the groundwater discharge to a lagoon with saltwater intrusion and distinct thermal regimes reveal that there are several challenges to overcome in order to obtain reliable estimates of submarine groundwater discharge. It points out the difficulties in choosing which tools to apply in the field, but it also gives insights to what method to use depending on which information of the study site is available beforehand.

The implementation of the two analytical solutions was only satisfactory for 18 points located closer to the shore and with a significant fresh groundwater component in the exchange flux. This has two implications: (1) Though the seepage meters allowed the measurement of the amount of exchanged water through the lagoon bed, it is not possible to infer the origin of this water unless other measurements are performed (e.g., EC); (2) it indicates that the thermal methods may not be well suited to study groundwater–surface water interactions if surface water recirculation processes are of importance due to the subsurface thermal regime (essentially providing unknown temperature boundary conditions).

Characterizing the saltwater-freshwater interaction in coastal areas is a key issue in the application of temperature measurements to estimate fluxes. The establishment of the areas dominated by saltwater or freshwater can be indicators of the limits of the applicability for the usual assumption of stable groundwater temperature. Also, the detection of the saltwater-freshwater interface is highly relevant due to the dominance of vertical-horizontal fluxes in the study area.

Choosing between the Bredehoeft and Papadopulos (BP) [26] or the Shan and Bodvarson (SB) [28] analytical solution to evaluate the subsurface temperature profiles depends mainly on the knowledge of the site. On the one hand, if groundwater is monitored in the surrounding area and thermal data is available, SB can be applied, allowing for a better constraint of the lower boundary condition. Also, if stratification of the subsurface is known or expected and it is possible to measure or infer the different thermal parameters for each layer, SB provides a good solution to better fit the thermal profiles. BP is better suited for environments with few or no prior data acquisition, as only the average groundwater temperature in the area and one value for the thermal conductivity are needed. SB reproduces more complex systems that are frequent in nature and BP only homogeneous scenarios. In both cases, it is necessary to address the implications that surface water recirculation processes have on the calculated exchange fluxes.

Nevertheless, the seepage meter and thermal methods have their own strengths and based on our study, it can be concluded that the thermal methods represent a fast way of getting the right order of magnitude of the exchange fluxes. The results (data/model) can be generated within one day and can be used to design an efficient seepage meter campaign (where and how much), which would likely give the most representative estimate of the exchange fluxes. Additionally, seepage meter measurements may allow observing the origin of the flux when taking samples for water chemistry.

Author Contributions: Conceptualization, J.T.-C., C.D and P.E.; methodology, J.T.-C., C.D. and P.E.; validation, J.T.-C., C.D. and P.E.; formal analysis, J.T.-C. and C.D.; investigation, J.T.-C., C.D., S.K. and P.E.; resources, P.E., C.D. and S.M.; writing—original draft preparation, J.T.-C.; writing—review and editing, J.T.-C., C.D., S.K., S.M. and P.E.; visualization, J.T.-C.; supervision, P.E. and C.D.; project administration, C.D. and P.E.; funding acquisition, P.E. and C.D.

Funding: This research has received funding from the European Union's Horizon 2020 Research and Innovation Programme under the Marie Sklodowska-Curie Grant Agreement No 722028 and the European Union Seventh Framework Programme FP7/2007–2013 under the Grant Agreement No 624496.

Acknowledgments: Comments from the editor Jörg Lewandoswki and from the two reviewers are gratefully acknowledged. We thank Iris Tobelaim and Christian Hansen for their help in the field.

Conflicts of Interest: The authors declare no conflicts of interest.

References

1. Winter, T.C.; Harvey, J.W.; Lehn Franke, O.; Alley, W. *Ground Water and Surface Water a Single Resource*; DIANE Publishing Inc.: Darby, PA, USA, 1998; Volume 1139.

2. Crawford, R.M.M. Eco-hydrology: Plants and Water in Terrestrial and Aquatic Environments. *J. Ecol.* **2000**, *88*, 1095–1096. [CrossRef]

3. Hayashi, M.; Rosenberry, D. Effects of Ground Water Exchange on the Hydrology and Ecology of Surface Water. *Ground Water* **2002**, *40*, 309–316. [CrossRef]

4. Brunner, P.; Therrien, R.; Renard, P.; Simmons, C.; Franssen, H.-J. Advances in understanding river—Groundwater interactions: River-groundwater interactions. *Rev. Geophys.* **2017**, *55*, 818–854. [CrossRef]

5. Rosenberry, D.O.; Lewandowski, J.; Meinikmann, K.; Nützmann, G. Groundwater—The disregarded component in lake water and nutrient budgets. Part 1: Effects of groundwater on hydrology. *Hydrol. Process.* **2015**, *29*, 2895–2921. [CrossRef]

6. Lewandowski, J.; Meinikmann, K.; Pöschke, F.; Nützmann, G.; Rosenberry, D.O. From submarine to lacustrine groundwater discharge. *Proc. Int. Assoc. Hydrol. Sci.* **2015**, *365*, 72–78. [CrossRef]

7. Duque, C.; Haider, K.; Sebok, E.; Sonnenborg, T.O.; Engesgaard, P. A conceptual model for groundwater discharge to a coastal brackish lagoon based on seepage measurements (Ringkøbing Fjord, Denmark). *Hydrol. Process.* **2018**, *32*, 3352–3364. [CrossRef]

8. Lee, D. A Device for Measuring Seepage-Flux in Lakes and Estuaries. *Limnol. Oceanogr.* **1977**, *22*, 140–147. [CrossRef]

9. Rosenberry, D.O. A seepage meter designed for use in flowing water. *J. Hydrol.* **2008**, *359*, 118–130. [CrossRef]

10. Burnett, W.C.; Taniguchi, M.; Oberdorfer, J. Measurement and significance of the direct discharge of groundwater into the coastal zone. *J. Sea Res.* **2001**, *46*, 109–116. [CrossRef]

11. Cable, J.; Burnett, W.C.; Chanton, J.P.; Corbett, D.; Cable, P.H. Field Evaluation of Seepage Meters in the Coastal Marine Environment. *Estuar. Coast. Shelf Sci.* **1997**, *45*, 367–375. [CrossRef]

12. Cable, J.; Martin, J.; Jaeger, J. Exonerating Bernoulli? On evaluating the physical and biological process affecting marine seepage meter measurements. *Limnol. Oceanogr. Methods* **2006**, *4*, 172–183. [CrossRef]

13. Rudnick, S.; Lewandowski, J.; Nutzmann, G. Investigating groundwater-lake interactions by hydraulic heads and a water balance. *Ground Water* **2015**, *53*, 227–237. [CrossRef]

14. Anibas, C.; Buis, K.; Verhoeven, R.; Meire, P.; Batelaan, O. A simple thermal mapping method for seasonal spatial patterns of groundwater–surface water interaction. *J. Hydrol.* **2011**, *397*, 93–104. [CrossRef]

15. Kazmierczak, J.; Müller, S.; Nilsson, B.; Postma, D.; Czekaj, J.; Sebok, E.; Jessen, S.; Karan, S.; Stenvig Jensen, C.; Edelvang, K.; et al. Groundwater flow and heterogeneous discharge into a seepage lake: Combined use of physical methods and hydrochemical tracers. *Water Resour. Res.* **2016**, *52*, 9109–9130. [CrossRef]

16. Sebok, E.; Duque, C.; Kazmierczak, J.; Engesgaard, P.; Nilsson, B.; Karan, S.; Frandsen, M. High-resolution distributed temperature sensing to detect seasonal groundwater discharge into Lake Væng, Denmark. *Water Resour. Res.* **2013**, *49*, 5355–5368. [CrossRef]

17. Burnett, W.C.; Aggarwal, P.K.; Aureli, A.; Bokuniewicz, H.; Cable, J.E.; Charette, M.A.; Kontar, E.; Krupa, S.; Kulkarni, K.M.; Loveless, A.; et al. Quantifying submarine groundwater discharge in the coastal zone via multiple methods. *Sci. Total Environ.* **2006**, *367*, 498–543. [CrossRef] [PubMed]

18. Befus, K.M.; Cardenas, M.B.; Erler, D.V.; Santos, I.R.; Eyre, B.D. Heat transport dynamics at a sandy intertidal zone. *Water Resour. Res.* **2013**, *49*, 3770–3786. [CrossRef]

19. Luce, C.; Tonina, D.; Gariglio, F.; Applebee, R. Solutions for the diurnally forced advection-diffusion equation to estimate bulk fluid velocity and diffusivity in streambeds from temperature time series. *Water Resour. Res.* **2013**, *49*, 488–506. [CrossRef]

20. Duque, C.; Müller, S.; Sebok, E.; Haider, K.; Engesgaard, P. Estimating groundwater discharge to surface waters using heat as a tracer in low flux environments: The role of thermal conductivity. *Hydrol. Process.* **2016**, *30*, 383–395. [CrossRef]

21. Land, L.; Paull, C. Thermal gradients as a tool for estimating groundwater advective rates in a coastal estuary: White Oak River, North Carolina, USA. *J. Hydrol.* **2001**, *248*, 198–215. [CrossRef]

22. Kirkegaard, C.; Sonnenborg, T.O.; Auken, E.; Jørgensen, F. Salinity Distribution in Heterogeneous Coastal Aquifers Mapped by Airborne Electromagnetics. *Vadose Zone J.* **2011**, *10*, 125. [CrossRef]

23. Haider, K.; Engesgaard, P.; Sonnenborg, T.O.; Kirkegaard, C. Numerical modeling of salinity distribution and submarine groundwater discharge to a coastal lagoon in Denmark based on airborne electromagnetic data. *Hydrogeol. J.* **2014**, *23*, 217–233. [CrossRef]

24. Müller, S.; Jessen, S.; Duque, C.; Sebök, É.; Neilson, B.; Engesgaard, P. Assessing seasonal flow dynamics at a lagoon saltwater–freshwater interface using a dual tracer approach. *J. Hydrol. Reg. Stud.* **2018**, *17*, 24–35. [CrossRef]

25. Holzbecher, E. *Modeling Density-Driven Flow in Porous Media: Principles, Numerics, Software*; Springer Science & Business Media: Berlin, Germany, 1998.

26. Bredehoeft, J.D.; Papaopulos, I.I. Rates of Vertical Groundwater Movement Estimated From the Earth's Thermal Profile. *Water Resour. Res.* **1965**, *1*, 325–328. [CrossRef]

27. Sebok, E.; Müller, S. The effect of sediment thermal conductivity on vertical groundwater flux estimates. *Hydrol. Earth Syst. Sci.* **2018**, in press. [CrossRef]

28. Shan, C.; Bodvarsson, G. An analytical solution for estimating percolation rate by fitting temperature profiles in the vadose zone. *J. Contam. Hydrol.* **2004**, *68*, 83–95. [CrossRef]

29. Kurylyk, B.; Irvine, D.; Carey, S.K.; Briggs, M.; Werkema, D.; Bonham, M. Heat as a groundwater tracer in shallow and deep heterogeneous media: Analytical solution, spreadsheet tool, and field applications. *Hydrol. Process.* **2017**, *31*, 2648–2661. [CrossRef] [PubMed]

30. Shinn, E.A.; Reich, C.; Donald Hickey, T. Seepage meters and Bernoulli's Revenge. *Estuaries* **2002**, *25*, 126–132. [CrossRef]

31. McBride, M.S.; Pfannkuch, H.O. Distribution of seepage within lakebeds. *J. Res. US Geol. Surv.* **1975**, *3*, 505–512.

32. Anderson, R.B.; Naftz, D.L.; Day-Lewis, F.D.; Henderson, R.D.; Rosenberry, D.O.; Stolp, B.J.; Jewell, P. Quantity and quality of groundwater discharge in a hypersaline lake environment. *J. Hydrol.* **2014**, *512*, 177–194. [CrossRef]

33. Belanger, T.; Mikutel, D.F.; Churchill, P.A. Groundwater seepage nutrient loading in a Florida Lake. *Water Res.* **1985**, *19*, 773–781. [CrossRef]

34. Rosenberry, D.; LaBaugh, J.W.; Hunt, R. Use of Monitoring Wells, Portable Piezometers, and Seepage Meters to Quantify Flow between Surface Water and Ground Water; In *Field Techniques for Estimating Water Fluxes Between Surface Water and Ground Water*; US Department of the Interior, US Geological Survey: Washington, DC, USA; Reston, VA, USA, 2008; pp. 39–70.

35. Rosenberry, D. Integrating seepage heterogeneity with the use of ganged seepage meters. *Limnol. Oceanogr. Methods* **2005**, *3*, 131–142. [CrossRef]

36. Shaw, R.D.; Prepas, E.E. Groundwater-Lake Interactions: I. Accuracy of Seepage Meter Estimates of Lake Seepage. *J. Hydrol.* **1990**, *119*, 105–120. [CrossRef]

37. Shaw, R.D.; Prepas, E.E. Groundwater Lake Interactions: 2. Nearshore Seepage Patterns and the Contribution of Ground-Water to Lakes in Central Alberta. *J. Hydrol.* **1990**, *119*, 121–136. [CrossRef]
38. Murdoch, L.C.; Kelly, S.E. Factors affecting the performance of conventional seepage meters. *Water Resour. Res.* **2003**, *39*. [CrossRef]

Article

Application of Stable Isotopes of Water to Study Coupled Submarine Groundwater Discharge and Nutrient Delivery

Carlos Duque [1,2,*], Søren Jessen [2], Joel Tirado-Conde [2], Sachin Karan [3] and Peter Engesgaard [2]

1   WATEC, Department of Geoscience, Aarhus University, 8000 Aarhus, Denmark
2   Department of Geosciences and Natural Resource Management, University of Copenhagen, 1350 Copenhagen, Denmark
3   Department of Geochemistry, Geological Survey of Denmark and Greenland (GEUS), 1350 Copenhagen, Denmark
*   Correspondence: cduque@geo.au.dk

Received: 12 August 2019; Accepted: 31 August 2019; Published: 4 September 2019

Abstract: Submarine groundwater discharge (SGD)—including terrestrial freshwater, density-driven flow at the saltwater–freshwater interface, and benthic exchange—can deliver nutrients to coastal areas, generating a negative effect in the quality of marine water bodies. It is recognized that water stable isotopes (^{18}O and ^{2}H) can be helpful tracers to identify different flow paths and origins of water. Here, we show that they can be also applied when assessing sources of nutrients to coastal areas. A field site near a lagoon (Ringkøbing Fjord, Denmark) has been monitored at a metric scale to test if stable isotopes of water can be used to achieve a better understanding of the hydrochemical processes taking place in coastal aquifers, where there is a transition from freshwater to saltwater. Results show that ^{18}O and ^{2}H differentiate the coastal aquifer into three zones: Freshwater, shallow, and deep saline zones, which corresponded well with zones having distinct concentrations of inorganic phosphorous. The explanation is associated with three mechanisms: (1) Differences in sediment composition, (2) chemical reactions triggered by mixing of different type of fluxes, and (3) biochemical and diffusive processes in the lagoon bed. The different behaviors of nutrients in Ringkøbing Fjord need to be considered in water quality management. PO_4 underneath the lagoon exceeds the groundwater concentration inland, thus demonstrating an intra-lagoon origin, while NO_3, higher inland due to anthropogenic activity, is denitrified in the study area before reaching the lagoon.

Keywords: submarine groundwater discharge; stable isotopes of water; nutrients; freshwater–saltwater interface; Ringkøbing Fjord

1. Introduction

Coastal regions are exposed to anthropogenic pressures frequently associated with high-density populations that require water supply [1]. Additionally, these regions are recipients for nutrients derived from farming, industrial activity, and sewage systems within the coastal region as well as their inland catchments [2]. Excess nutrient loads cause eutrophication [3–5] and algal blooms in coastal waters [6,7], and ultimately may have a global scale impact by altering marine food webs [2,8]. Another consequence is the effect on touristic activities such as fishing and leisure, [9] which are important drivers of economic activity.

The impact of groundwater-borne nutrients to the coastal regions was recognized by pioneering studies in the 1980s [10–12]. Subsequent investigations on submarine groundwater discharge have used seepage meters, radioactive tracers, or hydraulic methods to estimate the flux of groundwater [13–15]. Flux estimates combined with nutrient concentrations were used to calculate nutrient loads to coastal waters in several regions of the world [16–20].

Nevertheless, coastal aquifers present a series of hydrodynamic features that complicate the direct application of flux and nutrient concentration to estimate nutrient loads. The presence of a freshwater–saltwater interface separates two domains where flow is driven by different mechanisms, and where the hydrochemical properties of groundwater are distinct. Freshwater flows as a consequence of the higher hydraulic head inland after the recharge of aquifers. Saltwater fluxes are associated with density-driven flow [1,21]. The benthic fluxes, referred to here as the exchange between groundwater and surface water across the sea or lagoon bed [22], can be generated by different mechanisms including, among others, tides and waves [23]. Also, the mixing of water with different properties and ionic strengths can trigger multiple chemical transformations [24,25], making an accurate estimation of mass loads difficult. While the general trends are well-known, the complexity of coastal aquifers (i.e., spatial variability in hydraulic conductivity, flow paths, residence times) hinders the understanding of all the processes that are taking place, as well as their hydrochemical consequences. Recently, it was recognized that there are limited studies where there is a distinction between the different types of fluxes [26]. One reason for this is that it is required to differentiate between fluxes; however, that can be challenging due to the amount of field work required or the lack of appropriate methods to discern these fluxes. Despite the solid theoretical and modeling knowledge about hydrochemical processes [24,25], detailed field studies are required to link them to conditions that characterize aquifers, often with a high degree of complexity.

Using stable isotopes of water as a tracer is a well-established discipline in hydrogeological studies that allows the differentiation of water exposed to different physical processes relating to precipitation and evaporation [27]. Its use has been applied with multiple objectives [28]: To quantify recharge sources [29–31], for climate assessment [32], or to evaluate water origin in lakes [33] or hydrothermal systems [34,35]. The continuous improvement in analytical techniques has enabled researchers to quantify ^{18}O and ^{2}H with small volumes of water through a simple and direct sampling system. Since chemical processes do not have an effect on stable isotopes, they can be suitable tracers when combined with hydrochemical studies. If different sources of water are exposed to different physical processes, as naturally occurring in coastal areas, water stable isotopes can be an indicator of flow paths and the mixing of water, without being affected by chemical reactions.

The utility of water stable isotopes for studying the impact of the hydrogeological dynamics in coastal areas on nutrient discharge has been tested in this study. To this end, a field campaign was designed to combine the spatial distribution of nutrients and stable isotopes of water, where groundwater transitions from fresh to saline. Hence, the objective was to evaluate the use of stable isotopes of water, in combination with the spatial distribution of nutrients (PO_4 and NO_3) for characterizing submarine groundwater discharge (SGD) at the metric scale, including the challenges associated with field sampling (i.e., spatial variability, scattered samples, and outliers). The field site was established in a Danish lagoon (Ringkøbing Fjord), as a benchmark in which there are also major local concerns about water quality and potential eutrophication issues.

2. Materials and Methods

2.1. Study Area

Ringkøbing Fjord is a brackish coastal lagoon located adjacent to the North Sea near the west coast of Jutland in Denmark (Figure 1). The lagoon covers 300 km^2 with a mean water depth of 1.9 m (Ringkøbing Amt., 2004). Lagoon salinity ranges from 5 to 12 g/L during the year, due to the seasonal discharge from the Skjern River and in connection with the sea managed by a sluice. The region is characterized by a flat topography and low altitude hills. Annual mean precipitation and evaporation are 1050 mm and 630 mm, respectively [36].

The local hydrogeological setting consists of a shallow 12-m thick unconfined sandy aquifer, limited by a clay layer that was 5–10 m thick [37]. Other aquifer units below this range are out of the scope of this study, since their input to the lagoon in the near-shore environment would not be quantitatively relevant at the study area. Groundwater originates as recharge in agricultural and residential (summer housing) areas flowing from east to west, until discharging to the lagoon [37,38]. Nutrient transport could be generated by farming activity as NO_3 [39], and is associated with septic tanks and PO_4 [40].

Ringkøbing Fjord has been studied in the last decades, focusing on the status of the lagoon water and the in-fjord processes affecting water quality [41–43]. The management of the sluice changes the salinity in the lagoon, and the impact on the ecological systems was described by Petersen et al. [44]. However, the groundwater flow with the associated nutrient input to the lagoon has not been researched. Previous groundwater-related research focused exclusively on the physical processes that determine the flow of groundwater to the lagoon [22,37], and the development of methodologies to help quantify water fluxes [38,45].

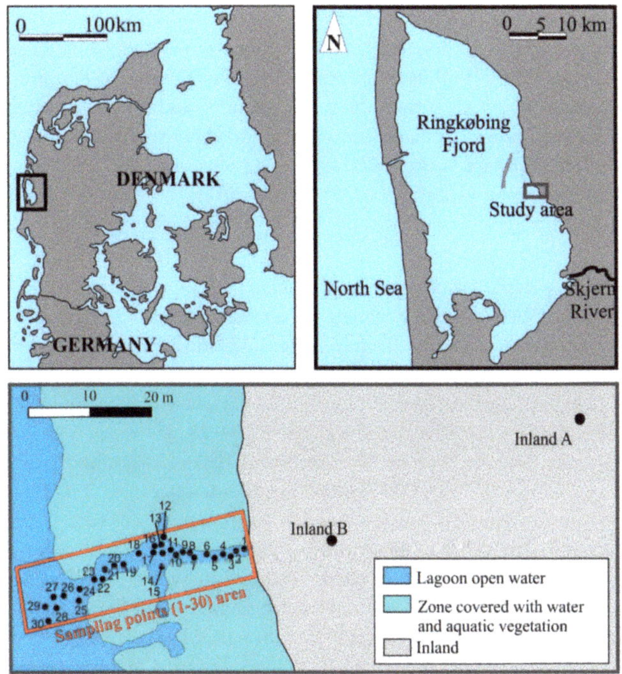

Figure 1. Study area location in Denmark and in Ringkøbing Fjord. Sampling points (1–30), inland wells, and general characteristics of the study area.

2.2. Field Campaigns

Field campaigns were conducted between the 14–19 August 2016, (Inland wells A and B) and between 7–14 July 2017 (sampling points 1–30). The data collection aimed to characterize the hydrogeological properties, fluxes, and hydrochemistry, including: freshwater aquifer inland (Inland wells A and B), shallow part of the aquifer underneath the lagoon (sampling points 1–30), and groundwater–lagoon water fluxes. The area was investigated in previous studies [22,38], allowing the optimization of the placement of sampling points 1–30 and the requirements to complete field work. Sampling points 1–30 extended along a transect from the shoreline up to 40 m offshore, with small differences in the water depth of the lagoon (water depth of 30 to 50 cm). The water depth of the lagoon did not exceed 50 cm for hundreds of meters offshore. Part of the transect was surrounded by submerged autochthonous vegetation; however, the field equipment was deployed along a corridor without vegetation, which serves as access to the lagoon for leisure and fishing. The first few meters of the lagoon from the shoreline were covered with organic material (degraded vegetation), followed by well-classified sand that extended to at least half a meter below the surface (tested by surficial core samples).

In 2016, groundwater samples were collected at two wells (Inland wells A and B) located inland 60 and 15 m from the lagoon shore, respectively (Figure 1). These wells were 8 and 9 m deep, respectively, and samples were collected at one-meter intervals.

In 2017, 30 measurement locations were established irregularly distributed along a transect (Figure 1). The distance between sampling points ranged between 1–3 m. This allowed for data to be obtained along the zone, with a transition from fresh to saline/brackish groundwater, gaining a representation of the spatial heterogeneity via multiple measurements at similar distances from the shore. At all locations, a set of measurements were completed both at the surface of the lagoon bed and at multiple depths to determine groundwater hydrochemistry (number of samples, n = 150), water isotopic characteristics (n = 150), aquifer hydraulic properties (n = 120), and groundwater seepage to the lagoon (n = 30). In four cases, it was not possible to pump up groundwater from piezometers due to local low hydraulic conductivity, or potential damage to the piezometer screens. Electrical conductivity (EC) data were presented as continuous in the study area, through the kriging interpolation method. The stage level in the lagoon and the water table in the Inland A and B wells were monitored regularly, showing steady conditions indicating that the results presented here correspond to a stable snapshot of the groundwater–lagoon water interaction. The sampling protocol and analytical methods (see below) were the same for both campaigns (August 2016 and July 2017).

2.3. Measurement of Fluxes

Seepage meters were deployed one week prior to the campaign in July 2017 to allow the flushing of the seepage meter chambers. Seepage meters had a classic Lee design [46], and were constructed from 200-L 58-cm diameter steel drums. To minimize the disturbance of the sediments in the study area, the fluxes of groundwater discharge were measured in the first two days of the campaign [47]. The seepage meter bags were thin-walled, durable plastic autoclave bags prefilled with approximately 1 L of freshwater, with known electrical conductivity (EC ~0.28 mS cm^{-1}). The measurements were repeated with 2–3 h measuring intervals on day 1 (round 1: 10:00–13:00, round 2: 13:00–15:00, round 3: 15:00–17:00) and day 2 (round 1: 9:00–11:00, round 2: 11:00–14:00, round 3: 14:00–16:00). After each interval, the sample bag was removed, and another prepared bag was immediately attached, to prevent inflow of lagoon water to the seepage meter during the bag exchange. Each bag had a valve that was closed during removal and while unattached. To approximate the simultaneous sampling of the seepage meters, the bag exchange was completed by two to three operators. The total time for sample collection and bag replacement never exceeded 20 min. The EC of seepage meter samples, corrected to 25 °C, was measured using a HACH IntelliCAL CDC401 Standard Conductivity probe, connected to a HACH HQ40d instrument (expected error less than 0.5% of the reading). The EC of measured flux was estimated with a mass balance:

$$\text{EC flux} = \frac{\left[\left(W_{off} - W_{bag}\right) \times \left(EC_{off}\right)\right] - \left[\left(W_{on} - W_{bag}\right) \times \left(EC_{on}\right)\right]}{W_{off} - W_{on}} \tag{1}$$

where W_{off}: Weight of the seepage meter bag after measuring, EC_{off}: EC of water in the seepage meter bag after measuring, W_{on}: Weight of the seepage meter bag before measuring, EC_{on}: EC of water in the seepage meter bag before measuring, and W_{bag}: Weight of the seepage meter bag and attachment accessories without water.

The proportion of freshwater and saltwater flux from each discharge sample was calculated by assuming an EC of 0.28 mS cm^{-1} for the freshwater endmember measured in Inland wells A and B, while the EC of the saline lagoon endmember was assumed to be equal to that measured for the lagoon water on the day of sampling (22 mS cm^{-1}). The results presented are the arithmetic mean of the set of six tests.

2.4. Wells

Two types of drilling methods were used depending on the desired depth. For shallow depths (up to 20 cm), a mini-piezometer was used (screen length 3 cm), to allow the sampling of small volumes with minor disturbance to the system, as the proximity to the lagoon bed could induce the vertical downward flow of lagoon water to the piezometer screen. Additionally, this sample could be a better indicator of the chemical characteristics of water flowing to the lagoon than deeper samples, as sampling was very close to the discharge point. In the same locations, a borehole was drilled with a $\frac{3}{4}$ inch steel pipe (screen length 10 cm) with a pneumatic hammer to depths of 1 m, 2 m, 3 m, and 4 m below the lagoon bed (sampling points 1–30), and for Inland wells A and B, every meter after reaching the water table (3–8 m or 1–9 m below terrain, respectively). For each location, sample collection was followed by slug tests, prior to pushing the well to the next depth. This system was used to facilitate the completion of hydraulic conductivity tests, and because it was difficult to reach depths below 1 m with the mini-piezometers.

2.5. Hydrochemistry Sampling and Chemical Analysis

In July 2017, groundwater samples were collected from the piezometers immediately after installation and purging (3 times the water volume contained in the well or after drying it) using a peristaltic pump. Water samples were collected in 60-mL polypropylene syringes. In total, 146 samples were collected, of which 69 were immediately passed through a 0.21-μm filter (Sartorius Minisart cellulose acetate) with a syringe, and directly into the collection containers. Filtration of the remaining 77 samples, of which most were saline, proved difficult and these were collected unfiltered into 100-mL acid washed amber glass bottles, completely filled to avoid headspace. These were filtered in the laboratory using a Mikrolab Aarhus A/S ML 303810 pneumatic syringe filtration unit. Upon collection in the field, the samples were immediately refrigerated to 5 °C. In either case, the filtrate was passed into one 10-mL polyethylene (PE) vial for analysis of major cations (Na^+, K^+, Ca^{2+}, Mg^{2+}) and ammonium (NH_4^+), another 10-mL PE vial for major anions (Cl^-, SO_4^{2-}, NO_3^-, Br^-), a 5-mL PE vial for PO_4^{3-} and Fe^{2+} analysis, a 100-mL amber glass flask for alkalinity titration, and a 2-mL septum glass for stable isotopes of water. In addition to filtration and refrigeration, samples for PO_4^{3-} and Fe^{2+} were conserved by 2 vol% 2 M H_2SO_4, samples for major cations and ammonium were acidified by 1 vol% 1 M HNO_3 (resulting in pH 2–2.1, depending on alkalinity), and samples for anions were stored frozen (−18 °C). In the August 2016 campaign, the procedure was similar; however, here a flow cell was used measuring O_2, pH, and EC directly in the field. A sample was taken when the readings were stable; then, they were filtered as described above.

Cations, anions, and ammonium were analyzed by ion chromatography using a Metrohm 820-IC Separation Center with 819-IC Detector. Alkalinity was measured by end point titration on a Metrohm autotitrator.

PO_4^{3-} were analyzed spectrophotometrically using the molybdate blue method of Murphy and Riley (1986). Total dissolved Fe, henceforth assumed to a direct proxy for the Fe^{2+} concentration, was measured on residual sample aliquots of the PO_4^{3-} samples after reduction to Fe^{2+} with hydroxylamine hydrochloride, using the spectrophotometric ferrozine method of Stookey (1970).

$\delta^{18}O$ and δD values, reported in ‰ units relative to the VSMOW/SLAP scale, were measured at the Geological Survey of Denmark and Greenland (GEUS) on a Picarro Cavity Ring-Down Spectrometer (CRDS) L2120-i. The standard deviation for each analysis was equal to or lower than 0.31‰ for $\delta^{18}O$ and 0.74‰ for δD.

Field measurements of pH and temperature-specific electrical conductivity (EC; corrected to 25 °C) and temperature were conducted using an IntelliCAL PHC101 Standard gel filled pH probe and an IntelliCAL CDC401 Standard Conductivity probe, both connected to a HACH HQ40d instrument. Field measurements of dissolved oxygen (O_2) were conducted using a WTW Optical IDS FDO® 925 sensor connected to a WTW oxi-3310 IDS Portable Dissolved Oxygen. The probes were in all cases immersed in a 0.3-L beaker flushed and freshly filled with sample water, delivered from the piezometer by the peristaltic pump.

2.6. Hydraulic Properties

The horizontal hydraulic conductivity at the well screen locations was estimated by the Hvorslev straight-line method, using specific software (AQTESOLV/Pro). It assumes a quasi-steady-state flow by omitting the storativity of the aquifer. After conversion into normalized response data (H/H_0), the Hvorslev solution for unconfined partially penetrating wells was used [48]. While developed for confined aquifers, the Hvorslev method can also be used for unconfined conditions, provided that the well screen is not too close to the water table [48]. The method assumes that the aquifer is homogeneous and of uniform thickness, and can be used in both cases of fully or partially penetrating wells. The test was repeated up to four times, depending on the time to get one slug test completed, and the results presented here are the arithmetic mean for each location.

3. Results

3.1. Physical Tracers of Flow (EC, Stable Isotopes of Water and Groundwater Seepage)

The data of sampling points 1 to 30 (Table S1) were projected into a cross-section (EC, D-excess, and hydraulic conductivity) to facilitate the interpretation and comparison. Some data are overlapped since they were located at the same distance from the shoreline; however, it is also a representation of the spatial variability in the study area.

Groundwater EC showed a consistent pattern starting from freshwater and transitioning to saltwater moving offshore (July 2017 campaign). The change from fresh to saltwater takes place in a mixing zone of 3–5 m width. Interpolation of the measurements at different depths and distances from the shore showed the shape of a classic saline wedge (in spite of representing only the top four meters of the aquifer), where higher density water intruded freshwater (Figure 2). The EC ranged from 0.15 to 20.20 mS cm^{-1}. The lowest EC corresponded to the fresh groundwater inland, and the highest corresponded approximately to the maximum EC of the lagoon during a year [38]. The measured profiles of EC in Inland wells A and B tapped the freshwater part of the aquifer (August 2016 campaign). EC fell in the range of 0.115–0.386 mS cm^{-1} (Inland well A) and 0.215–0.274 mS cm^{-1} (Inland well B), confirming that the inland water was fresh. The high density of measurements and the absence of outliers indicated that the observed patterns are apt for the further interpretation of chemical and hydrogeological processes.

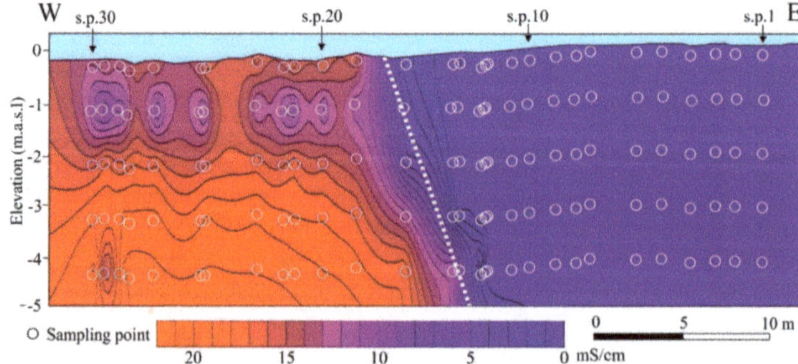

Figure 2. Interpolated electrical conductivity (EC) along the cross-section and sampling points (vertical scale × 2.5). The dashed line represents the transition from freshwater to saltwater. Sampling points are located with empty circles, and the locations of sampling points 1, 10, 20 and 30 are marked.

Seepage rates were spatially variable (Figure 3), due to changes in hydraulic properties and local hydrodynamics. There were a few points in the proximity to the shoreline and in the beginning of the saltwater area with low fluxes near zero. Still, in the majority of the locations, the seepage was from the aquifer to the lagoon. The observation of spatial variability of flux is frequent in seepage meters studies [49]. This was also confirmed in this study with fluxes ranging between ~0–10 cm/d in a 40-m transect. Within distances of 1–3 m, fluxes changed up to 10-fold. The EC that was estimated, with the mass balance of the seepage meter bags (Equation (1)) that were higher or lower than 10 mS/cm, correlated with measurements of saline/fresh groundwater below the seepage meters. Thus, a clear pattern of groundwater becoming more saline in the offshore direction was evident, from both seepage meter estimated EC and measured subsurface EC (Figure 3).

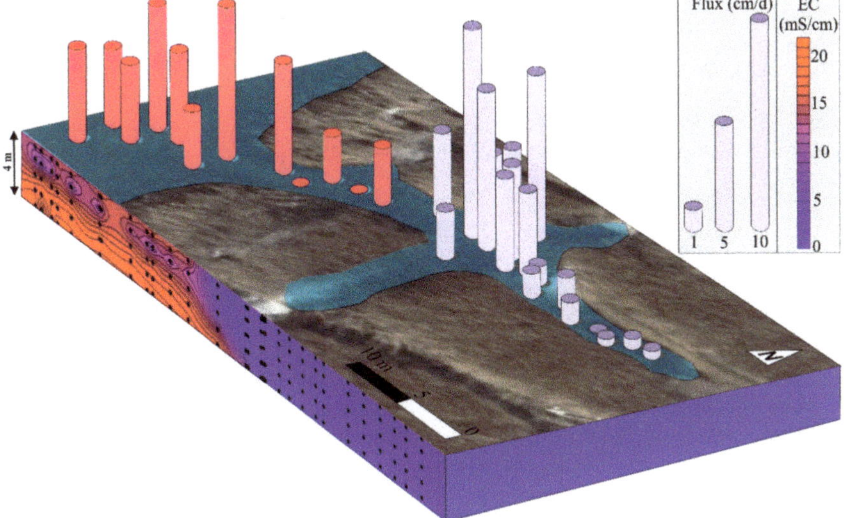

Figure 3. Spatial plot of discharging fluxes measured with seepage meters and an interpolation of measured EC in collected groundwater. The red/purple colors of the flux bars indicate if the mass balance of the seepage meters bags showed fresh (<10 mS/cm) or saline water (>10 mS/cm).

The deuterium excess (D-excess) (Equation (2)) was used as an indicator of evaporation:

$$D\text{-excess} = \delta D - 8\delta^{18}O \tag{2}$$

An independent spatial differentiation with three distinct zones was identified from the changes in D-excess that correlates with the EC distribution: a freshwater zone and a saline zone separated into shallow and deep zones (Figure 4). To probe the differences between the various zones, a two-tailed t-test was conducted between the D-excess measured in each zone. There was a significant difference (p-value < 0.1) in the D-excess content between freshwater and shallow saline samples (p-value < 0.001), and also between the samples from the two shallow and deep saline zones (p-value < 0.001). The freshwater zone was characterized by high D-excess and uniform mean and standard deviation values: −10.21 + 1.23‰, maximum: −13.09‰, n = 85). The samples from the shallow saline zone had a much lower D-excess (−4.76 + 1.61‰, maximum: −9.37‰, n = 32). In the deep saline zone, the D-excess was the lowest (−1.22‰ + 2.68‰, maximum: −6.13‰, n = 27). A dipping line marks the differences between the shallow and the deep saline zone (Figure 4). In the furthest offshore location, the shift took place at 2 m below the lagoon bed, while at the location of the saltwater–freshwater interface, this shift was at 4-m depth. The spatial distribution of the isotopic tracer was in good agreement with the EC distribution (Figure 2).

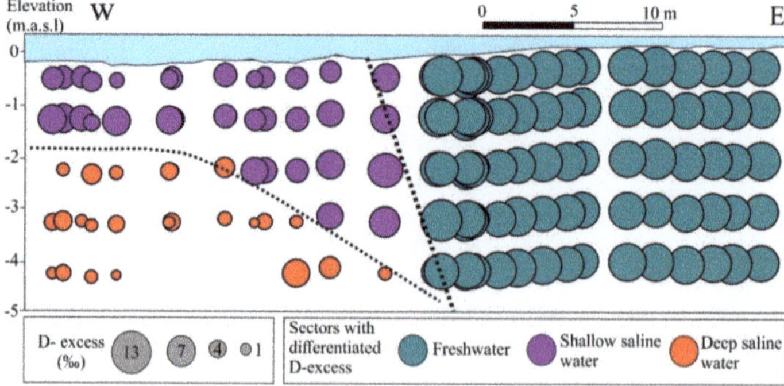

Figure 4. Deuterium excess along the cross-section with dashed lines dividing data into a freshwater zone, a shallow zone, and a deep saline zone, as explained in the text (vertical scale × 2.5).

Stable isotopes of water (δ^{18}O and δD) in groundwater showed alignment with the Global Meteoric Water Line (GMWL) and the Local Meteoric Water Line (LMWL) [50].

$$\text{LMWL} = 7.29 \times \delta^{18}\text{O} + 4.81‰ \tag{3}$$

The δ^{18}O values ranged between −7.69 and −2.75 ‰, while δD was −50.1 and −22.6‰ (Table S1). Many of the samples grouped around values near δ^{18}O = −7.6‰ and δD = −47‰, corresponding to freshwater (Figure 5). These values were similar to groundwater in other regions of Denmark [50–53]. There was no evidence of fractionation of freshwater samples, as all grouped together with no outliers. A conservative mixing line was established between the groundwater endmember and oceanic seawater (i.e., VSMOW standard), in which part of the samples aligned well. Additionally, an evaporation line was added, where the start and the slope were selected based on the last freshwater sample presence, and following the trend of the samples with lower δD, respectively (Figure 5). The enrichment in heavy isotopes for samples with higher EC was indicated by the alignment between the conservative mixing line with seawater and the evaporation line (Figure 5).

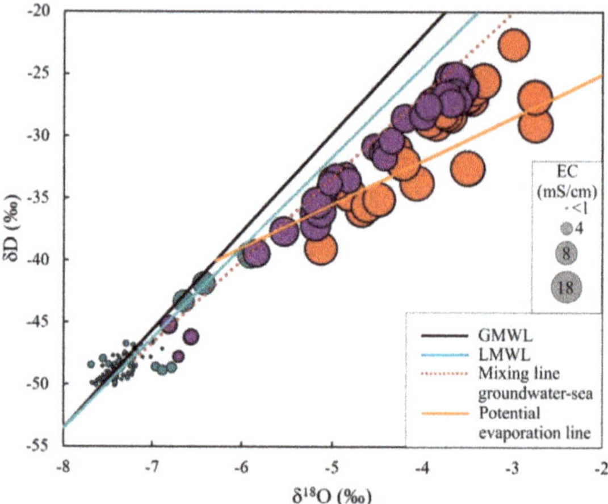

Figure 5. δ^{18}O and δD of groundwater samples with the Global Meteoric Water Line (GMWL), the Local Meteoric Water Line (LMWL), a mixing line between fresh groundwater in the study area, and the sea (δ^{18}O and δD: 0‰, 0‰), and a potential evaporation line. Colors indicate the sample categories from Figure 4.

Hydraulic conductivity (K) along the transect ranged between 0.02–28.0 m d^{-1}. The average values for each depth were similar: 9.3 m d^{-1} (1 m), 7.4 m d^{-1} (2 m), 9.3 m d^{-1} (3 m), and 9.8 m d^{-1} (4 m). In spite of the apparent homogeneity, there were spatial differences in K. At 1-m depth, K was relatively homogenous (standard deviation of 4.5 m d^{-1}) than at deeper locations (standard deviation at 2 m = 8.3 m d^{-1}, 3 m = 7.5 m d^{-1}, and 4 m = 10.9 m d^{-1}). At 2-m, 3-m, and 4-m depths, K values were more variable, especially in the zone dominated by saltwater (Figure 6). In some of the locations, measurements could not be completed, reflecting low K or field technical problems (Figure 6).

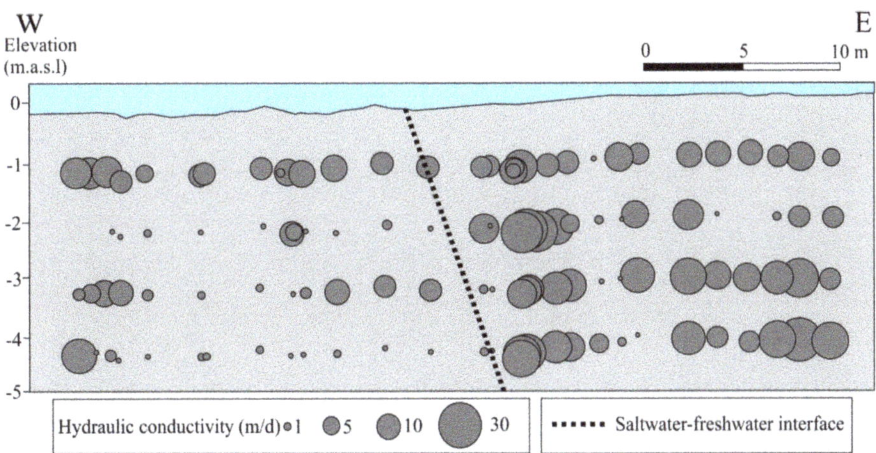

Figure 6. Hydraulic conductivity at the different depths measured, and location of the freshwater–saltwater transition in groundwater (vertical scale × 2.5).

3.2. Hydrochemistry of Groundwater

Inland wells A and B present different hydrochemical characteristics despite the short distance between them (50 m). In Inland well A, NO$_3$ concentrations reached a maximum of 772 µmol/L or 47.8 mg/L at a 6-m depth (Figure 7), indicating the presence of a NO$_3$ plume. The EC profile follows this trend; thus, the uncontaminated freshwater is probably reflected by the most surficial measurement (~0.115 mS/cm), showing infiltrating rain water near the coast. In Inland well B, no NO$_3$ is observed, except at the depth of 7 m (7 µmol/L). This indicates that between Inland well A to Inland well B, mixing and geochemical reactions occur, giving a relative uniform NO$_3$ profile at Inland well B. Alkalinity showed an increase from Inland well A to Inland well B, suggesting that denitrification with organic matter took place (notice that Inland well A only has three measurements). The alkalinity in Inland well B compared with the concentration observed at locations 1–30 (data not shown). Groundwater samples from the 2017 campaign (sampling points 1–30 at multiple depths) also showed no presence of NO$_3$ except for five samples that were below the detection limit.

For PO$_4$, Inland wells A and B had average values of 0.8 and 3.9 µmol/L, respectively, which is a notable increase in PO$_4$ moving toward the shore of the lagoon. In Inland well A, all the sampling points, except at 3-m depth, did not show any PO$_4$. In Inland well B, the highest concentrations were close to the topographic surface, decreasing to a minimum at 5–6 m, followed by an increase at deeper locations (Figure 7).

PO$_4$ distribution in groundwater along the cross-section under the lagoon showed differences that also allowed dividing the transect into three zones, by grouping the samples plus a fourth category (Figure 8) (Table S1). In the freshwater zone, PO$_4$ had an irregular distribution with the highest values isolated between much lower concentrations (Group I). The mean concentration for this zone (n = 85) was 1.33 µmol/L with a maximum of 33.28 µmol/L, and a standard deviation of 3.89 µmol/L. At the saltwater–freshwater interface, four samples showed a higher concentration, which were assigned to a different category (Group II) (arithmetic mean = 14.44 µmol/L, standard deviation = 5.86 for n = 4). In the saline zone, the arithmetic mean at variable depths showed marked differences (depth 0.2 m = 5.95 µmol/L, 1 m = 1.97 µmol/L, 2 m = 0.11 µmol/L, 3 m = 0.13 µmol/L, and 4 m = 0.09 µmol/L). Thus, higher PO_4 concentrations were measured at 0–1 m depth, and supported a subdivision of the saline zone into shallow and deep zones. The shallow saline zone (Group III) (Figure 8) had higher PO_4 concentrations (arithmetic mean 3.87 µmol/L for n = 21) with a maximum of 8.13 µmol/L, and a standard deviation of 2.52 µmol/L). For the deeper saline zone (Group IV) (Figure 8), the arithmetic mean, maximum, minimum, and standard deviation were much lower in comparison (0.11 µmol/L for n = 31, maximum 0.38 µmol/L, and standard deviation 0.10 µmol/L). The differences between groups were verified with a two-tailed t-test probing the differences in PO_4 concentrations in groundwater. Groups I and II showed significant differences ($p = 1.97 \times 10^{-2}$) as well as Groups II and III ($p = 3.41 \times 10^{-2}$) and Groups III and IV ($p = 1.20 \times 10^{-2}$).

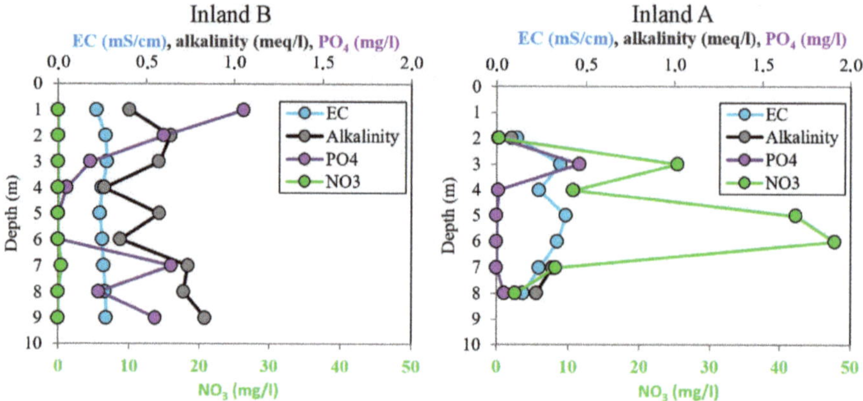

Figure 7. Profiles of measured EC, alkalinity, PO_4, and NO_3 for Inland wells A and B.

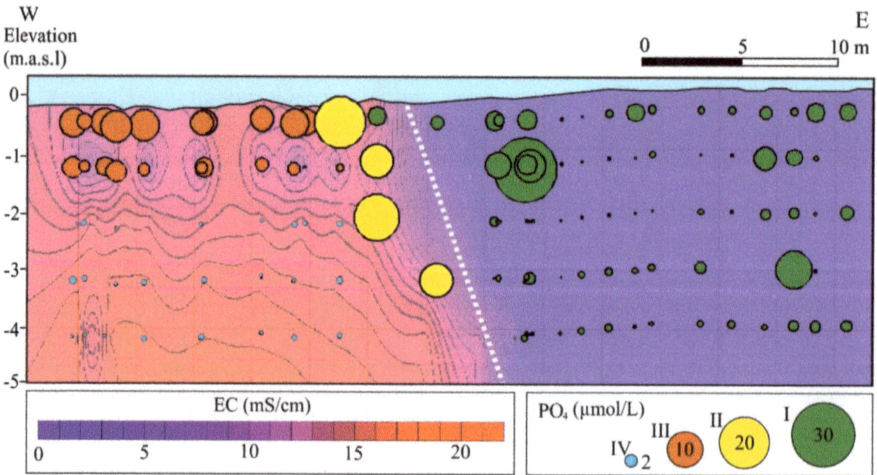

Figure 8. Bubble diagram of PO_4 concentrations measured in groundwater along the transect, together with interpolated EC. PO_4 concentrations have been plotted with different colors for the categories, as explained in the text (vertical scale × 2.5).

4. Discussion

4.1. Salinity and Drivers of Fluxes

Groundwater discharge to the lagoon takes place not only in areas dominated by freshwater at the subsurface, but also in areas dominated by saltwater at the subsurface. This was inferred from the patterns of the subsurface EC measurements (Figure 2) that corroborated with the seepage meter calculated EC (Figure 3), as well as groundwater stable isotopes' signals (Figure 4). Often, the conceptualization of groundwater discharge to lagoons includes only the freshwater component, as this can be estimated based on hydraulic gradient and estimated K value. In addition, it is frequently anticipated that the contamination sources originated inland [26]. However, these assumptions, based on the dataset presented in this work, can not alone explain our observations. There are fluxes that take place in the zone occupied by saltwater, which can have an important contribution to the chemical budgets of coastal areas and lagoons. The drivers of these fluxes are related to well-known variable density flow [21], and are also due to waves and water level oscillations of the surface water [23], as was demonstrated indirectly in several marine studies from the 1990s [15,54], and more specifically in recent studies [55–57].

In the study area, the differentiation between fresh and saline groundwater allowed the identification of different hydrochemical processes. While in the freshwater sector NO_3 concentration was decreasing, PO_4 was generated at different ratios, depending on the depth. Hence, the fresh SGD would have a minor impact on the nutrients discharge, while the saline SGD would be more relevant in this study area.

One major question is whether these results can be used to estimate nutrient budgets for the full lagoon by extrapolation. In natural environments, the spatial variability and the differences in nutrient sources require caution with this type of approach. In this study area, it can be assumed that the freshwater discharge is limited to a fringe of 10–20 m from the shoreline of the lagoon along all of the east coast, due to the similar hydrogeological setting (similar surrounding topography, hydrogeology, and precipitation regime). In terms of the saline discharge, it was measured in this study up to 40 m from the shore, and in principle, could be active even further offshore considering that: (i) The depth of the lagoon from the further offshore locations does not exceed 1.9 m [58], so similar conditions regarding depth can be assumed; (ii) the sediment characteristics at other offshore locations of the lagoon are unknown, though in small exploration campaigns a few hundred meters offshore, no changes in the sediment composition were observed; and, (iii) the presence of waves and lagoon level changes are common to the entire lagoon area. Under these assumptions, it is reasonable to assume that most of the lagoon, except for the 10–20 m wide shoreline fringe, is dominated by saline fluxes generated by processes not related to the inland hydraulic conditions. Other considerations, such as windows in deep confined aquifers [37], would be exceptions to this conceptualization.

The measured flux in seepage meters and estimated EC of groundwater that were used to characterize the lagoon and groundwater interactions agreed well, showing that one of the methods would be enough to inform about the differentiation between fresh–salt groundwater areas. Still, there are uncertainties when using seepage meters to estimate the EC of groundwater, due to the assumptions that are required: The seepage meter chamber has to be totally flushed out of groundwater before the first measurement, with no exchange between the chamber and the seepage meter bag during the measuring time, and where the groundwater seeping into the seepage meter mixes immediately in the chamber and can be collected in the seepage meter bag. Therefore, the variability in fluxes estimated with seepage meters, combined with the concentration of chemical components immediately below the lagoon bed (to estimate the mass load to Ringkøbing Fjord), should be considered with a range of variability that can only be provided if multiple seepage meters are used in field campaigns. The direct sampling of seepage meters can generate changes in the redox conditions of the samples [59].

4.2. Stable Isotopes of Water

Groundwater in coastal areas is exposed to different processes that affect its isotopic composition. Salinity (Figures 2 and 4) and isotopic signal (Figure 3) mapping along the cross-section supported the differentiation of fluxes, flow paths, and mixing of waters with different origin and physicochemical characteristics (i.e., salinity, oxygen, and content of nutrients). Overall, the LMWL matched the isotopic composition of groundwater. This is more evident for the freshwater samples, as the saline samples are likely affected by evaporation (Figure 5). Thus, the mixing line with saltwater fits well with part of the group of samples located in the shallow saline zone of the aquifer, while the other part indicated mixing between the lagoon water and groundwater. This mixing process can be explained by the benthic exchange occurring in the saline zone, as it affects only the shallow part of the aquifer where, for example, pressure oscillations (waves and surface water level changes) will have a reduced magnitude and would not encroach deeper into the aquifer [23].

The deviation in the plot of the samples for $\delta^{18}O$ and δD (Figure 5) can be explained by an evaporation process. Under the assumption that evaporation explains the isotopic composition of these samples, the reason for being located at a deeper location (deep saline water in Figure 4), can be associated to seasonal flow dynamics, due to the lagoon density changes (i.e., salinity in the lagoon is higher during the summer and would move deeper into the aquifer due to its higher density). In principle, evaporation would both increase the salinity as well as the concentrations of deuterium and $\delta^{18}O$, but in fact, each per-mille increase in $\delta^{18}O$ would result in only a few-percent increase in salinity.

Therefore, the use of water stable isotopes is an excellent method for identification of the different type of fluxes driving SGD. The main challenge is the high density of data required to determine the isotopic signature of each flux. At our field site, sampling at a meter scale was required, but in other natural systems, it might require an even higher density of data or a higher resolution at variable depths. The main advantage of water stable isotopes is the simplicity in the sampling, and the reliability of the results.

4.3. Nutrients

The distribution of PO_4, along the cross-section (Figure 8), allowed the distinction of four groundwater categories with different chemical composition. The spatial variability in the freshwater zone (category I), with no clear pattern, could be due to the natural spatial variability of the sediment composition by the dissolution of carbonates, which may contain adsorbed or co-precipitated phosphate. Unfortunately, there are no data on sediment composition to confirm this hypothesis. In the saline zone, the three categories seemed to be supported by the hydrogeological and isotopic measurements. The transition from freshwater to saltwater was characterized by an increase in the PO_4 content. This can be associated with the displacement by anions in seawater, and the release of P and N sorbed to Fe oxides from organic matter, as has been reported in oxic aquifers [60] and lab experiments [61].

The differentiation between a shallow and a deeper saline zone is also accompanied by changes in the PO_4 distribution (Figure 8). The coincidence between the chemical composition of PO_4 in groundwater and the isotopic/EC distributions could be an indicator of geochemical processes, which are triggered by the mixing of water with different oxic characteristics. Following this reasoning, the saltwater–freshwater interface would be characterized by the mixing of anoxic waters with different salinities, where the shallow part of the aquifer will be in contact with lagoon water enriched in oxygen, while the deeper zone will have more stable anoxic conditions. Additionally, changes in pH can also play a role. This is a complex environment, where the observations reported can be generated by different processes. We discuss three potential hypotheses that may explain the distribution of PO_4:

(1) Sediment composition differences. The horizontal layering of the changes observed in groundwater is concordant with the deposition of sediments. The reason that the most surficial sediments have a higher content of PO_4 would be associated with the changes that have taken

place in Denmark for the last century. The increase in the use of fertilizers and the sediment transport by the Skjern River that drains an agricultural catchment support the development of recent layers of sediments enriched in PO_4. This interpretation supports the observed changes in PO_4 in the top 1 m of the aquifer. Additionally, the hydraulic tests indicate a change in the characteristics of the sediments between 1–2 m depths (Figure 8). Nevertheless, this difference was not observed in the freshwater zone. Additional data about PO_4 sediment composition is required to confirm this hypothesis.

(2) Hydrodynamics of the coastal zone observed from the isotopic signature. In this case, the groundwater flow paths and the mixing of water with different properties would be responsible for the PO_4 distribution. The freshwater zone would be characterized by groundwater oxic conditions that do not enhance the mobility of PO_4 [24,25]. The shallow saline zone would be dominated by exchange processes where lagoon water with variable salinity, oxygen conditions, and organic matter generated by primary production would mix with saline, reducing groundwater. These processes can generate the microbial remineralization of organic matter [62]. The increase of PO_4 due to a high exchange of water has been documented in sand flats [63]. The agreement between PO_4 distribution and groundwater characteristics based on $\delta^{18}O$ and deuterium, as well as the measured seepage and the salinity changes, makes this possibility more feasible than sedimentological changes (Hypothesis 1). Groundwater hydrodynamics would control the hydrochemical distribution, rather than the sediment composition changes. The changes in hydraulic conductivity from the first meter to deeper zones can favor the sharp transition in isotopic signal associated to different hydrogeological processes.

(3) Diffusive and biological processes in the lagoon bed. A well-documented source of PO_4 is the release from sediments by diffusion [64]. Nevertheless, this has been highlighted as a process that only takes place in the top 20 cm [64], while here, the changes were observed further than 1-m depth. Another alternative is associated with the higher temperature due to the season (summer) when it was measured. In the lagoon bed, the high temperature would increase the biological activity, and the releasing of PO_4, as has been documented in lakes [65,66].

Most likely, the distribution of PO_4 monitored in the study area is a coalescence of the different processes discussed above. The contribution of each of these processes would require further specific research and sampling campaigns.

The presence of NO_3 in Inland well A showed contaminant sources in the upland area (Figure 7). The origin of NO_3 is probably associated to the septic tanks used in the summer houses (between 200–300), and the diffuse input due to fertilizer in agricultural fields. The difference with Inland well B (Figure 7) indicated that geochemical reactions can be responsible for the fast degradation of NO_3. Denitrification in the study area can be favored by the presence of organic matter in the sediments and the mixing of water with different redox conditions, as it has been shown in intertidal beaches [67]. The two wells are aligned along what would be expected to be a flow line, but an alternative hypothesis is that Inland wells A and B do not represent the same water, but different flow paths. Still, it is clear that NO_3 transported by groundwater is not reaching the lagoon, as only minor concentrations were measured (albeit below the detection limit) in a few of the 150 sampling points collected under the lagoon. Under the lagoon bed, the presence of organic matter that can contribute to the denitrification processes has been documented, especially in the proximity of the shoreline of the lagoon [68].

5. Conclusions

SGD at the field site shows a clear distinction between the hydrodynamic characteristics and processes taking place in the freshwater-dominated zone and the saline zone. The drivers of fluxes, the origin of discharging water, the chemical reactions that can take place, and the type of mixing of water must be considered in the analysis and methods for studying nutrients borne by SGD.

Stable isotopes of water can be used as a tracer of hydrogeological flow paths in coastal areas. The simplicity in sampling stable isotopes of water, and the relative low cost of the analysis, point to them as an efficient and useful tool for the differentiation of fluxes in SGD, which poses a persistent challenge in the study of coastal areas.

The spatial distribution of PO_4 detected in Ringkøbing Fjord can be explained by: (1) Changes in the sediment composition of the study area, (2) mixing and chemical reaction of water with different origin and composition, due to fluxes driven by variable sources (freshwater, density driven flow, benthic exchange), or (3) diffusive flux and biological activity in the lagoon bed. It may be that all processes are taking place simultaneously, but with different contributions.

Groundwater originating inland has low PO_4 content and high NO_3 content in the study area. The processes taking place in the proximity of the lagoon, as denitrification, reduce NO_3 concentrations before discharge to the lagoon, while PO_4 can be generated in the aquifer below the lagoon, even if it is not detected in the inland wells. These two processes revert the input of nutrients that could be inferred by simple measurements of groundwater inland.

Mixing of water with different salinities and origin can trigger reactions generating and/or degrading chemical components at multiple spatial and temporal scales in coastal aquifers. Addressing salinity distribution and the distinction of flow paths will help to develop better monitoring strategies, and more accurate models to improve the estimation of nutrient budgets due to SGD in coastal areas.

Supplementary Materials: The following are available online at http://www.mdpi.com/2073-4441/11/9/1842/s1, Table S1: In situ measurements, stable isotopes, nutrient and ion analysis for sampling points 1 to 30.

Author Contributions: Conceptualization, C.D. and P.E.; Formal analysis, C.D. and S.J.; Funding acquisition, C.D. and P.E.; Investigation, C.D., J.T.-C., S.K. and P.E.; Methodology, C.D., S.J. and P.E.; Project administration, C.D. and P.E.; Visualization, C.D. and S.J.; Writing—original draft, C.D.; Writing—review and editing, C.D., S.J., J.T.-C., S.K. and P.E.

Funding: The research leading to these results has received funding from the European Union Seventh Framework Programme FP7/2007-2013 under grant agreement number 624496, the European Union's Horizon 2020 Research and Innovation Programme under the Marie Sklodowska-Curie Grant Agreement number 722028 and the HOBE project (Hobe.dk) funded by the Villum Fundation. This work was carried out as part of the activities of the Aarhus University Centre for Water Technology, WATEC.

Acknowledgments: We thank Iris Tobelaim, Christian Hansen and students in the Hydrogeological field course (2016) for field and laboratory assistance.

Conflicts of Interest: The authors declare no conflict of interest.

References

1. Michael, H.A.; Post, V.E.A.; Wilson, A.M.; Werner, A.D. Science, society, and the coastal groundwater squeeze. *Water Resour. Res.* **2017**, *53*, 2610–2617. [CrossRef]

2. Valiela, I.; Foreman, K.; Lamontagne, M.; Hersh, D.; Costa, J.; Peckol, P.; DeMeo-Andreson, B.; Babione, M.; Sham, C.-H.; Brawley, J.; et al. Couplings of Watersheds and Coastal Waters: Sources and Consequences of Nutrient Enrichment in Waquoit Bay, Massachusetts. *Estuaries* **1992**, *15*, 443. [CrossRef]

3. Lee, V.; Olsen, S. Eutrophication and Management Initiatives for the Control of Nutrient Inputs to Rhode Island Coastal Lagoons. *Estuaries* **1985**, *8*, 191. [CrossRef]

4. Sinha, E.; Michalak, A.M.; Calvin, K.V.; Lawrence, P.J. Societal decisions about climate mitigation will have dramatic impacts on eutrophication in the 21st century. *Nat. Commun.* **2019**, *10*, 939. [CrossRef] [PubMed]

5. Vitousek, P.M.; Naylor, R.; Crews, T.; David, M.B.; Drinkwater, L.E.; Holland, E.; Johnes, P.J.; Katzenberger, J.; Martinelli, L.A.; Matson, P.A.; et al. Agriculture. Nutrient imbalances in agricultural development. *Science* **2009**, *324*, 1519–1520. [CrossRef]

6. Hallegraeff, G.M.; Hallegraeff, G. A review of harmful algal blooms and their apparent global increase. *Phycologia* **1993**, *32*, 79–99. [CrossRef]

7. Anderson, D.M.; Glibert, P.M.; Burkholder, J.M. Harmful algal blooms and eutrophication: Nutrient sources, composition, and consequences. *Estuaries* **2002**, *25*, 704–726. [CrossRef]

8. Lotze, H.K.; Milewski, I. Two centuries of multiple human impacts and successive changes in a north atlantic food web. *Ecol. Appl.* **2004**, *14*, 1428–1447. [CrossRef]

9. Niencheski, L.F.H.; Windom, H.L.; Moore, W.S.; Jahnke, R.A. Submarine groundwater discharge of nutrients to the ocean along a coastal lagoon barrier, Southern Brazil. *Mar. Chem.* **2007**, *106*, 546–561. [CrossRef]

10. Zimmermann, C.F.; Montgomery, J.R.; Carlson, P.R. Variability of Dissolved Reactive Phosphate Flux Rates in Nearshore Estuarine Sediments: Effects of Groundwater Flow. *Estuaries* **1985**, *8*, 228. [CrossRef]

11. Whiting, G.J.; Childers, D.L. Subtidal advective water flux as a potentially important nutrient input to southeastern U.S.A. Saltmarsh estuaries. *Estuar. Coast. Shelf Sci.* **1989**, *28*, 417–431. [CrossRef]

12. Valiela, I.; Costa, J.E. Eutrophication of Buttermilk Bay, a cape cod coastal embayment: Concentrations of nutrients and watershed nutrient budgets. *Environ. Manag.* **1988**, *12*, 539–553. [CrossRef]

13. Burnett, W.; Aggarwal, P.; Aureli, A.; Bokuniewicz, H.; Cable, J.; Charette, M.; Kontar, E.; Krupa, S.; Kulkarni, K.; Loveless, A.; et al. Quantifying submarine groundwater discharge in the coastal zone via multiple methods. *Sci. Total Environ.* **2006**, *367*, 498–543. [CrossRef] [PubMed]

14. Michael, H.A.; Mulligan, A.E.; Harvey, C.F. Seasonal oscillations in water exchange between aquifers and the coastal ocean. *Nature* **2005**, *436*, 1145–1148. [CrossRef] [PubMed]

15. Moore, W.S. Large groundwater inputs to coastal waters revealed by ^{226}Ra enrichments. *Nature* **1996**, *380*, 612–614. [CrossRef]

16. Andersen, M.S.; Baron, L.; Gudbjerg, J.; Gregersen, J.; Chapellier, D.; Jakobsen, R.; Postma, D. Discharge of nitrate-containing groundwater into a coastal marine environment. *J. Hydrol.* **2007**, *336*, 98–114. [CrossRef]

17. Bratton, J.F.; Böhlke, J.K.; Manheim, F.T.; Krantz, D.E. Ground Water beneath Coastal Bays of the Delmarva Peninsula: Ages and Nutrients. *Ground Water* **2004**, *42*, 1021–1034. [CrossRef]

18. Knee, K.L.; Street, J.H.; Grossman, E.E.; Boehm, A.B.; Paytan, A. Nutrient inputs to the coastal ocean from submarine groundwater discharge in a groundwater-dominated system: Relation to land use (Kona coast, Hawaii, U.S.A.). *Limnol. Oceanogr.* **2010**, *55*, 1105–1122. [CrossRef]

19. Povinec, P.; Burnett, W.; Beck, A.; Bokuniewicz, H.; Charette, M.; Gonneea, M.; Groening, M.; Ishitobi, T.; Kontar, E.; Kwong, L.L.W.; et al. Isotopic, geophysical and biogeochemical investigation of submarine groundwater discharge: IAEA-UNESCO intercomparison exercise at Mauritius Island. *J. Environ. Radioact.* **2012**, *104*, 24–45. [CrossRef]

20. Szymczycha, B.; Vogler, S.; Pempkowiak, J. Nutrient fluxes via submarine groundwater discharge to the Bay of Puck, southern Baltic Sea. *Sci. Total Environ.* **2012**, *438*, 86–93. [CrossRef]

21. Werner, A.D.; Bakker, M.; Post, V.E.; Vandenbohede, A.; Lu, C.; Ataie-Ashtiani, B.; Simmons, C.T.; Barry, D.; Barry, D. Seawater intrusion processes, investigation and management: Recent advances and future challenges. *Adv. Water Resour.* **2013**, *51*, 3–26. [CrossRef]

22. Duque, C.; Haider, K.; Sebok, E.; Sonnenborg, T.O.; Engesgaard, P. A conceptual model for groundwater discharge to a coastal brackish lagoon based on seepage measurements (Ringkøbing Fjord, Denmark). *Hydrol. Process.* **2018**, *32*, 3352–3364. [CrossRef]

23. Santos, I.R.; Eyre, B.D.; Huettel, M. The driving forces of porewater and groundwater flow in permeable coastal sediments: A review. *Estuar. Coast. Shelf Sci.* **2012**, *98*, 1–15. [CrossRef]

24. Slomp, C.P.; Van Cappellen, P. Nutrient inputs to the coastal ocean through submarine groundwater discharge: Controls and potential impact. *J. Hydrol.* **2004**, *295*, 64–86. [CrossRef]

25. Spiteri, C.; Slomp, C.P.; Tuncay, K.; Meile, C. Modeling biogeochemical processes in subterranean estuaries: Effect of flow dynamics and redox conditions on submarine groundwater discharge of nutrients. *Water Resour. Res.* **2008**, *44*. [CrossRef]

26. Rodellas, V.; Stieglitz, T.C.; Andrisoa, A.; Cook, P.G.; Raimbault, P.; Tamborski, J.J.; Van Beek, P.; Radakovitch, O. Groundwater-driven nutrient inputs to coastal lagoons: The relevance of lagoon water recirculation as a conveyor of dissolved nutrients. *Sci. Total Environ.* **2018**, *642*, 764–780. [CrossRef] [PubMed]

27. Dansgaard, W. Stable isotopes in precipitation. *Tellus* **1964**, *16*, 436–468. [CrossRef]

28. Barbieri, M. Isotopes in Hydrology and Hydrogeology. *Water* **2019**, *11*, 291. [CrossRef]

29. Blasch, K.W.; Bryson, J.R. Distinguishing Sources of Ground Water Recharge by Using δ^2H and δ^{18}O. *Ground Water* **2007**, *45*, 294–308. [CrossRef]

30. Duque, C.; Lopez-Chicano, M.; Calvache, M.L.; Martin-Rosales, W.; Gómez-Fontalva, J.M.; Crespo, F. Recharge sources and hydrogeological effects of irrigation and an influent river identified by stable isotopes in the Motril-Salobreña aquifer (Southern Spain). *Hydrol. Process.* **2011**, *25*, 2261–2274. [CrossRef]

31. Barbieri, M.; Nigro, A.; Petitta, M. Groundwater mixing in the discharge area of San Vittorino Plain (Central Italy): Geochemical characterization and implication for drinking uses. *Environ. Earth Sci.* **2017**, *76*. [CrossRef]

32. Jiang, W.; Wang, G.; Sheng, Y.; Shi, Z.; Zhang, H. Isotopes in groundwater (2H, 18O, 14C) revealed the climate and groundwater recharge in the Northern China. *Sci. Total Environ.* **2019**, *666*, 298–307. [CrossRef] [PubMed]

33. Mayr, C.; Lücke, A.; Stichler, W.; Trimborn, P.; Ercolano, B.; Oliva, G.; Ohlendorf, C.; Soto, J.; Fey, M.; Haberzettl, T.; et al. Precipitation origin and evaporation of lakes in semi-arid Patagonia (Argentina) inferred from stable isotopes (δ^{18}O, δ^2H). *J. Hydrol.* **2007**, *334*, 53–63. [CrossRef]

34. Battistel, M.; Hurwitz, S.; Evans, W.C.; Barbieri, M. The chemistry and isotopic composition of waters in the low-enthalpy geothermal system of Cimino-Vico Volcanic District, Italy. *J. Volcanol. Geotherm. Res.* **2016**, *328*, 222–229. [CrossRef]

35. Bolognesi, L.; D'Amore, F. Isotopic variation of the hydrothermal system on Vulcano Island, Italy. *Geochim. Cosmochim. Acta* **1993**, *57*, 2069–2082. [CrossRef]

36. Kirkegaard, C.; Sonnenborg, T.O.; Auken, E.; Jørgensen, F. Salinity Distribution in Heterogeneous Coastal Aquifers Mapped by Airborne Electromagnetics. *Vadose Zone J.* **2011**, *10*, 125. [CrossRef]

37. Haider, K.; Engesgaard, P.; Sonnenborg, T.O.; Kirkegaard, C. Numerical modeling of salinity distribution and submarine groundwater discharge to a coastal lagoon in Denmark based on airborne electromagnetic data. *Hydrogeol. J.* **2014**, *23*, 217–233. [CrossRef]

38. Müller, S.; Jessen, S.; Duque, C.; Sebok, E.; Neilson, B.; Engesgaard, P. Assessing seasonal flow dynamics at a lagoon saltwater–freshwater interface using a dual tracer approach. *J. Hydrol. Reg. Stud.* **2018**, *17*, 24–35. [CrossRef]

39. Fowler, D.; Coyle, M.; Skiba, U.; Sutton, M.A.; Cape, J.N.; Reis, S.; Sheppard, L.J.; Jenkins, A.; Grizzetti, B.; Galloway, J.N.; et al. The global nitrogen cycle in the twenty-first century. *Philos. Trans. R. Soc. B Boil. Sci.* **2013**, *368*, 20130164. [CrossRef]

40. Robertson, W.D. Development of steady-state phosphate concentrations in septic system plumes. *J. Contam. Hydrol.* **1995**, *19*, 289–305. [CrossRef]

41. Håkanson, L.; Bryhn, A.C. Modeling the foodweb in coastal areas: A case study of Ringkøbing Fjord, Denmark. *Ecol. Res.* **2007**, *23*, 421–444. [CrossRef]

42. Håkanson, L.; Bryhn, A.C.; Eklund, J.M. Modelling phosphorus and suspended particulate matter in Ringkøbing Fjord in order to understand regime shifts. *J. Mar. Syst.* **2007**, *68*, 65–90. [CrossRef]

43. Håkanson, L.; Bryhn, A.C. Goals and remedial strategies for water quality and wildlife management in a coastal lagoon—A case-study of Ringkøbing Fjord, Denmark. *J. Environ. Manag.* **2008**, *86*, 498–519. [CrossRef] [PubMed]

44. Petersen, J.K.; Hansen, J.W.; Laursen, M.B.; Clausen, P.; Carstensen, J.; Conley, D.J. Regime shift in a coastal marine ecosystem. *Ecol. Appl.* **2008**, *18*, 497–510. [CrossRef]

45. Tirado-Conde, J.; Engesgaard, P.; Karan, S.; Müller, S.; Duque, C. Evaluation of Temperature Profiling and Seepage Meter Methods for Quantifying Submarine Groundwater Discharge to Coastal Lagoons: Impacts of Saltwater Intrusion and the Associated Thermal Regime. *Water* **2019**, *11*, 1648. [CrossRef]

46. Lee, D.R. A device for measuring seepage flux in lakes and estuaries. *Limnol. Oceanogr.* **1977**, *22*, 140–147. [CrossRef]

47. Rosenberry, D.O.; Toran, L.; Nyquist, J.E. Effect of surficial disturbance on exchange between groundwater and surface water in nearshore margins. *Water Resour. Res.* **2010**, *46*. [CrossRef]

48. Butler, J., Jr. *The Design, Performance, and Analysis of Slug Tests*; Lewis Publication: Boca Raton, FL, USA, 1998.

49. Rosenberry, D.O. Integrating seepage heterogeneity with the use of ganged seepage meters. *Limnol. Oceanogr. Methods* **2005**, *3*, 131–142. [CrossRef]

50. Müller, S.; Stumpp, C.; Sørensen, J.H.; Jessen, S. Spatiotemporal variation of stable isotopic composition in precipitation: Post-condensational effects in a humid area. *Hydrol. Process.* **2017**, *31*, 3146–3159. [CrossRef]

51. Karan, S.; Kidmose, J.; Engesgaard, P.; Nilsson, B.; Frandsen, M.; Ommen, D.; Flindt, M.R.; Andersen, F. Østergaard; Pedersen, O. Role of a groundwater–lake interface in controlling seepage of water and nitrate. *J. Hydrol.* **2014**, *517*, 791–802. [CrossRef]

52. Kidmose, J.; Engesgaard, P.; Nilsson, B.; Laier, T.; Looms, M.C. Spatial Distribution of Seepage at a Flow-Through Lake: Lake Hampen, Western Denmark. *Vadose Zone J.* **2011**, *10*, 110. [CrossRef]

53. Hajati, M.-C.; Frandsen, M.; Pedersen, O.; Nilsson, B.; Duque, C.; Engesgaard, P. Flow reversals in groundwater–lake interactions: A natural tracer study using δ^{18}O. *Limnologica* **2018**, *68*, 26–35. [CrossRef]

54. Younger, P.L. Submarine groundwater discharge. *Nature* **1996**, *382*, 121–122. [CrossRef]

55. Kroeger, K.D.; Charette, M. Nitrogen Biogeochemistry of Submarine Groundwater Discharge. *Limnol. Oceanogr.* **2008**, *53*, 1025–1039. [CrossRef]

56. Sadat-Noori, M.; Santos, I.R.; Tait, D.R.; Maher, D.T. Fresh meteoric versus recirculated saline groundwater nutrient inputs into a subtropical estuary. *Sci. Total Environ.* **2016**, *566*, 1440–1453. [CrossRef] [PubMed]

57. Tamborski, J.J.; Cochran, J.K.; Bokuniewicz, H.J. Application of 224Ra and 222Rn for evaluating seawater residence times in a tidal subterranean estuary. *Mar. Chem.* **2017**, *189*, 32–45. [CrossRef]

58. Miljøstyrelsen. *Basisanalyse del I: Karakterisering af Vandforekomster og Opgørelse af Påvirkninger*; Miljøministeriet Departementet: Copenhagen, Denmark, 2004.

59. Belanger, T.; Mikutel, D.; Churchill, P. Groundwater seepage nutrient loading in a Florida Lake. *Water Res.* **1985**, *19*, 773–781. [CrossRef]

60. Nyvang, V. Redox Processes at the Salt-/Freshwater Interface in an Anaerobic Aquifer. Ph.D. Thesis, Technical University of Denmark, Lyngby, Denmark, 2003.

61. Price, R.M.; Savabi, M.R.; Jolicoeur, J.L.; Roy, S. Adsorption and desorption of phosphate on limestone in experiments simulating seawater intrusion. *Appl. Geochem.* **2010**, *25*, 1085–1091. [CrossRef]

62. Boehm, A.B.; Paytan, A.; Shellenbarger, G.G.; Davis, K.A. Composition and flux of groundwater from a California beach aquifer: Implications for nutrient supply to the surf zone. *Cont. Shelf Res.* **2006**, *26*, 269–282. [CrossRef]

63. Ullman, W.J.; Chang, B.; Miller, D.C.; Madsen, J.A. Groundwater mixing, nutrient diagenesis, and discharges across a sandy beachface, Cape Henlopen, Delaware (USA). *Estuar. Coast. Shelf Sci.* **2003**, *57*, 539–552. [CrossRef]

64. Søndergaard, M.; Jensen, J.P.; Jeppesen, E. Role of sediment and internal loading of phosphorus in shallow lakes. *Hydrobiologia* **2003**, *506*, 135–145. [CrossRef]

65. Daunys, D.; Petkuviene, J.; Zilius, M.; Bartoli, M. Sediment-water oxygen, ammonium and soluble reactive phosphorus fluxes in a turbid freshwater estuary (Curonian lagoon, Lithuania): Evidences of benthic microalgal activity. *J. Limnol.* **2012**, *71*, 33.

66. Petkuviene, J.; Zilius, M.; Lubiene, I.; Ruginis, T.; Giordani, G.; Razinkovas-Baziukas, A.; Bartoli, M. Phosphorus Cycling in a Freshwater Estuary Impacted by Cyanobacterial Blooms. *Chesap. Sci.* **2016**, *39*, 1386–1402. [CrossRef]

67. Heiss, J.W.; Michael, H.A. Saltwater-freshwater mixing dynamics in a sandy beach aquifer over tidal, spring-neap, and seasonal cycles. *Water Resour. Res.* **2014**, *50*, 6747–6766. [CrossRef]

68. Duque, C.; Müller, S.; Sebok, E.; Haider, K.; Engesgaard, P. Estimating groundwater discharge to surface waters using heat as a tracer in low flux environments: The role of thermal conductivity. *Hydrol. Process.* **2016**, *30*, 383–395. [CrossRef]

Article

A Hydrologic Landscapes Perspective on Groundwater Connectivity of Depressional Wetlands

Brian P. Neff [1,*], Donald O. Rosenberry [2], Scott G. Leibowitz [3], Dave M. Mushet [4], Heather E. Golden [5], Mark C. Rains [6], J. Renée Brooks [3] and Charles R. Lane [5]

[1] Former Post-Doctoral Research Hydrologist, National Research Program, U.S. Geological Survey, Lakewood, CO 80225, USA

[2] Hydro-Eco Interactions Branch, Water Mission Area, U.S. Geological Survey, Lakewood, CO 80225, USA; rosenber@usgs.gov

[3] Pacific Ecological Systems Division, Center for Public Health and Environmental Assessment, U.S. Environmental Protection Agency, Corvallis, OR 97333, USA; leibowitz.scott@epa.gov (S.G.L.); brooks.reneej@epa.gov (J.R.B.)

[4] Northern Prairie Wildlife Research Center, U.S. Geological Survey, Jamestown, ND 58401-7317, USA; dmushet@usgs.gov

[5] Center for Environmental Measurement and Modeling, Office of Research and Development, U.S. Environmental Protection Agency, Cincinnati, OH 45268, USA; golden.heather@epa.gov (H.E.G.); lane.charles@epa.gov (C.R.L.)

[6] School of Geosciences, University of South Florida, Tampa, FL 33620, USA; mrains@usf.edu

* Correspondence: bneff@uwaterloo.ca; Tel.: +1-303-747-4455

Received: 27 November 2019; Accepted: 17 December 2019; Published: 21 December 2019

Abstract: Research into processes governing the hydrologic connectivity of depressional wetlands has advanced rapidly in recent years. Nevertheless, a need persists for broadly applicable, non-site-specific guidance to facilitate further research. Here, we explicitly use the hydrologic landscapes theoretical framework to develop broadly applicable conceptual knowledge of depressional-wetland hydrologic connectivity. We used a numerical model to simulate the groundwater flow through five generic hydrologic landscapes. Next, we inserted depressional wetlands into the generic landscapes and repeated the modeling exercise. The results strongly characterize groundwater connectivity from uplands to lowlands as being predominantly indirect. Groundwater flowed from uplands and most of it was discharged to the surface at a concave-upward break in slope, possibly continuing as surface water to lowlands. Additionally, we found that groundwater connectivity of the depressional wetlands was primarily determined by the slope of the adjacent water table. However, we identified certain arrangements of landforms that caused the water table to fall sharply and not follow the surface contour. Finally, we synthesize our findings and provide guidance to practitioners and resource managers regarding the management significance of indirect groundwater discharge and the effect of depressional wetland groundwater connectivity on pond permanence and connectivity.

Keywords: hydrologic landscapes; landscape hydrology; depressional wetlands; geographically isolated wetlands; hydrologic connectivity; wetland connectivity; indirect groundwater connectivity; direct groundwater connectivity; VS2DI; groundwater-surface water interactions

---

1. Introduction

As their name implies, depressional wetlands are wetlands that form within depressions on the landscape where they have limited surface-water connectivity and are more dependent on atmospheric exchanges than other wetland types [1,2]. Interest in the hydrologic connectivity of depressional wetlands has accelerated in recent years, largely in response to the need for scientific information to guide wetland management.

A 2003 special issue of *Wetlands* provided a historical orientation to the legal and scientific context of depressional wetlands with limited surface-water connections [3], including a synthesis of the then-current scientific understanding and a guide to future research [4]. In the special issue, these wetlands were described as "geographically isolated." However, recent research has shown that even the surface-water connectivity of so-called "geographically isolated" wetlands has been underestimated [5,6], and the term has recently been described as being a misnomer [7]. A little over a decade after the special issue of *Wetlands*, the U.S. Environmental Protection Agency published a review and synthesis of over 1200 peer-reviewed scientific publications on the connectivity of headwater streams and wetlands [8]. This report identified connectivity of non-floodplain, primarily depressional, wetlands as a key gap in scientific understanding, particularly with regard to how connectivity varies over space and time. Lane, et al. [9] provided an updated review of the literature on non-floodplain wetlands.

Other studies are noteworthy for providing insights to the variable hydrologic regimes of depressional wetlands [10,11]. Intermittent connectivity to downstream waterbodies can be through surface water by 'fill and spill' [12–14] or 'fill and merge' [5] events. Groundwater can also flow between depressional wetlands and to downgradient waterbodies [15,16]. Hayashi, et al. [17] applied a water-balance perspective to simultaneously consider surface-water and groundwater connectivity and explain wetland permanence.

With regard to the effects of wetlands on other waterbodies, McLaughlin, et al. [18] modeled how depressional wetlands in Florida hydraulically moderate baseflow in downgradient waterbodies, even though no water passes from the wetland to the downstream waterbody. Golden, et al. [19] used a hybrid statistical and process-based modeling approach to quantify the cumulative effect of depressional wetlands on downgradient streamflow. Lane, et al. [9] provided a review of the hydrologic, physical, and chemical effects of non-floodplain wetland connectivity on downgradient aquatic systems. Rains, et al. [11] proposed that wide variability in wetland connectivity cumulatively creates a specific effect on the hydrologic functioning of downstream waterbodies.

The challenge remains to develop broad, perhaps universally applicable, insights into the hydrologic connectivity of depressional wetlands. Some studies have attempted to link landscape form, geology, and climate to conceptualize wetland persistence and sensitivity to land use [20,21]. Leibowitz, et al. [22] were notable for explicitly applying the hydrologic landscapes theoretical framework [23] to conceptualize the factors affecting hydrologic connectivity across a wide range of landscapes. The hydrologic landscapes theoretical framework was developed by Tom Winter in the 1980s and 1990s to provide a relatively simple framework to explain the flow of water across any given landscape, from highlands to lowlands, through groundwater or surface water [24–26]. Winter published two widely cited papers illustrating the hydrologic landscapes concept by explaining groundwater flow through six conceptual, two-dimensional landscapes [23,27]. Leibowitz, et al. [22] used these same landscapes to illustrate their framework for understanding connectivity. This approach is elegant in its understandability and use of a universally applicable theoretical framework with potential to conceptualize all aspects of hydrologic flow and connectivity in a landscape, including the interplay between surface water and groundwater.

While innovative and useful, the Leibowitz, et al. [22] study left important opportunities for future research. First, their study used the hydrologic landscapes framework to explain the relative magnitude of connectivity between floodplains and floodplain wetlands with rivers in diverse landscapes. They did not consider the connectivity of depressional wetlands existing within a landscape. Second, the focus of the Leibowitz study was the connectivity between headwaters and wetlands with rivers *downstream*. An opportunity exists to consider the importance of *upgradient* connectivity. Third, the Leibowitz paper effectively used the hydrologic landscapes framework to help explain diverse connectivity patterns reported in the literature but did not try to advance the hydrologic landscapes theoretical framework. This serves as the point of departure for the present study.

Our purpose in this study is to use the hydrologic landscapes perspective to develop insights into the hydrologic connectivity of depressional wetlands that are broadly, or universally, applicable. We define hydrologic connectivity to be the flow of water through various features of a landscape such as uplands, lowlands, wetlands and other surface water bodies and the ground. Our focus is primarily on groundwater flow and how groundwater interacts with surface waters. We numerically simulate hydrologic connectivity within five of the six generic landforms presented in Winter [23]: playa, plateau, mountain valley, riverine valley, and coastal terrain. We then insert depressional wetlands into these generic landforms and repeat the simulation exercise. We did not reproduce the sixth landform in Winter [23], hummocky terrain, largely because this landform is comprised of depressional wetlands. We do, however, return to the hummocky landform in our discussion and show how the lessons learned in our modeling apply to the patterns of groundwater flow in the hummocky landform in Winter [23].

In our numerical simulations, we evaluate the linkages between groundwater and surface water flows within the hydrologic-landscapes framework, then we evaluate how the presence of depressional wetlands affects hydrologic connectivity. Our evaluation has both qualitative and quantitative character. We assess the patterns and changes in flowpaths in the landscape both visually and by quantifying groundwater discharge to the surface and to depressional wetlands in the landscape.

The Hydrologic Landscapes Theoretical Framework

The hydrologic landscape theoretical framework describes the flow of water through *fundamental hydrologic landscape units* (FHLUs). Each FHLU has three components, a landform, climate, and geology [23,27]. The *landform* is the physical shape of the land and consists of an upland and a lowland connected by a slope. *Climate* is represented as the combined effect of precipitation and evapotranspiration (ET). *Geology* controls how water flows across or through the landscape in response to climate. When considered together, landform, climate, and geology comprise an FHLU (Figure 1).

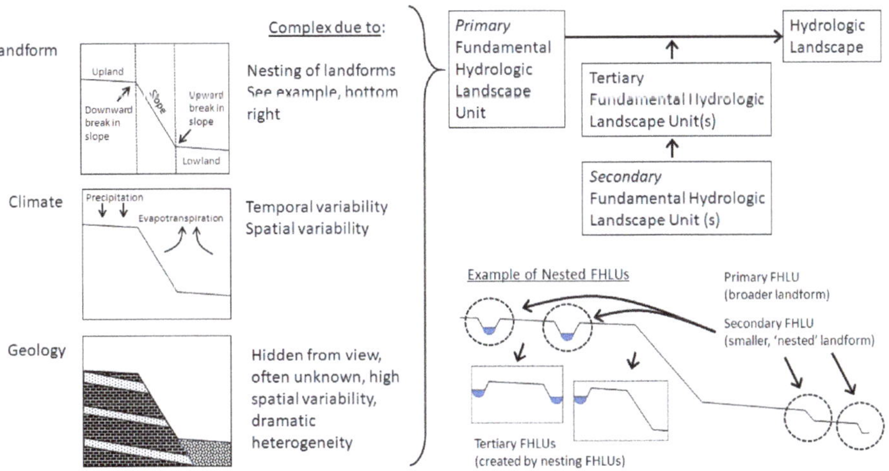

Figure 1. Conceptualization of a hydrologic landscape, consisting of one or more fundamental hydrologic landscape units (FHLUs). Each FHLU consists of a landform, climate, and geology. The right side of the figure depicts how FHLUs can become nested within a hydrologic landscape.

In practice, FHLUs exist at many scales. For example, a landform, defined as an upland and lowland divided by a slope, could be considered as the top of a plateau, an adjacent river, and the valley slope dividing the two. When combined with geology and climate this could be considered a plateau

FHLU. However, a landform could also be considered as being the upland area immediately above a depressional wetland, the slope of the depression, and wetland itself serving as the lowland. When this landform is combined with climate and geology it could be correctly considered a depressional wetland FHLU. Infinite possibilities of FHLUs exist. To further complicate matters, FHLUs are often embedded within one another. In the above examples, a depressional wetland FHLU could exist within a plateau FHLU. This exact scenario is considered in Figure 1 and later in our simulations.

The embedding of one FHLU within a relatively larger FHLU is described as *nesting* [23,27]. In the hydrologic landscape context, the nesting concept is unrelated to the concept of nested streamflow gauging stations or sampling sites and simply refers to one or more smaller FHLUs occurring within a larger FHLU. The configuration of a *hydrologic landscape* is defined as either a variant of a FHLU or as the sum of all nested FHLUs. For example, Winter [27] describes riverine hydrologic landscapes as having a landform "characterized by relatively broad lowlands that have smaller fundamental landscape units such as terraces nested within them." We expand here on the sparse hydrologic landscapes nomenclature to define the *primary FHLU* as a spatially larger unit, into which one or more *secondary FHLUs* are embedded, or nested. Depressional wetlands and terraces are common types of secondary FHLUs. The nesting of FHLUs often causes other, less obvious, landscape features with a highland and lowland divided by a slope, which we term *tertiary FHLUs* to draw attention to the nature of these features as being a byproduct of FHLU nesting. Tertiary FHLUs exist between or adjacent to secondary FHLUs such as depressional wetlands (Figure 1). A hydrologic landscape is the sum of all nested FHLUs.

In Figure 1, the broader, primary FHLU landform is depicted with a clear upland and lowland, divided by a slope. Four secondary FHLUs exist within the broader primary FHLU—two depressional wetlands and two terraces. The depressional wetland secondary FHLUs result in two areas that have their own highland and lowland, divided by a slope, shown in the two box insets labeled as tertiary FHLUs in Figure 1. In the leftmost tertiary FHLU, the two wetlands form the upland and lowland, and the slope actually rises and then falls in elevation between the two wetlands. In the rightmost tertiary FHLU in Figure 1, the wetland serves as the upland, the primary FHLU lowland serves as the lowland, and the slope again rises and falls between the two. When visually inspecting a landscape, it is convenient to first define the broad landform as the primary FHLU; second, to identify the embedded features such as depressional wetlands or terraces and define these as secondary FHLUs; and third, to focus on the less obvious tertiary landforms that are caused by the existence of secondary FHLUs.

2. Materials and Methods

The water table is crucial for determining hydrologic connectivity of wetlands. As a general rule, groundwater will not flow from a depressional wetland when surrounded by a mounded water table unless special circumstances exist [16,28,29]. The current study combined relatively simple geology and climate conditions to simulate groundwater flow and water table contour for five of the hydrologic landscapes presented in Winter [23]—playa, plateau, mountain valley, riverine valley, and coastal terrain. We chose these five landscapes to illustrate the hydrologic connectivity of depressional wetlands in a range of settings found throughout North America. In our simulations, we held the climate and geology variables as constant as possible while varying the landform to simplify our analysis of the effects of landform on connectivity of depressional wetlands. Later in the discussion section, we return to the consequence of variability in geology and climate.

Groundwater flow was simulated using the VS2DI model, described below, through each of the five domains presented in Winter [23,27]. Additional simulations were run using a second set of domains created by inserting depressional wetlands into various parts of each domain (Figure 2). There is an infinite combination of potential wetland locations and number. The location and number of depressional wetlands in the second set of domains were chosen to convey how a range of wetland locations affects the flow of water through each domain. Groundwater flow and connectivity were simulated in each domain.

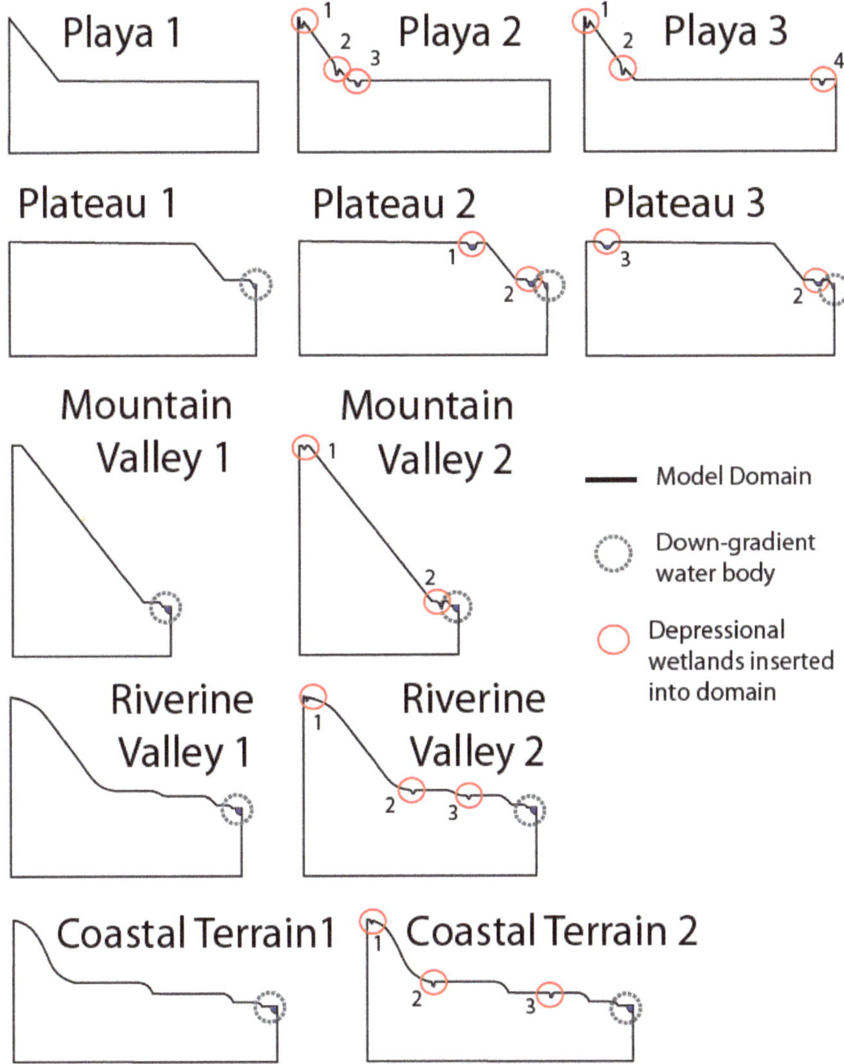

Figure 2. Model domains in this study. Depressional wetlands are numbered.

Tom Winter provided few details of his conceptual, generic landscapes. No specifics were provided regarding the dimensions of the domains, climatic conditions, or the geologic makeup of the domains other than being homogenous. Additionally, no detail was provided on how, or even if, the flow through each conceptual domain was numerically simulated. As a result, we used reasonable estimates of these variables in our simulations, and trial and error, to simulate a water table similar to that depicted in Winter [23,27]. With regard to the precise height and length dimensions of each model, we sought to create 'typical' dimensions for each landform relative to one another as depicted in Winter's work [23,27]. For example, mountains are typically taller and less expansive in the horizontal dimension than coastal landforms, and our chosen model dimensions reflect that. We chose moderately conductive, homogenous and isotropic geology and a net recharge (precipitation less evaporation) to represent a climatic condition within the range of what is commonly reported in North America. If our simulations reproduced a water table reasonably close to that reported in Winter [23,27], we stopped there.

The dimensions, climatic conditions, and geologic makeup of our model domains are provided in Table 1. The other decisions made regarding the model setup are provided in the following paragraphs. All models and simulation results presented here are archived and available for download at https://doi.org/10.23719/1504529. A supplemental to this article also contains our models, results, and instructions to rerun the simulations and reproduce our results (Supplementary Materials). Our purpose in this study was to draw broad lessons by conceptually modeling a range of hydrologic landscapes with and without embedded depressional wetlands, not to perfectly mimic the flows depicted in Winter [23,27]. In this context, we used the Winter [23,27] studies as a guide.

All simulations were performed using the Variably Saturated 2-Dimensional numerical model (VS2DI). VS2DI is a fully distributed model and uses a finite difference approach to solve the Richards equation for two-dimensional flow through unsaturated and saturated sediments [30,31]:

$$\frac{\partial(\theta(h)+sS_sH)}{\partial t} = \nabla \cdot K(h,T)\nabla H + q, \tag{1}$$

where $\theta(h)$ is volumetric soil water content, h is pressure head, s is saturation, S_s is specific storage, H is total head ($H = h - z$, with z being the positive downward vertical coordinate), $K(h,T)$ is the hydraulic conductivity tensor (assumed to be aligned with the coordinate axes and a function of temperature when heat transport is simulated), T is temperature, and q is the source/sink function. Healy and Essaid [30] also provided explanations of an alternate expression of the hydraulic conductivity tensor and how this equation has been modified from the original Richards equation to enable the simulation of the flow in both unsaturated and saturated zones.

Table 1. Cell sizes, net recharge, and geologic type used in each domain. Vert. is vertical, Horiz. is horizontal, Grad. is the regional topographic gradient. Parameters for generic geologic type followed by "CP" were documented in Carsel and Parrish [32], all others were described in Lappala, et al. [33]. Parameters were developed to replicate the water table and groundwater flow depicted in Winter [23] as closely as possible.

Domain	Dimensions (m)			Grid Cell Size (m)		Net Recharge (m/d)	Geology [1]
	Vert.	Horiz.	Grad.	Vert.	Horiz.		
Playa	1000	9500	0.11	10	63.3	2.00×10^{-4}	silty clay CP
Plateau	300	20,000	0.02	3	200	3.50×10^{-5}	silt-loam
Mountain Valley	300	7000	0.04	1.5	20	3.50×10^{-5}	silt-loam
Riverine Valley	250	10,000	0.03	0.75	20	3.50×10^{-5}	silt-loam
Coastal Terrain	200	11,500	0.02	1	20	3.50×10^{-5}	silt-loam

[1] van Genuchten Class.

The VS2DI model was selected for its ability to simulate flow in the unsaturated and saturated zones and allow the water table to adjust to fluxes in the model parameters. In addition, this program provides the ability to visually compare results between domains using a graphical user interface (GUI). Finally, the relative speed and ease of use of VS2DI allowed us to simulate many more scenarios than would have been possible with other programs.

For the geologic substrate, we assigned parameters using generic geologic types. In Table 1, the parameters for soil classes followed by "CP" were documented in Carsel and Parrish [32], and all others were described in Lappala, et al. [33]. We used silty clay CP for the geologic parameters for all playa simulations and silt-loam in simulations of other domains. The parameters for these geologic classes are available in the VS2DI model software and we encourage curious readers to download, inspect, and run our models for themselves.

The model grid cell sizes were selected by starting with relatively coarse cells and testing finer sizes until further reduction did not alter the groundwater flow patterns observed in the simulation results [34]. Ultimately, cell size is a tradeoff between model fidelity and computational demand, where model fidelity is the degree to which the actual attributes and processes of the system are accurately represented [35]. The cell sizes used in this study (Table 1) were deemed to be an appropriate compromise and sufficient for the study purpose of developing broadly applicable insight to wetland hydrologic connectivity.

The boundary conditions for all the model domains were assigned to produce results approximately consistent with Winter [23]. No-flow boundaries were assigned on the bottom and sides of each domain. Potential seepage faces [36] were assigned as needed to promote model stability at upward breaks in slope, adjacent to water bodies. In the case of the Plateau landscapes, promoting model stability while maintaining water table and flow patterns similar to those in Winter [23] required assigning a potential seepage face from the downward break in slope extending to the downgradient waterbody. Our use of seepage face boundary conditions near upward breaks in slope did not constrain seepage to these locations, as our model does permit seepage through specified flow (recharge) boundaries. However, use of the seepage face boundary conditions promoted model stability, and more realistically depicts seepage faces that are commonly observed at upward breaks in slope. The remaining portions of the top boundary were assigned a specified flux into the domain in a vertical direction, which is intended to represent net groundwater recharge (infiltration–evapotranspiration). The one exception to this is that the lower portions of the Playa domains were assigned a specified pressure head equal to zero to reproduce a thin unsaturated zone near the surface and to promote model stability.

The boundary condition for waterbodies was handled two ways. First, a specified total head boundary was assigned to low-lying water bodies (typically the modeled drain) and depressional wetlands extending downward to or below the depression-free FHLU water table. The specified total head was assigned to be equivalent to the surface elevation of ponded water in a depression. However, we found the specified head boundary condition to be unrealistic for ponded water in depressions existing above the normal water table. This condition caused an unrealistic water table to form, saturating up to tens of meters of the unsaturated zone with great and unsustainable (i.e., the wetland would soon become dry) amounts of downward seepage from the wetland. To produce more plausible results, we assigned a specified recharge boundary to these depressional wetland areas equal to one order of magnitude greater than that of the surrounding upland area.

Our decision to use a specified head boundary for low-lying water bodies necessitated the assumption that water seeping upward to the wetland is eliminated through ET, surface water outflow, possibly 'spill' surface water outflow, or groundwater outflow in the case of a flow-through wetland. All four of these conditions are features of real landscapes we regularly observe. This is not readily apparent in our 2D model figures and we acknowledge this here.

Our choice of a specified head boundary condition for low-lying wetlands caused the bending of groundwater flow from local and regional flow paths toward the wetland. Similar bending of groundwater flow toward depressional wetlands has been reported in field studies. For example, consider the water table elevation shown in Rains, et al. ([37] see Figure 4). This figure, and study, shows a water table which is warped towards the wetland complex at a surface-water outlet, where outflowing surface water depresses the wetland stage and therefore the local water table. The water table in this case resembles a cone of depression, but with an "outlet" on the downgradient side. This situation, in our experience, is typical in landscapes with any significant relief. The topographic slope in Rains, et al. [37] is approximately 0.02, the average slopes in our domains range from 0.015 to 0.105.

We considered an alternative boundary for low-lying wetlands—a seepage face boundary, where the model determines where the water table intersects the depression, a point at which we could declare the wetland pond elevation. A condition where the seepage-face boundary is likely better is where relief is very low, and the goal was to show seasonal oscillation between wetland recharge and discharge conditions driven by variability in runoff/precipitation and ET. We often see these switches between recharge–discharge conditions, with recharge–discharge neutrality at the crossovers and perhaps overall. Table 2 in Nilsson, et al. [38] described the case where wetlands are recharging local groundwater during the wet season and are receiving local groundwater discharge during the dry season. The landscape slope in the Nilsson case was close to zero. It would be difficult or impossible to model this oscillating condition and perhaps the best way to simulate it would be to simply let the wetland stage be set by the water table.

Table 2. Percentage of groundwater that discharges near the upward break(s) in slope of each landscape, within 500 m or as specified.

Domain and Break	Discharge	Domain and Break	Discharge
Playa 1	90.06%	Riverine Valley 2-1st break to wetland 2	100.00%
Playa 2-break to wetland #3	97.97%	Riverine Valley 2-2nd break to wetland 3	100.00%
Playa 3	91.99%	Riverine Valley 2-3rd break	48.12%
Plateau 1-600 m uphill to 500 m downhill of break	89.87%	Coastal Terrain 1-1st break	≫ 99.99%
Plateau 2-600 m uphill to 800 m downhill of break	90.15%	Coastal Terrain 1-2nd break	≫ 99.99%
Plateau 3-200 m uphill to 800 m downhill of break	87.76%	Coastal Terrain 1-3rd break	99.88%
Mountain Valley 1	99.95%	Coastal Terrain 2-1st break to wetland 2	100.00%
Mountain Valley 2-Break to wetland 2	98.27%	Coastal Terrain 2-2nd break to wetland 3	100.00%
		Coastal Terrain 2-3rd break	99.89%
Riverine Valley 1-1st break	99.98%		
Riverine Valley 1-2nd break	99.84%		
Riverine Valley 1-3rd break	77.54%		

Ultimately, both types of head boundaries for low-lying depressional wetlands have value and validity, provided they are interpreted correctly. Our purpose was to develop broadly applicable insight into the connectivity of depressional wetlands and we judged the constant-head boundary approach to better simulate the type of depressional wetland behavior we see in the field.

Simulations were run to steady state using homogenous isotropic geology. By maintaining uniform, relatively simple geology and climate conditions in our simulations, we intentionally isolated the effect of the landform on groundwater flow patterns. This approach is similar to that used in Winter [23,27], who depicted homogenous geology and steady-state conditions in developing the generalized hydrologic landscape figures. Furthermore, we used the same geologic and climatic conditions for all of the hydrologic landscapes, with the exception of the Playa landscapes. Geology and net recharge in the Playa landscapes were modified modestly to permit model stability while replicating the results in Winter [23,27] as described below.

For model domains without depressional wetlands, simulations were conducted using geologic substrate and net-recharge conditions suitable to replicate the water table and groundwater flow depicted in the generalized hydrologic landscape figures in Winter [23,27] as closely as possible. Depressional wetlands were inserted into the second set of model domains (Figure 2) and simulations were run using the same geology and climatic conditions. Readers can download and inspect our groundwater-flow models from a supplemental to this article or by accessing a separate publicly available archive at https://doi.org/10.23719/1504529.

3. Results

The results of our groundwater flow simulations are generally close to the water table contours and groundwater flow depicted in the iconic hydrologic landscapes in Winter [23]. Notable exceptions to this statement exist, such as the thin unsaturated zone reported in most of the Winter [23,27] landscapes and are explained in the Discussion section. The water table in our simulations closely matched those reported in Winter [23,27]; the water table declined in elevation at downward breaks in slope and rose to the top of the model domain at upward breaks in slope. However, in all simulations we found a large majority (range 48–100%, average 93.5%) of groundwater discharges to the surface at any particular concave-up break in slope (Table 2). Our use of seepage face boundary conditions near upward breaks in slope did not constrain seepage to these locations, as our model does permit seepage through specified flow (recharge) boundaries.

In all cases, the groundwater connectivity of depressional wetlands inserted into generic hydrologic landscapes depended on the surface elevation of ponded water in a wetland relative to the surrounding water table. However, a key aspect of our results is that in some cases, described below, the addition of one or more depressions to the model domain affects the elevation of the water table sufficiently to profoundly affect the groundwater and surface-water connectivity of the added depressional wetland(s).

3.1. Playas

The simulated groundwater flow through the generic Playa landscape (Playa 1; Figure 3) showed predictable flow paths from highlands to lowlands. However, 90% of recharge from upland areas discharged within 500 m of the upward break in slope. Discharge decreased rapidly as distance from the break in slope increased (Table 2). In fact, only 0.002% of all groundwater discharge occurred on the more distant half of the lowland. It is important to recognize that groundwater discharge at the upward break in slope either evaporates or continues flowing downgradient as surface water. As discussed further in the Discussion section, real-world playas are often arid and ET draws the water table well below land surface, causing many depressional wetlands to be recharge wetlands.

The Playa 2 landscape had three depressional wetlands inserted into the Playa 1, generic landscape, Wetland 1, 2, and 3. Wetland 1 was positioned near the top of the playa highland. This wetland was well above the water-table elevation at this location and recharged groundwater, causing the water table to rise somewhat relative to the Playa 1 landscape without wetlands. Wetland 3 was positioned slightly downgradient of the upward break in slope, with a water surface below the surrounding water table. This wetland received groundwater discharge from all sides and did not connect to downgradient waterbodies by groundwater. Wetland 2 is a more complex case. It is situated in a location that had a water table near or at ground level in the generic Playa 1 landscape. Thus, one might expect the water surface of Wetland 2 to be below the surrounding water table and to receive groundwater from all sides. However, the water table on the downgradient side of Wetland 2 does not rise with the land surface. It instead decreases in elevation to form a continuous downward slope away from Wetland 2, causing it to become a flow-through wetland by receiving groundwater discharge on the upgradient side and recharging groundwater on the downgradient side. With regard to surface water connectivity, it is possible that any of the three wetlands could fill and spill and connect to downgradient waterbodies by surface water. Wetland 3 is the most likely of the three wetlands to overfill its depression and spill for two main reasons. First, by virtue of the lower position of Wetland 3 on the landscape relative to Wetlands 1 or 2, Wetland 3 has a larger watershed and therefore is likely to receive more runoff. Second, Wetland 3 is the only wetland receiving groundwater from all sides and therefore is likely to receive more groundwater inflow.

The Playa 3 landscape is similar to the Playa 2 landscape, except Wetland 3 is moved to a position on the lowland far away from the break in slope and renamed Wetland 4. Wetland 4 exists below the surrounding water table and receives groundwater discharge on all sides. However, the presence of Wetland 4 created a groundwater flow divide near the center of the lowland portion of the Playa 3 landscape (Figure 3). This flow divide completely eliminates the groundwater flow from the playa upland to the far end of the lowland and Wetland 4. However, local groundwater flow to Wetland 4 from the nearby lowland was significant and vastly exceeded the exceptionally small groundwater flow in the far end of the Playa 1 and Playa 2 landscapes.

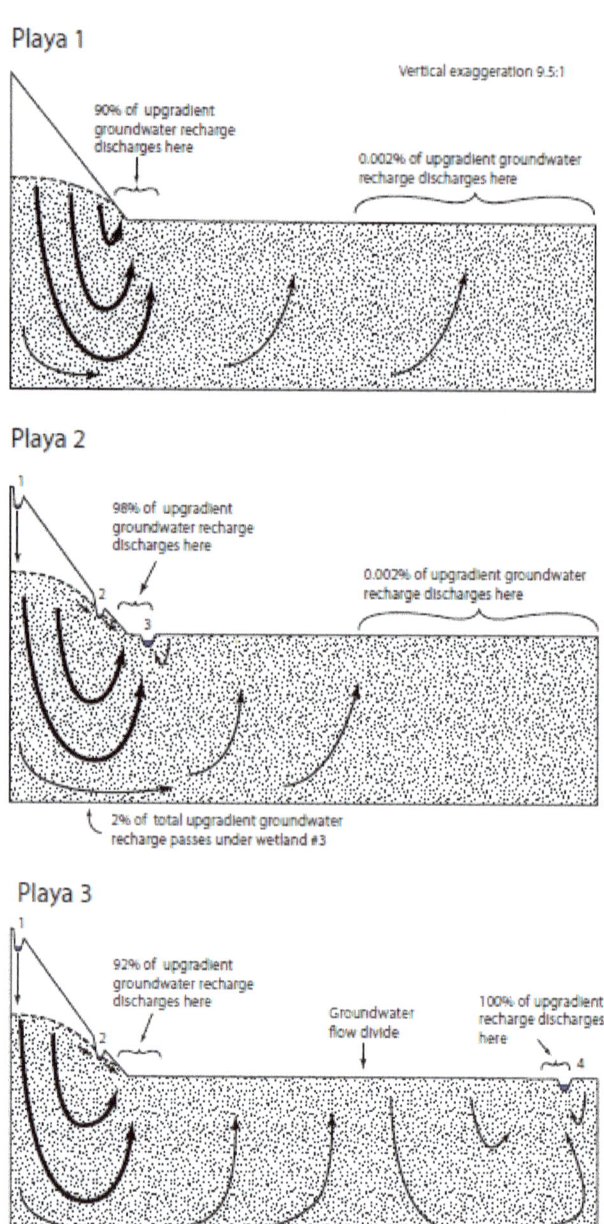

Figure 3. Simulated flow through the Playa landscapes.

3.2. Plateau

The Plateau landscapes (Figure 4) experienced large groundwater discharge near the upward break in slope. Similar to the Playa landscapes, 88–90% of the groundwater flow discharged near the upward break in slope and potentially continued toward the downgradient waterbody as surface water (Table 2). The seepage face on the uphill side of the break in slope was much wider in the Plateau 1 and Plateau 2 landscapes relative to the Plateau 3 landscapes. Discharge near the break in slope was roughly proportional to the extent of the seepage faces, as 61%, 63%, and 55% of groundwater discharge occurred uphill of the break in slope, respectively.

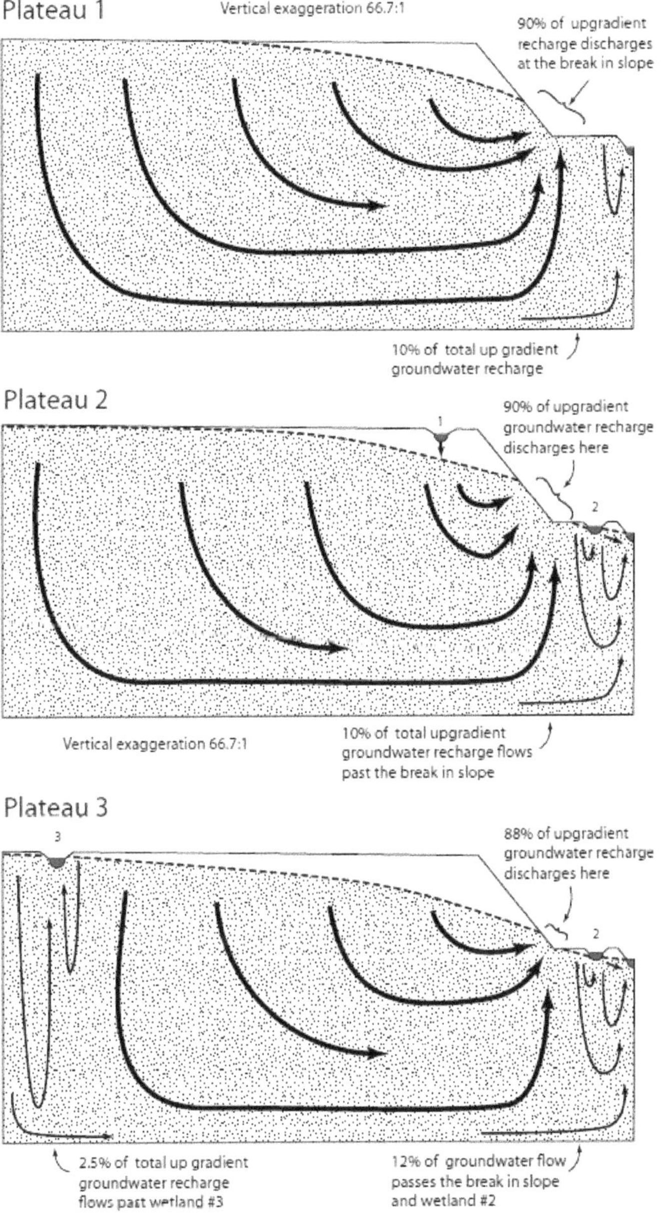

Figure 4. Simulated flow through the Plateau landscapes.

Insertion of depressional wetlands into the Plateau 2 landscape influenced groundwater flow much like in the Playa landscapes. In the Plateau 2 landscapes, Wetland 1 exists well above the water table and recharges groundwater. Wetland 2 exists in a groundwater discharge area and is in an area with a relatively high water table in the generic landscape. Nevertheless, the water table on the downgradient side of Wetland 2 did not form a water-table mound but instead sloped continuously downward, indicating flow toward the model drain. This caused Wetland 2 to be a flow-through wetland. In the Plateau 3 landscape, Wetland 1 was moved to the far-left side of the domain and renumbered Wetland 3. The simulated stage of Wetland 3 was lower than the surrounding water table and the wetland, and therefore received groundwater discharge from all sides. With regard to surface-water connectivity, it is possible for any of the wetlands in the Plateau landscapes to fill and spill.

3.3. Mountain Valley

The generic Mountain Valley 1 landscape (Figure 5) was similar to the generic Plateau and Playa landscapes, except nearly all water recharged in the uplands discharged at the upward break in slope, with only 0.05% of upland recharge continuing to the downgradient waterbody as groundwater (Table 2). In the Mountain Valley 2 landscape, Wetland 1 was inserted near the highest position in the landscape. This wetland was many tens of meters above the water table and simply recharged groundwater, causing a slight but largely inconsequential increase in the elevation of the water table. Inserting Wetland 2 at the upward break in slope provided an example of a depressional wetland where one might expect groundwater to discharge to the wetland from all sides. However, the water table on the downgradient side of Wetland 2 slopes continuously downward toward the model drain, allowing Wetland 2 to be a flow-through wetland. All other simulated patterns of groundwater flow were similar to the generic, wetland-free, Mountain Valley 1 landscape.

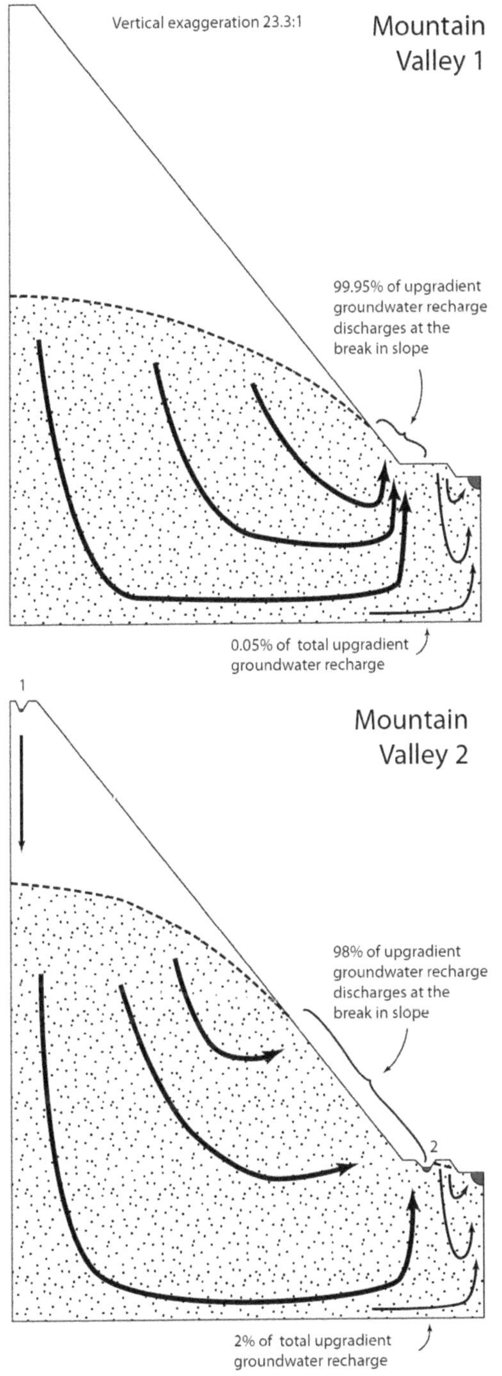

Figure 5. Simulated flow through the Mountain Valley landscapes.

3.4. Riverine Valley

The generic Riverine Valley 1 landscape (Figure 6) had three upward breaks in slope, causing relatively complex groundwater flow patterns. Nearly all of the groundwater discharged at the two highest upward breaks in slope (Table 2). In a technical sense, some small groundwater connection exists between the upland area and the downgradient water body. In a practical sense, the groundwater connection is likely insignificant. As with all other hydrologic landscapes, groundwater discharges to the surface and then either evaporates or continues its journey to downgradient water bodies as surface water. Adding depressional wetlands in the Riverine Valley 2 landscape created several groundwater-flow boundaries and relatively complex flow paths. Groundwater flow from upland areas to downgradient waterbodies is completely eliminated.

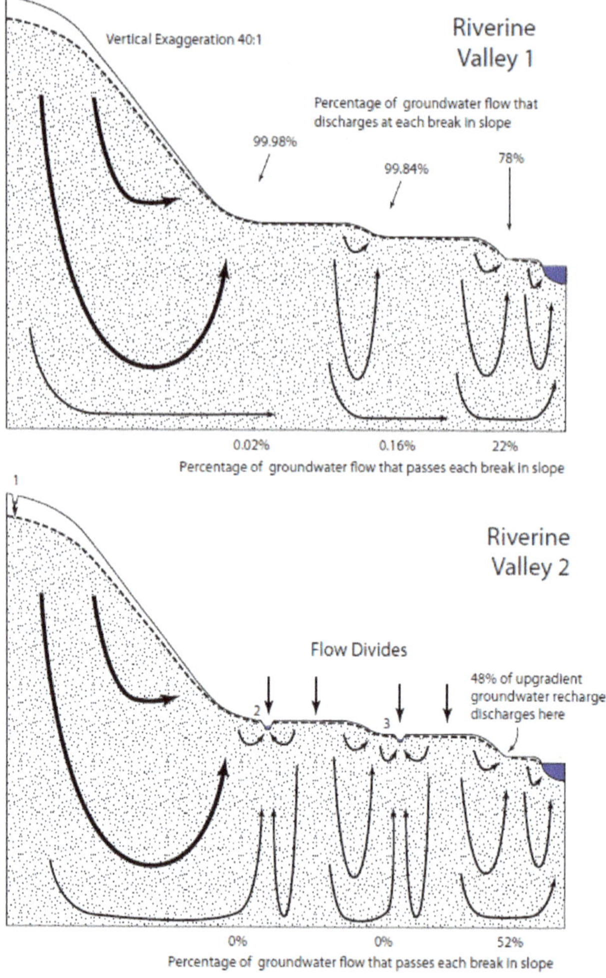

Figure 6. Simulated flow through the Riverine Valley landscapes.

3.5. Coastal Terrain

Groundwater flow patterns in the Coastal Terrain 1 and 2 landscapes (Figure 7) followed nearly the same patterns observed in the Riverine Valley 1 and 2 landscapes, respectively. Notably, the discharge at the upward breaks in slope was very nearly 100% (Table 2).

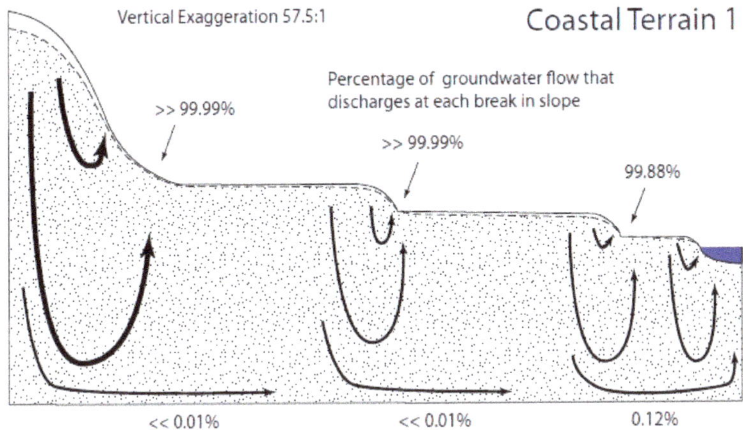

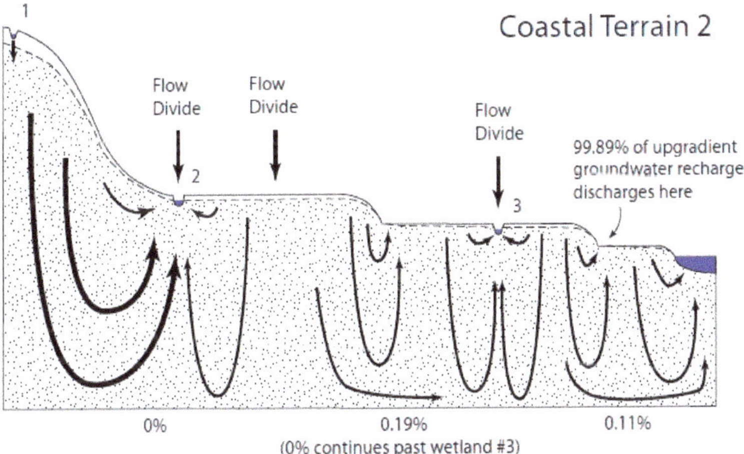

Figure 7. Simulated flow through the Coastal Terrain landscapes.

Readers can reproduce and manipulate the patterns we observed by accessing the publicly available archive of our groundwater-flow models at https://doi.org/10.23719/1504529 or in a supplemental to this article, running the models and then modifying the parameters in additional simulations. Modifying individual depressions to be deeper or shallower than the water table at a location is particularly illustrative.

4. Discussion

4.1. Groundwater Discharge at the Upward Break in Slope

The point where a downhill slope breaks, or begins to 'flatten out', is termed an upward break in slope (Figure 1). The existence of groundwater discharge at the upward break in slope has been widely acknowledged in the literature and is likely intuitive to many hydrologists [23,26,39,40]. However, to our knowledge, the proportion of groundwater flow that discharges at a break in slope has not explicitly been quantified or emphasized. In our simulations, a large majority of groundwater flow, between 48% and very nearly 100%, discharged to the surface at upward breaks in slope in the landscapes modeled in this study. Direct groundwater flow from an upland area in a landscape to a downgradient waterbody or drain still occurred, but it was very small (Table 2). Nested FHLUs (e.g., the terraces in Figures 6 and 7 and all wetland depressions) increased the number of upward breaks in slope and amplified this effect. Depressional wetlands added to the Riverine Valley and Coastal Terrain hydrologic landscapes, created flow divides, and caused all groundwater flow to discharge to the surface in the two-dimensional plane we simulated (Figures 6 and 7).

There exists an interesting question of what happens to the groundwater that rises toward the surface. At a basic level, the water rising to the surface either evaporates, is transpired, or becomes surface water and continues downgradient. Groundwater moving toward the surface may be intercepted by plants and transpired or may evaporate as it nears and reaches the surface. In many cases, ET lowers the water table and helps prevent groundwater discharge to the surface. In this case, groundwater may rise to the water table and flow laterally as groundwater until the water table intersects the surface. Finally, if the water table coincides with the surface, groundwater may discharge to the surface and become surface water. We believe the water table in the hydrologic landscape figures in Winter [23,27] reflects this process.

For example, the mountain valley landscape figure from Winter [23] is reproduced here as Figure 8. Note that a seepage face is depicted for one side of the valley but not the other. This feature of the mountain valley landscape figure is not directly addressed in either Winter [27] or Winter [23]. However, since the landform is identical on either side of the valley, we may implicitly interpret the difference in water table as being caused by either geology or climate conditions. Given the apparently homogenous geology throughout the landscape, climate conditions are the most likely suspect; in this case, riparian transpiration provides a plausible explanation for maintaining the water table beneath the surface. The seepage face on the left side of Figure 8 likely indicates diminished ET that is less than the quantity of groundwater moving toward the surface. On the right side of Figure 8, the water table is maintained beneath the surface and groundwater flow is forced to continue downgradient where it discharges directly to the waterbody at the landscape lowland.

Our models accommodated ET indirectly, by using a recharge boundary in upland areas equal to net recharge, defined as precipitation less ET. We did not simulate spatial heterogeneity in ET. Commonly, especially dense vegetation exists at the fringes of wetlands and causes a depression in the water table at the edge of the wetland pool, actually drawing water out of the wetland pool toward the root zone on the periphery of the wetland [41] (see especially Figure 7). Since we did not simulate this phenomenon, the seepage faces in our simulations may better be interpreted as areas where groundwater discharge to the wetland may occur, but not necessarily to the pool of water within the wetland. Hayashi, et al. ([17] see especially Figure 2) provided a useful explanation of the distinction between the spatial extent of a wetland and the extent of the pool of water within a wetland.

If groundwater upwelling near the upward break in slope manages to avoid being lost to ET and reaches the surface, it becomes surface water and continues downgradient. This surface water may also be lost to ET before flowing to a downgradient waterbody, or it may even return to the groundwater system as recharge. It is best to think of groundwater that discharges to the upward break in slope as *possibly* being connected to downgradient waterbodies. We interpret the groundwater that discharges to the surface in our models to represent the potential magnitude of connectivity through this route.

From a management perspective, this is worth considering. Groundwater discharge to upward breaks in slope has the potential to transform groundwater connectivity to downgradient waterbodies from a direct groundwater connection to an indirect connection via surface-water. Callahan, et al. [42] described this exact process in action, showing that much of the nitrogen-rich groundwater from alder-covered hillslopes is first discharged to and modified by toeslope wetlands in salmon-bearing headwater streams in Alaska.

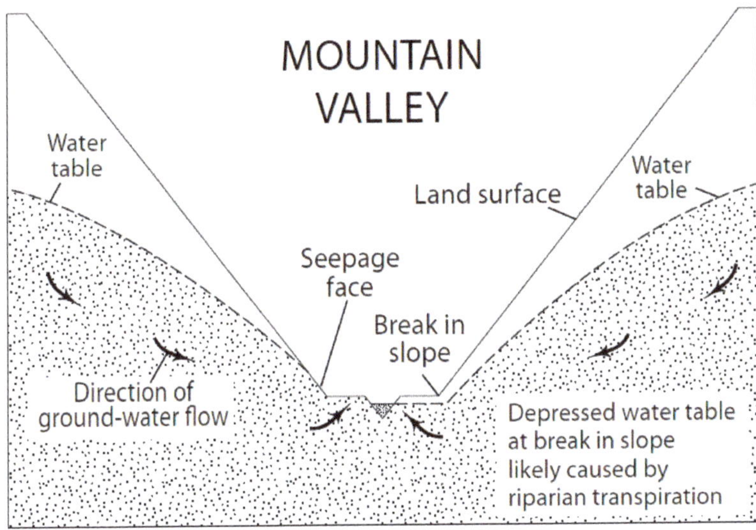

Figure 8. Mountain Valley hydrologic landscape modified from Winter [23].

4.2. Other Support for Study Findings

The conceptual modeling approach taken in this study has distinct limitations. However, alternate lines of reasoning support the conclusion that a large portion of groundwater may discharge to upward breaks in slope rather than flow directly to downgradient waterbodies. First, a wealth of literature documents groundwater discharge to the surface at areas near upward breaks in slope, far away from downgradient waterbodies. This type of groundwater discharge is responsible for stream baseflow, which can be a large portion of total streamflow [43–45]. The literature on this point is so abundant that our contribution here is merely to draw attention to the importance of breaks in slope and the potential magnitude of this type of hydrologic connection.

A second line of reasoning to support our conclusions with regard to groundwater discharge at upward breaks in slope is provided by observations of in situ hydrologic conditions. For example, we can observe that stream headwaters often occur at upward breaks in slope. Figure 9 provides an example of stream headwaters occurring at an upward break in slope, likely indicating locations of groundwater discharge. In this example, water flows from the Missouri Coteau highlands, at the left of the figure, past the Missouri Escarpment, which slopes downward from west to east, to the Drift Prairie lowlands at the right. Groundwater recharge likely occurs primarily on the Missouri Coteau, and primarily discharges near the upward break in slope on the Missouri Escarpment. Water that discharges near the break in slope often contributes to the headwaters of streams that flow to Pipestem Creek as surface-water flow, which we describe as indirect groundwater connectivity to Pipestem Creek. This process reduces the length of the relatively slow groundwater flowpath and introduces a much faster surface water flowpath, which can be expected to greatly reduce the time required for groundwater recharge from the Missouri Coteau to reach Pipestem Creek.

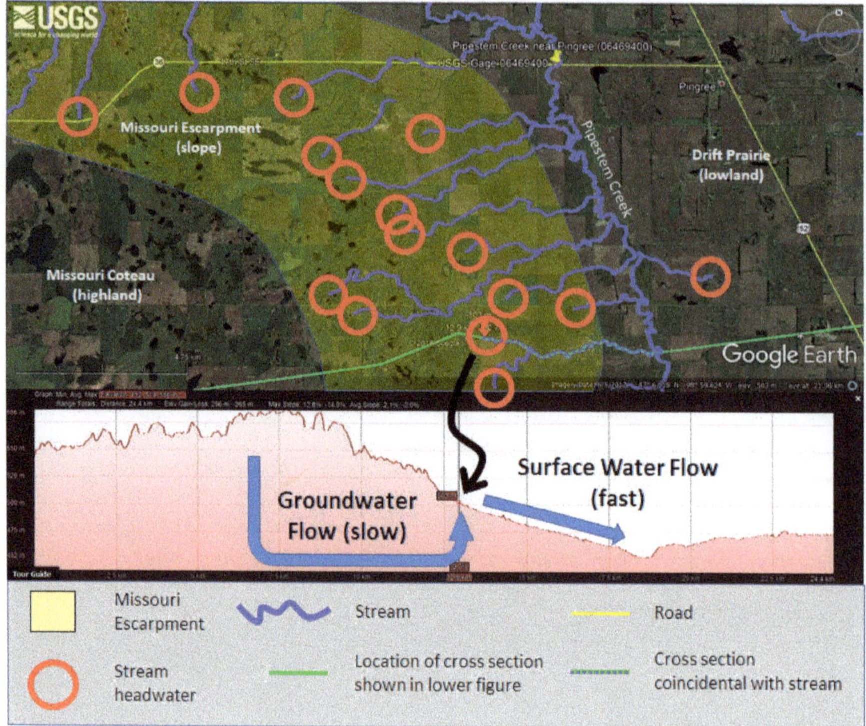

Figure 9. Annotated Google Earth screen capture illustrating the flow of water across a landscape near Pingree, ND. At bottom is an elevation profile of the cross section drawn through the image. The position of the headwaters of a tributary creek in the cross section is indicated on the cross section by the black arrow at a major upward break in slope.

Recently, Brooks, et al. [46] conducted a water isotope (δ^{18}O and δ^2H) study in this area that confirms the general assertions in the previous paragraph and refines the likely flowpaths in this particular setting. These authors found that the majority of water within Pipestem Creek originated from isotopically depleted groundwater. Isotopic variance within surface water throughout the watershed was related to the extent of evaporation of water residing in depressional wetlands, supporting the idea that direct and indirect pathways of groundwater dominated the flow in Pipestem Creek. The proportion of unevaporated water (groundwater with low residence time in the surface water system) increased in the downstream direction, indicating that wetland storage of groundwater was more important in the upper parts of the Pipestem watershed. The tributaries draining the Missouri Escarpment also showed the distinct groundwater signal, but they varied greatly in the degree of evaporation. However, they did not vary isotopically with precipitation inputs, indicating groundwater was the primary water source.

4.3. Consequences of Groundwater Discharge at the Upward Break in Slope

Groundwater discharging directly to tributary streams from adjacent uplands has profound consequences. This flow path maintains stream baseflow [45], moderates temperature [44,47], provides thermal refuge for aquatic species [48], and supplies nutrients to streams [42].

The magnitude of groundwater discharge at upward breaks in slope can enhance pond permanence of depressional wetlands and may support fill and spill or fill and merge surface water connectivity.

For example, springbrooks are floodplain wetlands that receive groundwater discharge, which then flows overland to a nearby stream [49]. Much of the groundwater discharge to springbrooks may originate as regional groundwater flowing from nearby uplands, often flowing over long distances [50,51].

The effect of groundwater discharge on downstream waterbodies is affected by the physical flow path taken to the downstream waterbody, as well as chemical and biological processes along that flow path [52]. Groundwater discharge to the upward break in slope often supports wetlands, headwaters of streams, and stream baseflow [44]. Farther downstream, the chemical and physical makeup of this water is attenuated by many processes that are notable. For example, evaporation affects groundwater that discharges to the surface and continues downgradient. This alters the isotopic signature of the groundwater source and Brooks, et al. [46] used this evaporation signal to estimate the importance of surface water storage on downstream flows.

The groundwater-discharge-supported surface flows from breaks in slope to downgradient waters provide pathways along which aquatic biota can move, thereby influencing a landscape's biotic connectivity via surface flows. However, while two-dimensional models like the ones we used in our simulations may be sufficient for elucidating the general aspects of water flows within a given FHLU, due to the fact that the movements of many organisms are not limited to aquatic pathways (e.g., adult amphibians, flying insects [53]), these models are overly simplistic in terms of identifying wetland connectivity that can result from the multidirectional movements of biota. However, such water flow models are needed to inform biotic connectivity models. Mushet, et al. [54] described what they called "freshwater ecosystem mosaics" to facilitate the marriage of models that consider movements of water, materials, and/or biota along specific water-flow pathways (e.g., streams, rivers), linking individual waterbodies (e.g., wetlands, lakes) with ecosystem models that include exchanges of energy, nutrients, materials and organisms between the aquatic features and surrounding upland areas. The knowledge gained from the use of two-dimensional FHLUs can be used to inform the transition to three-dimensional models that include the spatial distribution of water bodies, interconnecting flows, and the myriad influences of the upland areas between, i.e., all of the components of a complete freshwater ecosystem mosaic.

In addition to providing surface-water pathways along which biota can move and therefore support biotic connectivity, the substantial groundwater discharge that can occur at a break in slope can lead to unique communities occurring in these areas, thereby supporting biotic diversity [55]. For example, slope wetlands (e.g., fens) that do not pond water can occur at these slope breaks [2,56]. In places such as the prairie pothole region, slope wetlands and the plant and animal communities they support are much rarer than the depressional wetlands used in our simulations. The presence of slope wetlands created by the conversion of groundwater to surface flows at breaks in slope can add greatly to an area's overall biotic diversity [57].

Cumulatively, the magnitude of groundwater discharge at upward breaks in slope provides a potentially powerful connection from recharge areas, possibly including depressional wetlands, to downgradient waters. This point is easily lost when considering the hydrologic connectivity of wetlands to downgradient waterbodies as either a surface water (fill and spill) or groundwater (direct discharge) process.

4.4. General Patterns of Wetland Connectivity and Pond Permanence

The results illustrate that the elevation of the ponded-water surface of a depressional wetland relative to the elevation of the surrounding water table is the crucial variable that determines how the wetland pond is connected to other aquatic features on the landscape via groundwater. Where the surface of a wetland is above the surrounding water table, the wetland recharges the aquifer and groundwater flows away from it. Conversely, where the ponded-water surface of a wetland is beneath the surrounding water table, the wetland will normally receive groundwater discharge from all sides. Where the water table intersects the ponded-water surface of a wetland, the wetland will normally act as a flow-through wetland, both receiving and recharging groundwater (Table 3). These findings are consistent with the other literature [18,58]. Temporal variability in the balance between precipitation and evapotranspiration also causes many wetlands to transition between supplying groundwater recharge, receiving groundwater discharge, and serving as flow-through wetlands [59]. No exceptions to this general pattern were observed in this study, but Neff and Rosenberry [16] provided a summary of special conditions that can lead to exceptional cases.

Table 3. Summary of general patterns of wetland connectivity.

Wetland Surface Relative to Water Table (Dashed Line)	Connectivity in Upgradient Direction	Connectivity in Downgradient Direction	Likely to Spill?	Likely Pond Permanence	Likely Location(s)
Above	Surface water only	Groundwater; Surface water, if fill and spill occurs	No, unless depression is especially shallow	Low	Groundwater recharge areas, especially near a downward break in slope.
Below	Groundwater and surface water	Surface water, if fill and spill occurs	Yes, unless depression is especially deep	High	Groundwater discharge areas, especially if not near downward breaks in slope.
Same	Groundwater and surface water	Groundwater; Surface water, if fill and spill occurs	Possibly	Moderate	Slopes, also groundwater discharge areas if near a downward break in slope.

Whether a wetland receives or contributes to groundwater has profound implications for the nature of a depressional wetland's connectivity to the landscape and pond permanence. Wetlands that recharge groundwater (e.g., those with a ponded water level above the surrounding water table) are connected by groundwater to other parts of the landscape in the downgradient direction only. These wetlands also continually lose water to the groundwater system, are more likely to have low pond permanence [60,61], and may be less likely to fill and spill or fill and merge [5].

Depressional wetlands that receive groundwater discharge (e.g., those with a ponded water level below the surrounding water table) are connected by groundwater to other parts of the landscape in the upgradient direction only. These wetlands continually gain water from the groundwater system, are more likely to have high pond permanence, and are more likely to fill and spill or fill and merge.

Depressional wetlands that both receive and recharge groundwater are connected by groundwater to other parts of the landscape in both the upgradient and downgradient directions. Groundwater may not significantly affect the pond permanence or fill and spill and fill and merge behavior of these wetlands.

In concept, groundwater discharge at upward breaks in slope may also provide a more subtle source of water to help tip the water balance of a depressional wetland toward filling and spilling or merging. However, the current analysis was not intended to evaluate fill and spill or fill and merge. In particular, there are two limitations that make it difficult to examine this issue rigorously. First, fill and spill and fill and merge can be associated with high precipitation conditions in combination with high antecedent moisture or wet periods such as the spring snowmelt [13,62,63], conditions that can last for years due to possible regime shifts [5]. Since the present analysis used long-term average annual conditions, fill and spill and fill and merge that occur from shorter-term high precipitation events or periods are not included. Second, the current analysis has a limited ability to differentiate between fill and spill vs. fill and merge, since the scenarios do not include pairs of nearby wetlands. However, note that in both cases the existing model could be run in such a way in the future as to address these issues—first, by including higher frequency precipitation conditions and, second, by including scenarios that feature pairs of nearby wetlands and associated upstream contributions.

This would provide insight into which aspects of hydrologic landscapes have a greater groundwater influence on wetland fill and spill or fill and merge and the conditions leading to that occurrence.

4.5. Key Variables Affecting Connectivity

Our results provide insight to conditions and processes that are particularly impactful on the orientation of a given depressional wetland to the surrounding water table. The actual conditions on the landscape vary and are dependent on several key variables.

4.5.1. Climate and Geology

In the hydrologic landscapes theoretical framework, the water-table elevation of a given landform is a product of climate and geology. With regard to climate, areas with high groundwater recharge and low evapotranspiration facilitate a high water table relative to areas with low groundwater recharge and high evapotranspiration. With regard to geology, less permeable materials slow the gradient-driven flow of groundwater and raise the water table relative to coarse, more permeable materials. In this study, we held climate and geology as constant as possible in order to focus on the effect of landform on hydrologic flow through diverse landscapes; we recognize this intentional limitation.

Short-term timing of weather-based infiltration may cause the water table to fluctuate around a given depressional wetland. This may cause individual depressional wetlands to flip from recharge to flow-through to discharge wetlands, or possibly the reverse. This will alter the connectivity of *individual wetlands* with the landscape. However, it does not alter how each *type* of wetland is hydrologically connected to the landscape. In other words, the connectivity patterns of discharge, recharge, and flow-through wetlands will not change, even though individual depressional wetlands may shift between these states due to short-term fluctuations of the water table.

4.5.2. Depth of Wetland Basin

The depth of the depression in which a wetland sits affects hydrologic connectivity in two ways. First, the deeper the depression, the more likely it is that the bottom of the depression will be below the surrounding water table. When combined with ET from the wetland, this scenario causes groundwater to flow toward a wetland and isolates it from downgradient waterbodies with regard to groundwater flow. The constant discharge of groundwater to these wetlands enhances the permanence of their ponded water.

With regard to surface-water connectivity, deeper depressions are less likely to be connected to downgradient waterbodies by fill and spill, for two reasons. Deeper depressions are inherently larger in volume than shallow depressions of a similar width-to-height ratio and will require the addition of a greater volume of water before spilling occurs. This effectively reduces the likelihood that fill and spill will occur. It is notable that deeper depressions are more likely to extend below the surrounding water table, causing groundwater discharge to the wetland. This condition would likely increase the inflow to a given wetland. However, as the deep-depression wetland fills, the surface of the ponded water will likely eventually rise to the same elevation as the surrounding water table. Once this occurs, groundwater discharge to the wetland will cease. The combined effect of deeper depressions is to make a depressional wetland less likely to fill and spill.

4.6. Special Conditions that Affect Groundwater Connectivity

The water table is often considered to be a subdued replica of the surface. It is tempting to make this simplifying assumption when developing hypotheses of hydrologic connectivity, especially given the difficulty of measuring the water table directly. Unfortunately, the water table fluctuates independently of the surface under many conditions. It is important to recognize these conditions to avoid incorrect judgements of connectivity.

4.6.1. Tertiary FHLUs and the Water Table

Our simulations revealed certain arrangements of landforms cause the water table to fall and not follow the surface contour. Nesting of a depressional wetland within a broader hydrologic landscape often creates this specific arrangement. Figure 1 depicts two types of tertiary FHLUs, captured in inset boxes and labeled as tertiary FHLUs, both of which cause this unusual water table situation in our simulations. Note that the slope in both tertiary FHLUs is a concave-down land surface that actually rises from the wetland in the FHLU highland to the top of the wetland depression and then falls to the FHLU lowland. We refer to this slope as a 'rising then falling slope.'

If the water table beneath this type of tertiary FHLU falls far enough, the highland wetland will either recharge groundwater or become a flow-through wetland by receiving groundwater discharge on the upgradient side and recharging groundwater on the downgradient side. Examples of this type of flow-through connectivity include Wetland 2 in the Playa 2 and 3, Plateau 2 and 3, and Mountain Valley 2 landscapes (Figures 3–5).

Our simulations show the water table dropping beneath this type of tertiary FHLU is sometimes not sufficient to change the connectivity pattern of the wetlands. Examples of this case include our simulations in the Riverine Valley and Coastal Terrain landscapes. We believe the magnitude of water table decline is reduced in cases where (a) the width of the rising then falling slope is especially wide, (b) the climate is relatively wet, and (c) low-permeability geologic substrate is present. Rosenberry and Winter [59] provided a field study of hydrologic connectivity for two adjacent prairie pothole depressional wetlands separated by a rising then falling slope.

Evaluation of the flow depicted in the hummocky terrain landscape from Winter [23,27] provides an additional example of the significance of a tertiary FHLU with a 'rising then falling' slope (Figure 10). In this landscape, the primary FHLU has a highland at the right edge of the landscape and a lowland at the left edge of the landscape. The slope dividing the highland and lowland is interrupted by five depressional wetland secondary FHLUs. Four tertiary FHLUs exist in this landscape, between the depressional wetlands, as shown in Figure 10. All of these tertiary FHLUs have a rising then falling slope, and the water table is depressed to varying degrees. In some cases, the water table falls sufficiently to allow groundwater to flow away from a depressional wetland. In one case, the second depressional wetland from the left, the depression serves only as a lowland in the tertiary FHLUs and therefore receives groundwater discharge. The hydrologic consequence of having many depressional wetlands embedded on this landscape is to cause the water table to not represent a subdued version of the surface contour, which creates complex flow paths and groundwater connectivity.

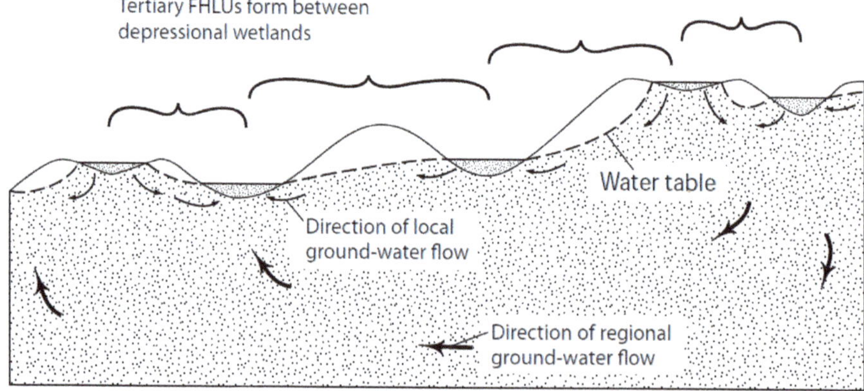

Figure 10. Groundwater flow through a hummocky terrain hydrologic landscape, tertiary FHLUs indicated by brackets. Modified from Winter [23,27].

4.6.2. Additional Situations Where the Water Table Contour Does Not Follow the Land Surface

The simulations done in this study reflect cases where the contour of the water table generally follows the land surface. However, there are situations where this does not hold true. Focused recharge, geologic heterogeneity, and riparian ET all can cause the water table to fall away from the land surface in specific locations [23]. Tile drains are often added to the landscape specifically to cause the water table to assume a shape that is the reverse from the surface contour and allow wet areas to become dry. Also, if the climate becomes especially dry, such as is the case with intense drought, reduced recharge and increased ET could lower the water table sufficiently to prevent the water table from following surface contours. Likewise, a sufficiently porous substrate could allow groundwater to drain and lower the water table sufficiently not to follow surface contours. Assuming the water table is a subdued replica of the surface in any of these cases can lead to the misinterpretation of groundwater flow and connectivity.

Hydrologically, these processes have two notable effects. First, breaks in slope on the surface may not reflect the contour of the water table below. This creates a case where groundwater flow must be judged by observing the water-table contours rather than using the landscape surface as a proxy for the water table. Second, these processes lower the water table and tend to dry out the 'watery patches' important for biodiversity (e.g., metapopulation migration moments, see [54]).

4.6.3. Situations Where Groundwater Flow Does Not Follow Water Table Contours

Both geologic heterogeneity and strong anisotropy can cause groundwater to flow from a wetland, even if a water-table mound is present, to downgradient waterbodies [16]. These factors are particularly important when the water-table mound is especially small, the slope of the water table on the downgradient side of the mound is relatively steep, or the elevation drop to a discharge point is especially great [28,29,64].

4.7. Guidance for Practitioners

A key dilemma for practitioners is to sufficiently assess both direct and indirect wetland groundwater connectivity with limited resources. Golden, et al. [35] made the case that measured, modeled, and hypothesized information can be creatively combined depending on resource availability to provide a defensible basis for decision making. Each form of information has its advantages, complements the other two, and can be pursued to varying degrees.

One contribution of this study is to improve the theory available to practitioners to develop a vision, or hypothesis, of groundwater connectivity on the landscape before employing measurements or modeling. This may constrain the information needed from measurements or modeling, or it may inform more cost-effective ways of employing those approaches. Figure 11 provides a simple workflow that uses hypothesized connectivity to efficiently build an understanding of wetland connectivity via groundwater.

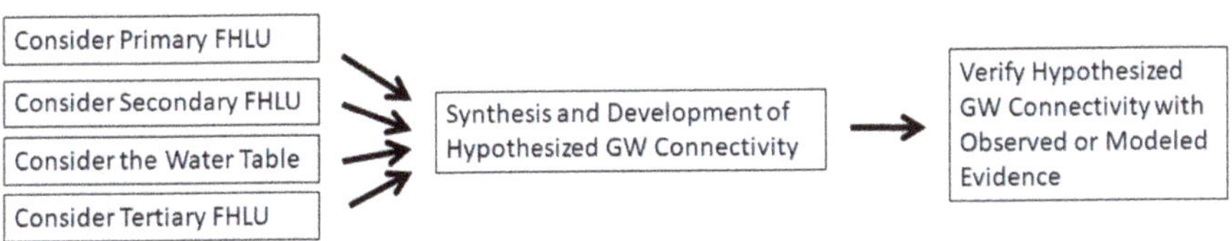

Figure 11. Flow chart to assess depressional wetland connectivity.

A good first step when developing a hypothesized vision of the groundwater connectivity of a landscape is to assess the primary FHLU. To do this, first identify the primary landform, without regard to small-scale local relief. It is useful to visualize the 'default' water-table contour. The tendency is for the water table to roughly follow the land surface in a subdued manner, with some exceptions. First, downward breaks in slope cause the water table to decline locally. This creates an opportunity for wetlands in these upland settings to serve as focal points for groundwater recharge (e.g., [58]). Second, upward breaks in slope cause the water table to rise toward or to the surface, where the large majority of groundwater discharge and contribution to surface-water flow occurs. In many scenarios, the key groundwater connection in a landscape is the indirect groundwater connection rather than the direct groundwater connection. Finally, geology and climate will further modify the water table and flow through a hydrologic landscape and existing knowledge of these parameters can be used to define the primary FHLU (Figure 1).

Identifying embedded, secondary FHLUs is the second step to develop a hypothesis of wetland groundwater connectivity. These can be relatively major landscape features, such as terraces, or relatively small features such as depressional wetlands. With regard to depressional wetlands, the size and depth of a depression are important to consider as key factors that affect groundwater connectivity (Table 3).

The third step in developing a hypothesis of wetland groundwater connectivity is to consider the predominant behavior of the water table (Table 3). Shallow well data tracking the water table elevation are ideal, but usually not readily available. Fortunately, the water table can be evaluated as creatively as funds, time, and available expertise allow and need not be expensive. In particular, a simple visual examination of the landscape features may prove sufficient to develop a hypothesis of the predominant water table behavior. For example, the presence of stream headwaters or springs often indicate groundwater discharge (Figure 9). The presence of biota adapted to highly mineralized waters often indicates strong regional groundwater inflow [65,66]. Historical aerial and satellite imagery, such as imagery available freely on Google Earth, can provide insight to pond permanence and response to seasonal or longer-term climate variability. Place names can also be relevant, examples include features such as Dry Creek or Great Sulfur Spring. The lack of, or flooding of, residential basements may provide information on water table behavior. Peculiar streamflow changes between wet and dry years may indicate a high degree of fill and spill connectivity, as described in Shaw [14] and Shaw, et al. [67].

Assessing the geology of the area is vital to understand the predominant water table behavior. Our results were obtained by assuming geological homogeneity and isotropy, to isolate the independent effects of the landform, especially when depressional wetlands are introduced. However, geological heterogeneity and anisotropy are the norm, and have profound effects on water-table shape, groundwater flow direction, and associated hydrologic connectivity [68]. For example, vernal pool wetlands of similar size, shape, and spatial arrangement, but formed in different geological settings, can have vastly different physical and chemical hydrological characteristics, with some connected to the landscape only by surface-water fill and spill and others being connected both by groundwater flow-through and surface-water fill and spill [69]. Fortunately, highly detailed geologic maps and literature are available in many areas. Well logs, observations of bedrock outcroppings, or the presence of sand or gravel mining can also provide insight to key aspects of the surrounding geology.

Similarly, evaluating the climate and its variability is essential to understand the predominant water table behavior. Our study evaluated steady state conditions with homogenous and constant infiltration. Real settings are dominated by spatial and temporal variability in precipitation and evaporation, which drives variability of infiltration. These processes greatly affect the water table. To aid in this evaluation, weather station data are often readily available and can provide much information about precipitation amount, type, and seasonality, as well as the evaporation rates in an area.

The fourth step to develop a hypothesis for wetland groundwater connectivity is to identify tertiary FHLUs and how they may cause exceptions to the predominant behavior and location of the water table. The presence of one or more depressional wetland creates an especially important type of tertiary FHLU, characterized by a slope between upland and lowland that rises to the top of the wetland depression then falls to the lowland, as depicted in the two tertiary FHLUs in Figure 1. In our simulations, this type of FHLU is especially likely to have a water table that does not follow the land surface contour and caused depressional wetlands in areas normally associated with groundwater discharge to the surface to act as flow-through wetlands with a groundwater connection to downstream waterbodies. Example of this type of connectivity include Wetland 2 in the Plateau 2, Plateau 3, and Mountain Valley 2 landscapes (Figures 4 and 5).

A synthesis of the FHLUs, climate, geology, and readily observed evidence of the water table should provide a hypothesis for how depressional wetlands are hydrologically connected to the landscape. From this point, hydrologic modeling and/or the addition of new observations, such as from well-placed piezometers, can confirm or refine the hypothesis of groundwater connectivity of depressional wetlands.

The value of evaluating the hydrologic landscape to develop a connectivity hypothesis is to quickly and inexpensively simplify, improve, and reduce our dependence on hydrologic modeling or installing new monitoring equipment for decision making. For example, a relatively large modeling effort, or extensive installation of monitoring equipment such as piezometers and streamflow gauging stations, may provide good enough information to confidently make land-use management decisions. However, these approaches may be relatively expensive, uncertain, time consuming, or demand expertise that is not readily available. By developing a high-quality hypothesis of connectivity, the modeling and collection of new measurements or observations can be done selectively and merely to confirm the hypothesis. In addition, a convergence between hypothesized, modeled, and observed information improves confidence that we correctly understand groundwater connectivity in the study area.

The authors used a less refined version of the process depicted in Figure 11 when siting piezometers and water quality sampling locations in a prior study [46]. In this example, the study area was in the prairie pothole region of North Dakota and included the area shown in Figure 9. One aspect of the example study was to evaluate the potential for groundwater connectivity between depressional, 'pothole' wetlands on the Missouri Coteau highlands and Pipestem Creek on the Drift Prairie lowland. The study area was conceptualized as being similar to a plateau hydrologic landform and two particular locations were targeted for piezometer nest installations and water quality sampling sites; the upward break in slope on the Missouri Escarpment and the bank of Pipestem Creek. The exploration of the study area revealed stream headwaters and one spring at the upward break in slope on the Missouri Escarpment, tentatively confirming the presence of groundwater discharge in this area. Piezometer nests were installed at prominent upward breaks in slope and near the bank of Pipestem Creek and monitored. A groundwater modeling exercise used these data and confirmed these locations as likely sites for groundwater discharge. Water quality analyses of groundwater samples at these locations verified the nature of discharging groundwater as having originated from recharge in upland areas and not pothole wetlands. In this example, the use of a groundwater connectivity hypothesis guided field instrumentation and modeling and the combination of all three sources of information provided greater confidence in the study conclusions.

5. Conclusions

Simulated flow through five generic hydrologic landscapes quantitatively shows that most groundwater flow from uplands to lowlands is indirect, discharging first to the surface at an upward break in slope and then continuing downgradient as surface-water flow. In essence, upward breaks in slope transform the nature of groundwater-connectivity across a landscape from a direct groundwater connection to, predominantly, an indirect groundwater connection. In addition, embedding depressional wetlands within a broader landscape introduces additional upward breaks in slope that serve to amplify groundwater flow toward the surface and restrict regional groundwater flow. The presence of depressional wetlands within a broader landscape also introduces additional downward breaks in slope that serve to lower the water table locally. In some cases, particularly on the downgradient side of a depressional wetland, a landform occurs that rises to the top of the depression, then falls to the lowland. The water table in this area is especially prone to not following the contour of the land surface and instead slopes monotonically away from a depressional wetland, causing it to be groundwater-connected to downgradient waterbodies. With regard to water-resource management, we explain how applying a hydrologic-landscapes approach to developing a hypothesis of groundwater connectivity can be used to improve practitioners' understanding of wetland connectivity and guide more resource-efficient measurement and modeling efforts to support decision making. Finally, this work provides a framework for using the hydrologic landscapes theory to understand wetland connectivity via groundwater.

In situ landscape complexity presents a key limitation of our analysis. In particular, our focus was on the role of landforms in determining groundwater connectivity and we intentionally restricted complexity related to climate and geology. Understanding all the nuances of landscape-scale connectivity will likely never be possible, but our analysis is an important step in that direction. One role of this research regarding the groundwater connectivity of wetlands is to provide a basis to develop tools and guidance needed by decision makers. Future research can help shrink the remaining gap.

Supplementary Materials: The following are available online at http://www.mdpi.com/2073-4441/12/1/50/s1: The 12 groundwater flow models used in this study, including all input and output files associated with each model, and instructions for readers to install the modeling software and run the models themselves. These files may be unzipped using the 7-zip file archiver software, available free of charge from https://www.7-zip.org/. (Note: Our use of proprietary software does not imply an endorsement of any specific software or company).

- Files related to each groundwater model are arranged in directories according to the basic hydrologic landscape (e.g., 1-Playa) and subdirectories according to the specific model domain used in this study (e.g., 1-Playa/Playa–No Wetland; 1-Playa\Playa-Wetland Added Close; 1-Playa\Playa-Wetland Added Far).
- Files within each directory include the executable file used to launch the model and various output and input files generated by the modeling software.
- Further description of the supplement content is provided in a ReadMe file within the supplement.

Author Contributions: D.O.R., B.P.N. and S.G.L. conceptualized the study and curated the data. D.O.R. and B.P.N. developed the methodology and performed the modeling and project administration. B.P.N. did the validation, wrote the paper, and created the visualizations. D.O.R. provided supervision. D.M.M. provided funding for publication fees. B.P.N., D.O.R., H.E.G., C.R.L., D.M.M., S.G.L., M.C.R. and J.R.B. provided formal analysis, comment, and interpretation of data through many manuscript reviews and revisions. All authors have read and agreed to the published version of the manuscript.

Funding: This manuscript greatly benefited from discussions held at the "North American Analysis and Synthesis on the Connectivity of 'Geographically Isolated Wetlands' to Downstream Waters" Working Group, supported by the John Wesley Powell Center for Analysis and Synthesis, funded by the US Geological Survey and the National Exposure Research Laboratory of the US Environmental Protection Agency's Office of Research and Development.

Acknowledgments: We thank Rick Healy for his advice and guidance on troubleshooting our groundwater models. Zeno Levy, Rose Kwok, and James Markwiese provided helpful reviews of the early manuscript. The manuscript was further improved by the work of two anonymous peer reviewers. The information in this document has been funded in part by the U.S. Environmental Protection Agency. Any opinions expressed in this manuscript are those of the authors and do not necessarily reflect the views of the U.S. Environmental Protection Agency. Protection Agency. Any use of trade, firm, or product names is for descriptive purposes only and does not imply endorsement by the U.S. Government.

Conflicts of Interest: The authors declare no conflict of interest.

References

1. NRCS. *Hydrogeomorphic Wetland Classification System: An Overview and Modification to Better Meet the Needs of the Natural Resources Conservation Service*; Technical Note No. 190-8-76; Natural Resources Conservation Service: Washington, DC, USA, 2008.

2. Smith, R.D.; Ammann, A.; Bartoldus, C.; Brinson, M.M. *An Approach for Assessing Wetland Functions Using Hydrogeomorphic Classification, Reference Wetlands, and Functional Indices*; Army Engineer Waterways Experiment Station: Vicksburg, MS, USA, 1995.

3. Nadeau, T.-L.; Leibowitz, S.G. Isolated wetlands: An introduction to the special issue. *Wetlands* **2003**, *23*, 471–474. [CrossRef]

4. Leibowitz, S.G.; Nadeau, T.-L. Isolated wetlands: State-of-the-science and future directions. *Wetlands* **2003**, *23*, 663–684. [CrossRef]

5. Leibowitz, S.G.; Mushet, D.M.; Newton, W.E. Intermittent surface water connectivity: Fill and spill vs. fill and merge dynamics. *Wetlands* **2016**, *36*, 323–342. [CrossRef]

6. Calhoun, A.J.; Mushet, D.M.; Alexander, L.C.; DeKeyser, E.S.; Fowler, L.; Lane, C.R.; Lang, M.W.; Rains, M.C.; Richter, S.C.; Walls, S.C. The significant surface-water connectivity of "geographically isolated wetlands". *Wetlands* **2017**, *37*, 801–806. [CrossRef] [PubMed]

7. Mushet, D.M.; Calhoun, A.J.; Alexander, L.C.; Cohen, M.J.; DeKeyser, E.S.; Fowler, L.; Lane, C.R.; Lang, M.W.; Rains, M.C.; Walls, S.C. Geographically isolated wetlands: Rethinking a misnomer. *Wetlands* **2015**, *35*, 423–431. [CrossRef]

8. EPA. *Connectivity of Streams & Wetlands to Downstream Waters: A Review & Synthesis of the Scientific Evidence*; EPA/600/R-14/475F; U.S. Environmental Protection Agency Office of Research and Development: Washington, DC, USA, 2015.

9. Lane, C.R.; Leibowitz, S.G.; Autrey, B.C.; LeDuc, S.D.; Alexander, L.C. Hydrological, physical, and chemical functions and connectivity of non-floodplain wetlands to downstream waters: A review. *J. Am. Water Resour. Assoc.* **2018**, *54*, 346–371. [CrossRef]

10. Euliss, N.H.; LaBaugh, J.W.; Fredrickson, L.H.; Mushet, D.M.; Laubhan, M.K.; Swanson, G.A.; Winter, T.C.; Rosenberry, D.O.; Nelson, R.D. The wetland continuum: A conceptual framework for interpreting biological studies. *Wetlands* **2004**, *24*, 448–458. [CrossRef]

11. Rains, M.; Leibowitz, S.; Cohen, M.; Creed, I.; Golden, H.; Jawitz, J.; Kalla, P.; Lane, C.; Lang, M.; McLaughlin, D. Geographically isolated wetlands are part of the hydrological landscape. *Hydrol. Proc.* **2016**, *30*, 153–160. [CrossRef]

12. Stitchling, W.; Blackwell, S. *Drainage Area as a Hydrologic Factor on the Glaciated Canadian Prairies*; Prairie Farm Rehabilitation Administration: Regina, SK, Canada, 1957.

13. Leibowitz, S.G.; Vining, K.C. Temporal connectivity in a prairie pothole complex. *Wetlands* **2003**, *23*, 13–25. [CrossRef]

14. Shaw, D.A. The Influence of Contributing Area on the Hydrology of the Prairie Pothole Region of North America. Ph.D. Thesis, University of Saskatchewan Saskatoon, Saskatoon, SK, Canada, 2009.

15. Winter, T.C.; LaBaugh, J.W. Hydrologic considerations in defining isolated wetlands. *Wetlands* **2003**, *23*, 532–540. [CrossRef]

16. Neff, B.P.; Rosenberry, D.O. Groundwater connectivity of upland-embedded wetlands in the Prairie Pothole Region. *Wetlands* **2018**, *38*, 51–63. [CrossRef]

17. Hayashi, M.; van der Kamp, G.; Rosenberry, D.O. Hydrology of prairie wetlands: understanding the integrated surface-water and groundwater processes. *Wetlands* **2016**, *36*, 237–254. [CrossRef]

18. McLaughlin, D.L.; Kaplan, D.A.; Cohen, M.J. A significant nexus: Geographically isolated wetlands influence landscape hydrology. *Water Resour. Res.* **2014**, *50*, 7153–7166. [CrossRef]

19. Golden, H.E.; Sander, H.A.; Lane, C.R.; Zhao, C.; Price, K.; D'Amico, E.; Christensen, J.R. Relative effects of geographically isolated wetlands on streamflow: A watershed-scale analysis. *Ecohydrology* **2016**, *9*, 21–38. [CrossRef]

20. Todd, A.; Buttle, J.; Taylor, C. Hydrologic dynamics and linkages in a wetland-dominated basin. *J. Hydrol.* **2006**, *319*, 15–35. [CrossRef]

21. Singh, M.; Tandon, S.K.; Sinha, R. Assessment of connectivity in a water-stressed wetland (Kaabar Tal) of Kosi-Gandak interfan, north Bihar Plains, India. *Earth Surf. Proc. Landf.* **2017**, *42*, 1982–1996. [CrossRef]

22. Leibowitz, S.G.; Wigington, P.J., Jr.; Schofield, K.A.; Alexander, L.C.; Vanderhoof, M.K.; Golden, H.E. Connectivity of streams and wetlands to downstream waters: An integrated systems framework. *J. Am. Water Resour. Assoc.* **2018**, *54*, 298–322. [CrossRef]

23. Winter, T.C. The concept of hydrologic landscapes. *J. Am. Water Resour. Assoc.* **2001**, *37*, 335–349. [CrossRef]

24. Winter, T.C. A conceptual framework for assessing cumulative impacts on the hydrology of nontidal wetlands. *Environ. Manag.* **1988**, *12*, 605–620. [CrossRef]

25. Winter, T.C. A physiographic and climatic framework for hydrologic studies of wetlands. In *Aquatic Ecosystems in Semi-Arid Regions: Implications for Resource Management*; Environment and Climate Change Canada: Ottawa, ON, Canada, 1992; pp. 127–148.

26. Winter, T.C. Landscape approach to identifying environments where ground water and surface water are closely interrelated. In Proceedings of the International Symposium on Groundwater Management, San Antonio, TX, USA, 14–16 August 1995; pp. 139–144.

27. Winter, T.C. The Vulnerability of wetlands to climate change: A hydrologic landscape perspective. *J. Am. Water Resour. Assoc.* **2000**, *36*, 305–311. [CrossRef]

28. Winter, T.C. Effects of water-table configuration on seepage through lakebeds. *Limnol. Oceanogr.* **1981**, *26*, 925–934. [CrossRef]

29. Winter, T.C. Numerical simulation of steady state three-dimensional groundwater flow near lakes. *Water Resour. Res.* **1978**, *14*, 245–254. [CrossRef]

30. Healy, R.W.; Essaid, H.I. VS2DI: Model use, calibration, and validation. *Trans. ASABE* **2012**, *55*, 1249–1260. [CrossRef]

31. Hsieh, P.A.; Wingle, W.; Healy, R.W. *VS2DI: A Graphical Software Package for Simulating Fluid Flow and Solute or Energy Transport in Variably Saturated Porous Media*; US Geological Survey Water-Resources Investigations Report 99-4130; US Geological Survey: Reston, VA, USA, 2000.

32. Carsel, R.F.; Parrish, R.S. Developing joint probability distributions of soil water retention characteristics. *Water Resour. Res.* **1988**, *24*, 755–769. [CrossRef]

33. Lappala, E.G.; Healy, R.W.; Weeks, E.P. *Documentation of Computer Program VS2D to Solve the Equations of Fluid Flow in Variably Saturated Porous Media*; Department of the Interior, US Geological Survey: Reston, VA, USA, 1987.

34. Barnett, B.; Townley, L.; Post, V.; Evans, R.; Hunt, R.; Peeters, L.; Richardson, S.; Werner, A.; Knapton, A.; Boronkay, A. *Australian Groundwater Modelling Guidelines*; Australian Government: Canberra, Australia, 2012.

35. Golden, H.; Creed, I.; Ali, G.; Basu, N.; Neff, B.; Rains, M.; McLaughlin, D.; Alexander, L.; Ameli, A.; Christensen, J.; et al. Integrating geographically isolated wetlands into land management decisions. *Front. Ecol. Environ.* **2017**, *15*, 319–327. [CrossRef]

36. Scudeler, C.; Paniconi, C.; Pasetto, D.; Putti, M. Examination of the seepage face boundary condition in subsurface and coupled surface/subsurface hydrological models. *Water Resour. Res.* **2017**, *53*, 1799–1819. [CrossRef]

37. Rains, M.C.; Fogg, G.E.; Harter, T.; Dahlgren, R.A.; Williamson, R.J. The role of perched aquifers in hydrological connectivity and biogeochemical processes in vernal pool landscapes, Central Valley, California. *Hydrol. Proc.* **2006**, *20*, 1157–1175. [CrossRef]

38. Nilsson, K.A.; Rains, M.C.; Lewis, D.B.; Trout, K.E. Hydrologic characterization of 56 geographically isolated wetlands in west-central Florida using a probabilistic method. *Wetl. Ecol. Manag.* **2013**, *21*, 1–14. [CrossRef]

39. Genereux, D.; Bandopadhyay, I. Numerical investigation of lake bed seepage patterns: Effects of porous medium and lake properties. *J. Hydrol.* **2001**, *241*, 286–303. [CrossRef]

40. Winter, T.C. Relation of streams, lakes, and wetlands to groundwater flow systems. *Hydrol. J.* **1999**, *7*, 28–45. [CrossRef]

41. Winter, T.C. *Ground Water and Surface Water: A Single Resource*; Geological Survey (USGS): Reston, VA, USA, 1998; Volume 1139.

42. Callahan, M.K.; Whigham, D.F.; Rains, M.C.; Rains, K.C.; King, R.S.; Walker, C.M.; Maurer, J.R.; Baird, S.J. Nitrogen subsidies from hillslope alder stands to streamside wetlands and headwater streams, Kenai Peninsula, Alaska. *J. Am. Water Resour. Assoc.* **2017**, *53*, 478–492. [CrossRef]

43. Holtschlag, D.J.; Nicholas, J. *Indirect Ground-Water Discharge to the Great Lakes*; Open-File Report 98-579; US Geological Survey: Reston, VA, USA, 1998.

44. Grannemann, N.G.; Hunt, R.J.; Nicholas, J.R.; Reilly, T.E.; Winter, T.C. *The Importance of Ground Water in the Great Lakes Region*; Water-Resources Investigations Report 00–4008; US Geological Survey: Reston, VA, USA, 2000.

45. Neff, B.P.; Day, S.M.; Piggott, A.R.; Fuller, L.M. *Baseflow in the Great Lakes Basin. Scientific Investigations Report 2005-5217*; US Geological Survey: Reston, VA, USA, 2005.

46. Brooks, J.R.; Mushet, D.M.; Vanderhoof, M.K.; Leibowitz, S.G.; Christensen, J.R.; Neff, B.P.; Rosenberry, D.O.; Rugh, W.D.; Alexander, L.C. Estimating wetland connectivity to streams in the Prairie Pothole Region: An isotopic and remote sensing approach. *Water Resour. Res.* **2018**, *54*, 955–977. [CrossRef] [PubMed]

47. Callahan, M.K.; Rains, M.C.; Bellino, J.C.; Walker, C.M.; Baird, S.J.; Whigham, D.F.; King, R.S. Controls on temperature in salmonid-bearing headwater streams in two common hydrogeologic settings, Kenai Peninsula, Alaska. *J. Am. Water Resour. Assoc.* **2015**, *51*, 84–98. [CrossRef]

48. Power, G.; Brown, R.; Imhof, J. Groundwater and fish—Insights from northern North America. *Hydrol. Proc.* **1999**, *13*, 401–422. [CrossRef]

49. Ward, J. Ecology of alpine streams. *Freshw. Biol.* **1994**, *32*, 277–294. [CrossRef]

50. Hammersmark, C.T.; Rains, M.C.; Mount, J.F. Quantifying the hydrological effects of stream restoration in a montane meadow, northern California, USA. *River Res. Appl.* **2008**, *24*, 735–753. [CrossRef]

51. Rains, M.C.; Mount, J.F. Origin of shallow ground water in an alluvial aquifer as determined by isotopic and chemical procedures. *Groundwater* **2002**, *40*, 552–563. [CrossRef]

52. Robinson, C. Review on groundwater as a source of nutrients to the Great Lakes and their tributaries. *J. Great Lakes Res.* **2015**, *41*, 941–950. [CrossRef]

53. Smith, L.L.; Subalusky, A.; Atkinson, C.L.; Earl, J.E.; Mushet, D.M.; Scott, D.E.; Lance, S.L.; Johnson, S.A. Biological connectivity of seasonally ponded wetlands across spatial and temporal scales. *J. Am. Water Resour. Assoc.* **2019**, *55*, 334–353. [CrossRef]

54. Mushet, D.M.; Alexander, L.C.; Bennett, M.; Schofield, K.; Christensen, J.R.; Ali, G.; Pollard, A.; Fritz, K.; Lang, M.W. Differing modes of biotic connectivity within freshwater ecosystem mosaics. *J. Am. Water Resour. Assoc.* **2019**, *55*, 307–317. [CrossRef]

55. Duval, T.P.; Waddington, J. Extreme variability of water table dynamics in temperate calcareous fens: Implications for biodiversity. *Hydrol. Proc.* **2011**, *25*, 3790–3802. [CrossRef]

56. Brinson, M.M. *A Hydrogeomorphic Classification for Wetlands*; Wetlands Research Program Technical Report WRP-DE-4; US Army Corps of Engineers: Washington, DC, USA, 1993.

57. Amon, J.P.; Thompson, C.A.; Carpenter, Q.J.; Miner, J. Temperate zone fens of the glaciated Midwestern USA. *Wetlands* **2002**, *22*, 301–317. [CrossRef]

58. Rains, M.C. Water sources and hydrodynamics of closed-basin depressions, Cook Inlet Region, Alaska. *Wetlands* **2011**, *31*, 377–387. [CrossRef]

59. Rosenberry, D.O.; Winter, T.C. Dynamics of water-table fluctuations in an upland between two prairie-pothole wetlands in North Dakota. *J. Hydrol.* **1997**, *191*, 266–289. [CrossRef]

60. Van der Kamp, G.; Hayashi, M. Groundwater-wetland ecosystem interaction in the semiarid glaciated plains of North America. *Hydrol. J.* **2009**, *17*, 203–214. [CrossRef]

61. LaBaugh, J.W.; Rosenberry, D.O.; Mushet, D.M.; Neff, B.P.; Nelson, R.D.; Euliss, N.H. Long-term changes in pond permanence, size, and salinity in Prairie Pothole Region wetlands: The role of groundwater-pond interaction. *J. Hydrol. Reg. Stud.* **2018**, *17*, 1–23. [CrossRef]

62. Wigington, P.; Moser, T.; Lindeman, D. Stream network expansion: A riparian water quality factor. *Hydrol. Proc.* **2005**, *19*, 1715–1721. [CrossRef]

63. Evenson, G.R.; Golden, H.E.; Lane, C.R.; McLaughlin, D.L.; D'Amico, E. Depressional wetlands affect watershed hydrological, biogeochemical, and ecological functions. *Ecol. Appl.* **2018**, *28*, 953–966. [CrossRef]

64. Winter, T.C. *Numerical Simulation Analysis of the Interaction of Lakes and Ground Water*; US Geololgical Survey Professional Paper 1001; US Government Printing Office: Washington, DC, USA, 1976; 45p.

65. Bridgham, S.D.; Pastor, J.; Janssens, J.A.; Chapin, C.; Malterer, T.J. Multiple limiting gradients in peatlands: A call for a new paradigm. *Wetlands* **1996**, *16*, 45–65. [CrossRef]

66. Quinton, W.; Hayashi, M.; Pietroniro, A. Connectivity and storage functions of channel fens and flat bogs in northern basins. *Hydrol. Proc.* **2003**, *17*, 3665–3684. [CrossRef]

67. Shaw, D.A.; Vanderkamp, G.; Conly, F.M.; Pietroniro, A.; Martz, L. The fill–spill hydrology of prairie wetland complexes during drought and deluge. *Hydrol. Proc.* **2012**, *26*, 3147–3156. [CrossRef]

68. Haitjema, H.M.; Mitchell-Bruker, S. Are water tables a subdued replica of the topography? *Groundwater* **2005**, *43*, 781–786. [CrossRef] [PubMed]

69. Rains, M.C.; Dahlgren, R.A.; Fogg, G.E.; Harter, T.; Williamson, R.J. Geological control of physical and chemical hydrology in California vernal pools. *Wetlands* **2008**, *28*, 347–362. [CrossRef]

 *water* 

Article

Evaluation of Stream and Wetland Restoration Using UAS-Based Thermal Infrared Mapping

Mark C. Harvey [1,*], Danielle K. Hare [1], Alex Hackman [2], Glorianna Davenport [3], Adam B. Haynes [1], Ashley Helton [1], John W. Lane Jr. [4] and Martin A. Briggs [4]

[1]   Department of Natural Resources and the Environment, Center for Environmental Sciences and Engineering, University of Connecticut, Storrs, CT 06269, USA
[2]   Massachusetts Division of Ecological Restoration, 251 Causeway Street, Suite 400, Boston, MA 02114, USA
[3]   The Living Observatory, 139 Bartlett Rd, Plymouth, MA 02360, USA
[4]   U.S. Geological Survey, Hydrogeophysics Branch, 11 Sherman Place, Unit 5015, Storrs, CT 06269, USA
*   Correspondence: mark.harvey@uconn.edu; Tel.: +1-860-798-9482

Received: 30 May 2019; Accepted: 16 July 2019; Published: 29 July 2019

Abstract: Large-scale wetland restoration often focuses on repairing the hydrologic connections degraded by anthropogenic modifications. Of these hydrologic connections, groundwater discharge is an important target, as these surface water ecosystem control points are important for thermal stability, among other ecosystem services. However, evaluating the effectiveness of the restoration activities on establishing groundwater discharge connection is often difficult over large areas and inaccessible terrain. Unoccupied aircraft systems (UAS) are now routinely used for collecting aerial imagery and creating digital surface models (DSM). Lightweight thermal infrared (TIR) sensors provide another payload option for generation of sub-meter-resolution aerial TIR orthophotos. This technology allows for the rapid and safe survey of groundwater discharge areas. Aerial TIR water-surface data were collected in March 2019 at Tidmarsh Farms, a former commercial cranberry peatland located in coastal Massachusetts, USA (41°54′17″ N 70°34′17″ W), where stream and wetland restoration actions were completed in 2016. Here, we present a 0.4 km^2 georeferenced, temperature-calibrated TIR orthophoto of the area. The image represents a mosaic of nearly 900 TIR images captured by UAS in a single morning with a total flight time of 36 min and is supported by a DSM derived from UAS-visible imagery. The survey was conducted in winter to maximize temperature contrast between relatively warm groundwater and colder ambient surface environment; lower-density groundwater rises above cool surface waters and thus can be imaged by a UAS. The resulting TIR orthomosaic shows fine detail of seepage distribution and downstream influence along the several restored channel forms, which was an objective of the ecological restoration design. The restored stream channel has increased connectivity to peatland groundwater discharge, reducing the ecosystem thermal stressors. Such aerial techniques can be used to guide ecological restoration design and assess post-restoration outcomes, especially in settings where ecosystem structure and function is governed by groundwater and surface water interaction.

Keywords: ecological restoration; wetlands; seepage; groundwater; springs; thermal; drone; UAS

---

1. Introduction

At temperate latitudes, cold water anomalies in summer and warm water anomalies in winter can indicate zones of spatially preferential groundwater discharge that can be mapped with ground-based thermal infrared (TIR) at high spatial resolution compared to more traditional groundwater discharge-characterization methodology [1]. Unoccupied aircraft systems (UAS) are now routinely used for collecting visible-light aerial imagery and creating digital surface models (DSM) of surface water-related features [2–4]. Lightweight TIR sensors provide another remotely sensed data type that can be used to generate high-resolution (<1 m) TIR orthophotos (e.g., [5]). However, the use of TIR-equipped UAS is relatively novel, only recently finding environmental and hydrological applications [1,5–10]. Examples include environmental monitoring of natural geothermal systems [5] and distinguishing sewer and stormflow discharges from groundwater based on characteristic temperatures [9]. In engineered peatlands, UAS TIR was used to guide water isotopic sampling for a similar goal of parsing groundwater discharge endmembers [10].

Recently, there has been a movement to better incorporate ecological services and functions into stream and wetland restoration projects [11]. Hydrological process-based wetland restoration requires understanding of site-wide hydrodynamics, which often involves scaling-up measurements from discrete physical sampling points to the square-kilometer scale [12]. A fundamental challenge to point-scale measurement upscaling is that wetland subsurface materials often exhibit strongly heterogeneous and anisotropic properties and preferential flow paths; traditional groundwater discharge characterization methods are often inadequate for representing system-scale dynamics [13]. In contrast, spatially distributed thermal methods (i.e., ground-based TIR and fiber-optic sensing) have shown great promise for the comprehensive mapping of preferential seepage processes in wetlands [14,15], although work is hindered by boggy terrain and vegetation. The use of spatially extensive thermal investigations to identify groundwater seepages could potentially improve restoration design and evaluation, particularly if deployed from a UAS platform. TIR-equipped UAS provides such an opportunity for developing effective and transferable methods to produce spatially contiguous and extensive groundwater seepage maps.

UAS-based imaging offers the potential to plan and evaluate process-based wetland restoration projects efficiently, with the combined collection of visible, DSM, and TIR data. However, the approach requires sufficient contrast between the temperature of the surface environment and groundwater temperature, enough to allow the TIR sensor to resolve the groundwater as relatively warm (winter) or cool (summer). Contrast is a function of the sensitivity and resolution of the TIR sensor (see Section 3.1) but also hydrodynamics, where turbulence may rapidly homogenize waters of different temperatures, making discreet seeps harder to identify. The use of TIR-equipped UAS to map high-temperature geothermal springs has been previously demonstrated [5], but imaging lower-temperature groundwater is more challenging, as temperature contrasts are much less. Further, relatively cold groundwater discharge often plunges in slow-flowing surface waters in summer, when most field work is conducted, reducing the surface expression of groundwater discharge zones [14]. Thermal images collected during winter may provide the best opportunity to explore the potential of the method to evaluate wetland restoration design and implementation, as groundwater is relatively warm and buoyant.

The objectives of the study were to evaluate (i) the ability of TIR-equipped UAS for identifying groundwater discharges and (ii) the usefulness of TIR-equipped UAS as a tool for validating thermal refugia goals of process-based restoration.

2. Site Background

2.1. Tidmarsh Farms Site Description

The study area, Tidmarsh Farms, is a former 2.5 square-km cranberry peatland that underwent a process-based restoration in coastal Massachusetts, USA (41°54′17.6″ N 70°34′17″ W) (Figure 1). The cranberry farm, through which the restored stream channel flows, was built on a series of smaller peat-filled kettle holes located in Manomet Village in Plymouth, Massachusetts. The Beaver Dam Brook Watershed has a small spatial extent (5 km^2), while the groundwater aquifer supplying this area (Plymouth–Carver–Kingston–Duxbury aquifer system) is 360 km^2, indicating that the groundwater flow paths are from a large hydrologic system [16], and the natural volumetric contributions to the system are predominately controlled by groundwater. The original stream channel, sometimes known as Beaver Dam Brook, was dammed in the early 1830s, forming an early version of what became known as Beaver Dam Pond. Beginning in the 1890s, cranberry production was conducted, for which water structures were built, maintained (including a perimeter ditch and lateral ditches), and peat covered with layers of sand. These anthropogenic modifications decreased the surface wetness and modified the location of the naturally flowing surface waters, which resulted in a decrease in groundwater connection to the main stem and thus a decrease in thermal stability, an important indicator of ecosystem health. In 2008, the owners of Tidmarsh Farms decided to take the farm out of production in order to ecologically restore and protect the area. In 2017, having completed the largest freshwater wetland restoration in Massachusetts, the property was purchased by the Massachusetts Audubon Society, who in 2018 opened the Tidmarsh Wildlife Sanctuary to the public. The area is covered in low-lying cranberry vegetation as well as a variety of sedges and cattails that are mostly adjacent to the central stream bank and marginal drainage ditches.

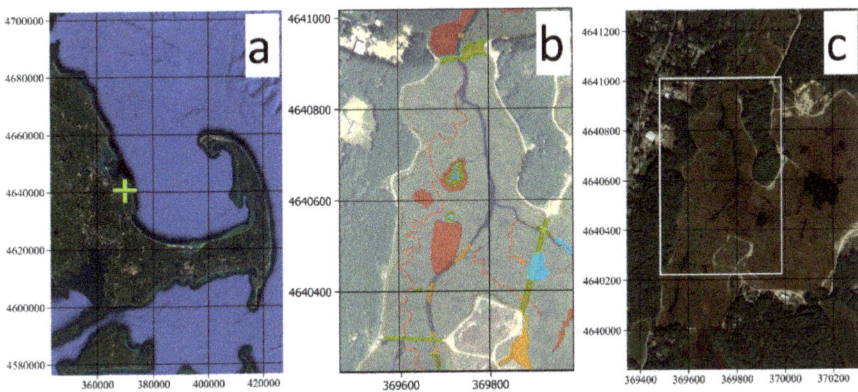

Figure 1. (**a**) Site location in Massachusetts (Google Earth), (**b**) feature layout, and (**c**) overview of survey area (Google Earth)—white box shows area covered by panel (**b**). Key for panel (**b**): Red lines are new channels to be built, blue lines are existing channels to remain, red polygons are no-disturbance zones, green polygons are dike removals, and orange polygons are fill.

2.2. Tidmarsh Farms Restoration Approach and Details

The Tidmarsh Farms restoration project used a "process-based" approach [17], which considers the underlying physical drivers of wetland and stream systems. The process-based approach identifies limiting factors to wetland recovery with the goal of reducing ecological stressors and encouraging site rejuvenation. Focusing on the natural movement and storage of water on the land, the project team identified legacy agricultural impacts that affected site hydrology. The three primary issues were (i) an anthropogenic sand layer placed during farming over native peatland soils that impacted water storage in the upper layer (~1 m), (ii) dams and water control structures that interrupted the movement of water, and (iii) physical simplification, including channel straightening, ditching, and peatland platform maintenance, that reduced water retention. The ecological restoration approaches developed to mitigate these impacts included ditch plugging, peatland surface roughening, dam and water control structure removal, stream channel reconstruction (5.6 km), and large wood addition (~3000 pieces) (Figure 1b). Collectively, these actions were intended to help increase hydrologic residence time on site and improve ecological services within the restored wetlands.

During the assessment and design phase, a key question for the project team focused on where to locate the reconstructed stream channel within the broad valley. Project designers hoped to replace highly modified (wide, straight, shallow) agricultural channels with more geomorphically appropriate (narrow, sinuous) channels that also provided cool/cold-water habitat by intercepting groundwater discharge [14]. As detailed in Hare et al. [16], each identified discharge was characterized as a low-flux or high-flux zone by scaling up vertical flux measurements based on a subset of thermal flux profiles

(~0.15 m/day, ~3 m/day, respectively) analyzed with 1DTempPro software [18]. These locations were compared to the underlying peat depth and subsurface peat basin structure, determined by ground penetrating radar (GPR) [16]. The results of the pre-restoration assessment demonstrated that both high- and low-flux seepage occurred along the margins of the peat surface, but high-flux preferential discharges also appeared within the peat surface interior (Figure 2a). No groundwater discharge was visible in the interior of these zones (Figure 2b), signaling an abrupt end to any groundwater input along the site edge. These interior preferential flow path discharges correlated with large increases in peat depth (high curvature) located on the upgradient side of the regional groundwater flow path as well as in areas of general peat thinning (Figure 2c); it was theorized that the high curvature of the subsurface peat induced high-seepage zones interior of where they were expected due to the abrupt change in hydraulic conductivity [16]. Therefore, the restoration design incorporated the observational theory and the site's interpolated peat depth, along with other observational and historical site data, to place the new stream channel to maximize groundwater input.

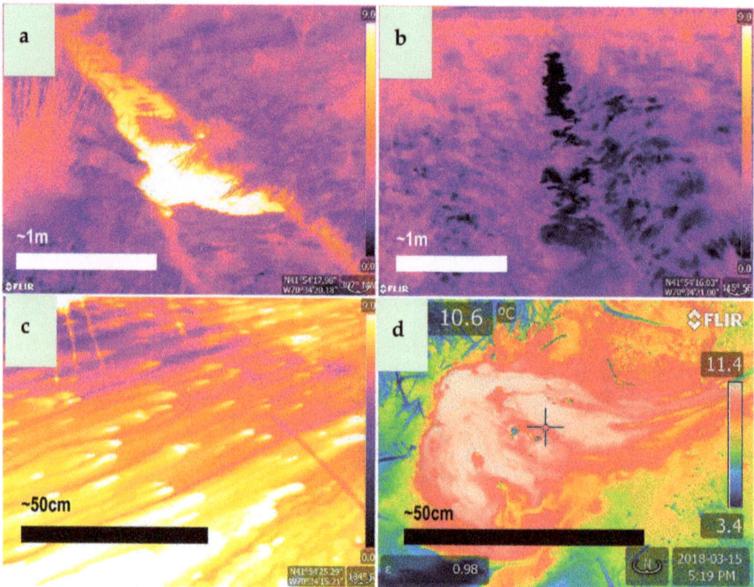

Figure 2. Hand-held thermal infrared (TIR) images collected before the restoration [16], along the relic cranberry peatland drainage ditches. Images show (**a**) natural groundwater discharge, (**b**) no discharge, (**c**) clustered centimeter-scale macropore discharge features, and (**d**) discrete discharge features observed after the restoration (March 2018) in similar locations to the pre-restoration period, indicating persistent subsurface hydrogeologic controls.

3. Methods

3.1. Aerial Imaging Field Methods

UAS TIR imagery was collected during the early morning of 19 March 2019 using a DJI Matrice 100 quadcopter fitted with an ICI 9640 640 × 480 uncooled TIR sensor (spectral response 7–14 μm, focal length 12.5 mm, temperature range −40 to 140 °C, and thermal sensitivity <50 mK) and automated image capture (ICI UAV module®) (ICI Cameras, Beaumont, TX, USA) that records raw sensor values (without calibration applied). Imaging was performed before sunrise, when most of the peatland surface was covered in frost (air temperature was approximately −3 °C) but flowing channel features were not frozen. In-situ temperature measurements were not taken during the TIR flight (a separate temperature calibration was conducted—Section 3.3). However, the objective of the study was to identify groundwater discharge locations, so it was sufficient to measure relative temperature; absolute temperature was not required to meet this objective. Regional groundwater temperature from a groundwater well in the Plymouth–Carver–Kingston–Duxbury aquifer system was reported to be 11.7 °C on 19th March 2019 [19]. Accordingly, available data indicated that groundwater discharge-influenced surface water could be relatively distinguished from other surface waters.

The flight plan was developed using UgCS® software (UgCS, Riga, Latvia). The UgCS photogrammetry tool was utilized with the following parameters: forward and side overlap 80%, flight speed 8 m/s, and ground resolution 16 cm (equates to a flight altitude of 120 m above ground level). The flight plan was then uploaded to the quadcopter's flight controller via a Samsung S4 smartphone running Android and the UgCS® for DJI application. Accordingly, both in-flight navigation and image capture were autonomous. Ground control points were not deployed at the time of data collection but were later identified in Google Earth for photogrammetry processing (see Section 3.2).

Visible imagery was collected from aircraft approximately one year prior (March 2018) using a Ricoh GRII camera (Ricoh Imaging Company, Ltd., Tokyo, Japan) when there was patchy snowpack over the land areas. Image stills were collected along multiple flight lines, with aircraft altitudes ranging approximately 90–120 m above ground level. Position of the aircraft was tracked by internal global positioning system (GPS). Additional detail of methodology for both TIR and visible data collection, including flight and photogrammetry processing parameters, is available from Harvey et al. [20].

3.2. Image Processing and Analysis

TIR images were processed using Agisoft Photoscan® commercial photogrammetry software, running on a computer equipped with an i7 processor and 32 GB RAM. Georeferencing was improved within Agisoft Photoscan using ground control points, which were identified during post-processing using Google Earth satellite imagery. This provided a horizontal position error of 1.6 m (root mean squared error) relative to Google Earth imagery. Agisoft Photoscan workflow parameters for TIR imagery were *Highest* (Align Photos) and *Medium* (Build Dense Cloud). The resulting TIR orthophoto was then post-processed using QGIS open-source desktop geographic information system (GIS). This involved the conversion of the raw 32-bit pixel values to calibrated temperature values (°C) (Section 3.3). All map coordinates shown in figures are WGS84 datum.

3.3. Temperature Calibration of TIR Images

Temperature calibration utilized aluminum trays filled with water of various temperatures recorded using a high-precision (0.01 °C) digital thermometer (Traceable Thermometer, Control Company, Friendswood, TX, USA). Five distributed temperature measurements were made inside each tray (each corner and center), while the UAS hovered (15 m above tray level) capturing TIR images (Figure 3a). The temperature probe was left to stabilize for 15 s before the temperature was recorded. In order to obtain the water surface skin temperature, the thermometer was placed in contact with the uppermost surface of the water.

The five temperature values were averaged to give a best estimate of the water surface temperature of each tray. Average temperature measurements were regressed against average raw pixel values from the TIR images of the trays during the flight (plus two additional sites on rivers with known temperature at time of image capture) (Figure 3b). A linear equation was fitted to the calibration scatter plot ($r^2 = 0.98$), which allowed temperature calibration of the TIR orthophoto (standard error of the regression = 1.9 °C). Note: the function is linear, so the conversion from raw values to temperature does not affect the appearance of the orthophoto, has no effect on the ability to detect groundwater seeps, and does not impact the conclusions of the study.

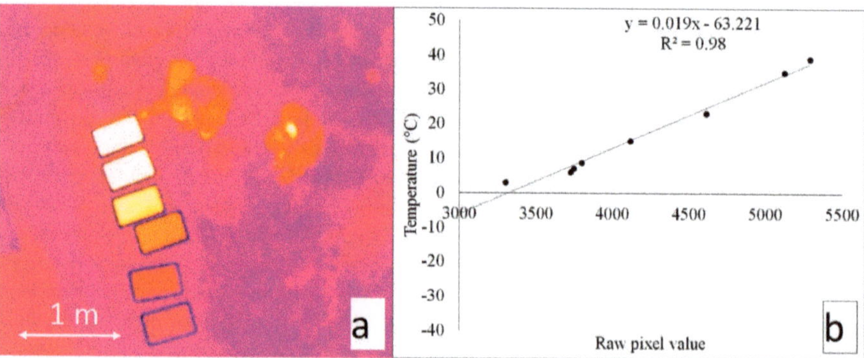

Figure 3. Temperature calibration of raw pixel values in TIR images: (**a**) TIR image capture by UAS hovering 15 m above trays and (**b**) linear calibration equation derived from average of pixel values in each tray and corresponding temperature measurements.

Temperature calibration was undertaken one week after the TIR flights at Tidmarsh Farm (25 March 2019), with warmer ambient air temperature (5 °C versus −3 °C). However, the effect of ambient temperature appeared to be minor; a previous calibration (same TIR sensor) undertaken with different ambient temperatures (13–20 °C) gave a similar linear relation (Table 1) [5], and standard error of the regression (2.3 °C).

Table 1. Comparison of calibrated temperatures from this study versus Harvey et al. [5].

Pixel Value	Calibrated Temp. (°C) (This Study)	Calibrated Temp. (°C) (Harvey et al., [5])
2800	−11.5	−10.0
3000	−7.3	−6.2
3200	−3.1	−2.4
3400	1.1	1.4
3600	5.2	5.2

4. Results and Discussion

4.1. Image Quality and Spatial Coverage

Results show the TIR-equipped UAS survey method has sufficient resolution and sensitivity to resolve subtle temperature contrasts presented by groundwater seeps in cold, temperate winter conditions. The UAS method can efficiently map large areas (i.e., 0.4 km² area captured in 36 min of flight) and is well suited for assessment of wetlands, both before and after restoration efforts.

We compared TIR, visible, and DSM imagery to evaluate the landscape forms and thermal signature of groundwater inflows (Figures 4 and 5). While the visible and DSM images show stream and pond morphology (Figure 5a,c,d), they provide no indication of groundwater input. However, the TIR image makes groundwater input obvious, and substantial inflows of warmer groundwater are visible along the restored channel (i.e., compared to other locations on the wetland platform) (Figure 5b). DSMs extracted from visible imagery (Figure 5c,d) provide an efficient method for mapping the restored channel forms (compared to ground-based surveys).

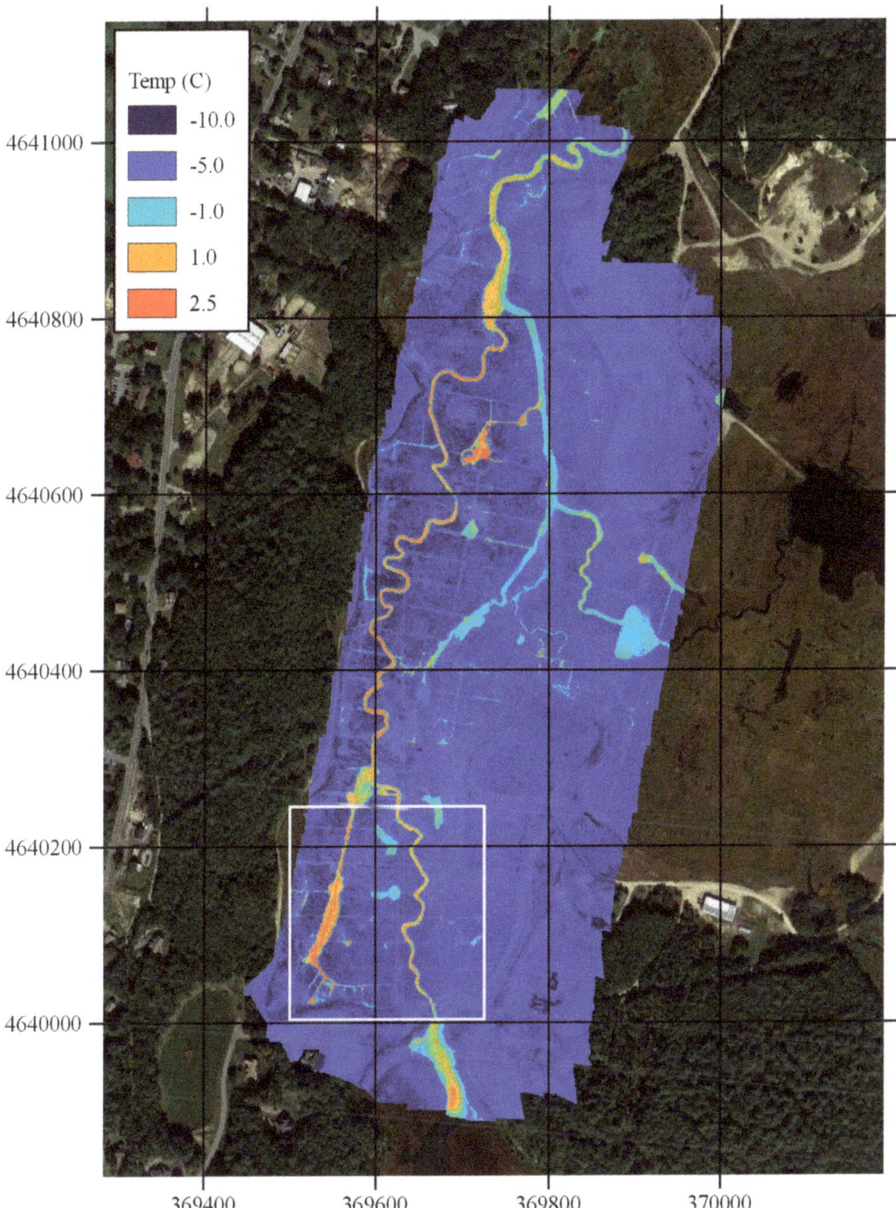

Figure 4. Calibrated TIR orthophoto of the Tidmarsh study area overlaid on Google Earth imagery. Note: white square detail is shown in Figure 5a–c. Map datum WGS84.

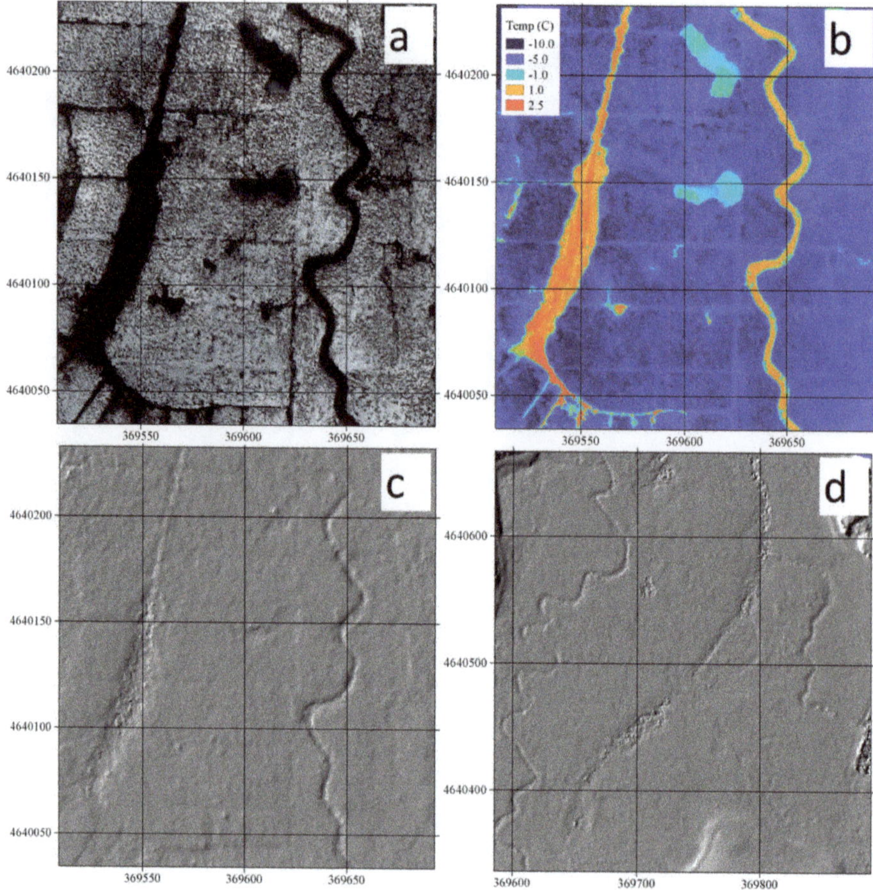

Figure 5. Comparison of winter UAS imagery showing (**a**) visible light and (**b**) TIR. (**c,d**) Digital surface models show complex channel forms that would be difficult to survey with ground-based methods (e.g., total station). Note: Only TIR shows both stream morphology and temperature. Panels (**a–c**) correspond to white square in Figure 4. Panel (d) shows area of complex channel form restoration in the central survey area.

Figure 6 provides close-up views of the TIR imagery, with colors adjusted to provide more contrast at the warmer end of the temperature scale (more clearly shows groundwater input). Note that (i) groundwater discharge (>1 °C) is clearly discernible from the relatively cooler surface waters (−1 °C) (Figure 6a), (ii) seepage zones are distributed along the reconstructed stream channel (Figure 6b), and (iii) discrete seeps within the stream channel are clearly identifiable (Figure 6c).

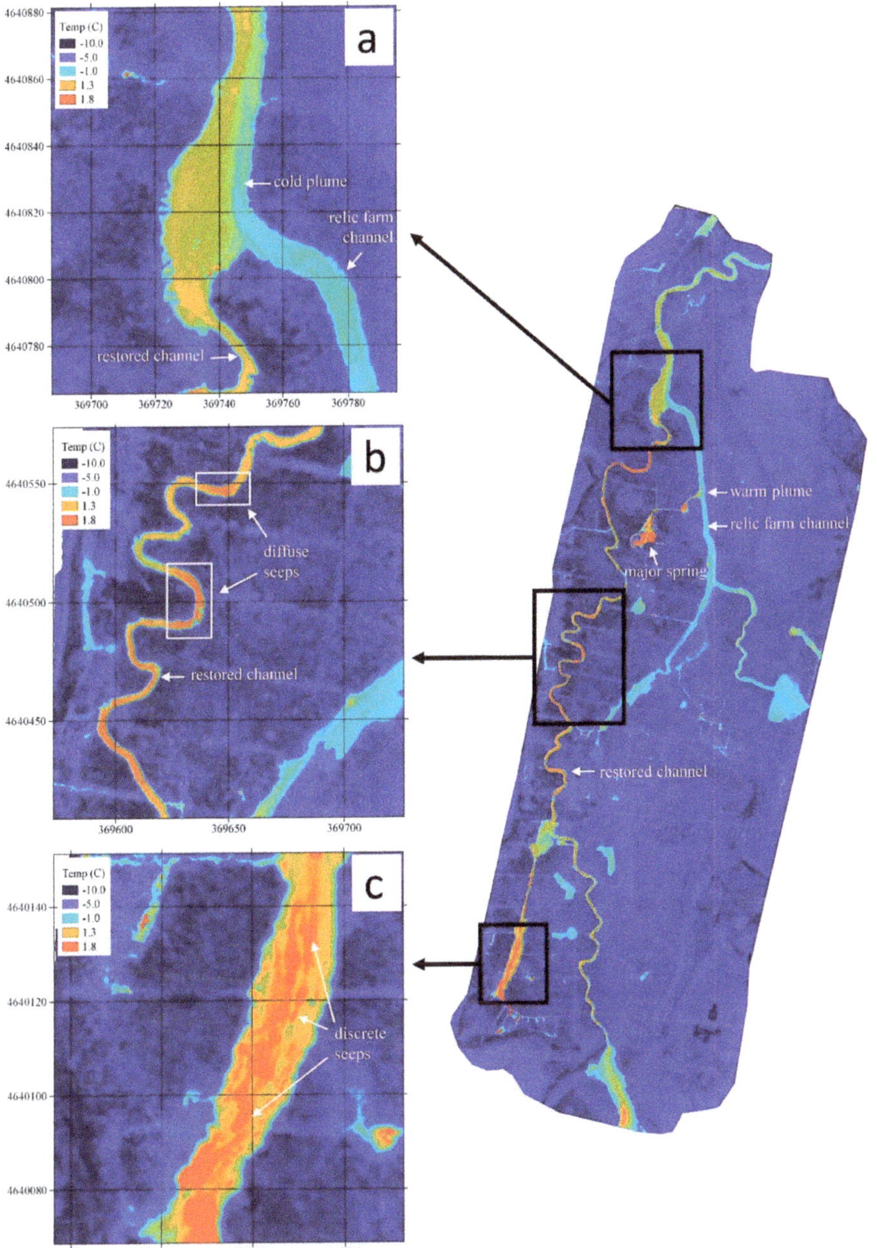

Figure 6. TIR imagery showing (**a**) confluence of cool surface waters and warmer groundwater, (**b**) diffuse seepage zones along reconstructed stream channel (white boxes), and (**c**) warm groundwater seeping into the stream channel. Note: water flows from south to north.

Before the restoration, ground-based TIR (Figure 2) showed that "wet areas", long known by farmers, were due to groundwater seeps, likely due to the thin peat in this area (rather than ponding of meteoric water in topographic depressions). This led to the seeps being utilized during the restoration process and demonstrated the value of TIR to inform the restoration design [16]; thermally stable pools and streams are important for ecological refugia, which are fundamental for successful ecological restoration [21]. Based on our results, we expect UAS TIR imagery would be equally useful in the pre-restoration design phase.

One of the objectives of the restored stream channel was to strengthen the connection between groundwater and surface water, encouraging groundwater discharge. To meet this objective, the main stream channel was moved through a process of filling portions of the relic farm channel and locating a newly dug restored channel according to underlying hydrologic drivers of peatland groundwater outflows (see Section 2, Figure 7). Our post-restoration TIR imagery confirms this approach was successful; both diffuse and focused groundwater seepage is visible along the reconstructed channel (Figure 6b,c), showing greater groundwater connection than in the relic channel. The process-based design resulted in a lateral thermal profile typical of a groundwater-dominated stream, which demonstrates the success of the stream placement based on the restoration objective.

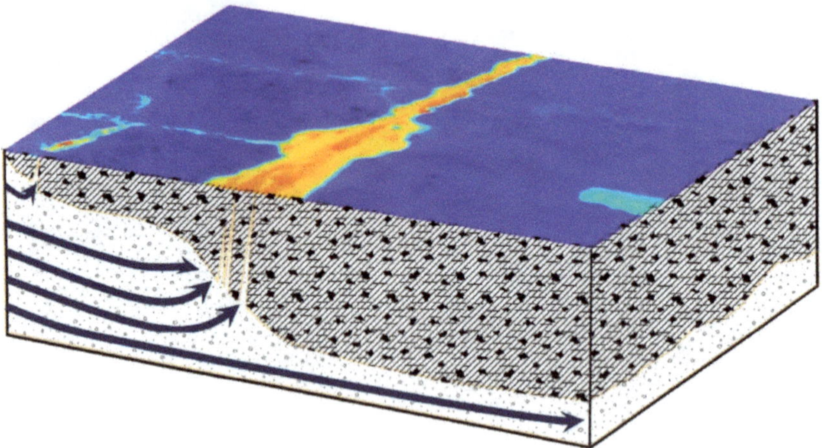

Figure 7. TIR imagery superimposed on conceptual model of subsurface. The restored channel was geolocated based in part on the conceptual model presented by Hare et al. [16], which predicted preferential groundwater discharge upward through the peatland platform based on the underlying topography of the sand–peat aquifer interface. Note: figure not to scale.

4.2. General Utility and Limitations of the Method in Wetlands

Here, we have successfully demonstrated the TIR-equipped UAS method in a continental climate during cold, winter conditions. The timing of the survey was intentional to provide maximum temperature contrast. The same method would not be as effective when applied in spring or autumn or in temperate to tropical climates; in these cases, the temperature contrast between surface environment and groundwater would be less and would be more quickly lost by surface mixing. Similarly, vegetation coverage is at a minimum in the winter, and this provided ideal conditions (TIR sensors cannot penetrate vegetation). In continental climates, surface waters may not be visible outside of winter because of foliage. In temperate/tropical climates, streams may be partly or completely covered by vegetation throughout the year.

Groundwater temperature is also a factor. The northeastern USA has a typical continental geothermal gradient [22], with shallow groundwater temperature approximating the mean annual surface temperature (~11 °C). Less commonly, groundwater may be much warmer due to magmatism and/or geologic faults that allow deeply circulating groundwater to reach the surface [5]. At Tidmarsh, it is possible that groundwater is warmed by decomposition of organic matter present in the peat soils.

Our results show how UAS-based TIR can be applied to mapping groundwater seeps in wetlands. This approach provides a viable alternative to ground-based seep identification methods (i.e., ground-based TIR or direct temperature measurements), particularly at the scale of this survey. Depending on survey size, UAS may provide a less labor-intensive solution, with spatial coverage that is unimpeded by site-access considerations. For instance, the hand-held TIR survey conducted by Hare et al. [16] took place over three days (two in the summer and one in the winter) for ~2 h each and was substantially more efficient than the weeks-long fiber-optic distributed temperature sensing survey [14,16]. The efficiency of UAS allows for repeat surveying, opening opportunities to conduct temporal thermal evaluations. More importantly, by using photogrammetry to combine images from a single flight, a complete landscape snap-shot-in-time image could be achieved, ensuring the entire area of interest is covered, including areas that would have been difficult to access by foot. Also, many wetland sites do not allow for direct foot access due to active farms, waterlogged areas, or protected areas, making UAS TIR the only viable approach for a temperature survey.

Here, we have demonstrated the UAS TIR method at Tidmarsh Farms, where groundwater flows are clearly large enough (i.e., compared to surface water flows) to be seen. In wetlands like Tidmarsh, the method will allow for more accurate and reproducible surface monitoring. In other wetland areas where groundwater inflows are relatively weak or in river environments, groundwater inflows may be greatly diluted and the thermal signature quickly lost, even in winter. In such areas, groundwater seeps may not be identified using TIR-equipped UAS, and the refugia potential would be similarly reduced. This remains to be tested in future studies.

5. Conclusions

A key question of the study was to determine if TIR-equipped UAS imaging would have sufficient resolution and sensitivity to resolve subtle temperature contrasts resulting from surface–groundwater connectivity and exchange. Our results confirm that UAS-based TIR imaging can delineate focused and diffuse seeps. The winter survey timing maximized temperature contrast between relatively warm groundwater and cooler surface waters (more equilibrated with air temperature) while minimizing vegetative cover that could obstruct the TIR sensor's nadir overhead view. Further, we targeted a winter day when there was no snow on the peatland platform and little ice on the surface water features. The same method would not be as effective during times of snow and deep freeze or in summer when relatively cold, dense groundwater tends to plunge below slow-flowing warmer surface water.

The primary objective of the study was to determine if the engineered stream channel at Tidmarsh Farms had successfully intercepted groundwater seepage compared to relic farm channels. The TIR-equipped UAS was able to image the entirety of the restored channel area in a single morning. Results show the reconstructed stream channel is warmer along its entire length, indicating spatially contiguous groundwater connectivity. Therefore, we conclude that the groundwater discharge "process-based" restoration design was successful and likely creates thermal refugia for aquatic habitats. Further, the DSM extracted from visible imagery was useful in efficiently mapping the restored channel forms (compared to ground-based surveys).

Our results show that TIR-equipped UAS can efficiently map wetland groundwater seeps. This approach provides a viable alternative to ground-based seep identification methods, including ground-based TIR, direct temperature measurements, and indirect methods (vegetation mapping), particularly at the scale of this survey. Depending on survey size, UAS may provide a less labor-intensive solution (hours versus days) with greater spatially continuous coverage than ground-based methods, particularly when site access is limited. Temperature calibration and georeferencing of imagery provides an accurate and reproducible approach for surface monitoring of wetlands, wetland restoration planning, post-restoration evaluation, and mapping thermal refugia. More generally, TIR imagery can be used as a base map for planning geochemical, geophysical, or ecological sampling campaigns and provides the high-resolution georeferenced TIR imagery necessary to identify, locate, and sample individual seeps.

Author Contributions: Conceptualization, M.A.B. and D.K.H.; methodology, M.C.H. and M.A.B.; validation, M.C.H.; formal analysis, M.C.H. and A.B.H.; investigation, M.C.H., M.A.B., A.B.H., D.K.H., A.H. (Alex Hackman), and A.B.H.; resources, M.C.H.; data curation, M.C.H.; writing—original draft preparation, M.C.H., D.K.H., A.H. (Alex Hackman), A.B.H., and G.D.; writing—review and editing, M.A.B., A.H. (Ashley Helton), and J.W.L.J.; visualization, M.C.H., D.K.H., and A.H. (Alex Hackman); supervision, M.A.B.; project administration, M.A.B.; funding acquisition, J.W.L.J., M.A.B., and A.H. (Ashley Helton).

Funding: Funding for this method development was provided by the National Science Foundation Division of Earth Sciences award (#1824820), U.S. Department of Energy grant DE-SC0016412, and the U.S. Geological Survey (USGS) Toxic Substances Hydrology Program.

Acknowledgments: We thank the Audubon Society for site access and logistical support. Any use of trade, firm, or product names is for descriptive purposes only and does not imply endorsement by the U.S. Government.

Conflicts of Interest: The authors declare no conflict of interest. The funders had no role in the design of the study; in the collection, analyses, or interpretation of data; in the writing of the manuscript; or in the decision to publish the results.

References

1. Briggs, M.A.; Hare, D.K. Explicit consideration of preferential groundwater discharges as surface water ecosystem control points. *Hydrol. Process.* **2018**. [CrossRef]

2. Briggs, M.A.; Dawson, C.B.; Holmquist-Johnson, C.L.; Williams, K.H.; Lane, J.W. Efficient hydrogeological characterization of remote stream corridors using drones. *Hydrol. Process.* **2018**, 1–4. [CrossRef]

3. Pai, H.; Malenda, H.F.; Briggs, M.A.; Singha, K.; González-Pinzón, R.; Gooseff, M.N.; Tyler, S.W. Potential for Small Unmanned Aircraft Systems Applications for Identifying Groundwater-Surface Water Exchange in a Meandering River Reach. *Geophys. Res. Lett.* **2017**, *44*. [CrossRef]

4. Woodget, A.S.; Carbonneau, P.E.; Visser, F.; Maddock, I.P. Quantifying submerged fluvial topography using hyperspatial resolution UAS imagery and structure from motion photogrammetry. *Earth Surf. Process. Landf.* **2015**, *40*, 47–64. [CrossRef]

5. Harvey, M.C.; Rowland, J.V.; Luketina, K.M. Drone with thermal infrared camera provides high resolution georeferenced imagery of the Waikite geothermal area, New Zealand. *J. Volcanol. Geotherm. Res.* **2016**, *325*, 61–69. [CrossRef]

6. Abolt, C.; Caldwell, T.; Wolaver, B.; Pai, H. Unmanned aerial vehicle-based monitoring of groundwater inputs to surface waters using an economical thermal infrared camera. *Opt. Eng.* **2018**, *57*, 053113. [CrossRef]

7. Dugdale, S.J. A practitioner's guide to thermal infrared remote sensing of rivers and streams: Recent advances, precautions and considerations. *Wiley Interdiscip. Rev. Water* **2016**, *3*, 251–268. [CrossRef]

8. Dugdale, S.J.; Kelleher, C.A.; Malcolm, I.A.; Caldwell, S.; Hannah, D.M. Assessing the potential of drone-based thermal infrared imagery for quantifying river temperature heterogeneity. *Hydrol. Process.* **2019**, *33*, 1152–1163. [CrossRef]

9. Fitch, K.; Kelleher, C.; Caldwell, S.; Joyce, I. Airborne Thermal Infrared Videography of Stream Temperature from a Small Unmanned Aerial System. *HPEYE* **2018**. [CrossRef]

10. Isokangas, E.; Davids, C.; Kujala, K.; Rauhala, A.; Ronkanen, A. Combining unmanned aerial vehicle-based remote sensing and stable water isotope analysis to monitor treatment peatlands of mining areas. *Ecol. Eng.* **2019**, *133*, 137–147. [CrossRef]

11. Hester, E.T.; Gooseff, M.N. Moving beyond the banks: Hyporheic restoration is fundamental to restoring ecological services and functions of streams. *Environ. Sci. Technol.* **2010**, *44*, 1521–1525. [CrossRef] [PubMed]

12. Grand-Clement, E.; Anderson, K.; Smith, D.; Luscombe, D.; Gatis, N.; Ross, M.; Brazier, R.E. Evaluating ecosystem goods and services after restoration of marginal upland peatlands in South-West England. *J. Appl. Ecol.* **2013**, *50*, 324–334. [CrossRef]

13. Hunt, R.J.; Krabbenhoft, D.P.; Anderson, M.P. Assessing hydrogeochemical heterogeneity in natural and constructed wetlands. *Biogeochemistry* **1997**, *39*, 271–293. [CrossRef]

14. Hare, D.K.; Briggs, M.A.; Rosenberry, D.O.; Boutt, D.F.; Lane, J.W. A comparison of thermal infrared to fiber-optic distributed temperature sensing for evaluation of groundwater discharge to surface water. *J. Hydrol.* **2015**, *530*, 153–166. [CrossRef]

15. Lowry, C.S.; Walker, J.F.; Hunt, R.J.; Anderson, M.P. Identifying spatial variability of groundwater discharge in a wetland stream using a distributed temperature sensor. *Water Resour. Res.* **2007**, *43*. [CrossRef]
16. Hare, D.K.; Boutt, D.F.; Clement, W.P.; Hatch, C.E.; Davenport, G.; Hackman, A. Hydrogeological controls on spatial patterns of groundwater discharge in peatlands. *Hydrol. Earth Syst. Sci.* **2017**, *21*, 6031–6048. [CrossRef]
17. Beechie, T.J.; Sear, D.A.; Olden, J.D.; Pess, G.R.; Buffington, J.M.; Moir, H.; Roni, P.; Pollock, M.M. Process-based principles for restoring river ecosystems. *BioScience* **2010**, *60*, 209–222. [CrossRef]
18. Koch, F.W.; Voytek, E.B.; Day-Lewis, F.D.; Healy, R.; Briggs, M.A.; Werkema, D.; Lane, J.W., Jr. 1DTempPro: A program for analysis of vertical one-dimensional (1D) temperature profiles v2.0: U.S. *Geol. Surv. Softw. Release* **2015**. [CrossRef]
19. U.S. Geological Survey. USGS 420316070433501 MA-D4W 79R Duxbury, MA, in USGS Water Data for the Nation: U.S. Geological Survey National Water Information System Database. 2019. Available online: https://waterdata.usgs.gov/nwis/uv?site_no=420316070433501 (accessed on 1 June 2019).
20. Harvey, M.; Briggs, M.A.; Dawson, C.B.; White, E.A.; Haynes, A.; Fosberg, D.; Moore, E. Thermal infrared and photogrammetric data collected by small unoccupied aircraft system for the evaluation of wetland restoration design at Tidmarsh Wildlife Sanctuary, Plymouth, MA. USA. *Geol. Surv. Public Data Release* **2019**. [CrossRef]
21. Lake, P.S.; Bond, N.; Reich, P. Linking ecological theory with stream restoration. *Freshw. Biol.* **2007**, *52*, 597–615. [CrossRef]
22. Blackwell, D.D.; Richards, M.C.; Frone, Z.S.; Batir, J.F.; Williams, M.A.; Ruzo, A.A.; Dingwall, R.K. SMU Geothermal Laboratory Heat Flow Map of the Conterminous United States. 2011. Available online: http://www.smu.edu/geothermal (accessed on 21 August 2015).

MDPI

St. Alban-Anlage 66

4052 Basel

Switzerland

Tel. +41 61 683 77 34

Fax +41 61 302 89 18

www.mdpi.com

Water Editorial Office

E-mail: water@mdpi.com

www.mdpi.com/journal/water